Mathematics and the Built Environment
Volume 9

Series Editors

Kim Williams, Kim Williams Books, Torino, Italy

Michael Ostwald ⓘ, Built Environment, University of New South Wales, Sydney, Australia

Throughout history a rich and complex relationship has developed between mathematics and the various disciplines that design, analyse, construct and maintain the built environment.

This book series seeks to highlight the multifaceted connections between the disciplines of mathematics and architecture, through the publication of monographs that develop classical and contemporary mathematical themes – geometry, algebra, calculation, modelling. These themes may be expanded in architecture of any era, culture or style, from Ancient Greek and Rome, through the Renaissance and Baroque, to Modernism and computational and parametric design. Selected aspects of urban design, architectural conservation and engineering design that are relevant for architecture may also be included in the series.

Regardless of whether books in this series are focused on specific architectural or mathematical themes, the intention is to support detailed and rigorous explorations of the history, theory and design of the mathematical aspects of built environment.

To achieve the highest research standards, each volume in this series undergoes a rigorous peer review process managed directly by the series editors.

Sotirios Kotsopoulos
Editor

Shape Computation

Fifty Years, 1972–2022

Editor
Sotirios Kotsopoulos
Department of Architecture
National Technical University of Athens
Athens, Greece

Department of Architecture
Massachusetts Institute of Technology
Cambridge, USA

ISSN 2512-157X ISSN 2512-1561 (electronic)
Mathematics and the Built Environment
ISBN 978-3-031-81622-2 ISBN 978-3-031-81623-9 (eBook)
https://doi.org/10.1007/978-3-031-81623-9

© The Editor(s) (if applicable) and The Author(s) 2025. This book is an open access publication.

Open Access This book is licensed under the terms of the Creative Commons Attribution-NonCommercial-NoDerivatives 4.0 International License (http://creativecommons.org/licenses/by-nc-nd/4.0/), which permits any noncommercial use, sharing, distribution and reproduction in any medium or format, as long as you give appropriate credit to the original author(s) and the source, provide a link to the Creative Commons license and indicate if you modified the licensed material. You do not have permission under this license to share adapted material derived from this book or parts of it.

The images or other third party material in this book are included in the book's Creative Commons license, unless indicated otherwise in a credit line to the material. If material is not included in the book's Creative Commons license and your intended use is not permitted by statutory regulation or exceeds the permitted use, you will need to obtain permission directly from the copyright holder.

This work is subject to copyright. All commercial rights are reserved by the author(s), whether the whole or part of the material is concerned, specifically the rights of translation, reprinting, reuse of illustrations, recitation, broadcasting, reproduction on microfilms or in any other physical way, and transmission or information storage and retrieval, electronic adaptation, computer software, or by similar or dissimilar methodology now known or hereafter developed. Regarding these commercial rights a non-exclusive license has been granted to the publisher.

The use of general descriptive names, registered names, trademarks, service marks, etc. in this publication does not imply, even in the absence of a specific statement, that such names are exempt from the relevant protective laws and regulations and therefore free for general use.

The publisher, the authors and the editors are safe to assume that the advice and information in this book are believed to be true and accurate at the date of publication. Neither the publisher nor the authors or the editors give a warranty, expressed or implied, with respect to the material contained herein or for any errors or omissions that may have been made. The publisher remains neutral with regard to jurisdictional claims in published maps and institutional affiliations.

Cover Art Image from G. Stiny (c. 2009), lecture notes. Shape grammars include generative grammars as a special case. Inscribed squares and quadrilaterals show why. These spatial relations define shape grammars and generative grammars alike, but only shape grammars allow for the yellow double-arrow and triangles, chevrons, etc.—in fact, whatever you see.

This book is published under the imprint Birkhäuser, www.birkhauser-science.com by the registered company Springer Nature Switzerland AG
The registered company address is: Gewerbestrasse 11, 6330 Cham, Switzerland

If disposing of this product, please recycle the paper.

Foreword by Nicholas de Monchaux

Fifty years ago, George Stiny and his collaborators began to publish a series of seminal papers that allowed us to look at the world in a new way. Drawing from mathematics, linguistics, and a deep craft and understanding of visual form, these authors created a new way to talk about and analyze the visual world. Most of all, they created a new way to see the world.

Over the past three decades, this conversation has been driven by the Design and Computation group of the Department of Architecture at MIT. The initial insight to deploy the framework of computation to engage the aesthetic led not only to new modes of computation but also to new possibilities for how computation could be taught and understood.

Spatial and graphic intersections are, of course, the essence of the original work on Shape Grammars, and essential to understanding its intellectual contributions today—from its influence on topology, transformations, computational morphology, to the intellectual framework it has contributed to design. But this strict sense of intersection is notable for its influence in a much broader domain, as well. For as is sometimes the case with new and unexpected tools, a particular way of seeing—in this case, synthesizing artistic expression, geometry, and mathematics—extended itself inexorably. Thus, beyond its original, geometric sense, Shape Computation's most transformative contribution at MIT has been its creative environment where intersections between the cultural and the calculated are the norm, not the exception. Vital to these intersections are the insights of the many hundreds of students from the Design and Computation group who occupy leading positions in design, technology, and education across the globe.

Against computation's all-too-common focus on optimization, this legacy affirms the most essential principle of the work this volume celebrates—that the world of computing and the world of aesthetics must be connected generatively, creatively, rigorously, and radically. For this legacy, as well as for the inventiveness and resilience of the underlying ideas, we owe a great debt to the legacy of Shape Computation at MIT—and beyond.

Nicholas de Monchaux
Professor and Head of Architecture
Massachusetts Institute of Technology
Cambridge, USA

Foreword by George Stiny

I worked out the key ideas for shape grammars in 1968, experimenting with painting and sculpture. The way this goes is strikingly simple. Start with a shape C and a rule A → B that aligns two shapes A and B in space, say, by drawing them on the same piece of paper. Then, look for (embed) a shape (part) that looks like A in C and replace it with a shape that looks like B. "Looking for" (embedding) and "looking like" animate this process. They both come down to seeing in new ways. Shapes are made up of finitely many points, lines, planes, and solids, and they're especially interesting whenever they're not entirely points. For example, rules elaborate Leon Battista Alberti's definition of architecture in the firm and graceful ordering of lines and angles, as they apply recursively to calculate. Architecture (design) adds to painting and sculpture (art) to highlight what shape grammars do.

The first published version of shape grammars was in 1972—it marked the official start of the subject. There were rules in a shape specification and rules in a material specification that worked in parallel for painting and sculpture. This mapped onto Alberti and architecture in the separation of shapes (form), and materials and making. (Today, weights put materials in shapes.) Jim Gips joined me to write the paper and brought in computers to calm my zeal for seeing/looking in art and design. The gap between computers and seeing opened three years later in our Ph.D. dissertations. (Gips did computer science at Stanford, and I was in engineering at UCLA.) Gips wrote the first computer program for shape grammars in terms of polygons given by points; they were the symbols/primitives—atoms and units, 0's and 1's—needed to calculate. (It's odd that atom and unit are two-syllable words.) Obviously, when two squares combine, there are two squares—and also a hexagon with collinear sides and a perpendicular diameter

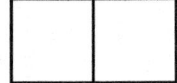

It's counting points 1, 2, 3. But nothing is obvious. Instead, I played around with embedding and fusing (each implies the other) and used them to calculate. Shapes are more than symbols—like pictures, there's too much in them to count. Without atoms or units or 0's and 1's, calculating goes a unique way. Combining (fusing) two squares is two squares and Gips's hexagon or a rectangle and a square that flips to and fro and other surprises—left and right end-brackets, links, and a middle I, or an E on its side and one on its arms, or ground plans with walls and doors and windows put in here and there, or two faces of a cube seen in a funny way

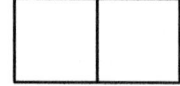

Shapes vary in vast excess. This is because lines fuse and divide endlessly into discrete and overlapping segments (lines) that appear and disappear at will. There's recursion in computers and shape grammars. But computers need atoms and units for recursive rules (code) to work—it's like this in linguistics for words. Computers don't do embedding for shapes; in computers, embedding is identity for symbols. Symbols are exactly how they're defined; there's no ambiguity or any tinge of doubt. But shapes are never what they seem to be; there's always another way to see. As a result, computers aren't very good at art and design. Art and design call for constant looking (embedding and fusing) in an open-ended process. Without this, there isn't any surprise—everything is clear from the start.

In 1975, I was positive that my kind of shape grammars would catch on right away and that many others would try to calculate in this way, without symbols in given vocabularies of atoms and units. Atoms and units—wherever they come from—are fixed before calculating begins. Seeing relies on embedding and doesn't care about this. How simple can it be? But it didn't happen. In fact, shape grammars are something of a mystery, a magical lure to avoid. Why bother with ambiguity in shapes when there are definite symbols that are secure in what they are? How to use symbols is what I learned in school—my teachers never let up on it. And Gips never tired of telling me that shape grammars were no more than a trivial toy for children to play with, to be replaced as they grew up by expert (altruistic) computer code in which everything worked in a mature way as it must in a neatly ordered, arithmetic world. There are heaps and gobs of data for this super utopia that can't be ignored; yep, it's right around the corner. Even so, I knew all along that the snag was ambiguity and embedding—they ask too much of computers and those who use them. My neighbor in Brookline is a computer scientist at a local university. Yet even with his musical bent ("Music is liquid architecture; Architecture is frozen music."), he refuses to take shapes and embedding seriously. Symbols and identity are required to calculate—end of story! After 50 years, I still haven't changed my mind about atoms and units—they aren't for me. Atoms and units combine in this way and that, and they let me count. But seeing is more than this. Lost and adrift, I'm never sure

Foreword by George Stiny

what I'll see next or if it will last. Art and design wane without this kind of aesthetic (perceptual) imagination. You can't take away the looking.

Lionel March asked me to write for *Environment and Planning B* in 1976. I like to think that what I did then influences the way I use shape grammars now. In "Two Exercises in Formal Composition" (1976), the first exercise is constructive, and the second one is analytic. This taxonomy is informal and seems intuitive beyond any test. But it fails if you poke at it in shape grammars. There's neither hierarchy nor division when I calculate with shapes and rules. The constructive exercise uses rules in the expression (schema)

$$x \to x + t(x)$$

In the right side of each rule, the shape x and a transformation of it t(x) add as one and fuse. This is common in architecture and in painting and elsewhere in art and design. It's an easy start and likely to go on, with changeable results. Shapes are liquid not frozen ("essentially fixed and dead")—parts appear and disappear seamlessly as rules are tried. Yet some make this "combinatoric" (Herbert Simon's word) in an atomic (Lucretian) swerve. My examples from 1976 are OK for atoms and not. In this example, the rule

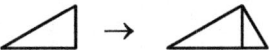

adds two similar right triangles, so that the second is larger than the first. The second triangle overlaps two sides of the triangle

in the left side of the rule. If I limit the rule's use to the largest, most recently added triangles (atoms) in a combinatoric process, I get the shapes

and others like Russian dolls, with two triangles, three, four, five, etc. (It's a snap with labels.) But beyond my limit, the rule goes for any right triangle I'm in the mood to see. Then, the number of triangles (they're no longer atoms) doubles less one to three, five, seven, nine, etc. Of course, it's a chore to use the rule one triangle at a time when multiple triangles are apt to be in sync—maybe three of the nine triangles in the rightmost shape are a line of school pennants, or four make three quadrilaterals in a pleated band. The summation schema

$$x \to \Sigma F(prt(x))$$

defines a single rule for any number of triangles I want. My two exercises are alike in key places. The second, analytic one marshals identity rules in the schema

$$x \rightarrow x$$

to divide shapes into parts in trees and other types of descriptions/representations—graphs and topologies, and sundry data structures in lists and so on. The identities in $x \rightarrow x$ are a subset of $x \rightarrow x + t(x)$. Adding shapes puts seeing first to tell where to add, whether this is more or not. Every rule is the same—no matter what it's for, seeing comes first. Identities rely on the summation schema to parse shapes in countless ways. For example, the identity for squares

and the identity for 2 x 1 rectangles

divide the shape

in twin trees

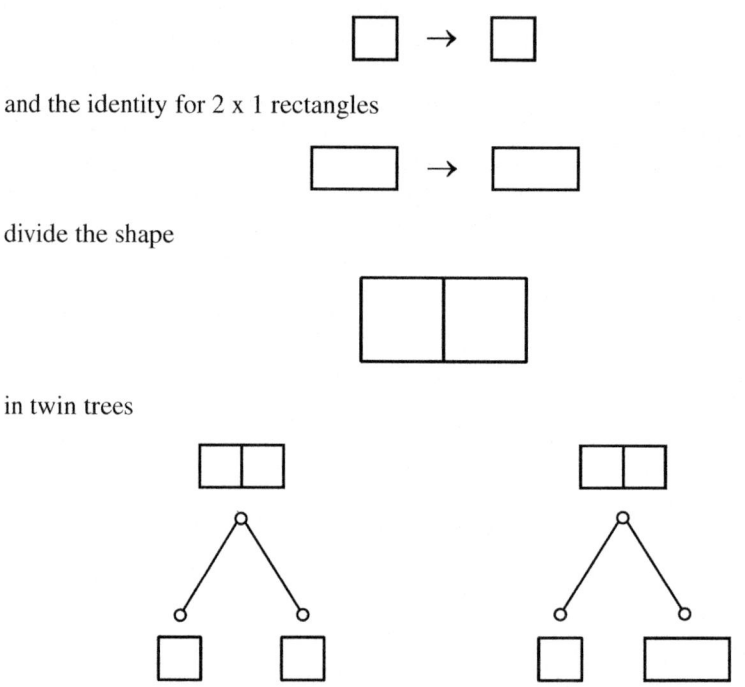

Two pairs of identities are used twice in the summation schema. In the left tree, adjacent squares share one side and in the right tree, the rectangle overlaps three sides of the square. Equally, identities divide partis (grids and such) in architecture; this runs through design from studio teaching to practice. Identities are useless in computers and indispensable in shape grammars. Both loop when they're tried, but identities in computers stick to prior atoms and units to recount what everyone already knows, while identities in shape grammars recreate what's there, to see for the first time. There's nothing to learn in computers and everything to learn in shape grammars. Initially, I thought the constructive exercise did more than the analytic one. I made a big deal about generative rules for vocabularies of shapes and their spatial relations—say, Alberti's lines and angles, the pair of overlapping triangles in the right side of my rule

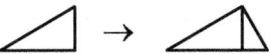

the undivided shapes in the rule's left and right sides, or a rectangle and a square that are resolved and composed in a tree. Now, the schema is

$$x \rightarrow x + t(x')$$

where x and x' are shapes in a given vocabulary, and x and t(x') form a spatial relation with specified properties. (The summation schema shows how this goes in a host of ways.) It's fun to get familiar designs and new ones from scratch. It's neat to see shapes grow, to gape at what you've done, and to add to style in lush detail. Still, seeing is vital in all rules—by itself, it generates (divides) shapes anew. Creativity, surprise, and style are easy to find in this, too. My two exercises are hard to tell apart. It's a fluke that pairs of polygons line up—adjacent triangles and adjacent squares and overlapping triangles and an overlapping rectangle and square. But in general, both exercises let vocabularies of atoms and units alter in terms of how rules (identities and not) apply. Beauty in aesthetic imagination is first before goodness in ethics that strives to make the world a better place (for Ezra Pound, "the arts provide data for ethics") and truth in logic and STEM. Beauty provides the basis (vocabulary) for the good and the true. Architecture puts this in the Vitruvian canon. Delight (beauty) paves the way for commodity (goodness in utility and use) and firmness (truth in building construction, physical performance, and technology). Shape grammars blur atoms and units in order to redefine them. Today, no one sees beauty in this strange and wonderful way if beauty is noticed at all. Our times are drab, dreary, and dull. With an agreed ABC for the atoms and units that computers combine, there's no reason to see—only counting what's given and known—and no need for art and design. Embedding and fusing in shape grammars are why calculating with shapes is open-ended and so, "essentially vital"—in motion, vibrant, and pulsing with life. Art and design are lodged fully in seeing.

This leaves out a lot—to start, a universal kind of embedding that's for more than geometric shapes made up of linear elements and such, for example, for Alberti's tree-trunks and clods of earth in *De statua* and for Leonardo da Vinci's blotches and stains on walls, etc. Reduction rules for embedding and fusing are next, for exotic lines, planes, and solids; algorithms to find where rules apply; the use of labels and weights to supplement shapes in a host of different ways; and description rules for designs (shapes, symbols, and words) in n-tuples, maybe for plans, sections, and elevations and for the enormous mix of architectural relationships that unfold in firmness, commodity, and delight. But this book fills in many of the blanks, as it surveys unmapped domains. The range is remarkable—going from personal tales of art, architecture, and calculating with shapes to shape grammars that bring in time in painting and making in craft to how and why shape grammars can't help but ignore data in ML and LLMs to the tricks of shape grammars in studio and practice to yet more and more. This shows the power and reach of seeing and doing in shape grammars at 50. It's a super start to another 50 years that may influence (re-create) what's been done, in the way embedding and fusing in shape grammars change familiar (prior) things in terms of uncertain (current) ones—what I know now is the result of what I see next. It's all fresh and new. With some luck, I'll revel in the sixth decade of this and the seventh; they promise boundless delight. Shape grammars swerve and weave in extravagant arcs and impossible curves to calculate "in the region of the many and variable"—their loftiest goal is surprise.

George Stiny
Massachusetts Institute of Technology
Cambridge, USA

References

Gips, J. 1975. *Shape grammars and their uses*. Basel: Birkhauser.
Stiny, G. 1975. *Pictorial and formal aspects of shape and shape grammars*. Basel: Birkhauser.
Stiny, G. 1976. Two exercises in formal composition. *Environment and Planning B* 3 187–210.
Stiny, G., and J. Gips. 1972. Shape grammars and the generative specification of painting and sculpture. In *Information processing*, vol. 71, ed. C. V. Frieman. Amsterdam: North-Holland.

Preface

In *Remarks on the Foundations of Mathematics 1937–1944*, Ludwig Wittgenstein touched on shape computation:

"An addition of shapes together, so that some of the edges fuse, plays a very small part in our life. As when

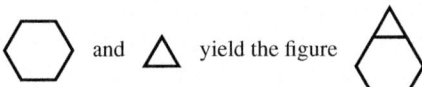

But if this were an *important* operation, our ordinary concept of arithmetical addition would perhaps be different."

Wittgenstein's 1939 lectures in Cambridge, attended by Alan Turing, challenged standard ideas of arithmetic. Turing's 1936 essay on computable numbers—introducing the concept of a Turing machine—set the standard for symbolic computation. The logicians Alonzo Church and Emil Post independently developed models based on the same general idea as Turing's. Computation and its applications introduced an algorithmic methodology to many scientific fields, propelling us into the digital age. However, in the late 1940s, the mathematician John von Neumann, one of the pioneers of the field, expressed his misgivings about the ability of symbolic descriptions to capture all of our visual experience. He argued that visual interpretation is idiosyncratic and often falls outside logical/symbolic representation in computers. But von Neumann did not explore visual computation any further. Using shapes for calculations seemed to have limited practical value in science. Nonetheless, it raised the question of the use of shape computation in art and design, where the whole often exceeds the sum of its parts. What would happen if Turing, Church, Post, and von Neumann took shape computation seriously?

In 1972, two Ph.D. students at UCLA and Stanford, George Stiny and James Gips, made a case for shape computation in their seminal essay, "Shape Grammars and the Generative Specification of Painting and Sculpture" (1972). This essay introduced a new approach to computation, showing that calculating directly with shapes in art and design is possible and useful. The essay pointed to a new frontier, especially for designers, that was soon valued as a discovery. Shape computation had yet to be explored. How could we use these ideas in creative fields? And what would general computing look like if shape computation was fully developed?

Stiny and Gips laid the foundations for shape computation through experiments in painting and sculpture. They introduced the concept of a rule for shapes that could be drawn or made without the use of symbols. Applying these rules defined a *language of designs*. Gips's work at Stanford (1975) demonstrated practical applications of shape grammars in computer vision, shape generation, and computer aesthetics, to showcase the relevance and potential of this research. Stiny's work at UCLA (1975) explored the mathematical properties of the shape grammar formalism in terms of an embedding relation, establishing that it is not the invariance of symbols that makes calculating possible.

In the following years, Stiny, mostly in the journal *Environment and Planning B: Planning and Design*, delineated the mathematics of shapes and how this works in shape computations. Shapes are made up of finitely many maximal elements—points, lines, planes, and solids—in a finite space. Shapes with elements of dimension $i \geq 0$ are manipulated in a space of dimension $j \geq i$. Shapes made of undivided points ($i = 0$) behave like abstract sets, while shapes made of elements of higher dimension ($i > 0$) behave visually—they have indefinitely many subparts embedded in them. This allows for a finite shape to be infinitely divisible. Shape computation is grounded rigorously in shape algebras U_{ij} (Stiny 1991). Moreover, these can be augmented with labels and weights to form new algebras, V_{ij} and W_{ij}, in which computations with multiple shape descriptions incorporating various properties, say, color or material, are defined (Stiny, 1992). Decompositions of shapes may also serve to analyze and explain calculations with shapes (Stiny, 1994). Topologies and related mappings show the continuity of shape computations.

The early research on shape grammars coincided with a period in which designers, architects, and visual artists attempted to adapt to new computational concepts, rationalizations, and procedures. In this context, Stiny proposed a youthful, liberating attitude, which was also formal and rigorous, to show that vision is a source of energy and wonder in spatial exploration, and that seeing informs calculation. Stiny's work introduced a new paradigm where the identification of shapes and their parts is unrestricted, liberating objects from the need to be represented in advance (Stiny, 1996). He placed our constant ability to discriminate and fuse parts into cohesive wholes at the core of visual computation, where seeing challenges logical constraints that often stifle artistic imagination. The potential for continued observation is boundless, always sparking new opportunities for further exploration.

I first came across shape grammars in 1994 at UCLA, during a graduate course given by Stiny. Before that, I had completed a 5-year diploma at the School of Architecture of the National Technical University of Athens (NTUA), which gave me a foundation in many technical, mathematical, and historical aspects of architecture and design. I was already familiar with Chomsky's *Syntactic Structures* (1957), Goodman's *Languages of Art* (1976), and March's *The Architecture of Form* (1976)—in particular, his "Boolean Description of a Class of Built Forms." Reflecting on my initial impression of shape grammars, I remember that it took a mere 15–20 minutes of Stiny's demonstrations to understand how the shape grammar formalism worked and to be impressed by how much it promised to do. Shape grammars offered more profound insights into design than any other theory I had come across—then and now. The strength of shape grammars lies in the combination of observation and calculation applied to designing, making, and seeing. Using shape algebras, embedding, and shape rules instead of symbols, counting, and symbolic instructions set shape grammars apart. Over the following years, Stiny's approach remained consistently unconventional and groundbreaking, while always being refreshing, elegant, and suggestive of future advances in shape computation. He further explored these in subsequent papers, establishing an original field of study at MIT.

Shape grammars stress the active role of the eye in how we engage with art and design. Forms, shapes, and symbols serve as the primary material of seeing by which we construct our visual configurations. Their appeal is predominantly aesthetic, based on perception rather than cognition—"In shape grammars, calculating relies on the eye (perceiving) to subsume mind (thinking) and hand (making). Without perception, thinking and making are idle, with scant to go on" (Stiny, 2022, xxv). Paintings, sculptures, and architectural works are visually broken down into parts to form relationships. They are put together again by the vigorous action of the eye: "Painting and drawing keep the eye alive, in motion, both inspired and vital" (Stiny, 2022, 157). Artworks invite dynamic participation where one identifies parts and relationships, interprets what they mean, how they do so, and based on what rules—"It seems to me that without new perception and a meaningful way to explain it, there's no cause to worry about art and design—there's nothing more to say about pictures and poems, and whatever they evoke to each of us" (Stiny, 2022, xx).

Shape grammars subsume Chomsky's formal syntax and March's Boolean representation of the built form—which are based on combining symbols—to calculate with elements of different spatial dimensions, whether points, lines, planes, or solids. Furthermore, Stiny's perspective differs from Goodman's, which ties pictorial representation to a conceptual framework. Goodman like Ernst Gombrich believes that perception relies heavily on conceptual schemata and that vision cannot occur independently of conceptualization, rejecting the idea of the "innocent eye" (Goodman, 1976, 7–8). Stiny, on the other hand, asserts that "art and design are for the eye and not the mind" (Stiny, 2022, 168). "The eye is the highest judge in visual calculating" (Stiny, 2022, xviii). Stiny contends that "The mind settles on descriptions that the eye intuitively ignores, to see in extravagant ways—proof once again that seeing supersedes thinking, as shapes supersede symbols" (Stiny, 2022, 25). Yet even so, open-ended seeing (perception) in the embed-fuse cycle may work in parallel with concepts (Stiny, 1981; 1990).

Leading academic institutions in the Americas, the UK, Continental Europe, and Asia have contributed in major ways to the theory and application of shape grammars. This widespread participation has provided a unique paradigm for computation and the generation and interpretation of designs. There are endless examples, but I can cite only a few, mostly from early contributors to shape computation. In particular, Terry Knight focused on the theory and practical use of shape grammars and their extensions, making significant contributions to design education at MIT. Her book *Transformations in Design* (1994) put style and stylistic change firmly in shape computation, while her later research extends to making—investigating how color, material, and sensory and temporal elements can be incorporated in shape grammars. Ulrich Flemming, at Carnegie Mellon University (CMU), researched generative design in architecture and engineering, refining the integration of shape grammars with established architectural practice in analysis and synthesis. Based at the Open University UK, Chris Earl explored the mathematical basis of shape grammars and design descriptions, particularly for computability. Also, at CMU, Ramesh Krishnamurti focused on computational design and emphasized the formal, semantic, and algorithmic aspects of shape grammars in computer programs. His work involved experimenting with shape grammar interpreters, pushing the boundaries of computer capabilities, to explore visual computation. Finally, Lionel March gave the young field of shape computation a home in his journal *Environment and Planning B: Planning and Design*, that fostered its growth and enabled it to thrive.

The place of shape grammars in the history of design, computation, and digital technologies has been showcased in numerous international conferences and exhibitions, including, for example, the SIGRAPH 2008 conference, the Centre Pompidou 2018 exhibition "Coder le monde," the Canadian Center for Architecture 2019 exhibition "Architecture Itself and Other Postmodernist Myths," and the Centre de Design de l'UQAM, Montreal 2021 exhibition "Vers un imaginaire numérique."

Shape grammars stand out as the only comprehensive theory of design that is formal and visual rather than speculative without rigorous foundations, or merely instrumental and ad hoc. It is an inclusive approach to calculating that considers how the mind processes existing knowledge, descriptions, and representations, as well as what the eye perceives, including things that may have been overlooked or that conflict with existing representations. The challenges posed by shape computation are ongoing. Currently, the global focus is on machines that learn from large datasets. Despite their practical efficacy, querying large datasets returns us to a brute-force approach and departs from actual observation. The seminal essay by Stiny and Gips sets things in another direction. Shape computation, using recursion and embedding, is indispensable for resourceful observation (seeing), action, and thinking.

The MIT Department of Architecture and the Scholarly Publications of the MIT Libraries are celebrating the 50th anniversary (1972–2022) of Stiny's and Gips's influential essay by publishing this special anniversary book on shape computation. Part of Springer's *Mathematics and the Built Environment* series, this book features contributions from a host of authors that explore the past, present, and future of shape computation and shape grammars. It includes scholarly articles and interdisciplinary investigations from art, architecture, design theory, shape studies, industrial design, computer implementation, design education, etc. The content addresses research practices and examples, combining theoretical reflection, retrospective assessment, applications, and speculation on the 50 years of shape computation.

I have divided the book into six parts, which contain 27 chapters in total. The first part showcases new research by pioneers of the field, including applications of the shape formalism in art, architecture, and shape studies, retrospective assessments, and memoirs after 50 years of shape grammar research. The second part presents new shape grammar research in art, craft, and engineering design, expanding shape grammars to consider time and embodied making that provide a fresh, holistic view of the subject. The third part focuses on the mathematics of shape computation and its applications. The fourth part highlights the computer implementation of shape grammars and presents original applications in design synthesis and analysis using a new shape grammar interpreter. It includes mathematical topics of shape modeling and applications relevant to real-world design challenges. The fifth part explores the connections between shape computation and artificial intelligence in architectural design and the creation of intelligent virtual worlds. The sixth part connects architectural practice and education, showcasing applications, methods, and tools for real-world scenarios. Applications of shape computation range far and wide, including the nature of computation itself in shape grammars. Shape computation is a computational lens in art, architecture, design theory, and more.

Sotirios Kotsopoulos
National Technical University of Athens
Athens, Greece

References

Chomsky, N. 1957. *Syntactic structures*. Berlin: Mouton & Co.
Gips, J. 1975. *Shape grammars and their uses*. Basel: Birkhauser.
Goodman, N. 1976. *Languages of art: An approach to a theory of symbols*. 2nd ed. Indianapolis: Hackett Publishing Company.
Knight, T. 1994. *Transformations in design: A formal approach to stylistic change and innovation in the visual arts*. New York: Cambridge University Press.
March, L., ed. 1976. *The architecture of form*. Cambridge: Cambridge University Press.
Stiny, G. 1975. *Pictorial and formal aspects of shape and shape grammars*. Basel: Birkhauser.
Stiny, G. 1981. A note on the description of designs. *Environment and Planning B: Planning and Design* 8: 257–267.
Stiny, G. 1990. What is a design? *Environment and Planning B: Planning and Design* 17: 97–103.
Stiny, G. 1991. The algebras of design. *Research in engineering design* 2:171–181.
Stiny, G. 1992. Weights. *Environment and Planning B: Planning and Design* 19: 413–430.
Stiny, G. 1994. Shape rules: closure, continuity and emergence. *Environment and Planning B: Planning and Design* 21: s49–s78.
Stiny, G. 1996. Useless rules. *Environment and Planning B: Planning and Design* 23: 235–237.
Stiny, G. 2006. *Shape: Talking about seeing and doing*. Cambridge: The MIT Press.
Stiny, G. 2022. *Shapes of imagination: Calculating in Coleridge's magical realm*. Cambridge: The MIT Press.
Stiny, G., and J. Gips. 1972. Shape grammars and the generative specification of painting and sculpture. In *Information processing*, vol. 71, ed. C. V. Frieman, 1460–1465. Amsterdam: North-Holland.
Turing, A.M. 1936. On computable numbers with an application to the *entscheidungsproblem* November 30 and December 23, 1936, of the *Proceedings of the London Mathematical Society*.
von Neumann, J. 1966. Theory and organization of complicated automata. In *Theory of self-reproducing automata*, ed. A.W. Burks, 46–47. Urbana: University of Illinois Press.
Wittgenstein, L. 1991. *Remarks on the foundations of mathematics, 1937–1944*. Revised edition. Translated by G.E.M. Anscombe, eds. G.H. von Wright, Rush Rhees, and G.E.M. Anscombe, 189e. USA: Blackwell Publishers.
Wittgenstein, L. 1976. *Wittgenstein's lectures on the foundations of mathematics, Cambridge 1939*, ed. Cora Diamond. Chicago: The University of Chicago Press.

Acknowledgments

I am grateful to the Massachusetts Institute of Technology and, specifically, the Department of Architecture. I want to express my thanks to Dean Hashim Sarkis and the Head of the Department, Nicholas de Monchaux, for their enthusiastic support and for recognizing the academic significance of this anniversary edition, which summarizes years of exciting research. Nicholas de Monchaux provided the necessary funding for the project and gave me useful advice on how to proceed. I also want to thank his administrative assistant, Alejandra Huete, for always finding time in his busy schedule to arrange meetings with him.

The Open-Access edition of this book was made possible through the generous funding of the MIT Libraries. Katharine H. Dunn and Katie Zimmerman managed the agreement process with Springer. I am deeply grateful to them and to the MIT Libraries for their essential support in this project. Furthermore, I would like to express my appreciation to the editorial team of the *Mathematics and the Built Environment* book series at Springer and Birkhäuser. The chief editors, Kim Williams and Michael Ostwald, warmly welcomed the volume into the series and followed its progress. My sincere thanks go to Frida Trotter, my editor and the series coordinator, for her exceptional support and guidance in bringing this book to publication. I'm also immensely thankful for the dedicated efforts of Ravi Vengadachalam, the production editor, and Bhagyalakkshme Sreenivasan, the production project manager, whose contributions were vital to the book's implementation.

I want to thank Marilyn Levine for her contribution to editing the chapters in this volume. She has diligently reviewed all 27 chapters and provided detailed feedback to authors to enhance the clarity and precision of their writing. Marilyn has also reached out to authors individually to discuss and explore potential avenues for improvement. Her efforts have significantly enriched the quality of this publication— I am truly thankful for her dedication and commitment. Finally, I am indebted to Chris Earl, Thanos Economou, Terry Knight, Ramesh Krishnamurti, Djordje Krstic,

Rob Woodbury, and George Stiny for their help and advice. This was invaluable to improving the form and content of this volume.

Sotirios Kotsopoulos

Contents

Shape Grammars from Art to Calculating

Scutum Fidei 3
George Stiny

Shape Grammars and the Architect 21
Ulrich Flemming

The Geometrical Art of Lionel March 33
Philip Steadman

Grids: Schemas, Rules and Cases 53
Chris Earl and Iestyn Jowers

Reflections on Interpreting Shape Grammars 73
Ramesh Krishnamurti

Shape Grammars in Art, Craft, and Engineering

***How Is That?* Computing the Temporality of Drawing** 99
Terry W. Knight

Computing with Abstract and *Material Shapes*: The Case of Triaxial Basket Weaving 125
Benay Gürsoy and Mine Özkar

[Un]making with____ 149
Dina El-Zanfaly

Preserving Brand Identity in Engineering Design 163
Alison McKay, Hau Hing Chau, and Alan de Pennington

The Science of Shape Grammar-Based Product Design 185
Jonathan Cagan

Shape Studies

Shape Rules: Their Algebras, Grammars and Decompositions 195
Djordje Krstic

Seeing and Drawing Shapes 221
Sotirios Kotsopoulos

Flows: Composite Shape Rules 261
Rudi Stouffs and Dan Hou

**Arrangements Containing Shapes: Mathematical Features
and Their Use in Visual Calculating** 289
Alexandros Haridis

Shape Computation Research at the University of Sydney 317
John Gero

Computer Implementation of Shape Grammars

Shape Machine: Shape-Based Search and Replace in CAD 337
Athanassios Economou

Design Space Exploration in the Shape Machine Interface 363
Robert Woodbury

**From Lines to Curves: Implementation of Shape Embedding
for Circular Arcs and Circles in 2D CAD Systems** 377
Tzu-Chieh Kurt Hong

Rewriting Shape Rules 407
Heather Ligler

**Automating the Archaeological Reconstruction of Classical Greek
and Roman Architecture in Shape Machine** 433
Myrsini Mamoli, Yichao Shi, and Violet Cerbone

Shape Grammars and Artificial Intelligence

**A Generative Grammar for the Design of Adaptive Intelligent
Virtual Worlds** 457
Ning Gu and Mary Lou Maher

**What Does Shape Grammar Implementation Tell Us About
Artificial Intelligence in Architectural Design?** 483
Thomas Wortmann

The *Shape* of Generative AI 493
Onur Yüce Gün

Shape Grammars in Design and Education

Computing Chinese Architecture 523
Andrew I-kang Li

Using Shape Grammar as an Analytical Tool for Shelters in Protracted Refugee Camps: The Zaatari Camp Grammar 537
Dima Abu-Aridah, José P. Duarte, and Rebecca L. Henn

Designing with Visual Calculation is Child's Play 567
Derek Ham

EthnoComputation: An Inductive-Deductive Shape Grammar on Toraja Glyph 575
Rizal Muslimin

Afterword 591

Index 601

Contributors

Dima Abu-Aridah College of Arts and Architecture, The Pennsylvania State University, University Park, PA, USA

Jonathan Cagan David and Susan Coulter Head of Mechanical Engineering, George Tallman and Florence Barrett Ladd Professor, Department of Mechanical Engineering, Carnegie Mellon University, Pittsburgh, PA, USA

Violet Cerbone School of Architecture, Georgia Institute of Technology, Atlanta, Georgia, GA, USA

Hau Hing Chau School of Mechanical Engineering, University of Leeds, Leeds, UK

Alan de Pennington School of Mechanical Engineering, University of Leeds, Leeds, UK

José P. Duarte College of Arts and Architecture, The Pennsylvania State University, University Park, PA, USA

Chris Earl School of Engineering and Innovation, The Open University, Milton Keynes, UK

Athanassios Economou School of Architecture, Georgia Institute of Technology, Atlanta, Georgia, GA, USA

Dina El-Zanfaly School of Design, Carnegie Mellon University, Pittsburgh, PA, USA

Ulrich Flemming School of Architecture and ICES, (Institute for Complex Engineered Systems), Carnegie Mellon University, Pittsburgh, PA, USA

John Gero Computer Science and Architecture, University of North Carolina at Charlotte, Charlotte, NC, USA

Ning Gu UniSA Creative, University of South Australia, Adelaide, SA, Australia

Onur Yüce Gün New Balance Athletics, Inc., Boston, MA, USA

Benay Gürsoy College of Arts and Architecture, The Pennsylvania State University, University Park, PA, USA

Derek Ham Entertainment Technology Center, Carnegie Mellon University, Pittsburgh, PA, USA

Alexandros Haridis School of Architecture and Planning, Massachusetts Institute of Technology, Cambridge, MA, USA

Rebecca L. Henn College of Arts and Architecture, The Pennsylvania State University, University Park, PA, USA

Tzu-Chieh Kurt Hong School of Architecture, University of Kansas, Lawrence, KS, USA

Dan Hou School of Architecture, Southwest Jiaotong University, Chengdu, Sichuan, China

Iestyn Jowers School of Engineering and Innovation, The Open University, Milton Keynes, UK

Terry W. Knight School of Architecture and Planning, Massachusetts Institute of Technology, Cambridge, MA, USA

Sotirios Kotsopoulos School of Architecture, National Technical University of Athens, Athens, GR, Greece

Ramesh Krishnamurti School of Architecture, Carnegie Mellon University, Pittsburgh, PA, USA

Djordje Krstic Calabasas, CA, USA

Andrew I-kang Li School of Design and Architecture, Kyoto Institute of Technology, Tokyo, Japan

Heather Ligler School of Architecture, Florida Atlantic University, Ft. Lauderdale, FL, USA

Mary Lou Maher College of Computing and Informatics, University of North Carolina Charlotte, Charlotte, NC, USA

Myrsini Mamoli School of Architecture, Georgia Institute of Technology, Atlanta, Georgia, GA, USA

Alison McKay School of Mechanical Engineering, University of Leeds, Leeds, UK

Rizal Muslimin School of Architecture, Design, and Planning, The University of Sydney, Darlington, NSW, Australia

Mine Özkar School of Architecture, Istanbul Technical University, Şişli, Istanbul, Turkey

Yichao Shi School of Architecture, Georgia Institute of Technology, Atlanta, Georgia, GA, USA

Philip Steadman Bartlett School of Energy, Environment and Resources, University College London, London, UK

George Stiny School of Architecture and Planning, Massachusetts Institute of Technology, Cambridge, MA, USA

Rudi Stouffs Department of Architecture, National University of Singapore, Singapore, Singapore

Robert Woodbury School of Interactive Arts and Technology, Simon Fraser University Surrey, Surrey, BC, Canada

Thomas Wortmann Department for Computing in Architecture, Institute for Computational Design and Construction (ICD/CA), University of Stuttgart, Stuttgart, Germany

Shape Grammars from Art to Calculating

Scutum Fidei

George Stiny

Abstract How identities x $\rightarrow$ x in the embed-fuse cycle distinguish the Father, Son, and Holy Spirit in God to crack the mystery of the Trinity.

Whenever I finish writing a book, there are always things that I wish I'd said and things that I regret. In this short essay, I'll concentrate on the former—better to add to than to take away, especially when you're going for a few extra words. It was the same for Samuel Taylor Coleridge when his *Biographia Literaria* turned out to be too short. I guess I'm on the same path to write more. No one's said my *Shapes of Imagination: Calculating in Coleridge's Magical Realm* (MIT Press, 2022) needs to be longer, but there are occasional spots where a couple of additional words might help; this may clear things up to make some of my ideas easier to grasp—time will tell. What I have to say assumes a passing familiarity with what I wrote originally. This isn't asking too much. My book can be downloaded in its entirety for free—

https://direct.mit.edu/books/oa-monograph/5489/Shapes-of-ImaginationCalculating-in-Coleridge-s

with its Creative Commons CC-BY-NC-ND license. I'm an enthusiastic fan of open access. Nonetheless, the paperback version is a lot prettier, and it comes with one of J. M. W. Turner's watercolors on the cover. Maybe it's worth the price, especially when Turner is unexpectedly tied to Coleridge and calculating. My goal is to be explicit in a handful of places that I'm not—still, there aren't any direct references to my book. They're implicit, so that you can use the index to thumb/scroll through its pages. To me, that's one of the true pleasures of reading, trying words from here and there in an uncertain order to see what happens as things pop up and fit together. The cool, new stuff in this essay suggests where to look. And if you're lost, that's OK, too—everything falls into place as you go on.

    Let's start with readymades, found objects (objets trouvés) whether they're made (manufactured) or not. According to Leon Battista Alberti, this is where art begins,

---

G. Stiny (✉)
Massachusetts Institute of Technology, Cambridge, USA
e-mail: stiny@mit.edu

modestly in tree-trunks and clods of earth—first observing and then adding to or taking away. This is how calculating works, as well, with symbols, numbers, and code, and in fact, it's Turing-complete. But observing, adding to, and taking away do more in visual calculating for shapes and rules in the embed-fuse cycle

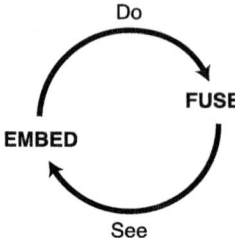

Of course, readymades extend past tree-trunks and clods of earth. There's an easy formula—readymades include whatever there is to see. This allows in smudges and stains, Rembrandts and Picassos, today's marvels of AI and machine learning that rely on gobs of data and labeled objects in vast training sets, and every stage of artistic work, no matter how stages are divided and measured. In painting, for example, a gessoed panel with nothing on it is OK, a single brush stroke or scant daub of paint, or a finished masterpiece on canvas. It takes imagination to make pictures and such; there's nothing before looking, that may involve adding to and taking away along with observing. Looking goes on and on in the embed-fuse cycle, where unfettered seeing begets unending surprise. The serious artist can't stop looking; it's a fulltime job in an ambiguous world of wonder and delight. (Ellsworth Kelly puts seeing first and foremost—"In my paintings I'm not inventing; my ideas come from constantly investigating how things look." David Hockney is this way, too, but maybe a little more relaxed—"I sit in the studio a lot, just taking in the pictures. I like being in here. A bed in the studio would suit me. It would be great. You need to do an awful lot of looking. I think unless you do that, you're not going to 'get' a lot of things.")

The relationship between Turing machines (computers) and shape grammars needs to be stated exactly once and for all. That Turing machines are a special case of shape grammars is easy to prove in a neat way, but the reverse of this, that shape grammars can be simulated in Turing machines, is mostly difficulties that need to be solved—some of them have nice solutions with new insights, but as many or more don't. The asymmetry between Turing machines (they're always 0-dimensional for points and symbols) and shape grammars (they're dimension $i \geq 0$ for points and symbols, and for lines, planes, and solids, etc.) shows the primacy of art including design and why shapes are a better choice than symbols. This is more of an aesthetic judgment than a strictly logical/rational one (Ockham's Razor, etc.), although logic needn't go for Turing machines every time. Even when the swerve to shapes seems effortless, it's not without crazy surprises; seeing isn't counting or consistent in the way numbers are. The result of calculating with shapes in the embed-fuse cycle needn't solve any given problem, even if it's spelled out in perfect detail, for example, x = 1 + 1 + 1 + 1. What happens if I add four triangles that move toward the common point on right angle axes, and then continue past it

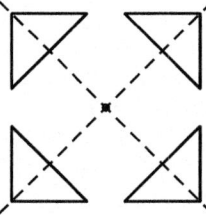

I may get four triangles or equally, two squares

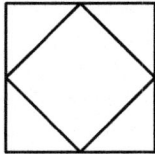

This looks right and feels wrong. Is there a hidden contradiction—do four triangles include two squares? Shapes may vary in number and kind, but no matter how I count, apples aren't oranges. (Aren't squares "unreachable," or is this Coleridge's thesis and antithesis in an indifferent synthesis?) And there's plenty more to see—pentagons, hexagons, big K's and little k's, and oodles of shapes without names. But let's go on, maybe to a four-bar trellis that looks one way diagonally and then another way horizontally/vertically. It's in a square in which the original triangles interact and are still easy to find

And there's a square and its diagonals that may form twin tetrahedra with the original triangles, like a Necker cube, or all the same, four little triangles that aren't hard to see, either going out of the page in a pyramid on a square base or into the page in perspective to vanish at infinity

Then, more is in view—a thin horizontal/vertical trellis crossing a wider diagonal one that's also horizontal and vertical triangles that exceed the added ones, or an eight-pointed star with two squares inside it, one inscribed in the other, where two

varieties of bowties/butterflies decorate the sides of the smaller square, or looking again, two kinds of bigger ones overlap its corners

And now, I can see a square and its diagonals in an alternative way and an octahedral spinning top twice

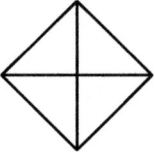

and if I look once quickly and decide to STOP, the one answer I've been waiting for—isn't this four triangles

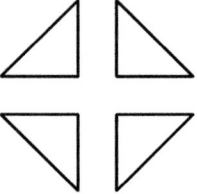

When I move four triangles parametrically on twin axes in a visual analogy (a spatial relation or the kind of "image" in Jean-Paul Sartre's *The Psychology of Imagination*), I expect four triangles in return. That's pretty clear, yet what I see and may try to count isn't always described this way. It isn't adding up triangles that are fixed in a word problem or BIM; it hinges not on what I've done but on how I go on, on the rule I try next in the embed-fuse cycle. That's why shapes are readymades. And that's why calculating with shapes seems so alive/vital—indifferent to what's gone before. It's how it is when things are strange and uncertain. No doubt, John von Neumann would put his faith in symbols (logic) before shapes (art). He assayed calculating this way during the Hixon Symposium at Caltech in 1948, hedging on things in domains where visual analogies are incomplete—for pictures and the Rorschach test that vary at will, swayed by personality and mood. Logic denies contradictions in visual analogies; it bars ambiguity in seeing to shield calculating from crazy sums like $1 + 1 + 1 + 1 = 2$ for triangles and squares in the second example in my series. Are visual analogies a must to calculate? Not in the embed-fuse cycle; it flouts visual analogies to use contradictions and ambiguity. There's less in visual analogies than meets the eye. Unless you do an awful lot of looking, you're not going to get a lot of things. That's for rules in the embed-fuse cycle, to get what visual analogies miss.

Shapes divide nonstop into parts (shapes) that appear and disappear without rhyme or reason—in triangles, squares, and a solitary line

---

John McCarthy tries to describe things on the phone in ways to make them—in a visual analogy or a parametric structure by filling in variables. The phone is quaint, and the idea isn't new. Giambattista Vico was the first to suggest that we understand what we can make. (And Richard Feynman 250 years later—"What I cannot create, I do not understand.") For McCarthy's AI and computers, understanding relies on descriptions (visual analogies) for making. Is this the same for data in DALL·E and 3-d printing? I guess some descriptions for making skip understanding, but which ones? No matter, descriptions for making omit a lot; they miss everything for von Neumann that's outside visual analogies in pictures and the Rorschach test, and for Oscar Wilde in a beautiful form, where one can put/read into it whatever one wishes, in "whispers of a thousand different things which were not present in the mind of him[/her] who carved the statue or painted the panel or graved the gem." Wilde's open-ended critical formula—to see things as in themselves they really are not—is a corollary of Coleridge's secondary imagination for aesthetic/poetic meaning and value. It's a keen way of talking about the embed-fuse cycle. Shapes can be made/remade simply by looking; seeing exceeds descriptions of underlying (parametric) structure in visual analogies—they inevitably fail. (A mate of mine at MIT told me to recant. Marvin Minsky was mad. He'd just read *Shape: Talking about Seeing and Doing* (MIT Press, 2006), and he didn't like it; it shows the limits of [symbolic] AI—"Stiny is telling us we're wrong. Speak to him as a friend, stop him before it's too late." *Shapes of Imagination* is apostasy twice.)

Some say Ivan Sutherland frames the difference between shapes that are dimension $i = 0$ and shapes of dimension $i > 0$ that's key in shape grammars. On most days, I'm OK with this and move on. But does Sutherland really go this far? He compares computer drawings and pencil and paper ones. The practical appeal of the former is that they're uniquely structured (represented) in visual analogies (Herbert Simon's constraints and hierarchies), and the downside of the latter is that they're unstructured without visual analogies to rely on, ambiguous in endless ways. Visual analogies are the root of CAD and today's BIM, but they do nothing for visual calculating. Sutherland's structured drawings in CAD are merely a special case of unstructured drawings with pencil and paper—it's visual analogies of dimension $i = 0$ in Turing machines vs shapes of dimension $i > 0$ in shape grammars. Sutherland doesn't try to calculate with pencil and paper drawings (shapes)—no one dares to. Nor does he notice that computer drawings are counted in pencil and paper ones. His target is traditional drafting. Nonetheless, drafting is related to computer drawings via visual analogies and shapes. Shape grammars unpack this relationship in terms of identity in dimension $i = 0$ and embedding in dimension $i > 0$. There's always a lot more to see in art and design—in open-ended calculating with shapes and rules in the embed-fuse cycle.

In the manner of I. A. Richards, I define a graph for a catchy children's song about counting stars and counting clouds to link God, Norbert Wiener's cybernetics, and my shape grammars—as unlikely a threesome as possible. My graph sets out the "differing types of interaction between parts of a total meaning"

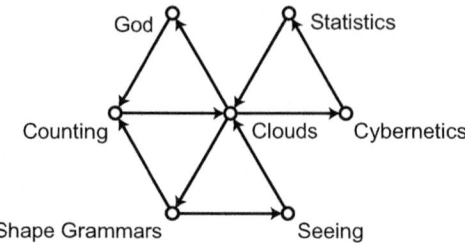

Many other dichotomies go for God (counting) and cybernetics (statistics)—today in computer science, for the now standard distinction between symbolic AI (axioms and logic/reasoning) and neural networks (machine learning with training sets)

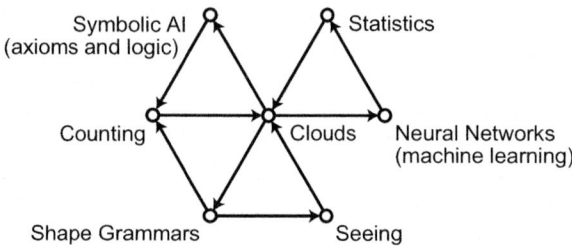

And in the way counting and statistics merge in representations, symbolic AI and neural networks interact jointly. Separate or not, logic and learning are equal in shape grammars, 0-dimensional throughout. (Total meanings aren't fixed in shape grammars; meaning alters freely in every pulse of the embed-fuse cycle.) But this isn't it for Richards. In *Principles of Literary Criticism*, he notes that "Coleridge's debt [to others] has been over-estimated. Such borrowings [thefts] as he made were more hampering to him than helpful" in his efforts to divide imagination and fancy. It's about the same for me. "Coleridge is very much on his own in the way he frames imagination—greater than anyone before him, expansive in the embed-fuse cycle." René Wellek doesn't think so; he "lists Coleridge's crimes in an indictment without exculpatory evidence or facts." No matter, the poet/critic has a ready reply—"the sense of Beauty overcomes every other consideration, or rather obliterates all consideration," as imagination "dissolves, diffuses, and dissipates [the facts], in order to re-create." One hundred years after Coleridge, R. Mutt answered a like charge of plagiarism in a similar way—imagination blurs the "useful significance" of ready-mades in new perception/thought. Today, Coleridge and Mutt have many ardent fans. In fact, Richards is Coleridge's biggest booster—once imagination catches on, "the order of our universes will have been changed." (Plagiarism is everywhere in art, for example, in Andy Warhol's campy Rorschach pictures. Each is a beautiful form, allegedly the result of painting half a canvas, folding it over vertically, and then opening it up before the paint sticks and dries. This one in gold indulges an over-the-top, simply divine baroque opulence

At first, Warhol assumed that patients made inkblots for psychiatrists and their avid disciples to interpret—"I thought that when you went to places like hospitals, they tell you to draw and make the Rorschach tests. I wish I'd known there was a [standard] set." Warhol's mistake was an artist's stroke of luck/genius, but it didn't tell him much about his own personality or anything else—

> I was trying to do these to actually read into them and write about them but I never really had the time to do that. So I was going to hire somebody to read into them, to pretend that it was me, so that they'd be a little more ... interesting. Because all I would see would be a dog's face or something like a tree or a bird or a flower. Somebody else could see a lot more.

And there really is a lot more to see—and a lot more to say. Looking may be key in making, but this is always incomplete, with countless things left unseen. That's why Warhol looks for a hired hand—a double who's an artist, or at least pretends to be. Are Warhol and his stand-in one and the same two-faced person/personality? Does this say anything about paintings and how to read into them? Is there an easy end to seeing and saying in pictures and poems that takes away the looking? This is the creed of AI, the Platonic myth of the eternal that visual analogies hold all experience, be it useful or aesthetic, in logical/statistical constraints and hierarchies in which there's nothing to learn. Or is it a combinatory ploy of the weak and squeamish, a generative dodge to avoid the anxiety of looking—to find the eternal in less than there is and vow that it's all that there is?)

Let's get back to graphs; they can elaborate a total meaning in myriad ways. What's one to make of the *Shield of the Trinity* (*Scutum Fidei*)? In Jerónimo Cosida's Renaissance rendition in the Cistercian Monastery of Tulebras in Navarra, Spain, a three-faced Trinity looks to its right, to its left, and straight ahead with four eyes shared in three ways, holding the shield in a common pair of outstretched hands—offering it, maybe as a ritual emblem to honor and remember the mystery of its unity in division and vice versa. And not to be outdone by Warhol and his stand-in who try for an amazing two in one, the Father, Son, and Holy Spirit trumpet their matchless world record of three in one in which they pop into view only to vanish, in and out of one another, together yet as in themselves they really are not—a unity, four eyes, three faces, two hands, one body, haunting in mystery and paradox. But words (visual analogies) aren't seeing

Nonetheless, the shield itself sidesteps the palpable wonders in God that I'm about to show. It's entirely abstract in the complete, planar graph K_4. Every pair of individually labeled vertices is connected by a single labeled edge; God occupies the center vertex of the graph, while the Father, Son, and Holy Spirit are each on a separate radius "est" (is) and are linked mutually by the opposite "non est" (is not). Saint Augustine describes this in *On Christian Doctrine*—

> The Trinity, one God, of whom are all things, through whom are all things, in whom are all things. Thus the Father and the Son and the Holy Spirit, and each of these by Himself, is God, and at the same time they are all one God; and each of them by Himself is a complete substance, and yet they are all one substance. The Father is not the Son nor the Holy Spirit; the Son is not the Father nor the Holy Spirit; the Holy Spirit is not the Father nor the Son: but the Father is only Father, the Son is only Son, and the Holy Spirit is only Holy Spirit. To all three belong the same eternity, the same unchangeableness, the same majesty, the same power. In the Father is unity, in the Son equality, in the Holy Spirit the harmony of unity and equality; and these three attributes are all one because of the Father, all equal because of the Son, and all harmonious because of the Holy Spirit.

There's a concrete visual analogue for this in which identities in the schema

$$x \rightarrow x$$

are used to alter how shapes look without changing them in any material way. After all, isn't God the aboriginal readymade—the "ground of being" in a clod of earth? (The Trinity defies equality to stump logicians and philosophers. It strikes me that the rapt and curious eye exceeds logic, reason, and thought. Who cares about counting and consistency if there's naught to see? Does anyone take logic seriously when seeing is believing without axioms and proof? Of course, if belief is too strong, it may take away the looking, but this doesn't last long—fast or slow, seeing alters belief.) God is a unity that appears in three ways. The Father, Son, and Holy Spirit are latent in a single image/shape that's undivided; they're ways of seeing, so that each is undeniable, and looks and acts its own way—not the three of them at once, but perceived separately one at a time, God's own duck/rabbit and 50% more, a kind of reversible figure as in Gestalt psychology. (It's easy to deny the Trinity—to believe in zero, one, or two, until Warhol hires someone to see three.) Ambiguity is two in one in reversible figures. Why—are they always similar to figure-ground relationships, maybe the Rubin vase that's twin silhouettes staring at one another across an empty divide, instead of the Rorschach test or a beautiful form? Is two in one proven science in Gestalt psychology that merits the respect of repeatability; an act of devotion to keep the majesty and power of the Trinity alone in the sacred number three; or prescriptive legislation to sway seeing, the visual etiquette and good manners favored by the cultivated few to educate rude yobs and shame the recently refined? (David Hume tries the last of these as a standard of taste and beauty.) The six relations in the *Shield of the Trinity* (4C2) unfold in visual analogies—in trees (hierarchies) like the ones linguists use to parse sentences into phrases and words—in the summation schema

$$x \to \sum F(prt(x))$$

The schema defines the identity for God, undivided and the same in the variable x and the sum of the terms $F(prt(x))$, as identities for $F(prt(x))$ individuate the Father, Son, and Holy Spirit. (Does this alter time and shift events? Are the secrets of seeing and looking in a new *Gospel of John*—in the beginning was God, and there were parts in God, and the parts were God.) As a readymade, God is a Rorschach test and a beautiful form, in a shape that's neither 0-dimensional nor empty. Could this be otherwise? I guess so, but odd quibbles and qualms are no excuse to pause; many shapes escape their taint. Let's try this one that may imply divinity

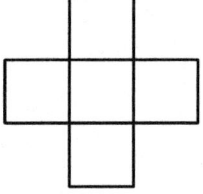

At least, it's an evocative shape for God—eight lines that are easy to draw with pencil and paper, and a T-square and a right triangle. For the polygonal identities

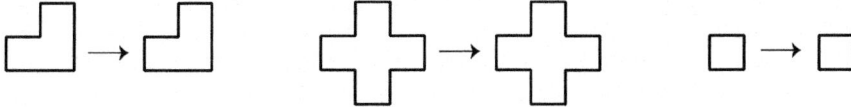

in the schema x → x, there are the three trees

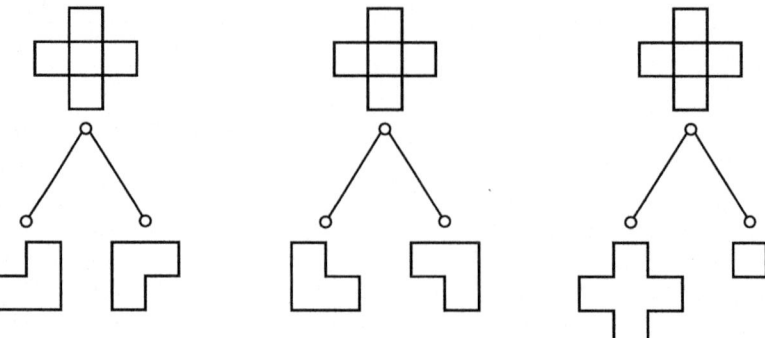

The Father is two L's on the left diagonal, the Son is two L's on the right diagonal, and the Holy Spirit is a Greek cross with a square inscribed at its center to align with its arms. (In pairs, trees are Coleridge's thesis/antithesis and synthesis/indifference). Then again, the two identities

put in a new similarity to define the Trinity in a π/2 clockwise rotation twice or with matching results, in a left diagonal reflection and in a right diagonal one

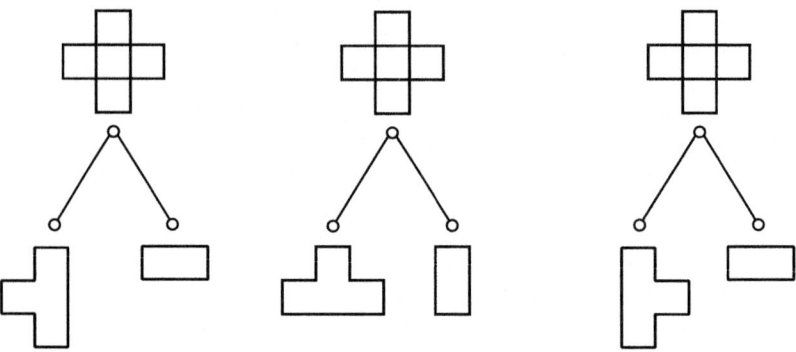

The Holy Spirit is the middle third, the Father on the left and the Son on the right—an intercalated version of the duck/rabbit that keeps the bilateral (left/right) symmetry of the Father and Son in my initial trees. But any trio of trees I choose—with symmetries and not—is a three-way duck/rabbit. It may take some work to see how trees structure shapes. Colors and such labels and weights help in identities to highlight parts, say, red for rectangles. The Father, Holy Spirit, and Son are similar yet distinct like the letters b, q, and d, and other triplets in b, d, p, and q

Scutum Fidei

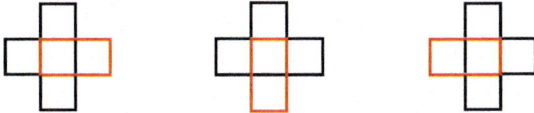

This series also describes the *Shield of the Trinity* counterclockwise looking head on, to show the bilateral symmetry of the Father and Son through the perpendicular bisector of its top edge that crosses the Holy Spirit—what do you see from behind the shield? Geometry in words is often evocative even if it's ambiguous or hard to grasp, incantatory in a sacred kind of magic, and entrancing beyond saying, so much more for rules to convey in the embed-fuse cycle. Geometry can be used to explain the Trinity, to validate clerical opinion and test it religiously. For example, maybe my series names the similar figures in Andrei Rublev's medieval *Troitsa*; the trio in this majestic icon can't not be on a circular arc—the bilaterally symmetric Father and Son measure God's reach on the longest chord, while the Holy Spirit in heavenly ascent tracks the highest point

There are six permutations in the Trinity (3!); five of them including mine show Rublev's *Troitsa* as in itself it really is not. Do permutations alter what I see? Yes, every time I look—give it a go and see for yourself. Still, this isn't for God who counts clouds and other fitful stuff (moving triangles) in Coleridge's primary imagination to ensure a faithful world in which things stack up the way they were to start, as in themselves they really are—constants in a parametric structure. At the center of the shield, God tallies permutations but for sure, this is all at once in an indifferent confluence that ignores Coleridge's secondary imagination and Wilde's critical formula. I guess God isn't an artist or a poet—and makes no effort to be. Primary imagination is how God keeps tabs on the world and rejects anything more. The following 4 × 6 array of shapes shows identities at work in the summation schema, as it's used to divide/resolve the Trinity all without dividing God

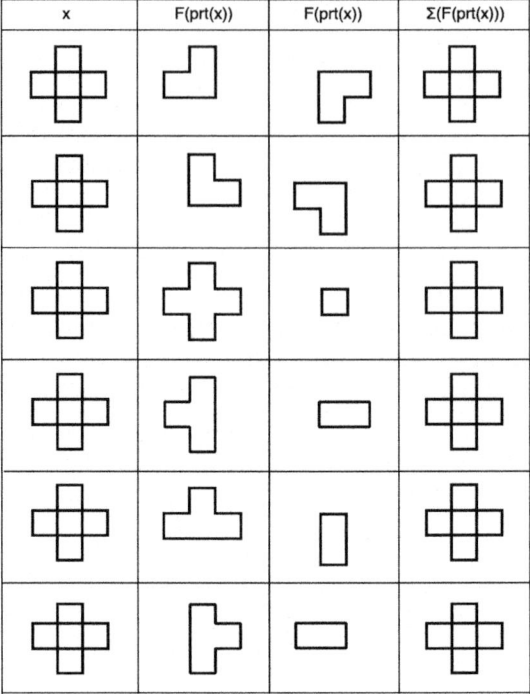

Every row is a tree—the end cells define the identity for God, and the middle cells show one of the Trinity. Rows begin and end with the same shape of God as parts vary. The Father, Son, and Holy Spirit appear in God and disappear in God. How God looks may alter in the world, but God doesn't change outside it—there's immanence and transcendence in the embed-fuse cycle. The two ways I show the Trinity are in the first three rows and in the last three rows—this is epiphanic (revealing/surprising) and influences how the eye feels in untold ways. (In fact, the summation schema does this recursively in two trees with four copies of God apiece—one copy is at the root and three copies are adjacent to it. Then, God divides in the Father, Son, and Holy Spirit.) The Father, Son, and Holy Spirit occur in numberless ways, no two alike. God is a Rorschach test and a beautiful form. (Does this vast variety in majesty and power finish off Saint Augustine's eternal and unchangeable? Even my array shows the Trinity in 3!(6C3) ways. Is this what God sees?) But the Trinity is a tentative start; there's less than three and more in four, five, six, etc.—are we in God like the Trinity, participating fully, everyone looking his/her own way? ("The Trinity, one God, of whom are all things, through whom are all things, in whom are all things.") This keeps the properties of the Trinity, except for the number three that's now a single value of an N-ity, as God absorbs, engages, and holds us all—one as many and many as one. Something in this vein also adds to the Father, Son, and Holy Spirit individually. (Shapes are all equally absorbing in one or two lines, a sketch, a Salvador Dali, or a computer DALL·E off the screen. There's the "infinite use of finite means" beyond Galileo's "discrete infinity" of atoms and words that tempts linguists and philosophers, and that may attract God. Alberti's tree-trunks and clods of earth with only identities to see exceed discrete infinities for recursive procedures in Turing machines, generative grammars, etc.—there's more in division and imagination than there is in combination and fancy. And the schema for parts and its inverse do as much, adding to with rules in $x \to \text{prt}^{-1}(x)$ or $x \to y + x$ and taking away with rules in $x \to \text{prt}(x)$ or $x \to y \cdot x$. These schemas intersect/meet in the identities $x \to x$, when $y = x$; and their dual compositions include all possible rules—for any rule $A \to B$, $A \to B \cdot (B + A)$ and $A \to B + (B \cdot A)$. Parts without atoms/words allow for recursion in the embed-fuse cycle to calculate visually, looking and maybe changing more. This quells the alarmist hype in AI. Like painting and photography, AI is a lush way of making. It leaves everything to see in readymades generated from verbal prompts and heaps of data—marks and stains to charm the eye. Of course, truth intrudes for pictures and shapes that don't yield to beauty in Wilde's critical formula. Then visual analogies are key, each with an eternal provenance. Beauty ignores origins, intentions, structure, and use "to wander in the region of the many and variable." For every shape, there's always a new visual analogy; personal experience never ends. AI uses words for pictures and shapes, and fails without them. That's for rules in the embed-fuse cycle, to see and do when there are no words. Aesthetic/poetic meaning and value of dimension $i > 0$ come before discrete infinities of dimension $i = 0$ in visual analogies and verbal prompts, scientific insight, routine language, and calm/cool reason.) Coleridge's esemplastic power and imagination blur old boundaries and divisions in order to re-create/re-divide/re-new; seeing is all magical surprises, as shapes (parts) fuse to embed different ones in an open-ended loop

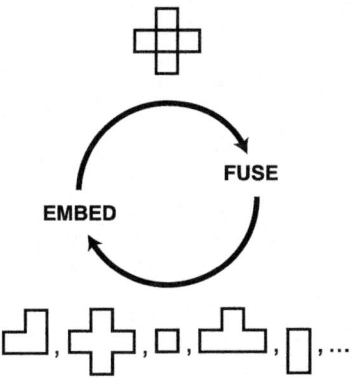

that's infinitely productive, trick after trick and surprise after surprise. Transformative experience in art (pictures and poems, etc.) and religion converge—if they were ever apart—coadunate in the embed-fuse cycle as imagination takes hold. Identities $x \to x$ in the summation schema trace an aesthetic (perceptual) arc. Wilde's *The Picture of Dorian Gray* is another example of this—not as a lurid Victorian tale ("stuff that were better unwritten") but as a history of how pictures change simply by looking (and how faces appear not to change by looking, too). The picture of Dorian Gray isn't a portrait or faithful description; it alters in sync with Dorian's transgressions and cruelties—not materially but aesthetically, as a beautiful form. There's a phantasmagoria of horror and mystery that ends only when imagination does; for whatever reason, Wilde's critical formula fails. Then, Dorian and his picture are seen as in themselves they really are, the latter a true portrait at last, immune to aesthetic interest and the uncharted risks of endless seeing, stuff to describe once and for all and to catalog only. For the first time, Dorian and his picture are the same—fixed and dead. This may be so, but the picture of Dorian Gray is somehow always a surprise, to recover brand-new whenever there's a chance to look, in motion, alive/vital. To invert Lord Henry in synonymic paraphrase, the visible is the true mystery of the world—this goes for Dorian Gray in a picture, for the Trinity in God, and for Andy Warhol and his stand-in, and then for the Rorschach test and beautiful forms, triangles and squares, and a single straight line. (Not everyone likes shape grammars. One big complaint that I hear all the time is that like art, shape grammars are useless, incorrigibly impractical—no more than a trivial child's toy. Yes, that's it precisely, that's where they're the most interesting in a kind of visual free play that's never structured and always open-ended; there's nothing like them anywhere I know. Practicality is a sure thing—it's the raison d'être for Turing machines, and shape grammars do them neatly as a special case, no problemo. Going from Turing machines to shape grammars to tackle real-world problems in alternative ways that are more efficient or easier to use may repay the effort; it's just not for me to re-frame what's already been done. Nor am I apt to explore new applications—I'm lost and at loose ends in the lush purposelessness of imagination, in a landscape forever in flux, in art for art's sake where meaning and value depend on seeing and personality. Is there anything wrong with this? Unless the impracticalities of seeing come first and foremost, shape grammars are a faint shadow of themselves, stripped of surprises and new intuitions—they're no more than Turing machines in disguise. And similar gripes abound. It's easy to worry about the embed-fuse cycle—it pops up in way too many ways—and to mistrust my visual tricks and their unintended results. They're self-indulgent, exploit perception without prediction, are hopelessly remote, extravagant, and far-out, run too many silly risks, are unprofessional, and rely too much on emotion from contempt to love and joy. I hesitate when it comes to emotion, not because I'm emotionless or because seeing and emotion aren't related—they are mutually, so that each influences the other. Rather it's because I don't have much to say about how and when seeing and emotion join together and interact. Isn't feeling the inevitable judge of seeing? Or is this better reversed? I'm sure it's meant in both ways at once, in a productive back and forth. Joy is a telling example; it begins and grows when I see afresh and at the same time, it's why I do. This is mostly ineffable—I'm apt to say something about how the eye feels and the tricks of imagination in the embed-fuse cycle, but I can't say everything. I do what I see—whatever I say is passing gossip, none of it lasts for long. Visual analogies are viciously incomplete.)

I've got some more to say/confess about plagiarism. I use the phrase "a wise teacher and an inculcator of good habits" to poke fun at pretentious (invidious) vocabularies of shapes (discrete infinities) that secretly censor seeing to encourage and improve hands-on education, social cohesion, mental health, and overall well-being, etc.—for children, there are Froebel's kindergarten blocks that combine in forms of life, knowledge, and beauty, and Lego bricks for resilient building in combinatorial play. The strict limits of vocabularies in everything that they exclude are what children, and artists, architects, and designers learn almost at once when what they see is more than they can do. No wonder I cut and glued Tinkertoy pieces as a child—impulsively most of the time but now and then, with a well-wrought plan clearly in mind. The words are Adrian Vermeule's, in "Beyond Originalism" in *The Atlantic*. I ignored the citation on purpose, to flout the law in a reckless act of adolescent rebellion. I'm pretty sure that no one is ever too old for this—it can be exhilarating and loads of mischievous fun. The

> law is parental, a wise teacher and an inculcator of good habits. Just authority in rulers can be exercised for the [common] good of subjects, if necessary even against the subjects' own perceptions of what is best for them—perceptions that may change over time anyway, as the law teaches, habituates, and re-forms them.

But this is entirely the wrong way around, something a fascist would advise—I speak to you as a dear friend, a benevolent father who knows what's best. "Just authority" in vocabularies and the rules they define just gets in the way. Vocabularies block seeing anything that can't be made combining what's in them; atoms, units, words, etc.—Coleridge's fixities and definites in fancy's counters that trace the "fixed and dead." New perception (seeing) re-forms the law soon enough, when vocabularies break down, collapse, and fail—when the embed-fuse cycle is needed to go on, for example, in the Trinity, to define, update, and revise vocabularies in retrospect. (Topologies and correlative structures in graphs and trees are involved in this, looking backwards. There's plenty of cool math, but I try hard not to show the details unless I'm asked repeatedly.) Most of the time vocabularies and rules are good habits to keep, sound commonsense to smooth out the expected bumps and jolts of daily life. Even so, rules are meant to break whenever Coleridge's secondary imagination kicks in, when the Greek spirit (aesthetics) and Wilde's beautiful forms hold sway despite the law. That's why there's an embed-fuse cycle, to make vocabularies and to break them as rules are tried—to replenish and refine vocabularies and to exceed whatever they allow in the flux of aesthetic/poetic experience. Imagination in the embed-fuse cycle re-forms fancy and its vocabularies of counters as it changes the order of our universes. Richards's opinion, however, comes as no surprise; in fact, the value of imagination was already certain beyond any reasonable doubt more than a century earlier in the judgment of the Honorable Percy Bysshe Shelley—"Poets [and artists] are the unacknowledged legislators of the world."

The opening lessons in Paul Klee's *Pedagogical Sketchbook* use drawings and words in heuristic schemas for art and design—in the manner of William Hogarth's serpentine and Johann Wolfgang Goethe's *Urpflanze*. My schemas for rules are similar in an algebra for shapes and their parts, transformations, and boundaries; at times, they define a lattice. Klee tests his schemas in the *Sketchbook*—even in the embed-fuse cycle. This may be a stretch, but Klee's schemas are worth trying as in themselves they really are not. Wilde's critical formula is best in lush fabrication.

It's the same for John Dewey in his incantatory logic of inquiry and experiment. (Dewey's *"definition* of Inquiry" in *Logic: The Theory of Inquiry* is closer to the embed-fuse cycle than his earlier riff on Coleridge in *Art as Experience.* In the latter, Dewey invokes esemplastic power "to characterize the work of imagination in art" as "a new and completely unified experience" in which there's "the welding together of all elements, no matter how diverse in ordinary experience." Super, but not Coleridge's imagination; elements and welds are left to see—nothing is fused to erase prior divisions and end the tedium of a stale vocabulary in order to embed anew. At most, this is "imagination" for William James, or a "total meaning" for Richards in a graph of "differing types of interaction [welds] between parts [elements]." There's Coleridge's fancy in dimension $i = 0$ without imagination in dimension $i > 0$—the embed-fuse cycle idles unengaged. I guess Coleridge's terms and "the vocabulary of his philosophic generation" sap Dewey's interest and his will to inquire. Esemplastic power is vital in art/calculating, but it's beyond Dewey's ken.) Klee begins lesson I.13 in the *Sketchbook* with "additive" ("stone upon stone") and "subtractive" ("chip from chip") schemas that mirror Alberti's adding to and taking away—for art in tree-trunks and clods of earth and for architecture in the firm and graceful ordering of lines and angles. (Use combines with firmness and grace in the Vitruvian canon.) Klee does more with this in the "receptive" and "productive" to and fro of the perceiving eye and the hand at work—"human action (genesis) is productive [fuses] as well as receptive [embeds]. It is continuity. ... The eye travels along the paths cut out for it in the [hand's] work," and the hand, no doubt, responds with silken skill to what the eye sees. The "creator controls whether what he[/she] has produced so far is good"—meaning and value inform seeing and doing. "Continuity" in receptive and productive "action" relies on rules, to see ("graze") and do in the embed-fuse cycle. This reads a lot into Klee, in the way Warhol's stand-in reads a lot into Warhol's pictures. My Klee may be more poetic (wishful) thinking than honest scholarship— and yes, this is how Coleridge marshals esemplastic power in imagination and fancy. My Klee avoids the bumps. In fact, Klee's two examples in I.13 aren't about seeing— they're 0-dimensional arithmetic, stacking building blocks and dividing triangles to add and subtract units in the successor function and the predecessor. Still, the nimble eye finds its way in drawings and sketches, seeing and doing in the artist's formula Eye and Hand. (Klee's "thinking eye" doesn't see; seeing/looking isn't thought in a cognitive process that's lodged in visual analogies in words, data, and parametric structures—this ends seeing and the tricks of new perception. Thought's limits in the rational mind blind the eye to take away the looking; the eye is for pictures that aren't cast in words—words only turn to stone.) It's a breeze to put Klee alongside Coleridge and Wilde in eddies of imagination and beautiful forms. And von Neumann is welcome, too, awed by pictures and the Rorschach test in excess of visual analogies. (Roger Penrose calls this creativity—"It's creativity if it's not acting within the scheme [vocabulary and visual analogies] we had previously.") Art and science are one culture—painting, poetry, and calculating are a single domain. An aesthetic/ creative trinity flashes in every direction; no two of the three are equal, yet they accord, coadunate in imagination

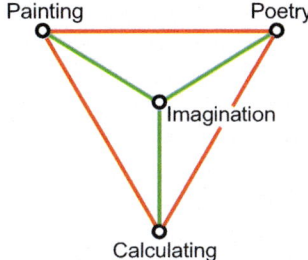

The vertices and edges in my graph make the four triangles in the *Shield of the Trinity*. To update this layout, I've put colors on edges for "est" and "non est"—keeping to best practice, there's green for "go" and red for "stop." For example, painting is calculating and vice versa, but not directly; it takes two steps that go first through imagination. Seeing and doing are alike in painting and calculating. Rules in the embed-fuse cycle are the receptive and productive impulse in both. (This is Saint Augustine in a secondary role—God replaced by imagination "of [which] are all things, through [which] are all things, in [which] are all things.") Many like to travel with companions to share the delicious rush of sights and sounds, and to see discomforts and calamities in funny ways. God already grounds a mysterious trio to fashion a common shield, while Klee, Coleridge, Wilde, and von Neumann are an odd gang of four—drifters and flâneurs, walking and talking for the fun of it, looking around aimlessly on an unending stroll for a breath of fresh air, alert with naught to know for good and all. Nothing blocks their way no matter what direction(s) they take—calculating in Coleridge's magical realm, free and at loose ends, prediction aside in an open-ended loop. Pathways and byways, divisions old and new, burst into view only to fade away—the eternal and unchangeable are long gone and forgotten. Everything is strange and wonderful in the unbounded region of the many and variable. This is how it is whenever imagination pulls us together in "sharp and eager observation." There are no fast and final goals looking and talking in this way; it's all about beauty—to see things as in themselves they really are not. There are shapes, and rules to try—adrift in the embed-fuse cycle.

Acknowledgements Maria Panagopoulou asked me about the Trinity on Easter Sunday (April 16, 2023). We agreed that it's a total mystery. But later, looking around for a third glass of wine, I thought about it again, this time in terms of visual analogues and how I could make them for God using identities in the summation schema. That was the origin of the main example in this essay and its Latin title—and of my answer to the mystery of the Trinity. No doubt, my answer isn't the eternal solution that God only knows, that's well-hidden out of sight—the way it is in computers for descriptions and representations (visual analogies) of pictures and drawings, etc. My answer relies on the tricks of strong imagination; it tests appearances and how the eye sees in endless ways. My Easter egg entered the festival's lists (τσούγκρισμα) hard-boiled, ready, and sure to win but secretly slipped away during the first round battered in tragic defeat, having clinked, knocked, and tapped for the last time. Still, rebirth is the rule at Easter. In a heartbeat, my egg joined Alberti's twosome to make an explicit threesome—a tree-trunk, a clod of earth, and a cracked egg. There's always a lot more to see; trinities are hard to ignore. Maria got it right—"mysteries are not to just be dismissed as such, but, like eggs, to be held, examined and, well, cracked!" Sotirios Kotsopoulos read the early drafts of this essay with good humor, a constant eye, and quick insights. He provided the occasion to talk to Maria and insisted I consider Samuel Taylor Coleridge in John Dewey's *Art as Experience*. Sotirios would be the first to agree that Coleridge's esemplastic power and his unique way of distinguishing imagination and fancy defy easy understanding outside the embed-fuse cycle. The embed-fuse cycle runs imagination through calculating with vast landscapes to explore and wild things to see.

Open Access This chapter is licensed under the terms of the Creative Commons Attribution-NonCommercial-NoDerivatives 4.0 International License (http://creativecommons.org/licenses/by-nc-nd/4.0/), which permits any noncommercial use, sharing, distribution and reproduction in any medium or format, as long as you give appropriate credit to the original author(s) and the source, provide a link to the Creative Commons license and indicate if you modified the licensed material. You do not have permission under this license to share adapted material derived from this chapter or parts of it.

The images or other third party material in this chapter are included in the chapter's Creative Commons license, unless indicated otherwise in a credit line to the material. If material is not included in the chapter's Creative Commons license and your intended use is not permitted by statutory regulation or exceeds the permitted use, you will need to obtain permission directly from the copyright holder.

Shape Grammars and the Architect

Ulrich Flemming

Abstract The architect in the title is the author, who traces how his perception of shape grammars evolved over 50 years. His particular focus is emergence and its role in the embed-fuse cycle. It allows us to describe, purely by graphical means, the way designers derive shapes from shapes. This is a great advantage because shapes cannot be described, at a sufficient level of detail, using words and as a consequence, the operations on shapes designers perform can also not be described effectively with words. The embed-fuse cycle pulls these operations out of the nebulous realm of "designer's intuition" and makes them intelligible as a form of reasoning on a par with reasoning using language—there is thinking outside of language. The cycle is entirely non-deterministic, which is one of its strengths. But this can turn into a hindrance, for example, when we try to develop a shape grammar that reproduces a given corpus of visually related designs in order to elucidate the underlying design logic *post festum*—emergence gets in the way. A related issue arises when we compare shape grammars with the traditional notion of an architectural grammar or language, as it has been employed by architects (notably F. L. Wright) and architectural historians alike. Central to these grammars is the concept of a limited vocabulary of shapes from which a design can be built, giving it formal coherence. Shape grammars don't usually describe such a vocabulary independently of its use, which has conceptual and practical implications.

U. Flemming (✉)
Carnegie Mellon University, Pittsburgh, USA
e-mail: ujf@andrew.cmu.edu

Seeing comes before words[1]

Two monographs introduced me to shape grammars in 1975.[2] At the time, I was finishing my dissertation on the generation of rectangular dissections, where I used recursive rewrite rules to generate these types of plans. Given such rules, it was easy to prove that they could, in fact, generate all possible configurations of a given number or rectangles (by induction on their number), an issue that had not been satisfactorily dealt with up to then.[3] I had been introduced to recursive rewrite rules through Chomsky's *Syntactic Structures*[4] and had come to appreciate them for the combination of power and elegance that characterizes them. I was, therefore, ready for shape grammars as a form of computation.

But I was also ready for them through my experience as an architect. I had observed numerous times how architects explain the logic behind a design through a sequence of sketches, in which a starting shape is subjected to a series of transformations or modifications, each plausible in its own right, leading to the final form. Shape grammars seemed to be able to capture this process. In addition, the form of the rules reminded me immediately of the process of design exploration I was familiar with, where tracing paper is used to copy parts of a design that was to be left unaltered (for the time being) and to add modifications indicating the intended change. Arranging these sketches in sequence produces a record of the underlying design process as a succession of transformations.[5]

I considered applying the shape grammar formalism to the work of an architect that was intuitively recognizable as his.[6] Stiny and Mitchell beat me to the punch,[7] which I took as encouragement. I demonstrated that the process of step-wise refinement they used to generate Palladian plans could be adapted to generate plans in the popular tradition.[8] And I found out that you could tell a detective story with shape grammars by dropping a profusion of clues leading to a grand denouement.[9]

---

[1] Berger et al. (1977, p. 7)

[2] Stiny (1975), Gips (1975). As a bow to George Stiny, I adopt in this essay some of his idiosyncrasies as a writer, a fondness for parentheses and lengthy footnotes, and a willingness to add bits, and sometimes whole chunks of autobiography to the simmering stew. (I use "essay" in the original sense of the term, derived from French *essayer* (to try) and denoting a presentation of ideas that does not claim to be the final word on the issues.)

[3] Flemming (1978).

[4] Chomsky (1957).

[5] To the present day, I'm puzzled by the fact that architects who object to the idea of shape grammars, on principle, fail to see these similarities.

[6] At the time, I knew of no female architect whom I could study in this way.

[7] Stiny and Mitchell (1978).

[8] Downing and Flemming (1981). It was crucial for this work that Stiny had extended the formalism to include parametric shape rules and labeled points in Stiny (1977).

[9] Flemming (1981). Note that at the time, Stiny had not yet extended the formalism to dimensions higher than 1. I therefore had to present the denouement, i.e. the rules that extrude a plan 3-dimensionally, in words. I never had the time to get back to the article and update it. But that may have been a good thing anyway because it may have diminished the surprise I had intended with the denouement.

Emergence was a problem in all this work. I found it easier to make a grammar do what I wanted it to do than to prevent it from doing what I did **not** want it to do. Quite a bit of formal machinery was needed to accomplish this. In a conversation I had about this with Stiny, he laughed and said, "You had to clobber it [emergence] wherever it raised its ugly head." Be this as it may, exercises like these can provide insights into the logic underlying a corpus of works that conventional descriptions (text and pictures) do not reveal.[10]

I learned to appreciate emergence from a different direction, which I want to introduce with two anecdotes. I once audited a course on machine learning at CMU, where we discussed in one lecture the problem of defining concepts symbolically, i.e. using words. Thus, when an elephant is grey with floppy ears, a trunk, four legs etc., what happens when we come across a white one, or one who has lost a leg? We start adding exceptions, and they soon pile up until we reach a stage where the exceptions outnumber the original description. It struck me that this is the wrong way of going about it from the start: Someone who has seen a single picture of an elephant will recognize one ever after, including all (or most) exceptions. Note that there is no training set involved, carefully designed or otherwise. A single image suffices to build an internal model of an elephant as a recognizable shape that is, at the same time, flexible enough to account for a large range of variations (at least that's what I think happens). The upshot for the present context is that shapes cannot be described sufficiently, or effectively, with words.

A second anecdote: a photographer friend once remarked that there is no thinking outside language (he was repeating what is a tenet in some schools of thought[11]). I protested by saying that when I lift my camera to shoot a scene that strikes me as promising, I don't do this after having reasoned, in words, why I think this is a good idea—that would take too long, and my response is instantaneous. Figure 1 shows an example: When I took the picture of a collapsing sail during a cruise on a sailing ship, I did this because the moment I saw it, Umberto Boccioni's sculpture "Unique Forms of Continuity in Space" had come to my mind.

I believe this type of mental process should not be dismissed as pure "intuition." This is even more true for what goes on in the head of designers when they derive a design through a series of transformations. I insist, emphatically, that this is a form of thinking. Nobody doubts that the mental operations by which statements are derived from other statements expressed in words (inferences) are a form of reasoning. Designers create designs by a sequence of mental operations involving shapes, and I see this also as a form of reasoning. This is not far-fetched when we realize that "spatial reasoning" is a category tested for in many aptitude tests. I see no reason why we should call one of these processes thinking, but not the other.

---

[10] I had made this a main criterion when I reviewed papers presenting a shape grammar that generates a corpus. I was always looking for at least one *Aha-Erlebnis*, a point where I would say, "Ah, that's how it works!" I have not been successful, though, when I tried to interest architectural historians in this type of analysis. It is clearly so far outside the paradigm they are working in that they cannot see the promise.

[11] See, for example, Langer (1976, p. 87).

Fig. 1 Boccioni's "Unique forms in space" and a photo inspired by it. The photo on the left is used under the Creative Commons CC0 1.0 Universal Public Domain Dedication. The image on the right was taken from the collection in Flemming (2012)

Am I just batting around definitions? I don't think so. If two things are the same, at some level of abstraction, they should be given the same name appropriate for that level. I have to admit, though, that I am guided here also by a strong voluntaristic impulse: I **want** design to be included in the "cogito" in "cogito ergo sum", that is, to be given the same existential status—call it "pride"!

I need not worry because one can arrive at the same conclusion methodically. Langer does this, based on her general theory of symbolism, which

> ... rather than restricting intelligence to discursive forms and relegating all other conception to some irrational realm of feeling and instinct, has the great advantage of assimilating all mental activity to reason, instead of grafting that strange product upon a fundamentally unintellectual organism.[12]

Where do shape grammars enter the discussion? If shapes cannot be described with words, we cannot describe with words what visual thinking consists of. It would be like describing a ballgame to a radio audience–listeners get a sense of what is going on, but cannot **see** the action and therefore cannot comment on or analyze it.[13] And

---

[12] Langer (1976, p. 143).

[13] Even a seemingly simple operation like translating a rectangle, expressed verbally like that, needs additional information, like the axis of transformation and the direction and distance of movement

this is where I finally paid attention to emergence, in the positive sense, as something to be embraced, not avoided. The genius of shape grammars lies (1) in the fact that they use recursive rewrite rules (in all their power and elegance) which, at the same time, are specified graphically, through shapes (no verbal description is needed); and (2) in the use of a representation of shapes that supports emergence. If we put the two together, we get the embed-fuse cycle, by which designers alternate between seeing and doing.[14]

Stiny hardly ever uses the term "visual thinking" in his writings. He calls the manipulations on shapes that designers perform "calculating." For a long time, I had problems with this because I just could not see any similarity between what my pocket calculator does (crunching numbers) and what I do while sketching. But I do not want to quibble with definitions and accept, therefore, Stiny's notion of calculating as the core of what designers do when they manipulate shapes. After all, "calculating" is recognized as a form of thinking when we talk about a "calculating" person (who may do too much of it). There is one proviso, though: if we accept that designers calculate, we also have to accept that poets calculate when they write or revise a poem, on paper or in their head (emergence is certainly involved). Be that as it may, in no case do I want to imply (nor would Stiny, I'm sure) that calculating with shapes is all there is to visual thinking.[15]

Shape grammars, then, are an important construct when we try to pull designing out of some nebulous notion of intuition and conceptualize at some level of detail what is going on. But they are meant to be more than a conceptual tool. They are meant to be of importance also at the practical level, i.e. assist designers in their work. For the rest of this essay, therefore, I would like to stretch them a little to see how far they cover the field of architectural design as I know it.[16]

Let me start with emergence. When I draw a rectangle inside another rectangle to represent a window in a wall, I immediately see eight additional rectangles created

---

along that axis. This is easier drawn than described in words. It's impossible to describe the majority of the rules of the published grammars in words.

[14] Stiny (2022, pp. 23–24). This was a remarkable leap of the imagination by the inventors of shape grammars. Ivan Sutherland, the father of computer graphics, had pointed out that a computable representation of a drawing needs to be structured (in terms of distinct geometric elements related to other such elements in an explicitly described topological structure), as opposed to a drawing, which has no inherent structure ("Pen and ink…only make dirty marks on paper") and can be interpreted by designers any way they want (Stiny 2006, p. 61). Shape grammars put the open-endedness back into the representation, but, in true dialectical fashion, at a new level, one that preserves computability.

[15] I have to get back to elephants because there remains one standing in the room, and that is the question, what is a shape? This is not an issue with shape grammarists because for them, a shape is a finite, unordered collection of maximal elements. But I doubt that we perceive the shapes we see around us in these terms. I could not review, for this essay, the vast literature on the subject, which spans several fields, from philosophy to psychology and neuroscience. I leave it at that because when I talk about shape grammars, I do not need to come up with a definition of my own.

[16] It goes without saying that shape grammars do not deal with the totality of architectural practice. A useful discussion of shape grammars has to stay within the realm of shape manipulation.

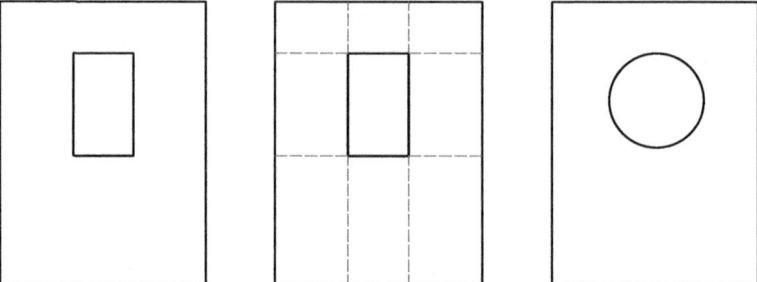

Fig. 2 Implied rectangles in a drawing

by the extensions of the four sides of the window (see Fig. 2).[17] I see these rectangles far less clearly than I see the emerging triangles in a star, but I see them nevertheless. Other architects seem to see them, too—at least that's the explanation I often have when I see them choose a circle instead of a rectangle for a window: It leaves the surrounding surface visually unstructured.

Shape grammars do not see these imaginary rectangles. They would if maximal lines were really maximal lines, as opposed to maximal line segments. But of course, I do not suggest that shape grammars should be extended to include these; that would open a can of worms better left unopened. I just want to point out that designers sometimes see things that are simply not there.

Except for design aids like LeCorbusier's "regulating lines" etc., shapes in an architectural drawing (or model) represent physical or spatial objects located in 3-dimensional space. When worked on in a 2-dimensional medium (like paper or a computer screen), they have to be projected on a picture plane. As a consequence, parts of the boundary of one object may merge with parts of the boundary of another object. In this case, we cannot erase parts of lines indiscriminately because we may remove more than we want. The reason is that

$$\dim(el) > \dim(em).$$

That is, the dimension of an element is greater than that of its embedding.[18] The reason for this, in turn, is that the shapes in the representation do not depict themselves, but physical objects. For the shape grammarist, a triangle in a shape **is** a triangle (rather than the representation of the mathematical concept "triangle"), or so it seems. But in architecture, we cannot neglect the distinction between an object and its representation.

Of course, we could avoid the problem by demanding that a projection show only the elements in the picture plane (which may be the intersection of a shape with that plane). This may work well when we work on an elevation, where I see

---

[17] It is, basically, the nine-square plan John Hejduk made famous through his assignments at the Cooper Union.

[18] See the discussion in Stiny (2006, pp. 77ff).

particularly promising opportunities for shape grammars. But it remains a crutch; it's like chopping the legs off a body to make it fit into a given bed.

Finally, I would like to discuss the concept of a vocabulary in the context of design. Stiny accepts it, barely, as something that can be distilled from a design in retrospect and may be useful for explanation and teaching, but sees no place for it otherwise.[19] His objections are two-fold and complementary. Following Langer, he sees no usefulness of the concept for an understanding of the visual arts because we cannot find constituent parts in artefacts from that realm ("Photography ... *has no vocabulary.*"[20]). This meshes neatly with the fact that the embed-fuse cycle does not accommodate such pre-given constituents—Fusion obliterates, immediately and relentlessly, the distinctions that appeared during emergence.[21] More generally, a pre-given vocabulary would get in the way when we "calculate in Coleridge's magical realm" and exercise our "negative capability"[22]—in short, it would stifle creativity.

I believe that we have to reconsider both objections for the case of architecture. Right away, we observe that a building, unlike a photograph, is prima facie parsable because it is made up of distinct, identifiable physical elements. The question is, **how** should we parse it? Consider Rietveld's Schröder House, an icon of the early modern movement.[23] We perceive it as being put together from a small set of element types: Rectangular panels placed horizontally or vertically to form floors, roofs, and walls; and a few types of linear "sticks" forming posts, railings, window mullions etc. These elements are 3-dimensional versions of the lines and rectangles in a Mondrian painting and taken from the aesthetic program of the de Stijl movement, with which both Rietveld and Mondrian were affiliated at certain times.

Rietveld described explicitly the rules he employed for placing the elements in the sketch shown in Fig. 3 (the drawing also shows the compositional rules for the Red-Blue Chair). They can be summarized as "Do not fuse!" That is, each element has to maintain its visual integrity.

I submit that the collection of element types and the rules for their composition can be viewed as visual equivalents, respectively, of the vocabulary and syntax of a natural language. It is an analogy that gives us a way to divide the elements we see into classes and, once we have the classes, to focus on the relations between the elements so classified (between panels and panels, sticks and sticks, and panels and sticks in the case of the Schröder House); that is, it gives us a way to parse a building when we look at it as a work of art.[24] Jaffé uses the analogy when he states, in his introduction to a monograph on the movement, "De Stijl ... redefined the vocabulary

---

[19] Stiny (2006, p. 19).

[20] Langer (1976, p. 95).

[21] See the discussion in Stiny (2006, pp. 89–98).

[22] Stiny (2022, p. xvi).

[23] Rietveld (1965).

[24] Note that the relations are perceptual and purely **visual**; they are not necessarily identical with the physical connections holding the pieces together. The distinction is nicely illustrated by the criticism often leveled against the Red-Blue Chair, that it defies practices of good furniture-making.

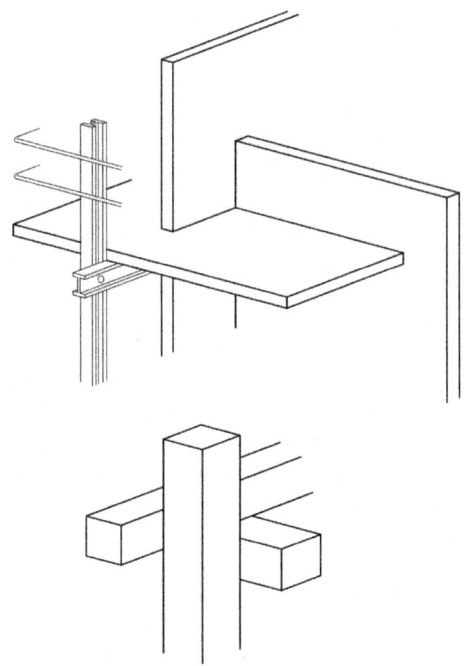

Fig. 3 Composition rules for Rietveld's Schröder House and Red-Blue Chair. Source: Rietveld (1965). Drawn by the author after Rietveld's sketch, which cannot be shown here for copyright reasons

and the grammar of the visual arts."[25] Do we want to chide Rietveld for lacking in creativity when he accepts this grammar as given and runs with it?

The Schröder House is an exception, both in the work of Rietveld and as the purest realization of De Stijl principles we have in architecture. But it is not an exception when it comes to demonstrating the usefulness of the grammar analogy for the conception and perception of architecture. Frank Lloyd Wright devotes an entire chapter to it in *The Natural House*. The chapter is titled "Grammar: The House as a Work of Art" and begins like this:

> Every house worth considering as a work of art must have a grammar of its own. "Grammar," in this sense, means the same thing in any construction—whether it be of words or of stone or wood. It is the shape-relationship between the various elements that enter into the constitution of the thing. The "grammar" of the house is its manifest articulation of all its parts.[26]

It is indeed the case that most of the houses Frank Lloyd Wright designed can be dated (plus or minus a few years) because the phase in which they were designed is characterized precisely by the vocabulary Wright used at the time, each with a concomitant set of composition rules.

---

[25] Jaffé (1982, p. 15).

[26] Wright (1954, p. 181). *The Natural House* is among the handful of books that impressed me most as a student. Perhaps my receptiveness to the notion of grammars in architecture stems from there.

When I visited the Schindler House in West Hollywood,[27] I was delighted when I realized that the architect designed the house as well as the furniture inside, using for both the same vocabulary and syntax—they were "talking to each other". I can think of no other expression that would express so succinctly the reason why I was so delighted (and the analogy has become a metaphor).

I am approaching a danger zone here. My phrase may suggest that an "architectural language" is a means of communication, which is a major function of a natural language. It is not.[28] This is not to say that a piece of architecture has no meaning; it is not "syntax without semantics". However, this is a topic outside the scope of this essay.

What I want to establish is the usefulness of the grammar analogy when we look at a building (or the design of a building) as a work of art.[29] It does not really matter if the grammar was given at the outset; if it emerged during design; or if the architect actually thought in such terms, i.e. if it is the result of an interpretation in retrospect. The analogy pops up time and time again and appears to be irrepressible. I have suggested elsewhere that there is a reason for this because there are deeper similarities between architecture and language.[30] Both are engaged in the production of **recognizable patterns**, sentences that can be understood because they follow rules that can be expressed in a grammar, on the one hand, and architectural compositions which can be recognized as non-arbitrary configurations of forms that are non-arbitrary precisely because we can recognize an underlying logic, expressible as a grammar with an associated vocabulary often enough to make the analogy useful.[31]

It should now be clear why I do not want to relegate the notion of a vocabulary to the margins of the discussion. One may point out, though, that all this talk about whether shape grammars can or cannot accommodate a vocabulary is for the birds and bees anyway because a shape grammar can definitely generate designs based on a vocabulary given at the outset—we can describe vocabulary elements in the right-hand sides of rules, which place them according to the associated syntax. In fact, shape grammars are uniquely apt at describing rules of composition because the rules deal with spatial or visual relations, which are easier to describe graphically than symbolically. The published shape grammars amply show this. This is true, but remains conceptually unsatisfactory, to me, because a shape grammar usually does

---

[27] Smith (1993).

[28] Scruton (1970, Chap. 7).

[29] Note that the use of "grammar" in "shape grammar" is not based on an analogy, but on the fact that is shares its basic computational mechanism with linguistic grammars like Chomsky's phrase structure grammar, both derived from Post production systems. See Gips and Stiny (1980).

[30] Flemming (1994).

[31] A deeper affinity between language and architecture emerges also when we realize that a language is not only a means of communications, but also a means of expression. The repertoire of forms artists use in their work is called in German their *Formensprache* (form language). The term implies that such a repertoire is more than just a catalog or kit of parts—it has expressive power when put to use.

not recognize a vocabulary of shapes as something that exists independently of its use; in fact, it decidedly resists doing this.[32]

This has also practical implications. Perhaps the best-known grammar I have published is the grammar of Queen Anne Houses.[33] I developed it to the point where the basic geometry of a house had been generated and was ready to be decorated by ornaments, which would turn the house into a Queen Anne House as we know it.[34] Essential elements would have to be the posts supporting the roofs of porches. They typically have a base, a shaft, and a top, each of which can be articulated in various ways. The concrete forms reflect the properties of the tools used in their fabrication (an example is a lathe that gives a round piece of wood a characteristic profile). It would be interesting to investigate the relationship between the forms and tools by developing a shape grammar that produces such posts—it would be fun.

In order to place such a post at all the desired locations, we would have to do something like marking these locations and then using these marks to trigger the rules that generate the description of a post, from scratch, at each mark. There is nothing wrong with this conceptually, but it is undesirable from a computational point of view. We would like to generate the description of a post once and then put it into a repository or library, from where it can be fetched and instantiated wherever it is needed. I do not see how this can be done with a shape grammar proper.[35] And with that, I have reached a point on the boundary I have been looking for.

References

Berger, J., et al. 1977. *Ways of seeing*. New York: Penguin.
Chomsky, N. 1957. *Syntactic structures*. The Hague: Mouton.
Downing, F., and U. Flemming. 1981. The bungalows of buffalo. *Environment and Planning B: Planning and Design* 8: 269–293.
Flemming, U. 1978. Wall representations of rectangular dissections and their use in automated space allocation. *Environment and Planning B: Planning and Design* 5: 215–232.
Flemming, U. 1981. The secret of the Casa Giuliani Frigerio. *Environment and Planning B: Planning and Design* 8: 87–96.
Flemming, U. 1987. More than the sum of parts: The grammar of Queen Anne houses. *Environment and Planning B: Planning and Design* 14: 323–250.

---

[32] As I look at this sentence, an interesting question emerges: **Should** we consider a vocabulary independently of its use? I cannot pursue this further here. Note also, as Stiny has pointed out to me, that 0-dimensional grammars can/do rely on antecedent vocabularies, for example, Froebel's building gifts.

[33] Flemming (1987).

[34] I had to stop there not because I wanted to, but because I had run out of funding. But I was thinking ahead.

[35] CAD systems have offered this capability for decades. For this and other reasons, I have always envisioned a shape grammar interpreter to sit on top of a CAD system providing two modes of computation, one symbolic, one graphical (i.e. visual), between which designers can alternate as they please.

Flemming, U. 1994. Get with the program: Common fallacies in critiques of computer-aided architectural design. *Environment and Planning B: Planning and Design* 21: s106–s116.
Flemming, U. 2012. *Found art around the world.* Available as iBook (ISBN 978-0-9852535-0-9) and Kindle Book (ISBN 978-0-9852535-1-6).
Gips, J. 1975. *Shape grammars and their uses.* Basel: Birkhäuser.
Gips, J., and G. Stiny. 1980. Production systems and grammars: A uniform characterization. *Environment and Planning B: Planning and Design* 7: 399–408.
Jaffé, H.L.C. 1982. Introduction. In *De Stijl 1917–1931. Visions of Utopia,* ed. M. Friedman, 11–15. New York: Abbeville.
Langer, S.K. 1976. *Philosophy in a new key,* 3rd ed. Cambridge: Harvard University Press.
Rietveld, G.T. 1965. Das Schröderhaus in Utrecht. *Bauen+Wohnen* 20 (11): 425–430.
Scruton, R. 1970. *The aesthetics of architecture.* Princeton: Princeton University Press.
Smith, K. 1993. The Schindler house. In *RM Schindler: Composition and construction,* eds. L. March and J. Sheine, 114–123. London: Academy Editions.
Stiny, G. 1975. *Pictorial and formal aspects of shape and shape grammars.* Basel: Birkhäuser.
Stiny, G. 1977. Ice-ray: A note on the generation of Chinese lattice designs. *Environment and Planning B: Planning and Design* 4: 89–98.
Stiny, G. 2006. *Shape. Talking about seeing and doing.* Cambridge: The MIT Press.
Stiny, G. 2022. *Shapes of imagination. Calculating in Coleridge's magical realm.* Cambridge: The MIT Press.
Stiny, G., and W.J. Mitchell. 1978. The Palladian grammar. *Environment and Planning B: Planning and Design* 5: 5–18.
Wright, F.L. 1954. *The natural house.* New York: Bramhall House.

Open Access This chapter is licensed under the terms of the Creative Commons Attribution-NonCommercial-NoDerivatives 4.0 International License (http://creativecommons.org/licenses/by-nc-nd/4.0/), which permits any noncommercial use, sharing, distribution and reproduction in any medium or format, as long as you give appropriate credit to the original author(s) and the source, provide a link to the Creative Commons license and indicate if you modified the licensed material. You do not have permission under this license to share adapted material derived from this chapter or parts of it.

The images or other third party material in this chapter are included in the chapter's Creative Commons license, unless indicated otherwise in a credit line to the material. If material is not included in the chapter's Creative Commons license and your intended use is not permitted by statutory regulation or exceeds the permitted use, you will need to obtain permission directly from the copyright holder.

The Geometrical Art of Lionel March

Philip Steadman

This paper is dedicated to the memory of my friend and colleague, Lionel March, in gratitude for all our many conversations about architecture, art, and mathematics.

1 Introduction

Lionel March (1934–2018) studied mathematics and architecture at Cambridge University, and devoted his career to mathematical and scientific research on buildings and cities. He was at the same time a historian, working on Frank Lloyd Wright, and on Renaissance arithmetic and proportional theory. In 1971 March and I wrote a book together on architecture and discrete mathematics, *The Geometry of Environment*.

I admire March for his many abilities and achievements, but for three things above all: first, his fundamental work in the 1960s with Leslie Martin on the relationship of density to the forms of buildings; second, his ability to enthuse and lead university research groups including the Centre for Land Use and Built Form Studies at Cambridge which took its name from the density work, and later groups at Waterloo in Canada, and the Open University; and third, I admire March for his paintings and digital artworks.

This essay is a celebration of March's art, which has not, I believe, received the critical attention that it deserves. He produced pictures in the tradition of geometrical abstraction at the two ends of his life: in England in the 1960s and in California when he retired from UCLA. The structure of these works is mathematical, inspired in part by serial music, and drawing on the combinatorial mathematics that we expounded

---

P. Steadman (✉)
University College London, London, UK
e-mail: j.p.steadman@ucl.ac.uk

in *The Geometry of Environment*, in particular planar symmetries and proportional systems.

I wish I had talked more with March about his pictures. I have tried here to infer some of the thinking and geometrical processes behind the conception and realisation of these works as mathematical "machines". March published one short article in *Architectural Design* magazine, but nothing more. I have made an effort to reverse engineer other works on the basis of their titles and observable regularities. I have pointed to the work of artists in the constructive tradition to whom March owed debts. But I am sure there is much more to be learned.

2 The Art of Lionel March

Lionel March devoted his research career to the study of geometry in architecture. In 1967 he and Leslie Martin founded the centre for Land Use and Built Form Studies at Cambridge University. In 1971 he and I published *The Geometry of Environment*, a book exploring applications of discrete mathematics to architectural plans and forms (March and Steadman 1975). At the Centre for Configurational Studies at the Open University in the 1980s, March worked with George Stiny and others on architectural shape grammars. At the University of California in the 1990s he studied Renaissance arithmetic and its role in the ideas and buildings of Alberti and Palladio. This research was brought together in his definitive book of 1998, *Architectonics of Humanism: Essays on Number in Architecture* (March 1997). March died in 2018.

At the beginning and end of his life March produced works of art that expressed and alluded to these geometrical and architectural interests. The pictures have not, I believe, received the attention they deserve. The work in the 1960s was in the broad tradition of geometric abstraction and following the principles of serial art. The pictures were large. The geometry was rectangular, organised in grids, often using just primary colours. Some were painted in acrylic, not always by March himself but by his assistant, Judith Oxley. Others were made with coloured adhesive tapes on white board. March had an exhibition of these works at the Institute for Contemporary Arts in London in 1962 (Fig. 1). The University of Hull commissioned paintings for their new Arts Building.

There was then a long gap in March's artistic activities, until he started to produce pictures again in the 1990s in California, this time using digital technology. The formal range of the work widened to incorporate circular elements, non-orthogonal geometries, and a more extended colour palette (Fig. 2). These digital works were shown by Lionel March on his website *The Museum of the Golden Ratio* but have not, I believe, been exhibited elsewhere. (The website has since been taken down.)

In all this work we can see or intuit certain common characteristics. There are basic shape elements—grid lines, rectangles, triangles, circles—that are repeated and organised into more complex arrangements. The process of construction involves combining and permuting these elementary shapes in systematic ways. We can guess at the presence of mathematical structures underlying the visual appearances. The

The Geometrical Art of Lionel March

Fig. 1 'ICA piece 2c', shown in an exhibition, 'Experiments in Serial Art', at the Institute of Contemporary Arts, London, 1962

Fig. 2 'MacMahon-Hay Suite, composition 3', digital print. Percy MacMahon was a twentieth century English mathematician specialising in number theory and combinatorics. I do not know who Hay was: perhaps the French-American mathematician Louise Hay?

movement and permutation of elements gives rise to sequences or families of related compositions. Beyond this point I am a little diffident about attempting to discuss March's art. I never talked to him at length about the work. He himself published only a few short explanations. I have not had the opportunity to study any of the notebooks that March left, which are now at the Canadian Centre for Architecture. Perhaps research in those archives will reveal in more detail what was in March's mind, if indeed he ever wrote it down. However, in the absence of other published criticism, I will try to explicate a small selection of works as best I can, using what little March did publish, plus some "reverse engineering" of my own. And I will try to make connections not just to larger movements in twentieth century painting, but to March's activities in architectural research, serial music, and the history of mathematics. There are hints of many of these connections in the titles he gave to

Fig. 3 'Rotations around a square', cover of *Architectural Design* magazine, February 1966

the works. But this is all—I must emphasise—partial and tentative. In 1966 March made a picture, or rather a proposal for a series of pictures, called 'Rotations Around a Square'. This was not a painting as such, but a printed design for the cover of an issue of *Architectural Design* magazine (Fig. 3) (March 1966). In a short note, Lionel March describes how this, and other similar works, are generated mathematically. The published design was to have been the first of a sequence of sixteen works, to be presented as a book (which never appeared).

The accompanying text explains that "In a work in which serial techniques are employed, there are five stages to the process of design: intention, selection, automatism, expression and interpretation." March conveys the broad *intention* of this particular work in a rough sketch, showing "A square impinged upon by events occurring around it" (Fig. 4). The colours are to be the subtractive primaries used in printing—magenta, yellow, and cyan—plus black. The colour plates are to be rotated to produce the sixteen different designs.

The second stage is the *selection* of elements. These are based on sequences of three numbers (a, b, c) representing lengths or 'rhythmic measures', for example (3, 4, 5). The order of the three numbers is permuted in all possible ways:

        a b c   b c a   c a b   b a c   a c b   c b a

        3 4 5   4 5 3   5 3 4   4 3 5   3 5 4   5 4 3

         A   B   C   Ɔ   ꓭ   Ɐ

The Geometrical Art of Lionel March 37

Fig. 4 The 'intention' of the work, 'Rotations Around a Square'

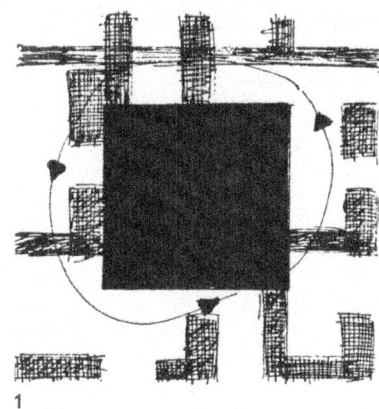

These triplets are referred to in brief as A, B, C, Ɔ, ᙠ, Ɐ. Notice that the last three triplets are mirror reflections of the first three triplets. The triplets are used to give dimensions to squares with black stripes on white backgrounds, in which the three numbers represent the width and position of the stripe (Fig. 5). The striped squares can be turned in two orientations, and the colouring reversed to give designs in white on black.

Pairs of these square designs are then combined using the Boolean operations of union and intersection (Fig. 6). The union of two designs combines *all* areas that are black in each design, while the intersection includes *only* areas that are black in both designs—that is to say, the extent of their overlap.

The result is a repertoire of small square designs that are placed in the picture in an ordered sequence around the edges of the large central square. As March says, this is a

Fig. 5 Divisions of squares into black stripes on white according to the numerical triplets, for 'Rotations Around a Square'

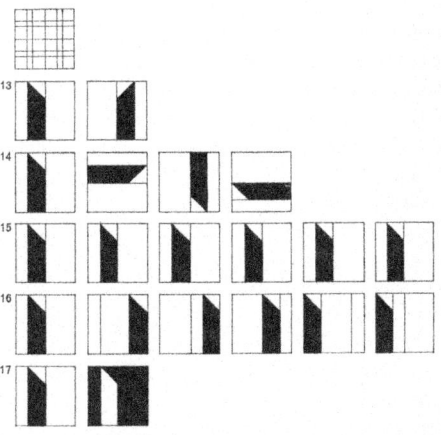

Fig. 6 Possible combinations of the square designs of Fig. 5 produced by the Boolean operations of union ∪ and intersection ∩

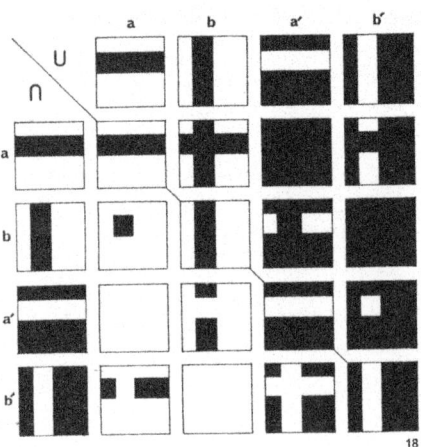

matter of 'arranging arrangements.' He has a schematic diagram showing the general principles, with twenty smaller squares around the large square (Fig. 7). The letters A, B, C, ᗡ, ꓭ, ∀. stand for the triplets of numbers, the different 'rhythmic measures'. The solid, dotted, and dashed lines show relationships of identity or symmetry among the small square designs. We are now at the stage of *automatism*, where the whole process becomes a kind of mathematical machine, and where varying the rules of arrangement of the small squares produces a variety of different overall compositions. The machine can throw up surprises and unexpected conjunctions.

Lastly come the *expression* and *interpretation* stages of the work, as March calls them. These are where the artist makes selections from the alternatives produced by the generative machine and develops their expressive possibilities. Part of this stage is the choice of colours, which can "change the appearance of the work without

Fig. 7 Analytical diagram showing the symmetry patterns of 'Rotations Around a Square'

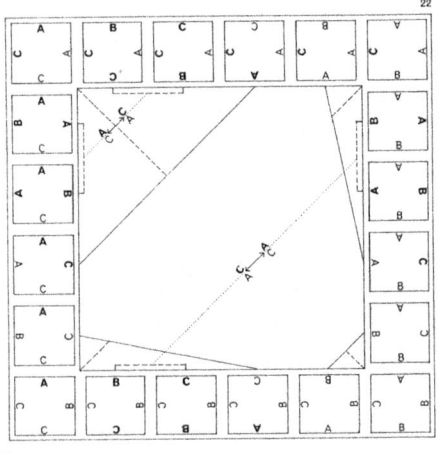

affecting the structure." In the work in question the black parts of the designs around the outside are printed in the three primary colours, as mentioned. Where the cyan, magenta, and yellow overlap, they create green, orange, purple, and black.

There are some larger points to make, beyond the minutiae of this particular piece, about the types of formal and mathematical tools which March is using. He starts with sequences of whole numbers, which have some affinities with the basic compositional units of serial music. He applies symmetry operations of translation, reflection, and rotation (in this case rotations through multiples of 90°) to elementary shapes. He produces composite designs by means of Boolean operations. All these are the topics from discrete mathematics which fascinated March throughout his life, and which were covered in *The Geometry of Environment*.

In the book we illustrated symmetries with examples from the work of Le Corbusier, Frank Lloyd Wright, and other architects. We did not at that early date anticipate all the connections with computer-aided design: but of course, symmetry operations in the plane are fundamental to the manipulation of shapes in 2D drafting systems; and the Boolean operators of union, intersection and subtraction are basic to certain types of 3D computer modelling systems. We repeated the diagram of Fig. 6 from the *Architectural Design* article in a section of *The Geometry of Environment* devoted to the representation of complex 2D shapes and 3D forms, in terms of the union and intersection of simpler geometrical elements.

March's interests in symmetries, permutations, and the enumeration of possibilities in architecture are all to be seen in a slightly eccentric diagram from *The Geometry of Environment*. This shows all possible ways in which two telephone boxes can be combined in plan such that their doors can open (Fig. 8). Of somewhat greater architectural interest is a complete catalogue, made more recently, of 126 possible plans for courtyard houses based on 3×3 squares, in which four of the nine spaces are courts (Fig. 9). These plans are not too far removed from some of the artworks. March devoted much effort in his architectural research to devising ways in which *all* plans and built forms of certain classes can be enumerated and catalogued, using methods of binary encoding.

If I am diffident when talking about March's pictures, I am extremely reluctant to venture into their possible relationship with serial music, a subject of which I know next to nothing (Bandur 2001). But links clearly exist with the methods of musical composition introduced by Arnold Schönberg in the 1920s and developed by Karlheinz Stockhausen and others after World War II. In serial music, from what I understand, basic elements are related one to another through measure and proportion. Changes in structure are used to create series of related movements or compositions. The formal principles of repetition, variation, contrast, and symmetry are employed, in combining the low-level structures of the basic elements into more complex superstructures. Within a given formal system, all possibilities of combination may be explored. If I have any of this right, then the parallels with March's method for 'Rotations Around a Square' are very clear. Stockhausen came to lecture at the Cambridge School of Architecture in the 1960s. Indeed, Stockhausen cited Le Corbusier's *Modulor* and its patterns of preferred dimensions derived from the Golden Section as a forerunner of serialism in music.

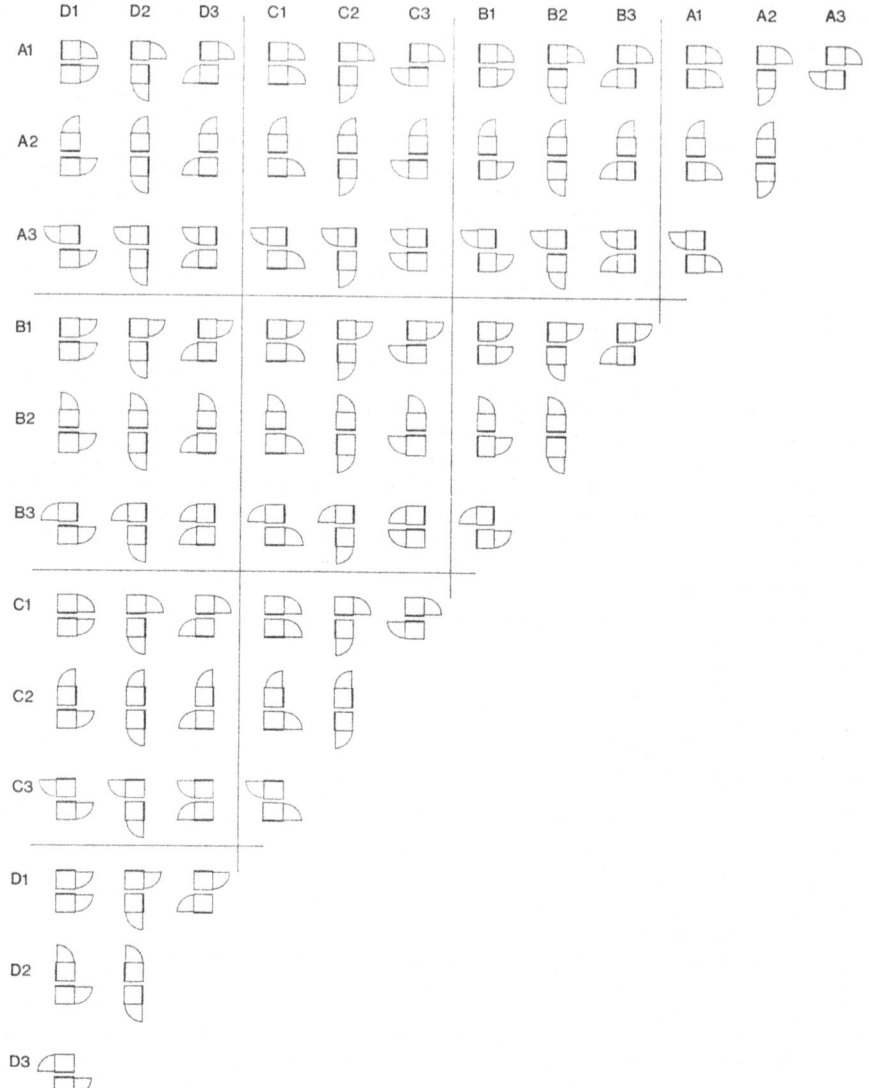

Fig. 8 Table showing 72 combinations of pairs of telephone boxes in plan, such that their doors can open: from *The geometry of environment*, 1971 p. 119

Going back to March's pictures from the 1960s: what was the wider historical context in which his art was situated, and which other artists and movements might have influenced his thinking and practice? He was certainly much interested in De Stijl, the Neoplastic movement of the 1920s, and the paintings of Piet Mondrian and Theo van Doesburg. I believe Leslie Martin knew Mondrian personally—certainly

The Geometrical Art of Lionel March

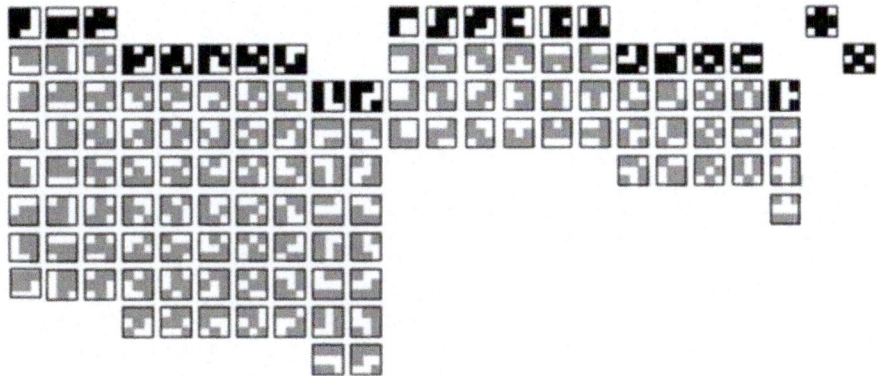

Fig. 9 Complete catalogue of 126 possible plans for courtyard houses based on 3 × 3 squares, in which 4 spaces are courts

he owned one of Mondrian's pictures. One series of digital works by March is called 'Counterplastic' (Fig. 10). These, according to his website, evoke Neoplastic compositions, but run 'counter to traditional principles'. By contrast with Mondrian's paintings, they are on black not white backgrounds; and they employ secondary colours, not the primary colours of De Stijl.

Of all the members of De Stijl, March was particularly attracted to the work of the Belgian artist George Vantongerloo, who believed that reality has a systematic order, and that the work of art is a model of that order. March made a group of digital works entitled 'Dialectic' which are explicitly intended, he says, to evoke the work of Vantongerloo (Fig. 11).

The Swiss artist and industrial designer Max Bill shared Vantongerloo's ambitions to create paintings and sculptures with underlying mathematical structure. Bill's

Fig. 10 Example from the 'Counterplastic' series 'evoking neoplastic compositions': digital print

Fig. 11 'Dialectic', evocative of the paintings of Georges Vantongerloo: digital print

'Fifteen Variations on a Theme' is probably his best-known exercise in serial art. March would have been familiar with Bill's graphic design and architectural work as well as his paintings. Other Swiss contemporaries of Bill who also crossed over between graphic design and painting were Karl Gerstner and Richard Paul Lohse. These Continental ideas in neoplastic, concrete, and serial art were only taken up seriously in Britain after the Second World War, by Kenneth Martin, Anthony Hill, and the Systems Group of artists who came together in 1969 and had a major travelling exhibition in 1972/73 (*Systems* 1972/73). Of the twelve members of the Systems Group, perhaps those whose work had the closest affinities with March were Jeffrey Steele, Malcolm Hughes, and Jean Spencer. All three artists worked with geometrical elements based on rhythmic sequences of numbers, which were repeated and subjected to symmetry transformations. A series of drawings and reliefs by Spencer, with notes on their mathematical basis, are shown in Fig. 12. March took part in a radio discussion of the Systems Group exhibition in 1972.

When March moved to the University of California he became deeply fascinated, as mentioned, with the relationship of Renaissance ideas about number to architectural composition in the work of Alberti, Leonardo, Serlio, and Palladio. His book *The Architectonics of Humanism* was conceived as a companion volume, fifty years on, to Rudolf Wittkower's *Architectural Principles in the Age of Humanism* (Wittkower 1949). The larger part of *The Architectonics* explains how Renaissance arithmetic had very different concerns from the modern idea of numbers as mere instruments for counting. Numbers, for ancient and Renaissance mathematicians, had distinctive individual characters: they were connected in complex networks of relationships; they had meaningful equivalents in the letters of the Hebrew and Roman alphabets; and above all they had geometrical connotations (there were triangular numbers, square numbers, pentagonal numbers and so on).

One might expect that March would have endorsed Wittkower's equation of geometrical proportions in architecture with musical harmonies; and he does indeed

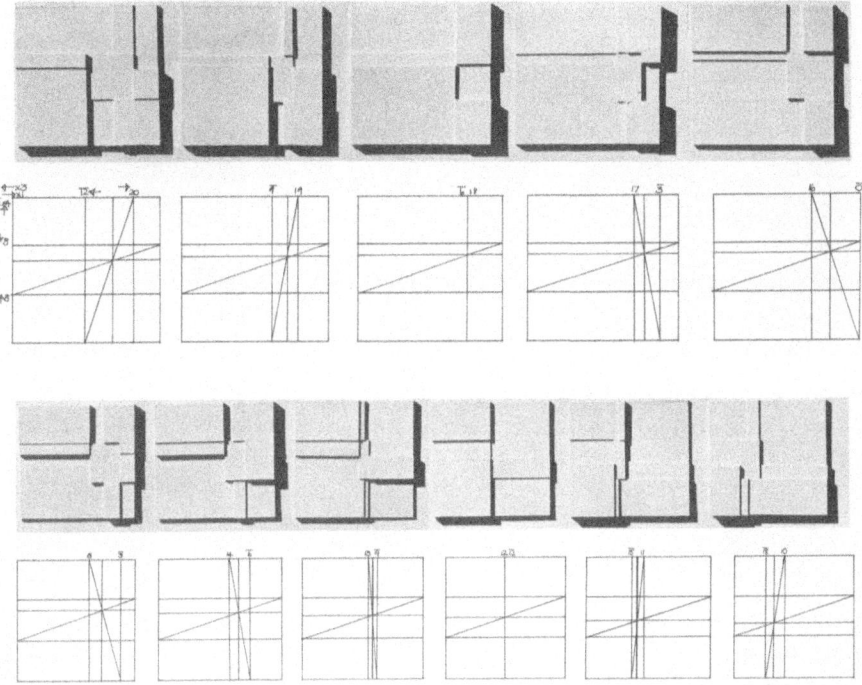

Fig. 12 Jean Spencer of the Systems Group, maquette and drawing for 'Relief 1', 1972: from the catalogue of the *Systems* exhibition, Arts Council, London 1972–73, pp. 49–50. Photographs by Graham Bishop, Corry Bevington

discuss the links. But March is sceptical of a simple correspondence and sees instead a common compositional basis for both arts in Renaissance arithmetic. As he points out, several ratios used in practice by Renaissance architects to determine the shapes of rooms or complete plans have no special musical significance. I will pick out and try to explain a few of the many Renaissance themes and allusions in March's digital artworks. Leonardo da Vinci was obsessed for a time with *lunulae*: the shapes of crescent moons (Fig. 13). He saw these as a possible route to solving the ancient problem of squaring the circle. Given a square, can one construct a circle of the same area, using just compass and straight edge? The problem was only shown to be incapable of solution in the nineteenth century. What attracted Leonardo to *lunulae*, was that it *was* possible to determine the combined area of two such shapes constructed on two sides of a given right-angled triangle (Fig. 14). March's series 'Da Vinci variations' pays homage to Leonardo's impossible ambition (Fig. 15).

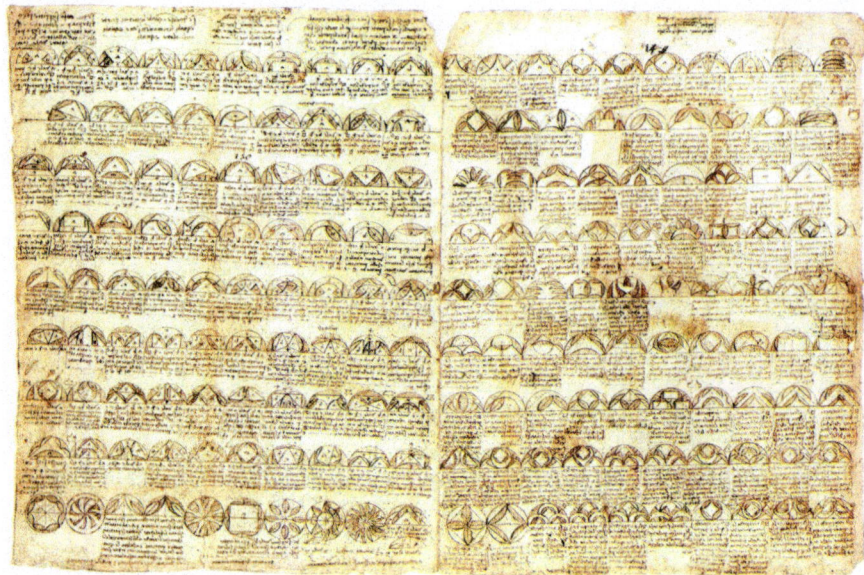

Fig. 13 Leonardo da Vinci, drawings of *lunulae* from the *Codex Atlanticus*, folio 455 recto

Fig. 14 Leonardo's construction showing *lunulae* set on the three sides of a right-angled triangle

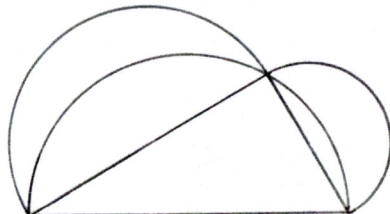

A further group of works by March refers indirectly to the five Platonic solids (Fig. 16): the tetrahedron, the cube, the octahedron, the dodecahedron, and the icosahedron. These, and the semi-regular polyhedra, were another of Leonardo's obsessions: he made some famous drawings of the solids in perspective for the mathematician Luca Pacioli's book *On Divine Proportion*.

March's interests in the work of Andrea Palladio date back to his time at the Open University (if not earlier), and to his work with George Stiny, Bill Mitchell, and Ramesh Krishnamurti on shape grammars for generating the plans of Palladian villas. In California he became more preoccupied with the specific numbers and ratios employed by Palladio for the shapes of rooms in the villas and churches. These are the points of reference for his 'Palladian septet' series (Fig. 17). The sequence of

The Geometrical Art of Lionel March

Fig. 15 'Da Vinci variations', digital print

Fig. 16 Example from the 'Bisshop Bradwardyne Cycle' based on transformations of the five Platonic solids: digital print. Thomas Bradwardyne was a fourteenth century English cleric and mathematician

works is based, March writes, "on the preferred room proportions cited by Andrea Palladio."

The pictures were made in 2008 to celebrate the 500th anniversary of the architect's birth. The ratios in question are, in sequence, 1:1 (orange), 4:3 (yellow), $\sqrt{2}$:1 (green), 3:2 (blue), 5:3 (indigo) and 2:1 (violet). The six rectangles are brought in turn to the fronts of the pictures. The circles provide a means for constructing the rectangles. (Notice the $\sqrt{2}$ proportion, which Palladio does indeed use, but which has no counterpart in musical harmony.) March provided a line diagram for the basis of all seven works (Fig. 18).

I have inferred a series of steps that could be taken in principle, using just compass and straightedge, to construct this diagram (Fig. 19). The steps are as follows:

Draw a circle of radius three units, whose centre is at A.

Raise a perpendicular on a horizontal diameter at point B, 1 unit from A.

Draw a right-angled triangle ABC, where AC is a radius of the circle.

Draw a second right-angled triangle ABD. Produce the edges of this triangle AD and BD.

Draw circles of radius 2 units, with centres at A, B and D.

Raise two more perpendiculars on the horizontal diameter of the original circle, 2 and 3 units from A.

The rectangles with Palladian proportions are now all found by the intersections of the straight lines with the circumferences of the circles (Fig. 20). The rectangles are added to the developing composition one at a time, and their colours then become muted as they recede into the background, and the new rectangle is superimposed. Why this specific construction works is another matter. Possibly March did not conceive the process in this way. But at least I hope to have demonstrated something of how the various geometrical elements of the compositions are related.

The Geometrical Art of Lionel March 47

Fig. 17 'Palladian Canon', digital prints

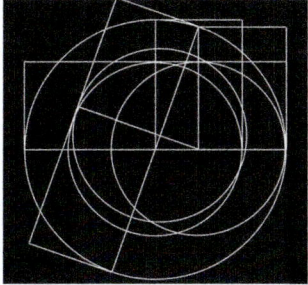

Fig. 18 Construction diagram for 'Palladian Canon'

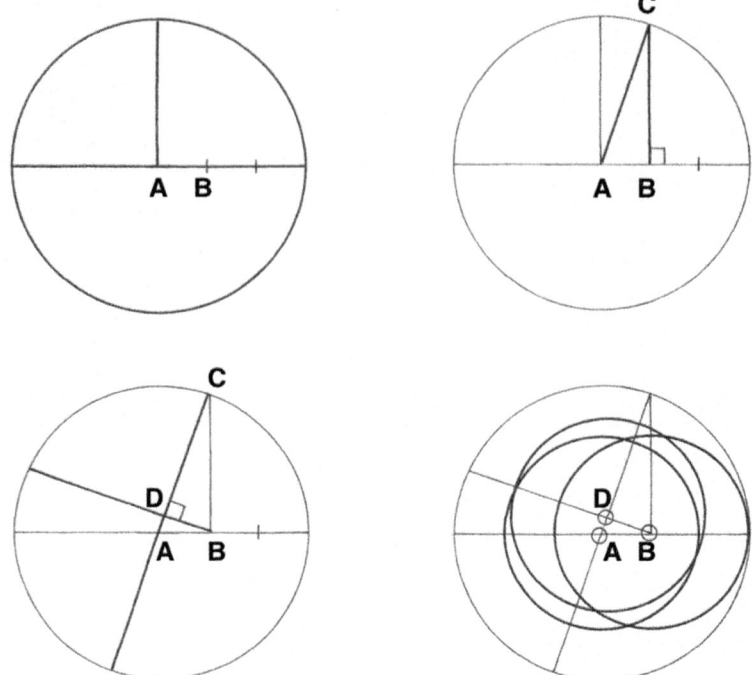

Fig. 19 Author's suggested geometrical construction by which Fig. 18 may be produced

I began with March's 1966 work 'Rotations Around a Square' and explained how he planned it as a programme for a series of sixteen pictures. More than fifty years later, in 2010, he completed the project as a set of digital prints, now with the more political title of 'Revolutions Around Red Square: Dusk to Dawn' (Fig. 21). In the accompanying notes he says that the black "varies in intensity from 10 to 80% in ten steps to 'midnight' and back again." I think this must refer to the central square and must mean tones of black over which a constant magenta is overlaid. Around the edges the smaller square patterns are rotated as before and printed in yellow, magenta, and cyan.

The work of a lifetime, condensed into a single night.

The Geometrical Art of Lionel March

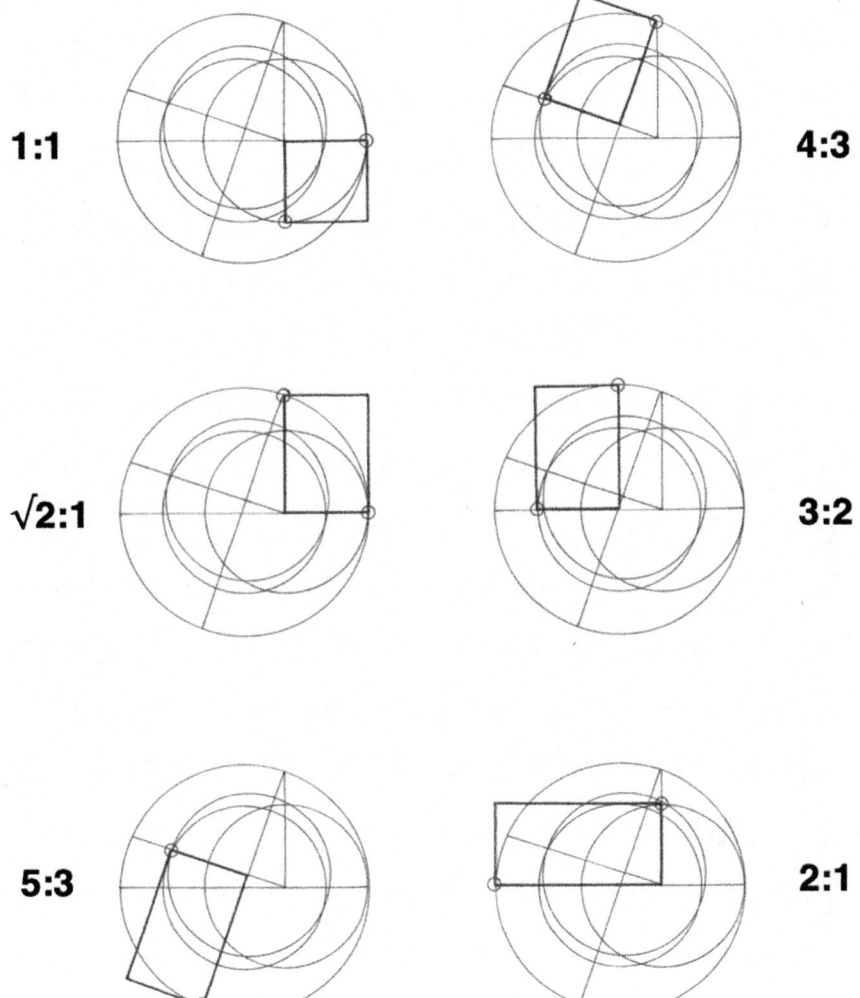

Fig. 20 Six rectangles of different proportions, derived from Fig. 19 and embodied in 'Palladian Canon'

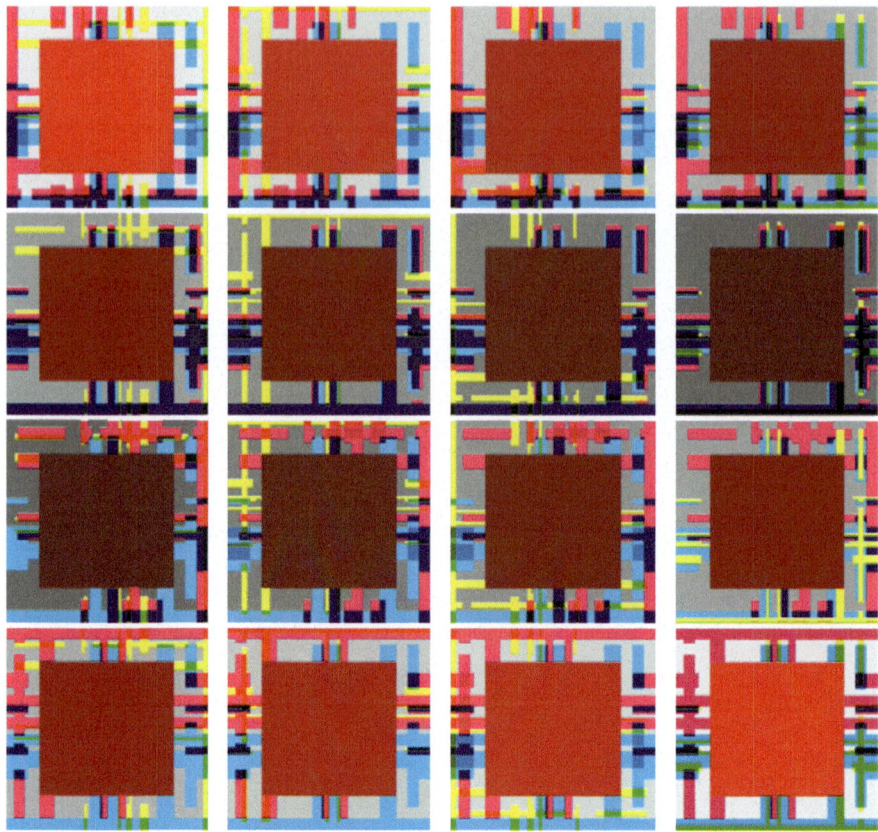

Fig. 21 'Revolutions Around Red Square: Dusk to Dawn', digital prints 2010

Acknowledgements I am most grateful to Candida March, Ben March, and Talitha Clift for all their help, and for allowing me to reproduce their father's works.

References

Bandur, M. 2001. *Aesthetics of total serialism: Contemporary research from music to architecture.* Basel: Birkhäuser.
March, L. 1966. Serial art: Notes on the cover design 'Rotations around a square'. *Architectural Design* 62–63.
March, L. 1997. *Architectonics of humanism: Essays on number in architecture.* Chichester: Academy Editions.
March, L., and P. Steadman. 1975. *The geometry of environment.* London: RIBA Publications. 1971: Paperback. Cambridge: The MIT Press.
Systems. 1972–73. London: Arts Council Travelling Exhibition, Whitechapel Gallery.
Wittkower, R. 1949. *Architectural principles in the age of humanism.* London: Warburg Institute.

Open Access This chapter is licensed under the terms of the Creative Commons Attribution-NonCommercial-NoDerivatives 4.0 International License (http://creativecommons.org/licenses/by-nc-nd/4.0/), which permits any noncommercial use, sharing, distribution and reproduction in any medium or format, as long as you give appropriate credit to the original author(s) and the source, provide a link to the Creative Commons license and indicate if you modified the licensed material. You do not have permission under this license to share adapted material derived from this chapter or parts of it.

The images or other third party material in this chapter are included in the chapter's Creative Commons license, unless indicated otherwise in a credit line to the material. If material is not included in the chapter's Creative Commons license and your intended use is not permitted by statutory regulation or exceeds the permitted use, you will need to obtain permission directly from the copyright holder.

Grids: Schemas, Rules and Cases

Chris Earl and Iestyn Jowers

Grids create a structural rhythm for spatial compositions. Paul Klee in his notebooks, *The Thinking Eye* (Spiller 1961), characterised these structural compositions as divisible, with (repeated) parts having measure and weight. He set the organic and individual, which could also have rhythm and structure, in contrast to the divisible. Shape computation reconciles these views. Shapes may be a rhythmic composition of parts but new parts, new shapes, emerge through the formalisms and mechanisms of shape computation. This paper illustrates two directions to exploit the structural rhythm of a grid. First, compositions develop a grid through new rules (the grid as generator). Second, new structural rhythms are generated from spatial relations among emergent parts of the grid.

1 Introduction

Grids are a common strategy, used in shape grammars to support the generation of designs, and can be viewed as elementary *schemas* on which plans and layouts are constructed. Grids have many forms and uses in design. In this chapter, we focus on orthogonal grids and their application in architecture and planning. Of particular interest is how a grid as a schema encapsulates spatial relations among sub-shapes in the grid. These include the orthogonality of lines and the incidence between the grid cells. Spatial relations define *rules* within the schema set down by the grid. The grid is the schema.

C. Earl · I. Jowers (✉)
The Open University, Milton Keynes, UK
e-mail: iestyn.jowers@open.ac.uk

C. Earl
e-mail: christopher.earl@open.ac.uk

While grids as schemas embody a variety of spatial relations which structure space, they can themselves be constructed according to schemas which pick out shapes such as areas bounded by lines (e.g. grid cells), and incidence relations between them. These schemas are primarily about shape elements and relations which are used to form *rules* to generate grids.

These two types of schema are complementary: the grid as a type of shape from which relations between sub-shapes are elicited, against the view of a grid as a construction from shapes (e.g. grid cells) and relations (e.g. edge incidence). These two senses of schema, extensional and intentional, both represent a wide class of shapes, encapsulating possibilities.

In designing, possibilities within shape schemas are explored by rules, choices made, and particular *cases* selected. Orthogonal grid schemas are just one example of this progression. Given their wide applications in architecture and planning, these grids provide a forceful illustration of how transitions from schemas, rules to cases, are used in design. However, transitions also happen in different sequences as another view of grids illustrates.

A grid can also be viewed less schematically, as a shape, with lines in shape algebra U_{12} and interstitial spaces or cells, bounded by the lines in the grid, in shape algebra U_{22} (Stiny 2006). This view of a grid, as a shape with well-defined selected parts, is common. Shape rules, based on the defined parts and their relations construct the shape and perhaps many others. Rules belong to more general rule schemas which generate a range of 'grid-like' shapes. This view represents a transition from case, to rule, to schema.

Returning to schemas. How do the grids as schemas fit with schemas as symbolic representations of shape rules? They are a middle ground between schemas and rules, a 'shapely symbol' of possible rules. Stiny (2006) represents a schema as a symbolic rule, $x \rightarrow y$, with x and y representing shapes. These schema take a variety of forms, including addition ($x \rightarrow x+y$), part ($x \rightarrow \text{prt}(x)$), division ($x \rightarrow \Sigma\text{prt}(x)$), transformation ($x \rightarrow t(x)$), boundary ($x \rightarrow b(x)$) and interior ($x \rightarrow b^{-1}(x)$). Schemas can be tied down through conditions on assignments to shapes x and y (including the spatial relation between them) which may result in parametric rules. However, some schemas such as boundary and interior are more categorical without scope for varying the spatial relations between x and y.

A grid schema might appear to pin down the grid in its 'shapely symbol'. But the schema is wide open to interpretation through constructions following rules based on spatial relations observed among emergent parts of the grid. How these rules can open up the initial idea of the grid will be examined in Sect. 4.

These views of grids exemplify three key features of design, namely: schema, rule and case. Designing may be seen more broadly as progressive transitions between and among these features. For example, Economou and Kotsopoulos (2014) analyse back and forth trajectories between schemas and rules, distinguishing the discursive nature of schemas from the visual character of shape rules.

We can look to history for a classical perspective. Views of grids, from schema, through rules, to a case, can be articulated in the historical and architectural context of Vitruvius' six principles (Vitruvius 2009). An analysis of these principles, presented

Fig. 1 Vitruvian principles in pairs, adapted from Economou (2018)

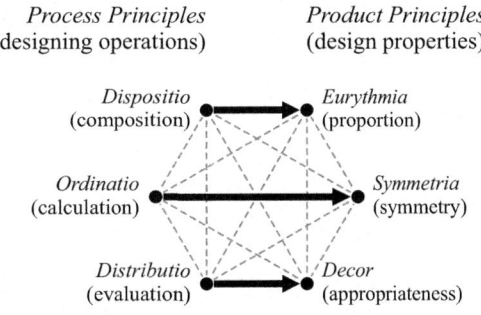

from the point of view of schemas and rules, by Economou (2018) divides the six principles into two categories, following Watzinger (1909) and Scranton (1974), corresponding to the processes of design and the characteristics of their productions. This work illustrates a neat parallelism, expressed through shape and shape rules, between classical concerns as adumbrated by Vitruvius and contemporary issues in the interplay between representations of process and product in design, as shown in Fig. 1.

The three process principles for Vitruvius correspond to designing operations: *dispositio, ordinatio, distributio* and the product principles correspond to design properties: *eurythmia, symmetria* and *décor*. With a rough and ready assignment of terms, based on the review of translations by Economou (2018), *eurythmia* (shapeliness or proportion) arises from the process of *dispotitio* (composition), *symmetria* (symmetry) from *ordinatio* (calculation) and *décor* (appropriateness) from *distributio* (evaluation). Economou (2018) offers a computational interpretation with reference to the concept of a Design Machine (Stiny and March 1981). The first Vitruvian pair corresponds to shape algebras and rules, with the second pair as application of rules (design algorithms) and the third pair as evaluation (criticism algorithms) to select a particular design. Economou emphasises how the Vitruvian pairs are applied recursively, reworking and constructing the design process itself.

A parallel might be drawn between the three Vitruvian pairs and the interpretation of grids as schema, rule and case, as shown in Fig. 2. The first pair *dispotitio-eurythmia* corresponds to conceptualising schemas, the *ordinatio-symmetria* pair corresponds to applying rules and finally the third pair *distributio-décor* corresponds to the selection of cases.

In this chapter, these themes from shape computation are drawn together in three steps. The first step examines examples of simple grids and how they can be generated with shape rules. This represents a {case → rule} activity. The rules may then generate a wider class of shapes which include the initial grids. The class is defined intensively. Next, refining the rules targets the production of specific grid shapes, or broadening them widens the scope of exploration. At this stage the exploration is restricted to shapes that are recognisable as simple grids. There appears limited scope to abstract from the rules to obtain schema although in principle this can be done freely yielding new concepts of grid schemas outside expectations. Overall, a combined

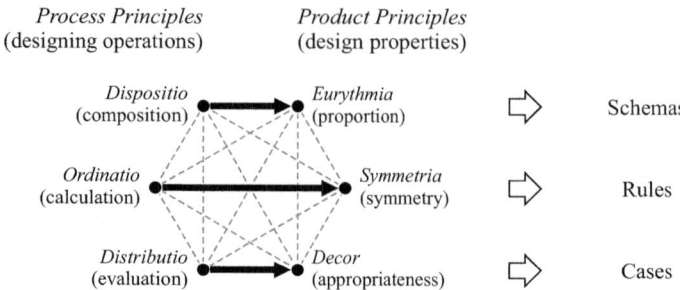

Fig. 2 Associating Vitruvian pairs and schemas, rules and cases

activity: {case → rule → schema} is possible. A detailed treatment of this sequence is developed by Harrison et al. (2015) and Harrison (2021) for kinematic designs. Starting from a case (a kinematic mechanism), rules for the case (and similarly behaving mechanisms) are constructed through making and experimentation; then schema abstract the rules. New rules consistent with the schema then generate new mechanisms far from the initial case.

A second step examines how a grid is a generator of designs, acting as the base for applying further shape rules. This draws on work by Leslie Martin and Lionel March on 'the grid as generator' (Martin and March 1972). This can represent both a {rule → case} activity and a {schema → rule} activity. In the former, rules are applied to the grid as an initial shape. In the latter, the grid is a schema for organising the placement of additional shapes using rules defined by spatial relations in the grid. In compositional terms these form a {schema → rule → case} activity.

Third, spatial relations are extracted from a grid schema and form the basis for new rules. Productions have grid-like properties, broadening the concept and scope of the initial grid schema. The wealth of variety in grid-like structures produced in this {schema → rule → schema} activity (Economou and Kotsopoulos 2014), is a target for exploration with computational tools such as the current developments in the Shape Machine (Economou et al. 2021; Hong and Economou 2023).

The applications to grids of the compositions of {schema, rule, case} examined in Sects. 2, 3 and 4, are examples of their use in computational design, in the iterative ways envisaged by Economou (2018) for the corresponding Vitruvian pairs. Speculatively, these compositions have implications for the logic of design (March 1976) in that the various transitions between schemas, rules and cases correspond to types of reasoning. The fluidity inherent in these compositions takes design away from hoarding parts and previous explorations, however intricate and complex these might be in their combinations, towards new schemas, new rules and new cases.

2 Simple Grids and Generative Rules

In some of the early work on shape grammar applications, such as Palladio's villas (Stiny and Mitchell 1978) and types of Chinese lattice (Stiny 1977), the first steps of design generation are to construct an underlying grid, with regularly spaced orthogonal lines. The grid for the Palladian plans forms the module for design in the sense that spaces with appropriate proportions are constructed by merging grid cells. The grid enables alignments and proportions.

How do rules in a grammar generate a simple grid of square cells with a rectangular boundary? The aim seems straightforward, but the implementation can be tricky, partly because the grids are not the final outcome. This means that generation, directed to producing labelled grids, is only the prelude to the next stage where further rules act on grid elements (labelled sub-shapes) such as the cells and their boundaries. For example, in Palladian plans grid cells are concatenated to form rooms, and in Chinese lattices grid cells are filled with motifs, oriented according to label positions.

This emphasises that grids are not only arrangements of lines but also include the cells bounded by the lines (and which don't themselves have any interior lines). We first examine simple orthogonal grids used in grammars for architectural compositions. These may have an overall rectangular boundary with the grid 'filling' the rectangle; a 6 × 4 example is shown in Fig. 3.

The rectangular grid is a shape, albeit a parametric shape, with conditions on the equality of spacing and parameters, which include grid spacing and numbers of lines in each orthogonal direction. In this sense, the rectangular grid is a shape schema. So, the question of how rules in a grammar generate grids might be refined a little: what rules generate instances of this shape schema?

One indicator for shape rules to generate grids are the spatial relations among sub-shapes in the grid. There are numerous spatial relations among many observed parts. First, relations between parallel lines of equal spacing and orthogonality, could give rules such as: add line (of same length) in parallel to existing lines, at given distance; add orthogonal lines between endpoints of two parallel lines. Formulated as shape rules, as in Fig. 4, do these generate grids? They can, but they also generate a lot more, especially when emergence plays a role, and rules are applied to embedded shapes, perhaps in ways initially not intended. For example, in Fig. 4, the first eight rule applications proceed to produce a small but acceptable 3 × 3 grid. Subsequent rule applications show the creative freedom that is available in shape rules, their 'rule bound unruliness' (March 1996), and gives some insight into what could happen if this simple grammar is allowed to generate shapes without direction.

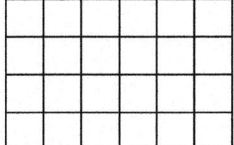

Fig. 3 Rectangular grid with evenly spaced lines and square cells

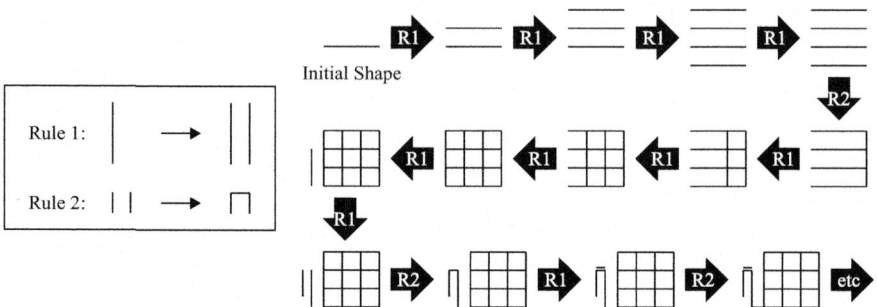

Fig. 4 A grid grammar, with orthogonal parallel lines

Clearly these rules generate more than grids, but the shapes produced are still defined within the rectangular grid schema. Labels at the endpoints of the lines might help to direct rule applications to always and only produce grids, but then rules will multiply to manage labels and to keep productions within a rectangular boundary as well as ensuring all the parts of the grid are present. A simple grammar that emphasises the spatial relations between lines in a grid seems to quickly become an exercise in book-keeping.

The simple grid, as defined by the rules in Fig. 4, is a shape with a declarative description of the constituent spatial relations based on equal spacing of parallel and orthogonal lines and a rectangular boundary. An alternative approach is to produce a square-celled arrangement within a rectangular boundary, so the rules must be engineered to effectively select (and label) the cells and intersections of grid lines. This is backed up by the examples where shape rules are used to construct underlying grids for Palladian Villa plans (Stiny and Mitchell 1978) and other layouts in a variety of architectural and design contexts, e.g. Queen Anne houses (Flemming 1986) and Malagueira houses (Duarte 2005).

This change in the schema from lines to lines + cells, forces consideration of other spatial relations in the simple grid, including those between the square cells. These are incident, edge to edge. However, cells in the middle of the grid are distinguished, in terms of their spatial relations to adjacent cells, from those at the edge or corners of the grid. This distinction is used to good effect in Stiny and Mitchell's rules for the starting grid for Palladian plans (Stiny and Mitchell 1978), before cells are combined to create the proportioned spaces. The trick here is to use the corner configuration of cells. The specific rules in the Palladian grammar are expressed with walls and discrete spaces which incidentally, while giving walls, also reduce scope for emergence. When abstracted to the simple grid they take the form shown in Fig. 5. Rule 1 adds an adjacent grid cell (a square area and its boundary) to another along its edge, and Rule 2 adds a square between two cells incident at a corner.

A label (as a shape marker), placed symmetrically on an edge, limits the size of the grid (when it is removed) but allows the addition rule (Rule 1) to act in a specified direction. An initial shape is a square cell with two labels to allow the grid to grow in two directions, and Rule 3 is applied twice to erase labels and stop generation.

Grids: Schemas, Rules and Cases

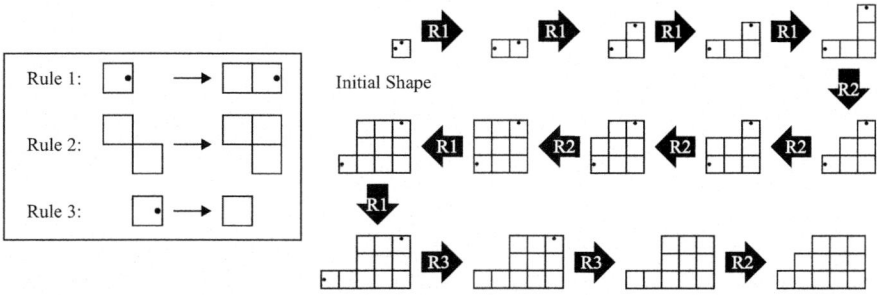

Fig. 5 A grid grammar, with repeating adjacent square cells

Rules and initial shape, amended by shifting both labels equal distances along the edges, also generate the grid. Applications of Rule 1, create threads of cells between corners of the grid. These threads minimally 'span' the grid. More economically, placing the label symmetrically in the corner, means that just one label in the initial shape is sufficient to generate the thread.

In Fig. 5, the first four applications of Rule 1 set up an L-configuration of cells between corners of the proposed grid. Then, four applications of Rule 2 fill in the grid. Application of Rule 1 to emergent cells, such as larger squares formed from four smaller squares, is avoided by using the placement of the label relative to the primary square cell. Infill with Rule 2 can start before labels fix the extent of the grid, but only grows the grid shape within the limits of the current thread. Rule 2 can be applied to emergent cells, but this either overwrites existing parts of the shape or is over-written by subsequent Rule 2 applications. The erasing rule used twice fixes the extent of the grid, otherwise the grid may grow 'tails', as shown, although the gaps can again be filled using Rule 2.

Using the spatial relations between cells (edge and corner incidence) yields a straightforward generative scheme. However, this is a device, to produce just the simple grid. It omits to place the grid in context as the grid is not the end game, but the opening. It is an underlying schematic arrangement which is then used as the initial shape in the next phase of rules. This is the sense that the grid is a generator of designs in Martin and March (1972), and the subject of the following section.

Perhaps the device is not wholly effective as any labels identifying grid cells have been removed. To use the grid as generator it would be necessary, at a subsequent stage, to re-identify the cells. However, another 'cell identification' label can be added by the rule, to indicate primary square cells, as shown in Fig. 6.

Fig. 6 Alternative rules with cell identification labels

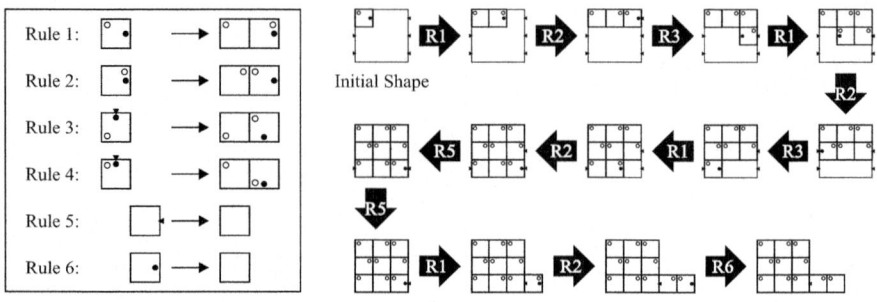

Fig. 7 Grid rules from the Chinese lattice grammar, from Stiny (1977)

The cell identification label is distinct from the grid generation label, and is distributed in a pattern across the grid, effectively orienting the cells. The grid, as shape schema, is now ready for interpretation and development with further rules.

What happens when the relative position of the labels in Rule 1 is altered or the spatial relations between the cell identifying labels and the generative labels are modified? In Fig. 5, shifting the labels in Rule 1 equal distances along the edges has little effect on generation. However, there are placements of the labels (on left and right sides of the rule) which radically limit the extent of the grid.

Different relative locations for the cell identifying labels, included in the alternative rules in Fig. 6, effectively give the cell an orientation. Patterns in the orientation of the cells across the grid act as the frame for subsequent development of the grid, such as in the shape grammar for a grid-based Chinese lattice (Stiny 1977), reproduced in Fig. 7. The first eight rule applications produce a labelled grid with alternating cells, and subsequent applications remove labels. However, ignoring triangular edge markers can result in the growth of a 'tail', as shown.

Exploring these varieties of rules and their productions through computational models, such as Shape Machine (Economou et al. 2021), allows different rule sets arising from the grid schema to be compared and particular designs (grids with emerging patterns of cell labels) to be selected.

Economou and Hong (2022) apply Shape Machine to good effect in generating patterns of cell labels which they elaborate with oriented patterns filling the grid cells. In particular, they illustrate rules producing geometrical conditions between the patterns in neighbouring grid cells. Such elaboration of the underlying grid is a prelude to the 'grid as generator'; the theme of the next section.

Another way to consider generations of simple grids draws on more recent developments in shape computation (Stiny 2022) where rule applications are not locked into the sequential step by step construction of shapes like the underlying grid. Rather, various rules apply in 'one-shot' multiple applications to produce a grid (and potentially more besides). This view reflects, in shape rule terms, the idea of the grid as a given, familiar shape which is immediately producible and readily interpretable. It can be made and seen in an instant, at a glance. Perhaps, other spatial relations in the grid will be more amenable to providing a straightforward generative description.

Grids: Schemas, Rules and Cases 61

This may be an enticing rabbit hole; tracking down rules to produce something that is visually familiar and readily interpreted (visually) in terms of parts, familiar in a particular design context.

All in all, it is surprising how the simple grids illustrate transitions between schema, rules and cases. However, by using a grid as the base for further generation, the 'grid as generator', we get a better picture of these transitions.

3 Grids as Generators

Leslie Martin's work on the grid as generator (Martin 1972) argues for both the conceptual and practical power of the grid in architecture and planning. Visual image and procedure in urban planning are distinguished in a way reminiscent of the distinction between processes of design and the characteristics of their productions, in the analysis of the Vitruvian principles. Martin's view that the grid can accommodate organic development through overlapping patterns of human activity mirrors the critical observation of Alexander (1965) that *"the city is not a tree"*. Such a combinatorial tree represents fixed parts in a nested structure, rather than overlapping parts (with their potential for merging and disaggregation) which form a combinatorial lattice. Martin remarks that *"organic growth without starting elements of some kind of framework is chaos.... understanding of that structural framework opens up a range of choices and opportunities for future development"* (Martin 1972).

The grid schema is a structural framework for possible design development (Martin and March 1972). Historical examples abound and Martin (1972) presents an analysis of how the New York grid layout accommodated changing patterns of activity and land use, by applying new rules. The main thrust of Martin and March's work is constructive rather than analytic, identifying general schemas and associated rules for designing with grids. One memorable example, which resonated strongly at the time, and continues to do so, was the comparison between pavilion and courtyard development of the grid cell in terms of densities, daylight exposures and connectivities, as shown in Fig. 8. These rules, although primarily intended for application to the grid cell, are equally applicable to other sub-shapes of the grid under scale transformations, as shown.

The grid as generator in urban planning is a frame for many different elements of the plan, such as boundaries, roads and blocks. The title of Martin's article is telling

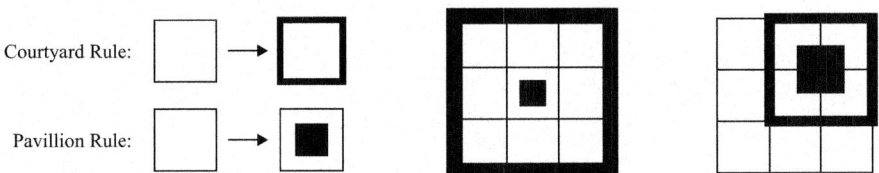

Fig. 8 Rules for the courtyard and pavilion development of a grid cell

as a precursor to subsequent research in generative systems employing shape rules in a particularly structured way, i.e. the shape grammar. Examples include designs for building and layout plans where shape rules form rooms and walls within the framework of the grid. One schema for these layouts is the division of a rectangular boundary into component rectangles (Earl 1977; Fleming 1978; Krishnamurti and Roe 1979, Steadman 1983). Schematically dissections are sub-shapes of a simple square grid. There are several views of generating these shapes including (a) removing lines from a square grid and (b) recursively dividing rectangles with lines. Rules for implementing (a) remove parts of grid lines and require a labelled grid to start, while implementing (b) requires adding lines across rectangles, a starting rectangle as well as an additional dissection rule, which introduces a pinwheel into a rectangle. The view (a) is illustrated in Fig. 9, with straightforward rules (shown here for a 4 × 3 initial shape) which aim to remove all the original grid labels. But this labelled grid will need generating in its own right. One issue with these rules is the possibility of 'hanging edges'. These will remain labelled (and hence invalid productions) if Rule 2 is modified so that the arrow-head label must be incident with a line segment for erasure to take place.

Grids as generators are not necessarily square grids. Each square grid is a schematic representation of multiple grids with various grid spacings. Examples are tartan grids with unequal but repeating sequences of grid spacings. A particular case is a grid with spacings *a* and *b* in the sequence *ababab* in both orthogonal directions, as represented in Tartan Rule 1 in Fig. 10. Cells are identified by shape labels in the corners and the sizes of these shape labels relative to the rest of the shape preclude application to emergent squares. Alternatively, in the spirit of developing a schematic square grid of lines, rules may apply to the boundary of the cell, as represented in Tartan Rule 2 in Fig. 10, which results in an *abababa* grid (the original grid is still shown to indicate how the tartan has been overlaid).

Cell subdivision rules, such as Tartan Rule 1, don't generate *abababa* grids easily. Taking a different approach, rules, similar to those used for the square grid, based on spatial relations between neighbouring cells including corner incidence, produce these grids ab initio, as shown in Fig. 11.

There is more to say about grids with sets of different spacings between horizontal and vertical lines. In the previous section we kept our attention to the regularly spaced grid with its square cells. Assigning different spacings is one way to develop the

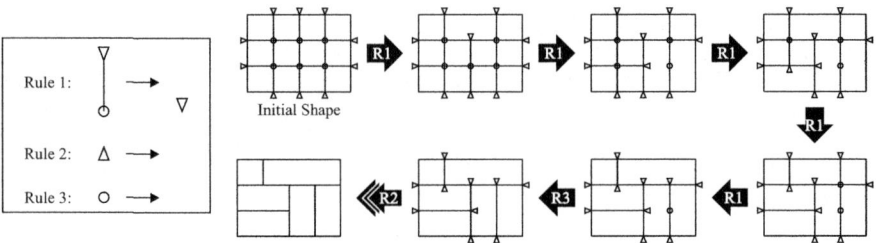

Fig. 9 Rules for developing a rectangular dissection

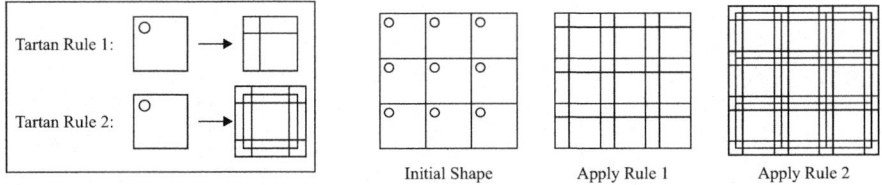

Fig. 10 Rules for developing tartan grids

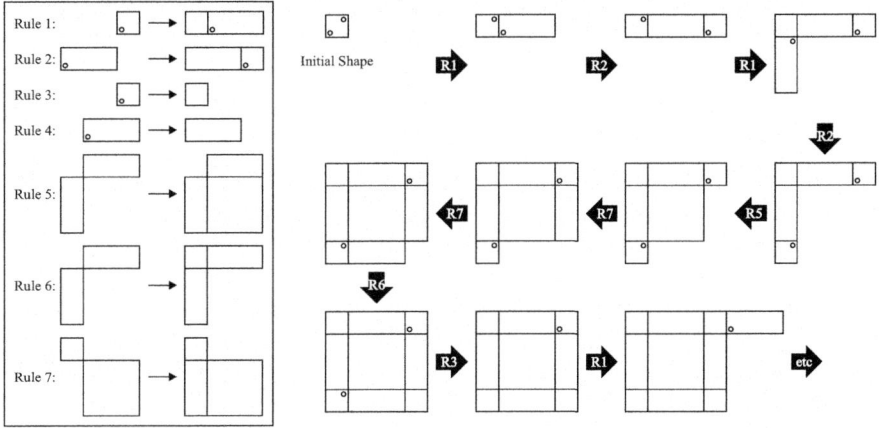

Fig. 11 Ab initio generation of *abababa* tartan grid

grid. But it is more than that because these spacings display a rhythm in how the grid structures space, i.e. the underlying purpose of the grid. This seems to have been the view of Bauhaus artists such as Paul Klee who in the *Thinking Eye* (Spiller 1961) discusses structure as 'dividual articulation' and develops measures and weights in ways reminiscent of the 'grid as generator' in Martin and March (1972).

Perhaps this is no surprise as Lionel March, in a paper on a class of proportioned grids (March 1981b), references how Klee's work *"prompted my own investigations"*. March presents a rhythm of these grids in terms of triples of spacings {a, b, c} permuted combinatorially as {a, b, c} {b, a, c} {b, c, a} {c, b, a} {c, a, b,} {a, c, b}, and overlaid, horizontally and vertically each with their own permutation order. Examples of March's proportioned grids are illustrated in Fig. 12, with additional measures and weights (colours).

Simple grids as schemas are straightforward to describe declaratively, but a little more complex generatively, based on spatial relations between the repeated boundary and cell sub-shapes of the grid. However, grids are rarely used as shapes in their own right but as frames for design development by shape rules. The examples above deliver specific geometric properties in a {schema → rule → case} transition. For example, the court-pavilion development of grids is a schema realised through

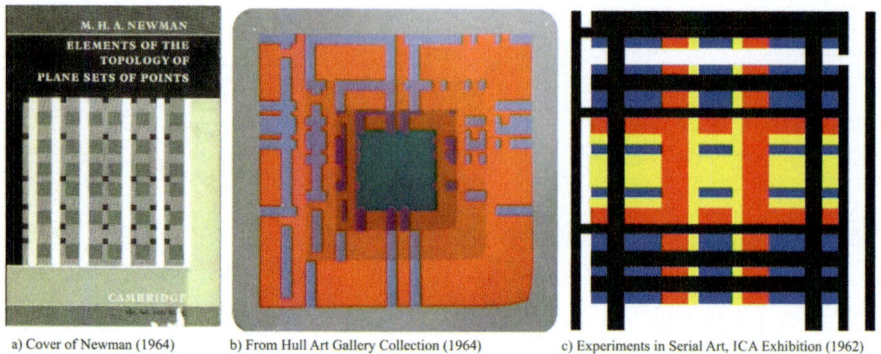

a) Cover of Newman (1964)   b) From Hull Art Gallery Collection (1964)   c) Experiments in Serial Art, ICA Exhibition (1962)

Fig. 12 Examples of grids by Lionel March

rules which present choices and opportunities for designers, architects and planners. Understanding choices and opportunities underlies the judgements of value and aesthetics made in constructing design cases (March 1976). The formal mechanisms of shape and shape rules come into their own.

The sequence from the grid schema to the rules of development to specific shape productions is fluid. Often the cases of specific grids serve to exemplify types or schema which are then realised through rules, giving a {case → schema → rule} transition. Other transitions are possible. Cases of developed grids such as Palladian villa plans, present a set of cases. Rules, constructed to produce these cases (and possibly more), can be generalised as schema. New rules consistent with this schema generate new cases as an evolution of the original plans. The transitions here are {case → rule → schema → rule → case}. The 'grid as generator' reveals how the varieties of transitions among schemas, rules and cases are identifying sequences for types of designs and designing.

Grids can reveal more. As suggested above, there are significant preconceptions about what a grid is or should be, often indicated by implicit or explicit cell labels. Turning off the cell labels opens up more spatial relations in grids and associated rules. The next section explores these rules along with their emerging choices and opportunities. Computational tools such as the Shape Machine (Economou et al. 2021) are invaluable for visualising how these wider choices and opportunities play out.

Perhaps grid schemas have been overly constrained. March (1981a) indicates how "*the restrictiveness of the grid is avoided*" through shape rules. He starts with the relation between grids and point sets, emphasising the topological view of cells and boundaries on a 'grating' examined by Newman (1964) in his book *Elements of the topology of plane sets of points*. March designed the cover for the paperback edition (Fig. 12a) and draws attention to parallels between Newman's treatment of point sets and grids with his own methods of using grids as generators of design. Two contemporary examples of Lionel March's work from 1964 (Fig. 12b, Hull Art

Gallery collection) and 1962 (Fig. 12c, ICA 1962 Exhibition, *Experiments in Serial Art*) are illustrated in Fig. 12.

March (1981a) then delineates the move to shape rules where, "*the grid is a product of the constructive process*", rather than an a priori structuring of spatial elements. Grids emerge from rules embodying spatial relations.

Up to this point, generation of grids uses the spatial relations of edge and corner incidence applied to labelled grid cells. Shape labels on the cells confine edge incidence to grid cells and prevent application to larger emergent squares. For corner incidence, rule applications for emergent squares are overwritten by subsequent applications (of corner incidence) to grid cells. What 'grids' emerge when spatial relations are relaxed?

4 Relaxing Spatial Relations in Grid Schemas

There are many varieties of rules associated with the grid schema. This includes abstracting spatial relations in the grid as new schemas with associated rules and productions (Stiny 2006). First the edge-to-edge relation between squares (as lines rather than areas) provides the underlying spatial relation for the square grid. The corresponding schema $x \to x + t(x)$: x is a square (in U_{12}) and $x + t(x)$ has a maximal line shared between x and $t(x)$; an example is shown in Fig. 13. The schema distinguishes two parts of $x + t(x)$, namely x and $t(x)$, in addition to $x + t(x)$. Cell interiors in U_{22} are identified by applying an interior schema $x \to b^{-1}(x)$ to each part. The interior $b^{-1}(x + t(x)) = b^{-1}(x) + b^{-1}(t(x))$ is an emergent rectangular area. The boundary of the interior, $b(b^{-1}(x) + b^{-1}(t(x)))$ is an emergent 2×1 rectangle in U_{12}. The operator bb^{-1} 'loses' the line dividing the cells U_{12}.

The schema can be applied to emergent squares and the interiors of these squares identified as grid cells by $x \to b^{-1}(x)$. The generation in Fig. 13 shows all applications of $x \to x + t(x)$ in 'one shot'. The next step includes applications to the emergent square in the centre. The result is a growing grid of squares with larger emergent squares accelerating growth. Placing a limit to growth forces the grid to fill available space.

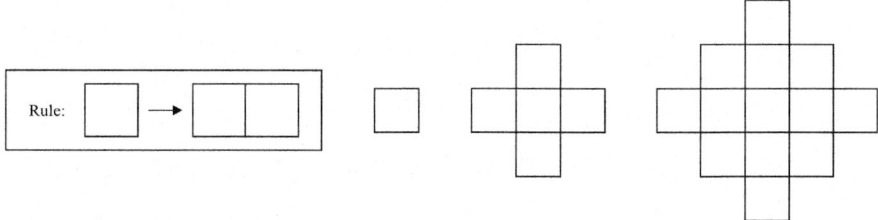

Fig. 13 Shape rule $x \to x + t(x)$, where x and $t(x)$ share a maximal line

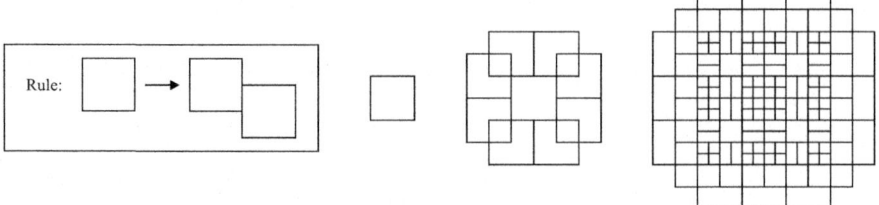

Fig. 14 Shape rule $x \rightarrow x + t(x)$, where x and $t(x)$ share a ½ edge line segment

Modifying the schema, maintaining alignment of the squares sharing a ½ edge line segment generates a 'grid' with increasingly smaller squares, which after n rule applications have an edge length of $(½)^n$, resulting in an intensification of the grid, as illustrated in Fig. 14. If the alignment shares a fraction a of the edge line, then defining $b = |1 - 2a|$, and as n increases, increasingly smaller squares emerge with side lengths of the general form $a^i b^j (a + b)^k$ with $i, j, k = 0, 1, 2, \ldots$, and $i + j + k \leq n$.

Alignment on a maximal line, but without sharing a line segment, and a ½ edge length distance between squares gives smaller emergent squares, which again after n rule applications have an edge length of $(½)^n$, as illustrated in Fig. 15. Alignment without incidence at distance, specified by the fraction a of the square edge length, also gives smaller and smaller emergent squares in the same sequences of decreasing sizes as the shared edge relation.

Alignment with two squares corner to corner (a spatial relation in the simple square grid) yields (consistent with $a = 0$ in the previous case) a square grid with emergent square 'holes', as shown in Fig. 16. If constructed within a boundary, this may be the most efficient way to produce an $n \times n$ square grid (when n is odd), even though nearly half of the square cells are not explicitly defined in the shape rule.

Generalising from specific grid spatial relations, to other corner and edge relations between square cells, Fig. 17 illustrates a schema $x \rightarrow x + t(x)$ with transformation t as translation and 45° rotation, maintaining incidence between edge and corner. These are reminiscent of experiments with Froebel's third gift (collection of cubes) discussed by MacCormac (1974) where symmetrical 2D arrangements of specific

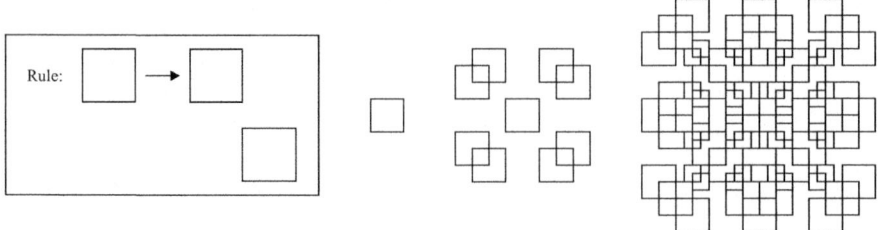

Fig. 15 Shape rule $x \rightarrow x + t(x)$, where x and $t(x)$ have colinear edges

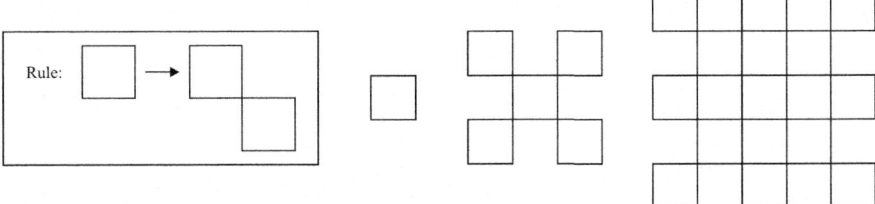

Fig. 16 Shape rule $x \rightarrow x + t(x)$, where x and $t(x)$ share a corner

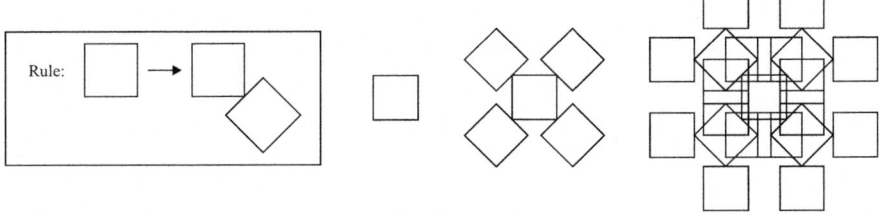

Fig. 17 Shape rule $x \rightarrow x + t(x)$, where x and $t(x)$ have incident edge and corner

spatial relations between squares are illustrated, and the formal treatment of 3D Froebel blocks by Stiny (1980). However, the blocks in Froebel gifts preclude the overlapping; a distinctive feature of the productions in Fig. 17, which are symmetrical with available rules applied in 'one shot'. Although not grids in the traditional sense, the resulting arrangements could be creatively employed as a generator.

Following the Froebel theme, the simple square grid displays a wealth of spatial relations among emergent shapes including rectangles. For a square x (as a line shape) and its neighbouring square t(x), application of boundary schema and its inverse x → b(b^{-1}(x) + b^{-1}(t(x)) gives an emergent 2 × 1 rectangle. Further, applying schema to the emergent rectangle, x → x + b^{-1}(x), adds the interior, namely b^{-1}(x), giving the 2D version of the block in the fourth Froebel gift. The spatial relation between 2 × 1 rectangles with the edges of rectangles aligned yields the shape rule in Fig. 18. The shape rule escapes the physicality of the Froebel blocks, allowing overlaps. Through repeated applications of the rule, square grid cells re-emerge, as shown in Fig. 18.

If only the longer and shorter edges are aligned in a symmetrical T, the spatial relation is again present in the square grid, although it uses emergent rectangles

Fig. 18 Shape rule $x \rightarrow x + t(x)$, applied to rectangles

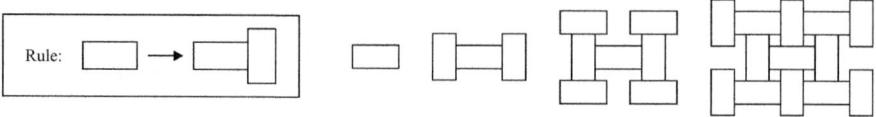

Fig. 19 Shape rule $x \rightarrow x + t(x)$, applied to rectangles aligned in a T

composed of 2 × 4 primary grid cells. This is represented in the shape rule in Fig. 19. The composition of these emergent rectangles is then reflected in the 'weave' patterns generated (also see MacCormac 1974), with the original grid cell recovered occasionally as a square 'hole', as illustrated in Fig. 19.

These are just a few illustrations which don't stray too far from a simple grid. Broader transformations, for example translating and shearing the square by an angle of 30° in the rule $x \rightarrow x + t(x)$, give rise to even more variation, as shown in Fig. 20, including a sheared grid when incident at an edge (Fig. 20, Rule 1) and a hexagonal tiling pattern when a reflection is added (Fig. 20, Rule 2). Following the development in square grid cells, if the spatial relation is partially incident at an edge, the grid is increasingly refined as generation progresses (Fig. 20, Rule 3).

Each of the spatial relations above follows schema $x \rightarrow x + t(x)$. Other possible spatial relations in the grid, such as rectangle and square incidence, as in Froebel's sixth gift, use schema $x \rightarrow x + t(y)$, constrained by: x is a rectangle (fractional ratio $a{:}b$), y is a square aligned with rectangle and incident to its long edge b, as shown in Fig. 21. Applying the rule parametrically eventually recreates a square grid with cell size a, although the example 'grid' in Fig. 21 is a long way from this endpoint.

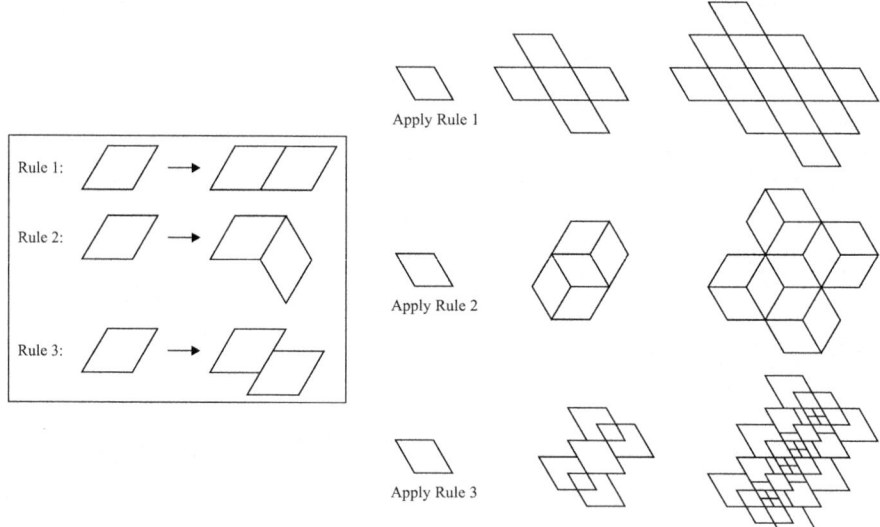

Fig. 20 Shape rules $x \rightarrow x + t(x)$, applied to sheared squares

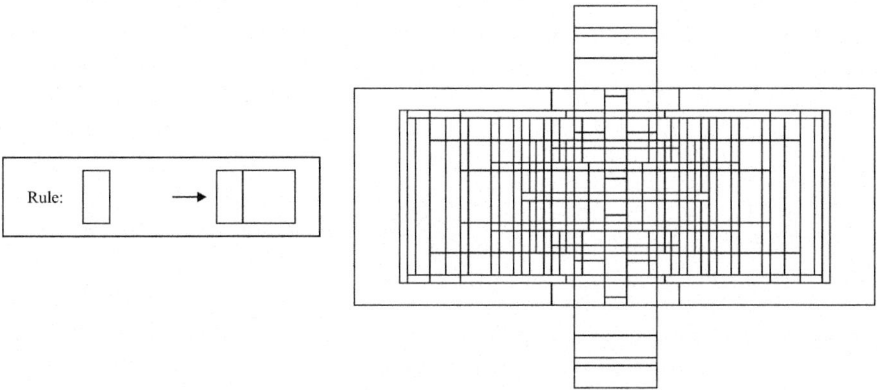

Fig. 21 Shape rule $x \rightarrow x + t(y)$, adds a square to the long side of a rectangle

Restricting application to a particular proportion closes down the generation quickly, except for proportion τ:1, the 'golden ratio'. The corresponding grids have spacings which from a first rough calculation appear as:

......τ + 1, τ, τ + 1, τ + 1, τ, τ + 1, τ, τ + 1, τ, τ + 1, τ, **1**, τ, τ + 1, τ, τ + 1, τ, τ + 1, τ, τ + 1, τ, τ + 1, τ + 1, τ, τ + 1 ......

in one direction (extending symmetrically from the starting rectangle) and in the other:

..... τ + 1, 2τ + 1, 2τ + 1, τ + 1, 2τ + 1, τ + 1, 2τ + 1, 2τ + 1, τ + 1, 2τ + 1, 2τ + 1, τ + 1, 2τ + 1, τ + 1, 2τ + 1, 2τ + 1, τ + 1, 2τ + 1, τ + 1, **τ**, τ + 1, 2τ + 1, τ + 1, 2τ + 1, 2τ + 1, τ + 1, 2τ + 1, τ + 1, 2τ + 1, 2τ + 1, τ + 1, 2τ + 1, 2τ + 1, τ + 1, 2τ + 1, τ + 1, 2τ + 1, 2τ + 1, τ + 1....

The sequences don't seem to be repeating periodically. The sequence of spacings τ, τ + 1, 2τ + 1, 3τ + 2, 5τ + 3, 8τ + 5, 13τ + 8, 21τ + 13, ... in a Fibonacci sequence will necessarily align with the spacings in these sequences as these are the proportions of emergent rectangles in the grid.

Continuing with the Fibonacci theme. Selecting different proportions of the rectangle from a Fibonacci sequence in the rule in Fig. 21 generates different grids. A set of proportions such as 1:2, 2:3, 3:5 appears to recover the complete square grid while applying the rule to just 1:2 and 2:3 proportioned rectangles seems to give a semi-tartan grid with spacings 2:1:2:1:2 in one direction and 1:1:1:1 in the other. These explorations are left to the reader.

5 Reflections

After a very enjoyable walk in the park, at least for the authors, we reflect on the views and avenues, drawing us to temples and pavilions that invite our attention—and more distant eye-catchers, which divert attention from the limits of the park.

In Alexander Pope's words, the park *"Surprises, varies and conceals the bounds"* (Pope 1963).

Perhaps most pleasant is to see the avenues in shape computation, mapped out by Lionel March and George Stiny nearly fifty years ago, holding their enduring appeal. The route of our particular walk was inspired by the way that March (1981a) invoked the grid as pivotal in interpreting the formalisms and methods of shape computation.

The parkland temples and pavilions in shape computation are bold insights that merit repeated views and perspectives. We notice how shape rules select parts and change them, often in an 'unruly' way (March 1996). Emergence, fusion and continuity in shapes (Stiny 1994, 2006) mean that surprise is around every corner. Reinventing rules through schemas, seeing particular cases through new rules that pick out parts and their spatial relations, and in a magical transposition, we are back to schemas and reinventing rules. We have seen this magic working on grids which are simple shapes, significant in architecture and planning.

A tour of the pavilions of shape computation must also mention the tools for putting rules into practice—the implementations and algorithms of Krishnamurti (1982), Krishnamurti and Giraud (1986) and Shape Machine (Economou et al. 2021). Excursions through rules, schemas and cases, such as our tour of grids, depend on these tools.

For history, context and reference, our walk in the park has had Vitruvius and his recent interpreters as guides. *Dispositio* (composition) and *Eurythmia* (shapeliness or proportion) set down an underlying grid with spacings and size, *Ordinatio* and *Symmetria* offer the means to develop the grid (the grid as generator) while *Distributio* and *Décor* provide for assessment and selection. Shape computation is the means to cycle through these steps recursively; elaborating, simplifying, and generalising, as eloquently outlined by Economou (2018).

What now; where next? Shape computation is a distinctive non-atomic view of computation (Stiny 2022). The possible spatial relations in the grid, for example, range endlessly over emergent sub-shapes of the grid. Rules are without bounds and the exercise started in Sect. 4 continues to surprise. At any stage, we can recapitulate, gather together productions, catalogue them and search ferociously. Perhaps nothing new emerges until we start again, using the grid as a generator of spatial relations, seeing new spatial relations and rules.

It really was a lovely walk, but what of the eye-catchers beyond the limits of the park, illusions of design intention, fancies without substance? Let's stay within an Arcadia of surprise and delight, laid out in fifty years of Shape Computation.

References

Alexander, C. 1965. A city is not a tree. *Architectural Forum* 122 (1): 58–62.
Duarte, J. 2005. A discursive grammar for customizing mass housing: The case of Siza's houses at Malagueira. *Automation in Construction* 14 (2): 265–275.

Earl, C.F. 1977. A note on the generation of rectangular dissections. *Environment and Planning B* 4 (2): 241–246.

Economou, A. 2018. The six Vitruvian principles of architectural design reframed within contemporary computational design discourse. In *Nexus conference: Relationships between architecture and mathematics 2018*, Pisa.

Economou, A., and T.C. Hong. 2022. Back to the drawing board: Shape calculations in shape machine. In *Design computing and cognition '22*, ed. J.S. Gero, 549–567. Springer.

Economou, A., and S. Kotsopoulos. 2014. From shape rules to rule schemata and back. In *Design computing and cognition DCC'14*, eds. J. S. Gero and S. Hanna, 419–438. Springer.

Economou, A., T.C. Hong, H. Ligler, and J. Park. 2021. Shape machine: A primer for visual computation. In *A new perspective of cultural DNA*, ed. J. H. Lee. KAIST research series. Singapore: Springer.

Flemming, U. 1978. Wall representations of rectangular dissections and their use in automated space allocation. *Environment and Planning B* 5 (2): 215–232.

Flemming, U. 1986. More than the sum of parts: The grammar of Queen Anne houses. *Environment and Planning B* 14 (3): 323–350.

Harrison, L. 2021. *Rules for making: Kinematic design, shape and structure*. Ph.D. dissertation. Milton Keynes: The Open University.

Harrison, L., C.F. Earl, and C. Eckert. 2015. Exploratory making: Shape, structure and motion. *Design Studies* 41 (A): 51–78.

Hong, T.-C.K., and A. Economou. 2023. Implementation of shape embedding in 2D CAD systems. *Automation in Construction* 146: 104640.

Krishnamurti, R. 1982. *SGI: A shape grammar interpreter, technical report, design discipline*. Milton Keynes: The Open University.

Krishnamurti, R., and C. Giraud. 1986. Towards a shape editor: The implementation of a shape generation system. *Environment and Planning B* 13 (4): 391–404.

Krishnamurti, R., and P. Roe. 1979. On the generation and enumeration of tessellation designs. *Environment and Planning B* 6 (2): 191–260.

MacCormac, R. 1974. Froebel's kindergarten gifts and the early work of Frank Lloyd Wright. *Environment and Planning B* 1 (1): 29–50.

March, L. 1976. The logic of design and the question of value. In *The architecture of form*, ed. L. March. Cambridge University Press.

March, L. 1981a. Editorial. *Environment and Planning B* 8 (3): 241–243.

March, L. 1981b. A class of grids. *Environment and Planning B* 8 (3): 325–332.

March, L. 1996. Rulebound unruliness. *Environment and Planning B* 23 (4): 391–399.

Martin, L. 1972. The grid as generator. In *Urban space and structures*, ed. L. Martin and L. March, 6–27. Cambridge University Press.

Martin, L., and L. March. 1972. *Urban space and structures*. Cambridge University Press.

Newman, M.H.A. 1964. *Elements of the topology of plane sets of points*. Cambridge University Press.

Pope, A. 1963. An epistle to the Right Honourable Richard Earl of Burlington: Occasion'd by his publishing Palladio's designs of the baths, arches, theatres, &c. of ancient Rome. In *The poems of Alexander Pope: A one-volume edition of the Twickenham text with selected annotations*, ed. Butt J. Yale.

Scranton, R.L. 1974. Vitruvius's arts of architecture. *Hesperia* XLIII: 494–499.

Spiller, J. 1961. *Paul Klee: The thinking eye*. Lund Humphries.

Steadman, P. 1983. *Architectural morphology*. Pion.

Stiny, G. 1977. Ice-ray: A note on the generation of Chinese lattice designs. *Environment and Planning B* 4 (1): 89–98.

Stiny, G. 1980. Kindergarten grammars: Designing with Froebel's building gifts. *Environment and Planning B* 7 (4): 409–462.

Stiny, G. 1994. Shape rules: Closure, continuity and emergence. *Environment and Planning B* 21 (7): s49–s78.

Stiny, G. 2006. *Shape: Talking about seeing and doing*. Cambridge: The MIT Press.
Stiny, G. 2022. *Shapes of imagination: Calculating in Coleridge's magical realm*. Cambridge: The MIT Press.
Stiny, G., and L. March. 1981. Design machines. *Environment and Planning B* 8 (3): 245–255.
Stiny, G., and W.J. Mitchell. 1978. The Palladian grammar. *Environment and Planning B* 5 (1): 5–18.
Vitruvius. 2009. *On architecture* (R. Schofield, Trans.). Penguin.
Watzinger, C. 1909. Vitruvstudien. *Rheinisches Museum für Philologie* 6202-223.

Open Access This chapter is licensed under the terms of the Creative Commons Attribution-NonCommercial-NoDerivatives 4.0 International License (http://creativecommons.org/licenses/by-nc-nd/4.0/), which permits any noncommercial use, sharing, distribution and reproduction in any medium or format, as long as you give appropriate credit to the original author(s) and the source, provide a link to the Creative Commons license and indicate if you modified the licensed material. You do not have permission under this license to share adapted material derived from this chapter or parts of it.

The images or other third party material in this chapter are included in the chapter's Creative Commons license, unless indicated otherwise in a credit line to the material. If material is not included in the chapter's Creative Commons license and your intended use is not permitted by statutory regulation or exceeds the permitted use, you will need to obtain permission directly from the copyright holder.

Reflections on Interpreting Shape Grammars

Ramesh Krishnamurti

Abstract This chapter, narrated mainly from a personal perspective, reflects on a journey in time and in topics, on aspects of shape grammar interpreters. The topics examined in this chapter move from a history of the early implementations, through shape algebra and shape representation to shape rule application and thence, onto implementation and search space interaction. The chapter concludes with a discussion on continuity of shape rule application.

1 Introduction

For his doctoral thesis, Jim Gips wrote the very first shape grammar interpreter—that is, an automaton, a program that takes a specification of shape and shape rule and provides the mechanism to produce new shapes by applying the rules appropriately (Gips 1975); I wrote the second, on a research project with George Stiny, funded by the UK Science and Engineering Research Council and The Open University (Krishnamurti 1982). However, Jim's interpreter did not incorporate George's embedding relation for shapes that allow for emergence and surprises in arbitrary parts or subshapes (Stiny 1975). As shape grammar interpreters go, mine took advantage of embedding and was thus the first to show shape grammars in their full generality. I wrote it again with Christian Giraud as an exercise in declarative programming as well as one exploring homogeneous coordinates to aesthetically simplify calculations (Krishnamurti and Giraud 1986). At the time neither Jim nor I had access to a windowing system, a graphic system, or even a structured programming language editor. Jim implemented his code in SAIL, an Algol-like functional programming language. I wrote my first version in FORTRAN, essentially a step up from assembly code, and my second in PROLOG, where graphics output was a side-effect. Our implementation efforts showed that a shape grammar could be regarded equivalently as a set of functions, a set of procedures and a collection of logical assertions.

R. Krishnamurti (✉)
Emeritus (Carnegie Mellon University), Pittsburgh, PA, USA
e-mail: ramesh@cmu.edu

There have been numerous implementations of shape grammar interpreters since. I will mention one which is especially satisfying to me, as it replicated my work thus validating it in the scientific sense. Scott Chase developed a shape grammar interpreter for his graduate thesis based on my algorithms (Chase 1987) and then went on to explore shape algebra for his doctoral thesis (Chase 1996). Chase wrote an elegant description of my algorithms (Chase 1989). He also gave, perhaps, the first extensive review of shape grammar interpreters at a Design Computing and Cognition workshop held in Delft (McKay et al. 2012). In their keynote article in a special issue of *AI EDAM* on shape grammars, Eloy, Pauwels and Economou provided an update to the list of interpreters (Eloy et al. 2018). Hong and Economou provide a more recent update (Hong and Economou 2022).

It is neither my purpose to add to this list nor to explore these implementations to any level of detail. Instead, I would like to offer a general reflection on the subject. Without attempting to provoke nor with any trace of hyperbole, I will assert that most implementations since have been essentially a variant of either Jim's or my interpreters with the added advantages afforded by advances in computing, representational capabilities, object-oriented programming techniques and libraries, graphics systems, geometry engines and plug-ins. Some of these interpreters target specific kinds of shapes, designs, languages, user-interactions, or are constrained to certain kinds and/or numbers of rules, for instance Piazzalunga and Fitzhorn (1998), Trescak et al. (2012), Tching et al. (2019), Li (2018). Some others consider shapes for a specific context or application, for example Chau (2002), Orsborn et al. (2006). Our interpreters namely, Jim's and mine, were not user-friendly and did not offer any facility beyond command line input; the subsequent interpreters were good attempts at making them so. Perhaps, the most user-friendly and easily impressively usable implementation is Kurt Hong's shape machine (Hong 2021; Economou et al. 2021). I have not yet seen at close hand Rudi Stouffs' interpreter (Stouffs 2022), which too is implemented on top of Rhino™ + Grasshopper™. Moreover, his interpreter deals explicitly with weighted shapes.

This is a personal account and so it includes a little history. Before I started work on my own interpreter, I spent several days as Jim's guest at his home, discussing aspects of his work. Those conversations and Jim's many subsequent encouragements were of immense help.

George and Jim had been interested in painting, composed of what I will refer to as 'patches' of color (Stiny and Gips 1972). These paintings comprised layers upon layers of patches of acrylic paint. Each patch could be regarded, two-dimensionally, as a blob with one or more holes. Back then, computationally, it was convenient to consider each patch as (a collection of) polygons. It was not until Iestyn Jowers' doctoral dissertation that shape computation with closed splines and curves were systematically studied (Jowers 2006; Jowers and Earl 2010, 2011). It would be remiss in this context not to mention Jay MacCormac's thesis where he employs curved shapes in the application of shape grammars to engineering product design (McCormack 2003; McCormack and Cagan 2003) (Fig. 1).

Gips implemented his interpreter on shapes as closed polygons, essentially as a necklace of points $<p_0, p_1, p_2, \ldots p_m, p_0>$ and shape rules as pairs of polygons.

Fig. 1 Acrylic paintings by George Stiny and Jim Gips (reproduced by permission of the author)

Although Gips restricted his shapes to a collection of simple polygons, it would have been possible to extend this notion to include self-intersecting polygons—that is, as a collection of a single outer polygon with several inner polygons. This is perfectly understandable, particularly from a painting standpoint, where his computations were essentially in U_2 (Stiny 1991), strictly speaking in $V_2 \times V_1$ as colored polygons arranged in layers. However, for practical reasons, Gips operated on his shapes in U_1 resulting in 'surprises' from shape rule application that gave rise to shapes made up of open or broken polygons. I have given an account of shape arithmetic in U_2 (Krishnamurti 1992a, b); Rudi Stouffs in his dissertation gives an elegant account of shape arithmetic in U_k, $k \geq 0$ (Stouffs 1994); together we have authored articles on computation in U_k, $k = 0, 1, 2, 3$ (Krishnamurti and Stouffs 2004; Stouffs and Krishnamurti 2006).

2 Shape

We take a shape to be a set of segments (spatial elements) x, each specified by a carrier shape $c(x)$ in which x is embedded and is of the same type (algebra), and a boundary shape $b(x)$, which represents the form of x and is a shape in a different type (algebra). A segment is a shape in itself. This model of shape algebra is independent of any underlying geometry. For any shape arithmetic operation $*$, the boundary of the resulting shape is a part of the sum of the boundaries of the two shapes (Krishnamurti and Stouffs 2004). That is, for shape operation $*$,

$$b(x * y) \leq b(x) + b(y) \tag{1}$$

where $\leq$ denotes the 'part' relationship. Interestingly, a variant of (1) is the start point for Chris Earl's thorough examination of shape boundaries and their properties (Earl 1997). More recently, Djordje Krstic has explored cross-algebra rule application involving shapes and their boundary shapes (Krstic 2023).

For implementation, expression (1) indicates that one never need inspect beyond the boundary of individual segments to obtain the boundary of a shape resulting from any shape arithmetical operation.

The boundary shape suggests various operations and relations. The boundary segments shared between x and y are specified by $S = b(x) \cdot b(y)$. Additionally, suppose $M = S \cdot b(x \cdot y)$ and $N = S \cdot b(x - y)$. Then, M and N are two mutually disjoint sets that sum to S. That is, $S = M + N$ and $M \cdot N = 0$, where 0 denotes the empty shape. We can specify relativised sets $I(x) = (b(x \cdot y) \cdot b(x)) - S = b(x \cdot y) \cdot b(y)$ and $O(x) = (b(x - y) \cdot b(x)) - S = b(x - y) - b(y)$. Then, $b(x) = I(x) + O(x) + S$ with $I(x) \cdot O(x) = 0$, $I(x) \cdot S = 0$ and $O(x) \cdot S = 0$. That is, given shapes x and y we can 'classify' the boundary of x into inner (I), outer (O) and shared (S) segments relative to y, and likewise, the boundary of y relative to x. Furthermore, we can 'construct' the boundary of any shape operation from their classified boundaries. That is,

$$b(x + y) = O(x) + O(y) + M$$
$$b(x \cdot y) = I(x) + I(y) + M$$
$$b(x - y) = O(x) + I(y) + N \qquad (2)$$

Classified boundary segments can be used to specify part or subshape relationship, namely, x is a part of y whenever $O(x)$, $I(y)$ and N are all empty. That is,

$$x \leq y \iff O(x) = 0 \text{ and } I(y) = 0 \text{ and } N = 0 \qquad (3)$$

Segments embedded in the same carrier shape can combine to form a single segment if they intersect or if their boundaries intersect. See Fig. 2. Disjoint segments are segments that cannot be so combined. The following proposition is the basis of the maximal segment representation of shapes: every shape is the sum of a unique finite set of disjoint segments with disjoint boundaries. That is, if $s = \{s_1, s_2, s_3, \ldots\}$ is a shape made of maximal segments, s_i, then,

$$s = s_1 + s_2 + s_3 + \ldots; \forall\, i \neq j, s_i \cdot s_j = 0 \text{ and } b(s_i) \cdot b(s_j) = 0 \qquad (4)$$

The concept of maximal shapes has its origins in George Stiny's doctoral dissertation which gives the first account for shape arithmetic in U_1 and where he introduces maximal lines (Stiny 1975). Stiny and Gips were the starting point for my work on a shape grammar interpreter. Much of the terminology I use in this article is adopted or adapted from Stiny (2008).

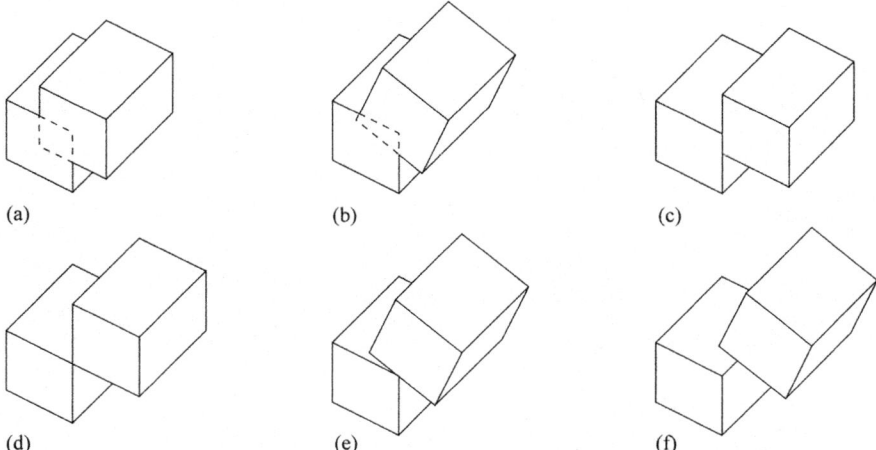

Fig. 2 The sum of two segments is a segment if the segments are not disjoint (cases **a**, **b**); or if their boundaries are not disjoint (case **c**); and is not a segment otherwise (cases **d**, **e**, **f**)

3 Spatial Change

Shape grammars encompass spatial change. The basic problem of shape change can be expressed as:

$$t = (s - a) + b \tag{5}$$

where s, t, a, and b are shapes, and the pair (a, b) represents a spatial relationship, the shape rule $a \to b$. It has the following interpretation: t is the shape that results from first taking a away from s, and then subsequently adding b. Clearly, the nature of the change depends on the relationship between s and a; the following cases arise:

(i) a and s have nothing in common in which case the difference $(s - a) = s$ and $(a - s) = a$.

(ii) a has everything in common with s, which we describe as the product $s \cdot a = a$, and $(s - a)$ has nothing in common with a. That is, $(s - a) \cdot a = 0$, the empty shape. Moreover, $(a - s) = 0$.

(iii) a has something but not everything in common with s, that is, $s \cdot a = x$, $x \neq a$, and the resulting shape $(s - a)$ has nothing in common with x. That is, $(s - a) \cdot x = 0$. Moreover, $(a - s) \neq 0$.

Shape grammars require that shape rules satisfy case ii: that is, it is required that $s \cdot a = a$, although, computationally, a shape algorithm will apply to all three cases. The pair (a, b) is equivalent to the pair $(a - b, b - a)$. That is, (5) can be rewritten to be the more efficient expression:

$$t = (s - (a - b)) + (b - a) \tag{6}$$

A related form of change is the notional 'undo' expression:

$$t = (s - b) + a \tag{7}$$

Here we subtract b from s before adding a. Clearly, the nature of the spatial change depends on b's relationship to both a and s. Using the equivalent shape rule pair, (5) can be rewritten as:

$$t = ((s - (b - a)) + (a - b)) + s \cdot a \cdot b \tag{8}$$

$s \cdot a \cdot b$ is a kind of hysteresis shape, the missing information that one needs to consider in an undo operation. Although this has only been marginally explored in Krishnamurti and Stouffs (1997) the 'undo' expression is worthy of further study.

It is instructive to consider (3) in the following way. If $s = \{s_1, s_2, s_3, \ldots\}$ are the maximal segments of s, then (5) can be written as for all s_i:

$$t_i = (s_i - a) + b \tag{9}$$

Moreover, $t = t_1 + t_2 + t_3 \cdots = \bigcup_i t_i$ and $\forall i \neq j, t_i \cdot t_j = 0$. That is, each shape t_i is made up of maximal segments, and the maximal representation of the shape is still preserved under sum (union) when shape rules are applied to the individual maximal segments.

4 Augmented Shapes

Shapes do not have to be simple. They can have properties, in fact, multiple properties. Stiny and Gips introduced colors into the discussion of shape grammars in the very first paper on shape grammars (Stiny and Gips 1972). Later, Terry Knight analyzed a variety of painting styles using 'colored' shapes and 'colored' shape rules (Knight 1989a, b, 1994). Stiny and Mitchell implicitly introduce texture in their shape grammar of Mughal gardens (Stiny and Mitchell 1980). Manuela Ruiz-Montiel introduces layers in her implementation of a shape grammar interpreter (Ruiz-Montiel et al. 2014). Stiny provides a rigorous treatment of weighted shapes for weights including colors, textures, and other properties (Stiny 1992). Expression (5) becomes:

$$w_t = (w_s - w_a) + w_b \tag{10}$$

where w_t, w_s, w_a and w_b are weighted shapes defined over a cross-product algebra with operations of $+, -, \cdot$ and part relation $\leq$ all defined separately in each algebra. Here $w_a \rightarrow w_b$ is the equivalent weighted shape rule. That is, for each shape s, there are attributes $a_s^1, a_s^2, a_s^3, \ldots$ to create a weighted shape expressed as the tuple, $w_s = (s, a_s^1, a_s^2, a_s^3, \ldots)$. Then, for any two weighted shapes w_s and w_t,

$$
\begin{aligned}
w_t + w_s &= (s + t, a_s^1 + a_t^1, a_s^2 + a_t^2, a_s^3 + a_t^3, \ldots \\
w_t + w_s &= (s + t, a_s^1 + a_t^1, a_s^2 + a_t^2, a_s^3 + a_t^3, \ldots \\
w_t \cdot w_s &= (s \cdot t, a_s^1 \cdot a_t^1, a_s^2 \cdot a_t^2, a_s^3 \cdot a_t^3, \ldots \\
w_t \leq w_s &= (s \leq t, a_s^1 \leq a_t^1, a_s^2 \leq a_t^2, a_s^3 \leq a_t^3, \ldots
\end{aligned}
\tag{11}
$$

We add the proviso that $w_s = 0$, the empty weighted shape when any of its constituent is 0 (or the equivalent 'empty' attribute) to ensure consistent weighted shape operations. We assume that $0 \leq w_s$, for any weighted shape w_s.

Weights offer an alternate interpretation to the notion of 'maximal' elements in these cross-product algebras. See Fig. 3 for illustration.

It should be clear from the figure, unweighted shapes that combine can be so weighted as to remain relatively maximal. s_1 and s_2 are overlapping lines that combine into a single line; let w_1 and w_2 denote their weighted counterpart. As is illustrated, overlapping thicknesses add, overlapping labels merge into a set, overlapping line

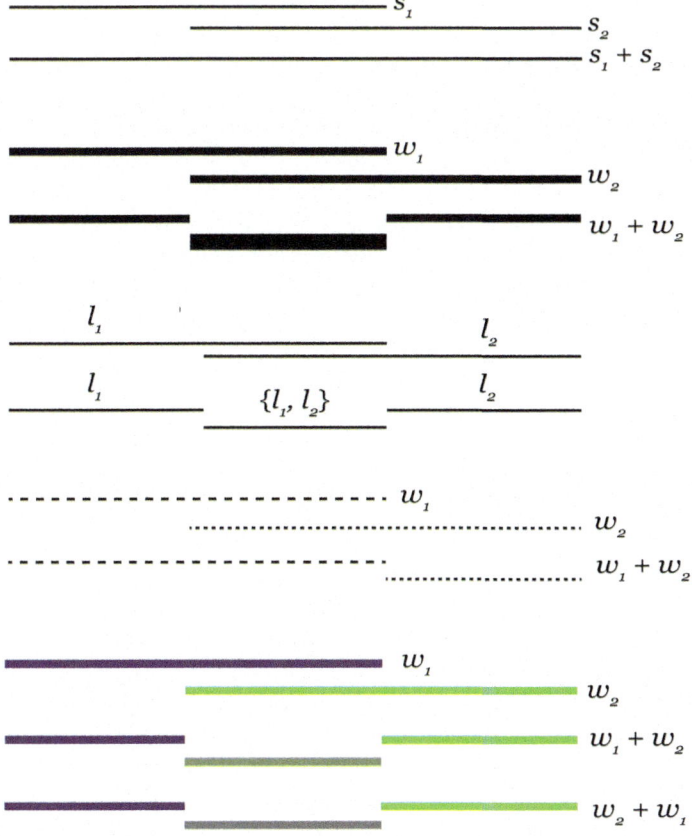

Fig. 3 Augmenting lines with various kinds of weights

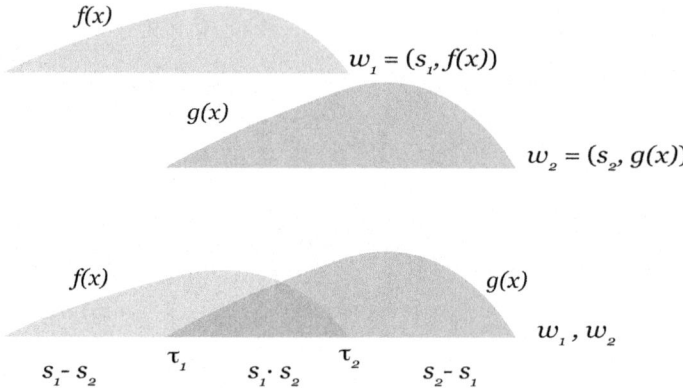

Fig. 4 Overlapping lines weighted by continuous functions

styles follow an order relation and overlapping colours may be treated as ordinals, as members of an ordered or enumerated set, or as is shown RGB values with an ordering on the lines (think of layers in Photoshop with a layering filter).

In this case, it is instructive to note that $+$ is not necessarily commutative in that $w_1 + w_2 \neq w_2 + w_1$. Neither is $\cdot$ necessarily commutative, $w_1 \cdot w_2 \neq w_2 \cdot w_1$. In the case of thickness, labels or colours, there are three resulting maximal weighted line segments: $w_1 - w_2$, $w_1 \cdot w_2$ (or $w_2 \cdot w_1$) and $w_2 - w_1$. In the case of colours, $w_1 \cdot w_2$ has a different colour depending on whether the sum is $w_1 + w_2$ or it is $w_2 + w_1$ (we assume that the first weight shape sits on top of the second weighted shape, much like Photoshop layers). Shapes can be augmented with properties that satisfy behaviors beyond those illustrated here. Stouffs and I have given a description of computing with shapes that have been augmented by different kinds of weights (Stouffs and Krishnamurti 2019).

It is important to note that the maximal representation, whether for weighted or unweighted shapes, works only when it is a definite description of a shape with indefinitely many parts. When lines are weighted as shown in Fig. 4 by continuous functions, then clever means are required to properly represent shapes with definite parts.[1] The figure poses a question that deserves an answer, namely, on how to specify operations on shapes. I offer here a hint of a solution, based on intervals.

Consider a function $f(x)$ defined over the range $\tau_1 \leq x \leq \tau_2$, which we represent as $[f(x), \tau_1 : \tau_2]$. This function can be normalized as $f((1-x)\tau_1 + x\tau_2)$ over the interval $0 \leq x \leq 1$. That is, $[f(x), \tau_1 : \tau_2] \equiv [f((1-x)\tau_1 + x\tau_2), 0 : 1]$.

Without loss in generality, we may suppose functions defined on the weighted lines are normalized. For the overlapping lines shown in Fig. 4 we can consider three separate line segments, $s_1 - s_2$, $s_1 \cdot s_2$ and $s_2 - s_1$. Let the ratio of $s_1 : s_2$ be λ. Then, $\lambda \tau_2 = (1 - \tau_1)$, and $[g(x), 0 : 1] \equiv [g(\lambda^{-1}x), 0 : \lambda]$.

---

[1] George Stiny in a private communication suggested that a general solution to the problem would be an interesting challenge :-).

Line segment $s_1 - s_2$ has a functional weight $[f(x), 0 : \tau_1]$ and segment $s_2 - s_1$ has a functional weight $[g(x), \tau_2 : 1]$. The common segment $s_1 \cdot s_2$ has two functional components $[f(x), \tau_1 : 1]$ and $[g(x), 0 : \tau_2]$ the contributions of which depend on the specific shape operation. Hence, for shape intersection,

$$\begin{aligned} w_3 &= w_1 \cdot w_2 \\ &\equiv (s_1 \cdot s_2, [min(f(x + \tau_1), g(\lambda^{(-1)}x))], 0 : 1 - \tau_1) \end{aligned} \quad (12)$$

For shape addition, there are three maximal weighted lines,

$$\begin{aligned} w_3 &= w_1 + w_2 \\ &\equiv \left(s_1, [f(\tau_1 x), 0 : 1]\right) \\ &+ (s_1 \cdot s_2, [(f(x + \tau_1) + g(\lambda^{(-1)}x)), 0 : 1 - \tau_1]) \\ &+ \left(s_2, [g(x), \tau_2 : 1]\right) \end{aligned} \quad (13)$$

For shape subtraction,

$$\begin{aligned} w_3 &= w_1 + w_2 \\ &\equiv \left(s_1 - s_2, [f(x), 0 : \tau_1]\right) \\ &+ \left(s_1 \cdot s_2, [max(f(x + \tau_1) - g(\lambda^{-1}x), 0)], 0 : 1 - \tau_1\right) \end{aligned} \quad (14)$$

In the case of subtraction, there may be segments that result with weight 0. That is, the segments that overlap, $s_1 \cdot s_2$, may yield zero weighted subsegments. These would need to be eliminated from further consideration. As one can see arithmetic on continuously weighted shapes, although doable and representable as definite descriptions, may be tricky. Figure 4 illustrates just one spatial case. There are other spatial situations to be considered. I look forward to reading a future publication that fully tackles the arithmetic of continuously weighted shapes.

It is interesting to note that Keles, Ozkar and Tari suggest an approach to shape recognition based on minimizing a cost function between a weight function for the shapes and a transformation of the weight functions of a given part (Keles et al. 2012). However, they did not employ continuous weights for rule application.

5 Computational Complexity

An important consideration for shape grammar interpreters is how tractable shape rule application can be, given its inherently combinatorial nature (Yue and Krishnamurti 2013). In 1993, Stouffs and I wrote an article on the computational complexity of the maximal representation of shapes in U_k, $k \geq 0$, which for various reasons remained unpublished (Stouffs and Krishnamurti 1993). Wortmann and Stouffs added to this analysis by considering the range of possible geometric transformations (Wortmann and Stouffs 2018).

At the core of the complexity analysis is the basic algorithm that takes any list of spatial segments and reduces it to its maximal representation. Segments of a shape are partitioned into equivalence classes based on their carrier shape. For shapes in U_k, $k \geq 0$, carriers have the same dimensionality as the segments they carry. Two segments can combine only when they belong to the same equivalence class; that is, for two c-equivalent segments, x and y, $c(x) = c(y)$. Each carrier has a unique descriptor, typically an equation, that specifies the class. For example, collinear lines have the same equation as the (infinite) line in which they are embedded. Likewise arcs on the same circle have the same carrier descriptor. By extension, this notion extends to higher dimensional spatial elements. For ease we assume that $c(\cdot)$ also denotes the carrier equation.

A segment x is described by the pair $(c(x), b(x))$, where $b(x)$ specifies the boundary of x. If the segment has dimensionality n, its boundary is an $(n - 1)$-dimensional shape. Segments of a shape are classified according to their carrier and their minimum boundary coordinate values and arranged as a list. As such, there is no strict order on the segments in a list for shapes in U_k, $k \geq 2$. The algorithms below determine the maximal representation from a given set of segments although straightforward, takes into consideration segment ordering. Notation follows that established in Cormen et al. (1980).

```
MAXIMAL (S)
1:  S ← SORT(S)
2:  M ← ∅
3:  if S ≠ ∅ then
4:    x ← POP(S)
5:    while S ≠ ∅ do
6:      y ← POP(S)
7:      if c(x) < c(y) then
8:        M ← M ∪ {x ← y}
9:      else
10:       C ← {x, y}
11:       while (S ≠ ∅) do
12:         y ← POP(S)
13:         if c(x) = c(y) then
14:           C ← C ∪ {y}
15:         else
16:           x ← y
17:           break
18:         end if
19:       end while
20:       M ← M ∪ REDUCE(C)
21:     end if
22:   end while
23:   M ← M ∪ {x}
24: end if
25: return M

REDUCE (C)
1:  R ← ∅
2:  if C ≠ ∅ then
3:    x ← POP(C)
4:    while C ≠ ∅ do
5:      y ← POP(C)
6:      if OVERLAP(x, y) then
7:        x ← COMBINE(x, y)
8:      else if SHARE-BOUNDARY(x, y) then
9:        x ← SHARE-COMBINE(x, y)
10:     else if DISJOINT(x, y) then
11:       if « C is in strict-order » then
12:         R ← R ∪ {x ← y}
13:       else
14:         T ← REDUCE({y} ∪ C)
15:         T ← SINGLE-REDUCE(x, T)
16:         R ← R ∪ T
17:         break
18:       end if
19:     end if
20:   end while
21:   R ← R ∪ {x}
22: end if
23: return R

SINGLE-REDUCE (x, C)
1:  if (C ≠ ∅) then
2:    v ← POP(C)
3:    if (OVERLAP(x, v)) then
4:      u ← COMBINE(x, y)
5:      R ← SINGLE-REDUCE(u, C)
6:    else if (SHARE-BOUNDARY(x, v)) then
7:      u ← SHARE-COMBINE(x, v)
8:      R ← SINGLE-REDUCE(u, C)
9:    else if (DISJOINT(x, v)) then
10:     R ← SINGLE-REDUCE(x, C)
11:     u ← POP(R)
12:     R ← {u, v} ∪ R
13:   end if
14: else
15:   R ← R ∪ {x}
16: end if
17: return R
```

Algorithm to convert a shape as a set of segments to its maximal segment representation.

The input list is sorted using a standard sorting algorithm (see MAXIMAL: 1). Union, '∪', can be accomplished in constant time, i.e., '$\Theta(1)$' (see MAXIMAL: 8, 15, 21, 24; and also REDUCE: 12, 15, 17, 22; and SINGLE-REDUCE: 12, 15). POP returns the 'next' element in a list (see MAXIMAL: 4, 6, 13 as well as REDUCE: 3, 5; and SINGLE-REDUCE: 2, 11). The call to REDUCE (in line 21) reduces the segments in a c-equivalence class to their maximal representation. MAXIMAL returns a lexicographically ordered maximal representation in M (see line 25).

REDUCE takes as input a list, C, of c-equivalent segments. Segments x and y are either disjoint or can combine to form a single segment—that is, when they overlap, share-boundary, or x contains y (y cannot contain x since x lexicographically precedes y in C). If they are disjoint, the class excluding x is reduced first to their maximal segments and then, x is reduced with respect to this maximal representation (see line 15) and described in procedure SINGLE-REDUCE shown on the lower left. It is possible to combine REDUCE and MAXIMAL into a single procedure (Krishnamurti 1980, 1992a). Figure 5 illustrates the possible spatial situations that arise, for shapes in U_2, before and after SINGLE-REDUCE has been invoked, when x and y are disjoint.

The algorithms apply to arbitrary shapes based on algebra-specific shape operations and relations, when applied to pairs of c-equivalent segments within the same algebra. For efficient algorithms (Krishnamurti 1992b), whenever two shapes are compared, their arithmetic operations and relations reduce to comparisons of c-equivalent segments as procedures MAXIMAL and REDUCE illustrate. Suppose the run time for each algebra-specific operation and relation on pairs of c-equivalent segments is asymptotically bound by some function $f(n)$, where n is a measure of the characteristic size of the boundary of the maximal segments in the shape. Then, the asymptotic upper bound on the run times for shape sum, difference, product,

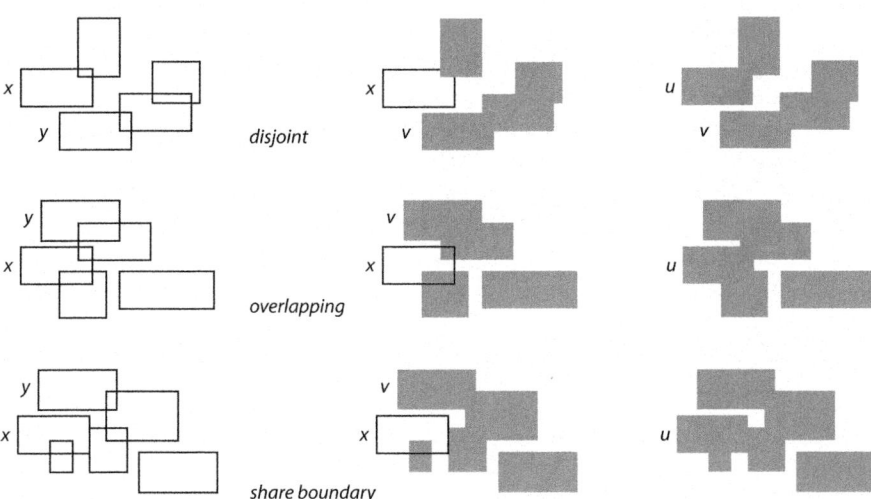

Fig. 5 Possible spatial situations that can arise before and after SINGLE-REDUCE is invoked when x and y are disjoint segments (see statement 12 within REDUCE). The leftmost column illustrates segments of the shape, the first two segments of which in lexicographical order are x and y respectively. The middle column describes the situation just before SINGLE-REDUCE is applied and indicates the element x and the maximal segments, formed by the remaining segments, whose first element is v. The rightmost column describes the situation just after SINGLE-REDUCE has been applied and indicates the maximal segments whose first element is u. The maximal segments of a shape are all shaded the same

Table 1 Time complexity for two c-equivalent segments and shapes

	$f(n)$	$\Theta(N^2 f(n))$	Lower bound
U_0	$\Theta(1)$	$\Theta(N)$	$\Omega(N)$
U_1	$\Theta(1)$	$\Theta(N)$	$\Omega(N)$
U_2	$O((m+n)\log n)$	$O((m_N + n_N)\log n_N)$	$\Omega((m_N + n_N)\log n_N)$
U_3	$O(Km + kn \log n)$	$O(K_N \, m_N + k_N \, n_N \log n_N)$	$\Omega((m_N + n_N)\log n_N)$

For two c-equivalent segments:
n: total number of their boundary segments
$m = O(kn)$: number of intersections of their boundary segments
k: number of c-equivalent classes and
K: number of simple boundaries

For two shapes:
$n_N = \sum_1^N n_i$: total number of boundary line segments
$m_N = \sum_1^N m_i = O(k_N \, n_N)$: total number of intersections of their boundary segments
$k_N = O(k\,N)$: total number of classes of boundary segments and
$K_N = O(K\,N)$: total number of resulting simple boundaries

and symmetric difference are quasi-linear, and at most quadratic in the number of maximal segments N. That is, $O(N f(n)) \leq \text{time} \leq O(N^2 f(n))$. Table 1 specifies $f(n)$ for each of the shape algebras U_k, $k \geq 0$, and their respective time complexities.

As suggested by Eqs. (2) and (3), shape operations and relations can be described by a pair of classification and construction algorithms, which are algebra specific. Classification in U_k, $k \geq 0$, typically involves a 'sweep' process which in U_0 is the identity matching, in U_1 is a comparison of boundary points, in U_2 a plane sweep to find polygon intersections (Shamos and Hoey 1976; Nievergelt and Preparata 1982), and in U_3, a space-sweep (Hertel et al. 1984; Tereshchenko et al. 2013) to find polyhedra intersections. Construction involves creating a topological representation of the boundary segments, classifying and extracting segments, and then reducing the extracted segments to their maximal representation. See Stouffs (1994), Stouffs and Krishnamurti (2006).

6 Recognizing Parts

In any spatial change, shapes a and b that specify a shape rule, $a \rightarrow b$ are representative. That is, a does not exactly occur in s, instead, a is 'alike' some part of s (Stiny 1975, 2008). In other words, there is a transformation g, such that $g(a)$ occurs in s, that is, $g(a) \leq s$. Then, expressions (5) and (10) respectively become:

$$t = (s - g(a)) + g(b) \text{ and } (s \cdot g(a) = g(a)) \tag{15}$$

$$w_t = (w_s - g(w_a)) + g(w_b) \text{ and } (w_s \cdot g(a) = g(a)) \tag{16}$$

For geometrical shapes specified on a coordinate system, g is a geometrical transformation; for parametric shapes, g additionally involves assignment of values to variables. Finding g is a significant aspect of any shape grammar interpreter. In every case, g is determined by a correspondence between 'distinguished' elements—that is, elements that remain invariant under a transformation and parameter assignment. Typically, one considers correspondences between points (or constructed points such as intersection points or feet of perpendiculars etc., labeled points, points weighted the same way and so on) that are invariant under the transformation.

In the case of general affine (linear) transformations, g, for shapes in U_n, g can be determined by a correspondence between two sets of $n + 1$ distinguished points (Krishnamurti 1981). For isometry and similarity (Martin 1982), this can be further reduced to a correspondence between two sets of n distinguished points. Others have considered other linear and projective transformations where additional correspondences are needed (Wortmann and Stouffs 2018; Hong and Economou 2022).

Once possible g's are determined, the subshape relation $g(a) \leq s$ (equivalently $g(a) \cdot s = g(a)$) can be tested to check whether g is a valid transformation. However, there may be an insufficient number of distinguished points, in which case there is an indeterminate number of possible transformations (Krishnamurti and Earl 1992; Krishnamurti and Stouffs 1993). Table 2 shows just the determinate cases for linear transformations of shapes in U_{kd}, $0 \leq k \leq 2$, $d \in \{2, 3\}$.

Table 2 Determinate cases for subshape recognition in U_{kd}, $0 \leq k \leq 2$, $d \in \{2, 3\}$

Algebra	Case	Determinate mapping
U_{02}, U_{12}	2 distinguished points	Third point is constructed in a direction perpendicular to the line formed by the points
U_{03}	3 distinguished points	Fourth point is constructed in a direction normal to the plane formed by the points
U_{13}	2 skew lines	Points on each line equidistant to the length of the common perpendicular from the end points of the common perpendicular
	3 coplanar lines not all concurrent nor parallel	Fourth point is constructed in a direction normal to the plane formed by the points of intersection of the lines
U_{23}	4 planes, not all parallel, and not all lines of intersection are parallel, concurrent or coincident	Two lines of intersection form a pair of skew lines. Use the construction above
	3 planes, and not all lines of intersection are parallel or coincident	The normal vectors of the planes are linearly independent—this reduces to case U_{13} above

Subshape recognition in $U_0 \times U_1 \times U_2 \times U_3$ is considered in Krishnamurti and Stouffs (1997). It is worth noting that subshape recognition in $U_0 \times U_1$, $U_0 \times U_2$ and $U_0 \times U_3$ is simply alternate ways of looking for subshapes in shapes with labeled points.

7 Representing Shapes

Interpretationally, expressions (5) (or (10)) are important in the following sense: s, t, a, and b are all represented in the same manner, or in shape grammar parlance, within the same algebra. A simple example will help. Mark Tapia in his dissertation developed a shape grammar interpreter using images (PICT files) to represent shapes (Tapia 1996, 1999). The expression involved a pre-processing of image to lines and post-processing of lines to image, and this was done consistently. The notion of pre-processing a shape description to apply the above spatial change expression is not unique, particularly for applications where other spatial and nonspatial augmentations hold.

Here are two particularly interesting exemplars to highlight the role that shape representation plays: the Chinese lattice and ice-ray grammars (Stiny 1977). They are among the more significant and oft cited examples of a shape grammar (Tapia 1992; Stouffs and Wieranga 2006; Yuan et al. 2011; Correia et al. 2016). Moreover, they are readily amenable to parametric shape grammar implementation (Stouffs 2018). They have been used as exemplars in their workshop on formal composition (Economou and Grasl 2018), in which they illustrate pedagogically motivated variations. Yu et al. employ the Chinese lattice rules along with weighting of the lines to denote mountain and valley folds as the basis for their origami shape grammar (Yu et al. 2021). As an aside, it is important to note that, computationally, origami structures require an alternative representation for shapes based on their fold structure and topology (O'Rourke 2011) rather than relying on one based strictly on coordinate values.

The lattice and ice-ray grammars are of paramount importance for the following reasons. Implicit in the rules for the Chinese lattice is the notion of aggregation to form configurations of connected cells. With appropriate choices for a seed cell and constraints these aggregations can become a grid or a tiling. It is straightforward to take the rules and adapt them to create a regular tiling of the plane. A further simple manipulation of the vertices can lead to a semi-regular tiling, which in turn can be further manipulated to create demi-regular and other tilings (Williams 1979). The Economou-Grasl workshop is particularly instructive here. With the help of constraints, aggregation can lead to designs that resemble structures with prescribed geometric properties, geometrical animals such as polyominoes, representations of real objects such as architectural floor plans, product designs and so on.

Implicit in the rules for ice-rays is the notion of subdivision to form new configurations from a given configuration. Cells are divided 'simply' into two, or even more complexly into many cells. In every case, new spatial elements are created.

Typical implementations of subdivision grammars consider s, t and a as collections of shape (polygonal) objects and a as a single object (polygon). Here, objects would be effectively pre- and post-processed as collections of lines. Shapes are indirectly represented and then translated to their equivalent rectilinear form albeit made up of lines, polygons, or polyhedral shapes. Shape rule application is essentially a form of polygon subdivision—a single polygon subdivided into two or more polygons. Structure sharing is a feature of most implementations; for instance, a polygon would have its own id, say p_{id}. An ice-ray design would be a collection of such polygon ids, each defined by a set of lines or a necklace of links to points. The points are maintained in a separate list along with their coordinate values.

The published grammars (far too numerous to cite) typically comprise rules that represent aggregation and/or rules that represent division. Most of these grammars tend to ignore emergence (Stiny 1994; Knight 2003). Nonetheless, the practical considerations in implementing grammars such as the lattice and ice-ray grammars become relevant. With Kui Yue, I suggested an approach based on implementing classes of shape grammar interpreters that rely on their intrinsic nature (Yue and Krishnamurti 2014).

The other more celebrated examples are shape rules like those shown in Fig. 6 which highlight emergence and visual calculation (Stiny 2022). These, incidentally, are the very shape rules that George Stiny asked me to demonstrate on my grammar interpreter. The power of a boundary-based shape representation such as the maximal line representation comes into its own here (Stiny 1986). Jowers and Earl examine such rules to explore an approach to implementation which exploits fixed structures inherent in shapes decompositions (Jowers and Earl 2018). Such fixed structures may not always be present; however, in the cases when they are, shapes can be formalized as a combinatorial word problem amenable to set grammar implementation (Jowers et al. 2019). Krstic likewise has explored shape decomposition in terms of algebraic lattices relating shape computation to symbolic computation and vice versa (Krstic 2010, 2017).

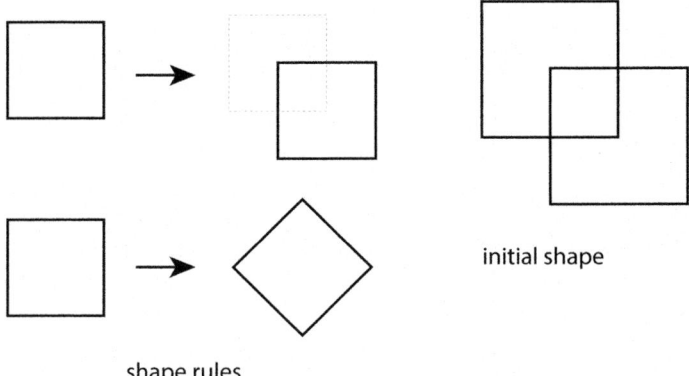

Fig. 6 Shape rules to demonstrate emergence

Shapes can be regarded as mechanical artifices, with Euclidean construction being the best-known system for creating shapes. Euclid's devices show means to produce points. From constructed points, shapes can be specified in terms of lines (and/or curves) that pass through these constructed points (Krishnamurti 2015). Analytically, we are more familiar with constructed points as corresponding to Cartesian coordinates. The coordinate system can be based on a range of possible number systems: integer (Z), rational (Q), Pythagorean (P), Euclidean (E), Vietean (V)[2] and real (R) where:

$$Z \subset Q \subset P \subset E \subset V \subset R$$

The number sets, apart from Z, are fields in the mathematical sense (Martin 1998), that is, in essence, they are closed under addition, subtraction, multiplication, and division. The first four number fields are Euclidean constructible, the fifth is paper folding (or origami)-constructible and the reals are neither. Shapes defined on a coordinate system based on a particular number field are closed in that field in the following sense. If F is a field, a 2-dimensional Cartesian coordinate system is defined by the cross product $F \times F = F^2$, then lines defined in F^2 all have carrier equations with coefficients in F. Moreover, geometrical transformations between shapes are also closed in F. Likewise, the field properties extend to higher-dimensional coordinate systems.

The Pythagorean, Euclidean, and Vietean numbers are expressible as radical expressions. However, reducing radicals for equality matching is neither intuitive nor straightforward. The best-known algorithms take exponential time (Landau 1992; Borodin et al. 1985). Consequently, currently, shape grammars can be defined exactly over the rational numbers, and only approximately over the other number sets.

Dealing with errors in calculation is thus important. When I implemented my shape grammar interpreter, I limited it to shapes defined over the rational numbers and included special-purpose routines to handle indefinite precision rational numbers to ensure exact computation. Most practical implementations tend to use built-in routines for numerical calculations (especially if they rely on an underlying geometry engine) and thus are at best approximations. As numbers are stored only to a particular precision in computer memory, accuracy of shape grammar computation relies on (i) the tolerance δ that is set and (ii) the degree to which the shape rule calculations compound errors. Essentially, points in memory are mapped to fixed-point numbers to within a given precision. See Fig. 7 illustrates the underlying grid representable in memory with coordinates indicated by open circles. The dark circle represents the real coordinate of interest. The δ's represent tolerances. Clearly, δ cannot be any lower than the precision afforded by memory and normally it is set to a value larger than the distance between pairs of grid lines. Clearly, the larger the value for δ the more the number of open circles that the dark circle is mappable to. Under geometrical transformations, the errors are likely to compound. Generally, if

---

[2] The numbers fall into two mutually disjoint sets: Dioclesian (D) numbers closed under cube root, and Glotin (G) numbers closed under trisection with $G \cap D = \emptyset$ and $V = G \cup D$.

Fig. 7 Precision error for different tolerance settings

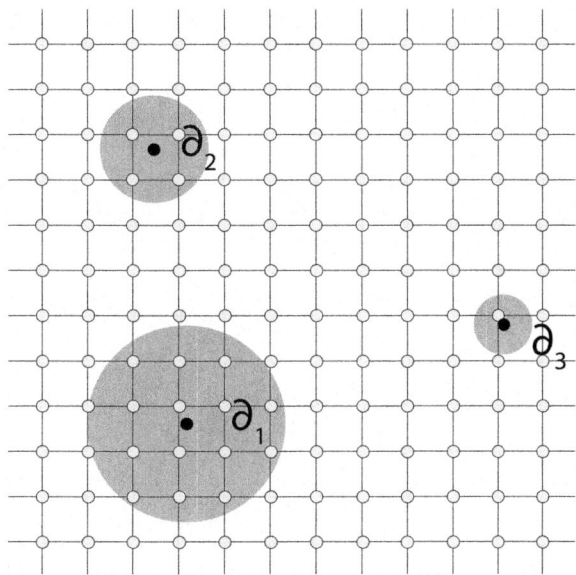

shape rule application is restricted to reflections about the horizontal, vertical, or 45° axes, and likewise rotations through multiples of 45°, the likelihood of errors percolating under smaller tolerances is low, even non-existent. As these seem to be the more common practical transformations for shape rule application, most interpreters should work just fine. In his dissertation, Kurt Hong describes in detail additional calibrations he employs to minimise precision errors (Hong 2021).

8 Search Space

Ultimately, users matter. To them, a shape grammar represents a world of possible designs, essentially, a combinatorial search space (Woodbury and Burrow 2003), to be explored as such. From the initial shape, different sequences of shape rule applications result in generally distinct designs. Each such sequence corresponds to a path from the initial shape in a tree of possibilities, more properly, a forest of possibilities. In an ideal world, the search space would be traversed—both forwards and backwards—in exploring possible designs and possible design choices (rule sequences). In an ideal world, the search space would be dynamically constructed as a (partial) gallery of choices made, explored, and periodically, revisited (Woodbury et al. 2017). In an ideal world, shape rules would be defined on-the-fly, and not necessarily, specified a priori. To be fair, Hong's Shape Machine performs splendidly in such user interactions (Economou et al. 2021).

With my then doctoral student Kui Yue I looked at ways to visually depict and traverse the search space and to explore user interaction with applicable shape rules

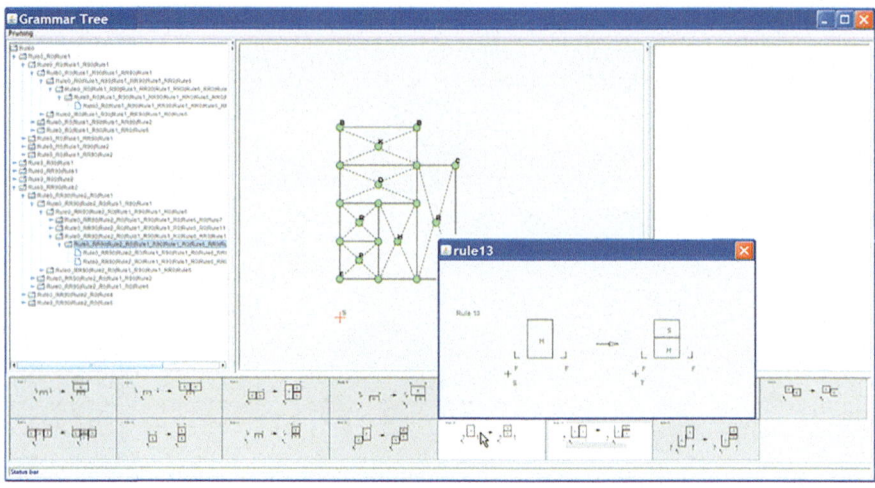

Fig. 8 Illustrating the search space for a Queen Anne grammar implementation

through various shape grammar implementations, each an experiment to test visual manipulation of the search space. Figure 8 depicts a screenshot of one such experiment for a subset of the Queen Anne grammar (Flemming 1987; Krishnamurti and Yue 2015). The window on the left depicts the entire search space, that is, all possible rule sequences. The highlighted rule sequence corresponds to a Queen Anne floor plan shown in the middle panel using a labelled graph-like notation. The bottom panel depicts the complete set of shape rules in the implemented grammar, with the two applicable shape rules highlighted. We subsequently used this implementation as the basis of grammar-based systems to predict the interior of Queen Anne houses in Pittsburgh and row houses in Baltimore from a specification of their external features (Yue et al. 2012).

9 Conclusion

The topics that are examined in this article from shape algebra move through shape representation to shape rule application and thence, to implementation and search space interaction. These are narrated mainly from a personal perspective and do not tell the whole story. They are suggestive of possible avenues of investigation, for instance, developing shape interpreters based on a constructive representation, rather than a coordinate-based representation of shapes. This would have important implications for foldable designs.

There remains one last topic of interest to implementation which features the notion of 'continuity' of shape rule computation, first described in Stiny (1994). Stiny basically shows that any shape grammar computation can be made continuous by retroactively constructing a topology for the shapes—a decomposition of the shapes into freely recognizable parts. By doing so, one achieves a 'kit-of-parts' approach to shape computation of shapes, as definite descriptions with definite parts—akin to computation on an object-oriented CAD system—one that side-steps issues posed by emergent shapes. Stouffs and I examined continuity to determine the conditions that must be satisfied a priori to ensure continuous computation (Krishnamurti and Stouffs 1997). Haridis and Stiny have subsequently explored a method to computing topologies algorithmically to effectively ensure continuity of shape rule application (Haridis and Stiny 2022).

This article reflects a journey in time and topics on the various aspects of shape grammar interpreters, though not chronologically nor in sequence. Whilst writing this article I reread familiar papers and discovered new articles on this subject by authors both familiar and less familiar to me. Although I could not possibly reference every one of these articles, the authors each deserve a nod of acknowledgment and their work merits praise. Shape grammars is fifty years old, and I have spent a sizable chunk of those years exploring the subject. To me it is heartwarming to see the continued interest in grammar interpreters and the depth of research shown by the more recent articles.

Competing Interests The author has no conflicts of interest to declare that are relevant to the content of this chapter.

References

Borodin, A., et al. 1985. Decreasing the nesting depth of expressions involving square roots. *Journal of Symbolic Computation* 1: 169–188.
Chase, S.C. 1987. *Computer implementation of shape grammars*. M.A. thesis. Los Angeles: University of California, Los Angeles.
Chase, S.C. 1989. Shapes and shape grammars: From mathematical model to computer implementation. *Environment and Planning B: Planning and Design* 16 (2): 215–242.
Chase, S.C. 1996. *Modeling designs with shape algebras and formal logic*. Ph.D. dissertation. Los Angeles: University of California, Los Angeles.
Chau, H.H. 2002. *Preserving brand identity in engineering design using a grammatical approach*. Ph.D. dissertation. Leeds: University of Leeds.
Cormen, T.H., C.E. Leiserson, and R.L. Rivest. 1980. *Introduction to algorithms*. Cambridge: The MIT Press.
Correia, R., A. Leitao, and J.P. Duarte. 2016. A generic shape grammar interpreter for discursive grammars. In *GA2016—XIX generative art conference*.
Earl, C.F. 1997. Shape boundaries. *Environment and Planning B: Planning and Design* 24 (5): 669–687.
Economou, A., and T. Grasl. 2018. Paperless grammars. In *Computational studies on cultural variation and heredity*, ed. J.-H. Lee, 139–160. Singapore: Springer.
Economou, A., T.-C.K. Hong, et al. 2021. Shape machine: A primer for visual computation. In *A new perspective of cultural DNA*, ed. J.H. Lee. Singapore: Springer.
Eloy, S., P. Pauwels, and A. Economou. 2018. AI EDAM special issue: Advances in implemented shape grammars: Solutions and applications. *Artificial Intelligence for Engineering Design, Analysis and Manufacturing* 32: 131–137.

Flemming, U. 1987. More than the sum of parts: The grammar of Queen Anne houses. *Environment and Planning B: Planning and Design* 14 (3): 323–350.

Gips, J. 1975. *Shape grammars and their uses*. Basel: Birkhauser Verlag.

Haridis, A., and G. Stiny. 2022. Analysis of shape grammars: Continuity of rules. *Environment and Planning B: Urban Analytics and City Science* 49 (7): 1929–1948.

Hertel, S., et al. 1984. Space sweep solves intersection of convex polyhedra elegantly. *Acta Informatica* 21: 501–519.

Hong, T.-C.K. 2021. *Shape machine: Shape embedding and rewriting in visual design*. Ph.D. dissertation. Atlanta: School of Architecture, Georgia Institute of Technology.

Hong, T.-C.K., and A. Economou. 2022. Five criteria for shape grammar interpreters. In *Design computing and cognition '20*, ed. J.S. Gero. Cham: Springer.

Jowers, I. 2006. *Computation with curved shapes: Towards freeform shape generation in design*. Ph.D. dissertation. Milton Keynes: The Open University.

Jowers, I., and C.F. Earl. 2010. The construction of curved shapes. *Environment and Planning B: Planning and Design* 37 (1): 42–58.

Jowers, I., and C.F. Earl. 2011. Implementation of curved shape grammars. *Environment and Planning B: Planning and Design* 38 (4): 616–635.

Jowers, I., and C.F. Earl. 2018. Visual structures of embedded shapes. In *Computational studies on cultural variation and heredity*, ed. J.-H. Lee, 175–187. Singapore: Springer.

Jowers, I., C.F. Earl, and G. Stiny. 2019. Shapes, structures and shape grammar implementation. *Computer-Aided Design* 111: 80–92.

Keles, H.Y., M. Özkar, and S. Tari. 2012. Weighted shapes for embedding perceived wholes. *Environment and Planning B: Planning and Design* 39 (2): 360–375.

Knight, T.W. 1989a. Color grammars: Designing with lines and colors. *Environment and Planning B: Planning and Design* 16 (4): 417–449.

Knight, T.W. 1989b. Transformations of De Stijl Art: The paintings of Georges Vantongerloo and Fritz Glarner. *Environment and Planning B: Planning and Design* 16 (1): 51–98.

Knight, T.W. 1994. Shape grammars and color grammars in design. *Environment and Planning B: Planning and Design* 21 (6): 705–735.

Knight, T.W. 2003. Computing with emergence. *Environment and Planning B: Planning and Design* 30 (1): 125–155.

Krishnamurti, R. 1980. The arithmetic of shapes. *Environment and Planning B: Planning and Design* 7 (4): 463–484.

Krishnamurti, R. 1981. The construction of shapes. *Environment and Planning B: Planning and Design* 8 (1): 5–40.

Krishnamurti, R. 1982. *SGI: An interpreter for shape grammars*. Technical report, 75 pp. Centre for Configurational Studies, The Open University.

Krishnamurti, R. 1992a. The arithmetic of maximal planes. *Environment and Planning B: Planning and Design* 19 (4): 431–464.

Krishnamurti, R. 1992b. The maximal representation of a shape. *Environment and Planning B: Planning and Design* 19 (3): 267–288.

Krishnamurti, R. 2015. Mulling over shapes, rules and numbers. *Nexus Network Journal* 17.

Krishnamurti, R., and C.F. Earl. 1992. Shape recognition in three dimensions. *Environment and Planning B: Planning and Design* 19 (5): 585–603.

Krishnamurti, R., and C. Giraud. 1986. Towards a shape editor: An implementation of a shape generation system. *Environment and Planning B: Planning and Design* 13 (4): 391–404.

Krishnamurti, R., and R. Stouffs. 1993. Spatial grammars: Motivation, comparison, and new results. In *CAAD futures '93*, ed. U. Flemming and S. Van Wyk, 57–75. Netherlands: Elsevier Science Publishers B.V.

Krishnamurti, R., and R. Stouffs. 1997. Spatial change: Continuity, reversibility, and emergent shapes. *Environment and Planning B: Planning and Design* 24 (3): 359–384.

Krishnamurti, R., and R. Stouffs. 2004. The boundary of a shapes and its classification. *Journal of Design Research* 4 (1): 75–101.

Krishnamurti, R., and K. Yue. 2015. Developing a tractable shape grammar. *Environment and Planning B: Planning and Design* 42 (6): 977–1002.

Krstic, D. 2010. Approximating shapes with hierarchies and topologies. *Artificial Intelligence for Engineering Design, Analysis and Manufacturing* 24 (2): 259–276.

Krstic, D. 2017. From shape computations to shape decompositions. In *Design computing and cognition '16*, ed. J. Gero. Cham: Springer.

Krstic, D. 2023. Notes on shape valued functions. In *Design computing and cognition '22*, ed. J.S. Gero. Cham: Springer.

Landau, S. 1992. Simplification of nested radicals. *SIAM Journal of Computing* 21 (1): 85–110.

Li, A. 2018. A whole-grammar implementation of shape grammars for designers. *Artificial Intelligence for Engineering Design, Analysis and Manufacturing* 32 (2): 200–207.

Martin, G.E. 1982. *Transformation geometry: An introduction to symmetry*. New York: Springer.

Martin, G.E. 1998. *Geometric constructions*. New York: Springer.

McCormack, J.P. 2003. *Implementing parametric shape grammars to capture and explore product languages*. Ph.D. dissertation. Pittsburgh: Carnegie Mellon University.

McCormack, J.P., and J. Cagan. 2003. Increasing the scope of implemented shape grammars: A shape grammar interpreter for curved shapes. In *Proceedings of the ASME 2003 international design engineering technical conferences and computers and information in engineering conference, vol. 3b. 15th international conference on design theory and methodology*, 475–484. Chicago: ASME.

McKay, A., et al. 2012. Spatial grammar implementation: From theory to useable software. *Artificial Intelligence for Engineering Design, Analysis and Manufacturing* 26: 143–159.

Nievergelt, J., and F.P. Preparata. 1982. Plane-sweep algorithms for intersecting geometric figures. *Communications of the ACM* 25 (10): 739–747.

O'Rourke, J. 2011. *How to fold it*. Cambridge University Press.

Orsborn, S., et al. 2006. Creating cross-over vehicles: Defining and combining vehicle classes using shape grammars. *Artificial Intelligence for Engineering Design, Analysis and Manufacturing* 20 (3): 217–246.

Piazzalunga, U., and P. Fitzhorn. 1998. Note on a three-dimensional shape grammar interpreter. *Environment and Planning B: Planning and Design* 25 (1): 11–30.

Ruiz-Montiel, M., et al. 2014. Layered shape grammars. *Computer-Aided Design* 56: 104–119.

Shamos, M.I., and D. Hoey. 1976. Geometric intersection problems. In *Symposium on foundations of computer science*, 208–215. New York: Institute of Electrical and Electronics Engineers.

Stiny, G. 1975. *Pictorial and formal aspects of shape and shape grammars*. Basel: Birkhauser Verlag.

Stiny, G. 1977. Ice-Ray: A note on the generation of Chinese lattice designs. *Environment and Planning B: Planning and Design* 4 (1): 89–98.

Stiny, G. 1986. A new line on drafting systems. *Design Computing* 1 (1): 5–19.

Stiny, G. 1991. The algebras of design. *Research in Engineering Design* 2: 171–181.

Stiny, G. 1992. Weights. *Environment and Planning B: Planning and Design* 19 (4): 413–430.

Stiny, G. 1994. Shape rules: Closure, continuity, and emergence. *Environment and Planning B: Planning and Design* 21 (7): S49–S78.

Stiny, G. 2008. *Shape: Talking about seeing and doing*. Cambridge: The MIT Press.

Stiny, G. 2022. *Shapes of imagination: Calculating in Coleridge's magical realm*. Cambridge: The MIT Press.

Stiny, G., and J. Gips. 1972. Shape grammars and the generative specification of painting and sculpture. In *Information processing*, vol. 71, ed. C.V. Frieman. Amsterdam: North-Holland.

Stiny, G., and W.J. Mitchell. 1980. The grammar of paradise: On the generation of Mughul gardens. *Environment and Planning B: Planning and Design* 7 (2): 209–226.

Stouffs, R. 1994. *The algebra of shapes*. Ph.D. dissertation. Pittsburgh: Carnegie Mellon University.

Stouffs, R. 2018. A practical shape grammar for Chinese ice-ray lattice designs. In *Computational studies on cultural variation and heredity*, ed. J.-H. Lee, 161–174. Singapore: Springer.

Stouffs, R. 2022. A multi-formalism shape grammar interpreter. In *Computer-aided architectural design. Design imperatives: The future is now. CAAD futures 2021*, ed. D. Gerber et al., vol. 1465. Communications in computer and information science, chapter 17. Singapore: Springer.

Stouffs, R., and R. Krishnamurti. 1993. *The complexity of the maximal representation of shapes*. Tech. rep. School of Architecture, Carnegie Mellon University.

Stouffs, R., and R. Krishnamurti. 2006. Algorithms for the classification and construction of the boundary of a shape. *Journal of Design Research* 5 (1): 54–95.

Stouffs, R., and R. Krishnamurti. 2019. A uniform characterization of augmented shapes. *Computer-Aided Design* 110: 37–49.

Stouffs, R., and M. Wieranga. 2006. The generation of Chinese ice-ray lattice designs on 3D surfaces. In *Communicating space(s), 24th eCAADe conference proceedings, eCAADe*, Volos, 316–319.

Tapia, M. 1992. Chinese lattice designs and parametric shape grammars. *The Visual Computer* 9: 47–56.

Tapia, M. 1996. *From shape to style. Shape grammars: Issues in representation and computation*. Ph.D. dissertation. Toronto: University of Toronto.

Tapia, M. 1999. A visual implementation of a shape grammar system. *Environment and Planning B: Planning and Design* 26 (1): 59–73.

Tching, J., J. Reis, and A. Paio. 2019. IM-sgi: An interface model for shape grammar implementations. *AI EDAM* 33 (1): 24–39.

Tereshchenko, V., S. Chevokin, and A. Fisunenko. 2013. Algorithm for finding the domain intersection of a set of polytopes. *Procedia Computer Science* 18: 459–464.

Trescak, T., M. Esteva, and I. Rodriguez. 2012. A shape grammar interpreter for rectilinear forms. *Computer Aided Design* 44 (7): 657–670.

Williams, R. 1979. *The geometrical foundation of natural structure: A source book of design*. Dover Publications.

Woodbury, R.F., and A.L. Burrow. 2003. Notes on the structure of design space. *International Journal of Architectural Computing* 1 (4): 517–532.

Woodbury, R.F., A. Mohiuddin, et al. 2017. Interactive design galleries: A general approach to interacting with design alternatives. *Design Studies* 52: 40–72.

Wortmann, T., and R. Stouffs. 2018. Algorithmic complexity of shape grammar implementation. *Artificial Intelligence for Engineering Design, Analysis and Manufacturing* 32 (2): 138–146.

Yu, Y., et al. 2021. Rethinking origami: A generative specification of origami patterns with shape grammars. *Computer-Aided Design* 137: 103029.

Yuan, X., J.-H. Lee, and Y. Wu. 2011. A new perspective to look at ice-ray grammar. In *Circuit bending, breaking and mending: Proceedings of the 16th international conference on computer-aided architectural design research in Asia*, 81–89.

Yue, K., and R. Krishnamurti. 2013. Tractable shape grammars. *Environment and Planning B: Planning and Design* 40 (4): 576–594.

Yue, K., and R. Krishnamurti. 2014. A paradigm for interpreting tractable shape grammars. *Environment and Planning B: Planning and Design* 41 (1): 110–137.

Yue, K., R. Krishnamurti, and F. Grobler. 2012. Estimating the interior layout of buildings using a shape grammar to capture building style. *Journal of Computing in Civil Engineering* 20 (1): 113–130.

Open Access This chapter is licensed under the terms of the Creative Commons Attribution-NonCommercial-NoDerivatives 4.0 International License (http://creativecommons.org/licenses/by-nc-nd/4.0/), which permits any noncommercial use, sharing, distribution and reproduction in any medium or format, as long as you give appropriate credit to the original author(s) and the source, provide a link to the Creative Commons license and indicate if you modified the licensed material. You do not have permission under this license to share adapted material derived from this chapter or parts of it.

The images or other third party material in this chapter are included in the chapter's Creative Commons license, unless indicated otherwise in a credit line to the material. If material is not included in the chapter's Creative Commons license and your intended use is not permitted by statutory regulation or exceeds the permitted use, you will need to obtain permission directly from the copyright holder.

Shape Grammars in Art, Craft, and Engineering

How Is That? Computing the Temporality of Drawing

Terry W. Knight

Abstract Process and time are key to studying, appreciating, designing, and making things. An essential question, then, for creative production is not just *What is that?* but *How is that?*—in other words, *How did that or, how can that, come to be?* As a process carried out over time, computation offers a means for rethinking, representing, and elevating the *How* in designing and making activities. Shape grammars are a special computational theory that have, from their inception 50 years ago, been directed toward the *How* of design. Building on the theory and applications of shape grammars, making grammars also prioritize the *How*, with a focus on the physical making of material things. Aspects of the time or temporality of making can be computed with making grammars too. A making grammar for the drawings of the artist Fritz Glarner is discussed here to demonstrate a computational framework for capturing some dimensions of time as it unfolds in creative processes.

> What is good is form-giving. What is bad is form. Form is the end, death, Form-giving is movement, action. Form-giving is life. These sentences constitute the gist of the elementary theory of creativity. (Klee 1973b, 269)
>
> The cardinal question: "How shall I give form to the movement from here to there?" embodies a time factor. (Klee 1973a, 31)
>
> Lines are the currency of architectural drawing. … the sequence of their making is, itself, another form of embedded time in the act of drawing. (Bryan and Grosman 2016, 101)

1 Introduction

The artist Paul Klee championed a philosophy of art and creativity that foregrounded becoming over being (Klee 1964, 307). He prioritized genesis and growth, and the path to form over form itself. As he put it, form is "an evil and dangerous

T. W. Knight (✉)
Massachusetts Institute of Technology, Cambridge, USA
e-mail: tknight@mit.edu

spectre" (Klee 1973b, 269). The path to form, by contrast, carries the essence of creativity—its vitality, its action, its movement. Klee's ideas influenced a host of prominent twentieth century philosophers—Walter Benjamin, Gilles Deleuze, Jacques Derrida, Michel Foucault, Martin Heidegger, Maurice Merleau-Ponty, among others—in their thinking on aesthetics and more. They continue to inspire and shape the work and worldviews of writers and scholars, as well as artists and designers, up to the present day. A key question, then, for the study and production of art and design, viewed through the lens of becoming or process, is not *What is that?* but *How is that?*—in other words, *How did that, or how can that, come to be?* Change, time, and the active, physical carrying out of a creative process are fundamental to answering the *How* question.

Computation is a process.[1] However, computational theories, methods, and technologies for art and design developed over the past decades have typically prioritized the *What*. They have been geared toward realizing particular results or ends, with whatever process or means works best. The *How* is often of technical interest and limited to the developers, and sometimes the users, of computational design techniques. However, a deeper or different focus on computational processes holds potential for realizing the broader significance and value of the *How* in art and design. This potential has mostly been neglected.

A singular exception is shape grammars and its recent adaptation, making grammars. Shape grammars and making grammars are generative, rule-based systems that are used to compute forms, either abstract or physical. However, a central aim of shape grammars from the start, and more recently that of making grammars, has been to expose the means, not just supply the ends. The aim has been to make visible the *How*, to pull back the curtains on the "process of giving form" (Klee 1973b, 269), whether that is a forward-looking creative process or backward-looking analytic one.[2] The unique visual and spatial nature of the rules and computations of shape and making grammars allow the *How* to come forward. Making grammars, in particular, were developed to express and understand the physical processes or paths to material forms, with rules that encode the knowledge and know-how of the maker (Knight and Stiny 2015).

Genesis, growth, becoming, and the path to form are processes carried out over time. So too is any computational process, including shape and making grammar computations. Making grammars express the *How* of creative making processes and can include the temporality of making, as demonstrated in a recent case study (Knight 2018). Making grammars also offer a possibility, not considered in that study, to answer a particular temporal aspect of the *How* question: *How long does this take to make?* Here, that possibility is explored through a study using drawing. Of all the arts, drawing has most often been singled out as expressive of time. It makes a particularly apt study. The drawings of the artist Fritz Glarner, which he made in preparation for his paintings, are the subject of the study.

---

[1] A 2011 ACM Ubiquity symposium on the question "What is Computation?" concluded with a first major point of agreement that computation is a process. See Denning (2011, 1).

[2] The idea of a reading creativity forward versus backward is taken up by Ingold (2009, 97).

2 Some Background on Process, Time, and Drawing

Computational methods for art, architecture, craft, and design have for the most part focused on productivity, on getting particular results. From the early algorithmic and computer art of the 1960s to art-making programs like AARON (Cohen, 1995) to evolutionary, genetic and cellular automata generated art and design to parametric design and AI techniques today, the interest has mainly been on outcomes. The underlying processes are mostly just that—underlying. Even among non-digital artists or designers who have used manual algorithms or instructions to create art, the instructions may have been foregrounded, but not necessarily the processes of carrying them out. Conceptual artists of the 1960s and 70s are a telling example. Sol LeWitt, a founder of conceptual art known for his instructions-produced Wall Drawings,[3] famously declared:

> In conceptual art the idea or concept is the most important aspect of the work. When an artist uses a conceptual form in art, it means that all of the planning and decisions are made beforehand and the execution is a perfunctory affair. The idea becomes a machine that makes the art. (LeWitt 1967, 79)

In other words, the concept as encoded by the instructions is what matters. The physical, embodied process of executing the instructions is of little account.

Some approaches within the art and design world, both computational and non-computational, have challenged this exclusive attention to the conceptual or physical thing. Some have drawn directly or indirectly from a philosophical tradition called "process philosophy." Alfred North Whitehead, the philosopher often associated with process philosophy, asserted the primacy of becoming over being, as Klee did (the two wrote at roughly the same time). For Whitehead, material things are always in the making and exist only as ongoing processes of becoming. There are no fixed things in the world: "apart from happenings there is nothing" (Whitehead 2004, 66). The Process Art movement of the 1960s advocated a similar point of view. Process artists critiqued the minimalist art and conceptual art focus on predetermined plans and final objects. They promoted an art practice grounded in process and events, and in making process visible as the primary content of a work of art. Robert Morris, one of the artists best known for this view, proposed an "anti form" philosophy in opposition to what he called "object-type" art. In his "Notes on the Phenomenology of Making," he wrote:

> Much attention has been focused on the analysis of the content of art making—its end images—but there has been little attention focused upon the significance of the means. … I believe there are "forms" to be found within the activity of making as much as within the end products. (Morris 1970, 62)

---

[3] For example, the instructions for LeWitt's Wall Drawing #118 at the SMFA Boston, 1971 are: "On a wall surface, any continuous stretch of wall, using a hard pencil, place fifty points at random. The points should be evenly distributed over the area of the wall. All of the points should be connected by straight lines."

Moreover, he attributed "the deep-seated tendency to separate ends and means within this culture to the simple fact that those who discuss art know almost nothing about how it gets made" (Morris 1970, 62).

Time is an essential dimension of process and it has naturally played a role in the work of process artists. Morris created his series of *Blind Time Drawings* to call attention to the embodied, the material and, importantly, the temporal nature of drawing. Recalling and perhaps referencing LeWitt's drawings, which were produced over roughly the same time period, Morris (and others he asked) made these drawings blindfolded, with their hands covered in graphite and according to timed sets of instructions that Morris predefined. But the execution of the drawings was hardly perfunctory; it was the essence of the work. In making these drawings, Morris diminished the sense of sight in favor of the tactile and material. He foregrounded the gestural quality of graphite strokes on paper, heightening the viewer's awareness of the moving hand and the traces it leaves as a record of time.

Morris's *Blind Time Drawings* were made over several decades from the early 1970s to the late 2000s, a time which saw the beginnings of a resurgence of interest and discourse on drawing as a medium in its own right (Kurczynski 2011). Since then, artists, architects, historians, and theorists alike have distinguished drawing, among art and design media, as an inherently active, movement-based, and temporal practice. In characterizing drawing, writers frequently echo the words of Whitehead and Klee. The art historian Norman Bryson writes: "If Painting presents Being, the drawn line presents Becoming" (2003, 150). In her introduction to a global survey of contemporary drawing, the artist Emma Dexter notes that "drawing forever describes its own making in its *becoming*" (2005, 6), and is "an unmediated record of an act" (2005, 7). And the art critic, artist, and writer, John Berger asks, "Isn't the act of drawing, as well as the drawing itself, about *becoming* rather than *being*?" (2005, 124).

Time is fundamental to becoming and is identified as a fundamental condition of drawing as well.[4] Tim Ingold, the anthropologist who has written extensively on lines, making, and more in relation to culture and life, points out that "drawing the line takes time. It cannot be reduced to a single instant" (2016, 62). Contrasting drawing to painting, the art historian Michael Newman remarks similarly that "drawing is closer to the movement of time, to lived temporality" (2003, 96) and that it "takes on the quality of an event" (2003, 105). And Dexter says of drawing that "it is a map of time recording the actions of the maker" (2005, 10). From a different perspective, psychologists and cognitive scientists have looked to the temporality of drawing as a way to understand the cognitive structure of the drawing activity, including the presence of perceptual "chunks" of active drawing times and pauses in between them (Cheng et al. 2001; Mao et al. 2020). Computer scientists interested in developing interactive graphics applications for sketching have turned to temporal features of

---

[4] See, for example, the description of drawing in the call for proposals for the Drawing Research Network Conference 2020: Temporal Drawing https://blog.lboro.ac.uk/tracey/drn2020-temporal-drawing/.

drawing—in particular, the pattern and order of drawing strokes over time—in order to facilitate sketch recognition (Arandjelovic and Sezgin 2011).

3 Time and Making Grammars

Activities, events, and processes in the world, including drawing, are continuous, ongoing, and multisensory. Computations, on the other hand, are abstract, formal, and discrete. However, a widely accepted notion in cognitive science is that the way people make sense of what is happening in the world, in other words, the way people perceive, understand, learn, predict, and act in the world, is by parsing (cognitively) the continuous stream of actions and events into discrete temporal parts or segments (Zacks and Tversky 2001). In computer science, AI, and robotics, this concept of temporal segmentation has been key to building computational models and robots to recognize, represent and generate human-like actions and gestures. (See, for example, the work of Spriggs et al. 2009.)

The idea behind making grammars intersects with these notions. A making grammar describes how things are made spatially and, at the same time, how things are made temporally. The rules express gestures or actions in time that correspond to temporal parts or segmentations of a continuous making process. In a making rule $M \rightarrow N$, the replacement operation $\rightarrow$ is interpreted as an action in time. That action might be a "doing" action (drawing, folding, painting, and so on) or a "sensing" action (seeing, touching, and so on). A making rule defines a temporal segmentation of the making of a thing when it is applied in a computation. In other words, when a making rule $M \rightarrow N$ is applied to a thing T to make another thing T′, it divides a continuous making process into a temporal segment with boundaries defined by T and T′.

To demonstrate this idea, a making grammar was developed recently to express the discrete temporal actions that comprise the craft activity of making *kolam* in India (Knight 2018). Kolam are made by laying down lines of rice powder on the ground to create intricate patterns. The making grammar encodes the maker's actions through rules that express both the *doing* actions of the maker—laying down or drawing lines of rice powder—and the *sensing* actions of the maker—seeing or looking ahead to determine the next step. The rules thus identify how this highly creative and more or less continuous activity might be understood temporally by decomposing it into distinct actions over time. The identification of temporal segments or actions in a making process is one aspect of understanding the temporality of making. Making grammars can be used to characterize another essential aspect of the temporality of making—the durations or times, either absolute or relative, of those actions. The following study explores this possibility through the drawings of the painter Fritz Glarner.

4 Timing the Drawings of Fritz Glarner

Fritz Glarner was a Swiss-born artist who emigrated to the US in the 1930s. He met and became close friends with the *de Stijl* luminary, Piet Mondrian, when Mondrian emigrated to the US in 1940. Around that time, Glarner's paintings switched from figurative and naturalistic themes to abstract ones. His early abstract paintings show the influence of Mondrian and *de Stijl*—they are composed of horizontal and vertical divisions of a canvas into rectangles. However, his work converged rapidly into his own unique style of form and color composition that he called Relational Painting. In his Relational Paintings, Glarner introduced a subtle (but radical in *de Stijl* terms) dynamic into an otherwise conventional rectangular division format by creating oblique divisions. For the next thirty years, until his death, Glarner experimented with and developed this style, producing progressively more complex and systematic paintings. The evolution of his Relational Paintings over thirty years can be divided into stages as described by Knight (1989). In that study, each stage is characterized with a shape grammar and the transformations from one stage to the next are characterized in terms of transformations of the rules of the grammars. Figure 1 shows a painting from 1957, which is typical of the paintings that Glarner produced in the stage between 1958 and 1962.

The painting exemplifies a very particular pattern of subdivisions that characterizes paintings from this stage. Figure 2 illustrates the subdivisions. They are sequenced in a way that likely corresponds to the actual process that Glarner used to make these paintings. First the canvas is divided into rectangles by orthogonal divisions, next the rectangles are divided into wedge-shapes by oblique divisions and, last, the wedge-shapes are divided at their wider ends to form borders. The oblique divisions follow an inviolable principle—the divisions always slant down right to left, either from the top to the bottom of a rectangle or from the right side of

Fig. 1 A Fritz Glarner painting typical of paintings he produced between 1958 and 1962. (*Photo Credit* The Art Institute of Chicago/Art Resource, NY Artist: Glarner, Fritz (1899–1972) Description: Relational Painting, No. 82. 1957. Oil on canvas, 111.5 × 81.3 cm (43 7/8 × 32 in.). Bequest of Richard S. Zeisler.)

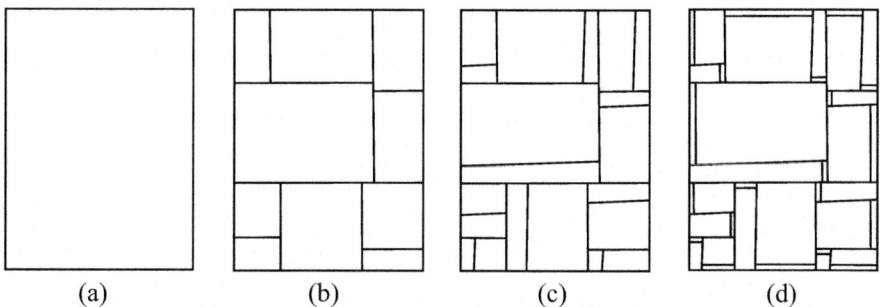

Fig. 2 The sequence of subdivisions underlying the painting in Fig. 1 and other paintings produced during the same time period. (**a**) The boundaries of the canvas. (**b**) Orthogonal divisions of the canvas into rectangles. (**c**) Oblique divisions of rectangles into wedge-shapes. (**d**) Divisions of wedge-shapes to produce borders of wedge-shapes

a rectangle to its left side. Additionally, when two rectangles form a larger rectangle, the oblique divisions are set at more or less right angles to one another (that is, not parallel). There are a few exceptions to the latter principle in some paintings from this stage. The subdivided areas of the canvas are painted in primary, black, and gray colors. The distribution of colors also follows rules though not with the same consistency as the subdivisions. The proportions of the subdivisions and the variations in the sizes of the subdivided areas vary from painting to painting and are less easy to characterize.

Glarner produced drawings in preparation for his Relational Paintings. Few can be found, though, perhaps because they were not saved or are in private collections. Figure 3 shows a drawing from 1963, the start of a stage that followed the stage exemplified by the painting in Fig. 1. There are additional subdivisions of wedge-shapes in this drawing that Glarner did not use in the previous stage.

Fig. 3 A Fritz Glarner drawing made in preparation for a painting (*Photo Credit* The Art Institute of Chicago/ Art Resource, NY Artist: Glarner, Fritz (1899–1972) Description: Color drawing for relational painting. Printed by Zigmunds Priede, published by Universal Limited Art Editions. 1963. Lithograph in black on white wove paper, image: 52.6 × 31.8 cm (20 3/4 × 12 9/16 in.)

This study focuses on the drawings that Glarner would have made in preparation for paintings like the one shown in Fig. 1. A making grammar that generates those drawings is outlined. It generates drawings for paintings Glarner actually produced as well as for hypothetical paintings. Like other making grammars, the rules of this grammar encode doing actions and sensing actions. Here, doing is drawing or erasing, and sensing is seeing. The rules also encode the time or duration of drawing or erasing strokes, as well as the time or duration of pauses for seeing and making decisions about next steps (like the pauses taken into account in cognitive science work mentioned previously). Time in this study is relative, not absolute, as absolute time is impossible to define or even guess at without some empirical information regarding Glarner's drawing process. Relative time, however, allows for comparisons between the times taken to produce different drawings, or the times taken to produce the same drawing developed in different ways. The grammar is not aimed at identifying temporal segments that distinguish different kinds of doing actions—an aim of the kolam making grammar (Knight 2018) and the knotting making grammar described by Knight and Stiny (2015). Here, the doing actions are all the same—the action of drawing or erasing a single, straight line.

The grammar is entirely speculative as there are no recordings or information available that show how Glarner actually made his drawings. It is an approximation of a possible drawing process. It is presented as a framework for a more detailed grammar based on "ground truth" or empirical data, if available. More generally, the grammar demonstrates a framework for computing the temporality of making activities beyond just drawing.

5 A Making Grammar for Glarner Drawings

The grammar is divided into five stages:

   I. Boundaries of a drawing
  II. Orthogonal divisions into rectangles
 III. Oblique divisions into wedge-shapes
  IV. Wedge border divisions
   V. Finish

The first four stages correspond to the stages shown in Fig. 2.

The rules of the grammar are either sensing rules that represent seeing, or doing rules that represent drawing or erasing. A sensing rule corresponds to seeing and identifying what part of a drawing to work on next, that is, where to make the next drawing stroke. The time needed to scan an entire drawing and decide where to do this is the key factor in calculating the time for a sensing rule. In other words, the time for a sensing rule is *decision time*.

A doing rule corresponds to making or erasing a line. The time needed to decide where and how to make or unmake a stroke within an identified portion of a drawing is one factor in calculating the time to implement a doing rule. The time needed to physically make or erase a stroke is another factor. In other words, the time for a doing rule includes *decision time* and physical, stroke or *gesture time*.

The rules and the drawings they compute are annotated with perceptual information or features of a drawing that an artist might monitor and track in making a drawing. These features control the progression from one stage to the next by indicating when a stage is complete and it is okay to move on to the next stage. The annotations are given by K [R, W, A] where

K =   a sensing (👁) or doing (✎) state.
R =   the number (density) of rectangles in a drawing.
W =   the number (density) of wedge-shapes in a drawing.
A =   the number of adjacent rectangles to check when making oblique divisions of rectangles in a drawing.

When a drawing is in a sensing state, then a sensing rule can be applied to see something in the drawing. When a drawing is in a drawing state, then a doing rule can be applied to make or erase a line. In order to apply a rule, the values for R, W, and A in the rules also have to match the values of R, W, and A in the current drawing.

The drawings in the rules are assumed to be parameterized, that is, the dimensions and proportions of rectangles and wedge-shapes are variable. The dimensions and proportions of shapes in Glarner's drawings for paintings (and in the paintings themselves) are certainly constrained in some ways, but conditions on parameters are not considered explicitly here and are not critical to this demonstration study. However, the existence of parameters is important and is considered in understanding and calculating the decision times for sensing and doing rules. Some liberties are taken with other technical details that would be possible to include in the rules, but those details are omitted here as they would obscure the main idea of the grammar. The grammar is defined to disallow dead ends, in other words, computations that would lead to an incorrect or incomplete drawing, something Glarner would likely not have done. Thus, no labeling devices or tricks are used that would allow for computations to end in designs with labels and, therefore, not in the language defined by the grammar (according to the original technical definition of shape grammars). Rules are defined to correspond to what an artist might think about, see, and do in making a drawing. There is one important exception—the rules preclude returning to a previous stage to rethink or change moves in a previous stage. In a real-world drawing scenario, returning or backtracking happens and is a necessity. The rules given here would need to be elaborated to do that.

5.1 Stage I Boundaries of Drawing

Figure 4 shows the initial shape of the grammar—a blank drawing represented by 0 and in a drawing state.

Initial shape

Fig. 4 The initial shape of the making grammar

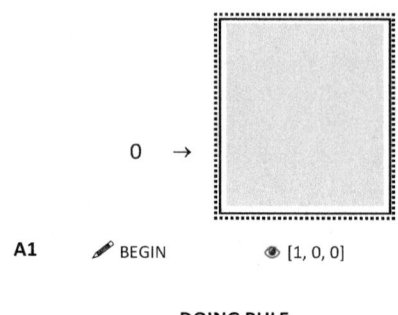

DOING RULE

Fig. 5 The Stage I rule to determine the boundaries and area of a drawing

Figure 5 shows the first rule **A1** of the grammar. It is a doing rule that applies to draw the boundaries of a drawing and determine its area. The left side of the rule is the initial shape, and the right side depicts the drawing's boundaries. Shading within the boundaries here, and in subsequent rules, represents rectangles perceived as planar areas. Shading also has a practical role in the rules, disambiguating rectangles from larger ones that might contain them. The dotted lines around the boundaries represent the "outside" of a drawing; they distinguish boundary lines from subdivision lines added subsequently within a drawing. A seeing state is associated with the right side. Thus, when the rule is applied, the state of a drawing changes from a drawing state to a seeing state.

The rule has a relative time or duration associated with its application. The time is *gesture time*, or the time to make drawing strokes. It is assumed here that longest (or maximal) lines are each drawn as a continuous stroke. The gesture time of rule **A1** is equal to the sum of the lengths of the boundary lines. This time may have no relation to the real time taken to draw the boundary lines. But as mentioned previously, the calculation of time here is not absolute, but relative, allowing for comparisons of the times taken to produce different drawings or different ways to produce the same drawing.

5.2 Stage II Orthogonal Divisions

Figure 6 shows the rules for dividing a drawing into rectangles. All of these rules are optional, though in practice Glarner made at least two divisions of a drawing (or painting) into rectangles. Rules **r1** and **r2** are sensing rules. They apply to see and identify within a current drawing which rectangle to subdivide or which rectangles to merge. Applications of these rules change the state of a drawing from a seeing state to a drawing state. Rules **R1** and **R2** are doing rules. They apply to draw or erase a dividing line. Applications of these rules change the state of a drawing from a drawing state back to a seeing state.

The perceptual features R, W, and A in the rules guide rule applications. The rules in this stage can only apply when there are no wedge-shapes yet defined (W = 0) and there are no adjacent rectangles to check (A = 0), which happen in subsequent stages. Thus, these rules can only apply in this stage, not in a subsequent stage. When either rule **r1** or rule **r2** applies, the values for R, W, and A remain the same. However, each time rule **R1** or rule **R2** is applied, the number of rectangles R is updated, either increased or decreased by 1.

Each of the rules in this stage has a relative time or duration associated with it. The time for rules **r1** and **r2** is *decision time*, the time taken to scan a current drawing, see, and decide which rectangle or rectangles to act on. In this demonstration study, the decision time for the two rules is based on the visual complexity of a drawing. In other words:

$$decision\ time = visual\ complexity.$$

Visual complexity is notoriously difficult to define. A multitude of ways have been proposed in the literature, mostly based on empirical studies. A common theme, however, is that visual complexity is multi-dimensional and that the amount of elements and the variety of elements in an image are key factors [for example, see

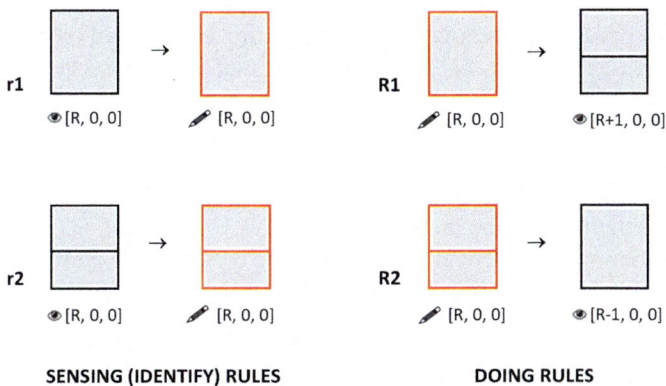

Fig. 6 Stage II rules for orthogonal divisions of a drawing into rectangles

Gartus and Leder (2017, 4) and Palumbo et al. (2014, 4)]. Accordingly, a simple but plausible measure of visual complexity is given here in terms of the *density* and the *diversity* of a drawing. The density of a drawing is equal to the number of rectangles in it. In other words:

$$density = R.$$

The diversity of a drawing is calculated as the difference between the areas of the largest and smallest rectangles in a drawing. In order to compare diversity across different drawings, the areas of rectangles are taken as the percentages of their areas in relation to the entire area of a drawing:

$$diversity = \%\,(area\ of\ largest\ rectangle - area\ of\ smallest\ rectangle)/$$
$$area\ of\ drawing$$

The visual complexity of a drawing is then calculated here simply as:

$$visual\ complexity = density \times diversity.$$

Certainly other factors are at play in measuring visual complexity. Arrangement is another factor that likely plays a role in visual complexity, but is not taken into account here. The measure here serves as a reasonable placeholder for a more nuanced or detailed measure.

Figure 7 shows visual complexity measures for three different drawings. The drawings in Fig. 7a and b have the same number of rectangles, but very different complexity values. The drawing in Fig. 7a has rectangles of more uniform size than those in the drawing in Fig. 7b, thus its complexity value is lower. The drawing in Fig. 7c has one more rectangle than the drawing in Fig. 7a, but because the sizes of the rectangles are more uniform overall, its complexity value is just slightly lower.

The time for rules **R1** and **R2** is *decision time* and *gesture time*. For rule **R1**, decision time is the time needed to decide where to make a division within an identified rectangle. The visual complexity of the surrounding context and other factors may play a role in this decision, but for the sake of simplicity, they are not taken into account here. Decision time is set arbitrarily here at a constant value of *30*, as any value would allow comparisons between paintings. For rule **R2**, decision time is set at a value of *0* as virtually no time is needed to decide where to erase a line already present in an identified rectangle. The gesture time for each of rules **R1** and **R2** is taken as the length of the line drawn or erased.

Figure 8 shows a hypothetical Glarner drawing typical of the period studied here. It is used to illustrate the application of the rules of the grammar. The dimensions of the orthogonal divisions and the areas of the wedge-shapes, including their borders (which combine to equal the areas of the rectangles they comprise), are indicated. These values are used for the calculations of the relative times of rule applications.

How Is That? Computing the Temporality of Drawing 111

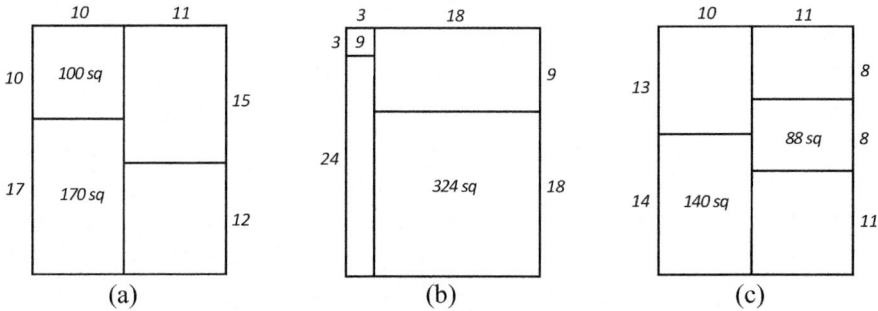

Fig. 7 The visual complexity values of three drawings. The dimensions of the drawings are given, without units of measurement. (**a**) Visual complexity = % (170 − 100/1718) × 4 = 16. (**b**) Visual complexity = % (324 − 9/1718) × 4 = 72. (**c**) Visual complexity = % (140 − 88/1718) × 5 = 15

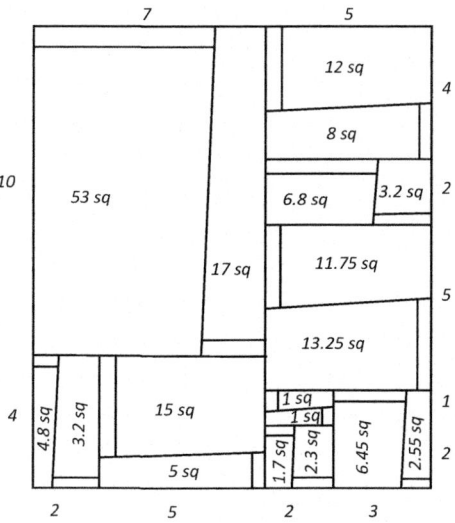

Fig. 8 A hypothetical Glarner drawing typical of the period studied here. It is used to illustrate the application of the rules of the grammar and the calculations of relative times for computations

The values are shown as unitless but could be understood as inches, for a 14″ × 12″ drawing.

Figure 9 shows a computation of the drawing through Stages I and II. Dimensions (without units) of rectangles are shown through the first four steps, along with the relative time calculations for these steps. The relative time for the entire computation, beginning with the initial shape, is *1490*.

For comparison, a different computation of the last drawing in Fig. 9 is shown in Fig. 10. In this computation, divisions are always made in the smallest possible rectangle in a current drawing, creating more visual complexity of the drawings at each step, compared to the computation in Fig. 9. The relative time for this computation is *2082*, much higher than the time for the computation in Fig. 9.

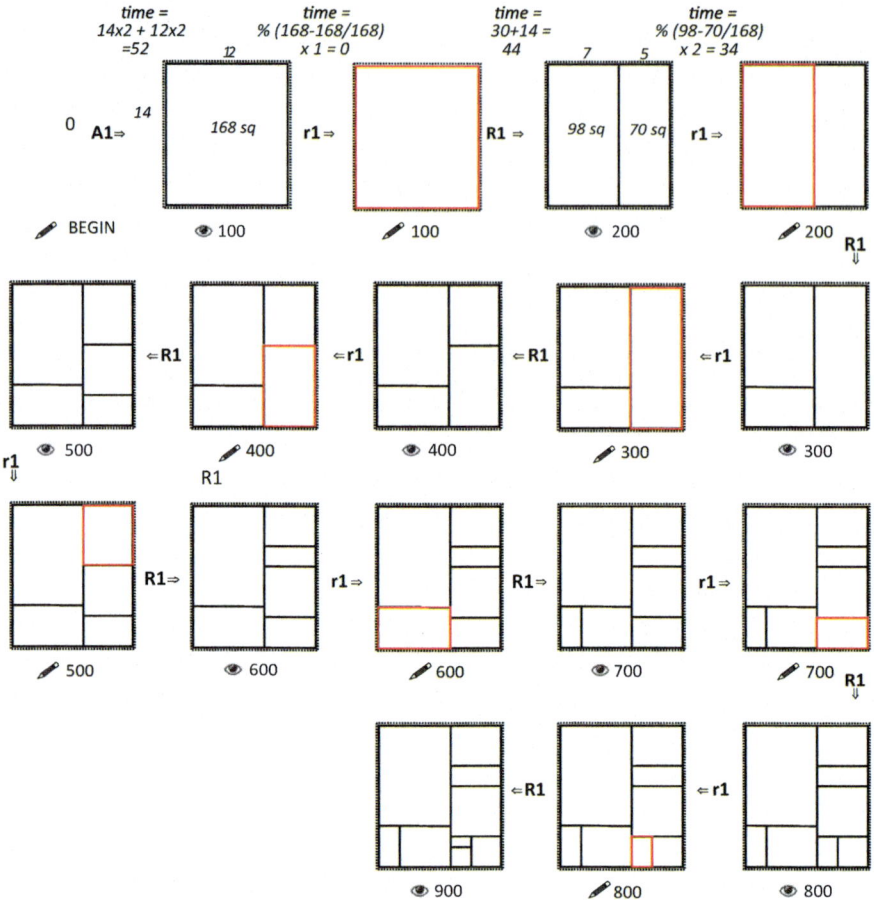

Fig. 9 A computation of a drawing beginning with the initial shape through Stage I and Stage II. Dimensions (without units) of rectangles are shown in the first four steps, along with the relative time calculations for those steps. The relative time for the entire computation is 1490. For simplicity of presentation, shading of rectangles (and later, wedge-shapes) as well as brackets around perceptual features are omitted here and in subsequent computations

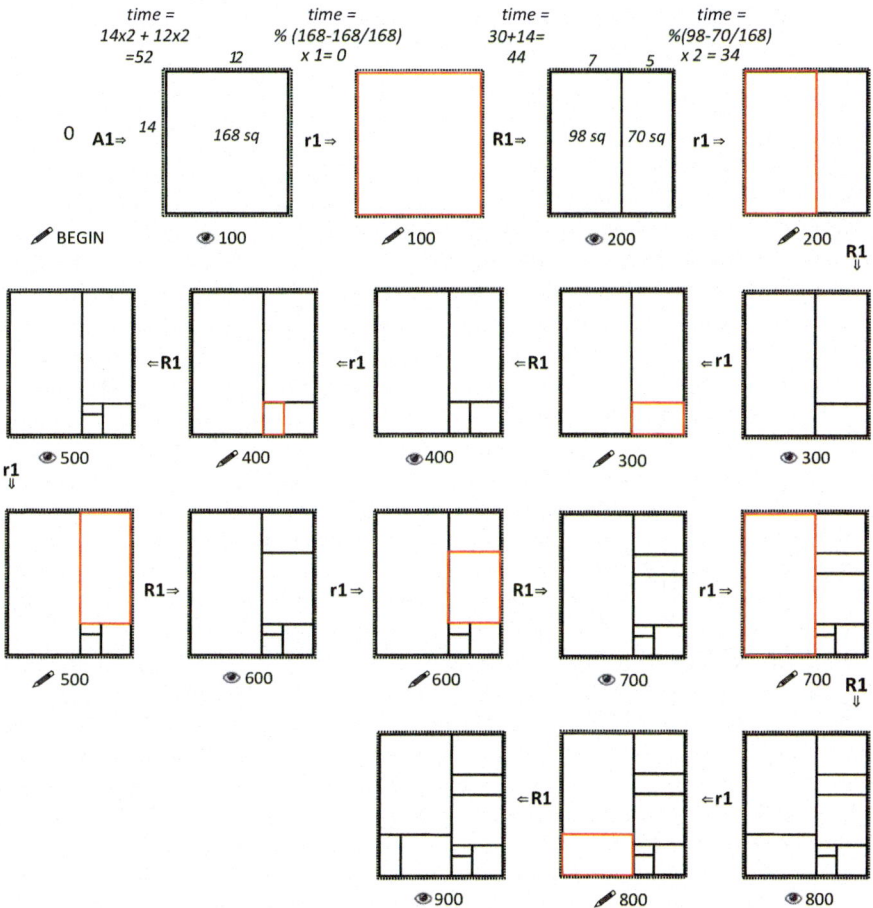

Fig. 10 An alternate computation of the last drawing in the computation in Fig. 9. Divisions are always made in the smallest possible rectangle in a current drawing, thus there is more visual complexity of the drawings in this computation compared to the one in Fig. 9 computation. The relative time for this computation is 2082, much higher than that for the computation in Fig. 9

5.3 Stage III Oblique Divisions

Figure 11 shows the rules for making mandatory oblique divisions of all rectangles in a drawing into wedge-shapes. As mentioned previously, oblique divisions always slant down right to left, either from the top to the bottom of a rectangle or from side to side. When two rectangles share a side to make a larger rectangle, the divisions within them are approximately orthogonal to one another. In an isolated rectangle (a rectangle not sharing a side with another rectangle), the oblique division can be from top to bottom or from side to side.

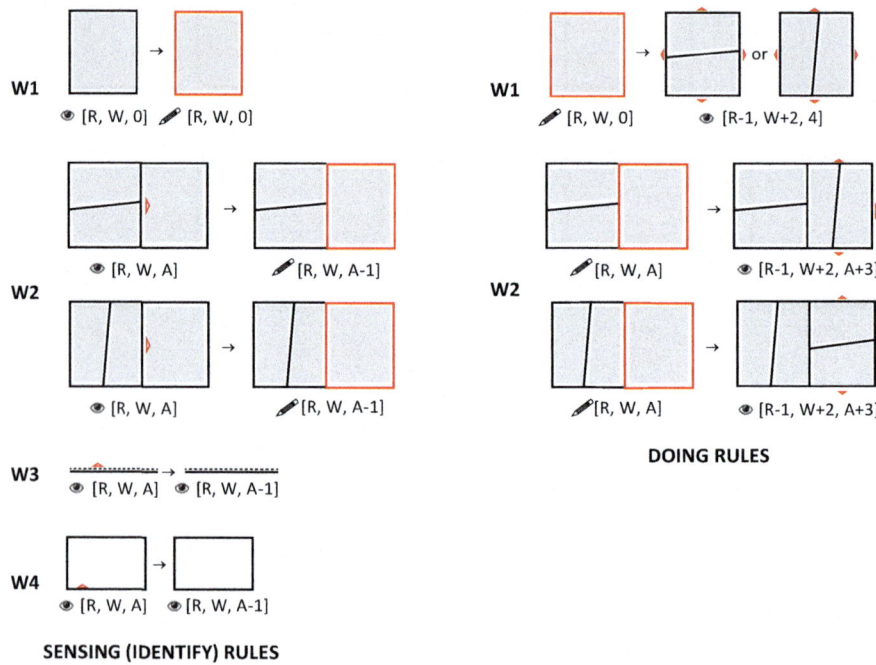

Fig. 11 Stage III rules for oblique divisions of rectangles into wedge-shapes

Rule **w1** is a sensing rule. It applies to see and identify an isolated rectangle in a drawing. It is the first rule to be applied in this stage, and it takes a drawing from a seeing state to a drawing state. This rule can only be applied when the perceptual feature A is equal to 0, which indicates that all previously divided rectangles have been checked for adjacent rectangles that share entire sides. The perceptual features R and W can be any value. Application of the rule does not change the values of R, W, and A.

Rule **w1** is followed by rule **W1**, a doing rule, which applies to draw an oblique dividing line side to side or top to bottom in a rectangle identified by rule **w1**. At the same time, a triangle or "eye" is added to each side of the rectangle that "sees" the length of the side it is on and is used to check for any rectangles adjoining it that share an entire side. The eye is uniquely located at the midpoint of the side and its width is a factor of the length of the side. When the rule **W1** is applied, the values of R, W, and A are updated accordingly—the number R of rectangles is decreased by 1, the number W of wedge-shapes is increased by 2, and the number A of sides to check is increased by 4. The state of a drawing changes from a drawing state back to a seeing state.

The **w2** rules are sensing rules that apply to see and identify a rectangle that adjoins a subdivided rectangle and forms a larger rectangle with it. The eye on the shared line is "deactivated" or erased, and the value of A is decreased by 1. The state of a drawing changes from a seeing state to a drawing state. These rules are followed by **W2** rules that apply to draw an oblique dividing line in a rectangle identified by a **w2** rule. At the same time, new eyes are added to the divided rectangle to check for any rectangles adjoining it that share an entire side. The value of R is decreased by 1, W is increased by 2, and A is increased by 3. The state of the drawing changes from a drawing state back to a seeing state. Because the orientation of the oblique dividing line is important, the **W2** rules apply under certain transformations that preserve the correct orientation: scale, translation, 180° rotation, and reflection + 90° rotation. Figure 12 illustrates all allowable **W2** rules between two rectangles that combine to form a larger rectangle.

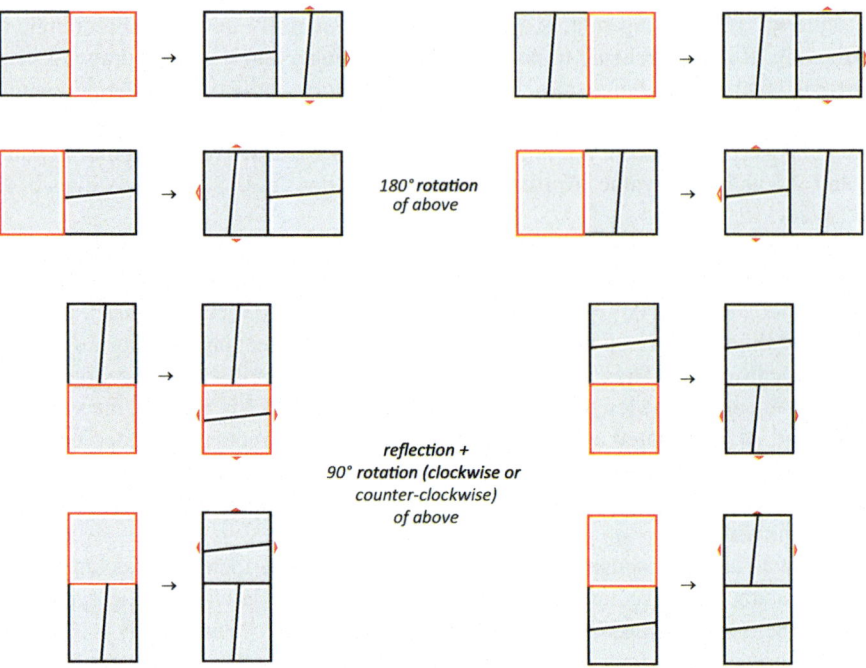

Fig. 12 All allowable **W2** rules between two rectangles that combine to form a larger rectangle. The top two rules are the two **W2** rules shown in Fig. 11

Rules **w3** and **w4** are sensing rules that deactivate or erase an eye on the side of a rectangle that does not form a larger rectangle with an adjacent rectangle. In rule **w3**, the eye is on a boundary line. In rule **w4**, the eye "sees" that it does not share an entire side with an adjacent rectangle because of its location and width in relation to the adjacent rectangle. When either rule **w3** or **w4** is applied, the values of R and W remain the same, and the value of A is decreased by 1. The drawing remains in a seeing state.

Each of the rules in this stage has a relative time or duration associated with it. The time for rule **w1** is *decision time*: the time taken to scan a current drawing, find, see, and decide which rectangle to act on. Here, the decision time is based on the visual complexity of a drawing, just as for rules **r1** and **r2**:

$$decision\ time = visual\ complexity.$$

However, there are fewer undivided rectangles to choose from with each application of the rule. Fewer options may or may not decrease decision time. For this study, the decrease in undivided rectangles is not taken into account. Visual complexity is measured in the same way as for rules **r1** and **r2**, using the density and diversity of isolated rectangles and wedge-shapes (that is, polygons) taken together in the calculations. The visual complexity of a drawing is thus:

$$visual\ complexity = density \times diversity$$

where

$$density = R + W$$

and

$$diversity = \%\,(area\ of\ largest\ polygon\ -\ area\ of\ smallest\ polygon)/\\ area\ of\ drawing.$$

The time for rule **W1** is *decision time* and *gesture time*. The decision time is the time needed to decide where to make an oblique division within an identified rectangle. It is calculated in the same way as for rule **R1**. It is set at a constant value of *30*, as any value would allow comparisons between paintings. The gesture time for rule **W1** is also calculated in the same way as rule **R1**. It is equal to the length of the line drawn.

The time for rule **w2** is *decision time*, the time needed to see a rectangle adjacent to an eye. The real time for this is likely short, almost instant, so is set here at a constant of *5*, much smaller in relation to the constant of *30* defined for rules **R1** and **W1**. The time for rule **W2**, which follows rule **w2**, is *decision time* and *gesture time*. Both are calculated in the same way as for rule **W1**. The times for rules **w3** and **w4** are decision times. They are both set in the same way as for rule **w2**, at a constant of *5*. Figure 13 shows a computation using Stage III rules, continuing from the last drawing of the computation in Fig. 9 (or Fig. 10). The relative time calculations are shown for the first two steps and the penultimate step. The total relative time for this computation is *1967*—more time than the Stage I and Stage II computation in Fig. 9. There are more steps, and more finding, seeing, and decision making involved.

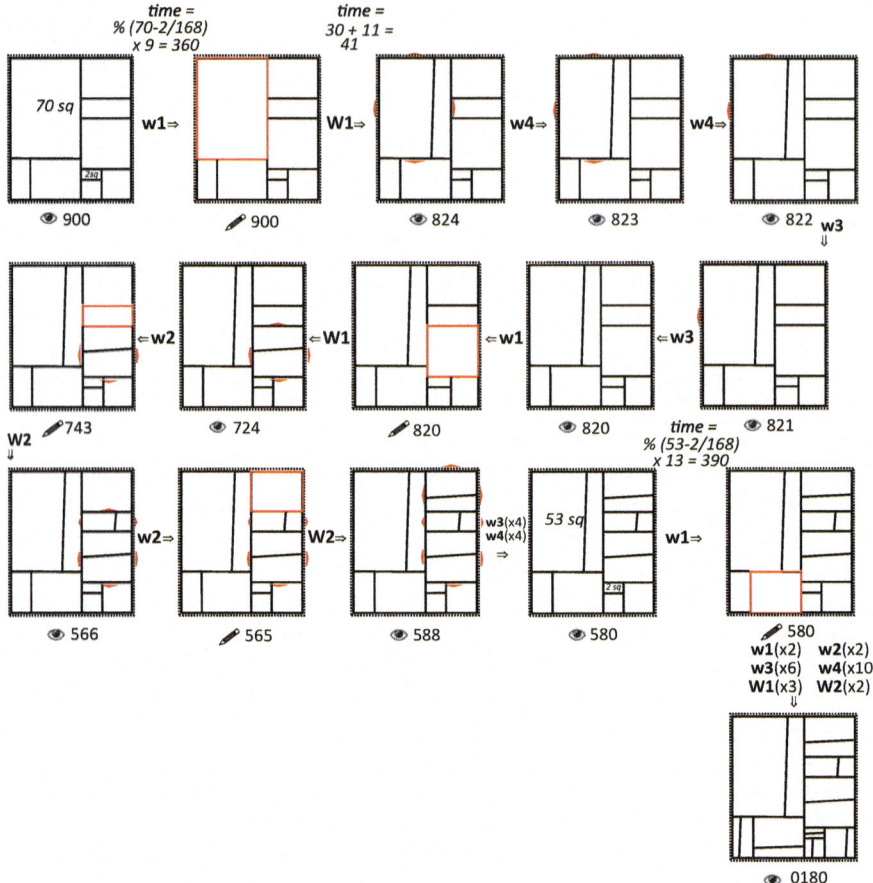

Fig. 13 A computation of a drawing using Stage III rules, continuing from the last drawing of the computation in Fig. 9 (or Fig. 10). The relative time calculations are shown for the first two steps and the penultimate step. The relative time for this computation is 1967—more time than the Stage I and Stage II computation in Fig. 9

5.4 Stage IV Wedge Border Divisions

Figure 14 shows the rules for making mandatory divisions of all wedge-shapes in a drawing to create borders at the wide ends of the wedge-shapes. Rule **b1** is a sensing rule. It applies to see and identify a wedge-shape in a drawing. It can only be applied when both the number of rectangles R and the number of adjacent rectangles A to check are equal to 0, in other words, when Stage III is complete. The values of R, W, and A remain the same when the rule is applied. The rule takes a drawing from a seeing state to a drawing state. Rule **B1** is a doing rule. It applies to draw a dividing line at the wide end of a wedge-shape identified by rule **b1**. It decreases W by 1 and changes the state of a drawing from a drawing state back to a seeing state. When W = 0, then no wedge-shapes remain to be subdivided.

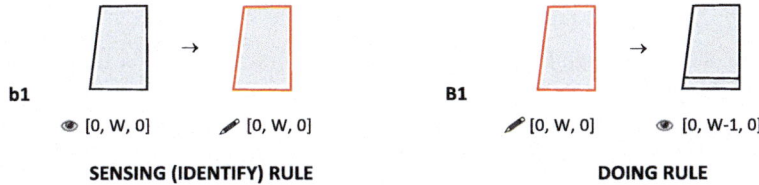

Fig. 14 Stage IV rules for dividing a wedge-shape to create a border at its wide end

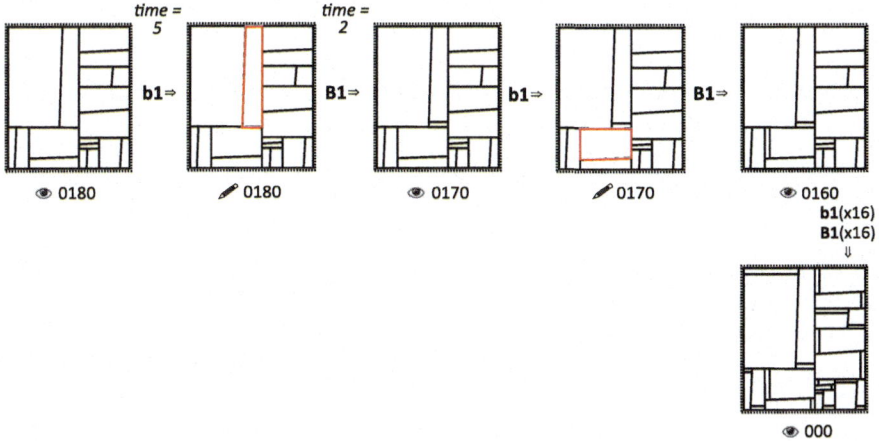

Fig. 15 A computation of a drawing using Stage IV rules, continuing from the last drawing in the computation in Fig. 13. The relative times for the first two steps are indicated. The relative time for the entire computation is 138—fast compared to the Stage II and III computations because little decision-making is involved

The time for rule **b1** is *decision time*. This rule must be applied everywhere possible and not in any particular order. It is easy and quick to find and see an unbordered wedge-shape to divide. The decision time is therefore set at a constant of 5. The time for rule **B1** is *gesture time*. The gesture time for the rule is calculated in the same way as for other subdividing rules. It is taken as the length of the line drawn. Decision time, or the time needed to decide where to make a division, is not counted because the widths of borders are more or less the same throughout a drawing (or painting). Deciding where to make the division is quick, if not immediate. Figure 15 illustrates a computation of a drawing using Stage IV rules, continuing from the last drawing of the computation in Fig. 13. The relative times for the first two steps are indicated. The total relative time for the computation is *138*—fast compared to the Stage II and III computations because little decision making is involved.

5.5 Stage V Finish

Figure 16 shows the mandatory last rule **F1** of the grammar. It is a sensing rule and can only be applied when the previous stage is complete, in other words, when W is equal to 0. (R and A are 0 on completion of Stage III.) The time for this rule is *decision time*, the time needed to see that all wedge-shapes have been divided. It is set at the same constant value of 5 as for the wedge finding/seeing rule **b1**.

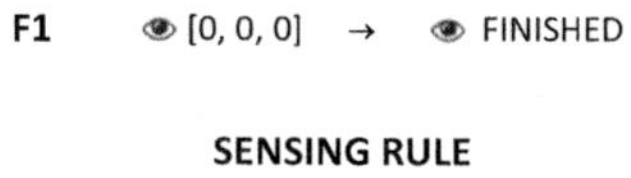

Fig. 16 The Stage V rule to finish a drawing

F1 👁 [0, 0, 0] → 👁 FINISHED

SENSING RULE

Figure 17 shows the last step of the computation of a drawing continuing from the last drawing of the computation in Fig. 15. The relative time for this step is 5.

Fig. 17 A computation of a drawing using the Stage V rule, continuing from the last drawing in the computation in Fig. 15. The relative time for this computation is 5

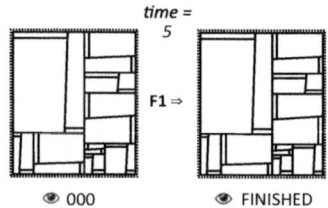

time = 5

F1 ⇒

👁 000   👁 FINISHED

The total relative time taken to compute, or draw, the last drawing beginning with the initial shape and continuing through the computations in Figs. 9, 13, and 15 is *3600*. The most time consuming aspect of the drawing is the stage for drawing oblique divisions. Different computations of the drawing—that is, different sequences of subdivisions—can result in different times to make the drawing. For example, if the computation in Fig. 10 is substituted for the one in Fig. 9, then the total time for making the drawing increases. The most time-consuming aspect of the drawing is then the stage for drawing orthogonal divisions. These differences in drawing times are the result of differences in decision times. With the rules given here, the total gesture times for any drawing computed in different ways will always be the same. Total gesture times (for drawing or erasing strokes) will only differ for different drawings. That might not be the case in a real-world scenario. The speed of drawing strokes might be affected by many factors, including perhaps visual complexity, which the rules here account only for decision making. Empirical studies would be needed to understand and describe actual times for making a drawing. The grammar here is hypothetical but provides a basis for more detailed, observation-based studies.

6 Reflections

The *How* of creative production, including both process and time, is key to studying, appreciating, designing, and making things. But it can be undervalued or neglected in comparison to the *What*, with its attention to the final, end thing. Time, in particular, is often sidelined, even though it may be essential to know about in order to make something. Digital fabrication is a case in point. Digital fabrication devices are touted as time savers, as accelerating processes of production. But novice users of personal fabrication devices or laser cutters can be surprised by how long it takes to make something with them, often longer than expected. Time is also critical to appreciating a made thing. The value we attribute to something, either handmade or machine-made, is often derived from its temporality—the amount of time it takes to make it.

Computation offers a means for rethinking, representing, and elevating the role of process and time in designing and making activities. Shape grammars are a special computational theory that have, from their inception 50 years ago, been directed toward the *How* of design. Building on the theory and applications of shape grammars, making grammars also prioritize the *How*, with a focus on the physical making of material things. Making grammars express—and compute—the sensory and action-centric nature of making through rules. Aspects of the time or temporality of making can also be computed. The Glarner study here demonstrates how duration might be computed, in particular, the durations of the sensory and gestural actions in drawing. Time has many other dimensions and understandings, from qualitative, subjective, and culturally determined ones to more quantitative and objective ones. It can be experienced and expressed in different ways, for example, through pace, tempo, and rhythm. The work here lays some new groundwork to capture the richness and significance of time as it unfolds in creative processes.

References

Arandjelovic, R., and T. Sezgin. 2011. Sketch recognition by fusion of temporal and image-based features. *Pattern Recognition* 44: 1225–1234. https://doi.org/10.1016/j.patcog.2010.11.006.

Berger, J. 2005. *Berger on drawing*, ed. J. Savage. Cork: Occasional Press.

Babak, B., and H. Grosman. 2016. Drawing in time: Processes of design and fabrication. *Architectural Design* 86 (1) (Special Issue: Architecture Timed: Designing With Time in Mind): 98–107.

Bryson, N. 2003. A walk for walk's sake. In *The stage of drawing: Gesture and act*, ed. Catherine de Zegher, Tate Gallery, and collaborator Avis Newman, 149–158. New York: Tate Publishing and The Drawing Center.

Cheng, P.C.H., J. McFadzean, and L. Copeland. 2001. Drawing out the temporal signature of induced perceptual chunks. In *Proceedings of the twenty-third annual conference of the cognitive science society*, 200–205.

Cohen, H. 1995. The further exploits of Aaron, painter. *Stanford Humanities Review* 4 (2): 141–158.

Denning, P.J. 2011. Ubiquity symposium: What have we said about computation? Closing statement. *Ubiquity* (April): 1–7.

Dexter, E. 2005. *Vitamin D: New perspectives in drawing*. New York: Phaidon Press.

Gartus, A., and H. Leder. 2017. Predicting perceived visual complexity of abstract patterns using computational measures: The influence of mirror symmetry on complexity perception. *PLoS ONE* 12 (11): e0185276. https://doi.org/10.1371/journal.pone.0185276.

Ingold, T. 2009. The textility of making. *Cambridge Journal of Economics* 34: 91–102. https://doi.org/10.1093/cje/bep042.

Ingold, T. 2016. *Lines: A brief history*. London: Routledge.

Klee, P. 1964. *The diaries of Paul Klee, 1898–1918*, ed. Felix Klee. University of California Press.

Klee, P. 1973a. Introduction. In *Notebooks Vol. 1 The thinking eye*, ed. Jürg Spiller. London: Lund Humphries.

Klee, P. 1973b. *Notebooks Vol. 2 The nature of nature*, ed. Jürg Spiller. London: Lund Humphries.

Knight, T.W. 1989. Transformations of De Stijl art: The paintings of Georges Vantongerloo and Fritz Glarner. *Environment and Planning B: Planning and Design* 16: 51–98.

Knight, T. 2018. Craft, performance, and grammars. In *Computational studies on cultural variation and heredity*, ed. Ji-Hyun Lee, 205–224. KAIST research series. Singapore: Springer. https://doi.org/10.1007/978-981-10-8189-7_16.

Knight, T., and G. Stiny. 2015. Making grammars: From computing with shapes to computing with things. *Design Studies* 41: 8–28. https://doi.org/10.1016/j.destud.2015.08.006.

Kurczynski, K. 2011. Drawing is the new painting. *Art Journal* 70 (1): 92–110.

LeWitt, S. 1967. Paragraphs on conceptual art. *Art Forum* 5 (10) (June): 79–83.

Mao, X., O. Galil, Q. Parrish, and C. Sen. 2020. Evidence of cognitive chunking in freehand sketching during design ideation. *Design Studies* 67: 1–26. https://doi.org/10.1016/j.destud.2019.11.009.

Morris, R. 1970. Paragraphs on conceptual art. *Art Forum* 8 (8) (April): 62–66.

Newman, M. 2003. The marks, traces, and gestures of drawing. In *The stage of drawing: Gesture and act*, ed. Catherine de Zegher, Tate Gallery, and collaborator Avis Newman, 93–108. New York: Tate Publishing and Drawing Center.

Palumbo, L., R. Ogden, A.D.J. Makin, and M. Bertamini. 2014. Examining visual complexity and its influence on perceived duration. *Journal of Vision* 14 (3): 1–18. https://doi.org/10.1167/14.14.3.

Spriggs, E.H., F. De LaTorre, and M. Hebert. 2009. Temporal segmentation and activity classification from first-person sensing. In *2009 IEEE computer society conference on computer vision and pattern recognition workshops, CVPR workshops 2009*, 17–24.

Whitehead, A.N. 2004. *The concept of nature*. New York: Prometheus.
Zacks, J.M., and B. Tversky. 2001. Event structure in perception and conception. *Psychological Bulletin* 127: 3–21.

Open Access This chapter is licensed under the terms of the Creative Commons Attribution-NonCommercial-NoDerivatives 4.0 International License (http://creativecommons.org/licenses/by-nc-nd/4.0/), which permits any noncommercial use, sharing, distribution and reproduction in any medium or format, as long as you give appropriate credit to the original author(s) and the source, provide a link to the Creative Commons license and indicate if you modified the licensed material. You do not have permission under this license to share adapted material derived from this chapter or parts of it.

The images or other third party material in this chapter are included in the chapter's Creative Commons license, unless indicated otherwise in a credit line to the material. If material is not included in the chapter's Creative Commons license and your intended use is not permitted by statutory regulation or exceeds the permitted use, you will need to obtain permission directly from the copyright holder.

Computing with Abstract and *Material Shapes*: The Case of Triaxial Basket Weaving

Benay Gürsoy and Mine Özkar

Abstract Visual rules and schemas play a crucial role in formally representing visual reasoning in design. The indeterminacy inherent in shapes within visual computation fosters creative discoveries, as shapes can be perceived differently in various contexts. Design, however, often transcends visual aspects, involving creative processes rooted in material engagement and making. This chapter explores the integration of visual and material reasoning in creative processes by employing *material shapes* in shape computation. We exemplify how *material shapes*, particularly those generated through basket weaving, contribute to a dynamic interplay between abstract representations and physical outcomes. Triaxial basket weaving serves as a case study, demonstrating basic rules and showcasing the use of abstract shapes and material shapes in computational making processes in tandem. The iterative nature of an "abstraction–materialization–abstraction" cycle is exemplified through the bending and twisting of thin linear strips of flexible material in basket weaving, highlighting the emergence of forms shaped by the physical system during the weaving process.

1 Introduction

The utilization of shape rules as analytical and creative tools has been extensively demonstrated in the shape grammar literature (Knight 1989). Visual rules and schemas are useful for formally representing visual reasoning in design (Stiny 2011). Creative processes may feed from making and engagement with materials (Knight and Stiny 2015). The indeterminacies in how physical things behave can also contribute to unexpected discoveries. Physical manipulations of materials can bring about an emergent shape or form. With the aim of exploring the possibility to

---

B. Gürsoy (✉)
Penn State University, 220 Stuckeman Family Building, University Park, PA 16802, USA
e-mail: bug61@psu.edu

M. Özkar
Istanbul Technical University, Harbiye Mahallesi, Taşkışla Cad. No: 2, Şişli, Istanbul 34367, Turkey
e-mail: ozkar@itu.edu.tr

computationally represent multisensory aspects of working with materials in design and focusing on the creative and emergent aspects of material engagement, we have previously introduced *"material shapes"* in computation as part of a dialogue between abstract representations and their material outcomes (Gürsoy and Özkar 2015). We looked at a particular *making for* case in which we cut regularly arranged incisions (referred to in the paper as the dukta technique) in craft paper, and physically manipulated the material samples in order to explore the emergent *material shapes*. While that exploration showed a need for *material shapes* to recognize emerging shapes as variations in texture, and the creative use of *material shapes* in design, it has been limited in the sense that it did not fully capture a continuous "abstraction–materialization–abstraction" cycle as part of spatial computation. In this chapter, we further exemplify the ways in which *material shapes* can be employed to integrate visual and material reasoning in creative processes in a way that they inform each other, continuously. We demonstrate the abstraction–materialization–abstraction cycles for a series of *material shapes* that are generated through basket weaving, namely bending and twisting of thin linear strips of flexible material. In basket weaving, the exact final form is rarely definable in advance. Rather than being imposed on the material, the form comes into existence during weaving: the strips perform as a physical system in shaping the final woven form. Although the final forms cannot be fully designed in advance, the formal relations of the physical system that anticipate these forms can be defined. With a specific focus on triaxial basket weaving, we exemplify a set of basic basket weaving rules and consequently demonstrate the use of *abstract shapes* and *material shapes* in tandem as part of a computation.

2 Background on Computing Making and *Material Shapes*

For visual design processes, shape grammars provide formal descriptions that give insight into processes of shape transformations and "a process model for the externalization of visual reasoning in design" (Oxman 2002, p. 134). This computational approach extends the understanding of computation beyond the mere use of computers to a more sensory and perceptual kind that is primarily based on visual reasoning. In visual computations, abstract shapes are not symbols, and they are indeterminate. The viewer can always see something new: "the way the shape was made and how it is divided into parts afterwards are independent" (Stiny 2001, p. 24). The indeterminacy of abstract shapes enables "embedding" in visual computations and is considered the basis for design emergence.

When materiality is prevalent in the design process, diverse sensory inputs operate in the course of design and making. Any haptic exploration such as bending, stretching, folding, and twisting a physical object requires sensory engagement and personal connection with the material. The need in the field to capture the full range of interactions that designers employ as they explore and discover the material properties of form has grown into a new research agenda with the increasing interest in making and craft practices over the past years. Recent studies that focus on filling this void have dwelt on the possibility to extend shape grammar formalism from shapes to "material things" with *making grammars* (Knight 2015). In a *making grammar*, the rules are "based on both the thing being made and a person's sensory interactions with that thing", rendering it a "theory of both the constructive and sensory aspects of a making activity" (Knight 2018). In *making grammars*, a making rule is either a *sensing rule* or a *doing rule*: while "a sensing rule represents a perceptual change in a person through the person's sensory actions with a thing", "a doing rule represents a physical change in a thing through a person's physical actions with the thing". Knight (2018) explored the use of this formalism to study the performative, time-based nature of craft practices, specifically focusing on structuring the temporal actions involved in the traditional *Kolam* pattern making in India.

There are other studies that use shape grammar formalism to analyze and represent the tacit design knowledge of making in various traditional craft practices. In one study, Muslimin (2014) looks at bamboo basket weaving and explores "methodologies to interpret tacit knowledge in traditional weaving" through a computational lens by developing a weaving grammar. The focus in his weaving grammar is on what the weaver sees and where the hands are placed during the weaving activity. In another related study, Muslimin (2017) explores the computability of an indigenous craft practice of engraving architectural ornaments from Indonesia called *Passura*. He presents a case study where he makes use of shape grammars to both "represent the logic behind the mental iteration in generating the design", and "record the making process" of *Passura*. Similarly, Noel (2015) explores the development of a formal methodology to document the traditional craft practice of wire-bending in Trinidad and Tobago. She investigates how the tacit knowledge of wire-bending can be made explicit through a shape grammar, the *Bailey-Derek Grammar*, that she named after the practice of well-known wire-benders from the Trinidad Carnival. More recently, Labrou and Kotsopoulos (2023) developed making grammars for a traditional bobbin lacemaking technique to capture the craftsperson's interaction with the physical tools and materials. These studies are significant not only for recording and documenting traditional craft practices, and communicating the making knowledge with broader communities, but also opening the way for novel explorations with the means and technologies available today. Along these lines, Hamzaoğlu and Özkar (2016, 2023) explore the integration of shape rules and digital fabrication methods in the reproduction of engraved Seljuk patterns that date back to the thirteenth century. They formally represent the underlying processes of the making of these patterns as shape rules, and fabricate the patterns by applying these rules in a CNC-milling process. Their methodology "unifies the components of making and design and expands the design space of any pattern with the factors of production". Özkar (2020) states that capturing and modeling physical information from built heritage through shapes can "increase [the] understanding of past architectural and building traditions". In a similar study that seeks formal descriptions for making activities using digital fabrication tools, MacLachlan (2018) investigates the use of shape grammar weights as "a way of modeling material properties alongside shape operations" for multi-material 3D printing. Vazquez et al. (2019) formalized the shape-changing behavior of 3D printed wood-based composite materials through a grammar that serves to compute their shape-change in response to variations in relative humidity. Making, however, is not only limited to the production of three-dimensional artifacts. Drawing and painting can also be considered as a form of making, as Knight and Stiny (2015) have previously pointed out. Gün (2017) explores the integration of digital and analog means in hand-drawing and watercolor painting that he calls "visual-making". He presents a "non-symbolic, open-ended and trace-based shape calculation system" where he makes use of visual schemas to computationally translate the hand-drawn and painted shapes for generative drawing processes that feed from "embodied actions and tactile material interactions". With the comparable aim of exploring embodied computational design knowledge, Smithwick and Sass (2014) present an action-based non-symbolic formal notation for physical actions that designers employ.

Following these examples, we are interested in formalizing emergent material behaviors and transformations in the making process as part of the computation. Unforeseen material aspects can be discovered through hands-on manipulations of the materials and physical things. These interactions introduce additional uncertainties to the creative design process. On the other hand, when material properties (i.e., stiffness, flexibility, density, color, texture, etc.) are assigned to abstract shapes, they rely on symbolic representations for material things, as in the case of *color grammars* (Knight 1993). Differently, *material shapes* (Gürsoy and Özkar 2015) are not symbolic representations for material things or material properties. They have a physical and tangible existence. They stand for themselves and can still be indeterminate: unexpected material behaviors and spatial configurations can emerge when they are physically manipulated in 3D space. The main difference between *material shapes* and "material things," as introduced by Knight and Stiny (2015), lies in how they are used in computations. *Material shapes* are not computed with directly, whereas material things are computed with directly. The approach presented in this chapter, and previously in Gürsoy and Özkar (2015), is distinctive as it allows splicing material manipulations into formal shape computations.

2.1 How to Compute with Material Shapes: Abstraction–Materialization–Abstraction Cycle

In spatial reasoning that involves making, abstract shapes are used together with *material shapes*. Transitions between "design through abstraction" and "design through materialization" (Aydın and Özkar 2015), and similarly between abstract shapes and *material shapes* in an "abstraction–materialization–abstraction" cycle are key to spatial reasoning and the computability of these processes. Abstractions in this cycle define temporary frameworks. Their natures change with regards to what the designer perceives in the course of making. Picking out certain parts that constitute the whole enables the designer to concentrate on a specific aspect of the making activity and to understand it deeply. Continuous transitions between abstract shapes

Computing with Abstract and *Material Shapes*: The Case of Triaxial ...

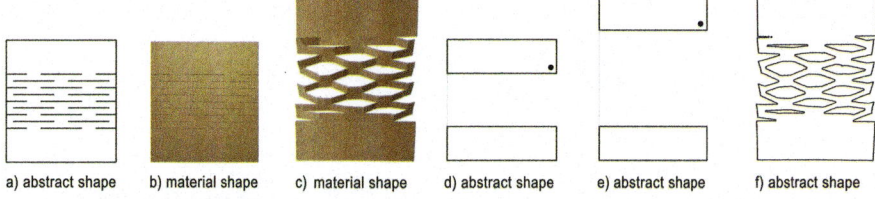

a) abstract shape b) material shape c) material shape d) abstract shape e) abstract shape f) abstract shape

Fig. 1 Transitions between abstract and *material shapes* in spatial reasoning

and *material shapes* are exemplified through a simple example in Fig. 1 (Gürsoy and Özkar 2015).

The example above shows a material exploration process that starts with the abstract shape at the far left (Fig. 1a). This abstract shape is generated through a series of visual transformations. The abstract shape is cut out from craft paper. This gives the *material shape* in Fig. 1b. When this material shape is physically manipulated (i.e., pulled from two edges), the incised areas in the middle change form (Fig. 1c). These changes cannot be predicted before physically manipulating the *material shape*. To understand, control, and *compute* these discoveries, the designer at this point can introduce a transition back to abstract shapes, as exemplified in Fig. 1d. Here, the rectangular uncut areas on the *material shape* in Fig. 1b are represented with two abstract rectangular shapes. This is only a temporary abstraction. It is unique to this process, and it defines a temporary framework for the explorations with that particular *material shape*. The abstract shape in Fig. 1e is a transformed version of the abstract shape in Fig. 1d. The rectangle on the top is translated vertically. In the temporary framework of the designer, it indicates *physically stretching the material shape*. The outcome of this physical manipulation is the *material shape* in Fig. 1c. The stretched incisions in the middle cannot be predicted before physically manipulating the *material shape* because the resulting *material shape* not only depends on the specifications of the initial abstract shape that defines the cut pattern, but also on the specifications of the material from which the *material shape* is cut. Different materials result in different configurations when they are stretched. There are also numerous ways for physical manipulations which all result in different spatial configurations. Hence, the abstract shape in Fig. 1e introduces constraints to physical manipulation and controls the emergence. Tracing the boundaries of the photographic representation of emerging *material shape* can yet be another translation from a *material shape* to an abstract shape (Fig. 1f). This way the material thing can become a design tool itself for the generation of new abstract shapes. This abstract shape can again be cut from craft paper, thereby translated to a *material shape* back again. The transitions from abstract shapes to *material shapes*, and from *material shapes* to abstract shapes can continue until it satisfies the designer's will.

The case-specific abstractions in design, as exemplified above, create natural stops in the continuous design process to perceive, think and reason. This cycle is in parallel with the "see-move-see" model for visual reasoning processes advocated by Schön and Wiggins (1992). Schön (1983) depicts the designer in a "reflective conversation with the situation" (p. 73) and in a visual interaction with abstract representations of the design problem. His interpretation has been widely accepted and has fostered much research on the externalization of design thinking in drawing and sketching.

2.2 Computations with Abstract Shapes and **Material Shapes**

Rules in shape computations are generally in the form A → B where A and B are shapes and the arrow → indicates that a shape A on the left side of the rule is to be replaced with a shape B on the right side. Both sides may be the same shape. A shape may as well be an "empty" (no) shape. Labels may be employed to limit the possible shape transformations and restrain the ambiguity. Figure 2 shows a basic shape rule.

The shape rule, above, prompts the user to look for the shape (A) in an existing abstract shape (C). The shape (A) in the shape (C) can be a spatially transformed (rotated, reflected, scaled, translated) version of the shape (A). Labels limit the possible transformations. If the user sees a shape (A) in the shape (C), then the shape rule asks the user to subtract it from the shape (C) and to replace it with the shape (B). Replacing (A) with (B) in the shape (C) gives a new shape, (C1). A shape grammar computation is the recursive application of shape rules. It starts with an initial abstract shape (C) and goes on with a sequence of shapes: (C1), (C2), (C3), etc. A computation, therefore, has the following form: (C) ⇒ (C1) ⇒ (C2) ⇒ (C3) ⇒... ⇒ (Cn). Each shape (Ci) is a shape computed from the previous shape (Ci − 1) and defines the next shape (Ci + 1). The double arrow (⇒) represents the application of a shape rule in a shape computation.

Spatial reasoning that involves physical making processes can be formalized as part of a shape computation in design. The computations then can occur within the "abstraction–materialization–abstraction" cycle and operate through the use of abstract shapes and *material shapes* in tandem. The abstract shapes are used in the computation to interpret the transformation as physical actions or manipulations of *material shapes*. Abstract shape rules, such as the one in Fig. 2, can specify formal actions (i.e., translations and rotations). These specified actions can be applied systematically and by-hand to physical things to gain insights into their material behaviors and generate novel, often unexpected, physical outcomes. The *material shape* in Fig. 1c is, for instance, the outcome of the computation with the abstract shape rule in Fig. 2. This abstract rule denotes the planar translation of the uncut parts of the *material shape* in Fig. 1b. Computational formalism, in return, serves

Fig. 2 A shape rule with abstract shapes

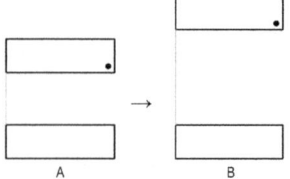

to establish the causal links between abstractions and physical manipulations. This way, the emergent outcomes can be repeated. Variations can also be introduced.

Note that *material shapes* in the computation are the physical things themselves, not just their representations as pictures. Broadening the scope of the computable in making from deterministic grounds to a more personal and perceptual approach opens new ways of looking at and understanding the material world: understanding not through imposed and absolute certainties but through personal and—possibly—temporary inferences. The transitions between *material shapes* and abstract shapes in a computation depend on the user and what the user perceives each time. The rules define temporary frameworks, and their natures change regarding what the designer perceives while making. As similarly pointed out by Knight (2018), the rules "like any finite description, can never capture all aspects of a making activity and can only approximate it." Picking out certain parts that constitute the whole enables the designer to concentrate on a specific aspect of the making activity momentarily. This is especially important for processes where the perceptions continually change between the parts and the wholes, as in the case of basket weaving. In basket weaving, the focus of the weaver continually shifts between individual strips to larger local nodes that emerge as voids between the strips, then back to the strips, and to the whole, etc. The whole is once again *more than the sum of its parts*. Different weavers can perceive different shapes while weaving and can formalize their process with different rules. Moreover, when the same rules are applied to different materials, different *material shapes* also emerge. Then the perceived shapes and the rule sets can change. A computational approach with abstract shapes and *material shapes* can reveal the temporality of making: the order in which new shapes are perceived. We can also formalize the body's interaction with things as part of a computation. While it is not the scope of this paper, formalizing these temporal and embodied aspects of making can enable the automation of production using digital and robotic fabrication tools.

3 Triaxial Basket Weaving Grammar

There are many different basket weaving patterns and basket weaving techniques. In this paper, our focus is on triaxial basket weaving (also called *hex weave*, or *kagome weave*). It has been practiced around the world for generations, from Asia to South America (LaPlantz 2016). Contemporary Japanese bamboo artists, including Morigami Jin, Yamaguchi Ryuun, Norie Hatakeyama, make use of triaxial basket weaving to create intricate non-figurative sculptures, some of which are exemplified in the books by Coffland (2000, 2006) and Earle (2008). The artist Alison Grace Martin (2015) creates woven structures with triaxial weaving technique "in order to explore how a basket weaving can be seen to represent a topological map of its structure". She also collaborates with architects and designers to construct large scale lightweight woven structures (Pineda et al. 2016; Ayres et al. 2018). These various weaving explorations offered a visual repository for this research.

In writing about what it means to make things, anthropologist Ingold (2013) distinguishes between two kinds of making: by blocks and by knots. While in the first, making proceeds "through the hierarchical assembly of preformed parts into larger wholes", the second corresponds to "weaving a pattern from ever unspooling threads that twist and loop around one another" where "there are initially no parts and wholes, and the form of a thing emerges from the process itself" (p. 26). Earlier Ingold (2000) wrote about the process of form generation in spiral basket weaving: "… the equable form of the spiral base of the basket does not follow the dictates of any design; it is not imposed upon the material but arises through the work itself. Indeed, the developing form acts as its own template, since each turn of the spiral is made by laying the longitudinal fibers along the edge formed by the preceding one" (p. 345). Similarly, in triaxial basket weaving, the form emerges during weaving, and cannot be fully defined in advance. Yet, the formal relations that anticipate the form can be defined. Basket weaving, crocheting, knitting, etc. all contain underlying rules. Following the same rules, one can replicate the forms generated. In a study on the computability of the craft of crochet, Çapunaman et al. (2017) pursue an analytical and systematic approach to discover the underlying rules of crocheting, and formally represent the deducted computational logic in the form of a computer algorithm. The algorithm is used to generate crocheting patterns of 3D modeled forms in the computer, which can then be crocheted by hand. In crocheting, the process of the generation of crocheted form demonstrates an "emergent behavior": several simple entities operate together in a systematic way to form more complex behaviors. Similarly, in triaxial basket weaving, the strands interwoven in a network all contribute equally to the overall strength, creating a *material system*. As noted by Çapunaman et al. (2017), "such systems are defined both by bottom-up and top-down relations between parts and wholes."

Previous studies that explore the computability of triaxial basket weaving mostly rely on top-down approaches and focus on recreating a given 3D modeled form in the computer as digitally woven surfaces. Among these are the works by Roelofs (2008, 2010), Mallos (2009, 2011, 2015), Akleman et al. (2009) and Baek et al. (2021). They approach weaving from a mathematical point of view, focusing on representing the "over-and-under" conditions in woven surfaces. The weaves generated through their algorithms are not straight strips when unrolled and flattened, they all have unique shapes. Recently, Ayres et al. (2018) published research where they explore the relation between the computational representation methods and the fabrication of interlaced lattice structures where the structures are made using straight strips of material.

Differently, this paper explores the computability of the hands-on triaxial basket weaving and seeks to include the feedback one gets from the materials. With a bottom-up approach, we identify the act of triaxial basket weaving (putting one strip across another) through visual rules and make use of *material shapes* to capture the emergent behaviors. There is also the possibility of visual emergence through the use of abstract shapes, and this can inform the creative making process. However, in this study our focus is on the material emergence that happens through the physical act of weaving, the process, the properties of the materials, and the emerging spatial relations.

3.1 Visual Rules for Triaxial Basket Weaving

Triaxial basket weaving is a 3-way weaving technique that is based on creating a semi-regular mesh of triangles and hexagons using straight strips of flexible material that spring back when released, such as bamboo, rattan, plastics, cardstock, etc. Callens and Zadpoor (2018) exemplify triaxial basket weaving as a technique that shows "lattice disclination": the insertion of a pentagon or a heptagon instead of a hexagon within the semi-regular mesh changes the surface curvature and creates "a type of topological defect that disrupts the orientational order of the lattice." These local nodes "cause the surfaces to deform out-of-plane" and "form concentrated sources of Gaussian curvature," as illustrated in Fig. 3. While this is a topological principle, it is also strictly dependent on the properties of the materials used and on the weaving itself, which creates a system of interwoven strands.

Introducing pentagons and heptagons in triaxial weaving is a technique that has been long employed by basket weavers to create variations in basket forms. Based on this topological principle, Martin (2015) physically models complex surfaces

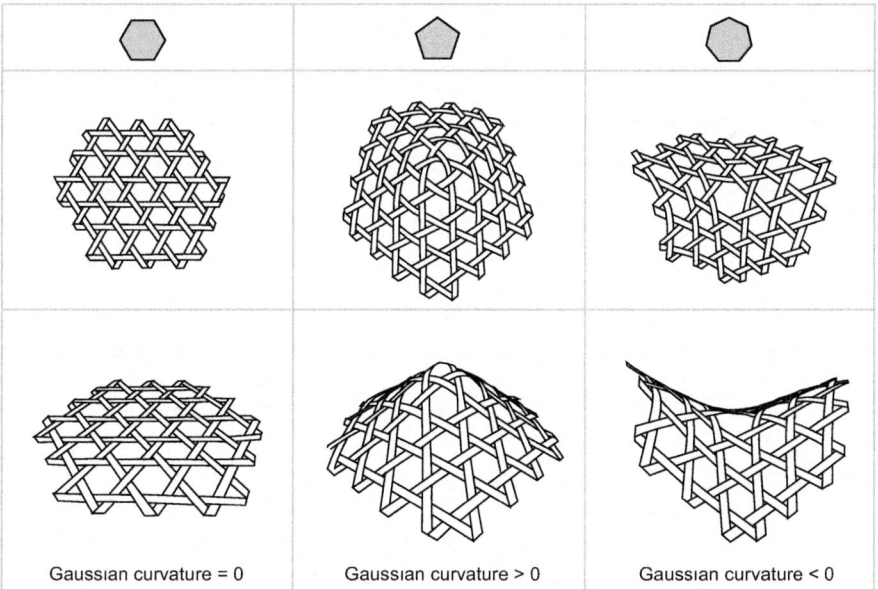

Fig. 3 A hexagonal weave is planar and flat. Pentagons in a hexagonal weave result in positive Gaussian curvature. Heptagons in the hexagonal weave result in negative Gaussian curvature. The drawings are based on the models by Callens and Zadpoor (2018)

through triaxial weaving, among which are various minimal surfaces (i.e., gyroid, Schwarz surfaces), saddle surfaces, hyperbolic paraboloids, etc. Analyzing various examples and after multiple hands-on trials of triaxial weaving, we have developed the following set of visual rules for triaxial weaving shown in Fig. 4 and formalized the triaxial weaving process using these rules. Emerging *material shapes* are an intrinsic part of the process. Note that these rules are subjective, not universal. The triaxial weaving process can be formalized with a completely different set of rules depending on what the user chooses to *see*. Yet we postulate that these rules serve to share and communicate our personal know-how of making with others, as we have tested through multiple student workshops.

The way the rules are presented in the table in Fig. 4 highlights their relations. Columns organize the rules according to the geometries to be woven: the first column shows the rules used in planar (hexagonal) weaving; the second column shows the rules used to create the pentagonal lattice disclinations; and the third column shows the rules for heptagonal lattice disclinations. These rules are related through the rows: rules that convey similar actions for introducing hexagons, pentagons, or heptagons in triaxial weaving are presented in the same row.

The rules in the first row (R1 group: R1, R1-P, R1-H) *add a strip* following the reference line indicated with a circular label and *shifts the label* to the next reference line in counter-clockwise direction. Notice that the strips added in pentagonal and heptagonal systems through the rules R1-P and R1-H are bent. The parts shown in red indicate the parts where spatial transformations emerge when the strips are bent during weaving. These parts are explored with *material shapes* during a computation and can be ignored when applying the subsequent rules. For this, a series of erase-label rules may be applied. The rules R1, R1-P, R1-H are parametric: the distance between the strips and the reference lines can change. This alters the density of the weave, as can be seen in the variations shown in Fig. 5. The width of the strips themselves can also be parametrically defined and altered. Materially, this can affect the bending performance of the strips, and the porosity of the surface. Since these rules are scalable, the preferred size for the strip can affect the bending behavior of the overall surface. Parametric relations can be defined as such and material parameters for the strips can introduce even more variations.

The rules in the second row (R2 group: R1, R2-P, R2-H) serve to *add reference lines*. Rules in the R1 group follow these reference lines to *add strips* and rules in the R7 group to *bend strips*. In a computation, rules in the R2 group are used to introduce pentagonal and heptagonal lattice disclinations, or to return to planar (hexagonal) weaving.

Rules R3 and R5 are both *identity rules*. "Identities are observational devices" (Stiny 2006) and identity rules display the exact same shape on either side of the arrow to indicate a recognition (Stiny 1996). Here, R3 and R5 are typically followed by the rules R4 and R6, respectively. With R3, the triangle emerging through the overlapping of three strips is *seen* and a red label is placed at one of the nodes to guide R4. Similarly, with R5, the pentagon emerging through the overlapping of five strips is identified and a red label is placed at one of the nodes to guide R6. Rules R4 and R6 change the overlap order (over-and-under) of the strips at the node indicated

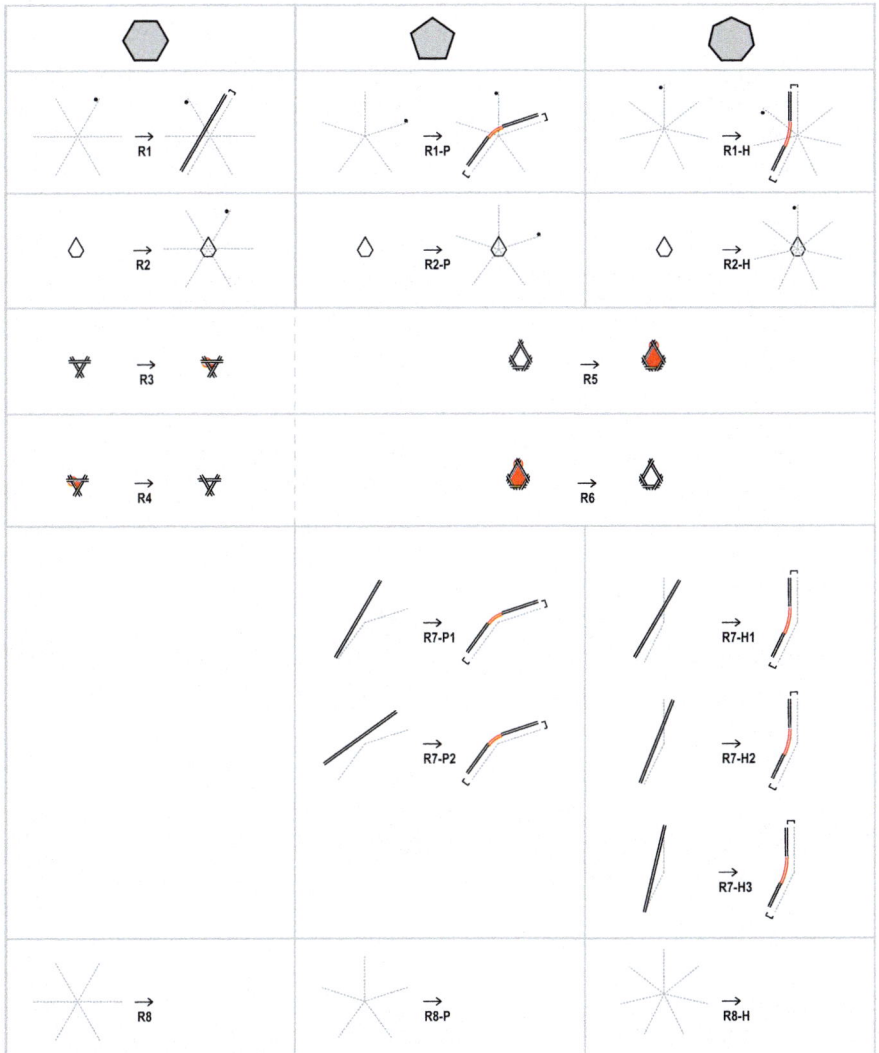

Fig. 4 Visual rules for triaxial basket weaving

by the red labels and erase the labels. When applied during the physical weaving process, these rules serve to physically *lock the strips at the nodes*. Rules R3 and R4 are used for creating hexagons, pentagons, or heptagons in triaxial weaving, while rules R5 and R6 are only used for pentagons and heptagons.

The rules in the R7 group serve to *bend the strips* on a woven surface following the reference lines for pentagons (R7-P1, R7-P2) or heptagons (R7-H1, R7-H2, R7-H3) in triaxial weaving. Like the rules in the R1 group, rules in the R7 group are parametric: the distance between the strips and the reference lines can change.

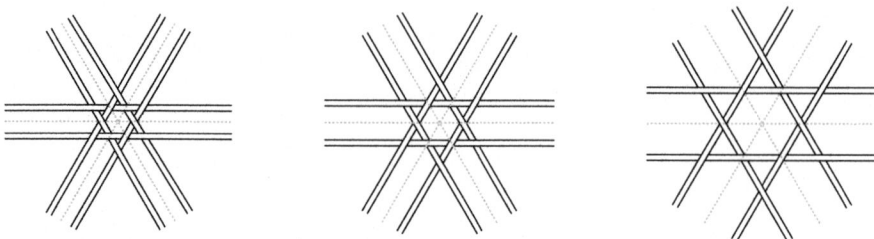

Fig. 5 Hexagonal weaves with different distance parameters used in Rule R1

The parts shown in red on the right-hand side of the rules again indicate the parts where bending takes place, resulting in three-dimensional displacement of the strips. Similar to the rules in the R1 group, the parts shown in red can be ignored during a computation and erased by a series of erase-label rules, if needed.

Finally, the rules in the R8 group (R8, R8-P, R8-H) *erase the reference lines*. They are typically preceded with the rules in the R2 group when a pentagonal and heptagonal lattice disclinations are introduced, or a return to planar (hexagonal) weaving is needed.

3.2 Planar Triaxial Weaving: Transformations of Abstract Shapes

As previously shown in Fig. 3, hexagonal triaxial weaves are planar and flat. Rules R1, R3, R4 are used in hexagonal triaxial weaving, and rules R2 and R8 can be used to introduce pentagons and heptagons. While material parameters, such as transparency, color, texture, can always affect and enrich the woven object, with rules R1, R3 and R4 the weaving process can be computed with abstract shapes and without a need for *material shapes*. Therefore, these rules can also be used for generating two-dimensional pattern designs in the form of drawings. In Fig. 6, we exemplify a computation to generate a planar triaxial hexagonal weave.

The process starts with adding three strips by applying rule R1 three times recursively, after which a triangle emerges between the three strips. In physical weaving, this triangle emerges as a void. Rule R3 is used to identify this triangle and place a red label. Rule R4 swaps the over-and-under order of the strips at the node where the red label is placed and erases the red label. The application of rules R1, R3 and R4 in this successive order constitutes one *cycle*. This *cycle* is repeated for a total of four times. This results in the placement of a total of six strips and the emergence of six triangles, as well as a hexagon at the center, all as voids while physical weaving. R3 and R4 are applied once more to identify the sixth triangle and flip the over-and-under order of the strips. This locks all the six strips at the nodes during physical weaving. Here is another way to describe the process:

Fig. 6 Computation to weave a planar hexagonal pattern

R1 – R1 –
(R1 – (R3 – R4)) –
(R1 – (R3 – R4)) –
(R1 – (R3 – R4)) –
(R1 – (R3 – R4)) –
(R3 – R4)

Moving forward, the (R1 – (R3 – R4)) cycle is repeated once. This time rule R1 is applied with a different distance parameter to create the second row of strips on both sides of the reference lines. After the first (R1 – (R3 – R4)) cycle in the second row, one cycle becomes (R1 – (R3 – R4)) – (R3 – R4). This is because the application of rule R1 in the second row results in the emergence of two triangles instead of one, and the over-and-under order of the strips at the nodes needs to be swapped twice. This cycle is repeated for a total of five times. This results in the placement of six additional strips and six additional hexagons. Rules R3 – R4 are applied once again to lock the new strips at the nodes while physical weaving. The process to weave the second row can also be described as follows:

(R1 – (R3 – R4)) –
(R1 – (R3 – R4)) – (R3 – R4) –
(R1 – (R3 – R4)) – (R3 – R4) –
(R1 – (R3 – R4)) – (R3 – R4) –
(R1 – (R3 – R4)) – (R3 – R4) –
(R1 – (R3 – R4)) – (R3 – R4) –
(R3 – R4)

Although the hexagonal weave shown in Fig. 6 only has two rows of strips on both sides of the reference lines, weaving of the subsequent rows would follow the same logic. Figure 7 shows planar triaxial weaves with a single, double and triple row of paper strips on both sides of the reference lines. With each new row of strips added, an additional (R3 – R4) is added to the *cycles*. Hence, the hexagonal weaving of the third row can be described as:

(R1 – (R3 – R4)) – (R3 – R4) –
(R1 – (R3 – R4)) – (R3 – R4) – (R3 – R4) –
(R1 – (R3 – R4)) – (R3 – R4) – (R3 – R4) –
(R1 – (R3 – R4)) – (R3 – R4) – (R3 – R4) –
(R1 – (R3 – R4)) – (R3 – R4) – (R3 – R4) –
(R1 – (R3 – R4)) – (R3 – R4) – (R3 – R4) –
(R3 – R4)

Fig. 7 Hand woven planar hexagonal weaves with paper

3.3 Pentagons and Heptagons in Triaxial Weaving: Introducing Material Shapes

The insertion of a pentagon or a heptagon within the planar triaxial weave causes the strips to bend and the surface to deform out-of-plane (Fig. 3). Therefore, we introduce *material shapes* in the computation, alongside abstract shapes. As mentioned before, *material shapes* are not symbolic or abstract representations for material things or material properties. They have a physical and tangible existence. They stand for themselves. In a shape computation, *material shapes* help capture the indeterminate material behaviors or spatial configurations that emerge when things are physically manipulated. In the case of computing triaxial weaving, we employ *material shapes* when abstract shapes fall short in representing how the strips are bent with the insertion of a pentagon or a heptagon.

Figures 8 and 9, and Figs. 10 and 11 show the two parts of the computations to generate triaxial weaves with a pentagon and heptagon (as voids) at their center, respectively. The last column in these figures show photographs of the *material shapes* side-by-side with the abstract shapes that emerge at the end of the computation in each row. It is possible to follow the computation in these figures both ways: through a step-by-step transition between *material shapes* in the vertical (top-down) direction, and through left-to-right and row-to-row as a step-by-step transition between abstract shapes, which are accompanied by a photo of the *material shape* at the end of each row.

Figures 8 and 10 show the initial steps to generate the first row of the weaves, and Figs. 9 and 11 show the subsequent steps to generate the second row of the weaves. Note that the initial shapes in the computations in Figs. 8 and 10 (that have five planar strips) are obtained by following the steps outlined previously in the first two rows of Fig. 6: *R1 − R1 − (R1 − (R3 − R4)) − (R1 − (R3 − R4)) − (R1 − (R3 − R4))*. The process takes a different turn after the irregular pentagons that emerge between the five strips are *seen* using Rule R5 and a label is placed. Rule R6 swaps the over-and-under order of the strips at the nodes where the red label is placed and erases the label. Rule R8 is then applied to erase the underlying hexagonal reference lines. These initial steps are common for generating pentagons and heptagons.

To create the pentagon, rule R2-P is applied to add new reference lines with five intersecting lines (Fig. 8). Moving forward, before adding strips, first the existing five strips are bent using rules R7-P1 and R7-P2 from the R7 group. Triangles emerge as voids, similar to the computation in Fig. 6 for a planar triaxial weave, *but this time on another plane.* This is where the *material shapes* come into play. These triangles that emerge on another plane are perceivable on the *material shapes* while physically bending the strips. *Seeing* the triangles on the *material shapes*, one can translate them to abstract shapes and apply rule R3 to place the labels. Rule R4 will then similarly swap the over-and-under order of the strips to physically lock them and erase the labels. Once all the five strips are bent and interlocked following the computation in Fig. 8, the first row of the triaxial weave with a pentagonal lattice disclination is

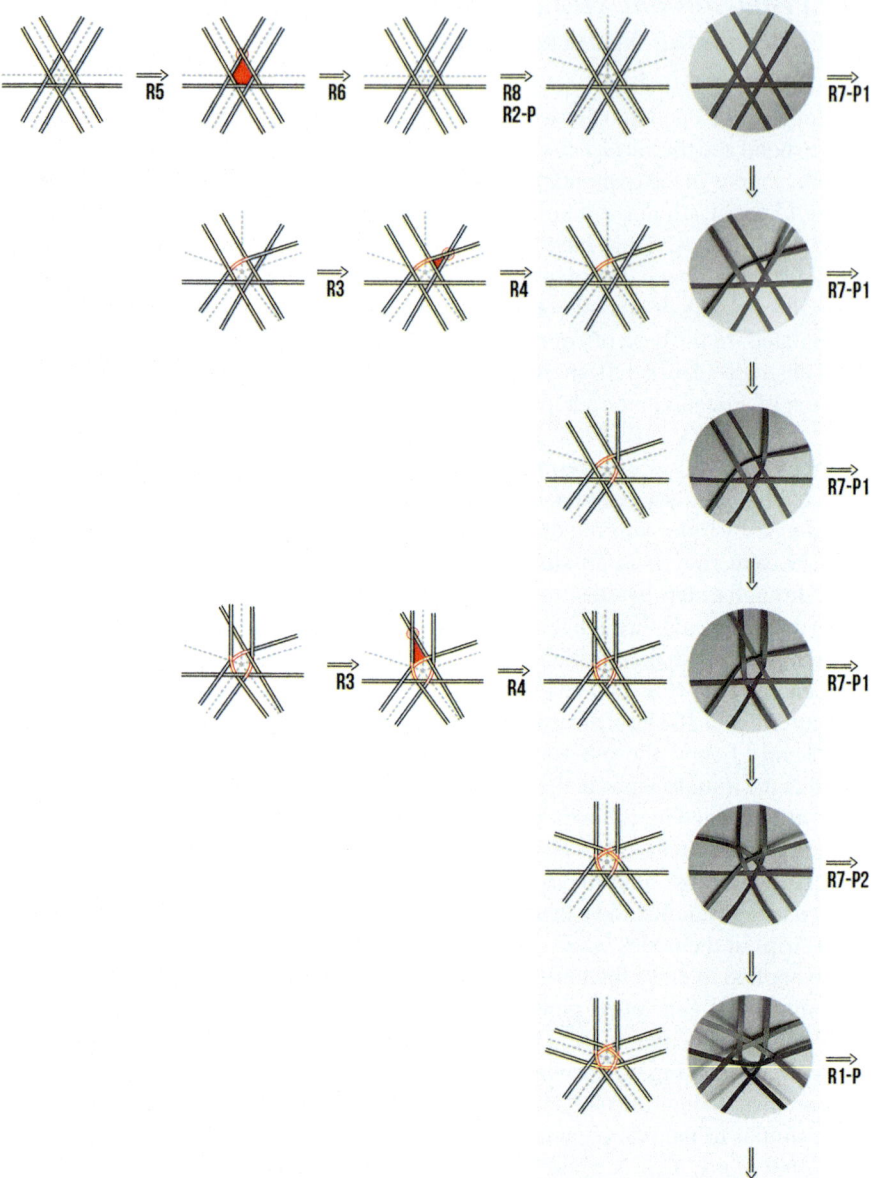

Fig. 8 Introduction of a pentagon in the triaxial weave

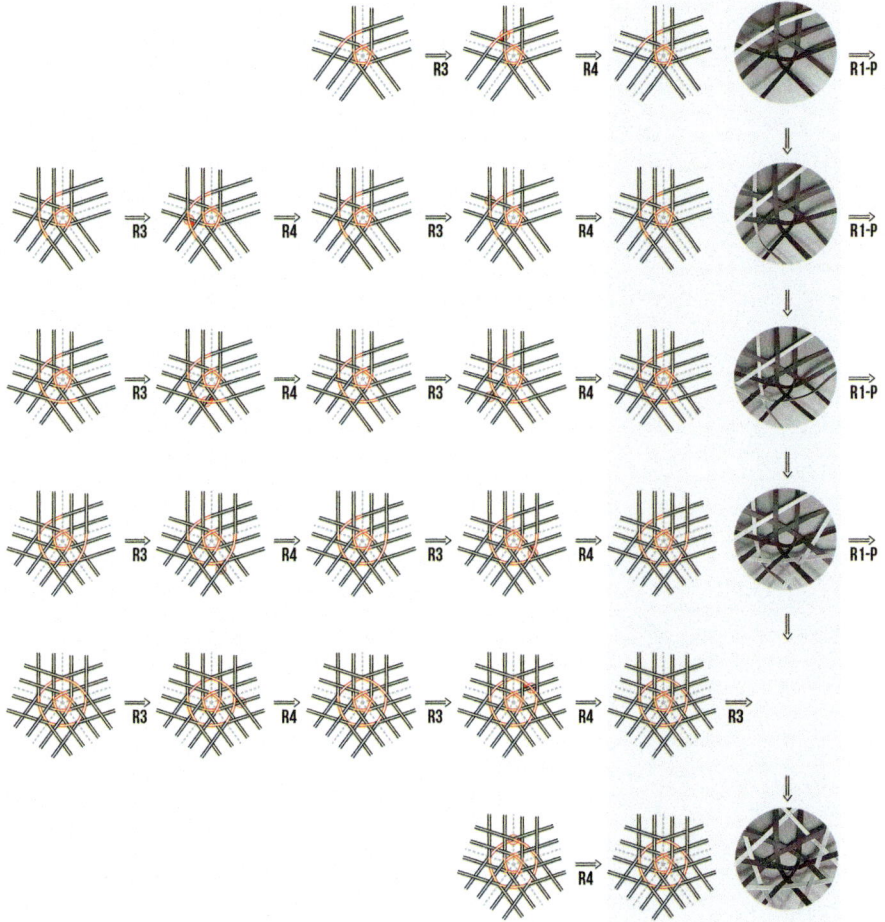

Fig. 9 Computation to generate the second row of the triaxial weave with a pentagonal lattice disclination

complete, resulting with the *material shape* (and the abstract shape) shown in the last row of Fig. 8.

The creation of the second row of the triaxial weave with a pentagonal lattice disclination requires adding five new strips with the rule R1-P. These strips are bent out-of-plane when they are added, as can be seen in the photos of the *material shapes* in the last column of Fig. 9. Red areas on the abstract shapes, in return, indicate the parts on the *material shapes* that go through bending. Similarly, triangles emerge as voids and are *seen* with rule R3, and the over-and-under order of the strips swapped with rule R4. In this computation, *material shapes* and abstract shapes work together in an abstraction-materialization-abstraction cycle. While the triangles emerge and are *seen in the material shapes* in the course of physically weaving the strips, abstract

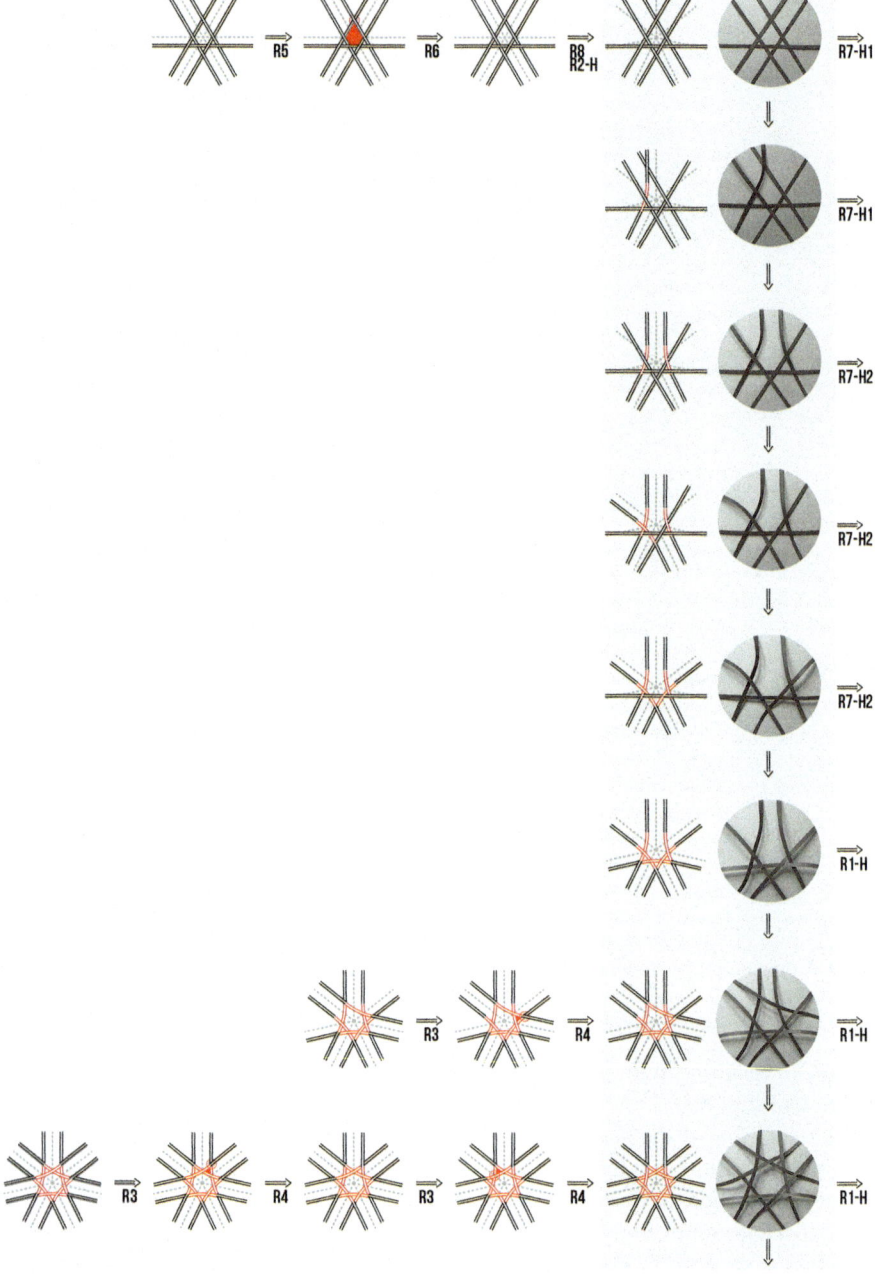

Fig. 10 Introduction of a heptagon in the triaxial weave

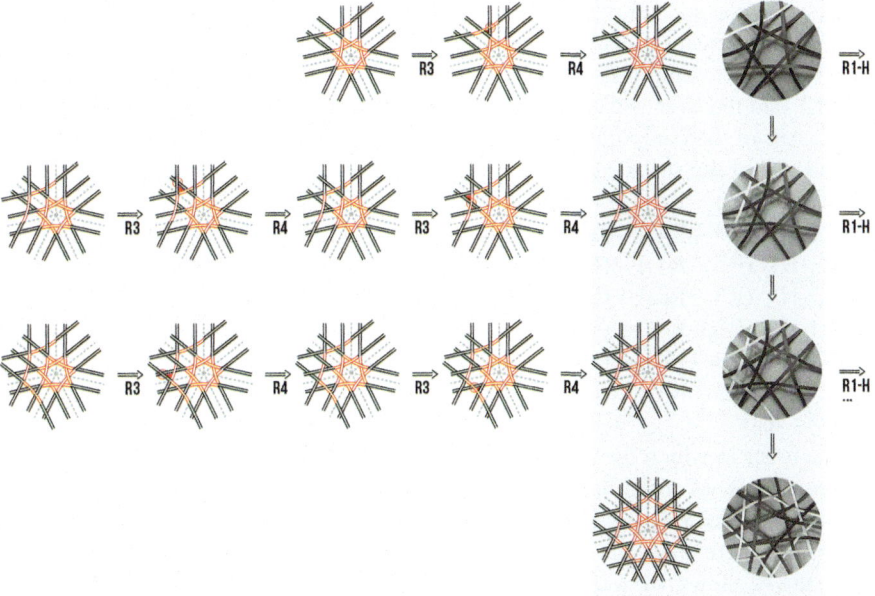

Fig. 11 Computation to generate the second row of the triaxial weave with a heptagonal lattice disclination

shapes help to guide the next steps, and keep track of the process. The process to weave the second row of the strips can also be described as follows:

(R1-P – (R3 – R4)) –
(R1-P – (R3 – R4)) – (R3 – R4) –
(R1-P – (R3 – R4)) – (R3 – R4) –
(R1-P – (R3 – R4)) – (R3 – R4) –
(R1-P – (R3 – R4)) – (R3 – R4) –
(R3 – R4)

The order in which the rules are applied follows the same process as in weaving the second row of strips in planar triaxial weaving, shown in Fig. 6. The rule R1 is replaced with the rule R1-P, and there is one less *(R1-P – (R3 – R4)) – (R3 –R4) cycle* since there are five strips to weave, instead of six.

To create the heptagon, rule R2-H is applied to add new reference lines with seven intersecting lines (Fig. 10). First, the existing five strips are bent using rules R7-H1, R7-H2 and R7-H3 from the R7 group. Then, two strips are added with the rule R1-H. When the strips are added, triangles emerge, again *on another plane* on the *material shapes*. Translating these to abstract shapes, rule R3 can be applied to place the labels and rule R4 to swap the over-and-under order of the strips to physically lock them and erase the labels. Once all the seven strips are placed and interlocked following the computation in Fig. 10, the first row of the triaxial weave with a heptagonal lattice disclination is complete.

To create the second row of the triaxial weave with a heptagonal lattice disclination, seven strips are added with the rule R1-H, which are all bent out-of-plane when added (Fig. 11). Triangles that emerge on the *material shapes* are again *seen* with rule R3, and the over-and-under order of the strips swapped with rule R4. Note that some steps of the computation are omitted in Fig. 11. The full process to weave the second row of the strips can be described as follows:

(R1-H – (R3 – R4)) –
(R1-H – (R3 – R4)) – (R3 – R4) –
(R1-H – (R3 – R4)) – (R3 – R4) –
(R1-H – (R3 – R4)) – (R3 – R4) –
(R1-H – (R3 – R4)) – (R3 – R4) –
(R1-H – (R3 – R4)) – (R3 – R4) –
(R1-H – (R3 – R4)) – (R3 – R4) –
(R3 – R4)

The order in which the rules are applied follows the same process as in weaving the second row of strips in planar triaxial weaving, shown in Fig. 6. The rule R1 is replaced with the rule R1-H, and there is one more *(R1-H – (R3 – R4)) – (R3 – R4) cycle* since there are seven strips to weave, instead of six.

Figure 12 shows two weaves, one with a pentagonal lattice disclination and another with a heptagonal lattice disclination. These weaves are generated by first following the computations shown in Figs. 8 + 9 and Figs. 10 + 11, respectively, and then continuing the process through adding additional rows of strips without changing the underlying reference lines. With each row of strip added, an (R3 – R4) sequence is added to the cycles. It is also possible to change the underlying reference lines along the way (with the rule groups R8 and R2) and introduce new pentagons and heptagons to create variations in the woven forms. The possibilities for form explorations are endless, as can be seen in the works of contemporary Japanese bamboo artists Morigami Jin and Norie Hatakeyama. Alison Grace Martin also uses this topological principle to weave paper sculptures with various levels of complexity, as well as to build large-scale bamboo installations.

Fig. 12 Triaxial weaves with a pentagonal and heptagonal lattice disclination

4 Discussion

When we introduced in an earlier work the *material shapes* in visual rules as part of a dialogue between abstract representations and the physical things in simple *dukta* examples, our sample design made the case for *material shapes* on computing with only one parameterized rule (Gürsoy and Özkar 2015). The *dukta* designs are straightforward and use a limited visual vocabulary to create great variety. In this chapter we considered the case of triaxial basket weaving that engages multiple rules. The rules are instrumental at local nodes of production, namely the triangles, hexagons, pentagons, and heptagons that emerge as strips are woven.

We use triaxial basket weaving as a case to exemplify the computability of a hands-on making process where the form emerges through a dialogue between the physical things and the rules applied to create certain geometries, where the parts are not predefined and can change depending on what the designer chooses to *see* in the process. Our initial assumption is that the person who engages with basket weaving perceives the production process differently. Instead of a top-down understanding, following the basket weaving process node by node (even rule by rule, or weave by weave) brings a bottom-up viewpoint where emerging situations can evoke and receive reaction. The weaver's perceptions continually change between the parts and the whole, so, their focus shifts between strips to larger local nodes that emerge as voids between the strips, then back to the strips and to the whole, etc. It is through the *material shapes* that the weaver can perceive emergent material features, which can then be extracted from the whole and abstracted as visual shapes to *compute*.

In line with the statement above, we show the process of weaving as computation where we pay attention not to numeric features, such as number of nodes, coordinates, placement, etc., but to properties of materials and spatial relations. The act of weaving is represented through visual rules, and *material shapes* are introduced to the computation where necessary. The variation in materials yields different *material shapes*. The rule set can also grow as other weaving techniques are introduced. With new techniques, for instance, where strips are stacked on top of one another, or that engage non-strip like materials, the need for *material shapes* will possibly increase because of the unpredictability of the spatial relations that will emerge as a result of applying simple visual rules, such as those we introduce in the paper. One can see such examples in the works of artists Yamaguchi Ryuun and Uematsu Chikuyu.

One challenge in triaxial basket weaving is to decipher the causal links between (1) the bottom-up rules of triaxial weaving that define the relationships between individual strips; (2) larger local nodes (i.e., triangles, hexagons, pentagons, heptagons) generated through the iterative application of the basic rules; and (3) the global distribution of the local nodes on the overall woven surface that define the final form of the artifact. This chapter focused on visualizing the basic bottom-up rules of triaxial weaving and how they can be used to formalize the triaxial basket weaving process, step-by-step. It has illustrated the effects of pentagons and heptagons on the global form of the woven system. Based on this, a formal representation of the deducted computational logic can be developed in the form of a computer algorithm, along the same lines of the work by Ayres et al. (2018). Also, a simple, two-dimensional visual notation system (as another level of abstraction) can be developed to represent the top-down distribution of the polygons (triangles, hexagons, pentagons, heptagons) on the surface, similar to the ones used in visualizing knitting patterns.

5 Conclusion

Any haptic exploration such as bending, stretching, folding, and twisting a physical object requires sensory engagement and personal connection with the material. The sensory and personal aspects often remain incommunicable. Explorations that involve material properties where parts perform within a physical system are crucial in shaping the designs. Formal representations of the process that cover as much of its full sensory breadth as possible would enable informed and resourceful conversations among designers. While quantifiable material properties such as flexibility and structural stability are more suitable for computer generated simulations, we can recognize emerging material properties (i.e., variations in texture, in light transmittance, in transparency, etc.) using *material shapes*. These properties are less likely to be perceived through computer simulations.

In basket weaving, the exact final form is rarely definable in advance. Rather than being *imposed on the material*, the form comes into existence during weaving: the strips perform as a physical system in shaping the final woven form. Although the final forms cannot be fully *designed* in advance, the formal relations of the physical system that anticipate these forms can be defined. We formalize triaxial basket weaving by developing a weaving grammar that uses *abstract shapes* and *material shapes* in tandem. The grammar is used to discover the underlying rules and relations in existing woven artifacts and can be used to generate new ones. *Material shapes* are the main objects of physical interaction in these explorations, and abstract shapes are created based on what one perceives in physical things.

Sometimes the design knowledge to convey to others is the know-how of a making process, and not just its outcome. Through a computational approach to making, this relative knowledge can become sharable. Computation enables us to understand how we make things, and in turn, we can share it with others. This approach is especially valuable for pedagogical contexts, such as design studios and ateliers, and for collaborative environments where the communication and transfer of knowledge is essential.

References

Akleman, E., J. Chen, Q. Xing, and J.L. Gross. 2009. Cyclic plain-weaving on polygonal mesh surfaces with graph rotation systems. *ACM Transactions on Graphics* 28 (3): 1–8.

Aydın, A., and M. Özkar. 2015. Material computability of indeterminate plaster behavior. In *Computer-aided architectural design: The next city—new technologies and the future of the built environment, CAAD futures 2015*, eds. G. Celani, D. Sperling, and J. Franco, 582–599. Berlin: Springer.

Ayres, P., A.G. Martin, and M. Zwierzycki. 2018. Beyond the Basket Case: A principled approach to the modelling of Kagome weave patterns for the fabrication of interlaced lattice structures using straight strips. In *AAG 2018: Advances in architectural geometry 2018*, 72–93. Chalmers University of Technology.

Baek, C., A.G. Martin, S. Poincloux, T. Chen, and P.M. Reis. 2021. Smooth triaxial weaving with naturally curved ribbons. *Physical Review Letters* 127 (10).

Callens, S.J.P., and A.A. Zadpoor. 2018. From flat sheets to curved geometries: Origami and kirigami approaches. *Materials Today* 21 (3): 241–264.

Çapunaman, Ö.B., C.K. Bingöl, and B. Gürsoy. 2017. Computing stitches and crocheting geometry. In *Computer-aided architectural design: Future trajectories, CAAD futures 2017*, eds. G. Çağdaş, M. Özkar, L.F. Gül, and E. Gürer, 289–305. Singapore: Springer.

Coffland, R. 2000. *Contemporary Japanese bamboo arts*. Art Media Resources.

Coffland, R. 2006. *Hin: The quiet beauty of Japanese bamboo art*. Art Media Resources.

Earle, J. 2008. *New bamboo: Contemporary Japanese masters*. New York: Japan Society.

Gün, O.Y. 2017. Computing with watercolor shapes: Developing and analyzing visual styles. In *Computer-aided architectural design: Future trajectories, CAAD futures 2017*, eds. G. Çağdaş, M. Özkar, L.F. Gül, and E. Gürer, 329–347. Singapore: Springer.

Gürsoy, B., and M. Özkar. 2015. Visualizing making: Shapes, materials and actions. *Design Studies* 41: 29–50.

Hamzaoğlu, B., and M. Özkar. 2016. Geometric patterns as material things: The making of Seljuk patterns on curved surfaces. In *Proceedings of Bridges: Mathematics, music, art, architecture, culture, 331–336*, eds. E. Torrence, B. Torrence, C.H. Séquin, D. McKenna, K. Fenyvesi, and R. Sarhangi, 331-336. Phoenix: Tessellations Publishing.

Hamzaoğlu, B., and M. Özkar. 2023. Rule-based milling of medieval stone patterns. *Nexus Network Journal*.

Ingold, T. 2000. Making culture and weaving the world. In *Matter, materiality, and modern culture*, ed. P. Graves-Brown. London: Routledge.

Ingold, T. 2013. Of blocks and knots. *The Architectural Review* 234 (1400): 26–27.

Knight, T. 1989. *Shape grammars in education and practice: History and prospects*. Internet Paper. http://www.mit.edu/~tknight/IJDC/.

Knight, T. 1993. Color grammars: The representation of form and color in designs. *Leonardo* 26 (2): 117–124.

Knight, T. 2015. Shapes and other things. *Nexus Network Journal* 17: 963–980.

Knight, T. 2018. Craft, performance and grammars. In *Computational studies on cultural variation and heredity*, ed. J.H. Lee, 205–224. KAIST research series. Singapore: Springer.

Knight, T., and G. Stiny. 2015. Making grammars: From computing with shapes to computing with things. *Design Studies* 41: 8–28.

Labrou, K., and S.D. Kotsopoulos. 2023. Making grammars for computational lacemaking. In *Design computing and cognition'22*, ed. J.S. Gero, 587–604. Cham: Springer.

LaPlantz, S. 2016. *The mad weave book: An Ancient form of triaxial basket weaving*. Dover Publications.

MacLachlan, L. 2018. *Making rules, making tools: How can shape grammar support creative making*. Unpublished Ph.D. dissertation, The Open University.

Mallos, J. 2009. How to weave a basket of arbitrary shape. In *Proceedings of ISAMA 2009: Eighth interdisciplinary conference of the international society of the arts, mathematics and architecture*, 13–19. Albany: The International Society of the Arts, Mathematics and Architecture.

Mallos, J. 2011. Extra ways to see: An artist's guide to map operations. In *Proceedings of ISAMA 2011: Tenth interdisciplinary conference of the international society of the arts, mathematics, and architecture*, 111–121. Chicago: The International Society of the Arts, Mathematics, and Architecture.

Mallos, J. 2015. Knotology baskets and topological maps. In *Proceedings of Bridges: Mathematics, music, art, architecture, culture*, eds. K. Delp, C.S. Kaplan, D. McKenna, and R. Sarhangi, 215–222. Phoenix: Tessellations Publishing.

Martin, A.G. 2015. A basketmaker's approach to structural morphology. In *Proceedings of the International Association for Shell and Spatial Structures Symposium: Future visions–computational design, IASS 2015*. Amsterdam: International Association for Shell and Spatial Structures.

Muslimin, R. 2014. *Ethnocomputation: On weaving grammars for architectural design*. Unpublished Ph.D. dissertation, Massachusetts Institute of Technology.

Muslimin, R. 2017. Ethnocomputation: An inductive shape grammar on Toraja Glyph. In *Computer-aided architectural design: Future trajectories, CAAD futures 2017*, eds. G. Çağdaş, M. Özkar, L.F. Gül, and E. Gürer, 329–347. Singapore: Springer.

Noel, V.A.A. 2015. The Bailey-Derek grammar: Recording the craft of wire-bending in the Trinidad Carnival. *Leonardo* 48 (4): 357–365.

Oxman, R. 2002. The thinking eye: Visual recognition in design emergence. *Design Studies* 23: 135–164.

Özkar, M. 2020. The matter of shape: A computational approach to making in architectural heritage (survey). In *Advances in mathematical sciences. Association for women in mathematics series*, ed. B. Acu, D. Danielli, M. Lewicka, A. Pati, R.V. Saraswathy, and M. Teboh-Ewungkem, 339–347. Springer.

Pineda, S., M. Arora, P.A. Williams, B.M. Kariuki, and K.D.M. Harris. 2016. The grammar of crystallographic expression. In *Proceedings of ACADIA 2016: Posthuman frontiers, data, designers and cognitive machines*, eds. K. Velikov, S. Manninger, M. del Campo, S. Ahlquist, and G. Thün. Ann Arbor: ACADIA Publishing Company.

Roelofs, R. 2008. Connected holes. In *Bridges conference proceedings*, 29–38.

Roelofs, R. 2010. About weaving and helical holes. In *Bridges conference proceedings*, 75–84.

Schön, D. 1983. *The reflective practitioner*. New York: Basic Books.

Schön, D., and G. Wiggins. 1992. Kinds of seeing and their functions in designing. *Design Studies* 13 (2): 135–156.

Smithwick, D., and L. Sass. 2014. Embodied design cognition: Action-based formalizations in architectural design. *International Journal of Architectural Computing* 12 (4): 399–418.

Stiny, G. 1996. Useless rules. *Environment and Planning B: Planning and Design* 23 (2): 235–237.

Stiny, G. 2001. How to calculate with shapes. In *Formal engineering design synthesis*, ed. E.K. Antonsson and J. Cagan. New York: Cambridge University Press.

Stiny, G. 2006. *Shape: Talking about seeing and doing*. Cambridge: The MIT Press.

Stiny, G. 2011. What rule(s) should I use? *Nexus Network Journal* 13 (1): 15–47.

Vazquez, E., B. Gürsoy, and J. Duarte. 2019. Formalizing shape-change: Three-dimensional printed shapes and hygroscopic material transformations. *International Journal of Architectural Computing* 18 (1): 1–17.

Open Access This chapter is licensed under the terms of the Creative Commons Attribution-NonCommercial-NoDerivatives 4.0 International License (http://creativecommons.org/licenses/by-nc-nd/4.0/), which permits any noncommercial use, sharing, distribution and reproduction in any medium or format, as long as you give appropriate credit to the original author(s) and the source, provide a link to the Creative Commons license and indicate if you modified the licensed material. You do not have permission under this license to share adapted material derived from this chapter or parts of it.

The images or other third party material in this chapter are included in the chapter's Creative Commons license, unless indicated otherwise in a credit line to the material. If material is not included in the chapter's Creative Commons license and your intended use is not permitted by statutory regulation or exceeds the permitted use, you will need to obtain permission directly from the copyright holder.

[Un]making with____

Dina El-Zanfaly

Abstract Despite the multiple ways of being and knowing, the design field has historically adopted a universal approach in which one size (or design) fits all. In this book chapter, I challenge this assumption in the computational and interaction design domains, arguing in favor of embracing plurality inherent in all human experiences. I introduce *[un]making with____* as a design approach that achieves inclusive interactions. I define "unmaking" here as a conscious method of deconstructing and iteratively remaking understandings, processes, and tools; in a sense, *unmaking* reveals new insights that make space for new ideas to emerge. The concept of *[un]making with____* emphasizes co-designing with people, rather than for them, and acknowledges that we design with tools, machines, and individual circumstances, allowing for multiple worldviews. Building on this, I focus on two levels of investigation: human sensory perception and context. I present three projects from my lab that exemplify the approach. Drawing from these three projects, I discuss how this approach emphasizes design as a web of relations and fosters mutual meaning-making.

1 Introduction

In this chapter, I present the concept of "[un]making with" as a co-design approach for bringing multiple views into the design process to achieve more inclusive interactions that respond to the plurality of human experiences, abilities, and perceptions. It is undeniable that *unmaking* is inherently rooted in the embed-fuse cycle (Stiny 2008) of Stiny's shape grammars, as the making, unmaking, and remaking of visual shapes. However, this chapter expands the concept of unmaking from being performed solely by the designer to being a collective act of resistance informed by a process of deconstructing and reconfiguring assumptions, understandings, and interactions during the design experience, leading to new insights and making space for the unexpected to emerge. By unmaking a single design, we can make room for the diverse interpretations that emerge from it. By unmaking how a computational design tool works,

---

D. El-Zanfaly (✉)
School of Design, Carnegie Mellon University, 5000 Forbes Avenue, Pittsburgh, PA 15213, USA
e-mail: delzanfa@andrew.cmu.edu

© The Author(s) 2025
S. Kotsopoulos (ed.), *Shape Computation*, Mathematics and the Built Environment 9, https://doi.org/10.1007/978-3-031-81623-9_8

we can investigate its biases and implications, and by engaging people in the act of undesigning assumptions, we are iteratively co-designing the tools, interfaces, and interactions as they emerge, rather than imposing a design on users and expecting them to respond in a preconceived way. This approach investigates what we can teach computational machines, and what we can learn from them. Rather than anthropomorphizing, my intention in this query is to reveal how interacting with machines and computational tools shapes our experiences and gets shaped by varying perceptions, experiences, and contexts. Through this lens, the question driving us as designers and scholars becomes: how can we accommodate differences and variations to reveal and inform the interactions between humans and machines? And how can we make these interactions more commonplace in systems, contexts, and cultures?

In the discussion that follows, I first discuss the theoretical and practical paradigms of the design process, as well as the emergence of "unmaking" as it has played out in computational and interaction design. I then present three projects, created with the students and collaborators in my lab, that came to embody my understanding of *[un]making with*, as driven by the sensory and contextual levels of human experiences. Finally, I reflect on the development of these ideas and the various ways they tangibly emerged in the process of making and then unmaking collective action and discovery within a highly interactive research environment.

2 [Un]making With and For

Even though we perceive the world differently based on individual sensory experience and physical abilities, culture, and context, the design field has historically adopted a universal approach in which one size fits all (Rosner 2018; Hendren 2020), in which designers have long assumed that any user can respond to the design as they intended it. However, this universal approach does not fully cover interactions with technology that require situated actions in environments (Suchman 2006). More recently, advances in computational tools and products such as Artificial Intelligence (AI) and mixed reality (MR) open new frontiers in design processes, services, and applications. Without shared conventions or contexts, interactions with these tools and products are naturally open to different interpretations—yet they are still often intangible experiences designed for a universal user, unable to adapt to different contexts and, potentially, unforeseen differences. For example, when manipulating a virtual object in a mixed reality application, one person may use a grasping hand gesture, while another may try pinching with their fingers. The function of selecting something in mixed reality headsets is performed by different gestures, depending on different devices (Fig. 1).

This book chapter challenges this approach in the computational and interaction design domains, arguing instead for embracing the plurality that is inherent in all human experiences. Over the years, there has been much support for multiple worldviews (Goodman 1978; Reddy 1993; Escobar 2018), emphasizing the notion of the "pluriverse" in design, a "world where different worlds exist" (Escobar 2018).

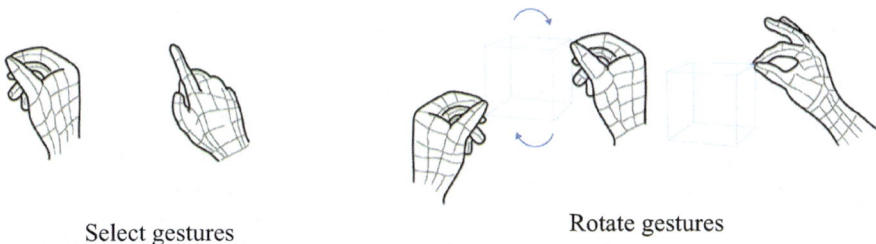

Fig. 1 Shows different gestures for two functions. On the left are select gestures in Apple Vision Pro and Hololens. On the right are rotate gestures in Apple Vision Pro and Hololens

Plurality emphasizes the importance of designing with an awareness of varied perceptions and interpretations to ensure our designs are inclusive, adaptive, and culturally sensitive. I build on the concept of plurality on two levels: sensory perception and context. On the sensory perception level, multiple interpretations go beyond just visual input to encapsulate the broader spectrum of how we engage our bodies to experience our environment and the world around us. For example, an artifact or space can be perceived in multiple ways based on our embodied abilities and sensory experiences—from motion and touch to sound. On the contextual experience level, a hand gesture can be used for different purposes based on culture or situation. For example, in some cultures, young people may use a peace sign to indicate an attitude or position, while displaying the similar gesture of a victory sign may be a way to look coquettish in a photo. And the gesture of OK in Western culture is considered an obscene gesture in Turkey and Brazil. Again, what is assumed is that the user is responding to an intuitively obvious, universal design, even when research has shown this not to be the case for many users with variable abilities, experiences, and cultural norms (Wennberg et al. 2018). During my research experience, I have come to embrace the process of "[un]making with" as a practice of resistance and, more tangibly, as a process of collective design that, I argue, enhances the potential for creative use and interpretations, and more inclusively reflects the variability of human perception, experience, and culture.

The idea of unmaking has itself been recognized since the fifteenth century, but its meaning and application have changed. Originally used to denote something destroyed, reversed, or undone (*New Oxford American Dictionary*), "unmaking" has been applied by the human–computer interaction field to denote the afterlife of products from a sustainability point of view (Lindström and Ståhl 2020; Song and Paulos 2021). Unmaking is also a form of physical reconfiguration with roots in architectural form, as when British architect and theorist Cedric Price introduced the concept of "anti-building" in his design of Fun Palace in 1961 (Steenson 2014). Price envisioned a flexible structure that is constructed, deconstructed, and reconstructed over time. On a more theoretical level, architect Christopher Alexander viewed design as an ensemble consisting of a form and its context, in which the form is a solution to a design problem, and the context defines that problem. Alexander proposed *decomposing* the relations in this ensemble to solve a design problem (Alexander 1964).

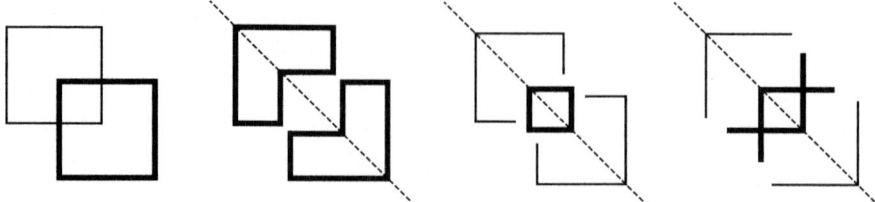

Fig. 2 Two intersecting squares can be perceived and interpreted differently. Redrawn based on George Stiny's originals (Stiny 2008)

The practice of "un" in the design field extends to other actions such as *uncrafting* (Murer et al. 2015), *unbuilding* (Wakkary 2021), and *undesigning* (Pierce 2014).

Along the same lines in computational design, George Stiny reminds us that we can perceive and interpret the same visual shape differently (Stiny 2008). For example, two intersecting squares can be interpreted as two "L" shapes, but they can alternatively be perceived as a central small square with lines around it (Fig. 2). These different visual interpretations are similar to the different ways of interpreting a hand gesture while interacting with an MR application.

As noted, it was during the development of three projects in my lab that I came to more fully recognize the need to approach design in a new, more adaptive and inclusive way. The research experience shifted from construction to reconstruction and deconstruction; we made sure our collective practice arose from the different ideas and varied visions of perceiving and making as a way to iteratively unmake the assumed methods and meanings.

2.1 Project One: Welding Craft and Computation

This project unfolded as an investigation into how craft-based practices, such as metalworking, inform how we design embodied interactions with computational tools, mainly mixed reality (MR) and AI tools. By looking at craft-based practices, we are able to create more inclusive interactions with computational tools, while also leveraging these tools for learning hands-on skills. A collaboration with a colleague, Daragh Byrne, an associate teaching professor at Carnegie Mellon's School of Architecture, the project partnered with the Industrial Arts Workshop (IAW), which trains teenagers in the underserved neighborhood of Hazelwood in Pittsburgh, Pa. Our aim was to aid the instructional process of the basics of metalworking by developing, testing, and deploying an inclusive in situ MR training system. Through co-designing with the IAW community (Fig. 3), instructors were incentivized to unpack their metalworking lessons, essentially leading to *unmaking* their instructional approaches. This included one-on-one instruction, using multisensory practices, to identify and test tangible methods of instruction that all users could follow. Over six months, we strove to unmake assumed instructor-student interactions that took into account situational

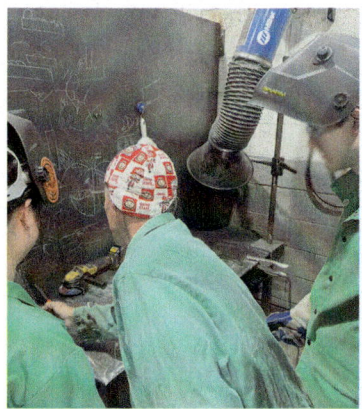 

Fig. 3 Two workshops with the IAW community: on the left, our team experiences welding training from a student's point of view; on the right, a co-design workshop where we introduced MR concepts and prototypes

and spatial aspects that we did not anticipate, such as the effect of welding noise on different students and using meditation to help students and instructors develop a sense of calm and body control.

The project engages with the challenge of designing inclusive MR interactions, including: (1) how to make tacit knowledge of a specific hands-on skill explicit (Collins 2012), as when the angle of the hand is based on the moving speed of a welding torch or when bending a metal piece; (2) how to transfer this tacit knowledge through a hybrid medium, as when one chooses the instruction modality in the MR application, and then co-designing the instruction to be clear to people with variable abilities and training. For example, ensuring that a green indicator illustration is perceived and understood by any user when they are holding a welding torch at the correct angle; (3) how to build a multisensory instructional application; and (4) how to respond to and accommodate the social and contextual aspects of the instructional processes of metalworking.

Our team took a full-day metalworking workshop (Fig. 3) to experience welding training from a student's point of view. This workshop helped us identify intervention opportunities and observe non-verbal interactions that are difficult to formalize, i.e., going to a student's welding booth to give feedback on improving their welding based solely on the welding sound they heard. We then organized a co-designing workshop (Fig. 3) with instructors and students who completed the training and became instructors themselves (Chen et al. 2024). In this workshop, we introduced the concept of mixed reality (MR) and the MR prototypes that we created, including an immersive interactive application that provided visual instructions to users to hold and move their welding torches according to the right angle and moving speed. In effect, we collaboratively *unmade with* instructors the MR interactions that we designed (Fig. 4). We created a clear acrylic screen and placed it in the welding

Fig. 4 [Un]making the MR interactions with the IAW that we created. We asked the IAW instructors to design the interactions on acrylic sheets

booth (Fig. 4). We then asked the instructors to imagine and draw the instructions they wanted to see and experience with the MR headset.

The process of collaboratively unmaking and remaking these interactions with the IAW community enabled us to weave the perceptual, social, and material aspects of craft-making and skill training into the MR system. On the level of sensory perception, it allowed us as researchers to include the welding instructors' different perspectives and understanding. On the level of contextual experience, using a physical mock-up of the interactions ensured the design and environmental aspects such as lighting are relevant to the situation. It also enabled us to realize the importance of the sonic cues of welding as part of the designed interactions. This approach is applicable beyond metalworking training, allowing us to consider the situatedness of many experiences, including audible cues, interactions with the instructor, and direct embodied interactions with students.

2.2 Project Two: Sand Playground

Generative AI applications have proliferated in creative industries, but these applications are often incomprehensible to users in terms of how they operate and what they produce. This makes it difficult to understand the capabilities and limits of these generative AI tools. Current work on Explainable AI (XAI) examines how an algorithm produces a result. However, the work I discuss here focuses on the "what," that is, what the generative AI does. Sand Playground, a human-AI tangible co-drawing interface, supports open-ended co-drawing on a sand canvas (Fig. 5). Our purpose initially was to enhance the co-creation experience of an intelligent system and a human being on a malleable medium. By understanding how a generative AI tool operates, and what its limitations and capabilities are, designers can leverage

Fig. 5 Sand playground, co-drawing with an AI agent on the sand

this information and thus make more informed design decisions (Ghajargar et al. 2022; El-Zanfaly et al. 2023; Nicenboim et al. 2022). This understanding generates increased trust in an AI system and transforms the processes and culture of creative practices. We investigated how we perceive the behavior of a generative AI agent when we co-create with it on an embodied interface. Additionally, we studied the role of increased understanding of an AI system within a co-creation experience.

More specifically, the AI-mediated interface consists of a sand bed and a metal ball that co-creates, depending on what users create using their hands. We started our *[un]making with__* approach by conducting a co-designing workshop with designers to imagine the co-creation scenarios between an intelligent system and a human user on a malleable medium such as sand (Fig. 6). We generated ten scenarios with workshop participants. We chose three co-drawing scenarios from the generated scenarios: artistic mimicry, zen gardening, and doodling. In artistic mimicry, the user leads the drawing sessions by directly drawing in the sand with one finger and then actuates the metal ball to be in the centrosymmetric position at a similar distance from the center from where the finger is. In zen gardening, the user initiates the drawing by placing a rock, or a group of rocks, on the sand. The system actuates the metal ball to draw a series of concentric contours around the rocks once it locates the placement of the rocks.

In doodling, our goal was to understand the mental models of users of how drawing-based generative AI works. More specifically, we wanted to investigate how people interpret the intention of a generative AI agent without any provided instructions. The doodling mode utilizes the open-source Sketch-RNN and the Quick Draw! Dataset (2022) to generate drawings based on the user's input. This is an improved technique compared to the previous two modes, where we only implemented AI to recognize user inputs. We used OpenCV and the Google MediaPipe library in Python to detect users' fingertip coordinates and input on the sand. Once a new session is initialized, Sand Playground selects ten random sketches from a pool generated from three pretrained Sketch-RNN models: "Catbus," "Elephantpig," and "Owl." When the user draws a stroke in the sand, the system records where their drawings end. The

 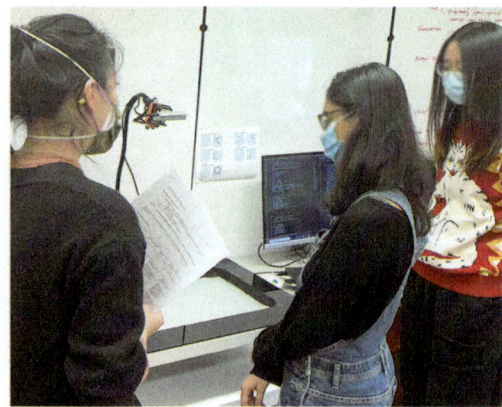

Fig. 6 Sand playground, co-drawing workshops

metal ball then moves to the position and adds to the user's drawings with a novel line from one generated sketch.

We then organized a workshop with ten designers to identify users' perceptions of a generative AI agent on Sand Playground. We utilized quantitative and qualitative approaches by asking these designers to fill out a survey with questions selected from the Creative Support Index (CSI) (Cherry and Latulipe 2014) and the USE Questionnaire (Lund 2001). We found that Sand Playground enabled participants to develop their approaches to understanding the intentionality of a generative AI agent by co-drawing freely and open-endedly in the sand in real time with the AI agent. The physicality of the AI agent, as the metal ball, enabled participants to treat the AI agent as a partner rather than a tool. Participants appreciated the imperfection of the agent's drawing and used it as part of their creative strategy.

Overall, the degree of collaborative interactions of the generative AI agent shifts from being a tool that, for example, mimics drawings to being a collaborator that detects what has been drawn and chooses lines as a response. This shift reconfigures the nature of interactions from an input to output experience to an evolving co-creation experience. The impermanence of the drawings on the sand canvas contradicts the assumption that an AI system is supposed to provide a precise finished product. On a sensory perception level, the sand acts as an open-ended, three-dimensional sensory canvas through which participants build their understanding. On a contextual experience level, any motions or gestures in the sand afforded by the interface can be used. As a result, participants constructed (made) and reconstructed (unmade) their preconceived notions about what the AI agent produces, taking the concept of AI co-creation from digital two-dimensional screens to a more accessible embodied interface.

2.3 Project Three: From I^3 to I^4: Learning to (Un)make and (Un)making to Learn

The third project focuses on challenging the separation of the design phase from the making phase, reinforced by computational design tools such as digital fabrication machines. This separation hinders the design and learning processes in both informal and formal design settings. This problem is deeply rooted in architecture as early as Alberti who adapted Aristotle's hylomorphic model of separating the design phase, *lineamenta*, from the execution phase, *structura* (Carpo 2011; Alberti 1992; Ingold 2011). In this case, design is a linear process that stops once the execution of the drawings is complete. Such an approach reiterates a separation between mind and body in creative practices. Consequently, the process separates design, the actions of the mind, from the making, the actions of the hand, putting design on a pedestal, compared to craft, which unifies the mind and body processes.

Similarly, digital fabrication machines were built in the mid-twentieth century to execute repetitive, autonomous tasks that respond to what was already designed (Cardoso Llach 2012). These machines were not built to be a creative design companion or a learning tool but rather as a universal computational tool, with the expectation of a single method of operation. When used, thinking and planning of the artifact must happen before the machine-driven execution and making phases. This universal approach hinders creative processes and ignores our diverse ways of knowing and being. In contrast, craft practices involve direct material manipulation and continuous reflection-in-action, fostering iterative exploration (Bamberger and Schön 1983; Schön 1983; Pye 1978). Consequently, learning how to operate a digital fabrication machine for novices is not enough to create and improvise with it. These machines have been integrated into formal and informal educational settings such as architecture schools and community-based fabrication labs for the past fifteen years. Many benefits to the learning process have been observed, such as precision, time-saving, iteration, and feedback between design and making (Özkar 2007; McCullough 1998; Carpo 2011). It has been observed that some users interact with these machines unexpectedly compared to what they are designed for, depending on embodied and situated dimensions (Landwehr Sydow et al. 2020). For example, research has shown that users of 3D printers develop tacit knowledge of printing through maintenance practices (Subbaraman and Peek 2023). Some researchers have tried to bridge the gap between planning and execution by introducing interactive fabrication (Fossdal et al. 2021; Willis et al. 2011).

By acknowledging that digital fabrication machines and their places need to support plurality in learning how to make, we can create more inclusive design practices and spaces. In 2014, I formalized I^3 as a multimodal interaction model that connects the design and making processes using digital fabrication machines for novice designers and makers. This multimodal interaction model challenges how computational tools are designed to be universal. The model focuses on iterative direct engagement and learning to design and make on several scales, from artifacts to small structures. I call it I^3 for its multi-layered approach of Imitation, Iteration,

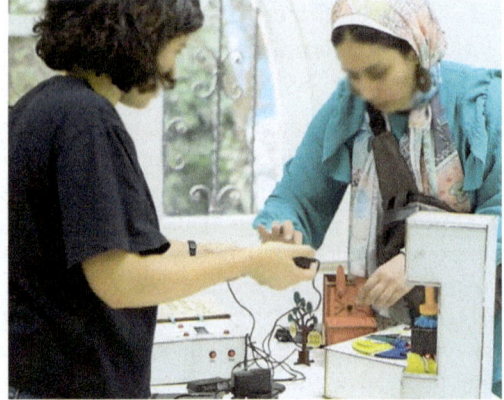

Fig. 7 [Un]making with the community as an approach for developing I³

and Improvisation. I³ builds the learner's sensory experience and spatial reasoning, enabling discovery and surprise (El-Zanfaly 2015, 2017, 2018).

I co-designed and implemented I³ through ethnographic work including observations, interviews, and post-workshop questionnaires with novice and experienced learners in different settings (Fig. 7). These include formal and informal educational settings in Turkey, Egypt, and the USA. The diversity of the contexts ensured that the model was adaptive and inclusive. The model was later adopted as a learning approach in other community-based fabrication labs in Italy and other European countries. In addition to ethnography, I³ is based on unmaking craft practices and studying how craft is learned and taught to inform how we interact with digital fabrication machines. This approach is similar to other research on how craft processes inform digital design tools and processes (coons and Ratto 2015). Along the lines of the first project, learning from craft to shape interactions with universally designed digital fabrication machines integrates both the sensory and contextual experiences of the learner.

3 Discussion

This chapter introduced a collaborative "*unmaking with__*" design approach that allows designers to re-imagine computational tools as a means of creating more inclusive interactions, and reflecting the plurality of multisensory experiences, contexts, and perceptions. While participatory design approaches have promoted workshops with users to solve a need or a problem they face, underlying this practice is the assumption that all users should collectively agree on the problem to solve. In counterpoint, the collaborative notion of *[un]making with____* enables participants to take apart existing interactions with computational tools in a way that explores divergent views and incorporates more inclusive and culturally appropriate capabilities and experiences into the design process.

All three projects presented in this chapter were conceived as experiments to develop my hypothesis that continuous making and unmaking with and among participants, designers, tools, and contexts will yield a more inclusive design approach, one that makes room for the rich, variable social and contextual experiences, cultures, and abilities of users. This model of iteratively making and unmaking with others reflects the diversity and multiplicity at the core of the human condition.

Drawing from these three projects, I suggest two aspects of *[un]making with___* that can be extrapolated from what we learned.

3.1 Unmaking as a Web of Relations

The concept of continuous making and unmaking with and among designers, participants, tools, and contexts acknowledges the web of relations between these entities in the design process. For example, both the I^4 and the metalworking projects emphasize the social and contextual aspects of learning to make with tools. It also expands the social dimensions of the computational interactions and tools being designed. The concept also helps mitigate the bias the designer may bring to the design, ensuring that marginalized voices are also heard. As seen in the three projects, [un]making as a web of relations ensures that the embodied and contextual aspects of the interactions are included.

3.2 Developing Mutual Meaning-Making

Put another way, the concept of *[un]making with___* presents a new paradigm of mutual meaning-making, where interactions are constructed and deconstructed with participants to define mutual objectives and produce shared understandings of how specific interactions with computational tools are perceived and used. This approach counters preconceived assumptions with fresh, variegated insights about the potential for computational tools to usher in a new understanding of the design process—one in which plurality, rather than universality, invigorates and democratizes a discipline that for so long has insisted on the uniformity of lived experience as the driver of design processes and solutions, and a predetermined use of computational tools.

References

Alberti, L.B. 1992. *On the art of building: In ten books.* Cambridge: The MIT Press.
Alexander, C. 1964. *Notes on the synthesis of form.* Cambridge: Harvard University Press.

Bamberger, J., and D.A. Schön. 1983. Learning as reflective conversation with materials: Notes from work in progress. *Art Education* 36 (2): 68. https://doi.org/10.2307/3192667.

Cardoso Llach, D. 2012. *Builders of the vision: Technology and the imagination of design*. Ph.D. dissertation. Cambridge: Massachusetts Institute of Technology. http://dspace.mit.edu/handle/1721.1/77775.

Carpo, M. 2011. *The alphabet and the algorithm*. Cambridge: The MIT Press. http://site.ebrary.com/id/10453040.

Chen, Z., T. Johnson, A. Knowles, A. Li, S. Yi, Y. Zhuang, D. Byrne, and D. El-Zanfaly. 2024. Augmenting embodied learning in welding training: The co-design of an XR- and tinyML-enabled welding system for creative arts and manufacturing training. In *Proceedings of the eighteenth international conference on tangible, embedded, and embodied interaction, TEI '24*, 1–14. New York: Association for Computing Machinery. https://doi.org/10.1145/3623509.3633398.

Cherry, E., and C. Latulipe. 2014. Quantifying the creativity support of digital tools through the creativity support index. *ACM Transactions on Computer-Human Interaction* 21 (4): 21:1–21:25. https://doi.org/10.1145/2617588.

Collins, H. 2012. *Tacit and explicit knowledge*. Chicago: University of Chicago Press.

Coons, G., and M. Ratto. 2015. Grease pencils and the persistence of individuality in computationally produced custom objects. *Design Studies*, Special Issue: Computational Making, 41 (November): 126–136. https://doi.org/10.1016/j.destud.2015.08.005.

El-Zanfaly, D. 2015. [I3] Imitation, iteration and improvisation: Embodied interaction in making and learning. *Design Studies*, Special Issue: Computational Making, 41, Part A (November): 79–109. https://doi.org/10.1016/j.destud.2015.09.002.

El-Zanfaly, D. 2017. A multisensory computational model for human-machine making and learning. In *ACADIA 2017: Disciplines and disruption [Proceedings of the 37th annual conference of the association for computer aided design in architecture (ACADIA)]*, Cambridge, 2–4 November, 2017, 238–247. CUMINCAD. ISBN 978-0-692-96506-1. http://papers.cumincad.org/cgi-bin/works/paper/acadia17_238.

El-Zanfaly, D. 2018. [I^3] Imitation, iteration and improvisation: Embodied interaction in computational making and learning. Ph. D. dissertation. Cambridge: Massachusetts Institute of Technology. https://dspace.mit.edu/handle/1721.1/118695.

El-Zanfaly, D., Y. Huang, and Y. Dong. 2023. Sand-in-the-loop: Investigating embodied co-creation for shared understandings of generative AI. In *Companion publication of the 2023 ACM designing interactive systems conference, DIS '23 Companion*, 256–260. New York: Association for Computing Machinery. https://doi.org/10.1145/3563703.3596652.

Escobar, A. 2018. *Designs for the pluriverse: Radical interdependence, autonomy, and the making of worlds*. Illustrated ed. Durham: Duke University Press Books.

Fossdal, F., R. Heldal, and N. Peek. 2021. Interactive digital fabrication machine control directly within a CAD environment. In *Symposium on computational fabrication*, 1–15. New York: Association for Computing Machinery. https://doi.org/10.1145/3485114.3485120.

Ghajargar, M., J. Bardzell, A.M. Smith-Renner, K. Höök, and P.G. Krogh. 2022. Graspable AI: Physical forms as explanation modality for explainable AI. In *Sixteenth international conference on tangible, embedded, and embodied interaction, TEI '22*, 1–4. New York: Association for Computing Machinery. https://doi.org/10.1145/3490149.3503666.

Goodman, N. 1978. *Ways of worldmaking*. Indianapolis: Hackett Publishing Company, Inc.

Hendren, S. 2020. *What can a body do?: How we meet the built world*. Riverhead Books.

Ingold, T. 2011. *The perception of the environment: Essays on livelihood, dwelling and skill*. London: Routledge.

Landwehr S., S. Martin Jonsson, and J. Tholander. 2020. Machine sensibility: Unpacking the embodied and situated dimensions of 3D printing. In *Proceedings of the 11th Nordic conference on human-computer interaction: Shaping experiences, shaping society, NordiCHI '20*, 1–13. New York: Association for Computing Machinery. https://doi.org/10.1145/3419249.3420166.

Lindström, K., and Å. Ståhl. 2020. Un/making in the aftermath of design. In *Proceedings of the 16th participatory design conference 2020—Participation(s) otherwise, PDC '20*, vol. 1, 12–21. New York: Association for Computing Machinery. https://doi.org/10.1145/3385010.3385012.

Lund, A. 2001. Measuring usability with the USE questionnaire. *Usability and User Experience Newsletter of the STC Usability SIG* 8 (January).

McCullough, M. 1998. *Abstracting craft: The practiced digital hand*. Cambridge: The MIT Press.

Murer, M., A. Vallgårda, M. Jacobsson, and M. Tscheligi. 2015. Un-crafting: Exploring tangible practices for deconstruction in interactive system design. In *Proceedings of the ninth international conference on tangible, embedded, and embodied interaction, TEI '15*, 469–472. New York: Association for Computing Machinery. https://doi.org/10.1145/2677199.2683582.

Nicenboim, I., E. Giaccardi, and J. Redström. 2022. From explanations to shared understandings of AI. *DRS Biennial Conference Series*. https://dl.designresearchsociety.org/drs-conference-papers/drs2022/researchpapers/293.

Özkar, M. 2007. Learning by doing in the age of design computation. In *Proceedings of the 12th international conference on computer aided architectural design futures, CAAD futures 07*, Sydney, 99–112. Singapore: Springer.

Pierce, J. 2014. Undesigning interaction. *Interactions* 21 (4): 36–39. https://doi.org/10.1145/2626373.

Pye, D. 1978. *The nature and art of workmanship*. Cambridge: Cambridge University Press.

Reddy, M.J. 1993. The conduit metaphor: A case of frame conflict in our language about language. In *Metaphor and thought*, 164–201. Cambridge University Press. https://doi.org/10.1017/CBO9781139173865.012.

Rosner, D.K. 2018. *Critical fabulations: Reworking the methods and margins of design*. Design thinking, design theory. Cambridge: The MIT Press.

Schön, D.A. 1983. *The reflective practitioner: How professionals think in action*. New York: Basic Books.

Song, K.W., and E. Paulos. 2021. Unmaking: Enabling and celebrating the creative material of failure, destruction, decay, and deformation. In *Proceedings of the 2021 CHI conference on human factors in computing systems, CHI '21*, 1–12. New York: Association for Computing Machinery. https://doi.org/10.1145/3411764.3445529.

Steenson, M.W. 2014. Architectures of information: Christopher Alexander, Cedric Price, and Nicholas Negroponte and MIT's Architecture Machine Group. Ph. D. dissertation. New Jersey: Princeton University.

Stiny, G. 2008. *Shape: Talking about seeing and doing*. Cambridge: The MIT Press.

Subbaraman, B., and N. Peek. 2023. 3D printers don't fix themselves: How maintenance is part of digital fabrication. In *Proceedings of the 2023 ACM designing interactive systems conference, DIS '23*, 2050–2065. New York: Association for Computing Machinery. https://doi.org/10.1145/3563657.3595991.

Suchman, L. 2006. *Human-machine reconfigurations: Plans and situated actions*, 2nd ed. Cambridge: Cambridge University Press.

Wakkary, R. 2021. *Things we could design: For more than human-centered worlds*. Cambridge: The MIT Press.

Wennberg, A., H. Åhman, and A. Hedman. 2018. The intuitive in HCI: A critical discourse analysis. In *Proceedings of the 10th Nordic conference on human-computer interaction, NordiCHI '18*, 505–514. New York: Association for Computing Machinery. https://doi.org/10.1145/3240167.3240202.

Willis, K.D. D., C. Xu, K.J. Wu, G. Levin, and M.D. Gross. 2011. Interactive fabrication: New interfaces for digital fabrication. In *Proceedings of the fifth international conference on tangible, embedded, and embodied interaction, TEI '11*, 69–72. New York: ACM.

Open Access This chapter is licensed under the terms of the Creative Commons Attribution-NonCommercial-NoDerivatives 4.0 International License (http://creativecommons.org/licenses/by-nc-nd/4.0/), which permits any noncommercial use, sharing, distribution and reproduction in any medium or format, as long as you give appropriate credit to the original author(s) and the source, provide a link to the Creative Commons license and indicate if you modified the licensed material. You do not have permission under this license to share adapted material derived from this chapter or parts of it.

The images or other third party material in this chapter are included in the chapter's Creative Commons license, unless indicated otherwise in a credit line to the material. If material is not included in the chapter's Creative Commons license and your intended use is not permitted by statutory regulation or exceeds the permitted use, you will need to obtain permission directly from the copyright holder.

Preserving Brand Identity in Engineering Design

Alison McKay, Hau Hing Chau, and Alan de Pennington

Abstract Brands are essential in maintaining the competitive advantage of many companies in the fast-moving goods sector. A challenge for such companies lies in preserving brand identity within a product range whilst developing new products and associated packaging designs that maintain and grow the brand. The authors' work in this area began c.2000 with H.H. Chau's PhD research which explored the preservation of brand identity in the design of packaging for high volume low value consumer goods such as personal hygiene products. This resulted in a U_{13} shape grammar implementation for use in the design of such products. Subsequently we have explored further applications and implementation issues for shape grammars: most recently in applying ideas from shape computation to design configuration. In this chapter, we explore the seven questions posed by G. Stiny in his recent book, Shapes of imagination: Calculating in Coleridge's magical realm, in the context of branded product design.

1 Introduction

Brand identity is a form of economic moat that is critical for companies in maintaining competitive advantage and so market share and profits. Unilever, with its Dove brand, is an example of such a quality company. The key characteristics of brands, reflected in the products that form a brand family, include shape, colour and graphics (Fischer et al. 2020). For example, a water droplet is used in the graphical communication of the Dove brand and has been used in successive generations of Dove products. There is continual pressure on brand owning organisations to develop brands through

A. McKay (✉) · H. H. Chau · A. de Pennington
School of Mechanical Engineering, University of Leeds, Woodhouse Lane, Leeds LS2 9JT, UK
e-mail: a.mckay@leeds.ac.uk

H. H. Chau
e-mail: h.h.chau@leeds.ac.uk

A. de Pennington
e-mail: a.depennington@leeds.ac.uk

the products that form their life cycles whilst also maintaining brand identity and so increasing market share. As an example of this, the Dove brand used a particular blue and white colour scheme c.2000 but this has now developed with different colours for different sub-brands (see Dove 2023). This chapter brings together examples of how shape grammars have been used to capture shape aspects of brand identity to support the design of branded products.

We have adapted Stiny's seven questions (Stiny 2022) to focus on the potential of shape grammars in the design of branded products. Our first question explores how shape aspects of brand style are described and communicated within product development teams to ensure the visual consistency of products within a given brand. In this context, our second and third questions consider the extent to which shape grammars might support the description of brand style and our rationale for beginning to explore shape grammars as a means of doing this. The field of shape grammars has evolved, and continues to do so, alongside our research. In answering our fourth and fifth questions, we consider how the application of shape grammars to branded products has evolved since we first began work in this area in the late 1990s, and where it might go next. The answers to our final two questions elaborate on this in two ways: by considering practical applications that already exist and how users of such applications, e.g., designers and design managers, might feel about adopting shape grammars as a design tool.

A table showing how our questions relate to Stiny's originals is provided in the Appendix. Before answering the questions, however, we introduce four cases that are used as examples in our responses. In conclusion, we consider how work in this area might develop further, particularly as to how the ambiguity that the mathematics of shape computation enables might be exploited in wider engineering design processes.

2 Cases Illustrating Applications of Shape Grammars for Style Preservation

Four cases that provide a chronology of developments in the use of shape grammars to capture style are discussed. The first is the widely used ice-ray grammar, which captures a style of ornamental windows developed in the nineteenth century. This is followed by Knight's grammar for Hepplewhite style chairs and then car grills, all in 2D, and finally, in 3D, the Dove brand. Our rationale for selecting these was that the Hepplewhite grammar begins with an initial shape that gives a geometric framework for applying parametric grammer shape rules involving 1D elements (straight lines) in a 2D space. This is followed by the Buick & Harley-Davidson cases where rules are defined parametrically in terms of key structures in the design of the product. A key difference between the Buick and Harley-Davidson grammars lies in the nature of the parameters: qualitative and quantitative respectively. Finally the Dove grammar is introduced where the rules operate on 1D elements in a 3D space and the initial shape is a point.

Case 1: Ice-Ray Grammar

The ice-ray grammar was created to capture a range of ornamental window or grille designs (Dye 1974). Here we look at a specific one: the grammar for Chengtu Szechwan 1880 AD (shown in Fig. 1) which has four rules. Rule 1 transforms a triangle into a smaller triangle and a quadrilateral; Rule 2 transforms a quadrilateral into a pentagon and a triangle; Rule 3 transforms a pentagon into two quadrilaterals; and Rule 4 transforms a pentagon into two quadrilaterals. By using these four rules, an artisan subdivides a larger area into smaller polygons recursively. The initial shape is the rectangle which forms the window frame. These four rules capture the style of ice-ray window designs which reflects both an aesthetic style and manufacturing constraints. For example, none of the rules bisect the shape on the left-hand side of the rule through a corner, thus preventing the generation of designs that cannot be made with the necessary structural integrity.

Case 2: Hepplewhite-Style Chair-Back Grammar

Knight (1980) proposed a grammar for Hepplewhite-style chair-back designs, shown in Fig. 2. Four parametric shape rules divide recursively a polygonal initial shape. In the original paper there are constraints on the range of parameter values that can be used, e.g., to ensure that the ends of new line segments are not too close to the ends of other lines. The final step to create the design of a chair back is to replace straight lines by curves following the steps provided in the paper; this is done after the structure of the chair back has been generated using the grammar.

Case 3: Buick and Harley-Davidson Grammars

Like many automotive manufacturers, Buick and Harley-Davidson use the external geometry of their vehicles to communicate their values to customers. A corpus of thirteen front-end designs, spanning 1939–2002, was used to create the Buick grammar (McCormack and Cagan 2003). It has sixty-three rules grouped according to the features of the design that can be seen in a front view of the car: grill, emblem, middle hood, centre, middle hood, outer hood, fender, hood flow line, roof line and headlight. A typical shape rule describes the geometry qualitatively, e.g., words "soft" and "sharp" are used as qualitative parameters to describe curves on the right-hand side of the rule (Fig. 3).

Similarly, Pugliese and Cagan (2002) propose a grammar for the design of Harley-Davidson motorcycles. Generated from an analysis of the anatomy of Harley-Davidson designs, the grammar has 45 rules grouped according to the features that can been seen in a side view of the motorcycle: low triangular cradle frame, teardrop fuel tank, triangular instrument panel, front fender, upper fork, lower fork, air filter, fuel cap, handlebar, headlight, taillight. Like the Buick grammar, this is a parametric but with quantitative parameters: wheel base, wheel size, tyre thickness, number of cylinders and their orientation (Fig. 4).

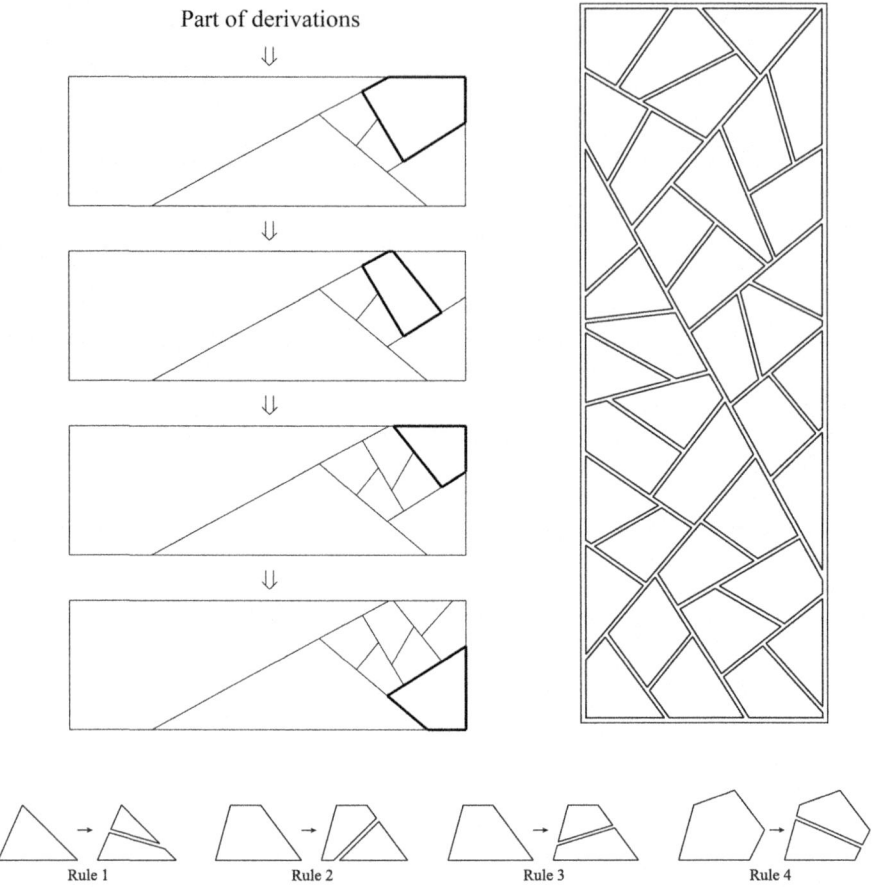

Fig. 1 Grammar-based derivation of shapes in the ice-ray style (reproduced from Stiny 1977)

Case 4: Dove Grammar

The Dove soap bar has a characteristic oval shape. Its top surface is ellipsoidal and its bottom surface is cylindrical. When Unilever introduced their Dove liquid soap, the overall geometry was mostly an oval shape. It had a doubly curved surface in the front that resembled an ellipsoidal surface, and a singly curved surface at the back that resembled a cylindrical surface. Subsequent generations of the Dove bottles gradually diverged from the geometry of the original soap bar. Nonetheless, the differences between consecutive generations were relatively small. The Dove grammar (Chau 2002) has 12 rules using a combination of surface and solid modelling to capture the Dove style. Some rules are used to generate the principal cross section, while others are used to produce the top and bottom surfaces. A final step of interpretation converts a surface model into a solid model.

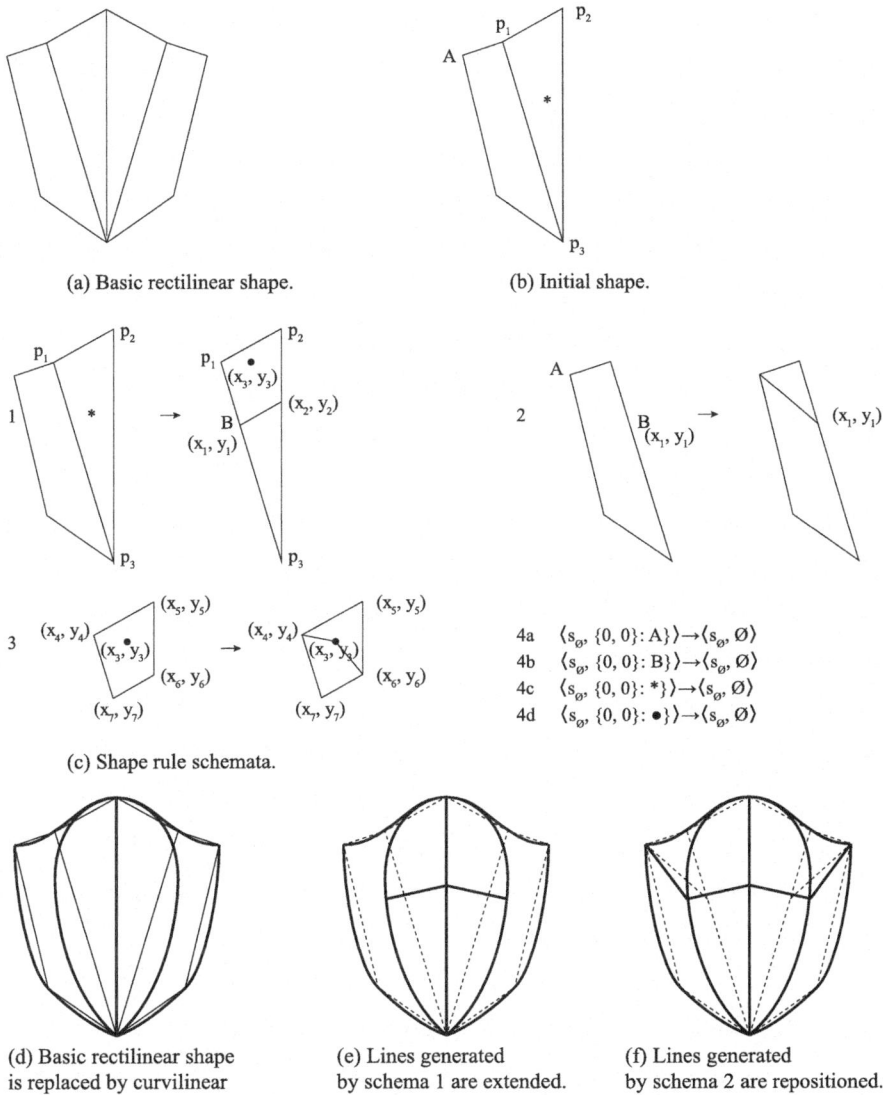

Fig. 2 Grammar-based derivation of shapes in the Hepplewhite chair-back style (reproduced from Knight 1980)

The sequence of rule application to develop a soap bar is shown in Fig. 5. In his thesis, Chau also includes an example generating a shower gel bottle (p.61 & 62). The structure of this development process, in terms of the sequence of rule applications, was mirrored in the relationship between the geometry of the Imperial Leather soap bar and its first liquid soap bottle. The wedge shape at the ends of

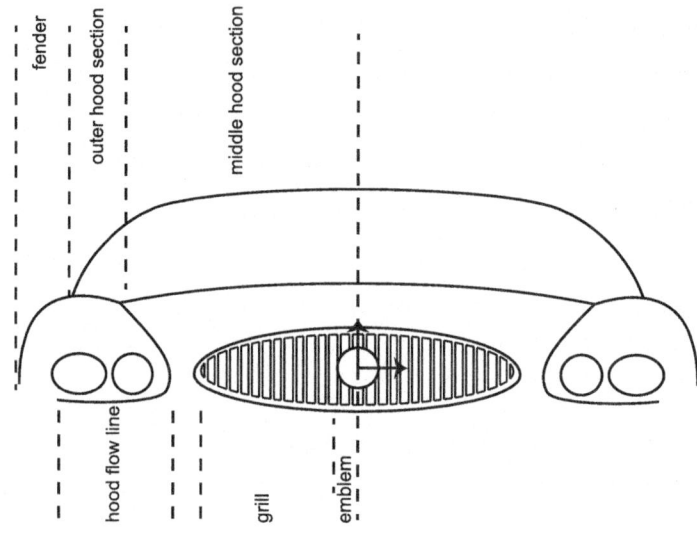

Fig. 3 Buick style (reproduced from McCormack et al. 2004)

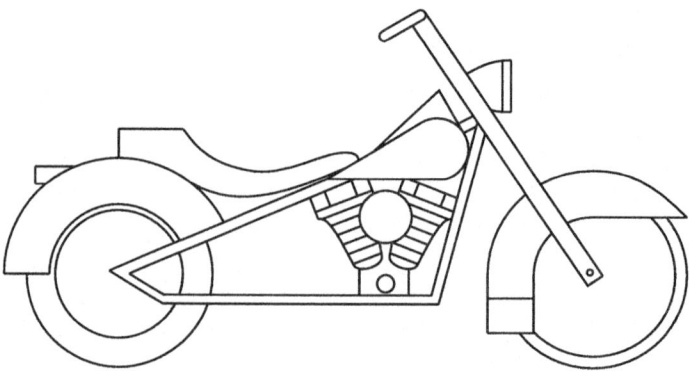

Fig. 4 Harley-Davidson style (reproduced from Pugliese and Cagan 2002)

the Imperial Leather soap bar are key shape elements of its brand identity and the squashed oval shape its principal cross section. Although its orientation was different from the Dove grammar, the overall principle remained the same. The initial shape is a point and, although the anatomy of the design can be seen, the grammar rules operate on shapes in a 3D space rather than explicit components of the design as in the Buick and Harley-Davidson grammars.

It is important to note that, as in the Hepplewhite chair grammar, the final interpretative step is not a part of the grammar. In essence, the grammar fixes the structure of a design in the style and geometric elements are then applied to this structure.

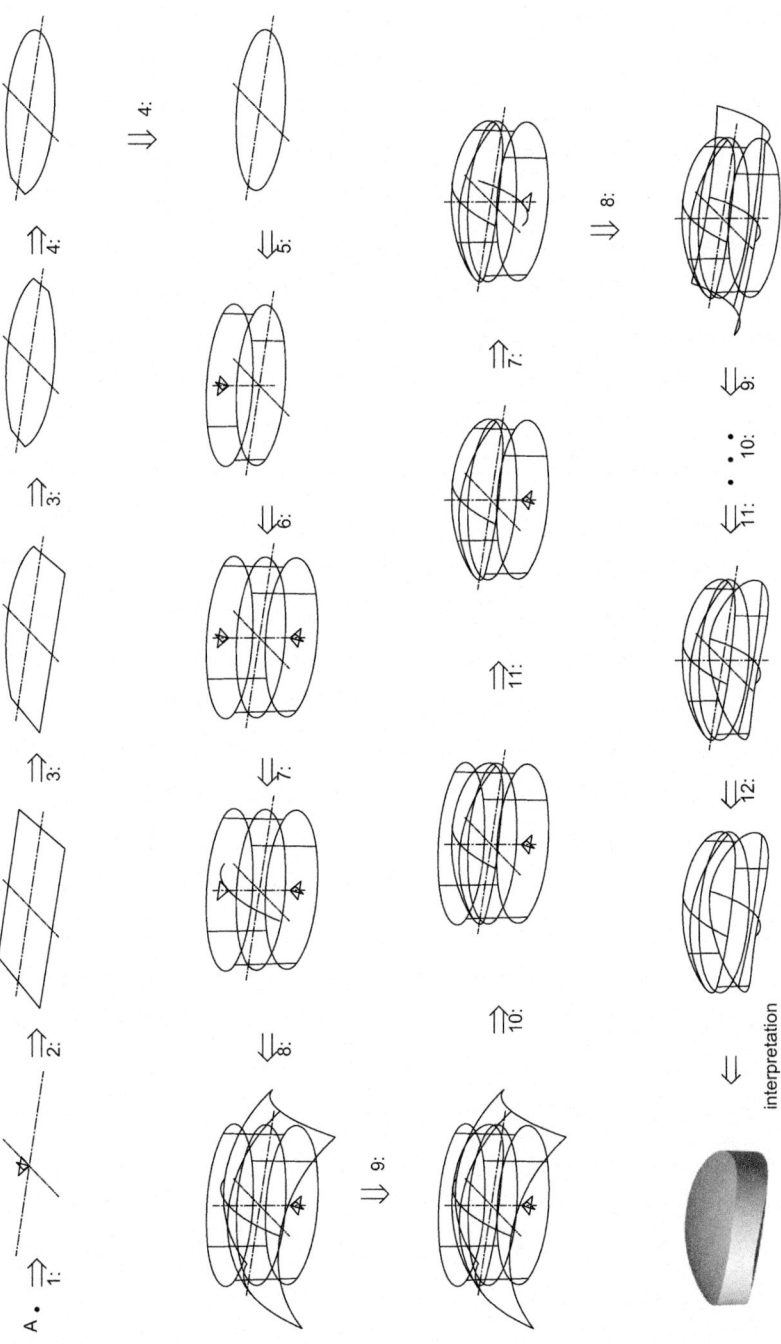

Fig. 5 Grammar-based derivation of shapes in the Dove style (reproduced from Chau 2002)

3 The Potential of Shape Grammars in the Design of Branded Products

The cases in the previous section are used here to explore the extent to which shape grammars can contribute to the design of branded products. We begin by exploring the notion of brand style in the context of consumer products and then the extent to which shape grammars can be used to describe brand style and our rationale for exploring this. From this basis we consider how the use of shape grammars to capture brand style has developed, both in research and practice, since its inception in the 1990s and how it might develop in the future.

Q1: What is a brand style in the consumer product sector? If you were to write one down on a piece of paper to show someone, what would it look like?
As technological distinctions between competing products diminish, the importance of brand grows. It is widely recognised that brands communicate with consumers through combinations of visual and other sensory cues (Batra 2019): so influencing purchasing decisions. Further, brands themselves develop and with this they develop cultural meaning (Krishna 2019). For example, as outlined by Zhang (2023), the Dove brand was launched circa 1955 as a beauty fragrance block and its advertising emphasised core product functions and properties: avoidance of the drying effects of soap on the skin (Unilever 2023). Since then the brand has developed to the point that adverts barely refer to the function of the products, instead focusing on psychological concerns of target users such as perceptions of beauty and feel-good factors.

One aspect of preserving brand identity of durable and fast-moving consumer goods is to maintain their visual appearance and so recognition. Given this, it is desirable to have a consistent brand identity across a range of products, which may or may not be in the same category. It also serves as a differentiation from competitors. Another dimension is temporal. Market pressure demands new styles every few years. New styles need to be different enough to be considered as new. However, in the interest of brand identity and brand loyalty, they cannot be too different or beyond recognition. Designers often use geometric form to deliver emotional and aesthetic experience. The appeal of shape grammar is to use an existing corpus of designs to define a set of shape rules. Then, apply them to produce new designs. If the process could be automated, it is appealing to produce a short list of prospective designs for human evaluation.

While both the brand and its cultural meaning have developed, e.g., the Dove brand now has its own web site and a number of sub-brands, the visual cues that consumers use, often subconsciously, to identify products continue to be reflected in the 3D designs of Dove products. Initially the brand included one product, a cleansing bar, that was characterised by its distinctive shape and blue and white colour scheme. By 2000, the range had increased to include other cleansing products, such as body wash, whose packaging reflected the characteristic shape and the blue and white colour scheme. Since then, the range has further extended as has the range of colours used. However, even as the brand message in advertising changed and the range of colours and logo designs have developed, the shapes of 3D products, such as cleansing bars and body wash bottles, continue to reflect the key shape characteristics of the original bar which is characterised by an upper ellipsoidal surface and a bottom surface that is cylindrical, joined by a flat side surface, as shown in Fig. 6.

Similarly, shape characteristics are used in many brands to establish visual associations within their product ranges, e.g., through the use of logos and visual cues in the shapes of products and their packaging. As a result, the geometric design of products and their packaging are critical tools in the realisation of brand identity. Our answer to the first question is provided in this context. For the purposes of this chapter and our research in shape grammars, brand style in the consumer product sector refers to shape aspects of the visual appearance of a product that relate it to other products in the same brand. Given this definition, writing a brand style down on a piece of paper involves drawings of shapes (in 2D, given the limitations of paper) and associated annotations. With reference to the cases in Sect. 2, in each case the brand style is communicated visually.

Q2: How far can shape grammars go in describing brand style? What are they good at, and what do they struggle with?

Fischer et al. (2020) highlight four aspects of branding (Historic Connection, Product Portfolio, Product Family and Competitor Brand) and, for products, four substructures: Layout, Shape, Color and Graphic. In the authors' research, and as reflected in the cases used in this chapter, shape grammars are particularly well suited to describing shape aspects of brand style. In this way, shape grammars can be used to describe brand characteristics of products in a portfolio and the information is available within them to gain insights into potential conflicts with competing brands. For example, given a grammar it becomes feasible to visualise historic connections

Fig. 6 A Dove soap bar and liquid soap bottle (reproduced from Chau 2002)

within a brand (e.g., see Chen et al. 2004). As highlighted in all four cases, with an appropriate implementation to compute a space within which shapes generated from a grammar might sit, shape grammars have the potential to underpin design tools that both evaluate designs with respect to their fit with the brand, and consider the extent to which they may encroach on competitor brands and vice versa.

In current practice, brand owners use visual language manuals to guide designers on the visual characteristics of given brands. Although these tend not to be in the public domain, as illustrated in the cases in Sect. 2, shape grammar rules can be used to describe shape aspects of a brand style and support the generation of new shapes in that style. All cases reflect structural aspects of the brand, some explicitly, e.g., the Buick and Harley-Davidson grammars refer explicitly to parts of the final product, others implicitly, e.g., the Hepplewhite and Dove grammars include shape elements that are key aspects of the visual identity of products in the brand.

In general, shape grammars are an effective way of capturing the anatomy of a style reflected in designs in a brand. However, the range of new shapes that can be generated depends on the generality of the rules used and how they are defined. For example, the Buick and Harley-Davidson grammars are constrained to the structure of current designs and their visual elements which reduces the risk of designs being generated that are not recognisable as part of the brand by limiting the size of potential solution spaces. On the other hand, the Hepplewhite and Dove grammars are less constrained geometrically, meaning larger potential solution spaces, but the risk of shapes not being recognisable within the brand increases, though this also increases the potential for surprising new shapes that are recognisable within the brand. Overall, there is a trade-off between the specificity of a given grammar, which makes it easier to implement, and the variety in the range of designs that can be generated.

While all four grammars in Sect. 2 capture visual aspects of brand identity, implementation that supports the generation of new designs remains a significant challenge. Different research groups are working on different fronts to address this problem with two recent implementations, both built on Rhino 3D, being shape machine (Economou et al. 2020) and Li and Stouffs' (2021) shape grammar implementation, that includes an API to integrate their SortalGI library with the Rhino 3D modelling environment. Further details on outstanding issues for shape grammar implementation, and so potential for widespread use, are provided in response to Question 4.

Q3: What inspired the exploration of the application of shape grammars to branded products from the late 1990s? What fields/interests was it coming from, and who was it for?

Engineering design research at Leeds began in the late 1970s with research on computational geometry and its application in design and manufacturing (McKay and de Pennington 2022). In the late 1990s, work on virtual packaging design (McKay et al. 2003) led to the identification of challenges related to the communication and evaluation of designs in a given brand style. Chau, through his PhD research (Chau 2002), responded by beginning to explore the application of shape grammars to the design of branded products. Building on time as a visiting scholar in Stiny's group at MIT,

Chau used a consumer goods case study, Dove, to explore the use of shape grammars as a means of maintaining brand identity in the design of consumer products. This led to a world first U_{13} implementation of shape grammars, demonstrated through application to the generation of designs in the style of the Dove brand (Chau et al. 2004).

This research led to the identification of the following questions that design practitioners (including design managers and both in-house and external designers) would value being able to answer.

(1) When creating a new design in the style, how well does it conform with the brand style?
(2) When selecting designs, e.g., from design suppliers, which designs fit best with the brand style?
(3) When creating a brand style, how likely is it that new designs encroach on competitor brands?
(4) For a given competitor design, to what extent does it encroach on our brand style?

Since then, design research at Leeds has continued to contribute to the application of shape grammars in the design of branded products. In addition, our research has explored the potential of using novel approaches (from computer vision and eye tracking) to support sub-shape detection in the implementation of shape grammar-based design systems. An important benefit of shape grammars lies in the mathematics of shape computation which enables the description of a given design in more than one way. While regarded as a disadvantage for manufacturing applications where unambiguous design descriptions are needed to maximise the efficiency of production systems and minimise waste, the benefits of ambiguity in early design processes are widely recognised.

Q4: How has research on the application of shape grammars to branded products evolved since?

From all four cases and our answers to the first three questions, and supported by findings reported by Manavis and Kyratsis (2021), there is strong evidence to support the assertion that shape grammars can support the description of shape aspects of brand style. For practitioners to benefit from this, however, they need to be able to generate new shape designs and evaluate existing ones, e.g., by answering the four questions for practitioners posed in Question 3. In essence there is a need to be able to compute with shapes. The theoretical frameworks of shape computation provide for this by supporting the generation of new shapes. However, they need to be implemented and achieving implementation requires the resolution of a series of technical challenges that are best explained with reference to Stiny's equation for the application of a shape rule:

$$C' = [C - t(A)] + t(B)$$

which asserts that the new designed shape C' is the original designed shape C with the left-hand side of the rule A under transformation t removed and replaced by the right-hand side of the rule B under the same transformation t.

Having demonstrated the potential of shape grammars, a key challenge for the community lies in finding ways to implement them so that they are accessible to design and other practitioners without the need for a detailed knowledge of the underlying shape computation methods. Working towards self-contained design applications for designers, who are not necessarily fluent in the technical details on how shape grammars work, raises a number of challenges though developments have been made on several fronts. The implementation of Stiny's equation requires resolution of the following technical challenges that are used to provide a structure for our answer to this question:

- ways to represent the curves and surfaces that act as the boundaries of shapes;
- means of finding shapes, i.e., the left-hand side of a rule A in a given shape C (often referred to as "sub-shape detection"); and
- given a shape that has been found, how to calculate the transformation t necessary for the removal of A from C and addition of B under the same transformation t.

The Representation of Shape Boundaries

The power of shape computation lies in the fact that it emulates the flexibility and fluidity of pencil and paper design processes where designers often switch between the interior of a shape and its boundary. For example, one can consider a square as a solid or empty area; and in either case it is equally valid to consider it as four edges that bound an area. Lim et al. (2008) report on experiments that explored the kinds of shape transformations that designers make when generating new shapes. In addition, shape computation facilitates design creativity through its treatment of emergent shapes that, through embedding, can be operated on as part of the design process.

For computer implementation, there is a need for representations of shape boundaries that accommodate these different perspectives. For grammars such as the ice-ray grammar that use only straight lines, computer implementation was largely resolved by Mark Tapia (1999) using a grid of horizontal and vertical lines. Significant progress lies in representing and operating on curved basic elements. Most notably, Jowers and Earl (2010) have a comprehensive treatment on the use of quadratic Bézier curves. A non-trivial task is on a complex manipulation of algebra equations. Another drawback to this method is that the curvature of a Bézier curve changes along its length. Therefore, sub-shape detection is required to be on the exact segment of the curve. An alternative to an exact analytical solution is an approximate approach which makes sense if we regard shape grammars as a way of formalising pencil and paper processes. In such cases, exact representation is not necessary; rather, it needs to be good enough for the eyes to see. For examples, the Design Synthesis & Shape Generation (DSSG) project used pixels to represent designed shapes and shape rules Jowers et al. (2010); Jowers and Earl (2010, 2011) used pixels to represent hand sketched curves; and Ensari and Özkar (2018) used NURBS curves for sub-shape matching.

It is a natural progression to support curves in design in additional straight lines. Jowers and Earl (2010, 2011) use Bézier curves in two-dimensions. Transformation and sub-shape detection are considered in all possible ways of the overlapping of two curve segments. Chau et al. (2018) use circular arcs represented by a rational quadratic B-spline curve. They use lattice representation to reduce a shape grammar to a set grammar at each step of shape computation thus reducing computational complexity. They solve the problem of an inexact numerical result of non-rational descriptors by introducing tolerances.

In a series of shape grammar implementations, Chau (2002) and Li (2009) borrowed the tolerances from Parasolid, a popular CAD kernel. Linear measurements are taken in metres and angular measurements are taken in radians. Any two points that are closer than 10^{-8} m are considered the same and snapped together. Any two angles that are closer than 10^{-11} are considered the same. This solves the problem of inexact representation of an irrational number in a floating point representation. The design cube is a 1km cube and accuracy is 0.01 μm which is sufficient for any product design applications. Pragmatically speaking, this is sufficient for emulating the pencil and paper process. There are for and against opinions on snapping. Nonetheless, this approach is possibly good enough for the time being.

In summary, the current state of the art for the representation of shape boundaries is in curves (including straight lines) and planar surfaces. This is reflected in the Hepplewhite chair and Dove cases where the introduction of more complex geometries is completed, after the structure of a design shape has been generated, through the final "interpretation" step. Further work is needed to include more complex surfaces and then interactions with lower order shapes, e.g., lines intersecting or sitting on surfaces.

Sub-shape Detection

Sub-shape detection is an essential capability that is needed to find the left-hand sides of rules in designed shapes: with respect to Stiny's equation, finding occurrences of A in C. Reis (2023) reports on an implementation of sub-shape detection for straight lines in a 2D space.

Krishnamurti and Earl (1992) provide a comprehensive process of shape recognition in three-dimensions for straight lines and planar surfaces with straight edges. The key idea is to identify three registration points or equivalent and then compute transformation matrices, t, that include translation, rotation, reflection and scaling.

Once a transformation, t, is chosen, shape computation can be performed. However, floating point representation of non-rational descriptors leads to drifting of coordinates due to numerical rounding error. For example, in Tapia's (1999) implementation, rotating a square through 45° a number times leads to a gradual drifting of the square due to rounding errors. A fundamental problem is how to represent and operate on maximal elements. The complexity varies hugely depending on which algebra the basic elements are operating in. Naturally, operations are less complex in algebra U_{12} compared to algebra U_{13}. Nonetheless, a generalisation from 2D to 3D could borrow the results from computer-aided design and computational geometries quite readily. A much harder question is what types of basic elements are being used. Parametric shape recognition in two- or three-dimension is a very challenging problem. For example, in the ice-ray grammar, to match the left-hand side of Rule 2, it is required to match a quadrilateral under transformation t. Since the quadrilateral could be in any proportion, it requires the use of topology before registration points could be identified. In the parametric form representation of curves and surfaces, each of the coordinates of a point on a curve is represented as an explicit function of an independent parameter (Piegl and Tiller 1997).

Calculation of Shape Transformation, t

Given a sub-shape A detected in a target design shape C a transformation matrix t that positions A in C is necessary for the application of Stiny's equation. The problem is illustrated in Fig. 7 which shows the transformations necessary for (and technical challenges involved in) the application of a rule where the left-hand side is a square, shown by the dashed lines, and the detected sub-shape is the square shown in solid lines. In summary, the problem is to superimpose a shape A with its local coordinate system (LCS) in the global coordinate system (GCS) of the design C in which it has been detected.

One-dimensional elements are straight lines and curves. A key issue lies in computing the potential transform matrices t for use in Stiny's equation. For example, for the square C in Fig. 7a, there are eight possible values for t, two are shown in Fig. 7c and d. The problem boils down to identifying sufficient registration points, e.g., the labelled corners in Fig. 7. The problem is solved most comprehensively by Economou et al. (2020) which includes many indeterminate cases. It is most useful when the initial shape is a single point (McCormack et al. 2003) or not much more

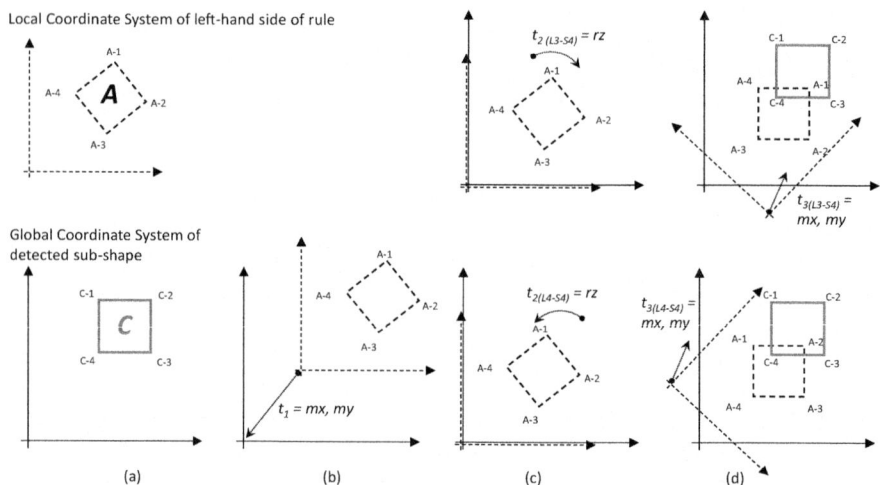

Fig. 7 Alternative derivation of transformation matrices in a rule application. (**a**) A match is found between the LHS of a rule A and a sub-shape C in the design shape; (**b**) The LCS of A is transformed to the GCS of C; (**c**) A is transformed (in the example rotated) match C. Often there is more than one option. For example, given its symmetry, the square can be matched in eight different orientations and two are shown in the figure; (**d**) Shape A is transformed to be superimposed on the sub-shape in C

than that. However, if we are willing to compromise and to require an initial shape to contain at least one straight line or curve segment, a real or virtual intersection and a third registration point which is not collinear to the above straight line, a 3D transformation matrix could be uniquely defined. The results produced by Krishnamurti and Earl (1992) in the 1980s are sufficient for straight lines. Chau et al. (2002, 2018) used parametric (parametric as in computational geometry not in parametric shape grammars) straight lines and circular arcs. There are no novelties in using parametric straight lines. In this context, exact representation of circular arcs are special cases of NURBS (non-uniform rational B-splines) where (a) quadratic rational B-splines are used and (b) the position and the weight (as in computational geometry, not in shape grammars) of the middle control vertex are automatically adjusted for exact representation of circular arc. The use of rational splines does increase complexity in representation but makes calculation easier. For examples, (1) a complete circle requires three spans of quadratic rational B-spline and there is no need to distinguish the start or the end; (2) a circular arc can slide along another circular arc of the same radius very much like a straight line can slide along another straight line; (3) when a circular arc degenerates into a straight line, there is no need to change the calculation or even to distinguish among them.

Q5: How is it continuing to develop? What is the future of shape grammars in the design of branded products?

We anticipate that future shape grammar-based design systems will allow designers to add and modify shape rules and create shape grammars visually, and so both explore new designs in existing brands and create grammars that describe new brands. Further, the design process could be done semi-autonomously where a short list of designs that conform to a given style is presented to a designer for evaluation and further development. The shape grammar implementations, discussed in our response to Question 2, are both built on top of commercially available computer-aided design systems. The use of a commercially available modelling environment increases the likelihood that grammatical design tools will become available to designers in a more realistic time scale.

Early shape grammar implementations tended to be written from the ground up. There are three notable exceptions. After the experience of extending Chau's implementation, Li et al. (2009) started their own implementation on the top of Auto CAD. More recent implementations were built on top of Rhino and Grasshopper. In addition to producing their own implementation, Stouffs and Li (2020) made the core part of their Python implementation a function library that offers potential for further work to build upon it. Relieved from the burden of building user interfaces allows implementations to focus on core capabilities of shape grammars which, in turn, could prove to be a more profitable strategy for faster development cycles.

Q6: Do you know of any examples of where shape grammars have been practically used in the design of branded consumer products?

We introduce two examples that show practical applications of shape grammars in the design of consumer products in a recognised style.

Martini Glass Grammar

As part of a Masters programme in product design, we used martini glasses in a research through design project (Frayling 1993) as a vehicle for students to consider issues when designing in a style (McKay 2014). In the project students generated designs of martini glasses and these were then analysed (in a workshop involving the students, authors and Stiny) to generate the martini glass grammar shown in Fig. 8. An important benefit of this analysis lay in the analysis of students' designs and how they manipulated shapes when generating new designs. For example, all students decomposed the glass into three parts (the body, bowl and stem) but only one changed the shapes of these aspects rather than reconfiguring them. Through this activity, students were able to identify ways in which they might enhance their design generation processes and so their creativity. In addition, students were able to analyse their own design generation styles and those of colleagues which, again, resulted in the generation of a wider range of potential designed shapes.

Chocolate Bar Grammar

In her PhD thesis, Chen (2005) developed a grammar based on the Kit Kat chocolate bar grammar. The grammar was constructed through analysis of the range of Kit Kat branded products available in UK supermarkets at the time and used to generate new designs in the brand. The new designs were used in a consumer survey to identify those that were recognised as being in the Kit Kat brand. In this way, Chen demonstrated how shape grammars coupled with consumer research methods might contribute to answering the first two of the four questions posed in our response

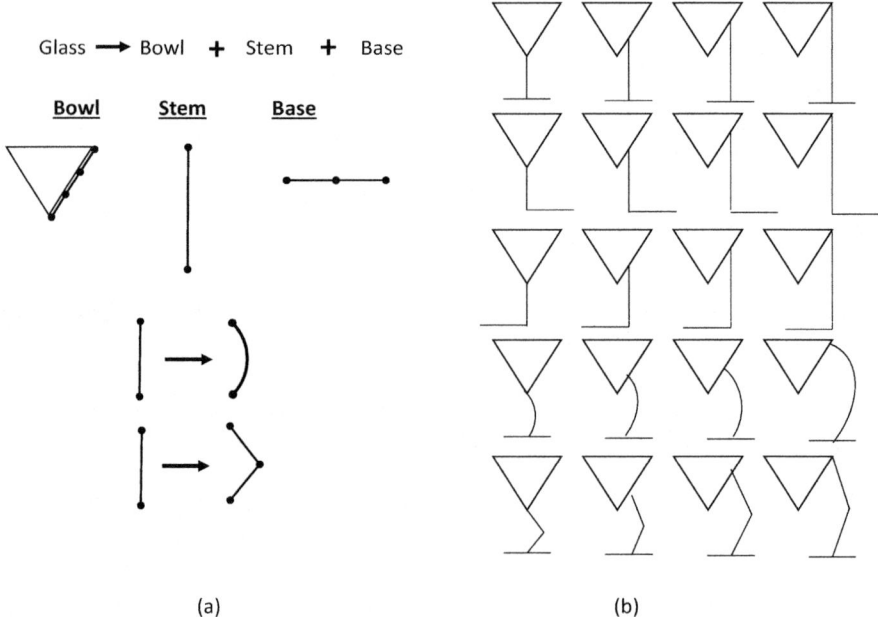

Fig. 8 Martini glass design: (**a**) examples of rules and (**b**) potential designs

to Question 3. For the last two, related to competitor brands, Chen does not report results but the surveys could have addressed these questions by including designs of competitor products in addition to the target brand.

Q7: How do you think designers and design managers might react to the idea that brand styles can be defined using shape grammars? How might they feel about the possibilities it creates?

Like any new innovation, the answer to questions such as this depends on how the innovation is pitched to potential users and whether the perceived benefits outweigh the costs of use. For example, shape grammars could be used as a generation tool to inspire new design directions, so enhancing design creativity and reducing the time needed to generate new designs, or as an analysis tool to support the selection of design concepts to be taken forward for further development. However, to achieve such benefits, like CAD packages that implement theories from computational geometry, shape grammar implementations need to be easy to use by users who are not familiar with the underlying technology.

A result of the DSSG project was a model for using shape grammar-based design systems to support design synthesis. The model, reproduced in Fig. 9, had two independent cycles – a designer designing [Practice] and a shape grammar-based system generating shapes [Technology] – linked by a third cycle where designers were influenced by shapes that had been computed and the design system was influenced by decisions made by the designer. How this third cycle operated was not elaborated but it could be controlled in different ways. For example, for students designing martini glasses where the goal might be to increase the range of concepts generated, this third cycle might present to the designer designs generated through underused rules. On the other hand, there may be other situations where the computer system might apply rules automatically when design pathways are identified, e.g., using emerging artificial intelligence technologies.

4 Conclusions

The importance of being able to design products that are recognised by consumers as belonging to a given brand, and to which they have emotional attachment, is a critical business capability for many brand owning organisations. In this chapter we have explored how shape grammars might contribute to achieving this. Through the four cases and our responses to Questions 2 and 6, we have demonstrated how shape grammars can be used to describe shape aspects of a brand and generate shape designs that conform to the brand both mathematically and in terms of brand recognition by consumers. However, as highlighted in our answers to Questions 3–5, while significant progress is being made, there remain technical issues that must be addressed if shape grammars are to be integrated into product development processes and the computer-based tools that support such processes (McKay et al. 2012). In addition, as outlined in our response to Question 7, it will be important

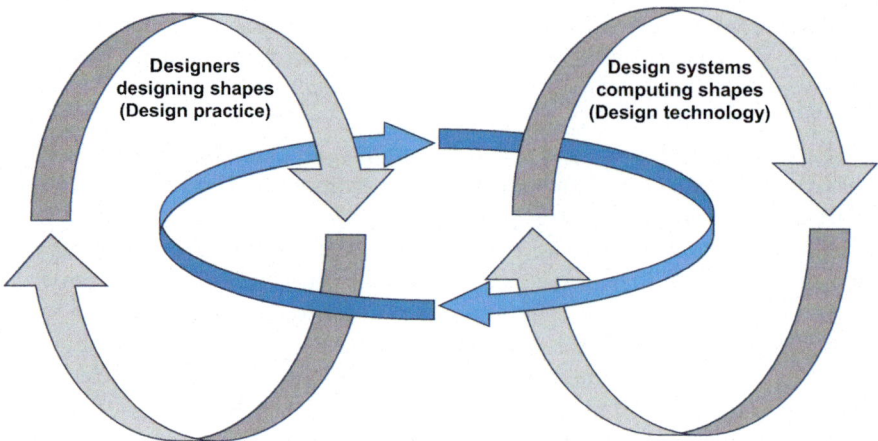

Fig. 9 Cycles of design generation and design computation. Reproduced from McKay et al. (2009)

to engage target users (both designers and design managers) in the development of such systems to ensure that their use does not detract from, and ideally improves, the key performance indicators for product development processes: namely, time, cost and quality. Thus, in addition to addressing technical challenges associated with the design of branded products, solutions will ideally result in better quality designs (i.e., designs that are more likely to be recognised by consumers but also can be produced efficiently in large volumes) faster and using fewer resources.

The discussion in this chapter has focused on the development of shape grammar-based design systems through the implementation of the mathematics of shape computation. While these issues need to be resolved, developments in computer science are creating new opportunities for the design of branded products. For example, Chen et al. (2023) consider the integration of style into shape design processes and Ghasemi et al. (2023) report on the potential of generative adversarial networks (GANs) in concept design. More widely, the ambiguity provided through the mathematics of shape computation has the potential to enhance the design tools that underpin today's product development processes. In particular, the benefits of computational geometry, which ensure unambiguous descriptions of shapes that are essential in downstream applications such as manufacturing, are limiting for activities such as shape synthesis where ambiguity is an essential part of creative processes. Early work on how current computer-aided design tools might be improved through the integration of shape computation and computational geometry is explored in (McKay and de Pennington 2022) and (Hong and Economou 2022). More widely, the increasing availability of shape grammar implementations is enabling the application of shape grammars in new areas. For example, Orynek et al. (2023) report progress in the use of grammars to relate design computation with craft-based design of woven textiles with a view to enabling the exploration of craft practice in design research.

Acknowledgements Our research in this area has been funded by the UK Engineering and Physical Sciences Research Council (EPSRC), most recently through Grant: EP/S016406/1, "Assuring the quality of design descriptions through the use of design configuration spaces". Earlier projects such as Design Synthesis & Shape Generation (DSSG, UK Arts & Humanities Research Council) and Designing with Vision (The Leverhulme Trust) were critical in the development of our work on shape grammars and their application to the design of branded fast-moving consumer goods.

Appendix: Recast questions for this chapter compared with Stiny's originals

	Question from Stiny (2022)	Question for this chapter
1	What is a shape grammar? If you were to write one down on a piece of paper to show someone, what would it look like?	What is a brand style in consumer products? If you were to write one down on a piece of paper to show someone, what would it look like?
2	How far can shape grammars go in describing forms – can they describe anything? What are they good at, and what do they struggle with?	How far can shape grammars go in describing brand style? What are they good at, and what do they struggle with?
3	What inspired your seminal research into shape grammars in the early 1970s? What fields/interests was it coming from, and who was it for?	What inspired the exploration of the application of shape grammars to branded products from the late 1990s? What fields/interests was it coming from, and who was it for?
4	How has research into shape grammars evolved since?	How has research on the application of shape grammars to branded products evolved since?
5	How is it continuing to develop? What is the future of shape grammars?	How is it continuing to develop? What is the future of shape grammars in the design of branded products?
6	Do you know of any examples of where shape grammars have been practically used in architecture and design?	Do you know of any examples of where shape grammars have been practically used in the design of branded consumer products?
7	What is the typical reaction of designers to the idea that their forms can be broken into grammars? How do they feel about this possibility?	How do you think designers and design managers might react to the idea that brand styles can be defined using shape grammars? How might they feel about the possibilities it creates?

References

Batra, R. 2019. Creating brand meaning: A review and research agenda. *Journal of Consumer Psychology* 29 (3): 517–518. https://doi.org/10.1002/jcpy.1122.

Chau, H.H. 2002. *Preserving brand identity in engineering design using a grammatical approach.* Ph.D. dissertation. Leeds: University of Leeds.

Chau, H.H., X. Chen, A. McKay, and A. de Pennington. 2004. Evaluation of a 3D shape grammar implementation. In *First international conference on design computing and cognition (DCC'04),* ed. J.S. Gero, 357–376. Dordrecht: Kluwer.

Chau, H.H., A. McKay, C.F. Earl, A. K. Behera, and A. de Pennington. 2018. Exploiting lattice structures in shape grammar implementations. *Artificial Intelligence for Engineering Design, Analysis and Manufacturing* 32 (s2): 147–161.

Chen, X. 2005. *Relationships between product form and brand: A shape grammatical approach.* Ph. D. dissertation. Leeds: University of Leeds.

Chen, X., A. McKay, A. de Pennington, and H.H. Chau. 2004. Package shape: Design principles to support brand identity. In *14th IAPRI world conference on packaging (IAPRI04),* Stockholm, Sweden, June 2004.

Chen, Y., L.B. Kara, and J. Cagan. 2023. *Automating style analysis and visualization with explainable AI—Case studies on brand recognition.* https://arxiv.org/abs/2306.03021. Accessed 17 Aug 2023.

Dove. 2023. *About Dove.* https://www.dove.com/uk/stories/about-dove.html. Accessed 16 Aug 2023.

Dye, D.S. 1974. *Chinese lattice designs.* New York: Dover Publications.

Economou, A., K. Hong, H. Ligler, and J. Park. 2020. Shape machine: A primer in visual composition. In *A new perspective of cultural DNA,* ed. J.H. Lee, 65–92. KAIST research series. Singapore: Springer.

Ensari, E., and O. Özkar. 2018. Shape computations with NURB curves. *Artificial Intelligence for Engineering Design, Analysis and Manufacturing* 32 (3): 282–294.

Fischer, M.S., D. Holder, and T. Maier. 2020. Evaluating similarities in visual product appearance for brand affiliation. In *Advances in affective and pleasurable design,* ed. S. Fukuda, 2–12. *Proceedings of the AHFE 2019 international conference on affective and pleasurable design,* July 24–28, 2019, Washington D.C., USA. Advances in intelligent systems and computing (book series AISC, vol. 952). https://doi.org/10.1007/978-3-030-20441-9_1.

Frayling, C. 1993. Research in art and design. *Royal College of Art Research Papers* 1 (1): 1993/4. London: Royal College of Art. ISBN 1874175551.

Ghasemi, P., C. Yuan, T. Marion, and M. Moghaddam. 2023. Are generative adversarial networks capable of generating novel and diverse design concepts? An experimental analysis of performance. In *DS 123: Proceedings of the design society: 24th international conference on engineering design (ICED23).* https://doi.org/10.1017/pds.2023.64.

Hong, T.-C.K., and A. Economou. 2022. What shape grammars do that CAD should: The 14 cases of shape embedding. *Artificial Intelligence for Engineering Design, Analysis and Manufacturing* 36 (e4): 1–20.

Jowers, I., and C.F. Earl. 2010. The construction of curved shape. *Environment and Planning B: Planning and Design* 37 (1): 42–58.

Jowers, I., and C.F. Earl. 2011. Implementation of curved shape grammars. *Environment and Planning B: Planning and Design,* 38 (4): 616–636.

Jowers, I., D.C. Hogg, A. McKay, H.H. Chau, and A. de Pennington. 2010. Shape detection with vision: Implementing shape grammars in conceptual design. *Research in Engineering Design* 21 (4): 235–247. https://doi.org/10.1007/s00163-010-0088-z.

Knight, T. 1980. The generation of Hepplewhite-style chair-back designs. *Environment and Planning B: Planning and Design* 7 (2): 227–238.

Krishna, A. 2019. How brands acquire cultural meaning: Introduction. *Journal of Consumer Psychology* 29 (3): 517–518.

Krishnamurti, R., and C. F. Earl. 1992. Shape recognition in three dimensions. *Environment and Planning B: Planning and Design* 19 (5): 585–603.

Li, A., and R. Stouffs. 2021. Towards a useful grammar implementation: Beginning to learn what designers want. In *A new perspective of cultural DNA*, ed. J. H. Lee, 55–64. KAIST research series. Singapore: Springer.

Li, A., H.H. Chau, L. Chen, and Y. Wang. 2009. A prototype system for developing two- and three-dimensional shape grammars. In *CAADRIA 2009: Proceedings of the 14th international conference on computer-aided architecture design research in Asia*, ed. T.-W. Chang, E. Champion, S.-F. Chien, and S.-C. Chiou, 717–726. Touliu, Taiwan: Department of Digital Media Design, National Yunlin University of Science and Technology.

Lim, S., M. Prats, I. Jowers, S. Chase, S. Garner, and A. McKay. 2008. Shape exploration in design: Formalising and supporting a transformational process. *International Journal of Architectural Computing* 6 (4): 415–433.

Manavis, A., and P. Kyratsis. 2021. Computational study on product shape generation to support brand identity. *International Journal of Modern Manufacturing Technologies* XIII (1): 115–122. ISSN 2067-3604.

McCormack, J.P., J. Cagan, and C.M. Vogel. 2003. Speaking the Buick language: Capturing, understanding, and exploring brand identity with shape grammars. *Design Studies* 25 (1): 1–29.

McKay, A. 2014. Examples of design research and their implications for design and designing. In *The Routledge companion to design research*, eds. P. Rodgers and J. Yee, 422–430. London: Routledge.

McKay, A., and A. de Pennington. 2022. Shape embedding: A means of superimposing alternative design descriptions on shape models. *Computer-Aided Design* 152: Article 103366.

McKay, A., S. Chase, S.W. Garner, I. Jowers, M. Prats, D.C. Hogg, H.H. Chau, A. de Pennington, C.F. Earl, and S. Lim. 2009. Design synthesis and shape generation. In *Designing for the 21st century. Volume II: Interdisciplinary methods and findings*, ed. T. Inns, 304–321. London: Routledge. ISBN 1409402401.

McKay, A., S. Chase, K. Shea, and H.H. Chau. 2012. Spatial grammar implementation: From theory to usable software. *Artificial Intelligence for Engineering Design, Analysis and Manufacturing* 26 (2): 143–169.

Orynek, S., B. Thomas, and A. McKay. 2023. Prototyping a novel visual computation framework for craft-led textile design. In *Proceedings of the international conference 2023 of the design research society special interest group on experiential knowledge (EKSIG), EKSIG 2023: From abstractness to concreteness—experiential knowledge and the role of prototypes in design research*, 19–20 June 2023, eds. S. Ferraris, V. Rognoli, and N. Nimkulrat, 380–397. Milano: Politecnico di Milano.

Piegl, L., and W. Tiller. 1997. *The NURBS book. Monographs in visual communication*, 2nd ed. Berlin: Springer.

Pugliese, M.J., and J. Cagan. 2002. Capturing a Rebel: Modeling the Harley Davidson brand through a motorcycle shape grammar. *Research in Engineering Design* 13 (3): 139–156.

Reis, J. 2023. Supporting creativity with emergent shapes in shape grammars. In *18th Iberian conference on information systems and technologies (CISTI)*, 20–23 June 2023, Aveiro, Portugal. ISBN 978-989-33-4792-8.

Stiny, G. 1977. Ice-ray: A note on the generation of Chinese lattice designs. *Environment and Planning B: Planning and Design* 4 (1): 89–98.

Stiny, G. 2022. *Shapes of imagination: Calculating in Coleridge's magical realm*. Cambridge: The MIT Press.

Stouffs, R., and A. Li. 2020. Learning from users and their interaction with a dual-interface shape-grammar implementation. In *RE: Anthropocene, design in the age of humans. Proceedings of the 25th international conference of the association for computer-aided architectural design research in asia (CAADRIA) 2020*, Bangkok, Thailand, 5–6 August 2020, eds. D. Holzer, W. Nakapan, A. Globa, and I. Koh, vol. 2, 153–162. Hong Kong: Association for Computer-Aided Architectural Design Research in Asia (CAADRIA).

Tapia, M. 1999. A visual implementation of a shape grammar system. *Environment and Planning B: Planning and Design* 26 (1): 59–73.

Unilever. 2023. *1950–1980—Building our brands.* https://www.unilever.com/our-company/our-history-and-archives/1950-1980/. Accessed 16 Aug 2023.

Zhang, X. 2023. Research on the history and innovation path of the Dove brand. Advances in economics and management research. In *2023 international symposium on economic development and management engineering (ISEDME 2023)*, ed. J. Feng, vol. 5, 337–369, June 10, 2023, Haikou, China.

Open Access This chapter is licensed under the terms of the Creative Commons Attribution-NonCommercial-NoDerivatives 4.0 International License (http://creativecommons.org/licenses/by-nc-nd/4.0/), which permits any noncommercial use, sharing, distribution and reproduction in any medium or format, as long as you give appropriate credit to the original author(s) and the source, provide a link to the Creative Commons license and indicate if you modified the licensed material. You do not have permission under this license to share adapted material derived from this chapter or parts of it.

The images or other third party material in this chapter are included in the chapter's Creative Commons license, unless indicated otherwise in a credit line to the material. If material is not included in the chapter's Creative Commons license and your intended use is not permitted by statutory regulation or exceeds the permitted use, you will need to obtain permission directly from the copyright holder.

The Science of Shape Grammar-Based Product Design

Jonathan Cagan

Abstract Herb Simon's *The Sciences of the Artificial*, published in 1969, argued for a scientific approach to the design or creation of new solutions to problems. Shortly after, Stiny and Gips introduced shape grammars as a mathematical approach to representing and reasoning about design. This work discusses the field and impact of shape grammars from the perspective of Simon's attributes within the lens of product design.

1 Introduction

In 1969, Herb Simon discussed the science of design in *The Sciences of the Artificial* (Simon 1969). Two years later, Stiny and Gips introduced shape grammars as a mathematical approach to representing and reasoning about (architectural) design (Stiny and Gips 1972), with many deep explorations by Stiny and others to follow, including Stiny's beautiful presentation of *Shape* (Stiny 2008). This short discussion presents the field and impact of shape grammars from the perspective of Simon's attributes in "The Science of Design: Creating the Artificial" (Chap. 5, from his seminal book) from the lens of product design, and from my own work over more than three decades.

2 The Science of the Design

Simon (Simon 1969) laid out five main attributes of cognitive and computational design:

J. Cagan (✉)
David and Susan Coulter Head of Mechanical Engineering, George Tallman and Florence Barrett Ladd Professor, Department of Mechanical Engineering, Carnegie Mellon University, Pittsburgh, USA
e-mail: cagan@cmu.edu

Representation: "How representations are created and how they contribute to the solutions of problems will become an essential component of the future theory of design."

Memory: "Systems must have some means of storing in its memory information about states of the world—afferent, or sensory, information—and information about actions—efferent, or motor, information."

Configuration: "Problem-solving systems and design procedures in the real world do not merely assemble problem solutions from components but must search for appropriate assemblies."

Complexity: "Complex systems might be expected to be constructed in a hierarchy of levels."

Search: "Only in trivial cases is the computation of the optimum alternative an easy matter."

Arguably, that chapter from the *Sciences of the Artificial*, started, or at least set the stage for, the field of design research today.

Shape grammars provide the means to address each of these attributes:

Representation: Shape grammars are, by their creation and application, a means to represent the design space, and their execution results in a design concept. Importantly they capture the logic of the human designers, uncovering the cognitive basis for design.

Memory: The efficiency of an elegant shape grammar provides a means to trace and recall states of a design process. They result in explainability of decisions that are made.

Configuration: Shape grammars provide a bottom up means to assemble a design. In the most raw form a set grammar (a shape grammar without emergence) allows for combination, resulting in configuration design. Emergence allows for sharing of features (such as function sharing) that can result in efficiency or novelty of a design.

Complexity: The language, from a well-formed shape grammar, is rich. A concise set of rules can describe an infinite space of paths and solutions. Emergence of new, unexpected shapes upon which rules can apply adds to the complexity of designs that can be generated.

Search: Exploration of a language of shape through rule application provides a means to search for a solution. Evaluation results in assessment of the appropriateness of the assemblies generated.

Shape grammars are a way to view the physical, human made and envisioned world. Although created in the context of architectural design, shape grammars have found application in the modeling and creation of products as well. In 1998, Manish Agarwal and I presented the language of coffee makers in *Environment and Planning B* (Agarwal and Cagan 1998). In doing so the logic of how to build a coffee maker through its shape but based on its function was described. Going further, by identifying certain shape rules and parameter value ranges, specific coffee makers that can be associated with different brands could be generated.

As with the style of Frank Lloyd Wright (Koning and Eizenberg 1981), Palladian (Stiny and Mitchell 1978) and more in architectural design, the coffee maker work highlighted the potential for shape grammars to capture the branding of consumer and other products. Next, Michael Pugliese and I presented the motorcycle grammar (Pugliese and Cagan 2002). After an assessment of the history of Harley bikes, mathematical ranges were specified within which the design generated would capture the brand identity of Harley Davidson (as verified through a survey of Harley owners). Next Jay McCormack, Craig Vogel and I analyzed the history of the Buick brand, creating a shape grammar that generates Buick vehicle designs including transition rules between design eras (McCormack et al. 2004). Such thinking has now been applied in a variety of fields and industries.

An algorithmic language, shape grammars began as a means for humans to uncover and "explain" the primarily architectural configurations and styles that other humans created through design. The basis for shape grammars naturally presented a computer language as well, one that could enable machines to generate designs in a given style, and allow for exploration of what could be generated within that language. There were two problems: (1) how to implement such a language?; (2) how to control its search?

3 Implementation

Implementation had been hard coded; each grammar was coded rule-by-rule into a computer production system. The goal was, and is, a more general means to go from "paper" to code. This goal requires the ability of a system to identify shapes upon which rules can apply. Work on shape interpreters, from the earliest days of shape grammars, have sought to address this capability. The word "interpreter" is particularly relevant in that shapes must be interpreted in order to find a left-hand-side of a rule that can be applied. Work by Ramesh Krishnamurti (1982) and Krishnamurti and Chris Earl (1992), as well as others, explored the range from foundations for such computational representations to means to implement systems able to readily take lines or curves as left-hand- and right-hand-sides of rules and result in the implemented and usable systems. My work with Jay McCormack (2002, 2006) addressed implementation needs for product-focused shape grammars, in particular.

However, even with this ability, the basic grammar rules themselves still need to be uncovered by the human user. For many purposes, that is not only acceptable but also desirable, in that the process results in explainability and human learning. People can now understand how Frank Lloyd Wright, Palladio, Siza (Duarte 2005), the architects of Queen Anne Houses (Flemming 1987), and many others, reasoned about the designs they created. But in new domains, or for the purpose of creating explainable AI that can apply in new applications where generative systems are desired, a major challenge remains: how a computer system can look at the pixels of representative images, and derive a set of shape rules that can generate those images and other images with a similar style and purpose. This requires the induction of rules.

Induction has been a core part of machine learning for decades. The challenge here is that the data is based on shape, and, in its raw sense, only pixels. Humans can parse what those pixels mean, holistically, and what features, within, form the building block of those designs. Computers, however, do not yet have that understanding of the world and its decomposition. Induction is necessary if shape grammars are going to be a basis for codifying the artificial and the natural world with no or minimal human input. With a focus on product design, several efforts include Seth Orsborn, Peter Boatwright and my work using Principal Component Analysis (PCA) to derive primary clusters of similar and non-similar lines and curves of shapes into chunks that are then codified within shape rules (Orsborn et al. 2008a, b); and Mark Whiting, Phil LeDuc and my (Whiting et al. 2018) work using un-coded structured datasets from which objects are extracted and then expanded into rules based on frequency, where upon rule similarities reduce the language through consolidation or abstraction.

4 Search

Although implemented shape grammars allow for human exploration of the design space, if the goal is generative design, such search needs to be automated. Bill Mitchell and I took on this challenge by merging stochastic optimization methods with shape grammars. Due to the discrete nature of a possibly infinite design space, one where emergence could change the course of the design mid-stream, gradient or other local optimization methods would not enable search with general shape grammars. Instead, global search is required. Our idea was that a non-deterministic search method, allowing for the identification and application of a left-hand-side rule, would be feasible to merge with optimizing search to select a rule, apply it, evaluate it, and then allow for the design to unfold in a goal-directed way. We chose simulated annealing (Kirkpatrick et al. 1983), as our search method because it emulated problem solving of humans (Cagan and Kotovsky 1997), and readily followed the discrete sequential application of shape grammars. This is a method that, conceptually, numerically mimicked the annealing of metals from high energy states to low energy states. The result, *shape annealing* (Cagan and Mitchell 1993), has been shown to be a powerful way to generate designs with shape grammars. Other stochastic search methods, such as genetic algorithms, have since been applied with shape grammars, but the original idea is the same. Shape annealing opened up the opportunity for automated search with shape grammars, an early approach to generative design. Shape annealing was been used for structural (truss) design by Kristi Shea (e.g., Shea and Cagan 1997) and others (Reddy and Cagan 1995), manufacturing process planning by Ken Brown (Brown and Cagan 1997), as well as other applications in the synthetic-natural worlds (discussed below). But the algorithm needs to be directed with feedback on its decisions, and the ability to identify a good, if not best, design solution.

Evaluation, then, becomes the last piece of the generative design puzzle. In the world of engineering, physics can often be used to tell you just how good a design is. For example, given a connected truss with dimensions and loads, its weight and structural integrity (including how close it is to failure) can be assessed. Still, for incomplete designs, strategies and means, for example heuristic based, need to be developed to determine whether the solutions are heading toward a quality configuration, or even toward a coherent configuration that can then be refined and optimized once developed.

For many products, however, the physical evaluation is only part of the overall assessment of a design. Your favorite consumer product needs to work, but it also needs to look good and excite you as the user. Any assessments can be included in the objective function in shape annealing (or any optimization method), but such qualitative assessments need to be quantified. Seth Orsborn, Peter Boatwright and I introduced an approach to codify preferences through utility functions from human users, and then to use that approach to generate preferred solutions for form for that user from shape grammars (Orsborn et al. 2008a, b). To merge the qualitative preference with the physical assessment, Ian Tseng, Ken Kotovsky and I employed a neural network to learn preferences for different parametric ranges, and then coupled that qualitative assessment with a computational aerodynamics analysis to determine vehicle forms that were efficient, while stylistically preferred, to its users (Tseng et al. 2012).

Herb Simon states, "Only in trivial cases is the computation of the optimum alternative an easy matter" (Simon 1969). Shape annealing and the means to assess the whole of a product are steps to find that optimum. By modeling and evaluating user preference and product performance, the process is a "course of action aimed at changing existing situations into preferred ones" (Simon 1969).

5 Applications in the Natural World

Finally, what has been interesting to me over recent years is a step toward using shape grammars to understand the biological world, and to generate synthetic biological systems. Aspects of how the natural world looks and works remains mysterious. Still, by uncovering and representing nature's languages through shape grammars, new insights into how nature works can result. These same insights can be used for generating synthetic biological systems and manipulating natural systems to achieve desirable capabilities, or even diagnosing diseases.

Paul Egan, Chris Schunn, Phil LeDuc and I modeled myosin, the building block of muscle, based on the logic of shape grammars (Egan et al. 2013). By then integrating that language into an agent system, myosin with particular performance capabilities could be specified and built (Egan et al. 2017). More recently, Tito Babatunde, Sebastian Arias, Bex Taylor and I articulated simple shape rules that enable the formation of structures out of DNA, the building block of life (Babatunde et al. 2021). These DNA origami structures can be used in applications ranging from drug delivery to biosensing. Using shape annealing, such DNA structures that efficiently fill or coat a mesh base with DNA are generated.

Ryan Yeh, Ken Nischal, Phil LeDuc and I modeled the generation of vascular networks in the eye with shape grammars (Yeh et al. 2020). By manipulating the parameter ranges of the generation of the networks, ophthalmologists can identify at what point such vascular systems become diseased.

Finally, by using inductive methods, my work by Whiting and LeDuc induced shape grammar rules from MRI scans of vascular networks in the brain (Whiting et al. 2022). Given the resulting rule sets, new brain scans can be broken down to the rules from which they are generated. It turns out that certain rules are more probably found under certain diseased conditions. Thus, this reverse engineering of the scan using shape grammars can indicate, through the identification of a rule application, possible diseases that can then be highlighted to a physician.

6 The Science of Shape Grammar-Based Product Design

After 50 years of shape grammars, how design researchers and designers look at and seek to understand the world is more crisp. The insight of Stiny and Gips and their influence on design science *a la* Simon have impacted the mathematics of CAD systems, the way products are envisioned, and how brands are understood. This also provides new insights as technologies emerge that uncover the natural and synthetic world around us. It would be interesting to know what this book will look like in another 50 years.

Acknowledgements This work was supported by the National Science Foundation (NSF) under Award CMMI-2113301.

References

Agarwal, M., and J. Cagan. 1998. A blend of different tastes: The language of coffee makers. *Environment and Planning B: Planning and Design* 25 (2): 205–226.
Babatunde, B., D.S. Arias, J. Cagan, and R.E. Taylor. 2021. Generating DNA origami nanostructures through shape annealing. *Applied Sciences* Article 2950: 21.
Brown, K.N., and J. Cagan. 1997. Optimized process planning by generative simulated annealing. *Artificial Intelligence in Engineering Design, Analysis and Manufacturing* 11: 219–235.
Cagan, J., and K. Kotovsky. 1997. Simulated annealing and the generation of the objective function: A model of learning during problem solving. *Computational Intelligence* 13 (4): 534–581.
Cagan, J., and W.J. Mitchell. 1993. Optimally directed shape generation by shape annealing. *Environment and Planning B* 20: 5–12.

Duarte, J.P. 2005. Towards the mass customization of housing: The grammar of Siza's houses at Malagueira. *Environment and Planning B: Planning and Design* 32 (3): 347–380.

Egan, P., J. Cagan, C. Schunn, and P.R. LeDuc. 2013. Design of complex biologically-based nanoscale systems using multi-agent simulations and structure-behavior-function representations. *ASME Journal of Mechanical Design* 135 (6): 061005-1-12.

Egan, P., J.R. Moore, A.J. Ehrlicher, D.A. Weitz, C.D. Schunn, J. Cagan, and P.R. LeDuc. 2017. Robust mechanobiological behavior emerges in heterogeneous myosin systems. In *Proceedings of the National Academy of Sciences (PNAS)* 114 (39): E8147–E8154.

Flemming, U. 1987. More than the sum of the parts: The grammar of Queen Anne houses. *Environment and Planning B* 14: 323–350.

Kirkpatrick, S., C.D. Gelatt, and M. P. Vecchi. 1983. Optimization by simulated annealing. *Science* 220 (4598): 671–680.

Krishnamurti, R. 1982. SGI: An Interpreter for Shape Grammars. Working Paper. Milton Keynes, UK: Centre for Configurational Studies, The Open University.

Koning, H., and J. Eizenberg. 1981. The language of the Prairie: Frank Lloyd Wright's Prairie houses. *Environment and Planning B* 8: 295–323.

Krishnamurti, R. 1980. The arithmetic of shapes. *Environment and Planning B: Planning and Design* 7: 463–484.

Krishnamurti, R., and C.F. Earl. 1992. Shape recognition in three dimensions. *Environment and Planning B: Planning and Design* 19: 585–603.

McCormack, J.P., and J. Cagan. 2002. Supporting designer's hierarchies through parametric shape recognition. *Environment and Planning B: Planning and Design* 29: 913–931.

McCormack, J.P., and J. Cagan. 2006. Curve-based shape matching—Supporting designers' hierarchies through parametric shape recognition of arbitrary geometry. *Environment and Planning B* 33 (4): 523–540.

McCormack, J.P., J. Cagan, and C.M. Vogel. 2004. Speaking the buick language: Capturing, understanding, and exploring brand identity with shape grammars. *Design Studies* 25: 1–29.

Orsborn, S., P. Boatwright, and J. Cagan. 2008a. Identifying product shape relationships using principal component analysis. *Research in Engineering Design* 18 (4): 181–196.

Orsborn, S., J. Cagan, and P. Boatwright. 2008b. A methodology for creating a statistically derived shape grammar composed of non-obvious shape Chunks. *Research in Engineering Design* 18 (4): 163–180.

Pugliese, M., and J. Cagan. 2002. Capturing a Rebel: Modeling the Harley-Davidson brand through a motorcycle shape grammar. *Research in Engineering Design* 13 (3): 139–156.

Reddy, G., and J. Cagan. 1995. Optimally directed truss topology generation using shape annealing. *ASME Journal of Mechanical Design* 117 (1): 206–209.

Shea, K., and J. Cagan. 1997. Innovative dome design: Applying geodesic patterns with shape annealing. *Artificial Intelligence in Engineering Design, Analysis and Manufacturing* 11: 379–394.

Simon, H.A. 1969. *The sciences of the artificial*. Cambridge: The MIT Press.

Stiny, G. 2008. *Shape—Talking about seeing and doing*. Cambridge: The MIT Press.

Stiny, G., and J. Gips. 1972. Shape grammars and the generative specification of painting and sculpture. In *Proceedings of IFIP congress '71*. Amsterdam: North Holland Publishing Co. (also in *The best computer papers of 1971*, ed. O.R. Petrocelli. Philadelphia: Auerbach).

Stiny, G., and W.J. Mitchell. 1978. The Palladian grammar. *Environment and Planning B* 5: 5–18.

Tseng, I., J. Cagan, and K. Kotovsky. 2012. Concurrent optimization of computationally learned stylistic form and functional goals. *ASME Journal of Mechanical Design* 134 (1): 111016-1-11.

Whiting, M.E., J. Cagan, and P.R. LeDuc. 2018. Efficient probabilistic grammar induction for design. *Artificial Intelligence in Engineering Design, Analysis and Manufacturing* 32 (2): 177–188.

Whiting, M.E., J. Mettenburg, E.M. Novelli, T. Santini, T. Martins, T.S. Ibrahim, P.R. LeDuc, and J. Cagan. 2022. Inducing vascular grammars for anomaly classification in brain angiograms. *ASME Journal of Engineering and Science in Medical Diagnostics and Therapy* 5 (2): 021002.

Yeh, R., K. Nischal, P.R. LeDuc, and J. Cagan. 2020. Written in blood: Applying shape grammars to retinal vasculatures. *Translational Vision, Science and Technology* 9(9): 1–10, article 36.

Open Access This chapter is licensed under the terms of the Creative Commons Attribution-NonCommercial-NoDerivatives 4.0 International License (http://creativecommons.org/licenses/by-nc-nd/4.0/), which permits any noncommercial use, sharing, distribution and reproduction in any medium or format, as long as you give appropriate credit to the original author(s) and the source, provide a link to the Creative Commons license and indicate if you modified the licensed material. You do not have permission under this license to share adapted material derived from this chapter or parts of it.

The images or other third party material in this chapter are included in the chapter's Creative Commons license, unless indicated otherwise in a credit line to the material. If material is not included in the chapter's Creative Commons license and your intended use is not permitted by statutory regulation or exceeds the permitted use, you will need to obtain permission directly from the copyright holder.

Shape Studies

Shape Rules: Their Algebras, Grammars and Decompositions

Djordje Krstic

> I have great love for narrative, clearly told, because that is where ambiguity could actually emerge. That you make something with such clarity that it can't be vague, yet it can still have multiple meanings. That's the richest place to play.
> James Gray[*]

Abstract The paper contributes to the shape grammars based formal design theory by first providing a formal definition of shape rules and shape rule algebras, and then utilizing this new tools to: construct new shape rules from the old ones, define meta shape grammars that generate new shape grammars by modifying the rules of the existing ones, provide a formal representation of Stiny's rule schemas which allows for their manipulation, define a shape rule decomposition as a set of shape rules that sums to a shape rule, and analyze calculations with shapes via shape rule decompositions.

This paper is in the field of formal design theory, which relies on shape grammars to generate design languages for analysis/synthesis in the areas of design, engineering design, architecture/urbanism, archeology, art, and (recently) making/manufacturing. The theory started in 1972 with the introduction of shape grammars by Stiny and Gips (1972), and it has evolved since. Numerous papers have been published in the field, with books (Stiny 2006, 2022) featuring a recent account of the theory, while also revealing its intersections with philosophy, literature and poetry.

A shape grammar is a set of shape rules that includes one distinguished initial rule: a rule can act on a given shape by replacing one of its parts with a new shape. A sequence of rule actions can gradually change a shape to produce a new one. If all

---

[*]We could not resist quoting director James Gray talking about the movie narrative. It is an accurate description of what rules of shape grammars do when, with perfect mathematical clarity, they create ambiguous and (often) unexpected outcomes. Indeed, "the richest place to play."

D. Krstic (✉)
22415 Kearny St, Calabasas, CA 91302, USA
e-mail: gkrstich@aol.com

rules in such a sequence belong to the same grammar, and the sequence starts with the initial rule, then the resulting sequence of shapes is a derivation of the last shape in the sequence. The latter shape is an instance of the language being specified by the grammar.

Shape rules are central to shape grammars and, as the title suggests, we will mostly deal with them in this paper. A shape rule is simply a pair of shapes separated by an arrow (*shape* → *shape*). The first shape represents the part that needs to be replaced, while the second one replaces it. To do the replacement, we first need to be able to rotate, translate, scale, i.e., transform the first shape so that it becomes a part of the given shape. This is not always possible, but if the right transformation is found, the rule action may proceed. The transformed first shape is subtracted from the given shape, and an equally transformed second shape is added–in effect changing the given shape. To do the calculations, we need an algebra of shapes featuring subtraction and addition operations, as well as a partial order relation. The latter is needed to be able to determine shape parts. The algebra also needs to be closed under geometric transformations that can act on shapes to move and scale them.

A more formal review of shapes, algebras of shapes, shape derivations with grammars and shape decompositions is given in the following section. The review sets the stage for the next sections featuring (formal) introduction of rules, their algebras, their construction and generation, as well as their usage for generating and analyzing shapes.

1 Shapes, Algebras of Shapes, Shape Derivations with Grammars, and Shape Decompositions

A shape is represented by a finite set of basic geometric elements which are maximal and confined to a finite chunk of space. The elements could be 0, 1, 2, or 3-dimensional i.e., points, lines, planes, and solids, and be situated in the space of the same or greater dimension. Thus, each shape is characterized by i the dimension of its basic geometric elements, and j the dimension of the space it is situated in, such that $i \leq j$. Shapes are not limited to the set structures of their representations as they may have parts beyond the subsets. The only exception are shapes made of points, which do behave like sets. Unless stated differently, these will not be considered in the text. A part of a shape or a *subshape* is a shape with elements that are parts of the maximal elements of the original shape. This allows the otherwise finite shape to contain an infinite number of parts, i.e., to be infinitely divisible.

For example, in Fig. 1, triangle (a) is represented by the set of 3 maximal lines (b) where its subshape (part) (c) has two maximal lines that are parts of the two oblique maximal lines of the triangle.

The set of all shapes characterized by dimensions i and j are partially ordered by the subshape relation ($\leq$) and there is a smallest shape, namely the empty set of basic elements of dimension i in space of dimension j, , but there is no greatest

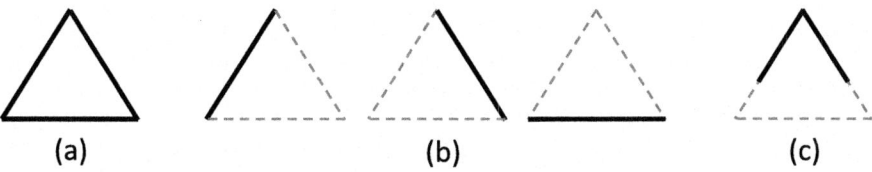

Fig. 1 Triangle **a** represented by set of maximal lines **b**, its subshape **c** with two maximal lines

shape. Partial order relation allows for meet and join operations to be defined on the set of shapes effectively turning it into an algebra, or more specifically, a lattice. The meet of two shapes is the greatest shape that is part of both shapes, while the join is the smallest shape that has both shapes as parts. Meet and join dub as the respective Boolean operations of product ($\cdot$) and sum ($+$) for shapes. To support rule actions, we also need a difference operation. Typically, Boolean complements are used to define the difference; however, these are not available because there is no greatest shape. In contrast, there are complements relative to certain shapes that could be used instead. If a and b are shapes, then difference $a - b$ is the complement of b relative to $a + b$, alternatively it is the complement of $a \cdot b$ relative to a. In both cases $a - b$ is the greatest shape which has parts of a, but no part of b and nothing else. There is another Boolean operation, symmetric difference ($\oplus$), that is used with shapes. The latter is defined as $a \oplus b = (a - b) + (b - a) = (a + b) - (a \cdot b)$, or the greatest shape that has only the parts of a and b that they do not share and nothing else.

For example, in Fig. 2, two overlapping squares (a) have the product (b), sum (c), differences (d) and (e), and symmetric difference (f).

Set U_{ij} of all shapes of dimension i situated in the finite chunk of space of dimension j closed under the above operations is an algebra for shapes U_{ij} (Stiny 1991). It is a generalized Boolean algebra i.e., a relatively complemented distributive lattice with the smallest element but not the greatest one (Birkhoff 1993). Set U_{ij} has also to be closed under similarity transformations in order for shapes to be moved around, reflected and scaled (Stiny 1991). These may be external to the algebra or integrated into its structure. The latter is achieved by treating transformations as an infinite set of operators and adding them to the signature of the algebra rendering U_{ij} a generalized Boolean algebra with operators (Krstic 1999, 2014; Stiny 2006). In contrast, set T_{ij} of transformations may be added to the carrier making it two-sorted so that U_{ij} becomes a two-sorted algebra operating on both shapes and transformations (Krstic 1999, 2014). Group operations of composition ($\circ$), inverse ($^{-1}$) and identity (ι) are

Fig. 2 Two overlapping squares (**a**) have the product (**b**), sum (**c**), differences (**d**) and (**e**), and symmetric difference (**f**)

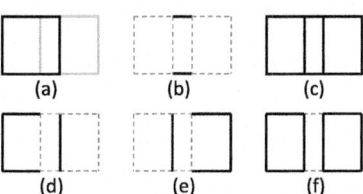

added to the signature of U_{ij} to be able to combine transformations. It appears that U_{ij} is the two algebras in one. It has a Boolean part, which is a generalized Boolean algebra operating on shapes, and a group part, which is a group operating on transformations. The two are combined with the group action operation (()) which takes a shape and a transformation and yields the transformed shape, or $t(a)$, where a is a shape and t a transformation.

Algebras for shapes could be combined to operate on *compound shapes*. For example, the direct product of an algebra for solids and one for planes, or $U_{33} \times U_{23}$, may be used to pair solid shapes with their planar boundaries. Direct product $U_{33} \times U_{23}$ has ordered pairs of shapes as well as ordered pairs of transformations as elements so its carrier consists of two Cartesian products $U_{ij} \times U_{kl}$ and $T_{ij} \times T_{kl}$. Operations are done componentwise. For example, $(a, b) + (u, v) = (a+u, b+v)$, or $(t_1, t_2)((a, b)) = (t_1(a), t_2(b))$, where $a, u, t_1 \in U_{ij}$ and $b, v, t_2 \in U_{kl}$. There are also subdirect products which are certain subalgebras of a direct product algebra. They retain all of the elements of the component algebras but not all of their combinations. A subdirect product of U_{ij} and U_{kj} which retains all of the combinations of shapes, but allows only ordered pairs of transformations with both components the same, is the sum of the two algebras, or $U_{ij} + U_{kj}$ (ibid.). Although sum $U_{ij} + U_{kj}$ is a subalgebra of product $U_{ij} \times U_{kl}$ only their group parts are so related. Their Boolean parts are equal. Sums of algebras guarantee the integrity of compound shapes by transforming their components—which belong to different algebras–in the same way. There is no such guarantee with direct products, however, this is the price to pay for more freedom in choosing the components.

As mentioned earlier, derivations of shapes by grammars are sequences of shapes produced by the sequences of rule actions. An action of rule $a \to b$ on shape c may take place if there is a transformation t such that condition

$$t(a) \le c \tag{1}$$

is satisfied. The action results in a new shape $c\prime$ and is given by

$$c' = (c - t(a)) + t(b) \tag{2}$$

Consequently, a derivation of shape c_n may be seen as a single calculation, a sequence of n rule actions given recursively by

$$\begin{aligned}
c_1 &= (0 - t_1(0)) + t_1(b_1) = t_1(b_1) \\
c_2 &= (c_1 - t_2(a_2)) + t_2(b_2) \\
&\ldots \\
&\ldots \\
&\ldots \\
c_n &= (c_{n-1} - t_n(a_n)) + t_n(b_n),
\end{aligned} \tag{3}$$

where $0 \to b_1$ is the initial rule, $a_i \to b_i$ ($i > 1$), the rule acting in the i-th step of the derivation, and transformations t_i are given by the sequence of $n - 1$ conditions

$$t_2(a_2) \leq c_1$$
$$\ldots$$
$$\ldots \qquad (4)$$
$$\ldots$$
$$t_n(a_n) \leq c_{n-1}.$$

Note that there is no condition related to t_1, as this transformation may be freely chosen given that the left-hand side of the initial rule is 0. Also, rules $a_i \to b_i$ and $a_j \to b_j$, for $i, j > 1$ and $i \neq j$, need not be different.

We will next take a closer look at decompositions of shapes and how they are used to analyze/explain calculations with shapes.

Any finite set of shapes is a decomposition of its sum, or inversely any finite set of parts of a shape that sum-up to that shape is its decomposition. We are considering only finite decompositions because U_{ij} algebras are not closed under infinite sums. However, some infinite decompositions do exist and an interesting one is a set of all parts of a shape. The latter is a proper subalgebra of U_{ij} closed under Boolean operations as well as under the symmetry group of the shape. Decompositions are important as they explain shapes in terms of their certain parts (properties) while neglecting the other (infinitely many) parts. As such they are often used in place of shapes as shape approximations. Whenever we use words to explain a shape, we are decomposing it by mentioning its certain parts and omitting the rest. Even naming a shape is a decomposition with one element, the shape itself (Krstic 2005). Computers and possibly our brains store shapes as shape decompositions.

Decompositions of shapes may also serve to analyze/explain calculations with shapes. Stiny (1994) and recently Haridis and Stiny (2022) used decompositions structured as topologies and related continuous mappings to examine continuity of shape derivations. We will follow a different approach here in which *argument lattices* and *argument decompositions* are introduced to analyze calculations with shapes (Krstic 1996 pp 133–135, 2017).

Any calculation with shapes implies that its arguments are decomposed in some way. Moreover, any shape related to a calculation is analyzed in terms of parts that emerge as combinations of the arguments. These are argument decompositions and their upper bound is the argument lattice. An argument lattice depends only on arguments and the relations between the arguments and not on particular calculations because its elements are the results of all possible calculations with the arguments. Consequently, argument lattices are also the upper bounds for topologies resulting from calculations examining continuity of shape derivations.

More formally, given finite set $X = \{x_k\}_{k=1,\ldots n}$ of n shapes from an U_{ij} algebra and a set R of relations between them, which includes $\sum X = 1$,[1] a Boolean algebra

---

[1] 1 stands for a Boolean unit.

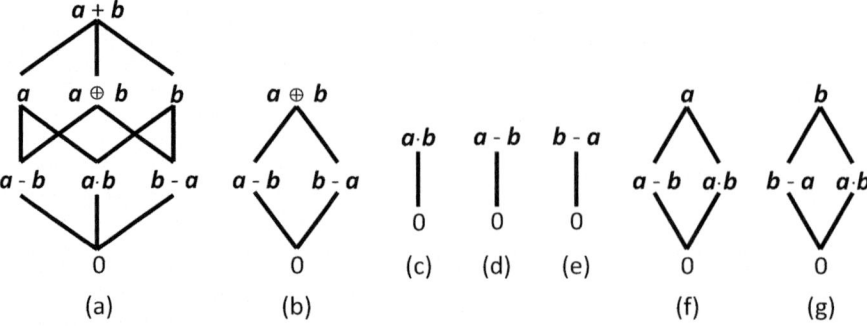

Fig. 3 Argument lattice with shapes a and b as arguments (**a**) and argument decompositions for $a+b, a \oplus b, a \cdot b, a-b, b-a, a$, and b (**a–g**), respectively

$Ar(X)$ may be generated by X and R as respective, generators and relations–in a standard universal algebraic way (Birkhoff 1935; Vickers 1989, pp. 39; Krstic 1996, pp. 15). Because values of all n-ary polynomials with X as arguments defined in the Boolean part of U_{ij} are elements of $Ar(X)$ the latter is the argument lattice for all of the calculations with X as arguments. For any shape $u \in Ar(X)$ principal ideal $\mathbf{I}(u)$ generated by u, or $\mathbf{I}(u) = \{y \in Ar(X) | y \leq u\}$, is the argument decomposition of u with respect to all the calculations with X as arguments. The argument lattice itself is the argument decomposition of $\sum X$.

For example, an argument lattice with shapes a and b as arguments and relation $a+b=1$ is given in Fig. 3a. It is also an argument decomposition of calculation $a+b$. Argument decompositions $\mathbf{I}(a \oplus b)$, $\mathbf{I}(a \cdot b)$, $\mathbf{I}(a-b)$ and $\mathbf{I}(b-a)$ for other calculations are given in Fig. 3b-e, respectively. Even the argument decomposition of a in Fig. 3f may be seen as representing calculations $a + a = a \cdot a = a$. The same holds for the argument decomposition of b in Fig. 3g, while $\mathbf{I}(0) = \{0\}$ represents calculations $a - a = b - b = a \oplus a = b \oplus b = 0$.

Derivation of a shape by a shape grammar can, as mentioned earlier, be seen as a single calculation with shapes that make the rules as arguments. This creates an opportunity to explain/describe derivation via argument lattice and argument decompositions. Shapes $c_1, c_2, \ldots c_n$ appearing at each step of the derivation as results of rule actions are, in accordance with (3), results of calculations with shapes that make rules, transformed in accordance with relations (4), as arguments. Consequently, the argument lattice of the derivation is generated by set

$$C_n = \{0, t_1(b_1), t_2(a_2), t_2(b_2), \ldots t_n(a_n), t_n(b_n)\} \qquad (5)$$

of arguments as well as relations (4) augmented by relation

$$\sum C_n = 1. \qquad (6)$$

Note that we assumed here that argument lattices are finite, however, that is not guaranteed. There are calculations with shapes which lead to infinite argument lattices (Krstic 1996, pp. 167; Jowers et al. 2019). In such a lattice all of the principal ideals are infinite as well. They do not qualify as argument decompositions which by definition are finite.

2 Rules and Rule Algebras

From a mathematical point of view shape rule $a \to b$ is ordered pair (a, b) which is a compound shape belonging to the direct product of two U_{ij} algebras, or U_{ij}^2. Though, in the course of a rule action both a and b are transformed in the same way–in accordance with (2)–which is how components of shapes belonging to sums of algebras are transformed. Consequently, rule $a \to b$ could be seen as an element of the sum $U_{ij} + U_{ij}$, or $2U_{ij}$, which is smaller and better suited algebra than U_{ij}^2. Because any pair of shapes amounts to a rule, compound shapes of $2U_{ij}$ are rules, rendering $2U_{ij}$ as a *shape rule algebra*. This makes partial order, Boolean operations, and transformations readily available for rules.

The partial order of compound shapes of $2U_{ij}$ is the partial order for rules defined componentwise as $u \to v \leq a \to b$ if $u \leq a$ and $v \leq b$. Rule $u \to v$ is a *subrule* of $a \to b$. The same works for sum, product difference and symmetric difference operations, or $u \to v * a \to b = (u*a) \to (v*b)$ where $*$ stands for $+, \cdot, -,$ and $\oplus$, respectively. Group action operation in $2U_{ij}$ is defined as $(t, t)((a, b)) = (t(a), t(b))$. Because group parts of $2U_{ij}$ and U_{ij} algebras are isomorphic,[2] we may write t in place of (t, t) so that the group action operation for rules is $t(a \to b) = t(a) \to t(b)$ There is the smallest rule, which is the *empty rule* defined as $0 \to 0$, but a greatest rule does not exist.

Note that rule action described by (1) and (2) cannot be carried out in $2U_{ij}$ as it requires interaction between shapes a and b which are elements of U_{ij}. Operations of $2U_{ij}$ are done componentwise prohibiting interactions between components of compound shapes. To support rule actions i.e., to check relation (1) and calculate formula (2), we need shapes a and b extracted from ordered pair $(a, b) \in 2U_{ij}$ which represents rule $a \to b$. Direct products and subdirect products like $2U_{ij}$ have Cartesian products of shapes as their carriers. Consequently, set-theoretic projection operators may be used to extract components from the elements of Cartesian products. For rule $r = (a, b) \in 2U_{ij}$ projection operator pr_L extracts a and pr_R extracts b. These are defined as $pr_L(r) = \cup(\cap r) \in U_{ij}$ and $pr_R(r) = \cup(\Delta r) \in U_{ij}$, where Δ denotes the symmetric difference for sets.[3] We need another pair of projections for the group part of $2U_{ij}$. However, as both components of a shape from $2U_{ij}$ are

---

[2] Each element t of U_{ij} is mapped to element (t, t) of $2U_{ij}$ and vice versa.
[3] $pr_L(r) = \cup(\cap r) = \cup(\cap(a, b)) = \cup(\cap\{a, \{a, b\}\}) = \cup\{a\} = a$ and $pr_R(r) = \cup(\Delta r) = \cup(\Delta(a, b)) = \cup(\Delta\{a, \{a, b\}\}) = \cup\{b\} = b$, where indexes L and R denote, respectively, the left-hand and the right-hand sides of the rule.

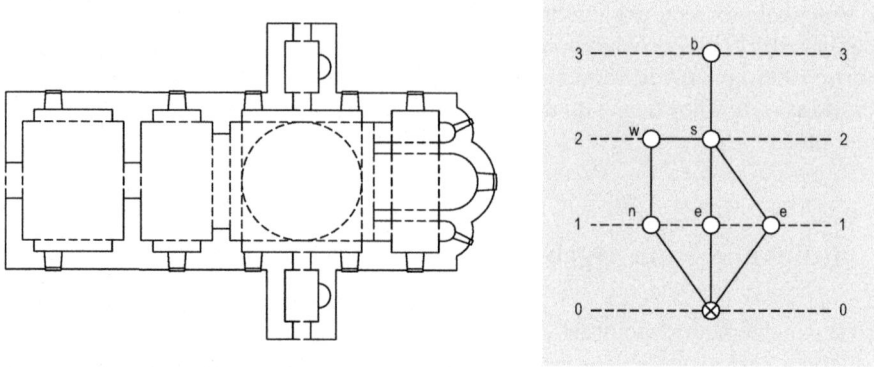

Fig. 4 The plan and the reflected ceiling plan of the church of the Virgin in Studenica, Serbia (on the left) and justified permeability graph of the plan (right)

transformed in the same way one projection is sufficient. For $t' = (t,t) \in 2U_{ij}$ projection operator is given by $pr_t(t') = \cup(\cup t')$.[4] Rule action expressions (1) and (2)—if precision takes over simplicity—may be written as $pr_t(t')(pr_L(r)) \leq c$ and $c' = (c - pr_t(t')(pr_L(r))) + pr_t(t')(pr_R(r))$, respectively. We will keep the projections implicit–for simplicity–in the remainder of this paper.

The approach, above, may be extended to *compound rules* that act on compound shapes of *parallel grammars*. The latter are grammars that generate several shapes in parallel, i.e., all floors of a high-rise building. Because compound shapes span several algebras, compound rules have to follow. Whenever a compound rule acts on a compound shape each of the component shapes is acted upon by a matching component rule.

For example, if compound shape c is defined in a direct product of n algebras $U_{i_1 j_1} \times U_{i_2 j_2} \times \ldots U_{i_n j_n}$, then compound rule r that can act on it has to be defined in $2(U_{i_1 j_1} \times U_{i_2 j_2} \times \ldots U_{i_n j_n})$ algebra. Shape c is ordered n-tuple $(c_{i_1 j_1}, c_{i_2 j_2}, \ldots c_{i_n j_n})$ and r is ordered pair $((a_{i_1 j_1}, \ldots a_{i_n j_n}) \to (b_{i_1 j_1}, \ldots b_{i_n j_n}))$. For r to act on c, relation $t_k(a_{i_k j_k}) \leq c_{i_k j_k}$ has to hold for each component $k \in \{1, 2, \ldots n\}$. Then n calculations $c'_{i_k j_k} = (c_{i_k j_k} - t_k(a_{i_k j_k})) + t_k(b_{i_k j_k})$ can take place in parallel to produce shape $c' = (c'_{i_1 j_1}, c'_{i_2 j_2}, \ldots c'_{i_n j_n})$.

Compound shapes may have more complicated structures with components belonging to sums and direct products of shape algebras as, for example, compound shapes in the grammar of Serbian medieval churches (Krstic 2015), each of which includes the plan, reflected ceiling plan and a justified permeability graph of the plan (see Fig. 4). Because both plans need to be transformed in the same way they belong to a sum of two U_{12} algebras. In contrast, the graph is transformed in a different way than the plans are so that the U_{12} algebra it belongs to is combined in the direct product with the (sum) algebra of plans. The resulting algebra is $2U_{12} \times U_{12}$ and the rules of the grammar are defined in $2(2U_{12} \times U_{12})$ shape rule algebra.

---

[4] $pr_t(t') = \cup(\cup(t')) = \cup(\cup(t,t)) = \cup(\cup\{t, \{t\}\}) = t$.

This section could be summarized by–and seen as the proof of–the following two propositions:

Proposition 1: Compound shapes belonging to $2(U_{i_1j_1} \times U_{i_2j_2} \times \ldots U_{i_nj_n})$ algebra are (compound) shape rules and the algebra is a shape rule algebra.

Proposition 2: Finite nonempty subsets of a shape rule algebra are shape grammars provided that they have one and only one rule of the form $r = ((0, \ldots 0) \to (b_{i_1j_1}, \ldots b_{i_nj_n}))$, where $b_{i_kj_k} \neq 0, k \in \{1,2, \ldots n\}$. Rule r is the initial rule.

Note that for $n = 1$ compound shape rules from proposition 1 are just shape rules as we have $2U_{ij}$ as the shape rule algebra.

3 Constructing Rules

Having algebras for rules provides a formal background for combining rules to construct new ones. Rules could be combined in a Boolean fashion via operations of sum, product, difference and symmetric difference and they can also be moved around or scaled and reflected via similarity transformations. All of these options are already available in $2U_{ij}$ algebras and we shall proceed to test them.

In Fig. 5a we have two rules defined in $2U_{12}$ rule algebra. Both rules replace a square by its translated version and in both rules the square is translated by 1/3 of its side. In the top rule the square is moved to the right and in the bottom one is moved to the left. New five rules are constructed by combining the original two rules in a Boolean fashion. The rule in Fig. 5b is the sum of the original two rules carried on in $2U_{12}$ rule algebra. It has been done componentwise so that the sum of the left-hand sides of the original rules carried on in U_{12} algebra is the left-hand side of the new rule while its right-hand side is the sum of the right-hand sides of the original rules. The other rules in Fig. 5 are the results of the following calculations with the original two rules as arguments: product (c), difference, where the bottom rule is subtracted from the top one (d), symmetric difference (e), and difference, where the top rule is subtracted from the bottom one (f).

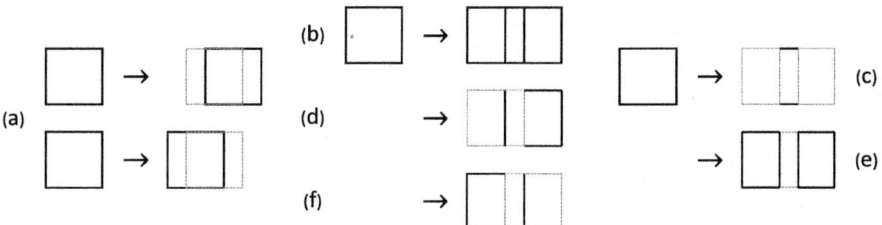

Fig. 5 Two rules translating the square defined in $2U_{12}$ rule algebra (**a**), are arguments in calculations creating five new rules as sum (**b**), product (**c**), difference between the top and bottom rules (**d**), symmetric difference (**e**) and difference between the bottom and top rules (**f**)

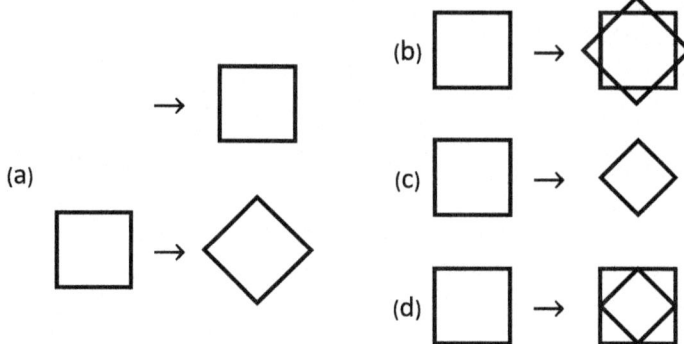

Fig. 6 The initial rule and rule rotating square (**a**), are arguments in calculations creating three new rules. Rule (**b**) is the sum of the arguments, while rule (**c**) is a transformed bottom rule (**a**) via transformation $(\iota, s_{1/\sqrt{2}})$. Rule (**d**) is the sum of the initial rule and rule (**c**)

In a second example, we shall take a look at the use of transformations together with Boolean operations in constructing rules.

Two rules of a small shape grammar are given in Fig. 6a. The first one is the initial rule which introduces a square to begin a shape derivation. The second one replaces a square with its rotated version, where the angle of rotation is 45° counterclockwise and the center of the rotation is the centroid of the square. The rule in Fig. 6b, which combines the original square with its rotated version, is the sum of the original two rules. The rule in Fig. 6c is the original square rotating rule but with a scaled right-hand side by factor $1/\sqrt{2}$. To construct it, the original rule should be transformed via a compound transformation which preserves its left-hand side and scales its right-hand side. Such transformation is $(\iota, s_{1/\sqrt{2}})$, where ι is the identity (transformation) and s_u is scaling by factor u. Unfortunately, this cannot be carried out in $2\boldsymbol{U}_{12}$ algebra because it requires that both components of a transformation are equal. However, nothing is lost as we can do it in direct product algebra $\boldsymbol{U}_{12}^2$, which does not place restrictions on transformations. The newly constructed rule is summed with the initial rule to get the rule in Fig. 6d, which inscribes the scaled and rotated square into the square appearing on its left-hand side.

In the next section we will go a step further in constructing rules: we will generate them.

4 Meta Shape Rules and Grammars

Algebras $\boldsymbol{U}_{ij}$ support calculations with shapes and provide the framework for shape grammars. Similarly, $2\boldsymbol{U}_{ij}$ algebras support calculations with compound shapes and provide a framework for shape grammars which generate compound shapes. As it was shown before, compound shapes of $2\boldsymbol{U}_{ij}$ may be interpreted as shape rules so that grammars defined in $2U_{ij}$ generate shape rules. The rules of such a grammar act on shape rules to produce new shape rules in the same way shape rules act on shapes to produce new shapes. This renders the rules of such grammar *meta shape rules* and the grammar itself a *meta shape grammar*. The language that a shape grammar generates is a set of shapes in the same style while the language a meta shape grammar generates is a shape grammar. Meta shape rules are compound shapes of compound shapes, i.e., compound shapes which have compound shapes as components. They are defined in a sum of sums algebra $2(2(U_{ij}))$. An element of $2(2(U_{ij}))$ algebra is represented as $(a \rightarrow b) \rightarrow (c \rightarrow d)$, where $a, b, c, d \in U_{ij}$, when interpreted as a meta shape rule, and as compound shape of compound shapes $((a, b), (c, d))$ with no such interpretation. The same algebra is a meta shape rule algebra in the first case and a compound shape of compound shapes algebra in the second. Table 1 shows the hierarchy of shapes, and rules as well their interpretation, representations, and membership in different algebras and grammars.

Note that the hierarchy in Table 1 is actually three hierarchies in one. All three hierarchies have three levels with shapes belonging to a different algebra at each level: U_{ij}, $2U_{ij}$, and $2(2U_{ij})$ on levels 1, 2, and 3, respectively. The difference between the hierarchies is in how they see/interpret shapes. All three share four shapes from U_{ij} on the first level, and all three see them simply as shapes with no (other) interpretation involved. One of the hierarchies continues in the same fashion with compound shapes on level 2 and a compound shape of compound shapes on level 3. The other one does the same on level 2, but interprets the shape on level 3 as a compound shape rule. The third hierarchy interprets compound shapes at level 2 as shape rules, and the compound shape of compound shapes at level 3 as a meta shape rule.

We will use several examples to show how the new (meta) rules work.

For example, the meta rule in Fig. 7a transforms the lower rule from Fig. 6a into the rule in Fig. 6c. It matches the original rule, erases it, and adds the transformed rule in its place. The meta rule in Fig. 7b does the same transformation, but it does it parsimoniously. Because the original and the new rule have the same right-hand sides, it only replaces the left-hand side.

Table 1 Hierarchy of shapes, and rules, and their interpretation, representation, and membership in different algebras and grammars

Level	Hierarchy	Interpretation	Representation and membership	Algebra supports
3	meta shape rule	yes	(shape → shape) → (shape → shape) $\in$ meta shape grammar $\subset 2(2U_{ij})$	N/A
2	shape rule shape rule	yes	shape → shape $\in$ shape grmmar $\subset 2U_{ij}$	meta shape grammars
1	shape shape shape shape	no	shape $\in U_{ij}$	shape grammars
2	compound shape compound shape	no	(shape, shape) $\in 2U_{ij}$	compound shape grammars
3	compound shape rule	yes	(shape, shape) → (shape, shape) $\in$ compound shape grammar $\subset 2(2U_{ij})$	N/A
3	compound shape of compound shapes	no	((shape, shape), (shape, shape)) $\in 2(2U_{ij})$	N/A

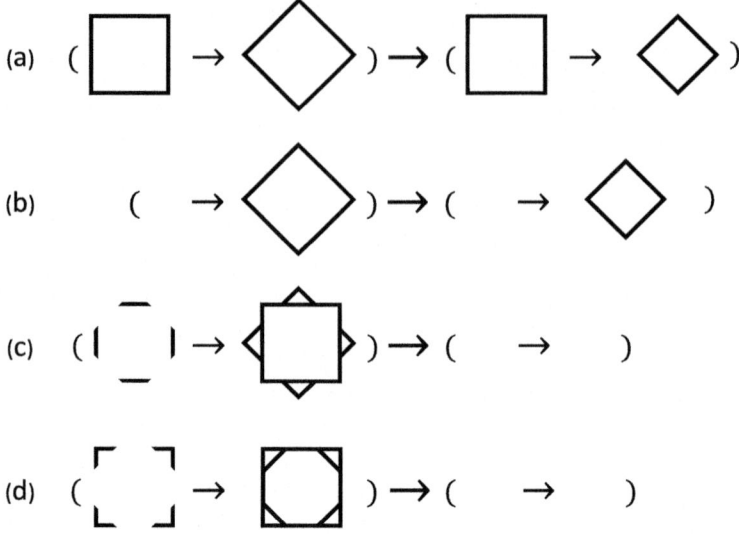

Fig. 7 Meta rules (**a**) and (**b**) convert rule from Fig. 6a into rule from Fig. 6c. Meta rules (**c**) and (**d**) transform rule in Fig. 6b by erasing certain parts

Both meta rules in Fig. 7c and d act on rule in Fig. 6b. They transform the rule by erasing parts of its both sides. The actions of these meta rules are shown in Fig. 8a. The original rule is in the middle and the new rules resulting from the actions of the meta rules in Fig. 7c and d are right and left of it, respectively. The shape rule on the right is an "octagon rule" capable of transforming a square into an inscribed octagon, while the one on the left is a "star rule" which turns a square into an 8-pointed star. Two derivations of shapes starting with a shape consisting of two overlapping squares and carried out with the octagon and star rules are shown in Fig. 8b.

Note that adding or taking the symmetric difference of an octagon and a star rule in the $2(2U_{ij})$ algebra results in a rule in Fig. 6a that rotates a square.

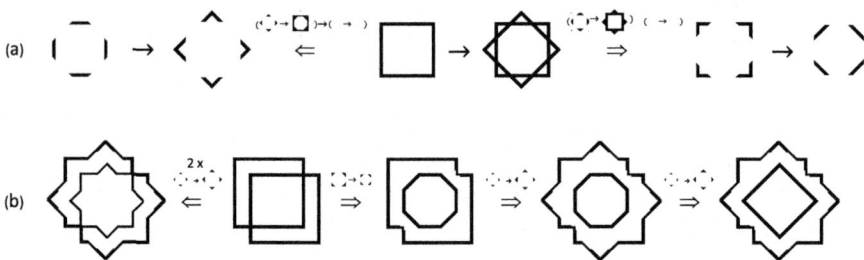

Fig. 8 The actions of rule in Fig. 7c and d on rule in Fig. 6b, producing "star rule" and "octagon rule," respectively (**a**). The latter rules acting on a shape consisting of two overlapping squares (**b**)

In the last example, meta shape grammar is used to transform a grammar that generates abstract floor plans, where walls are represented by single lines and positions of the doors/passages are marked by labels, into a more elaborate grammar for floor plans. One may start with the simple abstract grammar and generate several abstract representations of floor plans to check different layouts. If these are satisfactory the simple grammar may be converted into a more elaborate grammar generating floor plans. This can be done by using a meta grammar as a conversion tool. A meta rule that transforms the initial rule of the original abstract grammar into the initial rule of the floor plan grammar is shown in Fig. 9a. The former initial rule appears on the left-hand side while the latter one is on the right-hand side of the meta rule. An abstract floor plan grammar rule that adds a new space to an existing one is transformed into two different rules of the floor plan grammar via meta shape rules in Fig. 9b and c. To fully transform this rule i.e., cover all of the possible cases, one needs three more meta rules. Thus, the rule maps to five different rules of the floor plan grammar. The mapping between the abstract and more elaborate floor plan grammars is one to many. Note that the rules of the abstract grammar are defined in the $2U_{12}$ algebra while the rules of the floor plan grammar are defined in the $2(U_{12} \times U_{22})$ algebra so that the meta rules are in the $2U_{12} + 2(U_{12} \times U_{22})$ algebra.

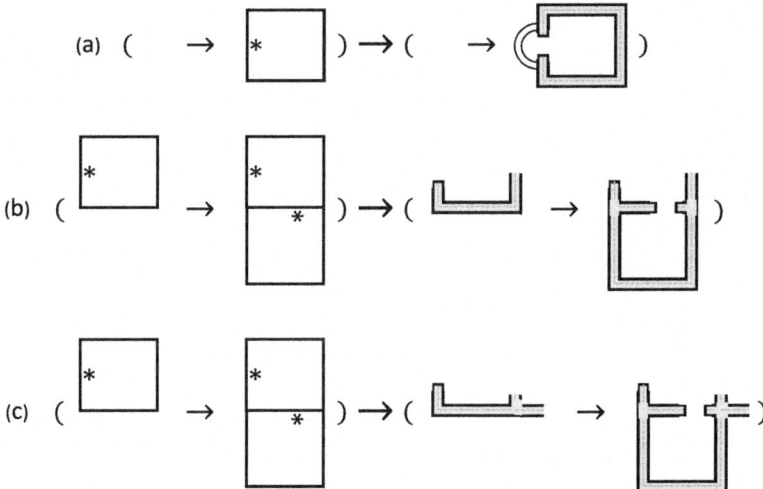

Fig. 9 Transforming a grammar that creates abstract floor plans into a more elaborate floor plan grammar. Transforming the initial rule (**a**) and a rule that adds a new space to an existing one (**b**) and (**c**)

5 Rule Schemas

Although shapes a and b of rule $a \to b$ could be arbitrary, it is more common that there is some kind of relation between them. There are certainly relations between them once the rule is applied and shapes c and c' as well as transformation t get in the picture. Because the rule acts on shape c under transformation t to produce shape c', in accordance with (1) and (2), we can with certainty say that $t(a) \le c, t(b) \le c'$, $c - t(a) \le c'$ and derive other relations—such as $t(a \cdot b)$ being part of both c and c'. There are also relations between shapes a and c' that are there on purpose, in order for the rule to be able to accomplish certain tasks. For example, $a \to t(a)$ will replace shape a with its transformed copy. This rule, depending on the transformation, could accomplish many tasks in moving and scaling a. Such relations are described in Stiny's rule schemas which resemble shape rules but have symbols in place of shapes, transformations, and other operators (Stiny 2011). Schemas distill infinite numbers of shape rules into simple expressions (with symbols) to examine/show how the rules work. A schema becomes a rule when symbols are initiated with shapes, transformations, and operators and the subsequent calculations are done.

For example, In Fig. 10 we have schema $x \to t(x)$ (a), its initiation, where x becomes a square and t a 30° rotation around the square's centroid ($t = rot_{30}$) (b), and the final rule after calculating $rot_{30}(x)$ (c).

Although U_{ij} algebras provide the framework for rule actions–i.e., calculating (1) and (2)–they do not have representation for rules, and by extension for Stiny's rule schemas. In the previous work we used shape valued functions as such representations (Krstic 2023). These are *rule generating* functions $f : U_{ij} \to U_{ij}$ defined in the framework of U_{ij} algebras. If there is schema $x \to y$ and function $f : U_{ij} \to U_{ij}$ given by $f(x) = y$, then the schema may be written as $x \to f(x)$ and function f is its representation in U_{ij}.

For example, schema $x \to t(x)$ is represented by function $f_t(x) = t(x)$ and the inverse of the schema $t(x) \to x$, which is equivalent to $x \to t^{-1}(x)$, by $f_t^{-1}(x) = t^{-1}(x)$. Although functions f_t and f_t^{-1} appear to be neat representations of the schemas, above, they are not without problems. Stiny remarks that schema $x \to t^{-1}(x)$ is superfluous because both t and t^{-1} are transformations so that schema $x \to t(x)$ is equivalent to its inverse (Stiny 2011). This does not apply to f_t and f_t^{-1} which are two different functions.

Another of Stiny's pair of schemas $x \to x + a$ and inverse $x + a \to x$ does not fare as well when represented by rule generating functions. The function representing the

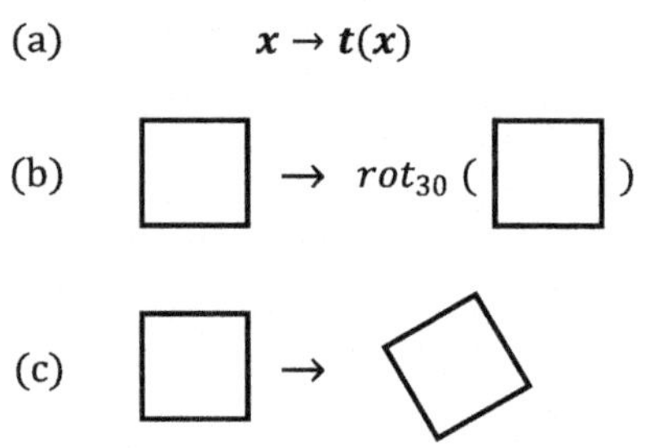

Fig. 10 Schema $x \to t(x)$ (**a**), its initiation with x a square and t a 30° rotation (**b**), and the final rule after calculating $rot_{30}(x)$ (**c**)

first schema is $f_a(x) = x+a$, but there is no function to represent the inverse schema. Because shapes fuse in sums all we can get is $f_a^{-1}(f_a(x)) = f_a(x) - a = (x+a) - a = x - a$, which is at best equal to x but usually smaller than it. Then there is a schema $x \to prt(x)$, meaning "x goes to one of its parts" (or $x \to y \leq x$) and its inverse $x \to prt^{-1}(x)$ meaning "x goes to a shape with x as a part" (or $x \to y \geq x$) (ibid.). Neither schema nor its inverse could be represented by a rule generating function as $prt(x)$ and $prt^{-1}(x)$ are not shapes but infinite sets of shapes.

It could be concluded that rule generating functions are not equivalent to rule schemas but could, in some cases, be their representations. We will check next how the elements of rule algebras $2U_{ij}$, from the previous sections, fare as rule schema representations.

Representation of a rule schema in $2U_{ij}$ follows the representation of a rule.

For example, rule schema $x \to t(x)$ is represented as an ordered pair $(x, t(x)) \in 2U_{ij}$ and the inverse schema as $(t(x), x) \in 2U_{ij}$, which can be transformed using t^{-1} to get an equivalent schema, or $t^{-1}((t(x), x)) = (x, t^{-1}(x))$. Note that we made use of the extension of equivalence for rules to rule schemas.[5]

Unlike with rule generating functions representation of schemas $x \to x + a$ and $x + a \to x$ is effortless in $2U_{ij}$: they are $(x, x+a)$ and $(x+a, x)$, respectively. The same is true for schemas $x \to prt(x)$ and $x \to prt^{-1}(x)$ which are represented by $(x, y \leq x)$ and $(x, y \geq x)$, respectively.

Important rule schemas are related to the boundaries of shapes. Stiny defines two such schemas $x \to b(x)$ and $x \to x + b(x)$, where b is a boundary operator which takes a shape to its boundary (ibid.). Shapes and their boundaries belong to different algebras, as boundaries are shapes one dimension lower than the shapes they delineate. Because $x \in U_{i+1j}$ and $b(x) \in U_{ij}$, shape $x + b(x)$–appearing on the right-hand side of the second schema above–is a compound shape belonging to the sum $U_{i+1j} + U_{ij}$ so that + in the schema is not Boolean sum but sum of algebras. Consequently, shape x–on the lef-thand side of the schema–has to belong to the same algebra so that the schema becomes $(x, 0) \to (x, b(x))$. Similarly, schema $x \to b(x)$ becomes $(x, 0) \to (0, b(x))$. The schemas themselves belong to sum of sums algebra $2(U_{i+1j}+U_{ij})$. Although representing the two schemas in the framework of $2(U_{i+1j}+U_{ij})$ algebra is reasonable, it is not without problems. Sum $U_{i+1j}+U_{ij}$ is a subalgebra of a direct product algebra so that its operations are done componentwise. This does not allow for operators, like the boundary one, that bridge the algebras. There is another sum of algebras for shapes, the commutative sum, that makes the definition of the boundary operator possible. Commutative sum $U_{i+1j}+_c U_{ij}$ has two-element sets such as $\{x, y\} = \{y, x\}$, where $x \in U_{i+1j}$ and $y \in U_{ij}$, instead of the ordered pairs of the standard (noncommutative) sum algebras. With elements instead of components, operations are done exhaustively instead of componentwise. This allows for bridging the algebras that are summed. The definition of commutative

---

[5] **Definition**: Two rules are equivalent if there is no shape that one rule can act on and another cannot (a) and the results of each rule acting on the same shape are the same (b). **Proposition**: A rule and the rule constructed by transforming both sides of the original rule are equivalent (Krstic 2014, 2023).

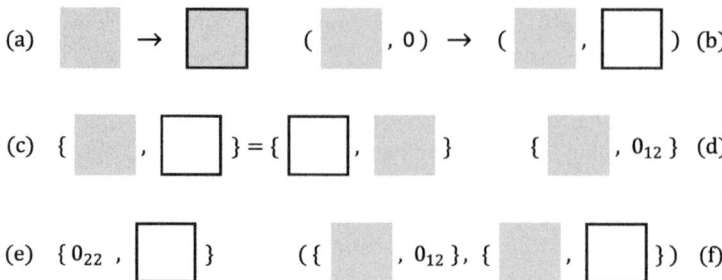

Fig. 11 A rule based on schema $x \to x + b(x)$ shown in its pictorial form (**a**) and as defined in $2(U_{22}+U_{12})$ algebra (**b**); shapes $x+b(x), x$ and $b(x)$, respectively, shown as defined in commutative sum algebra $U_{22}+_c U_{12}$ (**c-e**), while the schema itself is defined in $2(U_{22}+_c U_{12})$ (**f**)

sum algebras and their operations is given in (Krstic 2014), and details on how the boundary operator is defined in such an algebra are given in (Krstic 2023).

Finally, schemas $x \to b(x)$ and $x \to x + b(x)$ when defined in the framework of the sum of commutative sums algebra $2(U_{i+1j}+_c U_{ij})$ are $(\{x, 0\}, \{0, b(x)\})$ and $(\{x, 0\}, \{x, b(x)\})$ respectively.

For example, in Fig. 11, a rule based on schema $x \to x + b(x)$ is shown in its pictorial form (a) and as defined in (noncommutative) sum of sums algebra $2(U_{22} + U_{12})$ algebra (b). Although this algebra allows for the initialization of the schema, it does not support the boundary operator. In contrast, the commutative sum algebra allows for both. Shapes $x + b(x)$, x and $b(x)$ from the schema are initialized in commutative sum algebra $U_{22}+_c U_{12}$ and shown in Fig. 11c–e, respectively. The schema itself is initialized in the sum of commutative sums algebra $2(U_{22}+_c U_{12})$ and shown in Fig. 11f.

Note that elements 0_{22} and 0_{12} appearing in shapes in Fig. 11d–f, are empty shapes from U_{22} and U_{12}, respectively. Commutative sum algebras differentiate between empty shapes belonging to different algebras. This was omitted in the text above to avoid clutter in the formulas, but appears in the figure. Also note that the shape in Fig. 11c is depicted in two equivalent ways to highlight the commutativity of the algebra.

6 Analyzing Calculations with Shapes via Rule Decompositions

We have seen that shape decompositions can serve as shape approximations which is especially handy when analyzing calculations with shapes. Elements of such decompositions are shapes which are recognized as a consequence of calculations involving their sums. We used argument lattices and argument decompositions to analyze calculations with shapes including ones which are consequences of rule actions. The latter ones are analyzed in (Stiny 1994) and (Haridis and Stiny 2022) via decompositions structured as topologies and continuous functions, which approximate rule actions. In both approaches the analysis is done after a shape derivation has taken place. In the first case the argument lattice is calculated when all $t(a)$-s and $t(b)$-s are known while in the second case one starts with the final shape and works backwards creating decompositions for which the chosen functions representing rule actions are continuous.

We will take a new approach here. Shapes will be analyzed at the same time they are generated. The same tools (rule actions) that were used for shape generation will be used for the analysis. We will keep topologies but instead of continuous functions each action of a rule will be approximated with a series of actions of rules belonging to a decomposition of the original rule. This is a straightforward approach where in order to understand an action of a rule one decomposes the rule into a set of simpler rules and examines shapes that are generated by each of the new rules.

We have established that a shape rule is a compound shape belonging to a $2U_{ij}$ algebra so that rule decompositions are nothing more than shape decompositions. A rule could be decomposed into sets of rules that sum up to it and, like with shapes, we will consider only finite decompositions of rules.

For example, set $\{a \to 0, 0 \to b\}$ is a discrete decomposition of generic rule $a \to b$, because $a \to 0 \cdot 0 \to b = 0 \to 0$ and $a \to 0 + 0 \to b = a \to b$, respectively. It is the *natural decomposition* of $a \to b$ as it only highlights the minimum that the rule does i.e., removes a and includes b. The smallest decomposition is singleton $\{a \to b\}$, while the greatest one $d_{max}(a \to b)$, which includes all of the possible decompositions of rule $a \to b$ obtained in combinations of shapes a and b, is the set of all rules $x \to y$, where $x \in \mathbf{I}(a)$ and $y \in \mathbf{I}(b)$. Recall that $\mathbf{I}(a)$ and $\mathbf{I}(b)$ are argument decompositions of a and b depicted in Fig. 3f and g, respectively. Elements of $d_{max}(a \to b)$ are calculated in Table 2 while the decomposition itself is given in Fig. 12.

In our next example, we will analyze a shape derivation using natural decompositions of the rules involved.

Let c_k be a shape emerging in the k-th step of the derivation as a consequence of the action of rule $r_k = a_k \to b_k$ under transformation t_k, i.e., $c_k = (c_{k-1} - t_k(a_k)) + t_k(b_k)$–in accordance with (3). Further, let $d'_{top}(c_{k-1})$ be a topological decomposition of c_{k-1} which is a consequence of calculations up to and including the previous step in the derivation. The latter is the initial decomposition for the k-th step of the derivation. Finally, let $d_{nat}(r_k) = \{a_k \to 0, 0 \to b_k\}$ be the natural decomposition of rule r_k and let $cl_{\cdot,+}(A)$ be the closure of set of shapes A under finite products and

Table 2 Calculating elements of the greatest decomposition of rule $a \to b$

$\mathbf{I}(a) \quad \mathbf{I}(b)$	b	$b-a$	$a \cdot b$	0
a	$a \to b$	$a \to b-a$	$a \to a \cdot b$	$a \to 0$
$a-b$	$a-b \to b$	$a-b \to b-a$	$a-b \to a \cdot b$	$a-b \to 0$
$a \cdot b$	$a \cdot b \to b$	$a \cdot b \to b-a$	$a \cdot b \to a \cdot b$	$a \cdot b \to 0$
0	$0 \to b$	$0 \to b-a$	$0 \to a \cdot b$	$0 \to 0$

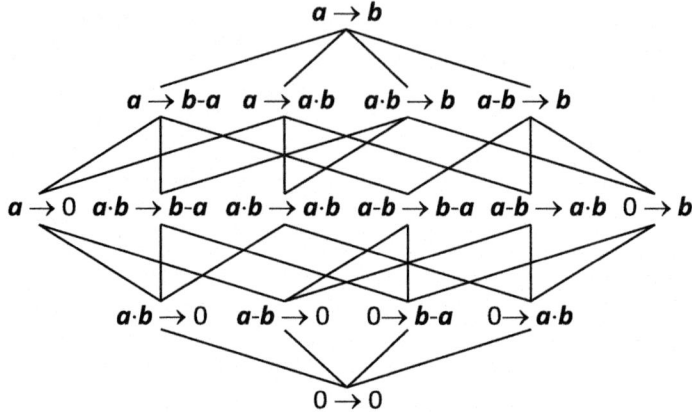

Fig. 12 The greatest decomposition of rule $a \to b$

sums. The following two-step procedure yields the final topological decomposition of shape c_{k-1} or $d_{top}(c_{k-1})$ and also the initial topological decomposition $d'_{top}(c_{k-1})$ for the $(k+1)$-th step of the derivation.

Procedure 1:

Step 1. In accordance with (3), we let rule $a_k \to 0 \in d_{nat}(r_k)$ act on decomposition $d'_{top}(c_{k-1})$ under transformation t_k. The result is decomposition $d(c_{k-1} - t_k(a_k)) = \{x - t_k(a_k) | x \in d'_{top}(c_{k-1})\}$. Because $c_{k-1} - t_k(a_k) \leq c_{k-1}$ all of the elements of $d(c_{k-1} - t_k(a_k))$ should be included in the final topological decomposition of c_{k-1} so that $d_{top}(c_{k-1}) = cl_{.,+}\left(d'_{top}(c_{k-1}) \cup d(c_{k-1} - t_k(a_k))\right)$.

Step 2. Again following (3), we let rule $0 \to b_k \in d_{nat}(r_k)$ act on $d(c_{k-1} - t_k(a_k))$ under t_k to produce $d(c_k) = \{x + t_k(b_k) | x \in d(c_{k-1} - t_k(a_k))\}$. Because $c_{k-1} - t_k(a_k) \leq c_k$ all of the elements of $d(c_{k-1} - t_k(a_k))$ should be included in $d'_{top}(c_k)$ so that it becomes $d'_{top}(c_k) = cl_{.,+}(d(c_k) \cup d(c_{k-1} - t_k(a_k)))$. If c_k is the final shape in the derivation i.e., no other rules are applied, then $d'_{top}(c_k)$ is the final topological decomposition of c_k, or $d_{top}(c_k) = d'_{top}(c_k)$. Otherwise, we go to the step 1 of the procedure and repeat it for the $(k+1)$-th step of the derivation.

Note that rule r acting on shape u to generate shape v generates a decomposition $d(v)$ of v when it acts on decomposition $d(u)$ of u. Each element of $d(v)$ is the result of r acting on some element of $d(u)$.

For example, in Fig. 13 we have a shape grammar with 3 rules: the initial rule which introduces a shape resembling two B letters facing each other, a rule that translates letter B, and a rule that translates letter C (a) and a 3-step shape derivation using the latter rules to turn shape BB into shape DK (b). The example is adopted from (Krstic 2017). We will analyze derivation (b) via natural rule decompositions and the above procedure.

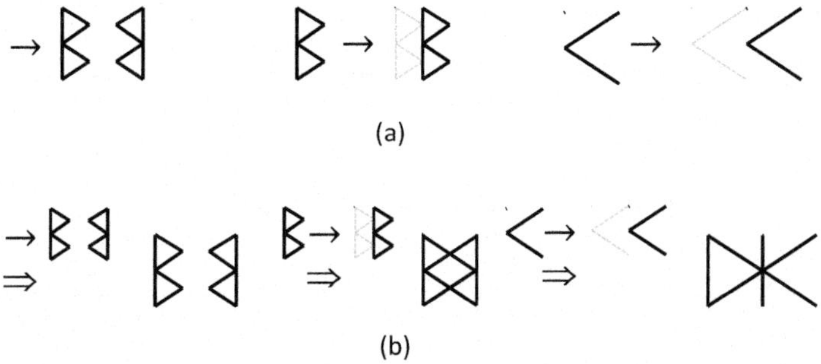

Shape Rules: Their Algebras, Grammars and Decompositions

Fig. 13 A shape grammar with three rules (**a**) and a 3-step derivation using the rules (**b**)

We start with the initial rule and its natural decomposition:

$$d_{nat}(r_1) = d_{nat}(\to \text{B}\text{$\lhd$}) = \{0 \to 0, \to \text{B}\text{$\lhd$}\}$$

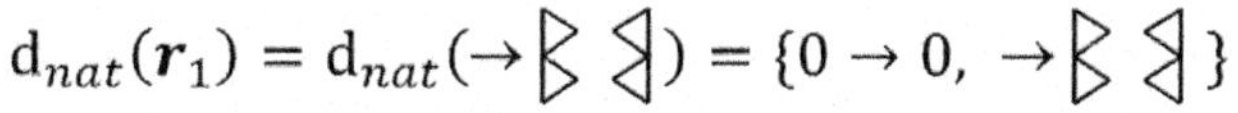

Calculations involving actions of rules that are elements of $d_{nat}(r_1)$ are done in accordance with Procedure 1 and given in Fig. 14. It starts with the empty shape and yields $d'_{top}(c_1)$.

The next step in the derivation involves the rule that translates the B shape. Its natural decomposition is:

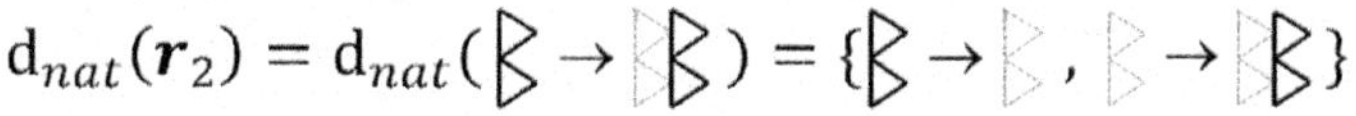

$$d_{nat}(r_2) = d_{nat}(\text{B} \to \text{B}) = \{\text{B} \to \text{ }, \text{ } \to \text{B}\}$$

The calculation, which starts with $d'_{top}(c_1)$–from the previous step–and uses $d_{nat}(r_2)$ to follow Procedure 1 resulting in $d'_{top}(c_2)$ and $d_{top}(c_1)$ is shown in Fig. 15. In the final step, rule translating C shape produces the final DK shape. The calculation follows Procedure 1 by utilizing natural decomposition:

$$d_{nat}(r_3) = d_{nat}(\text{<} \to \text{<<}) = \{\text{<} \to \text{ }, \text{ } \to \text{<<}\}$$

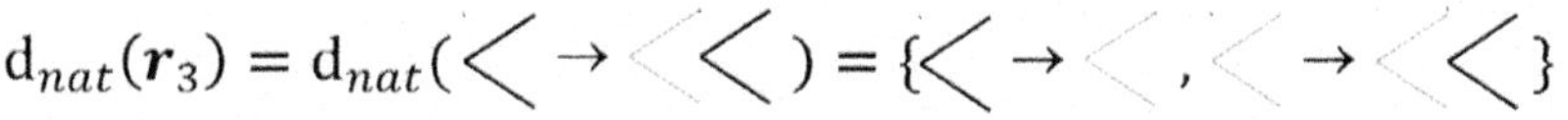

Fig. 14 Calculation involving actions of the rules which are elements of $d_{nat}(r_1)$ yielding $d'_{top}(c_1)$

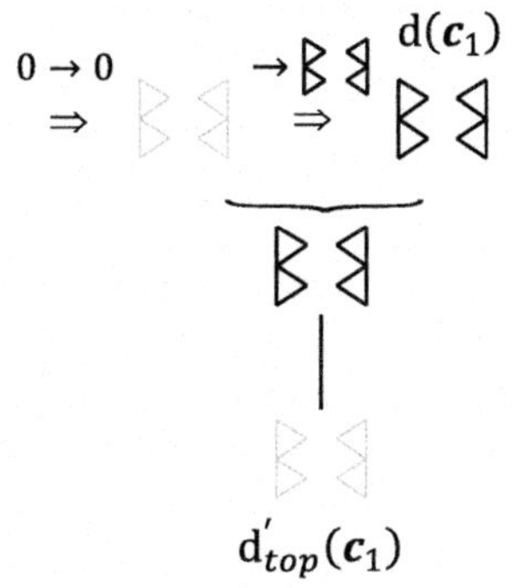

Fig. 15 Calculation involving actions of the rules that are elements of $d_{nat}(r_2)$ yielding $d'_{top}(c_2)$ and $d_{top}(c_1)$

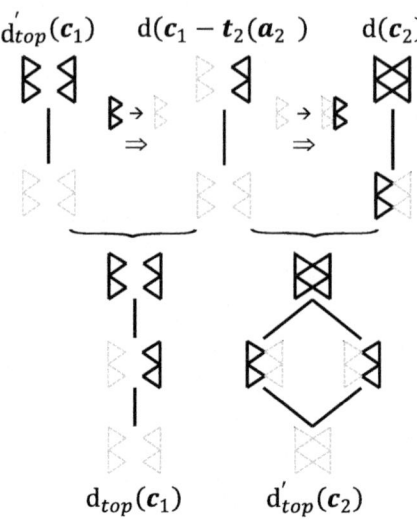

It starts with $d'_{top}(c_2)$ and results in decompositions $d_{top}(c_2)$ and $d'_{top}(c_3)$ as shown in Fig. 16. The latter becomes the final decomposition $d_{top}(c_3) = d'_{top}(c_3)$ because no new rules are applied.

The final sequence of shape decompositions describing the derivation when the rules are represented by their natural decompositions is depicted in Fig. 17.

Note that $d_{top}(c_3)$ is a topology but also has the structure of a Boolean algebra. The other two topologies $d_{top}(c_1)$ and $d_{top}(c_2)$ lack such a structure. Moreover, $d_{top}(c_3)$ is the argument decomposition of shape c_3 with respect to this derivation, or $d_{top}(c_3) = \mathbf{I}(c_3)$. It is the greatest decomposition of c_3 that can be produced with shapes (5), and relations (4) and (6). Also note that topologies above do not recognize shapes $t_k(a_k)_{k=1,2,3}$ although a_k is a half of each rule $a_k \to b_k$. Taking some bigger decompositions of rule $a_k \to b_k$ like the set of atoms $d_{at}(a_k \to b_k) = \{a_k \cdot b_k \to 0, a_k - b_k \to 0, 0 \to b_k - a_k, 0 \to a_k \cdot b_k\}$ of $d_{max}(a_k \to b_k)$ would add some new shapes to the resulting topologies, but shapes $t_k(a_k)$ would still be missing. The reason for this is that in Procedure 1 formulas (3) define rule actions and according to (3) $t_k(a_k)$ is removed from c_{k-1}. However, formulas (3) are not alone in defining rule actions, there are also inequalities (4) which recognize $t_k(a_k)$ as a part of c_{k-1}. Inequalities (4) could be turned into certain rules i.e., *useless rules* (Stiny 1996). These are of the form $x \to x$ and, in accordance with the name, do not change the shapes they act on, but recognize x as a part of every such shape instead. We can add a useless rule $a_k \to a_k$ before every rule $a_k \to b_k$ in the sequence without changing the outcome of the derivation. Only the topologies $d_{top}(c_k)$ will change. They will now include shapes $t_k(a_k)$ and their parts. The derivation will have twice as many steps as the original one. After doing all of the calculations using Procedure 1 we will only take into account every other decomposition. To avoid enlarging the derivation we can take the union of the natural decompositions of the useless and the original rule,

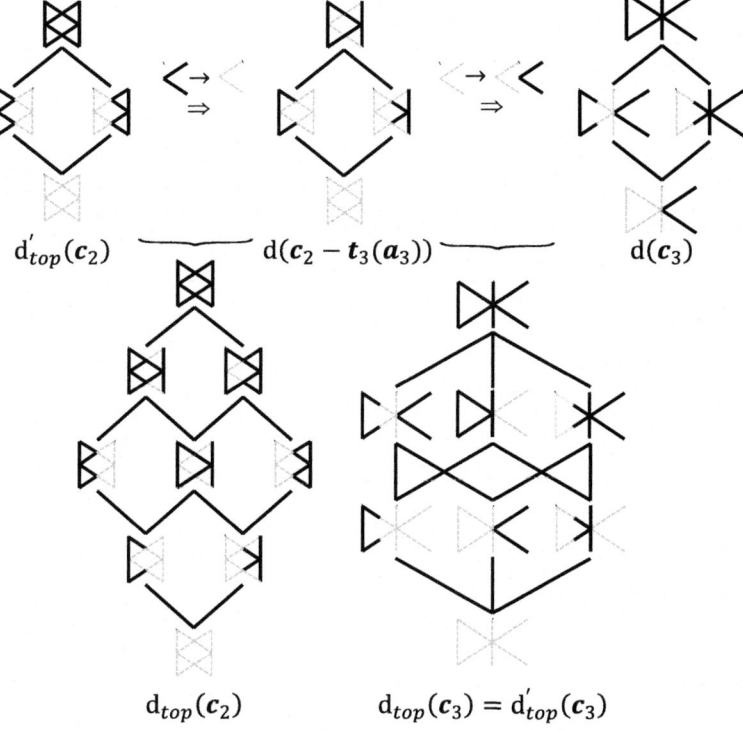

Fig. 16 Calculation involving actions of the rules that are elements of $d_{nat}(r_3)$ resulting in $d_{top}(c_2)$ and the final decomposition $d_{top}(c_3) = d'_{top}(c_3)$

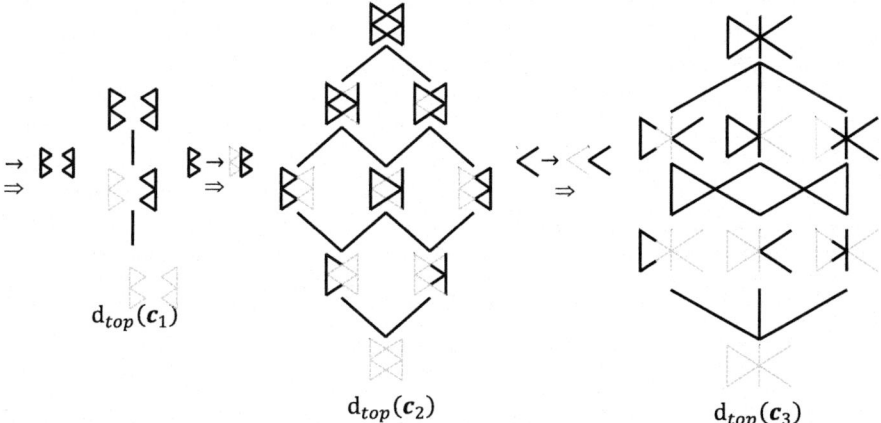

Fig. 17 The final sequence of decompositions describing the derivation when the rules are represented by their natural decompositions

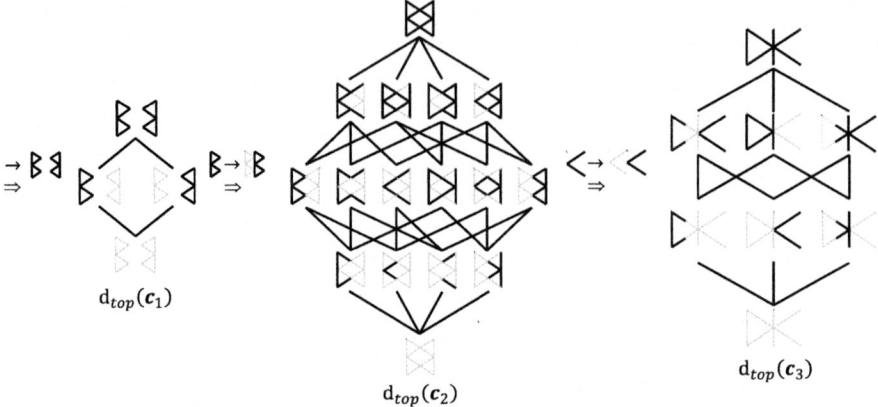

Fig. 18 The final sequence of decompositions describing the derivation when the rules are represented by their natural decompositions and the appropriate useless rules are inserted

or $\{a_k \to 0, 0 \to a_k\} \cup \{a_k \to 0, 0 \to b_k\} = \{a_k \to 0, 0 \to a_k, 0 \to b_k\}$ in effect fusing the two rule actions into one. The result is a decomposition of $a_k \to a_k + b_k$ so that the procedure needs to be modified to ignore parts of $c_k + t_k(a_k)$ that do not belong to c_{k-1} or c_k.

The useless rules to be added to the sequence of rules in our example are

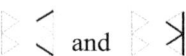

that go ahead of rules r_2 and r_3, respectively, as well as (trivially) rule $0 \to 0$ to start with. The resulting sequence of decompositions describing the derivation is given in Fig. 18.

Note that shape $t_2(a_2)$ is now an element of $\mathrm{d}_{top}(c_1)$ and also that shapes $t_2(a_2)$ and $t_3(a_3)$ as well as some of their parts are now elements of $\mathrm{d}_{top}(c_2)$. Decomposition $\mathrm{d}_{top}(c_3)$ is the same as in the previous example because there is no $t_4(a_4)$. All three topologies in this example are Boolean algebras. Topology $\mathrm{d}_{top}(c_3)$ is an argument decomposition, as before, but is now accompanied by $\mathrm{d}_{top}(c_2)$, or $\mathrm{d}_{top}(c_2) = \mathbf{I}(c_2)$ and $\mathrm{d}_{top}(c_3) = \mathbf{I}(c_3)$. Topology $\mathrm{d}_{top}(c_1)$, although a Boolean algebra, is not an argument decomposition. The argument decomposition of c_1, or $\mathbf{I}(c_1)$, is given in Fig. 19.

Note that $\mathbf{I}(c_1)$ like $\mathrm{d}_{top}(c_1)$, has $t_2(a_2)$, but it also has shapes.

which come as the result of the action of rule r_3. The first one is a part of $t_3(a_3)$ while the second one is $(c_1 - t_2(a_2)) - t_3(a_3)$ which is a part of $c_2 - t_3(a_3)$.

This shows the difference between the method that uses rule decompositions to analyze shape grammar derivations and the one that uses argument lattices and argument decompositions (Krstic 2017).

Fig. 19 The argument decompositions of shape c_1, or $\mathbf{I}(c_1)$

The first method creates decompositions in parallel with the derivation. It provides a local view at the derivation, where a shape decomposition is influenced by the actions of two rules: the one that makes the shape and the one that breaks it to make a new shape.

The second method is global in the sense that all of the rule actions influence every argument decomposition of the shapes involved in the derivation. The argument lattice, that contains all the argument decompositions, is created after the derivation is done. All the rules and transformations are known at that time and are used to generate the lattice, via set (5) and relations (4) and (6).

Note that the method that utilizes topologies and continuous functions (Stiny 1994; Haridis and Stiny 2022), that we mentioned earlier, is in the first category. It requires that the derivation be completed in order to calculate the topologies, by starting with the final shape and going back to the initial one. However, calculations are done by using local information. The topology of each shape depends on the topology of the next shape in the derivation and the given function connecting the two shapes, i.e., depends on the rule that breaks the shape to make a new one.

7 Conclusion

For the last 50 years, shapes have been generated by shape grammars using operations from algebras of shapes, which took their modern form as U_{ij} algebras more than 30 years ago.[6] In contrast, rules, despite being the building blocks of shape grammars, were left alone for most of that time. Only relatively recently has Stiny started manipulating rule schemas, which are abstract representations of rules (Stiny 2011). He was adding, subtracting and transforming rule schemas to create new ones in the way shapes have been manipulated in the framework of U_{ij} algebras.

---

[6] It has been more than 50 years since the publishing of (Stiny and Gips 1972) and more than 30 years since the publishing of (Stiny 1991).

In this paper, we attempted to provide the framework for such manipulations, which entails a formal definition of shape rules as well as that of shape rule algebras. Much to our surprise, it proved to be an easy task, requiring only a simple reinterpretation of what was already there. Compound shapes belonging to the sum of two U_{ij} algebras were interpreted as shape rules, and their algebras took the role of rule algebras. It appears that all rules are shapes, but that not all shapes are rules, or *rules* ⊂ *shapes*. Going from seeing the compound shapes as rules, i.e., the "rule interpretation," to seeing them as shapes, i.e., the "shape interpretation" and back allowed for effortless development of other interesting constructs.

For example, when rules are seen as compound shapes, rule algebras become shape algebras, which serve as the framework for shape grammars. The latter generate compound shapes, but when we go back to the "rule interpretation," these grammars appear to generate shape rules. Their building blocks are meta shape rules that replace rules with rules and the grammars are meta shape grammars which generate shape grammars. The examples given in this paper seem to only scratch the surface of possibilities such grammars afford. They can be valuable tools for analyzing transformations in design, featured in Terry Knight's early work (1994), or generating grammars defining different phases of design, i.e., transforming the sketching grammar into the conceptual design grammar and then transforming the latter into the final design grammar.

Rule algebras worked well as the framework for calculations with rule schemas. They did much better than the shape valued functions we used as representations for schemas in our previous work. Although we did not try it here, it is clear that rule algebras allow for schemas to be added, subtracted, multiplied, and transformed in order to create new schemas from the old ones. It is also possible to reason about the relation between the two schemas, that is, whether one schema is a part of another or if one is the inverse of another. Whenever the outcome of the latter case is positive, the sum of the two schemas—or the sum of the schemas equivalent to the two–is a useless (rule) schema.

For example, "octagon rule" and "star rule" from Fig. 8a are related so that one is the inverse of the other. This becomes obvious when examining a rule equivalent to the "star rule" which is obtained by rotating the latter by 45°, Fig. 20a. When this rule is summed with the "octagon rule" we get the useless rule in Fig. 20b and c. If we go from rules to schemas, as in (Economou, Kotsopoulos 2015), "octagon rule" yields $prt(x) \to t(x - prt(x))$ schema while "star rule" yields $x - prt(x) \to t^{-1}(prt(x))$. The latter schema is, via t, transformed into an equivalent schema $t(x - prt(x)) \to prt(x)$ which when summed with the "octagon rule" schema yields useless schema $prt(x) + t(x - prt(x))$ ↻.[7]

Finally, meta shape grammars could be seen as rule schema grammars that generate new schemas, which would be interesting to explore further.

In the last section of the paper, we used the "shape interpretation" of rules to get to compound shapes and compound shape decompositions. The latter become rule decompositions when we get back to the "rule interpretation." Rule decompositions

---

[7] We take (x) as fixed throughout these calculations.

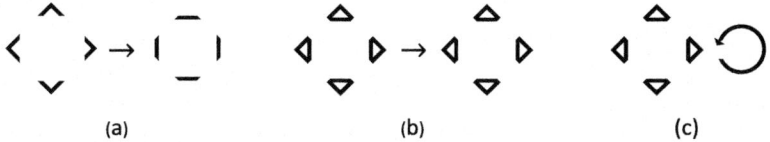

Fig. 20 A rule equivalent to the "star rule" (**a**), and its sum with the "octagon rule" (**b**), or shown in a compact form (**c**)

are seen as alternative tools for analyzing derivations of shapes with shape grammars. We used simple natural decompositions of rules and represented shapes in the derivation by their topological decompositions. Using other types of rule decompositions and differently structured shape decompositions is possible and worth exploring.

Traditionally, a shape grammar is a set of rules which requires an initial shape to start a shape derivation. Here, and in a previous paper (Krstic 2014), we opted for an initial rule, $0 \rightarrow b_1$, in place of the initial shape. The advantage of this approach is twofold. First, because 0 is part of any shape, we can start the derivation with any shape including 0. For example, a grammar defining a certain building style can start a derivation with a shape representing the building site. Second, because any transformation of 0 is 0, we can freely pick the first transformation t_1 to put shape b_1 into the desired place, and make it of the desired size and orientation. This is not possible with the initial shape, which occupies a fixed place in space and is of fixed orientation and size. Finally, an initial rule under "shape interpretation" becomes an initial shape, which blurs the difference between the two approaches.

References

Birkhoff, G. 1935. On the structure of abstract algebras. *Mathematical Proceedings of the Cambridge Philosophical Society* 31: 433–454.
Birkhoff, G. 1993. *Lattice theory*, 3rd ed. Providence: American Mathematical Society.
Economou, A., and S. Kotsopoulos. 2015. From shape rules to rule schemata and back. In *Design computing and cognition DCC'14*, ed. J.S. Gero and S. Hanna, 383–399. Springer Science + Business Media B.V.
Haridis, A., and G. Stiny. 2022. Analysis of shape grammars: Continuity of rules. *Environment and Planning B: Urban Analytics and City Science* 49 (7).
Jowers, I., C.F. Earl, and G. Stiny. 2019. Shapes, structures and shape grammar implementation. *Computer-Aided Design* 111: 80–92.
Knight, T.W. 1994. *Transformations in design: A formal approach to stylistic change and innovation in visual arts*. Cambridge: Cambridge University Press.
Krstic, D. 1996. *Decompositions of shapes*. Ph.D. dissertation. Los Angeles: University of California, Los Angeles.
Krstic, D. 1999. Constructing algebras of design. *Environment and Planning B: Planning and Design* 26: 45–57.
Krstic, D. 2005. Shape decompositions and their algebras. *Artificial Intelligence for Engineering Design, Analysis and Manufacturing* 19: 261–276.

Krstic, D. 2014. Algebras of shapes revisited. *Design computing and cognition DCC'12*, ed. J.S. Gero, 361–376. Springer Science + Business Media B.V.
Krstic, D. 2015. Language of the Rascian school: Analyzing Rascian church plans via parallel shape grammar. In *Design computing and cognition DCC'14*, ed. J.S. Gero and S. Hanna, 421–436. Springer Science + Business Media B.V.
Krstic, D. 2017. From shape computations to shape decompositions. *Design computing and cognition DCC'16*, ed. J. S. Gero, 249–266. Springer Science + Business Media B.V.
Krstic, D. 2023. Notes on shape valued functions. In *Design computing and cognition DCC'22*, ed. J.S. Gero, 569–586. Springer Science + Business Media B.V.
Stiny, G. 1991. The algebras of design. *Research in Engineering Design* 2: 171–181.
Stiny, G. 1994. Shape rules: Closure, continuity and emergence. *Environment and Planning B: Planning and Design* 21: s49–s78.
Stiny, G. 1996. Useless rules. *Environment and Planning B: Planning and Design* 23: 235–237.
Stiny, G. 2006. *Shape: Talking about seeing and doing*. Cambridge: The MIT Press.
Stiny, G. 2011. What rules should I use? *Nexus Network Journal* 13: 15–47.
Stiny, G. 2022. *Shapes of imagination: Calculating in Coleridge's magical realm*. Cambridge: The MIT Press.
Stiny, G., and J. Gips. 1972. Shape grammars and the generative specification of painting and sculpture. *Information processing '71*, ed. C.V. Frieman, 1460–1465. Amsterdam: North-Holland.
Vickers, S. 1989. *Topology via logic*. London, New York, New Rochelle, Melbourne, Sydney: Cambridge University Press.

Open Access This chapter is licensed under the terms of the Creative Commons Attribution-NonCommercial-NoDerivatives 4.0 International License (http://creativecommons.org/licenses/by-nc-nd/4.0/), which permits any noncommercial use, sharing, distribution and reproduction in any medium or format, as long as you give appropriate credit to the original author(s) and the source, provide a link to the Creative Commons license and indicate if you modified the licensed material. You do not have permission under this license to share adapted material derived from this chapter or parts of it.

The images or other third party material in this chapter are included in the chapter's Creative Commons license, unless indicated otherwise in a credit line to the material. If material is not included in the chapter's Creative Commons license and your intended use is not permitted by statutory regulation or exceeds the permitted use, you will need to obtain permission directly from the copyright holder.

Seeing and Drawing Shapes

Sotirios Kotsopoulos

Abstract Observing a pattern allows us to appreciate its visual characteristics and how it was created. A likely construction process is enacted in the perceiver's mind, suggesting how to explain and implement the pattern. We see the pattern and understand its construction, but we explain it in one way and produce it in another. Analyzing the form of the pattern and how it was created corresponds to two different ways of seeing. This research contributes to formally representing the production of two-dimensional patterns in both a visual and physical sense. Central to this study is shape grammar theory, which uses shape rules and algebras to formalize the manipulation of shapes. The production of a pattern can be explained *figuratively*, where we identify its parts and transformations to produce the pattern conceptually. It can also be performed *physically*, using a set of parts and rules to generate the pattern based on a given apparatus. The differences between *figurative* and *physical* construction inform how we perceive and implement designs. Three examples of ornaments, including straight lines, arcs, and combinations of lines and arcs, demonstrate how designs are explained and drawn using ink. Decompositions with structures such as Boolean algebras, topologies, and hierarchies capture the parts and the order of execution. Shape rules provide instructions on how and where to draw. A notable feature of drafting hierarchies is that the sets of their components are diagonal shape decompositions. Lastly, it is shown how a *design machine* models the figurative and physical construction of designs.

S. Kotsopoulos (✉)
National Technical University of Athens, Patission 42, 10682 Athens, GR, Greece
e-mail: skotsopoulos@arch.ntua.gr

© The Author(s) 2025
S. Kotsopoulos (ed.), *Shape Computation*, Mathematics and the Built Environment 9, https://doi.org/10.1007/978-3-031-81623-9_12

1 Introduction

Before you know it, the eye furnishes multiple interpretations for shapes, tracks geometric features and proportions, and supplies explanatory constructions. A good representation is one that makes specific attributes explicit. For example, according to Birkhoff and Eysenck, the aesthetic value of a shape is evaluated based on the visual association of order (O) and complexity (C). Birkhoff (1933) introduced a representation for polygons and simple rectilinear ornaments, considering factors such as the complexity of contour (C), symmetry along the vertical axis (V), optical equilibrium (E), rotational symmetry (R), relation to a horizontal-vertical network of shapes (HV), similarity (S), and unsatisfactory form (F). He then calculated the aesthetic value as $\beta = f(O/C)$, where $O = V + E + R + HV + S–F$. Eysenck (1949) later modified Birkhoff's formula to $\beta = f(O*C)$. On a different note, Leyton (1992) suggests a psychological relationship between shape and time. According to him, our minds assign shapes a causal history to explain their formation. Shape transformations play a crucial role in remembering the past, with symmetry influencing how shape is transformed into memory. Leeuwenberg and van der Helm (2013) distinguish between perception and cognition, arguing that perception follows the simplicity principle, favoring the minimum number of steps to figuratively construct a shape. The number of required transformations dictates visual interpretation, with shape complexity measured by the minimum number of steps.

Shape grammar theory, as proposed by Stiny, views shapes as unanalyzed entities. Stiny (1996) suggested that shape is simply a visual appearance without inherent aesthetic value, structure, or complexity, which are all products of description. In essence, the eye keeps wondering (θαυμάζειν) captured by the thousand shapes and labels, knowing that the world is a whole of amazement, wonder (θαύμα), and surprise. Shape interpretation leads to shape decomposition and description, producing a set of shape parts that highlight specific properties of the whole and ignoring others. These identified parts account for the entire shape, which is their sum (Krstic 2010).

Two common reasons for interpreting a shape are to explain it and to construct it. When we explain a shape, we take a series of figurative steps to understand how it was created. In physical construction, apparatus and form suggest a productive technique comprising breaking the shape into parts, ordering the parts, and executing a series of productive steps to build it. Technique coincides with the Aristotelian techne (τέχνη), encompassing both interpretive (see) and performative (do) characteristics. According to Aristotle, mastering a technique involves the crucial element of repetition, practicing specific actions repeatedly. Similarly, Wittgenstein (1939) suggests that creating a technique requires establishing a pattern of behavior and teaching others how to replicate it. For example, Wittgenstein cites arithmetic as a technique children learn using an abacus. He argues that what we learned when we were taught arithmetic was simply a technique, a way of writing things down and that we continue to use this technique. There is no right or wrong result before the calculation is invented and the technique is established.

Designers spontaneously observe how something is presented and made but may interpret it one way and execute it in another. Judging form (what is presented) and implementation (how it is made) are distinct practices of seeing and assigning structure. Visual recognition, figurative, and physical construction are different ways of perceiving. Recognition involves recalling familiar visual features and impressions, figurative construction involves generative explanation, and physical construction involves actual making. All three suggest plausible visual structures, but perceiving them feels effortless and unlike choosing among structures.

A diverse range of figural and abstract features may describe a shape (Leeuwenberg and van der Helm, 2013). Figural features like parallel, vertical, or intersecting lines, orthogonal, acute, or obtuse angles, and abstract features like symmetry, proportion, and congruity may uniquely characterize a shape but lack generative power. *Figurative construction* is a generative, explanatory process that produces a shape by copying, translating, or rotating certain parts. Physical construction requires expertise in handling tools and materials and an appropriate decomposition acknowledging the equipment. Execution rules are then used to streamline the physical assembly.

This study focuses on the formal representation of the figurative and physical construction of two-dimensional patterns. Central is shape grammar theory, which formalizes calculations with shapes. Three examples of ornamental designs featuring straight lines, arcs, and composite forms, including lines and arcs, show how visual structure relates to constructing a pattern, figuratively and physically. Shape rules, transformations, and decompositions—having the structure of Boolean algebras, topologies, and hierarchies—capture how the designs are interpreted, generated, and drawn. Stiny (2022) explicates the temporary nature of structure in the embed-fuse cycle, which also captures the interaction between seeing and doing. In his words: "Visual calculating in shape grammars unfolds in a metabolic process; it isn't mechanical repetition with fixed components and standard parts, or recursion with atoms and words…This lets art and design all the way in, so that seeing (drawing, painting, etc.) overtakes thinking."

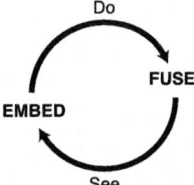

Knight and Stiny (2015) proposed a modification of shape grammars that has implications for the creation of physical objects in craft. Their approach redefines shape rules and algebras to reflect the craftsperson's focus and use of materials. This study underscores the necessity of having a specific technique in order to physically produce design instances within a design language. This technique integrates a set of assumptions about the design language and the production apparatus, effectively guiding the maker's performance. The examples of designs from three ornamental languages serve as tangible proof of the technique's effectiveness, demonstrating figurative and physical constructions using ink on paper. Drafting is examined as a physical creation process with three phases. First, the designs are classified based on their figural and abstract features. Second, select subshapes and transformations are used to create figurative constructions. Third, a technique is specified to capture the drafting steps, determining a kit of parts (decomposition), an execution order, and a sequence of shape rules.

In ink drafting, specific hierarchies of shape parts are used. The sets of their atoms are diagonal shape decompositions (Kristic 2022). When a maximal line or a sequence of collinear lines is drawn to a shape, it becomes a component of the diagonal decomposition of the shape. Collinear lines are drawn in the minimum space where a line can be defined, a U_{11} algebra. This effectively captures the focus of the drafter, who concentrates on one line or line sequence at a time. Therefore, the decompositions used in drafting divide shapes into their diagonal elements. Finally, it is demonstrated that a particular type of *design machine* (Stiny and March 1981) encompasses the figurative and physical construction of designs.

2 Background

Shape grammars consider shapes as finite arrangements of basic elements, such as points, lines, planes, or solids. Shape algebras with operations, addition (+), subtraction (−), multiplication (•), and an empty shape as the zero element enable the definition of addition, subtraction, and meet with shapes. In an algebra U_{ij}, every shape is a finite set of basic elements of dimension i (where $i = 0, 1, 2, 3$) manipulated in dimension j (where $j = 0, 1, 2, 3$), and $i \leq j$. New shapes are produced with shape rules. A shape rule A → B, involving two shapes A and B, is applied to shape C in two steps: First, a transformation of shape A, found on the left side of the rule, is matched with shape C—the embed phase of the embed-fuse cycle. Second, the same transformation of shape B, found on the right side of the rule, is added in place of shape A—the fuse phase of the embed-fuse cycle—as shape B fuses in C to produce a new shape C'. In short, $C \Rightarrow [C - t(A)] + t(B)$. The rule is recursively applied to C to produce the shapes C, C', C'', C''', and so on, with each new shape resulting from applying the shape rule to the previous shape.

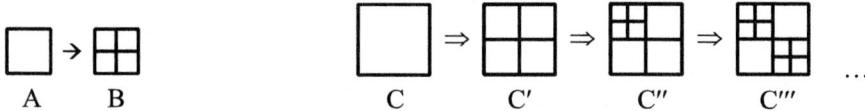

Rule schemata represent rules using an expression containing syntactical variables, casting large rule families. In a rule x → y, associating x and y to define rule instances linking the input and output shapes on the rule's left and right sides is useful. For example, if y = x, the rule becomes x → x, which is an identity. Identities capture the act of observation, distinguishing parts in a shape as one sees them without modifying them (Stiny 1976, 1996). If y is determined to be a similarity transformation t of x, the rule is rewritten x → t(x). If y is determined to be part of x, then the rule is rewritten x → prt(x). Finally, if y is determined to be the boundary of x, then the rule is rewritten as x → b(x).

For basic elements of dimension i > 0, the sets defining shapes are organized by an equivalence relation. The relation operates as follows: A set S is included in the set T when every basic element embedded in a basic element in S has a basic element embedded in it that's also embedded in a basic element in T. The sets S and T are equivalent if and only if each is included in the other. The smallest set in the equivalence class determined by S contains only *maximal* elements. For i > 0, basic elements of the same dimension are maximal with respect to one another when they are separated (a), or touch and are not embedded in a common element (b). For i = 0, points are atomic and always maximal. The same applies to dashed, composite, and dotted lines; their constituents are atomic and do not interact (c).

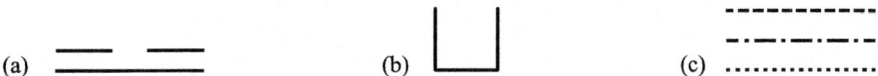

We draw shapes with a finite number of strokes within a finite space. However, a shape can be divided into an infinite number of parts. Every maximal element of a shape has an indefinite number of parts of the same dimension embedded within it. Lines could be parts of lines, planes parts of planes, and solids parts of solids. A set of lines is changed into a set of maximal lines if: (a) one line is embedded in the other; (b) two lines overlap, but neither is embedded in the other or (c) two lines are collinear and discrete and share an endpoint; they are replaced by the line that starts at the leftmost endpoint and ends at the rightmost endpoint of the two lines, that is the longest line containing both.

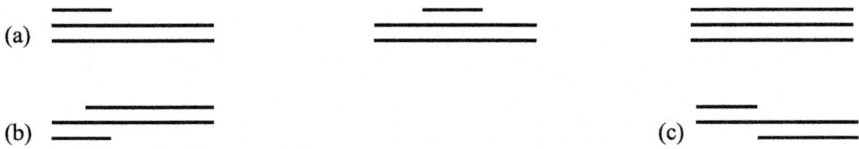

A shape M is part of a shape S (M ≤ S) if every maximal element of M is embedded in the maximal elements of S. For an element of M to be part of an element of S, they both must be of the same dimension and defined in the same space. The relation ≤ is a partial order on shapes. It is reflexive, antisymmetric, and transitive, and with the operations of sum (+), difference (−), product (•), and the empty shape, shape decompositions can be defined. Decompositions are defined as sets of shapes that show how their sums are divided into parts of specific kinds (Stiny 1991). Three examples of decompositions of a square as lattices, including maximal lines, are presented in Fig. 1. Decompositions can perpetually be reconfigured by embedding parts on the shape that appears at the lattice top. The lattices (a) and (b) have structures of Boolean algebras. They include the whole shape, the empty shape, the maximal lines, and their combinations. Meets and joins appear on all leaves of a lattice. Meets producing new shapes occupy the upper leaves, and joins producing new shapes occupy the lower leaves.

In (a), the meet of a line | and 0 is 0, and their join is the line |. Both occupy the lower leaves. The meet and join of shapes ⊔ and □ occupy the upper leaves. The lattices (b) and (c) present decompositions that are proper subsets of (a). They present parts with specific figural features: (b) presents parallel lines, and (c) presents right angles. Parallel maximal lines are discrete; they combine disjunctively without a conjunctive possibility, while right angles combine conjunctively, though in (c), the lattice does not recognize this. It is not a Boolean algebra but a lattice with meets and joins that do not coincide with products and sums. For example, the meet of the right angles ⌊ and ⌋, as presented in (c), is 0, which is smaller than their product, that is, a horizontal line —. Their join is the square on the lattice top, which is greater than their sum ⊔.

The decomposition presented in Fig. 2—a proper subset of the one in Fig. 1a—is a hierarchy including the whole shape, select parts, and the empty shape (Krstic 2010). This allows for uniform characterizations of decompositions which are lattices, topologies, Boolean algebras and hierarchies. The hierarchy presents the shape sensibly decomposed into parts based on the production apparatus. The parts are nested explicitly to frame the drafting steps, which appear on the left branch of the lattice. At the bottom, the empty shape represents the empty page and completes the algebra; in the middle row of the lattice, the outermost left shape presents the top horizontal line of the square, which is the first line to draw; second in the row is the bottom horizontal line, the second line to draw; the two lines in combination

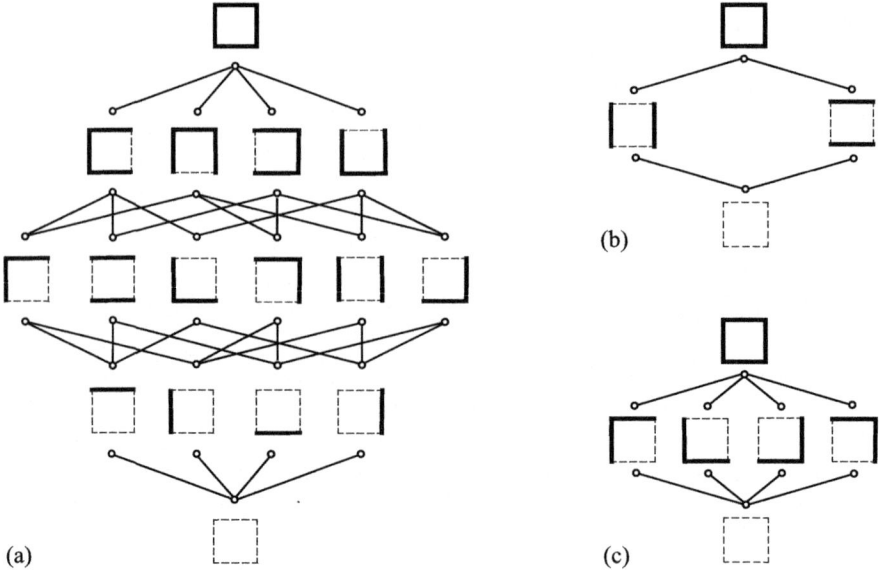

Fig. 1 Three decompositions of a square as lattices, including maximal lines

Fig. 2 The hierarchy presents the maximal parts of a square in drafting order

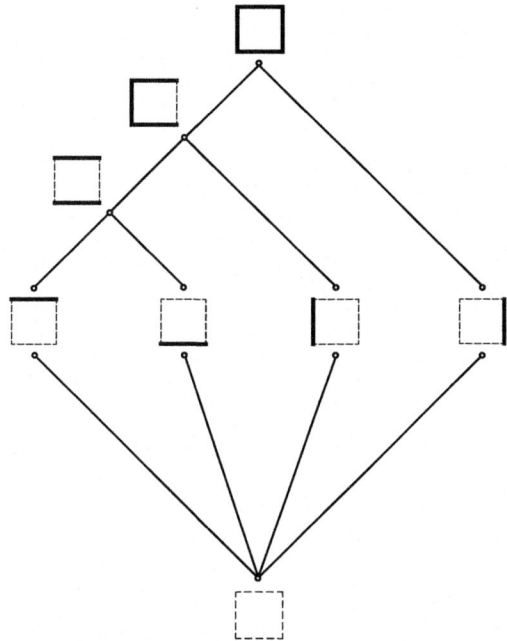

occupy the first leaf on the left branch of the lattice. The vertical lines follow; the left vertical line of the square is third in the row, and the right vertical line is fourth, completing the square at the top. The four lines in the hierarchy's middle row are a natural decomposition of the square since they contain singleton shapes of maximal lines that define the shape (Stiny 2006). They are also a diagonal decomposition of the shape since they present a discrete decomposition in which each element is a diagonal shape, and no two of them share the same diagonal space, namely, the same U_{11} algebra (Kristic 2022).

3 Seeing and Drawing

Stiny (2006) considers the concept of maximal representation of shapes with a reference to drafting. Therefore, "maximal lines" are the longest lines a skilled draftsperson would draw. The shape of the example can be divided in numerous ways. For instance, it can be decomposed into medium-sized triangles or represented as a large square with bisectors and diagonals.

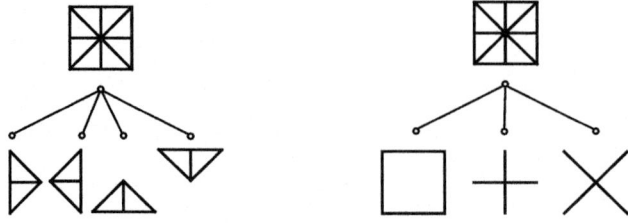

To draw the shape, it's best to draw the smallest set of the longest discrete lines, which are the maximal lines. Stiny shows how to draw eight maximal lines in sequence, starting with the horizontal, then the vertical, and finally the diagonal lines.

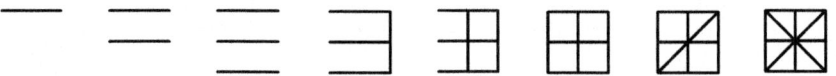

The diagram in Fig. 3 displays the drafting order with slight adjustments based on Stiny's work from 2006. This hierarchy represents a natural and diagonal decomposition of the shape, including individual shapes consisting of eight maximal lines, with no two distinct lines sharing the same diagonal space. The first line drawn is the one in the bottom-left corner, and the remaining lines are drawn from left to right in order. The steps are shown on the left side of the diagram, and the complete shape is displayed at the top. The empty shape is not included.

Notice that the relation between the shapes C and C′ at each decomposition node is C < C′, which reveals that additive rules are applied as in (a). In a different medium, like in sculpture, the process could be subtractive (C′ < C), as in (b).

The order of steps in this nested structure may be crucial or optional for the execution, depending on the equipment used. The hierarchy described here is specifically relevant for ink drafting by right-handed draftspersons as part of a technique involving additive rules of the general form, $x \rightarrow x + t(x)$, for drawing straight lines. These rules are not included here. The hierarchy and rules cover guidelines for handling drafting tools and materials. The T-square moves from top to bottom, and the triangle moves from left to right. When drawing a line, the hand moves from left to right. First, draw the horizontal lines from left to right; second, draw the vertical lines from left to right, with the hand moving from bottom to top; third, draw lines of other angles with the hand moving left to right. Following this order ensures no time is wasted but allows adequate drying time before retracing the same line to avoid blotting and spoiling the work, which is a sign of a weak draftsperson.

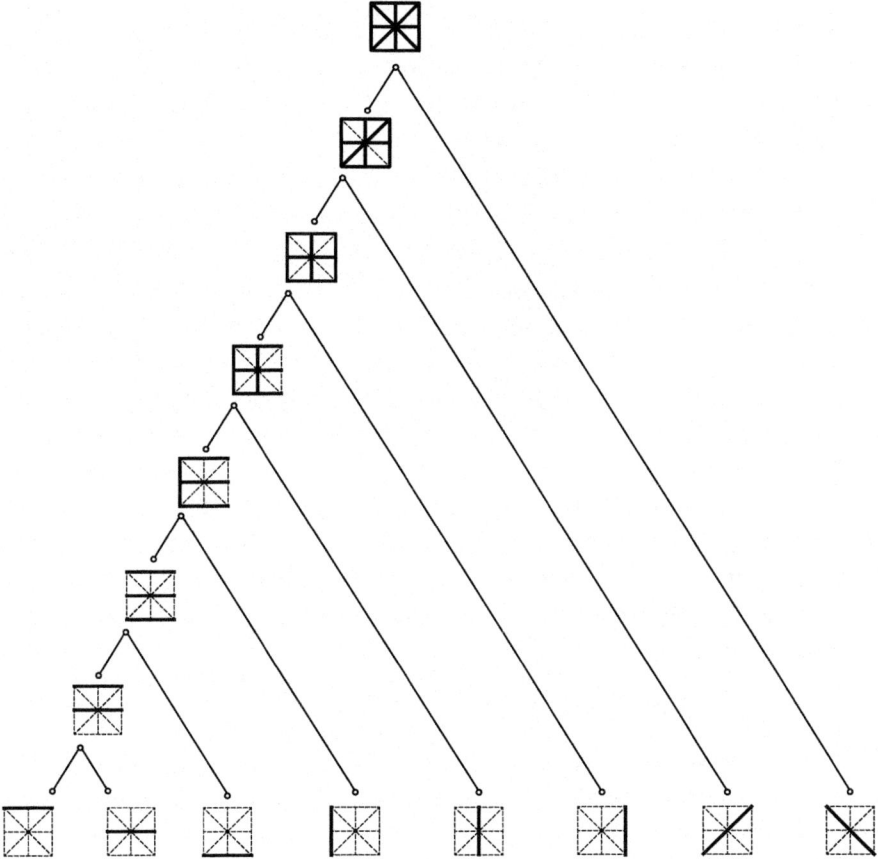

Fig. 3 The hierarchy presents the maximal lines in drafting order

Harmony, derived from the Greek word αρμονία, is attained through the sensible treatment of joints (αρμοί). Select joints that contribute to the aesthetic presentation of the work are highlighted and kept visible, while the remaining joints should be concealed to hide the parts, the construction logic, and the assembly order. An advanced technique deceives the eye, making the form generation appear effortless and seamless. But how is a harmonious pattern created? Harmonious continuity is a characteristic of two-dimensional ornamental pattern languages, such as Greek meanders and Indian Kolams. Greek meanders, or lines, can be single or appear in friezes with repeated motifs. Meanders with circular parts are referred to as waves, locks, or braids (πλοχμοί in Greek). Knight (1986) introduced a shape grammar to produce simple and multi-story meander friezes of the Late Geometric period (circa 900–700). Knight's grammar explains the generative process without specifying a production apparatus. A formal technique for drawing meanders with ink, which expands on Knight's grammar, is presented here. Two grammars for two

Fig. 4 Menelaos and hector fight over euphorbos's body. Credit: Jastrow (2006), Public Domain, https://commons.wikimedia.org/w/index.php?curid=1405480

distinct languages of meander designs are presented: a meander frieze incorporating a swastika—an ancient symbolic form with many connotations—and a wave meander including circle patterns. Figure 4 depicts a Rhodian plate (circa 600 BC) with a swastika and a wave. The scene portrays a duel from the Iliad, with Menelaos and Hector fighting over the dead body of Euphorbos.

The figural features of the meander include general visual impressions. The lines are continuous, and the swastika is repeated. The frieze extends infinitely in one direction. Horizontal and vertical lines are either parallel or perpendicular. The pattern shows figure-ground qualities. The swastika is represented in a darker color, such as black, blue, or red. The ground consists of a sequence of rotating L-shapes in either white or brown-orange terracotta colors. Sometimes, the figure and the ground are inverted, with a light color applied to the figure and a dark color to the ground.

Our perception of abstract features, such as symmetry and congruence, relies on our prior knowledge of the geometric properties of shapes. This knowledge allows us to make classificatory distinctions. The frieze belongs to the simplest line symmetry group, $t2$, which consists of a twofold (180°) rotation around a center and translation. Successive translations of twofold centers generate it. The swastika is a shape in point group 4, with a fourfold (90°) rotational symmetry. In the frieze, its symmetry is reduced to two. The visual interplay between the symmetry of the individual swastika and the symmetry of the frieze increases the ambiguity of the pattern.

Presenting the motif in gradient color tones helps identify it as part of the frieze. There are multiple ways to generate the pattern figuratively.

Figure 5 presents two alternative decompositions (a) and (b) with the structure of Boolean algebras. The instruction for generating the whole from the parts is: "Copy and rotate by 180°".

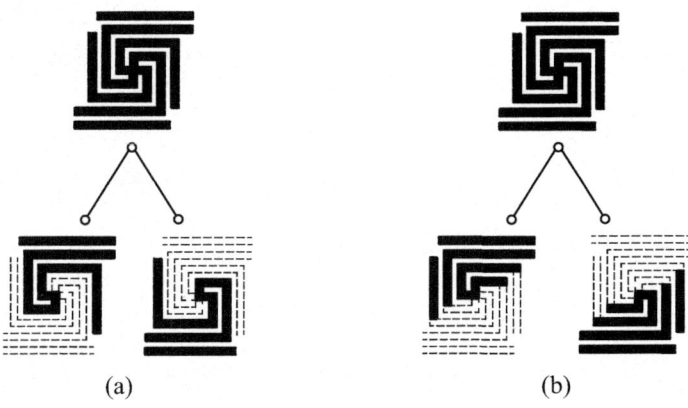

(a)   (b)

Fig. 5 Two swastika decompositions with the structure of Boolean algebras

Figurative construction is straightforward when it generates a pattern in a few steps with a few parts. Rules organized into shape grammars, like Knight (1986), may explain and generate an entire design language. A four-step process is presented here for the figurative construction of the frieze. Rule schemata describe the steps: (a) identification of the individual meander [x → x], (b) identification of a meander part [x → prt(x)], (c) copying and rotating the part by 180° [x → x + t(x)], (d) copying the part and translating it horizontally without erasing what was drawn before [x → x + t(pt(x))]. The steps (c) and (d) repeat to produce the frieze.

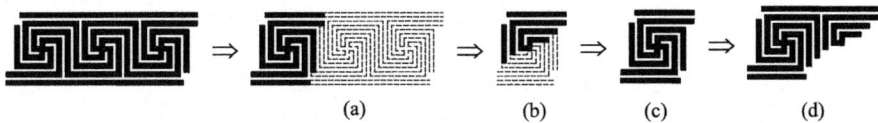

Digital software lets us draw with repeatable parts and shape primitives, blocks, and transformations. While the above parts and transformations help us explain and generate the patterns, they are not useful for physical production. Breaking the swastika into nine horizontal and seven vertical maximal lines, as shown on the next lattice with the structure of a Boolean algebra, Fig. 6a, aids in drafting. It's important

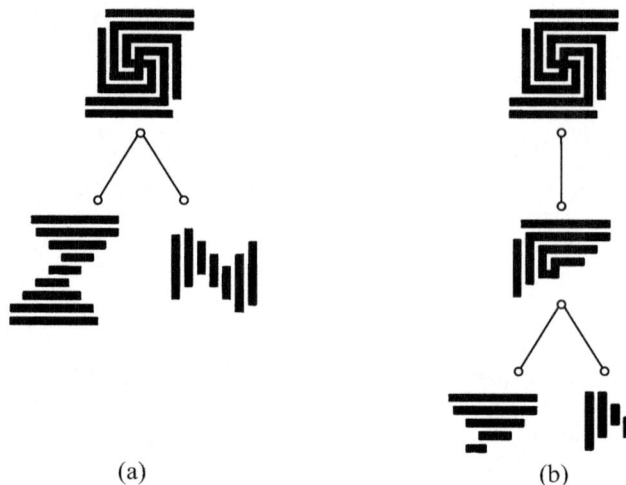

Fig. 6 **a** A decomposition with the structure of a Boolean algebra, including nine horizontal and seven vertical maximal lines. **b** Topology of a selected subshape and its maximal lines

to note that the pattern of these lines does not resemble the meander. The maximal lines have limited explanatory value and do not aid in recalling the pattern. The drafter draws line by line in a finite number of steps and evaluates the results at the end by observing the entire drawing. Figure 6b displays the topology of a selected subshape and its maximal lines while disregarding symmetrical subshapes. The explanatory value of this representation is low at the lower structure and higher at the higher structure. Notice that the maximal lines in the middle node of this lattice are not maximal in the context of the entire meander.

The hierarchy in Fig. 7 displays the maximal lines of a single meander arranged in their drafting order. The sixteen lines in the bottom row of the lattice represent a natural and diagonal decomposition of the shape. The line in the bottom left leaf is drawn first, followed by the rest in sequential order from left to right. The drafting progress is depicted on the leaves of the left branch, and the final shape is visible at the top of the lattice.

What remains to be determined is the execution rules. Meanders and waves are designed as decompositions of grids, which may be rectangular or circular. Shape rules determine which parts to select to create a visible pattern. This specific meander design is based on a grid of u × v square cells, where a specific number of cells are filled with color to create each line. Earl and Jowers (in this volume) explore the practical uses of grids as design generators and how new structural rhythms are created from the spatial relationships among emerging parts of the grids. Colored lines can be drawn on top of the grid lines, or the contour lines of the pattern can be drawn (on the left), and then color can be filled in at the end to create the shading effect. For this demonstration, thick black lines are used (on the right) for simplicity.

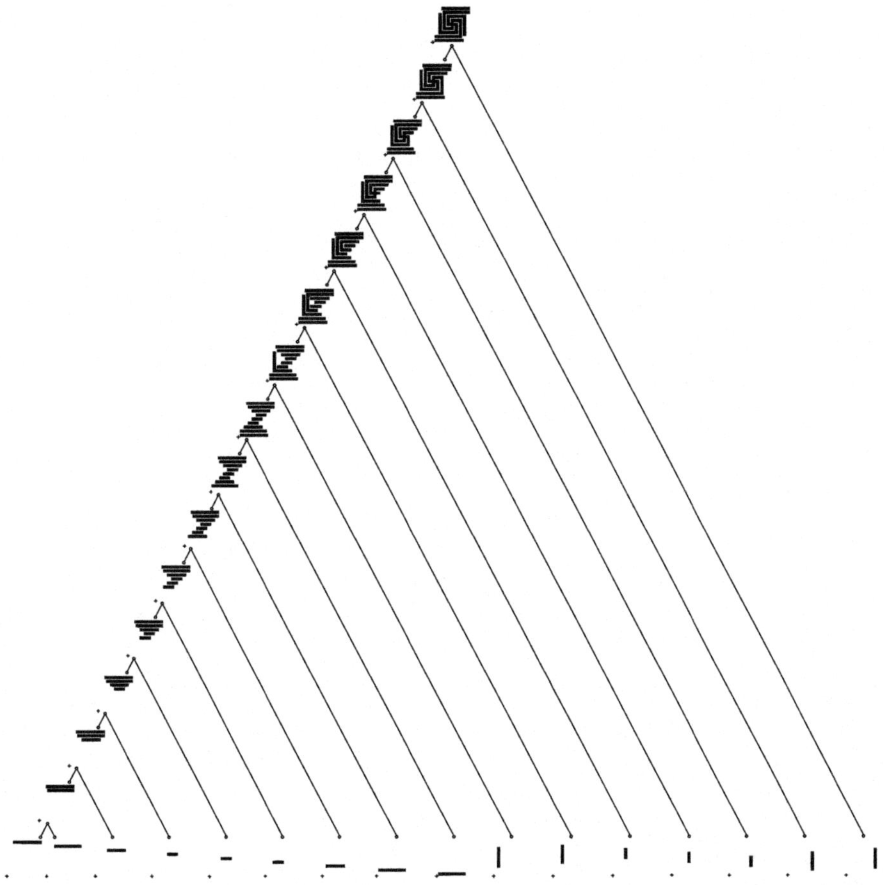

Fig. 7 The meander hierarchy of nine horizontal and seven vertical maximal lines in drafting order. The first line to execute is at the bottom left leaf

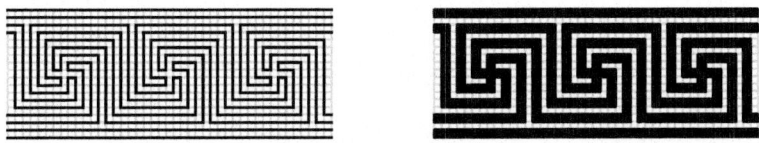

Some practical guidelines are held here. Start by drawing horizontal lines from left to right. Then, draw vertical lines from bottom to top, moving from left to right. Shape rules specify where to begin and finish a line of length L at specific locations on the grid. In a grid with u x v cells, if the meander's height is 17 rows (u = 17), the number of columns v can vary. For a single meander, v = 15, and in this example of two meanders, v = 31. To construct a meander, you will need a set containing three horizontal lines of lengths L_h and three vertical lines of lengths L_v. The visual explanation of the form precedes the information of line lengths. It remains crucial to explain the pattern visually before considering the metrical information. The horizontal lines should be drawn starting from the top left end of the grid. A small triangular execution arrow indicates the top left cell and the movement direction of the drafting tool. The end of each row is marked with a dot that serves as a counter. To generate the meander, horizontal lines are drawn on every other row, specifically on rows 1, 3, 5, 7, 9, 11, 13, 15, and 17.

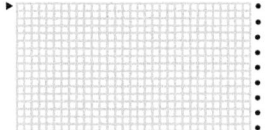

The grid includes horizontal lines at the top and bottom rows, spanning the entire grid length. In the remaining rows, there are various horizontal lines of different lengths. The horizontal lines (L_h) can have lengths of 13, 9, or 5 cells. The vertical lines (L_v) can have lengths of 11, 9, or 5 cells.

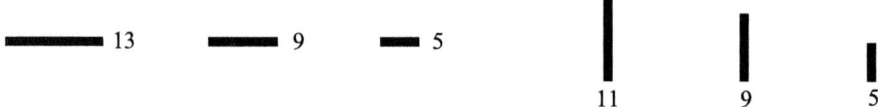

Shape rules focus and act locally, like the draftsperson's eye and hand. Each line has a specific length and may leave a certain number of empty cells. There are three options for the number of empty cells $e = 9, 5,$ or 1. As the drawing moves to the right end of the grid, the row is considered complete if there are not enough free cells to fit another line in the row. Specifically, if the number of remaining empty grid cells f in the row is less than the sum $L_h + e$ cells, the specific row is considered complete. There are nine rules 1, 2, …9 for drawing horizontal lines in nine rows. Rules 1 and 9 are exceptional; they draw two horizontal lines from the beginning to the end of the grid's top and bottom rows. Rule 1(a) draws the first line from the beginning to the end of the top row, and 1(b) moves to the third row to execute the following line in sequence.

Rules 2–8 consist of three steps each. They generate collinear horizontal lines of 13, 9, or 5 cells following the rule schema $x \rightarrow x + t(pt(x))$, where the transformation t is a translation. Rule 2(a) is applied once at the row indicated by the arrow, starting from the left end of the third row. It leaves four grid cells empty and draws a line of 13 cells. Rule 2(b) repeats the same line multiple times by tracking the end tip of the previously drawn line, leaving one cell empty (t = 1 cell), and drawing a line of length 13. When there are not enough free cells left to draw a line of 13 cells, rule

2(c) is applied. This rule erases the counter and moves the execution arrow to the fifth row.

2(a) ▸▭▭▭▭▭▭ . → ▸▭▭▭▭▭▭ .

2(b) ▸▬▭▭▭▭▭ . → ▸▬▬▬▬▭ .

2(c) ▸▬▬▬▬▭ : → ▸▬▬▬▬▬ .

Rule 3(a) leaves five cells empty at the left end of the fifth row and draws a line of 9 cells. Rule 3(b) then tracks the end tip of this line, leaves five cells empty (t = 5 cells), and repeats a line of 9 cells. Finally, rule 3(c) erases the counter and moves the execution arrow to the seventh row.

3(a). ▸▭▬▬▬▬▭ . → ▸▭▬▬▬▬▭ .

3(b). ▸▬▭▭▭▭▭ . → ▸▬▭▭▬▬▭ .

3(c). ▸▬▭▬▬▬▭ : → ▸▬▭▬▬▬▭ .

Rule 4(a) leaves eight empty grid cells starting from the left end of the seventh row and draws a line of 5 cells. Then, 4(b) tracks the end tip of this line, leaves nine cells empty (t = 9 cells), and repeats a line of length 5. Finally, 4(c) erases the counter and moves the execution arrow to the ninth row.

4(a). ▸▭▬▬▭▭ . → ▸▭▬▬▭▭ .

4(b). ▸▬▭▭▭▭▭ . → ▸▬▭▭▭▬▭ .

4(c). ▸▭▭▬▭▭ : → ▸▭▭▬▭▭ .

Rule 5(a) leaves six cells empty at the left end of the ninth row and draws a line of 5 cells. Rule 5(b) tracks the end tip of this line, leaves nine cells empty (t = 9 cells), and repeats a line of length 5. Rule 5(c) erases the counter and moves the execution to the eleventh row.

5(a). ▸▭▬▬▭▭ . → ▸▭▬▬▭▭ .

5(b). ▸▬▭▭▭▭▭ . → ▸▬▭▭▭▬▭ .

5(c). ▸▭▬▬▭▭ : → ▸▭▬▬▭▭ .

Rule 6(a) leaves four cells empty from the left end of the eleventh row and draws a line of 5 cells. Rule 6(b) then tracks the end tip of this line, leaves nine cells empty (t = 9 cells), and repeats a line of 5 cells. Finally, rule 6(c) erases the counter and moves the execution to the thirteenth row.

6(a).

6(b).

6(c).

Rule 7(a) leaves two cells empty at the left end of the thirteenth row and draws a line of 9 cells. Rule 7(b) tracks the end tip of this line, leaves five cells empty (t = 5 cells), and then repeats a line of length 5. Rule 7(c) erases the counter and moves the execution arrow to the fifteenth row.

7(a).

7(b).

7(c).

Rule 8(a) draws a line of 13 cells starting from the left end of the fifteenth row. Rule 8(b) then follows the end point of this line, leaves one cell empty (t = 1 cell), and repeats a line of 13 cells. Rule 8(c) erases the counter and moves the execution arrow to the seventeenth row.

8(a).

8(b).

8(c).

Rule 9(a) draws a continuous line from the beginning to the end of the seventeenth row. Rule 9(b) tracks the top and bottom lines, erases the counter, moves the execution arrow to the bottom left position, and adds counters in position for the vertical lines.

9(a).

9(b).

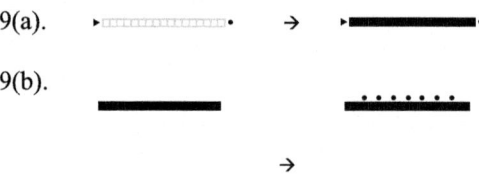

Seeing and Drawing Shapes

The rules for vertical lines involve tracking the end cells of horizontal lines and drawing from one end cell to the other. Vertical lines are drawn in every other column. Specifically, columns 3, 5, 7, 9, 11, 13, 15, 17, 19, 21, 23, 25, 27, and 29 receive lines. There are seven rules (10–17) of the form $x \rightarrow x + t(pt(x))$. Each rule entails drawing a single vertical line at the column indicated by the arrow, then erasing the counter and moving the arrow to the next position on the right.

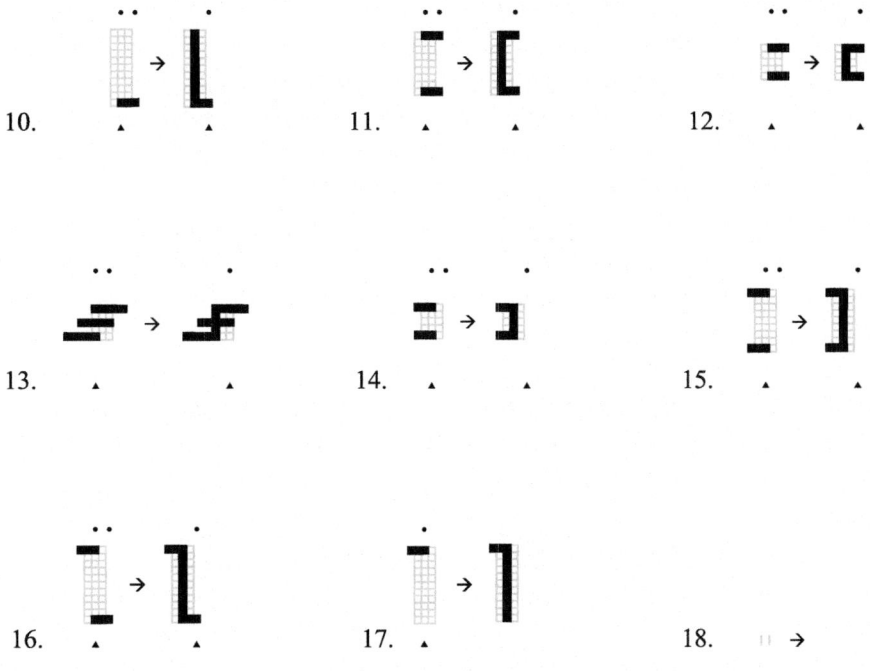

A derivation of horizontal lines is presented in Fig. 8. Collinear lines are drawn in a single pass and appear in a single derivation step. The drafting hierarchy is not included due to space constraints. The decomposition is diagonal, meaning each discrete element is a diagonal shape and no two elements share the same diagonal space. However, it is not natural because each hierarchy node does not contain singleton maximal lines.

Vertical lines are drawn after the horizontals. The derivation is presented in Fig. 9. Rules 10 and 17 apply once to draw the first and last vertical lines. Rules 11–16 apply twice —to complete the first and the second swastika. Rule 18 is a termination rule. The execution order is this:

10, [11, 12, 13, 14, 15, 16], [16, 11, 12, 13, 14, 15], 17, 18.

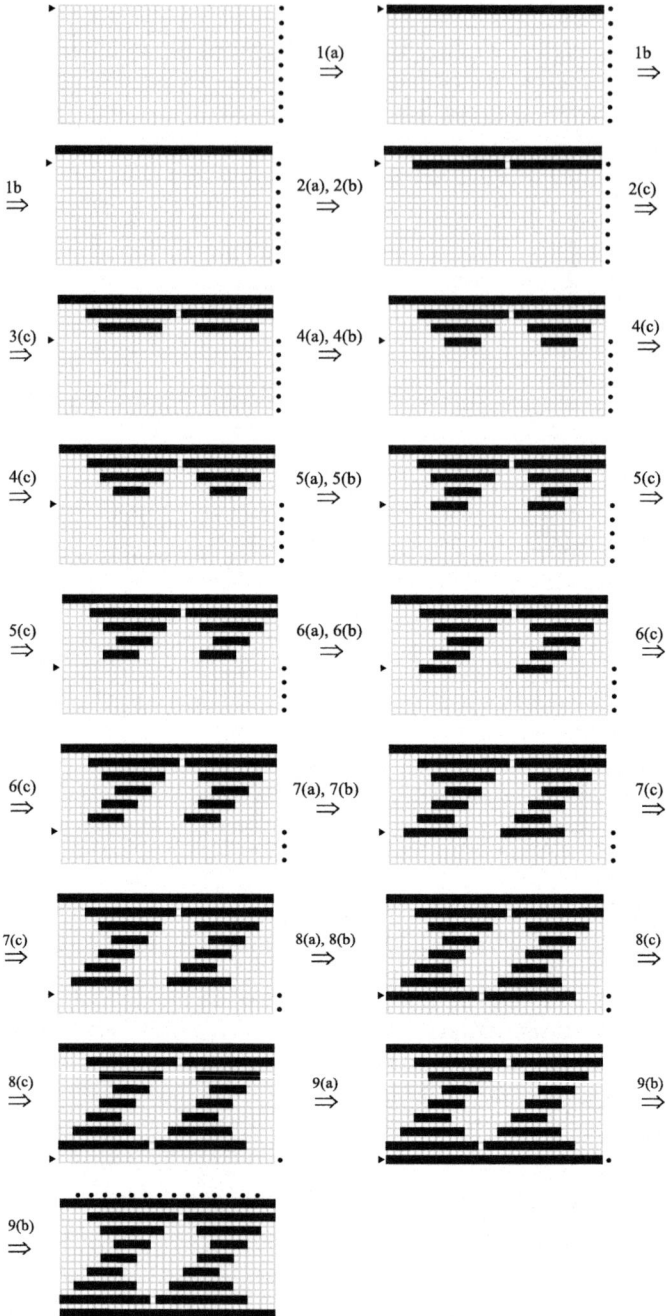

Fig. 8 Derivation of horizontal lines for a meander frieze with two swastikas

Seeing and Drawing Shapes

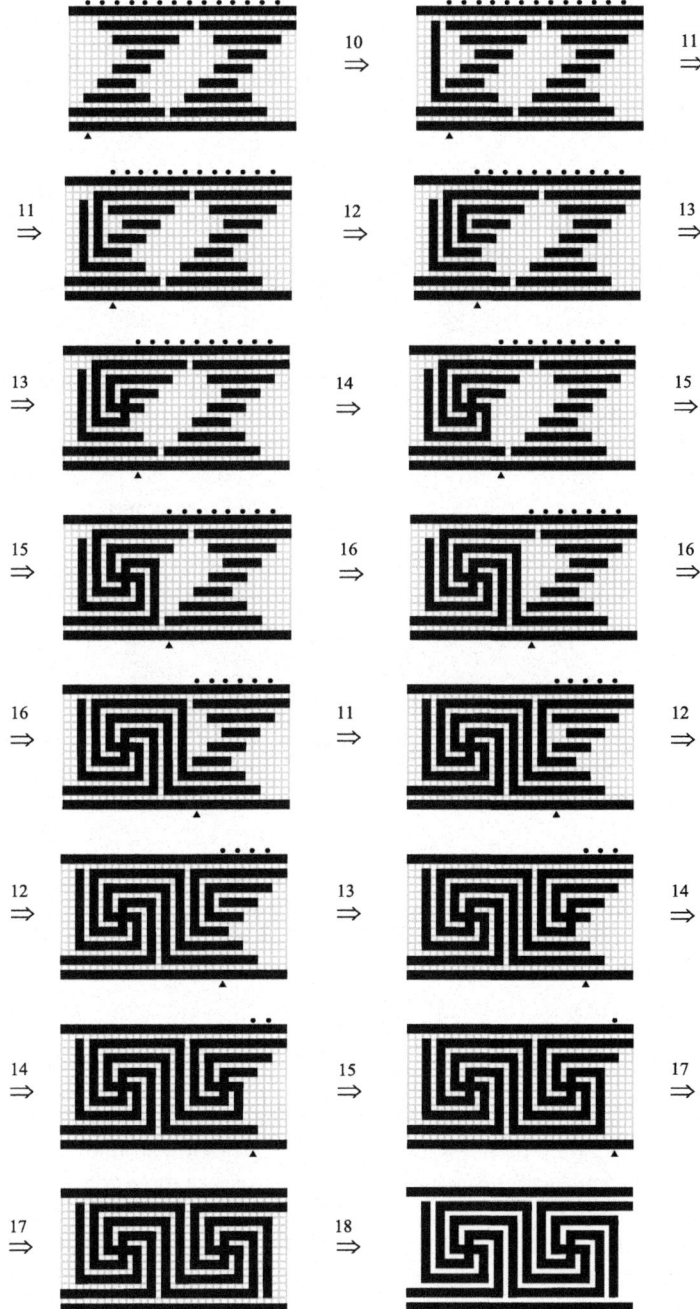

Fig. 9 Derivation of vertical lines for a meander frieze with two swastikas

Shape decompositions specify the parts and the execution order. Shape rules dictate the execution steps, such as how and where to draw a line. Drafting guidelines inform all three aspects: parts, order, and rules. This process is further explained in the following example. Ornamental friezes like waves, locks, or braids feature arcs based on a grid of concentric circles with no straight lines. Related shape studies involving parametric curves, algorithms for shape operations, and embedding have been presented by Jowers and Earl (2010, 2011).

The pattern prominently features circles and circular arcs. The arcs are continuous, and the pattern repeats by translating specific subshapes. The pattern can extend indefinitely in one dimension but has a distinct start and end. It has figure-ground qualities, with the figure typically in darker colors such as black, blue, or red and the ground in white or terracotta. In some variations, the figure and ground colors are reversed (Fig. 10).

Fig. 10 Decoration on a fragment of a raking sima circa 550 BC. Credit: Acropolis Museum, Athens, Greece (*Source* Personal archive)

Perceiving the wave as a series of overlapping concentric disks reveals its repetitive nature. Distinct disks mark the start and end of the frieze. The following figure presents the disks in gradient gray tones to make them more distinguishable.

A distinct abstract feature of the pattern is that it falls under the line group of symmetry *t2*. This group includes a twofold (180°) rotation around a center and the translation of concentric arcs. Each typical disk consists of three concentric arcs with varying radii, all rotated 180° around their center, and a small circle (see Fig. 11a). The frieze is made up of several of these disks. The starting disk includes a circle and three concentric arcs different from those found in the typical disk (see Fig. 11b). The starting and ending disks are the same, just rotated 180°.

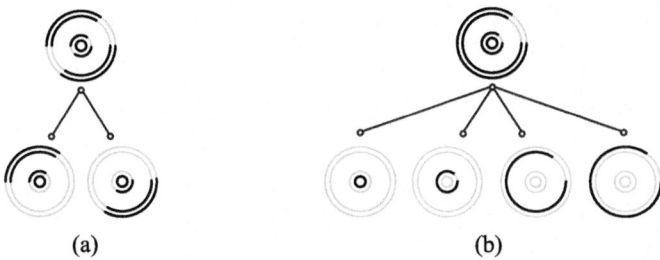

Fig. 11 The parts of the typical disk and the start–end disk

A six-step process generates the frieze figuratively:
Identify the typical disk [x → x].
Identify a part of the disk [x → prt(x) or x → x • y].
Construct the typical disk by rotating the part 180 degrees [x → x + t(x)].
Translate the disk part horizontally [x → x + t(pt(x))].
Repeat steps (d) and (c) to create the pattern body.
Complete the front end of the pattern by adding the start disk [x → prt^{-1}(x)].
Produce the end disk by rotating and translating the start disk [x → x + t(x)].

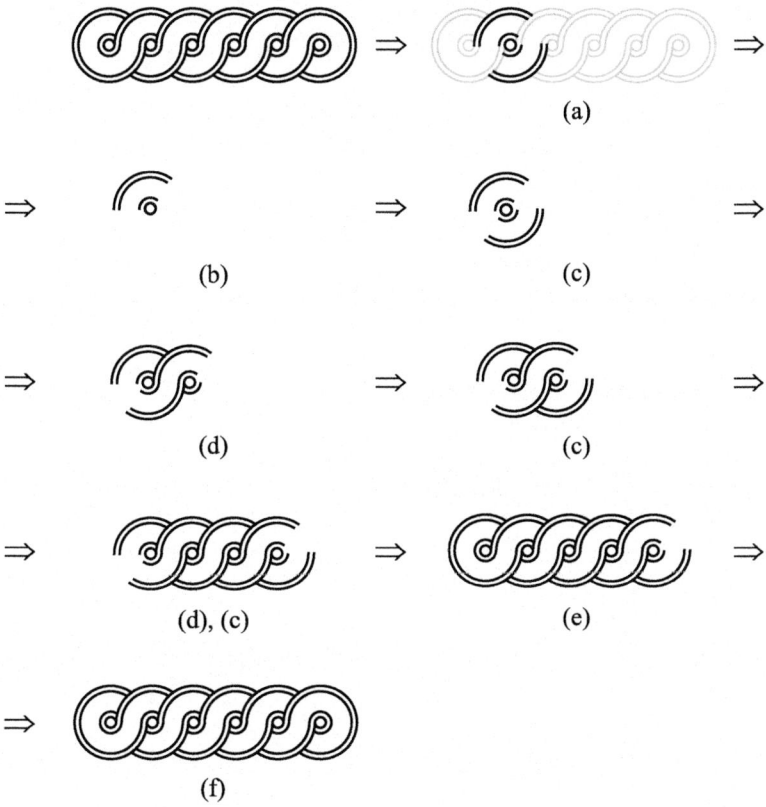

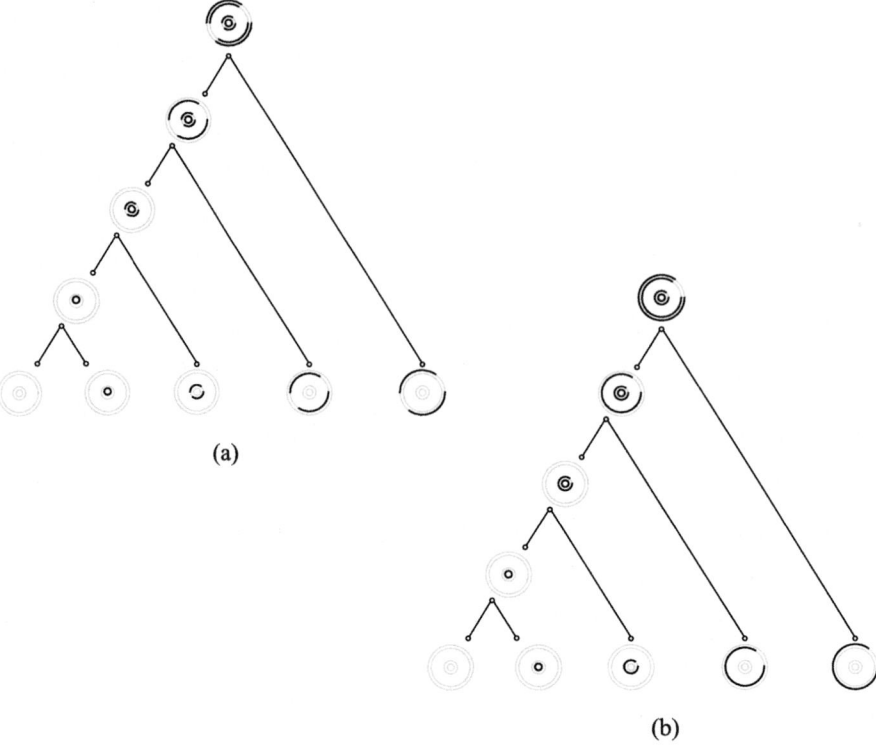

Fig. 12 Hierarchies of the drafting order for the typical (**a**) and the start–end (**b**) disks

The two hierarchies shown in Fig. 12a and b illustrate the drafting order of the typical and start–end disks. The arcs at the bottom of each lattice are drawn from left to right in sequence. The drafting process is depicted at the leaves of the left branch.

Figure 12a and b display both diagonal and natural decompositions of the shapes. They consist of singleton shapes with maximal arcs that define each disk. Additionally, they present a discrete decomposition in which each element is a diagonal shape, and no two share the same diagonal space. Moving on to Fig. 13, it shows a frieze with four disks. This frieze includes six arcs of various degrees, radii, and a small circle, forming a kit of parts. The hierarchy is organized so arcs with the same radii are drawn in a single pass of the compass. Although the hierarchy presents

Seeing and Drawing Shapes

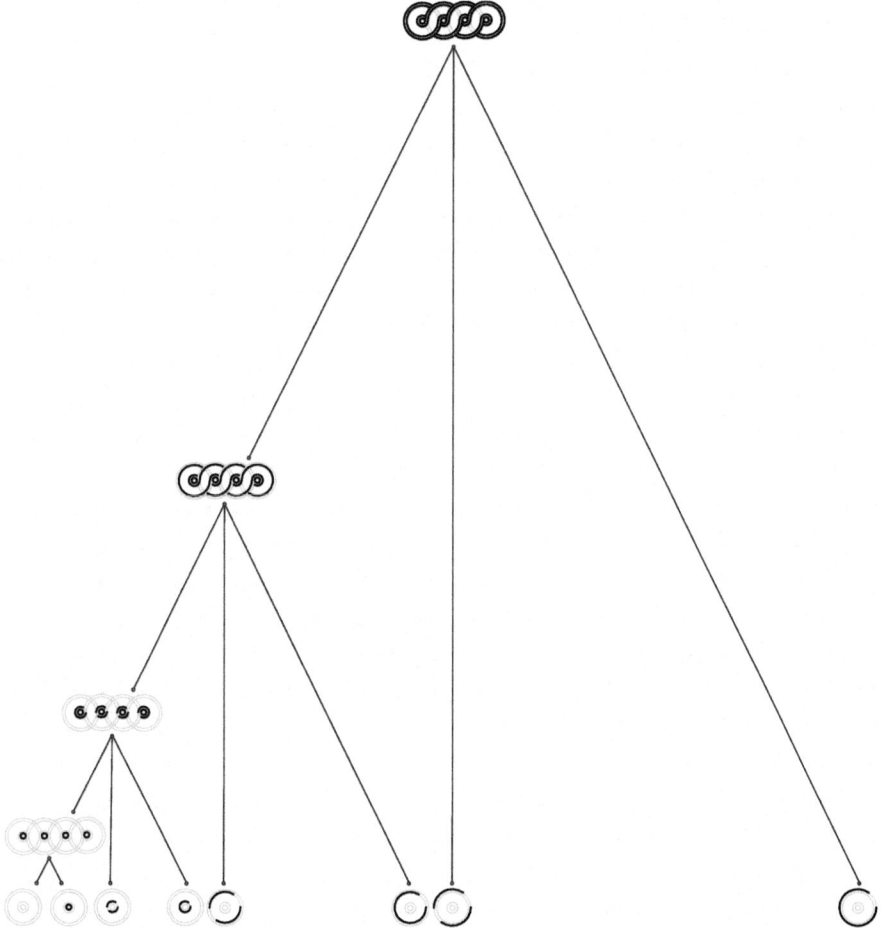

Fig. 13 The drafting hierarchy of a frieze with four disks

a diagonal set of atoms, it is not a natural decomposition, as each node does not include singleton maximal shapes. Regarding physical construction, the circle in the second bottom leaf from the left is drawn first, and the remaining arcs follow from left to right. It is important to note that the physical construction of the pattern differs significantly from its figurative construction.

Seven additive rules are applied in an ordered sequence from 1 to 7 to generate the pattern, with rule 8 erasing the grid.

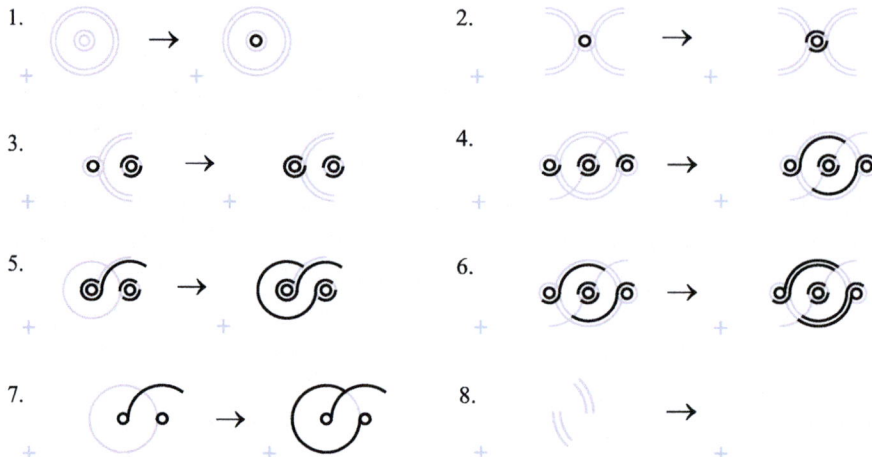

The initial shape is a grid of intersecting circles. The intersections determine where each arc starts and ends. The right-handed drafter draws the arcs from left to right and from the inside out. This approach allows for better handling of minor drafting discrepancies that may occur during the execution. Arcs with the same radii are drawn in a single passing of the compass to minimize readjusting the instrument.

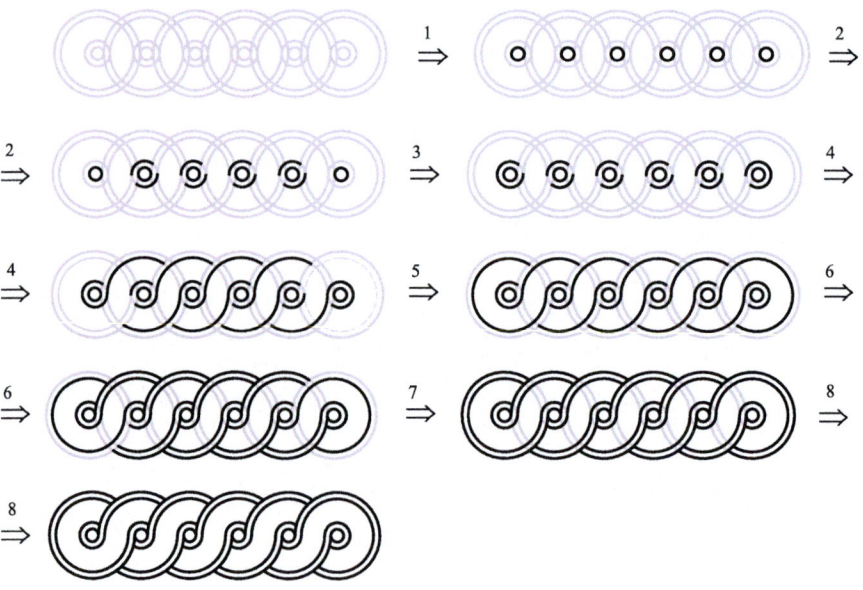

Pulli Kolam is an art form in South India that features geometric patterns with straight lines, circles, and arcs. Women create these patterns on the thresholds of their homes from memory. In traditional Kolam, the lines are made using coarse rice powder trickled between the fingers in a thin stream onto the ground or charcoal. Grids of dots (pullis) help the artist remember and accurately create the designs. Different design spaces can be created by arranging the dots in specific ways. There are various sublanguages of Kolam, including Sikku Kolam, which consists of only curved lines; Pulli Kolam, involving grids, arcs, and straight lines; Woda Pulli Kolam, which includes loops with hexagonally packed dots; Ner Pulli Kolam, featuring loops with square-packed dots; and Kambi Kolam, which includes intricate symmetry patterns with straight lines. Four examples of Pulli Kolam on different Pulli grids are shown below. The first design is also known as the Buddhist "knot of eternity" or the "knot of meditation," drawn with one unending line.

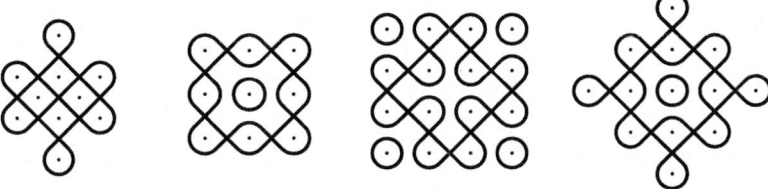

Gopalan (2023) outlines three constraints defining a Pulli Kolam: (i) all dots are circumscribed; (ii) two linear segments cannot overlap; and (iii) all line orbits are closed, meaning no loose line ending tips exist. Pulli Kolam features line continuity, with all straight lines angled at 45° and only four arcs (of the same radius) used: 360°, 270°, 180°, or 90°. These figurative features are approximated when the patterns are made by hand. Symmetry is an abstract feature of Kolam. Patterns can exhibit rotational or reflective symmetry across two perpendicular or diagonal axes. 3 × 3 Kolams are classified into one of the eight-point symmetry groups. The $4mm_d$ point group presents four-fold rotational symmetry with 90° rotation and four intersecting mirrors, two perpendicular and two diagonal. Bilateral symmetry, noted by m, involves two mirrors parallel to the edges of the pattern. Two mirrors along the diagonals are noted as m_d. Patterns with two-fold rotational symmetry (180°) are noted with 2. Lastly, all patterns have a onefold rotation by 360°.

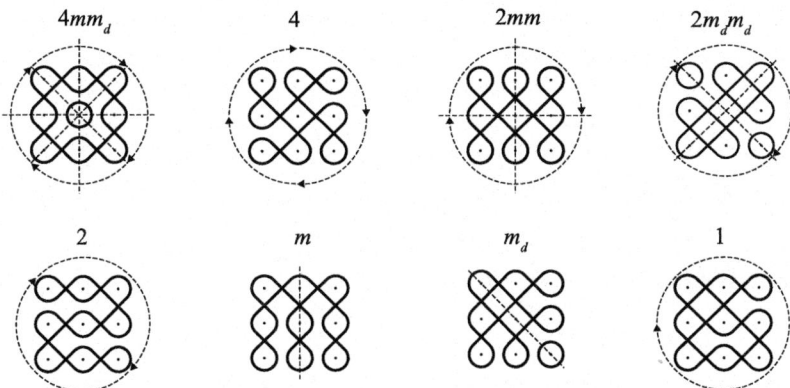

Tile sets allow Kolam patterns to be created using a vocabulary of shapes depicted on the tiles. While the tile sets are fun, they can make seeing the continuous looped line difficult. The designs can be decomposed in different ways. For example, in Fig. 14a, the pattern is decomposed using the logic of tiles, recognizing the pattern's rotational symmetry and having the structure of a Boolean algebra. In Fig. 14b, the decomposition acknowledges shape emergence.

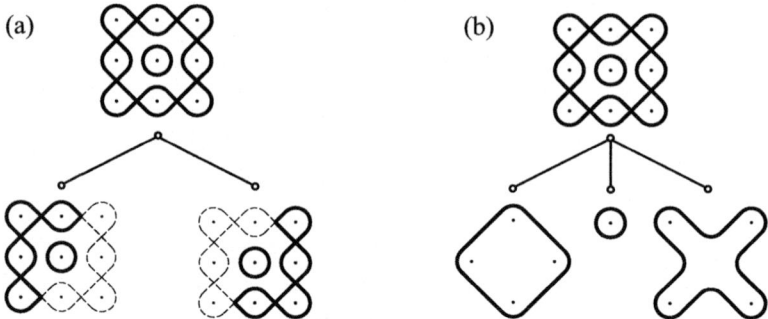

Fig. 14 Two possible decompositions of a Kolam pattern with symmetry $4mm_d$

A figurative construction involving a two-step computation is as follows:
Select the appropriate subshape using the schema $x \rightarrow prt(x)$.
Copy and rotate the subshape by 180° using the schema $x \rightarrow x + t(x)$.

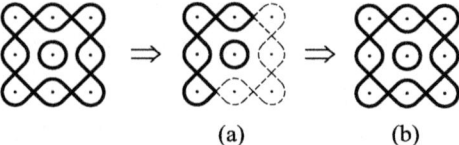

In recent years, several studies have explored the mathematical aspects of Kolam, particularly as a combinatorial game. Gopalan (2023) uses tiles to categorize symmetry and enumerate square Kolams, while Waring (2012) employs gestural lexicons. Knight (2018) introduces three shape grammars: a *modular* grammar using tiles, a *mirror* curve grammar that mimics the process of making Kolam, and a *making* grammar that aims to capture the real-time creation of Kolam as an act of seeing and doing. These three grammars address the figurative and physical construction of Kolam. The *making* grammar focuses on the physical technique of using coarse rice powder. In contrast, the technique presented here treats Pulli Kolams as graphic compositions drawn with ink, following the three Kolam constraints and the ink drafting conventions. When using tiles, two of the three Kolam constraints are automatically satisfied: dots are always circumscribed, and two lines cannot have common parts. The "no loose ends" constraint is achieved by generating the appropriate tile combinations. Below are the six tiles with their shapes. Each tile has a Pulli black dot at its center, and the gray lines represent the tile boundary. Loose ends are indicated by red markers, and there are 0, 1, 2, 2, 3, and 4 loose ends.

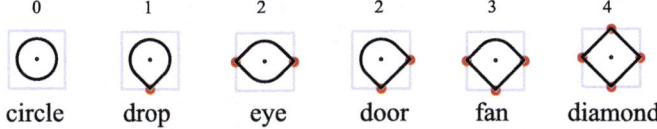

| 0 | 1 | 2 | 2 | 3 | 4 |
| circle | drop | eye | door | fan | diamond |

All six shapes are decompositions of a single shape: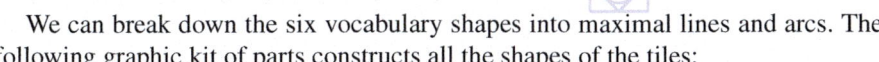

We can break down the six vocabulary shapes into maximal lines and arcs. The following graphic kit of parts constructs all the shapes of the tiles:

A precedence rule applies to drawing shapes with curves and straight lines: curves, circles, or arcs are drawn first, and maximal straight lines second (Fig. 15). Straight lines should begin and end at the tips of arcs and should be continuous with them. This order makes it easier to extend the tip of an existing arc with a straight line. All arcs are parts of a circle with a preset radius, so the compass is set once to draw all arcs.

(a) Circle: 360° arc.

(b) Drop: ¾ circle, 270° arc.

(c) Diamond: square.

(d) Eye: 2 × ¼ circle, 90° arc.

(e) Door: ½ circle, 180° arc.

(f) Fan: ¼ circle, 90° arc.

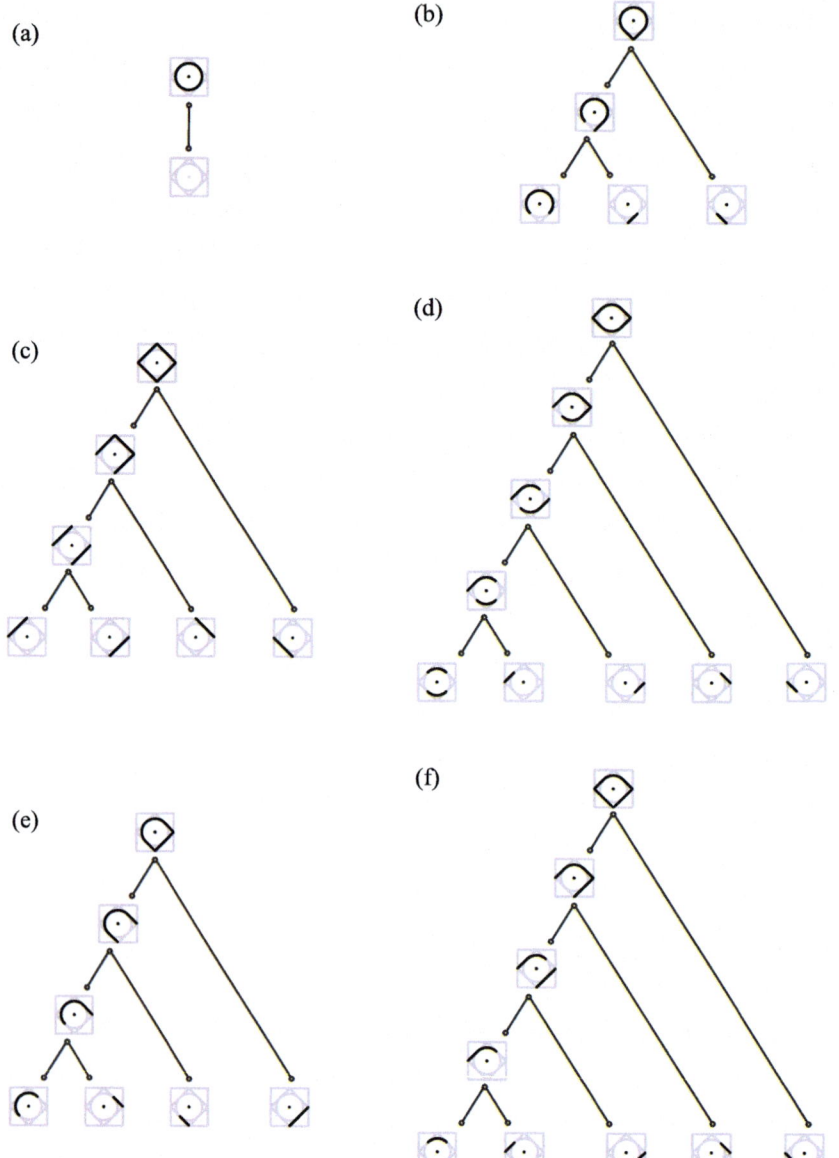

Fig. 15 The drafting hierarchies for the six shapes of the Kolam vocabulary contain maximal arcs and lines, presenting natural and diagonal decompositions

Seeing and Drawing Shapes

We can create all Kolam designs as decompositions of grids based on the schema $x \to \Sigma F[prt(x)]$ and draw them as configurations of maximal arcs and straight lines. The illustrations present 3×3, 4×4, and a cross-shaped 5×5 grids.

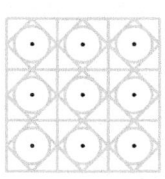 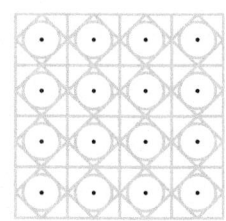 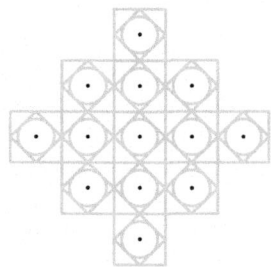

Not all grid decompositions are Kolam designs. A shape grammar ensures that only Kolam designs are produced. The grammar's novelty lies in its ability to generate Kolams as ink drawings by simultaneously satisfying the Kolam and drafting constraints. In the drawings, Pullis are represented by black dots, and loose-ending tips are marked with red dots as labels that do not interact with each other. Four distinct labels are used, with 1, 2, 3, or 4 red dots.

⊙ full circle, 360° arc, no loose ends, no red dots.

◌ ¾ circle, 270° arc, one loose end, one red dot.

◌ ½ circle, 180° arc, two loose ends, two red dots.

◌ ¼ circle, 90° arc, three loose ends, three red dots

◌ 2 x ¼ circle, 90° arc, two loose ends, two red dots

◌ Diamond, zero arcs, four loose ends, four red dots.

Arc-generating rules draw 360°, 270°, 180°, or 90° arcs and apply in five groups in this order: [1–7], [8], [9–18], [19], [20–23]. The rules in each group do not apply sequentially. The symbol "|" on a rule's right-side means "exclusive or" (XOR), indicating alternative shape outputs that a shape input may generate. The designer chooses which one. Blue lines mark the perimetric boundary. Rules 1–7 apply first.

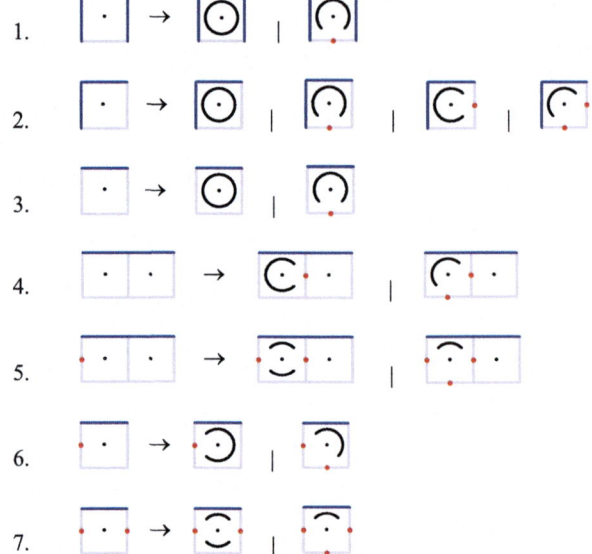

Rule 8 shifts the focus from the perimeter to the center, which may include 1 or 4 cells. It applies to 4 × 4, 5 × 5, or larger grids. In 3 × 3 grids, it simply erases the blue lines.

Rules 9–18 pertain to the grid center. The designer has the option to select which rules to use, ensuring that no Pulli are left uncovered. These rules are intended for 4 × 4, 5 × 5, or larger grids. For a 3 × 3 grid, the designer should refer to rules 20–23.

Seeing and Drawing Shapes

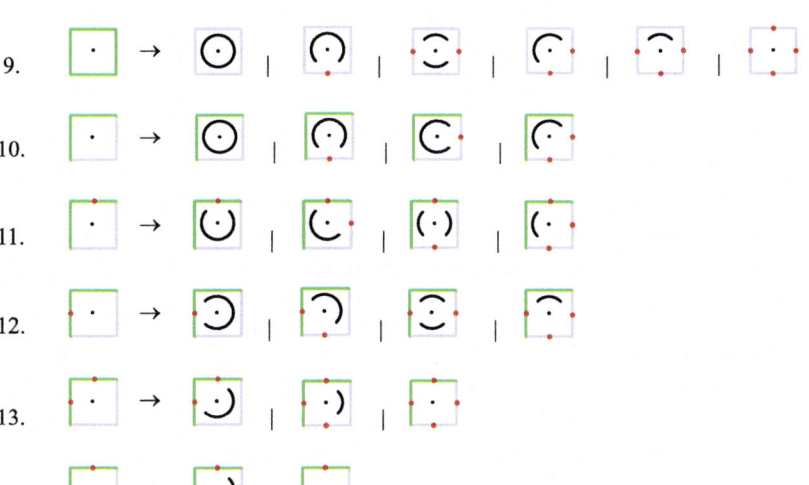

Rule 19 erases the green lines that mark the center of the grid. Rules 20–23 apply to any remaining empty grid cells.

Maximal straight lines are drawn at a 45° angle, passing over the red marks without crossing any arcs. For grids of sizes 3 × 3, 4 × 4, and 5 × 5, a straight line may pass from one to six red marks (n = 1, 2, ..., 6). The ending tips of arcs are considered tiny, straight lines. Collinear straight-line segments are drawn in a single pass of the drafting instrument.

Rules 24–29 draw lines with the 45° drafting triangle moving from left to right and the inking instrument moving from bottom to top. Rules 24 m–29 m are a mirror-reflection version of rules 24–29, used to draw lines in the opposite direction, with the triangle moving from right to left and the ink tool from top to bottom. The hand always moves from left to right. Notice that the continuity of straight lines and arcs is restored with these rules —the atomic elements of the arcs fuse when the lines are added. Maximal straight lines are drawn across the grid to be seamless with the arcs. Rules 24–29 and 24–29 m are listed based on the number of red marks their lines pass. Due to space constraints, only rules 24 and 24 m, 25 and 25 m are illustrated.

Rules 24 and 24 $_m$ the line passes over one mark.
Rules 25 and 25 $_m$ the line passes over two marks.
Rules 26 and 26 $_m$ the line passes over three marks.
Rules 27 and 27 $_m$ the line passes over four marks.
Rules 28 and 28 $_m$ the line passes over five marks.
Rules 29 and 29 $_m$ the line passes over six marks.

... etc.

Rule 30 ends the process by clearing the grid and removing the red dot markers.

30.

A Kolam is derived on a 5 × 5 cross-shaped grid. Rules 1 and 2 apply to the perimetric cells, and rule 8 to the center. Rule 9 erases the green lines, and Rule 21 finishes the arcs. Rules 24, 25, 26, 27, and 24 m, 25 m, 26 m, 27 m add the straight lines.

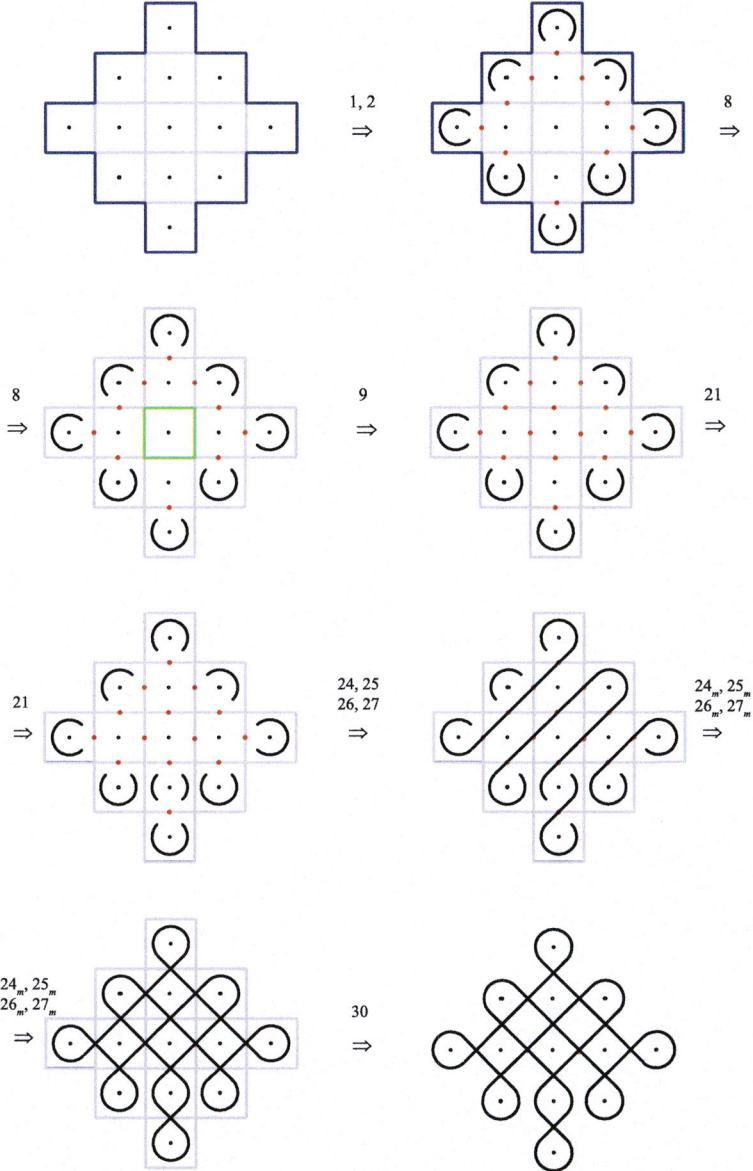

The drafting hierarchy of the Kolam pattern is summarized in Fig. 16. The initial shape is the Pulli grid on the first bottom left leaf. The parts on the bottom leaves are added from left to right: arcs first, straight lines second. Collinear straight-line segments are drawn in a single step. If expanded, this hierarchy presents a diagonal set of atoms but not a natural decomposition.

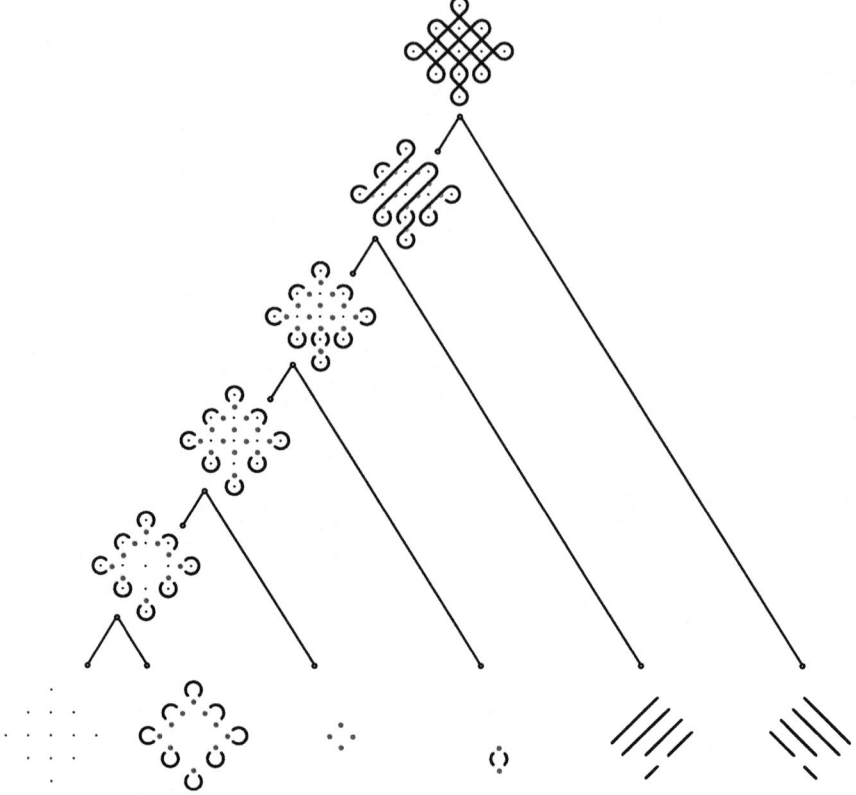

Fig. 16 The drafting hierarchy of the derived Kolam pattern

4 Discussion

This paper explored the creation of two-dimensional patterns, focusing at their figurative and physical construction. Each construction type has unique formal properties that relate to different ways of perceiving and ascribing structure. Figurative construction is driven by theory and aims to explain a design through parts and transformations that generate a design figuratively. Physical construction is practical and involves using a kit of parts and rules to physically produce a design, considering the tools and materials used. Designers consider both the form and the process of making a design but often find themselves explaining it in one way and implementing it in another.

There were several sources of inspiration for this research. One of the key influences was John Ruskin's *Elements of Drawing*, a seminal work published in 1853. In this treatise, Ruskin discusses the techniques of free-hand drawing and painting, offering theoretical and practical insights. Even today, it remains one of the most sensible books on how to draw. Ruskin emphasized the importance of perception in art, stating that "the excellence of an artist as such depends entirely on the refinement of perception." This statement forms the basis of this study and highlights the significance of perception in art and design. Ruskin further posits that "the subtlety of sight and the delicacy of drawing" are subjects of art education. In this study, the refinement of perception begins by distinguishing a broad range of figural and abstract visual features in two-dimensional designs.

Figural features include parallel, vertical, or intersecting lines and orthogonal, acute, and obtuse angles. Abstract features such as symmetry and congruence are also considered. These features are unique to each design but do not have the power to construct (Leeuwenberg and van der Helm 2013). However, the perceiver engages in a figurative construction, which explains and generates a pattern in a few steps with a few parts. On the other hand, physical construction is associated with a specific production apparatus. This process involves a 'kit of parts', explicit syntax, and rules governing production by integrating practical and material guidelines. This physical construction process, or *technique*, is supplied for any design language or apparatus. Like any repetitive behavior, a *technique* is performed unconsciously after sufficient practice. Modifying a technique by revising the parts, syntax, or apparatus marks improvisation.

Shape grammars are effective in capturing stylistic consistency and change. Recent research in computational design by Knight and Stiny (2015) and Krstic (2018) has explored the making of physical objects, providing additional inspiration. This study establishes a technique within specific design languages to address the creation of two-dimensional design patterns in a particular medium. A production technique is never abstract but incorporates assumptions based on prior knowledge of tools and materials. It involves rules of decomposition and assembly. It combines perception and reasoning: perception deals with designs freely and without logical limitations, while reasoning is goal-oriented, producing descriptions that can be ordered according to a certain logic the artisan applies, rooted in established norms of production.

To formalize a technique, it better be a familiar one. For this study, analogue drafting was the obvious choice. Designers have been using drafting for centuries as an accurate and efficient medium of expression. Three design examples featuring straight lines, arcs, and composite forms, with lines and arcs, were used for demonstration. Shape decompositions and rules determine what, when, where, and how to draw.

The design parts were nested in hierarchies acknowledging the requirements of ink drafting. Treating hierarchies as lattices allows their classification as Boolean algebras, topologies, or lattices. Boolean algebras are Figs. 1a and b, 5a and b, 6a, 11a and 14a. Topologies are Fig. 6b and all the Boolean algebras above. Latices are all the above and the rest of the hierarchies in the paper. A distinctive formal feature of drafting hierarchies is that the sets of their atoms are diagonal shape decompositions (Kristic 2022). The algebras appearing diagonally in the table of U_{ij} algebras enable shape fusion—lines fuse with lines, planes with planes, and solids with solids. Therefore, fusing happens in the diagonal U_{ii} component. Drafting decompositions adhere to a definition of shape algebra as an infinite sum of diagonal algebras, in which shapes are represented by their diagonal decompositions. Although somewhat esoteric, this definition emphasizes the spatial character of shape calculation. Diagonal shape algebras are the only ones where shapes can fuse to produce new shapes.

Shapes are finite sets of maximal elements that can be partitioned into subsets belonging to different diagonal algebras, forming their diagonal decompositions. An algebra U_{ij} is an infinite sum of diagonal algebras. When drawing a line or a sequence of collinear lines in a single pass of the drafting instrument, the line or line sequence becomes an element of the diagonal decomposition of the shape. The lines are executed in the *minimum space*—in terms of dimensionality—where a single line can be defined: a U_{11} algebra. The visual attention of the draftsperson is also reduced to U_{11}, focusing on the space of a line. When the sequence of collinear lines is completed to inspect the results, the draftsperson perceives the whole in U_{12}. Perceptual transitions like these are captured with rules. The hierarchies presented in Figs. 2, 3, 7, 12a and b and 15 contain sets of atoms that are both diagonal and natural decompositions. Natural decompositions follow Stiny's standard definition of maximal representation of shapes. Their elements are shapes. Each element of the natural decomposition of a shape is a shape represented by the singleton set containing one of the maximal basic elements of the original shape. Being a singleton, it has two subsets: itself and the empty set, since the empty set is part of every shape. The hierarchy of Fig. 13 presents a diagonal set of atoms but not a natural decomposition. The same holds for the hierarchy of Fig. 16 (if expanded), presenting the drafting of a Kolam pattern and the production of a meander frieze (Figs. 8 and 9), the hierarchy of which is not presented.

The third source of inspiration was *Shapes of Imagination* (2022), where Stiny explores a computational theory that emphasizes the importance of visual perception in the process of calculation. The embed-fuse cycle captures how seeing defies logic that would like to extinguish imagination, silencing our best aspirations. One can remain free to see in an open-ended computation cycle in which nothing prior is given, aiming in the direction of new things, which endlessly rekindles the possibility of seeing more. The *embed-fuse* cycle encompasses both interpretation (seeing) and generation (doing). This study adapts the concept of embed-fuse to incorporate the framing of figurative and physical construction. Accordingly, the external, broader cycle of the diagram captures figurative construction. The internal, narrow cycle captures the course of physical construction: seeing—breaking into parts to acknowledge the apparatus—and assembling that fuses the parts into new wholes (Fig. 17a). Both figurative and physical construction are reviewed through seeing.

The cycle can be defined by a *design machine* (Stiny and March 1981) consisting of four parts: (1) a receptor for seeing, (2) an effector for doing, (3) a language of designs, and (4) a theory that determines the fit between designs, receptor, and effector. In the diagram of Fig. 17b, description occupies the center of the see-do cycle, introducing an internal loop that engages goal-oriented embedding for physical production. A technique encompasses both a kit of parts (description) and a set of instructions (rules) directing the course of the effector (do). The design context is shaped by the receptor and effector. The receptor characterizes objects based on

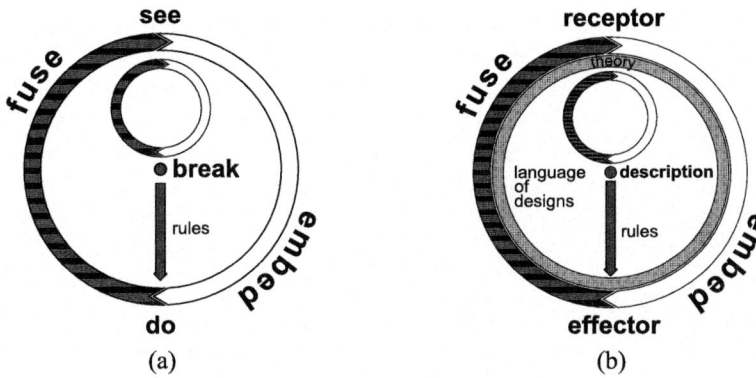

Fig. 17 a A modified version of the embed-fuse cycle that frames figurative and physical construction. **b** A mapping of this cycle to a design machine

specific criteria, while the effector uses design descriptions and a production apparatus to fabricate physical objects. A unique form of reasoning mediates between seeing and doing, the two ends of the embed-fuse cycle, leading to appropriate design descriptions that match the design context. Design interpretation involves explanation and production. The differences and agreements between figurative and physical formations inform the seeing-doing cycle of the design machine. Explanatory descriptions interpret a design language figuratively by illustrating how to generate design instances, while descriptions of physical production provide parts and rules based on specifications of physical making. Both are connected by a theory of design that acts as a conduit between them. The theory compares the visual perception of designs with the handling of the apparatus by reasoning about the fit between designs and context. Seeing and doing are perpetually reconfigurable based on the receptor and effector. Theory bridges seeing and doing.

Finally, the broad fascination with Generative AI, served as a fourth motivation, since it reminds the AI Assistants of the 60 s, like the *Advice Taker* of John McCarthy (1959) and the *Computer-Aided Design* system of Steven Coons (1963). These systems were inspired by engineering success and potential commercial applications. Ivan Sutherland (1963), a student of Coons, valued computer-generated graphics for transforming drawings from dirty marks on paper into hard-coded data, upholding that in the future, "designers will [have to] be vitally concerned with the structure of the design description in computer memory." Coons envisioned a Computer-Aided Design (CAD) system as a versatile design assistant, even though he recognized its limitations in exhibiting the interconnections of design ideas. Despite this, he believed in CAD's ability to revolutionize intellectual work, stating that "When the computer replaces pencil and paper, it will bring about a truly miraculous change in man's intellectual potential."

Generative AI systems, also known as Large Language Models (LLMs), are designed to perform creative tasks through learning. However, they have limitations. Yan Lecun (2023), a pioneer in this field, points out that humans learn through direct observation of the world, which current AI systems are unable to do spontaneously. Browning and Lecun (2022) note that symbolic languages excel at expressing discrete objects, properties, and abstract relationships but struggle in open-ended world interaction. They emphasize: "Abandoning the view that all knowledge is linguistic permits us to realize how much of our knowledge is nonlinguistic. While books contain a lot of information we can decompress and use, so do many other objects: IKEA instructions don't even bother writing out instructions alongside its drawings."

Throughout history, designers have used drawings, diagrams, and physical models as primary visual aids. The bias of CAD and AI toward perception was to shift it to a symbolic form and treat it as logical problem-solving. Today, there is a statistical shift as machines have become increasingly adept at identifying patterns from large datasets through intensive processing. While neural networks and backpropagation work well in artificial worlds like games, they struggle to function effectively in the real world. Although successful in simulating certain phenomena, these statistical models fail to provide insight. In contrast, humans navigate the world without relying on vast data or energy-intensive processing. The visual examples in this paper demonstrated how we perceive and construct patterns figuratively and physically by using different modes of seeing, reasoning, and configuring their topologies. For this to happen, calculating with spatial elements of all dimensions, recursion, and embedding is vital to enable impulsive and resourceful seeing, thinking, and doing. Many of us think Stiny got it right; calculating with lines, planes, or solids represents a profound shift in how we view and represent knowledge and the world—and solve problems—underscoring the role of seeing.

The recognition that much of human intelligence is nonlinguistic, although this is already well-known to those working with shape grammars, is promising. Despite its efficacy, quarrying into enormous data sets takes us back to narrow intelligence and abandoning seeing. This raises the question of whether visual presentation has finally been devalued by information processing, as Sutherland and Coons yearned for sixty years ago. It is time to decide whether we want to train computers to see like designers or designers to see like computers.

References

Aristotle. 2009. *Nichomachean ethics*. Trans. David Ross. Revised with an Introduction and notes by Lesley Brown, xvi, 1, 23, 24, 43, 44, 104, 105–110, 238 n1140b. Oxford University Press.
Birkhoff, G.D. 1933. *Aesthetic measure*. Cambridge: Harvard Univ. Press.
Browning, J., and Y. LeCun. 2022. *AI and the limits of language, Noema*. The Berggruen Institute. https://www.noemamag.com/ai-and-the-limits-of-language/.
Coons, A.S. 1963. An outline of the requirements for a computer-aided design system. In *Proceedings, spring joint computer conference*, 299–304.
Earl, C.F., and I. Jowers. 2024. *Grids: Schemas, rules and cases, shape computation: Fifty years 1972–2022 (this volume)*. Springer.
Eysenck, H.J. 1942. The experimental study of 'Good Gestalt'—A new approach. *Psychological Review* 49: 344–364.

Gopalan, V. 2023. *Symmetry classification and enumeration of square tile sikku kolams.* arXiv: 2304.14134v1.
Jowers, I., and C. Earl. 2010. The construction of curved shapes. *Environment and Planning B: Planning and Design* 37: 42–58.
Jowers, I., and C. Earl. 2011. Implementation of curved shape grammars. *Environment and Planning B: Planning and Design* 38: 616–635.
Knight, T.W. 1986. Transformations of the meander motif on Greek geometric pottery. *Design Computing* 1 (1): 29–67.
Knight, T. 2018. Craft, performance, and grammars. *Computational studies on cultural variation and heredity*, ed. J.H. Lee. KAIST research series. Singapore: Springer.
Knight, T., and G. Stiny. 2015. Making grammars: From computing with shapes to computing with things. *Design Studies* 41, 8–28.
Krstic, D. 2010. Approximating shapes with hierarchies and topologies. *Artificial Intelligence for Engineering Design, Analysis and Manufacturing* 24: 259–276.
Krstic, D. 2018. Grammars for making revisited. *Design computing and cognition DCC'18*, ed. John S. Gero, 519–538. Springer.
Kristic, D. 2022. Diagonal decompositions of shapes and their algebras. *Artificial Intelligence for Engineering Design, Analysis and Manufacturing* 36: e10.
LeCun, Y. 2023. *The impact of ChatGPT and other large language models on physics research and education.* Keynote address, Department of Physics, MIT. https://www.youtube.com/watch?v=vyqXLJsmsrk.
Leeuwenberg, E., and P. Van der Helm. 2013. *Structural information theory: The simplicity of visual form*, 25, 34. Cambridge: Cambridge University Press.
Leyton, M. 1992. *Symmetry, causality, mind*, 1–7. Cambridge: The MIT Press.
McCarthy, J. 1959. Programs with common sense. In *The proceedings of the symposium on the mechanization of thought processes*, National Physical Laboratory, Teddlngton, Middlesex, 24–27 November 1958.
Ruskin, J. 1853. *The elements of drawing in three letters to beginners*, xi, xv, 21, 22. Aquitaine Media Corp.
Stiny, G. 1976. Two exercises in formal composition. *Environment & Planning B: Planning & Design* 3: 187–210.
Stiny, G. 1991. The algebras of design. *Research in Engineering Design* 2: 171–181.
Stiny, G. 1996. Useless rules. *Environment & Planning B: Planning & Design* 23: 235–237.
Stiny, G. 2006. *Shape: Talking about seeing and doing*, 3, 81, 90, 182, 184. Cambridge: The MIT Press.
Stiny, G. 2022. *Shapes of imagination: Calculating in Coleridge's magical realm*, xi, 24. Cambridge: The MIT Press.
Stiny, G., and L. March. 1981. Design machines. *Environment and Planning B* 8: 245–255.
Sutherland, I. 1963. Structure in drawings and the hidden-surface problem. In *Reflections on computer aids to design and architecture*, vol. 1975, ed. N. Negroponte, 73–77. New York: Petrocelli-Charter.
Waring, T.M. 2012. Sequential encoding of Tamil kolam patterns. In *Society for science on form*, Japan, 83–92.
Wittgenstein, L. 1976. *Wittgenstein's lectures on the foundations of mathematics, Cambridge 1939*, ed. Cora Diamond, 38, 42, 61, 62, 77, 95, 275. Chicago: The University of Chicago Press.

Open Access This chapter is licensed under the terms of the Creative Commons Attribution-NonCommercial-NoDerivatives 4.0 International License (http://creativecommons.org/licenses/by-nc-nd/4.0/), which permits any noncommercial use, sharing, distribution and reproduction in any medium or format, as long as you give appropriate credit to the original author(s) and the source, provide a link to the Creative Commons license and indicate if you modified the licensed material. You do not have permission under this license to share adapted material derived from this chapter or parts of it.

The images or other third party material in this chapter are included in the chapter's Creative Commons license, unless indicated otherwise in a credit line to the material. If material is not included in the chapter's Creative Commons license and your intended use is not permitted by statutory regulation or exceeds the permitted use, you will need to obtain permission directly from the copyright holder.

Flows: Composite Shape Rules

Rudi Stouffs and Dan Hou

Abstract Generally, non-terminal symbols such as labeled points are used to constrain rule application and, thereby, guide rule selection in the application of shape grammars. However, distinguishing between salient rules that offer the user design choices and deterministic rules that together, and in a certain order, (mechanically) complete a specific design transformation may require other means of guiding rule selection that better reflect on the logic of the rule derivation process. We present a concept of composite shape rules embedding algorithmic patterns for rule automation. We denote these composite shape rules *flows* and adapt the notation from regular expressions. In this paper, we describe the context that led to the conception of this approach, describe the sequencing mechanisms, and present a case study. We conclude with a discussion disclosing additional potential of the notation.

1 Introduction

Shape grammars are a formal rewriting system for producing languages of shapes (Stiny 1980). At a minimum, a shape grammar consists of a set of productions, or shape rules, operating over a vocabulary of spatial elements, e.g., line or plane segments, optionally augmented with qualitative attributes, e.g., line thicknesses or colors. Then, a shape is defined as any composition of (augmented) spatial elements, and a shape rule as any combination of a left-hand-side shape and a right-hand-side shape, where the former cannot be empty. A shape rule applies to a shape if a transformation can be determined under which the left-hand-side of the shape rule is a part of the given shape. Rule application involves replacing this part with the right-hand-side of the shape rule under the same transformation.

R. Stouffs (✉)
Department of Architecture, National University of Singapore, 4 Architecture Drive, Singapore 117566, Singapore
e-mail: stouffs@nus.edu.sg

D. Hou
School of Architecture, Southwest Jiaotong University, West High-Tech Zone, Chengdu 611756, Sichuan, China

Traditionally, a distinction is made between the terminal vocabulary of (augmented) spatial elements and a non-terminal vocabulary of symbols or markers, e.g., labeled points (Stiny 1980; Yue and Krishnamurti 2013). In this case, a shape and, by extension, both sides of a shape rule may combine both spatial elements and symbols from the respective vocabularies. A shape grammar conceived in this way generally includes an initial shape as the starting point in the productive (generative) process. The language defined by such a shape grammar is the set of shapes generated by the grammar from the initial shape that do not contain any non-terminal symbols.

While this traditional approach may seem overly formalistic in comparison to simply "calculating with shapes" (Stiny 2006), most shape grammars presented in literature do adopt a notion of non-terminal symbols. Often, shape grammars emanate from an analysis of a particular body of architecture or are designed to generate a particular type or style of building. Any restrictions posed on the generative language generally necessitate constraining the application of rules. Non-terminal symbols are commonly used to constrain rule application and, thereby, guide rule selection. In addition, a shape grammar may need to specify many rules, further exacerbating the problem of constraining rule applications. While rules may be collected into stages, stages often are defined to rely on distinct non-terminal symbols.

The downside of using non-terminals is that they clutter the shape rule and make both the rule and its role in the derivation process more difficult to understand. Such an understanding must necessarily include the role of these non-terminals. Unfortunately, there is usually little relationship between the specification of non-terminals (including their naming) and the logic of the rule derivation process, beyond the identification of stages.

Our motivation for this study comes from an active development of a design grammar using railway station design as a demonstration study. Design grammars, also termed "grammars for designing" (Beirão et al. 2009), denote grammars that are progressively developed by designers for a new design context. They are distinct from analytical grammars that are developed from a specific body of designs, e.g., similar buildings by the same architect, and are constrained to only generate designs from that body or designs that can be assessed as belonging to the same body. Developing an analytical grammar involves systematically determining all possible rule variations and encoding these into a grammar. Rule variations are necessarily finite, and the encoding is done by the developer of the grammar, not by the user. Rule complexity is therefore not much of an issue and non-terminal symbols can be used to guide rule selection and derivation. For a design grammar, however, the designer is both the developer and the user of the rules; rule development becomes an important issue and reducing the need for using non-terminal symbols in the specification of shape rules a critical objective.

In this paper, we consider an algorithmic approach to rule sequencing, defining composite rules composed of shape rules or other composite rules and including sequencing directives. Firstly, an overview of the literature on rule sequencing is presented. Subsequently, the selected sequencing mechanisms are described and explained. Next, two examples and a case study are presented. Finally, a discussion discloses additional potential of the notation.

2 Rule Sequencing

The literature on sequencing of shape rules is not very extensive. Knight (1999) defines a deterministic grammar as a grammar imposing a restriction on rule ordering, without any restriction on rule format. She considers at least two different ways to define a deterministic grammar. Firstly, a function can be associated with any grammar to determine for each step of a derivation which rule to apply next (and under which transformation). Secondly, rules are distinguished (e.g., using non-terminal symbols) in such a way that at most one rule is applicable (under only one transformation) in each step of a derivation (Knight 1999).

Liew (2004) similarly considers deterministic rules as rules that together and in a certain order (mechanically) complete a specific design transformation. He distinguishes deterministic rules from salient rules, which offer the user design choices. Specifically, Liew (2004) conceives of an explicit sequence of rules, denoted a "directive" or "macro". This directive takes the form of an if-then-else control structure where the condition is a rule application that either succeeds (true) or fails (false) and the consequence (then or else) is another rule application, possibly applying the same rule. Either the then (success) or else (failure) part of the control structure can be omitted, allowing the rule sequence to end. The first rule in the rule sequence is denoted the primary rule, the others secondary rules. It is understood that primary rules are salient rules, whereas secondary rules belong to the deterministic category. Whereas Knight deems deterministic as emergent from the fact that in each step of the derivation at most one rule applies under a single transformation, Liew considers a prescriptive, algorithmic approach. While the latter also uniquely prescribes which rule to apply, there is no guarantee that the rule applies under a single transformation. If multiple transformations apply, one needs to be selected (randomly) for the algorithm to proceed automatically.

Grasl and Economou (2014) present different approaches to automating rule selection using rule selection agents that implement a sensor-logic-actuator mechanism. The sensor allows an agent to inquire about the shape at hand or apply "control" rules to determine whether a specific condition is met. These control rules do not make any changes to the shape, that is, their left-hand-side and right-hand-side are identical. The actuator allows an agent to apply a specific rule or undo the previous rule application. Grasl and Economou (2014) consider different agents implementing different rule selection approaches. For example, one agent performs a depth-first search over a set of rules to enumerate all possible derivations. Another agent randomly picks a rule, though its performance may be improved using sequencing or weighted randomness. Variant approaches include a genetic algorithm approach and a rule-based approach. In the latter case, a backward chaining technique is used to find the appropriate rule sequence, using control rules to decide which action to take next (Grasl and Economou 2014).

Grasl and Economou do not distinguish salient from deterministic rules, assuming that all rules are salient to some extent. This is very much in line with Stiny's (2006) concept of calculating with shapes, which amounts to having a grammar with a limited number of rules that can generate an unlimited number of designs. Unfortunately, shape grammars that support the generation of architectural designs often require many rules and, while the creative process of salient rule development and selection is obviously of primary interest to the designer, deterministic rules may need to complement salient rules to achieve a desired result. As such, this work is motivated by the desire to be able to automate the application of deterministic rules following the selection and application of a salient rule. That is, we recognize that while design grammars are developed by the user of the grammar, thereby necessitating simpler rules, parts of a design grammar may need to be automated or semi-automated. Instead of suggesting the use of non-terminals to guide automated rule selection, we propose the use of composite shape rules embedding algorithmic patterns for rule automation.

3 Flows

We denote composite shape rules embedding algorithmic patterns as *flows*, to clearly distinguish shape rules from composite shape rules, where necessary. At the basic level, we support sequence, iteration, and selection as algorithmic patterns for *flows*. Given a rule r specified in the form $lhs \rightarrow rhs$, with lhs and rhs denoting the left-hand-side and right-hand-side of the rule, respectively, rule r applies to a shape s if there exists a transformation $t \in T$ such that $t(lhs) \leq s$. The set T is generally considered to contain all similarity transformations, that is any composition of translation, rotation, reflection, and uniform scaling. Then, the application of rule r to s under transformation t yields the shape $s - t(lhs) + t(rhs)$.

Two rules r_1 ($lhs_1 \rightarrow rhs_1$) and r_2 ($lhs_2 \rightarrow rhs_2$) apply in sequence if upon a successful application of r_1, r_2 is applied to the shape resulting from the application of rule r_1. Considering Liew's (2004) "directive", this can be graphically represented as in Fig. 1(left). Algorithmically, this may be written in the form:

```
if ∃ t₁∈ T: t₁(lhs₁) ≤ s then
    s ← s - t₁(lhs₁)+t₁(rhs₁)
    if ∃ t₂ ∈ T: t₂(lhs₂) ≤ s then
        s ←s - t₂(lhs₂)+t₂(rhs₂)
    end
end
```

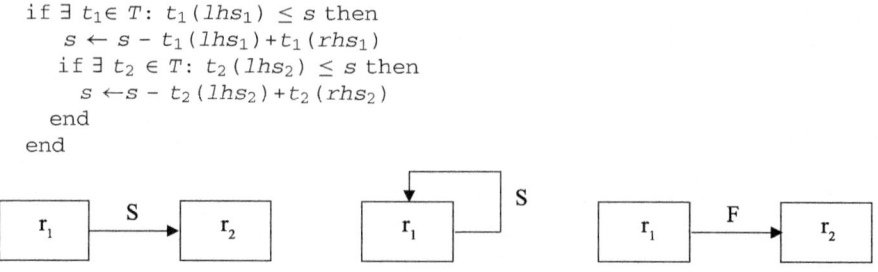

Fig. 1 Sequence (left), iteration (center) and selection (right), graphically represented using Liew's (2004) "directive". Rule application, represented as a rectangle, may be successful or not. Depending on success or failure, the next rule to be applied is indicated by an arrow marked "S" or "F", respectively. Such an arrow is omitted if no subsequent rule is available

A single rule r_1 can be applied iteratively if upon every successful application of r_1, a new application is attempted (Fig. 1(center)). Algorithmically, this can be expressed using a while-do construct:

```
while ∃ t₁ ∈ T: t₁(lhs₁) ≤ s do
  s ← s - t₁(lhs₁) +t₁(rhs₁)
end
```

Finally, selection specifies two (or more) alternative rules. These are attempted to be applied in order and as soon as one rule applies, the remaining rules are ignored (Fig. 1 (right)). Algorithmically, this can be written as follows:

```
if ∃ t₁ ∈ T: t₁(lhs₁) ≤ s then
  s ← s - t₁(lhs₁)+t₁(rhs₁)
else if ∃ t₂ ∈ T: t₂(lhs₂) ≤ s then
  s ← s - t₂(lhs₂) + t₂(rhs₂)
end
```

3.1 Backtracking

The sequential application of two rules may assume the successful application of both rules. Both Fig. 1 and the algorithm above only require rule r_1 to be successful. If rule r_2 subsequently fails, rule r_1 will remain applied. However, in many cases, we may want rule r_1 to apply only if r_2 subsequently applies as well, that is, rules r_1 and r_2 are considered as (part of) a sequence of rules that all apply or none. This is difficult to express algorithmically using the constructs above (if–then–else and do-while), as we would need to combine the two conditions, whether rules r_1 and r_2 apply, as well as the intermediate calculation of the result of rule r_1, all within the condition of a single if–then structure before actually applying both rules. If the sequence contains more than two rules, the algorithmic expression will only be more complicated. However, such is simply an application of backtracking which can be achieved using a recursive algorithmic structure. More importantly, Liew's (2004) "directive" does not support backtracking, as such, we must adopt a different way of expressing a *flow* (composite shape rule).

Instead, we adapt the notation from regular expressions. Regular expressions are patterns that are used to match strings by string searching algorithms. Regular expressions are composed of tokens that are combined in a prescribed order, with some variation built into the expression, to match a goal string. Similarly, *flows* are composed of shape rules that are combined in a prescribed order, with some algorithmic variation, to produce a valid final shape. In both cases, partial matches may not lead to any final result, requiring backtracking to undo the partial match and attempt a different match. The main difference lies in the fact that in the case of regular expressions, the goal string is given and can guide the matching process. In the case of composite shape rules, the goal is to arrive at any valid final shape. As such, there is no guide but the algorithmic expression of rules, and a purely trial-and-error approach must be adopted.

Another difference lies in the vocabulary of terminals. In the case of regular expressions, these are any character that can be represented in a string. These characters are finite and ordered. As such, regular expressions generally use shortcuts such as "." to match any single character, "a-z" to indicate any letter between "a" and "z", "a" and "z" included, and "[^abc]" to match any character other than "a", "b" or "c". The latter may be useful when explicitly searching for combinations of "a", "b", and "c", treating any other characters simply as separation characters, thereby ignoring the specific characters used for separation, or vice versa. In the case of *flows*, we also have a finite set of rules. However, they are usually unordered, or only partially ordered, and we tend to be only interested in a very limited subset of rules at any one time. As such, we require rules to be always explicitly enumerated, rather than be identified as a group or by exclusion.

3.2 A Notation from Regular Expressions

Let us revisit the algorithmic structures of sequence, selection, and iteration. In a regular expression, a sequence of characters can be literally explicated as such, e.g., "abc" matches the substring "abc". In the case of rules, we use rule names to identify individual rules and separate these rule names with spaces. Note that rule names can be required to be identifiers, that is, any combination of letters, digits, or the underscore symbol ("_"), excluding any spaces or other special characters that could be misinterpreted. Thus, "r_1 r_2" (or "r1 r2") is a representation for the sequence of two rules r_1 and r_2.

In terms of selection, matching one from a series of alternative tokens, regular expressions commonly offer two variant notations. The first one, which we have touched upon before, uses square brackets to collect alternative tokens. We adopt the same notation for the selection and application of a single rule from several alternative rules, separating the rule names with spaces and enclosing them within square brackets. Thus, "[r_1 r_2]" (or "[r1 r2]") is a representation for the selection of one from two rules r_1 and r_2. The second notation commonly used within regular expressions is to separate the tokens (or sequences or groups of tokens) with vertical bars. We omit this notation for *flows*. Instead, as is possible within regular expressions as well, we allow for the grouping of a sequence of rules within parentheses. Thus, "[r_1 (r_2 r_3)]" (or "[r1 (r2 r3)]") is a representation for the selection between rule r_1 and the sequence of rules r_2 and r_3. A grouping of rules can be considered to specify a composite rule; both square brackets and parentheses group rules into composite rules. Much of the notation below applies equally to rules and composite rules; we write "(composite) rule" when we want to ignore the distinction between a single rule or a group of rules. Otherwise, we use the term sub-*flow* to denote a grouping of rules as distinct from a single rule.

In a regular expression, a quantifier after a character (or token) specifies how often the character (or token) is allowed to occur. We allow the same quantifiers to be used in the expression of *flows*. These quantifiers are distinguished (Table 1):

- A question mark ("?") indicates at most one (zero or one) application of the preceding (composite) rule.
- An asterisk ("*") indicates any number (zero or more) of applications of the preceding (composite) rule.
- A plus sign ("+") indicates any number, excluding zero, (one or more) of applications of the preceding (composite) rule.
- Any positive number n within curly brackets ("$\{n\}$") requires the preceding (composite) rule to be applied exactly n times.
- Any positive number n followed by a comma, within curly brackets ("$\{n,\}$") requires the preceding (composite) rule to be applied minimally n times.
- Any two positive numbers n and m, separated by a comma and within curly brackets ("$\{n, m\}$") requires the preceding (composite) rule to be applied minimally n times, but no more than m times.

3.3 A Greedy Algorithmic Approach

In the case of iterating a (composite) rule, the quantifier only specifies how many times the (composite) rule may be applied. For example, in the case of the asterisk, the preceding (composite) rule may apply zero, one or more times. We may interpret this to mean as many times as possible, however, we may as well opt to not apply the (composite) rule at all. In a regular expression, how many times a token, when followed by an asterisk, is matched usually is dependent on the string that is being matched. For example, matching the regular expression "a*b" to the string "aaab" will normally force the token "a" to be matched three times such that the token "b" can match the letter "b". Similarly, given a *flow* "r_1* r_2", there may be only one solution for the number of times rule r_1 must be applied for the application of rule r_2 to succeed.

However, often, there may be multiple solutions involving a different number of applications for rule r_1 leading to a successful application of rule r_2. For example, in Stiny and Mitchell's (1978a) Palladian grammar, in stage 1 (grid definition), rule 5 allows for the extension of the grid by adding two additional grid columns, one on either side of the existing grid columns. When applying the grammar to generate the plan for the Villa Malcontenta, rule 5 is applied exactly once, yielding a grid that is five columns wide. While most of Palladio's villa plans are based on a five by three grid, Palladio also presents a villa ground plan in Book 2 of his Quattro Libri (Palladio 1965) based on a three-by-three grid (Stiny and Mitchell 1978b). As such, we might choose to use the quantifier "?" for rule 5, allowing rule 5 to be applied either zero or one time. Nevertheless, since both zero and one time are equally

Table 1 Overview of the regular expression-inspired notation for *flows* (and sub-*flows*), under a greedy approach. The metacharacter "!" is adopted as logical negation

Metacharacter	Description
␣	A space separates two (composite) rules in a **sequence** or **selection**. In a sequence, if either rule fails to apply, the entire sequence fails to apply. In a selection, only one rule needs to succeed for the selection to succeed
(…)	Parentheses enclose a **sequence** of (composite) rules. Rules are attempted to be applied one after the other, each time on the result of the previous application, in the order specified. If one of the rules fails, backtracking will occur. A sequence specifies a *sub-flow*
[…]	Square brackets enclose a **selection** of alternative (composite) rules. Alternative rules are attempted to be applied in the order specified. As soon as one application succeeds, subsequent rules are skipped. If no alternative applies, backtracking will occur. A selection specifies a *sub-flow*
[*…]	Square brackets enclose a **selection** of alternative (composite) rules. When the first character within square brackets is an asterisk, the alternative rules are attempted to be applied in a random order instead of in the order specified. If no alternative applies, backtracking will occur. A selection specifies a *sub-flow*
!	Success and failure of the succeeding (composite) rule are inverted. If the rule fails, the application succeeds, whereas if the rule succeeds, backtracking occurs
?	The preceding (composite) rule may apply zero or one time. A single application is attempted. Success or failure, no backtracking occurs, unless backtracking arrives from a later point to this rule and all alternatives within the application of this (composite) rule have been exhausted
*	The preceding (composite) rule may apply zero, one or more times. The iteration proceeds until the rule fails to apply. No backtracking occurs, unless backtracking arrives from a later point to this rule and all alternatives within the application of this iterative (composite) rule have been exhausted
+	The preceding (composite) rule may apply one or more times. The iteration proceeds until the rule fails to apply. Backtracking only occurs if the rule fails at the very first time, unless backtracking arrives from a later point to this rule and all alternatives within the application of this iterative (composite) rule have been exhausted
{n}	The preceding (composite) rule may apply exactly n times. The iteration proceeds until n consecutive applications or until the rule fails to apply. Backtracking occurs if fewer than n consecutive applications succeed or, upon backtracking from a later point to this rule, if all alternatives within the application of this iterative (composite) rule have been exhausted
{$n,$}	The preceding (composite) rule may apply n or more times. The iteration proceeds until the rule fails to apply. Backtracking occurs if fewer than n consecutive applications succeed or, upon backtracking from a later point to this rule, if all alternatives within the application of this iterative (composite) rule have been exhausted
{n,m}	The preceding (composite) rule may apply any number of times between n and m. The iteration proceeds until m consecutive applications or until the rule fails to apply. Backtracking occurs if fewer than n consecutive applications succeed or, upon backtracking from a later point to this rule, if all alternatives within the application of this iterative (composite) rule have been exhausted

applicable in generating valid designs, how should the algorithm select between these two options?

In fact, the same issue applies to regular expressions, as in some cases there might seem to be multiple valid options at first. For example, matching the regular expression "a*.b" to the string "aaab" should force the token "a" to be matched twice, with the token "." matched to the third letter "a", such that the token "b" can match the last letter "b". However, the string-matching algorithm may first attempt to match the token "a" three times, followed by the token "." (matching the letter "b"), before realizing that the regular expression cannot be matched in this way as there is no letter "b" remaining. It may then backtrack and reduce the number of matches of the token "a" by one, leading to a complete match. Alternatively, it can first try skipping the token "a", backtrack and match the token "a" once, and backtrack once again before matching the token "a" twice to successfully lead to a complete match. The first approach is denoted greedy matching, the second one lazy matching. Greedy matching is usually the default behavior for regular expressions.

Similarly, we apply a greedy approach to the application of *flows*, by default, but allow for other approaches to be specified as well. Firstly, we allow for a lazy approach for iterations, such that the number of iterations is kept as low as possible. The specification of a lazy approach is achieved by adding a question mark ("?") as a suffix to the iteration specification. Note that backtracking expresses a different behavior in the case of a lazy approach than in the case of a greedy approach. In the latter, as an iteration is applied as many times as possible, backtracking may undo the last application in the iteration and try the remainder of the *flow* again with one fewer application. In the case of a lazy approach, instead, the iteration is applied as few times as possible, and backtracking may add another application to this iteration. In both cases, backtracking may eventually lead to the entire iteration being undone, backtracking to the rule or sub-*flow* preceding the iteration.

Next to greedy and lazy matching, regular expressions commonly also allow for a possessive matching, which is a form of greedy matching, but disallows backtracking. That is, in the example of matching the regular expression "a*.b" to the string "aaab", a possessive matching would fail as the algorithm will match the token "a" three times, followed by the token "." (matching the letter "b"), and being unable to backtrack. Similarly, in rule application, sometimes it is appropriate to disable backtracking of an iteration. In a possessive approach, the iterated (composite) rule will be applied as many times as possible, until rule application fails, or the maximum number of iterations has been reached. If, upon backtracking from a later point, the focus of application returns to the possessive iteration, no alternative number of applications within this iteration will be attempted and, instead, backjumping will occur to a rule or sub-*flow* preceding the iteration. The specification of a possessive approach is achieved by adding a plus sign ("+") as a suffix to the iteration specification.

In addition, next to greedy, lazy, and possessive approaches, we also consider a probabilistic approach to an iteration. Let us assume that we allow a (composite) rule to be applied any number of times between some minimum and maximum number of times. Rather than trying to iterate the rule as many times as possible and allowed (greedy and possessive) or iterating the (composite) rule as few times as allowed (lazy), the algorithm could randomly select any value between the minimum and maximum values as the specific number of times to try to iterate the (composite) rule. To clarify, in the example of the width of the grid for Palladio's villa plans, a greedy algorithm would initially search for grids that are five columns wide, a lazy algorithm for grids that are three columns wide, while a probabilistic matching algorithm would randomly choose between a five-column wide grid and a three-column wide grid to start. Where the question mark and plus sign are used to specify lazy and possessive matching, respectively, we adopt the asterisk ("*") as a suffix to the iteration specification to indicate a probabilistic approach (Table 2).

3.4 Selecting from Multiple Rules and Valid Applications

Other issues that must be dealt with, different from regular expressions, are the order of selection from among alternative (composite) rules, and the order of selection from among multiple valid applications of a same rule.

In the case of selection, the alternative (composite) rules are necessarily specified in some order. If we are interested in all possible derivations, the order is of little concern, but if we are exploring only a few derivations, we might not want these derivations to attempt to apply the (composite) rules each time in the same order. Therefore, we consider two alternative orders. The first is the order of the (composite) rules in the selection group. This is the default. The second is a random reordering of the (composite) rules in the selection group for the purpose of (composite) rule selection and application. To achieve this random reordering, we precede the selection by an asterisk ("*"), that is, the asterisk is the first symbol within the square brackets identifying the selection. This is in line with the choice for the asterisk as a suffix to identify a probabilistic approach to iteration. For example, in the case of "[*r_1 r_2]" (or "[*r1 r2]"), the two rules r_1 and r_2 will be attempted to be applied in any order. Obviously, as soon as one application succeeds, any other (composite) rules that have not been tried yet are skipped.

In the case of multiple valid applications of the same rule, one application must necessarily be selected. Any ordering of valid alternative applications will be the result of the matching algorithm and may not be obvious to the user. For this reason, we consider a random selection of a single application from among multiple valid applications of the same rule. Only in a very few cases may a fixed ordering be appropriate. For example, in the case of developing and testing a *flow*, when some debugging is necessary, it may be appropriate for the application of a rule to always yield the same result, to be able to compare the *flow* result before and after debugging. For this, we do not provide any notation, but include a parameter within the implementation that can be set to apply to any *flow* in its entirety. Considering these selection mechanisms, note that *flows* are not exactly deterministic and generally include a probabilistic aspect.

In addition, the fact that a single rule may allow for multiple valid applications, in contrast to regular expressions, means that backtracking must be extended to any

Table 2 Overview of the extended regular expression-inspired notation for *flows* (and sub-*flows*) under a possessive approach (" + "), a lazy approach ("?") and a probabilistic approach ("*"). The last approach does not apply to regular expressions

Metacharacter	Description
?+	The preceding (composite) rule may apply zero or one time. No backtracking occurs. When backtracking arrives from a later point to this rule, rather than backtracking within this (composite) rule, backjumping takes place to a point before this rule
*+	The preceding (composite) rule may apply zero, one or more times. The iteration proceeds until the rule fails to apply. No backtracking occurs. When backtracking arrives from a later point to this rule, rather than backtracking within this iterative (composite) rule, backjumping takes place to a point before this rule
++	The preceding (composite) rule may apply one or more times. The iteration proceeds until the rule fails to apply. Backtracking only occurs if the rule fails at the very first time. When backtracking arrives from a later point to this rule, rather than backtracking within this iterative (composite) rule, backjumping takes place to a point before this rule
{n}+	The preceding (composite) rule may apply exactly n times. The iteration proceeds until n consecutive applications or until the rule fails to apply. Backtracking occurs if fewer than n consecutive applications succeed. When backtracking arrives from a later point to this rule, rather than backtracking within this iterative (composite) rule, backjumping takes place to a point before this rule
{n,}+	The preceding (composite) rule may apply n or more times. The iteration proceeds until the rule fails to apply. Backtracking occurs if fewer than n consecutive applications succeed. When backtracking arrives from a later point to this rule, rather than backtracking within this iterative (composite) rule, backjumping takes place to a point before this rule
{n,m}+	The preceding (composite) rule may apply any number of times between n and m. The iteration proceeds until m consecutive applications or until the rule fails to apply. Backtracking occurs if fewer than n consecutive applications succeed. When backtracking arrives from a later point to this rule, rather than backtracking within this iterative (composite) rule, backjumping takes place to a point before this rule
??	The preceding (composite) rule may apply zero or one time. Application is skipped at first. When backtracking arrives from a later point to this rule, then application is tried. Backtracking occurs if all alternatives within this iterative (composite) rule have been exhausted
*?	The preceding (composite) rule may apply zero, one or more times. Application is skipped at first. When backtracking arrives from a later point to this rule, then application is tried or, eventually, repeated. Backtracking occurs if all alternatives within this iterative (composite) rule have been exhausted
+?	The preceding (composite) rule may apply one or more times. A single application is tried at first. If successful, additional applications may be tried, but only upon backtracking from a later point to this rule. Backtracking occurs if the single application fails or if all alternatives within this iterative (composite) rule have been exhausted

(continued)

Table 2 (continued)

Metacharacter	Description
{n}?	The preceding (composite) rule may apply exactly *n* times. The iteration proceeds until *n* consecutive applications or until the rule fails to apply. Backtracking occurs if fewer than *n* consecutive applications succeed., or upon backtracking from a later point to this rule, if all alternatives within this iterative (composite) rule have been exhausted
{n,}?	The preceding (composite) rule may apply *n* or more times. The iteration proceeds until *n* consecutive applications or until the rule fails to apply. If successful, additional applications may be tried, but only upon backtracking from a later point to this rule. Backtracking occurs if fewer than n consecutive applications succeed or if all alternatives within this iterative (composite) rule have been exhausted
{n,m}?	The preceding (composite) rule may apply any number of times between *n* and *m*. The iteration proceeds until *n* consecutive applications or until the rule fails to apply. If successful, additional applications may be tried, but only upon backtracking from a later point to this rule, and never more than *m*. Backtracking occurs if fewer than *n* consecutive applications succeed or if all alternatives within this iterative (composite) rule have been exhausted
?*	The preceding (composite) rule may apply zero or one time. The one application may be skipped randomly. Backtracking only occurs upon backtracking from a later point to this rule and all alternatives within this (composite) rule have been exhausted
**	The preceding (composite) rule may apply zero, one or more times. The iteration proceeds a random number of times or until the rule fails to apply. When backtracking arrives from a later point to this rule, fewer or additional applications may be tried as well, in this order. Then, backtracking occurs if all alternatives within this iterative (composite) rule have been exhausted
+*	The preceding (composite) rule may apply one or more times. The iteration proceeds a random number of times (at least one) or until the rule fails to apply. When backtracking arrives from a later point to this rule, fewer or additional applications may be tried as well, in this order. Backtracking occurs if a single application already fails or if all alternatives within this iterative (composite) rule have been exhausted
{n}*	The preceding (composite) rule may apply exactly *n* times. The iteration proceeds until *n* consecutive applications or until the rule fails to apply. Backtracking occurs if fewer than *n* consecutive applications succeed., or upon backtracking from a later point to this rule, if all alternatives within this iterative (composite) rule have been exhausted

(continued)

Table 2 (continued)

Metacharacter	Description
{n,}*	The preceding (composite) rule may apply n or more times. The iteration proceeds a random number of times (at least n) or until the rule fails to apply. When backtracking arrives from a later point to this rule, fewer or additional applications may be tried as well, in this order. Backtracking occurs if fewer than n applications succeed or if all alternatives within this iterative (composite) rule have been exhausted
{n,m}*	The preceding (composite) rule may apply any number of times between n and m. The iteration proceeds a random number of times (at least n and at most m) or until the rule fails to apply. When backtracking arrives from a later point to this rule, fewer or additional applications may be tried as well, in this order. Backtracking occurs if fewer than n applications succeed or if all alternatives within this iterative (composite) rule have been exhausted

rule, even in the absence of iteration. That is, when, upon backtracking from a later point, the focus of application returns to a single rule that allowed for multiple valid applications, an alternative application should be selected from among these upon which the *flow* application process can once again move forward. Only when all valid applications have been attempted, one at a time, then backtracking will proceed to the preceding (composite) rule. We can extend the application of the possessive approach to individual rules that allow for multiple valid applications, to disallow backjumping to such rules. As a single plus sign is necessarily interpreted as an iteration, we suggest adopting the suffix "{1}+ " to a rule to indicate a possessive approach without iteration of this rule. Note that while the same suffix can be applied to a sub-*flow*, it only indicates the disallowance of backtracking or backjumping to this sub-*flow*, not within the sub-*flow*.

4 Examples: Restricting Applications

A shape rule combines a specification of recognition and manipulation, of search and replace. The left-hand-side of the shape rule specifies the pattern to be recognized, while the right-hand-side specifies the replacement pattern, relative to the left-hand-side. That is, rule application is only constrained by the left-hand-side and, specifically, by what is present in the left-hand-side and thus must be present in the shape under investigation. Additional constraints can be specified within description rules that accompany the shape rules (Stouffs 2018a). Descriptions are textual specifications that are intended to complement the geometric specifications of shapes. They were first proposed by Stiny (1981) to construct the verbal descriptions of design but have been extensively used to specify constraints on the application of shape rules. Shape descriptions are parametric in nature, that is, when constituting the left-hand-side of a description rule, it generally contains one or more parameters that can be matched onto parts of the description accompanying the shape under investigation. We refer to Stouffs (2018b) for a variety of applications adopting description rules to augment and constrain shape rules.

Despite the applicability of description rules to constrain rule applications, various conditions can be envisioned that cannot simply be expressed as a combination of spatial elements and accompanying descriptions. A simple example is the explicit absence of certain spatial elements. Shape rules only specify what spatial elements should be present and recognized, not what should be absent. Liew (2004) proposes the concept of a void descriptor that serves to specify an area being devoid of any spatial elements. For example, if the left-hand-side of a shape rule specifies a square and the void descriptor is used to denote the interior of the square to be void, this rule would only apply to any empty squares, not any other squares containing interior line segments or other spatial elements. Figure 2 shows an example presented by Raynard Yu, inspired by the façade of Lane 189 by UN Studio (2017). Identifying two different relationships among adjacent rhombi, he proposed two rules to explore compositions of rhombi. Rule 1 (Fig. 2, left bottom) adds a mirrored rhombus to one of the sides, rule 2 (Fig. 2, left above), instead, adds a 180°-rotated rhombus to one of the sides. Applying first rule 1 to any side of the initial rhombus, followed by rule 2, offers eight variations, two of which have overlapping rhombi (Fig. 2, right). To avoid such overlap, the void descriptor could be used to identify whether a new rhombus overlaps with any existing rhombus. However, since the void descriptor constrains the recognition part of rule application, an additional rule 3 would be needed to check if the most recently added rhombus is void inside or not. Identifying the most recently added rhombus can easily be achieved using a temporary label assigned to the rhombus. If the new void rule succeeds it can remove the temporary label at the same time. However, if rule 3 fails, backtracking must necessarily occur such that the addition of the rhombus is undone. This requires a simple flow sequence of two rules, rule 2 and rule 3. This flow sequence will only result in an application of rule 2 that adds a rhombus not overlapping an existing rhombus. The entire sequence may take the form *rule1 rule2 rule3*.

The following example includes iteration and negation in a three-step process of activation, assessment, and application. The three steps correspond to three rules (Fig. 3, top left). The initial shape is composed of four squares composing a larger square, with four points, one near each corner of the larger square (Fig. 3, bottom left). The *activate_rule* finds any square with a point near one of the corners and "activates" the square by adding a second point near the opposite corner point. Considering the specific distance of the point to the corner of the square, the rule only applies to each of the smaller squares and adds the second point near the center of the larger square. The rule is intended to be applied twice to activate two of the four squares. Subsequently, the *constrain_rule* searches for two adjacent points divided by the stem of a T-shape and removes it. In fact, this rule is never intended to be applied. In the flow, the rule is negated such that the negated rule succeeds when the rule itself fails. Hereto, we adopt the exclamation mark ("!") as logical negation (Tables 1 and 5). Success in this case means that no two adjacent points can be found, and that the activation rule has been applied to two diagonal squares.

Fig. 2 Two rules exploring compositions of rhombi. Applying the first rule to any side of the initial rhombus, followed by the second rule, offers eight variations, two of which have overlapping rhombi

If the *constrain_rule* were to succeed because two adjacent squares have been activated, the negated rule would fail, and backtracking would occur undoing one of the activations. Together, applying the *activate_rule* twice followed by the negation of the *constrain_rule* results in the flow *activate_rule*{2} !*constrain_rule*. Finally, the third rule finds an activated square and replaces it with another shape, in this case a triangle. Based on the symmetry of the left-hand-side shape with respect to the right-hand-side shape of the *triangle_rule*, there are two possible outcomes, each looking like a folded corner, where the corner is either folded open or closed (Fig. 3, right). The *triangle_rule* is also applied twice, to both activated squares. All possible outcomes are generated for a single activation outcome. To achieve this, a possessive approach is adopted for the activation part. The entire flow is written as.

(*activate_rule*{2} !*constrain_rule*){1}+ *triangle_rule*{2}.

Note that each unique result is generated twice, as the first application of the *triangle_rule* may be applied to either activated square.

Fig. 3 Three rules applied in a three-step process of activation, assessment, and application. All eight outcomes of the flow (*activate_rule*{2} !*constrain_rule*){1}+ *triangle_rule*{2} are shown, only four of which are unique

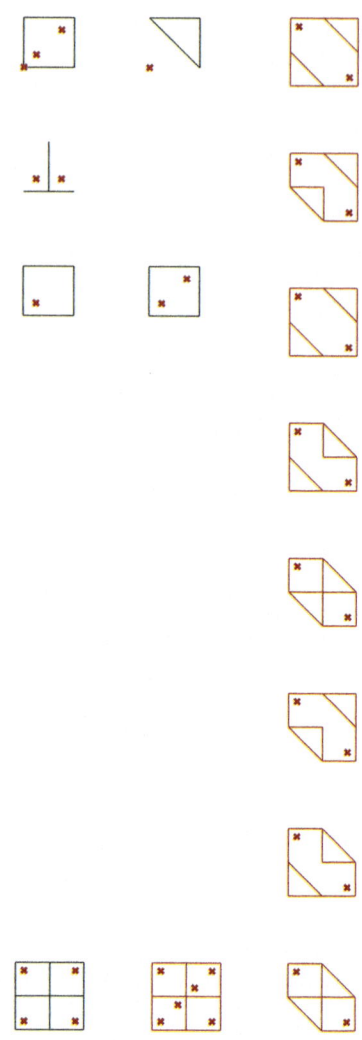

5 Study: Layout Generation Problem

For a more elaborative example, we consider an abstracted case study from the railway station design grammar previously mentioned (Hou and Stouffs 2018). It concerns a layout generation problem for which we prefer a semi-automated solution. We consider a grid composed of 25 cells with two predefined, adjacent spaces (Fig. 4 (above)). We consider a set of activities ("B1" through "B7") to occupy these cells. Each activity identifies how many times it should be made available and how much space must be provided (Table 3).

Flows: Composite Shape Rules

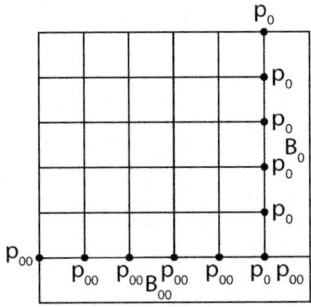

activities: {`("B_1", 1, 12, 2)`, `("B_2", 2, 5, 2)`, `("B_3", 1, 18, 5)`, `("B_4", 2, 8, 0)`, `("B_5", 1, 10, 3)`, `("B_6", 1, 4, 1)`, `("B_7", 1, 7, 1)`}

adjacencies: {`("B_0", "B_3")`, `("B_{00}", "B_1")`, `("B_1", "B_{00}")`, `("B_1", "B_7")`, `("B_2", "B_3")`, `("B_2", "B_5")`, `("B_3", "B_0")`, `("B_3", "B_2")`, `("B_3", "B_5")`, `("B_3", "B_6")`, `("B_5", "B_2")`, `("B_5", "B_3")`, `("B_6", "B_3")`, `("B_7", "B_1")`}

assigned: {`("p_0", "B_0", 20, 1)`, `("p_{00}", "B_{00}", 24, 1)`}

unassignedCount: {`9`}

Fig. 4 The initial shape and descriptions for the layout generation problem: (above) the grid of 25 cells, with two predefined, adjacent spaces, denoted "B0" and "B00"; (below) the initial descriptions encompassing all information from Table 3

Table 3 Activities to be assigned to the grid cells. For each activity, the number of spaces is specified and the minimum size of each space (a grid cell has an area of 4 units) as well as the adjacency requirements for each space

Activity	Count	Minimum area	Adjacency requirements							
			B0	B00	B1	B2	B3	B4	B5	B6	B7
B1	1	12		1							1
B2	2	5					1		1		
B3	1	18	1			1			1	1	
B4	2	8									
B5	1	10				1	1				
B6	1	4					1				
B7	1	7			1						

We adopt descriptions to encode all non-geometric information. Static information is encoded in the descriptions "activities" and "adjacencies". The results of the activity assignment process are encoded in the descriptions "assigned" and "unassignedCount". In addition, we consider a few descriptions to contain temporary information. These are "adjacenciesUse", "adjacenciesNece", "adjacenciesFound", "assignment", "surrounding" and "candidates" (Hou and Stouffs 2018). Finally, each geometric data type, i.e., points, line segments and plane segments, has a description attribute, respectively, "ptD", "lnD" and "plD". Figure 4 presents the initial shape and descriptions; Table 4 shows the rules for the layout generation problem.

Figure 5 offers an abstract, graphical illustration of the *flow*, presenting the groups of rules and their roles in the *flow* and including intermediary results from the derivation. Figure 6 offers an expanded graphical illustration of the *flow* distinguishing individual rules, using Liew's (2004) "directive".

There are 14 rules, collected into five groups, Activation (Ac), Assignment (As), Extension (Ex), Filling (Fi) and Checking (Ch). The Activation group contains a single rule (Ac_1), which randomly selects a grid cell and activates it by marking the grid as well as its four vertices (Fig. 5). As soon as a grid cell is activated, the derivation continues with rules from the Assignment group (Fig. 6). A first rule (As_1) detects an activity surrounding the cell and stores it in the "surrounding" description. This rule is repeatedly applied until all such activities have been identified (zero, one or more iterations). A second rule (As_2a) is then used to find an activity to be assigned (with count greater than one) that needs to be adjacent to one of these surrounding activities. If none is found, instead, rule As_2b finds any activity to be assigned (with count greater than one). Thus, As_2a and As_2b are alternative rules, with As_2a to be tried first. Then, rule As_3 assigns the activity found to the grid cell, using a transitional "assignment" description. It also initializes the as yet unsatisfied area as the difference between the activity's minimum area and the grid cell's area (identified as "planeSeg2D.area") If this assignment satisfies any adjacency requirements, this information is updated in rule As_4. As such, rule As_4 is repeated zero, one or more times. As_4 also clears the "surrounding" description with respect to these adjacent activities. As the last rule in the Assignment group, rule As_5 clears the information collected on other surrounding activities (zero, one or more iterations). This derivation, so far, can be described by the *flow*.

Ac_1 As_1* [As_2a As_2b] As_3 As_4* As_5*

Before a new grid cell can be activated, two tasks remain relating to the current assignment. The first task is to fulfill the area requirement, specifically, to reduce the area value in the "assignment" description to zero (or negative) as it represents the yet unsatisfied area. While this value remains positive, rule Ex_1 of the Extension group finds an adjacent grid cell and assigns the current activity to it, while updating the area value (zero, one or more iterations). The second task is to finalize the assignment, moving the assignment information from the transitional "assignment" description into the resulting "assigned" description, while also updating the activity count (how many times the activity should go through the assignment process). Rule Ex_2 performs this in a single application. At this stage, the process can now be repeated with a new activation, leading to the following *flow*:

(Ac_1 As_1* [As_2a As_2b] As_3 As_4* As_5* Ex_1* Ex_2)*

Although all activities should now have been assigned, this does not necessarily result in the grid being filled. The Filling group contains a single rule that is very similar to rule Ex_1 and takes a single cell and assigns any adjacent activity to it. At the same time, it updates the area value in the respective "assigned" description. At the very end of the process, the (negative) area value is a measure of the excessive space that has been assigned to the activity. After rule Ex_1 has been repeatedly

Flows: Composite Shape Rules

Table 4 The grammar rules addressing the layout generation problem. Thick lines represent line segments, thin lines denote a boundary, and a hatched area represents a plane segment. Description parameters are italicized. The predicate "void" indicates the area should be devoid of any geometry

Rule	Shape component and description components
Ac_1	Activate a grid cell "p".unassignedCount.*j* void → "AB" "p".unassignedCount.*j* unassignedCount: {`(j?>0)`} → {`(j−1)`}
As_1	Detect surrounding activities "AB" *y* → "AB" *y* assigned: {`(index?=ptD.x, name?=pID.y, area, count)`} → {`(index, name, area, count)`} surrounding: ø → {`(assigned.index, assigned.name)`}
As_2a	Find an activity with count greater than one that needs to be adjacent to one of the surrounding activities surrounding: {`(index, name)`} → {`(index, name)`} activities: {`(name, count?>0, area, adj_count)`} → {`(name, count, area, adj_count)`} adjacencies: {`(name_c?=surrounding.name, name_c?= activities.name)`} → {`(name_r, name_c)`} candidates: ø → {`(activities.name, activities.area, activities.adj_count)`}
As_2b	Find an arbitrary activity with count greater than one activities: {`(name, count?>0, area, adj_count)`} → {`(name, count, area, adj_count)`} candidates: ø → {`(activities.name, activities.area, activities.adj_count)`}
As_3	Assign the activity to the grid cell "AB" → ──── candidates.*name* candidates: {`(name, area, adj_count)`} → ø assignment: ø → {`(pID.x, candidates.name, candidates.area − planeSeg2D.area, candidates.adj_count)`}

(continued)

Table 4 (continued)

As_4	Update information on satisfied adjacency relations
	surrounding: {`(i, name)`} → ø
	assigned: {`( i?=surrounding.i, name?=surrounding.name, area, count)`}
	  → {`i, name, area, adj_cnt−1)`}
	assignment: {`(i, name, area, adj_cnt)`} → {`( i, name, area, adj_cnt−1)`}
	adjacenciesUse: ø
	  → {`(assigned.i, assignment.i)`, `(assignment.i, assigned.i)` }

As_5	Clear the information collected on surrounding activities surrounding: {`(index, name)`} → ø
Ex_1	Extend the assigned space if the area constraint is not yet satisfied 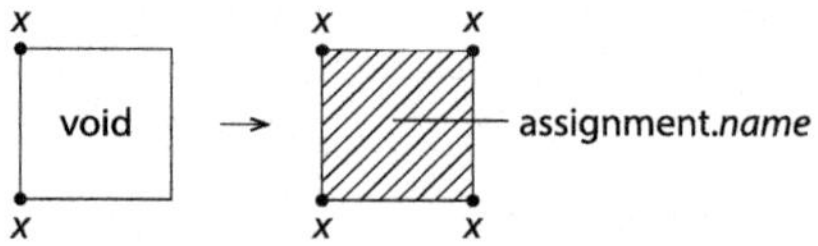 [diagram: void → assignment.name] assignment: {`(index?=pID.x, name, area?>0, adj_count)`} → {`(index, name, area − planeSeg2D.area, adj_count)`}

Ex_2	Finalize the assignment
	
	assignment: {`(index, name, area, adj_count)`} → ø
	activities: {`(name?= assignment.name, count, area, adj_count)`}
	  → {`(name, count−1, area, adj_count)`}
	assigned: ø → {`(assignment.index, assignment.name,`
	      `assignment.area, assignment.adj_count)`}

Fi_1	Assign any unassigned grid cells with the activity of an adjacent cell 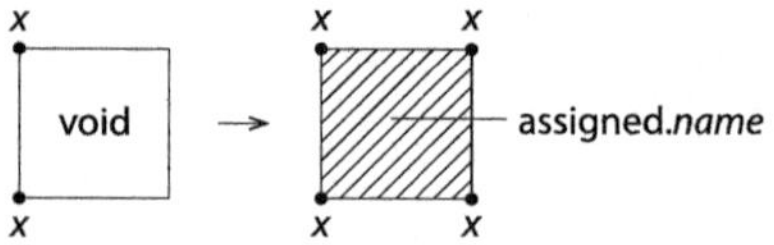 [diagram: void → assigned.name] assigned: {`(index?=pID.x, name, area?>0, adj_count)`} → {`(index, name, area − planeSeg2D.area, adj_count)`}

Ch_1	Identify any adjacencies that failed to be satisfied
	
	adjacencies: {`(name₁, name₂)`} → {`(name₁, name₂)`}
	assigned: {`(i₁, name₁?= adjacencies.name₁, area₁, adj_cnt₁?>0)`,
	      `(i₂, name₂?= adjacencies.name₂, area₂, adj_cnt₂?>0)`}
	  → {`(i₁, name₁, area₁, adj_cnt₁)`, `(i₂, name₂, area₂, adj_cnt₂)`}
	adjacenciesNece: ø → {`(assigned.i₁, assigned.i₂)`}

Ch_2	Check this information against adjacency relations marked as satisfied adjacenciesUse: {`(i, i)`} → {`(i, i)`} adjacenciesNece: {`(i₁?= adjacenciesUse.i, i_c?= adjacenciesUse.i_c)`} → ø

(continued)

Table 4 (continued)

Ch_3	Check this information against adjacency relations satisfied in the grid $\begin{matrix} x_1\,x_2 & & x_1\,x_2 \\ \bullet & & \bullet \\ \big| & \rightarrow & \big| \\ \bullet & & \bullet \\ x_1\,x_2 & & x_1\,x_2 \end{matrix}$ adjacenciesNece: $\{`(i_{r1}?=ptD.x_1, i_{c1}?=ptD.x_2)`, `(i_{r2}?=ptD.x_2, i_{c2}?=ptD.x_1)`\} \rightarrow \emptyset$ adjacenciesFound: $\emptyset \rightarrow \{`(ptD.x_1, ptD.x_2)`\}$
Ch_4	Update the information on related activities adjacenciesFound: $\{`(index_1, index_2)`\} \rightarrow \emptyset$ assigned: $\{`(index_1?=adjacenciesFound.index_1, name_1, area_1, cnt_1)`,$ $`(index_2?=adjacenciesFound.index_2, name_2, area_2, cnt_2)`\}$ $\rightarrow \{`(index_1, name_1, area_1, cnt_1-1)`, `(index_2, name_2, area_2, cnt_2-1)`\}$

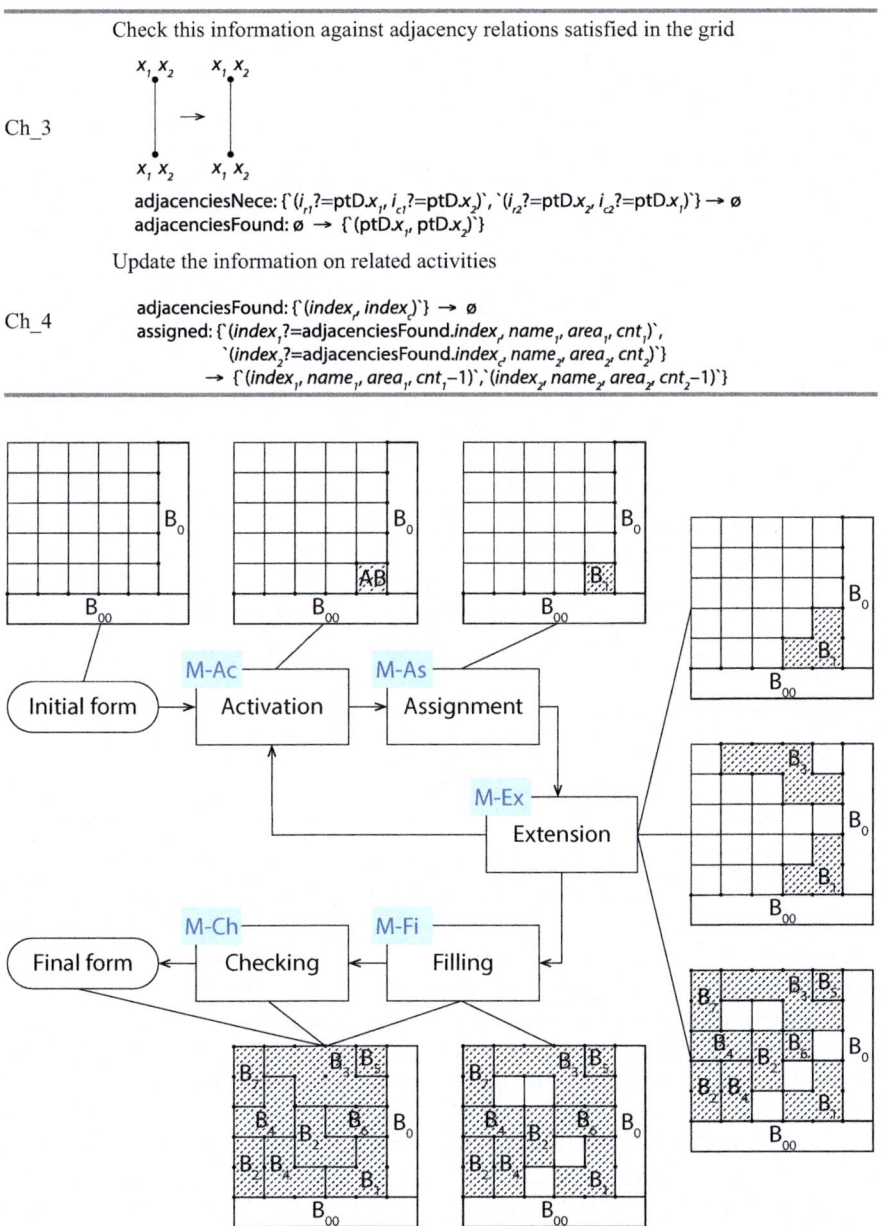

Fig. 5 A flowchart showing the groups of rules and their roles in the *flow*. Results from a derivation are also shown to illustrate the flowchart. Note that labels to points are omitted for the sake of clarity

Fig. 6 The complete *flow* graphically represented using Liew's (2004) "directive", with sub-flows identified by module (labeled, dashed boxes). Note that this visualization omits backtracking

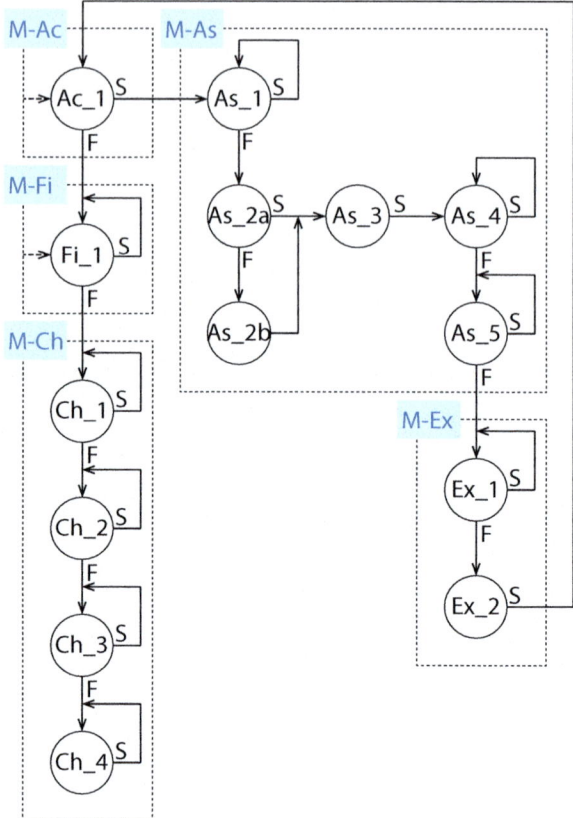

applied until all grid cells have been assigned an activity, the derivation continues by updating all adjacency relationship information. It is entirely possible that due to the application of rule Ex_1, additional adjacency constraints have been satisfied. It is important to capture this information as it serves to evaluate the success of the solution resulting from this single derivation. The Checking group of rules ensures the correctness of this information. The first step (rule Ch_1) identifies which adjacency pairs failed to be satisfied, by checking the adjacency count value in the "assigned" descriptions. However, it is possible that some adjacency count values are still positive even though these adjacency constraints are satisfied. Therefore, rule Ch_2 removes any "failed" adjacencies if these have already been captured in the "adjacenciesUse" description. Subsequently, rule Ch_3 checks if an adjacency constraint that is still marked as unsatisfied is fulfilled in the drawing. Finally, the information of related activities is updated by rule Ch_4. Each of rules Ch_1, Ch_2, Ch_3 and Ch_4 iterates zero, one or more times. The final *flow* becomes:

(Ac_1 As_1* [As_2a As_2b] As_3 As_4* As_5* Ex_1* Ex_2)* Fi_1* Ch_1* Ch_2* Ch_3* Ch_4*

6 Flow Patterns

The above case study demonstrates the applicability of shape grammars—using *flows*—to solve a design problem in an algorithmic way, organizing rules into sequences, selections, and iterations. While doing so may not be the primary value of a shape grammar, we argue it is rather unavoidable when designing with rules. Nevertheless, the case study only demonstrates the outcome of the *flow* development process and not the possible interaction of the user with the process. Obviously, the process of developing such a *flow* is not straightforward and involves trial and error. We envision assisting the designer in this process by providing him or her with a toolbox of simple patterns of rules or *flows*. For example, rule Ch-1 presents an example of requirement checking, iterating over the list of required adjacencies, and creating a new description for every failed adjacency requirement. The number of descriptions created in this way serves as a measure for success or failure. In particular cases, this search process may need to be reversed and extended, and involve making a copy of existing descriptions before removing any descriptions that actually meet the requirement. In an extension of the case study (Hou and Stouffs 2019) we explore different deductive and heuristic strategies as well, resulting in different *flows* (and a few different rules). An important aspect of the automation process is checking whether certain requirements (such as area and adjacency constraints) have been met. Various checking patterns can be identified that are more generally applicable than only in the case study and that can serve as templates for designers to develop their own *flows*. Obvious patterns are "conjunction" and "disjunction" that directly correspond to sequences and selections, respectively.

A "conjunction" applies when multiple requirements must all be satisfied and it prescribes a rule sequence where a violation of a single requirement requires a backjumping to an alternative design candidate (e.g., the rhombi example in Sect. 4). The conjunction pattern may require the application of the possessive approach, as a failure of one of the rules necessarily requires a backjumping to a point before the conjunction which may result in an alternative design candidate (e.g., the activate-constrain-apply example in Sect. 4). Especially noteworthy is the fact that the possessive approach, in this case, is not combined with an iteration, requiring the addition of the quantifier "{1}" to be able to specify the possessive approach (see Table 5). A "disjunction" applies when only one of several requirements must be satisfied and it prescribes a rule selection with implicit backtracking to identify a single requirement that is not violated.

An "enumeration" is a basic iteration, but one where the order of application within the iteration doesn"t matter. The "enumeration" pattern enumerates a set of occurrences and collects a description of them. This can often be achieved with a single rule, although an initialization rule may be necessary. The single rule will search for all occurrences and select a single occurrence that has not yet been described. As every occurrence will be matched once and the order has no effect on the outcome, the iteration should apply under a possessive approach, such that backjumping to the enumeration never takes place.

Table 5 Common flow patterns

Pattern	Exemplar flow
Conjunction	(r_cond1 r_cond2 ...){1}+ r_action
Disjunction	[r_cond1 r_cond2 ...] r_action
Enumeration	r_enum++
Negation	(r_copy* r_remove* !r_find){1}+
Iteration	r_init r_iterate* r_final

A less obvious pattern is "negation", that applies when the non-existence of something must be proven. Demonstrating a negative statement in a reasoning system may not be straightforward. A "negation" takes a contrary method to demonstrate that a description contains anything but the target data using, at a minimum, a sequence of three rules: the first rule iteratively makes a copy of the description; the second rule iteratively removes any other data from the copy; and the third rule checks for an empty description, to prove the negation. However, a rule specifies a left-hand-side that must match something. To check for nothing, we expect the rule to fail while allowing the *flow* application to proceed at the same time.

Stouffs and Hou (2017) implicitly present an iterative pattern, one that includes an initialization rule, an iteration rule, and a termination rule (Table 5). Specifically, they consider the generation of building forms whose height and number of floors is guided by a specified Gross Plot Ratio or Floor Area Ratio (see also, Janssen et al. 2016). This means that the number of subdivisions or the number of floors is initially unknown and thus cannot be incorporated into the shape rule. While constructing alternative shape rules for different numbers of floors would be undesirable, it is possible to iteratively add any number of floors using this iterative pattern. Here, the initialization rule determines the number of parts (e.g., floors), n, by comparing the (average) length or height, $(l_1 + l_2)/2$, to a target length l_t: $n = \lfloor l_1 + l_2 + 0.5 \rfloor / 2l_t$. Additionally, it may create the first division (e.g., floor) by adding a single line (or plane) segment at the appropriate distance(s) from the starting segment. The description "n − 2" can be added to the new segment, both to identify the segment that will be the subject of the iteration rule and to constrain the number of times the iteration rule can be applied. The iteration rule can read the required distance(s) from the previous set of parallel segments or from a description, add another segment and move the descriptions one segment over while, at the same time, reducing the number by one. The iteration rule must be constrained to apply only as long as $n > 0$; alternatively, the termination rule can be constrained to apply only when n equals zero, however, in this case a lazy approach should be adopted instead of the default greedy approach. Finally, the termination rule removes the descriptions.

A (possibly obvious) pattern that cannot be dealt with, with this notation, is the deterministic iteration of a single rule (or a set of rules). Consider a user that prefers to decide upon rule application only when there are multiple options, either multiple rules that apply all at once, or a single rule that matches multiple shapes. However, when only a single rule applies multiple consecutive times, or a set of rules applies in a single pattern, he or she may prefer to wait until the entire series of rule applications has completed automatically. This distinction between purely deterministic rule application and a (probabilistic) *flow* derivation is not explicitly captured in the notation. If the rule or rule set has been conceived to result in a purely deterministic derivation then the difference doesn"t matter, as the probabilistic character of the *flow* will never arise under these circumstances. However, if the (partial) deterministic derivation is simply a result from the interaction between the rule (or rules) and the shape under rule application, then the automation of this deterministic derivation is simply a convenience to the user and does not, in any way, constitute the automatic application of a composite rule or *flow*. A *flow* either applies in its entirety or not at all, whereas a deterministic derivation, unless explicitly designed for, may be only partial.

7 Discussion and Conclusion

We have presented a concept of composite shape rules embedding algorithmic patterns for rule automation. We have denoted these composite shape rules *flows* and adopted a notation from regular expressions. More than an algorithmic notation for automated rule selection, we consider *flows* as an alternative approach to rule specification. Next to rules of the form $r: lhs \rightarrow rhs$, we consider rules of the form $f: [r_1 \ (r_2 \ r_3)]$, allowing rules to be composed (or decomposed) at will.

The notation we adapt from regular expressions. We have adopted most of the operations to construct regular expressions as operations to construct *flows*, including quantification, grouping and the square bracket expression for selection. We have omitted a few metacharacters that are rather irrelevant to *flows*, such as the starting and ending position of the string, or matching any single character. We have also omitted the choice operator, as we have opted for the square bracket expression to be used instead. At the same time, we have extended the notation with an asterisk preceding a square bracket extension, allowing for a random selection of a single (composite) rule from the collection of (composite) rules identified as alternatives within the square bracket expression. We have also adopted different matching mechanisms, allowing for a greedy, a lazy, a possessive and even a probabilistic approach. Note that a probabilistic matching is not applicable to regular expressions.

Even though there are important differences between regular expressions and composite shape rules, most notably the fact that regular expressions work towards a target while composite shape rules are intended to be used exploratory or enumerative, we must conclude that the notation for regular expressions fits composite shape rules remarkably well. The notation is also very compact, simplifying the specification of a *flow*. Although Liew's (2004) graphical approach may seem much more readable, such visual readability would be entirely lost if we attempted to explicate the ability for backtracking.

The validation of the presented approach lies in its ability to automate rule sequencing and application, especially when deterministic rules are concerned.

Acknowledgements This work received some funding support from Singapore MOE's AcRF start-up grant, WBS R-295-000-129-133. The second author also benefited from a China Scholarship Council grant. We want to thank Bui Do Phuong Tung and Bianchi Dy for their work on the SortalGI shape grammar interpreter and API. The first author also wants to acknowledge the students in his Shape Computation course for their experimentation with *flows* and Raynard Yu for the example included. This paper extends a paper presented at the Eighth International Conference on Design Computing and Cognition (DCC"18) (Stouffs and Hou, 2019).

References

Beirão, J.N., J.P. Duarte, and R. Stouffs. 2009. Grammars of designs and grammars for designing–grammar based patterns for urban design. In *Joining Languages, Cultures and Visions*, eds. T. Tidafi and T. Dorta. Montreal: Université de Montréal.

Grasl, T., and A. Economou. 2014. Towards controlled grammars. In *Fusion*, ed. E. M. Thompson, vol. 2, 357–363. Brussels: eCAADe.

Hou, D., and R. Stouffs. 2018. An algorithmic design grammar for problem solving. *Automation in Construction* 94: 417–437. https://doi.org/10.1016/j.autcon.2018.07.013.

Hou, D., and R. Stouffs. 2019. An algorithmic design grammar embedded with heuristics. *Automation in Construction* 102: 308–331. https://doi.org/10.1016/j.autcon.2019.01.024.

Janssen, P., R. Stouffs, A. Mohanty, E. Tan, and R. Li. 2016. Parametric modelling with GIS. In *Complexity & Simplicity*, eds. A. Herneoja, T. Österlund, and P. Markkanen, vol. 2, 59–68. Brussels: eCAADe.

Knight, T.W. 1999. Shape grammars: Six types. *Environment and Planning B: Planning and Design* 26 (1): 15–31.

Liew, H. (2004). *SGML: a meta-language for shape grammar*. Ph.D. dissertation. Cambridge: Massachusetts Institute of Technology.

Palladio, A. 1965. *The Four Books of Architecture*. New York: Dover Reprinted from the 1738 translation by Isaac Ware of I Quattro Libri dell"Architettura.

Stiny, G. 1980. Introduction to shape and shape grammars. *Environment and Planning B: Planning and Design* 7 (3): 343–351.

Stiny, G. 1981. A note on the description of designs. *Environment and Planning B: Planning and Design* 8 (3): 257–267.

Stiny, G. 2006. *Shape: Talking about seeing and doing*. Cambridge: The MIT Press.

Stiny, G., and W.J. Mitchell. 1978a. The Palladian grammar. *Environment and Planning B: Planning and Design* 5 (1): 5–18.

Stiny, G., and W.J. Mitchell. 1978b. Counting Palladian plans. *Environment and Planning B: Planning and Design* 5 (2): 189–198.

Stouffs, R. 2018a. Description grammars: A general notation. *Environment and Planning B: Urban Analytics and City Science* 45 (1): 106–123.

Stouffs, R. 2018b. Description grammars: Precedents revisited. *Environment and Planning B: Urban Analytics and City Science* 45 (1): 124–144.

Stouffs, R., and D. Hou. 2017. The complexity of formulating design(ing) grammars. In *Proceedings of education and research in computer aided architectural design in Europe, eCAADe 2017*, eds. Fioravanti, A., S. Cursi, S. Elahmar, S. Gargaro, G. Loffreda, G. Novembri, and A. Trento. vol. 2, 443–452. Brussels: eCAADe.

Stouffs, R., and D. Hou. 2019. Composite shape rules. In *Design computing and cognition'18*, ed. J.S. Gero, 439–457. Switzerland: Springer Nature.

UN Studio. 2017. *Lane 189*. https://www.unstudio.com/en/page/11768/lane-189; https://www.archdaily.com/874598/lane-189-unstudio.

Yue, K., and R. Krishnamurti. 2013. Tractable shape grammars. *Environment and Planning B: Planning and Design* 40 (4): 576–594.

Open Access This chapter is licensed under the terms of the Creative Commons Attribution-NonCommercial-NoDerivatives 4.0 International License (http://creativecommons.org/licenses/by-nc-nd/4.0/), which permits any noncommercial use, sharing, distribution and reproduction in any medium or format, as long as you give appropriate credit to the original author(s) and the source, provide a link to the Creative Commons license and indicate if you modified the licensed material. You do not have permission under this license to share adapted material derived from this chapter or parts of it.

The images or other third party material in this chapter are included in the chapter's Creative Commons license, unless indicated otherwise in a credit line to the material. If material is not included in the chapter's Creative Commons license and your intended use is not permitted by statutory regulation or exceeds the permitted use, you will need to obtain permission directly from the copyright holder.

Arrangements Containing Shapes: Mathematical Features and Their Use in Visual Calculating

Alexandros Haridis

Abstract Construction lines and registration marks in shape grammars ground the appearance of shapes to provide an algorithmic approach to visual calculating. Here, the two concepts are studied in unity as a new object called *point-line arrangement*. This chapter develops such topics as the comparison and classification of shapes by their arrangements, the characterization of the "geometries" that arrangements give rise to, the algebra of arrangements, and others. In this way, it provides a more holistic view of construction lines and registration marks, beyond the usual roles they receive in algorithmic implementations of shape grammars.

1 Introduction

Mathematics of Shape Grammars

The "mathematics of shape grammars" began with the introduction of the formalism in the 1970s. Among the various topics of research that have emerged over the past five decades, the following three relate to the work I present in this paper:

(a) The *algebras of shapes* U_i, defined according to the dimensionality i of the basic elements that form a shape ($i = 0$ for points, $i = 1$ for line segments, $i = 2$ for planar regions, and $i = 3$ for solids).
(b) *Structural descriptions* defined in terms of embedded parts that are related in special ways, such as finite topologies and other decompositions (e.g., Haridis 2020a; Haridis and Stiny 2022; Krstic 1996).
(c) *Algorithmic features of visual calculating* with shapes (e.g., shape embedding with registration marks) emphasizing the features that distinguish it from other formal models of computation.

A. Haridis (✉)
Department of Architecture, Massachusetts Institute of Technology, 77 Massachusetts Avenue, Cambridge, MA 02139, USA
e-mail: charidis@mit.edu

Table 1 Summary of key properties of shapes made with basic elements of different dimensions

Basic element	Dimension	Content	Part relation	Shape topology	Topological boundary
Point	0	None	Identity	Finite	Point-set boundary
Line	1	Length	Partial order	Finite or infinite	Empty
Plane	2	Area	Partial order	Finite or infinite	Empty
Solid	3	Volume	Partial order	Finite or infinite	Empty

Table 1 summarizes some of the key properties of shapes related to the topics (a) and (b).[1]

Work on the mathematics of shape grammars illustrates a broad methodological direction for linking the design disciplines with mathematics and computing, whereby seeing and spatial intuition in the former inform developments in the latter. This methodological direction continues to this day to reveal unique connections between visual calculating with shapes and some key issues studied in a variety of other fields, including philosophy, design, artificial intelligence, logic, linguistics, aesthetics, and literary studies (Stiny 2022).

"Construction Lines" and "Registration Marks"

In this paper, I present a mathematical study of "construction lines" and "registration marks," two key geometric concepts traditionally linked with algorithmic implementations of shape grammars. The two concepts appear under various disguises in the literature, manifested more or less independently.

Registration marks, or "points of intersection," were introduced in Stiny (1975) to show that rewrite rules in shape grammars are algorithmic (i.e., that they can be simulated with a Turing machine). The registration marks are used to derive transformation matrices to geometrically match or embed a shape into another shape. The basic idea is illustrated pictorially in Fig. 1. This algorithmic approach to shape embedding that employs the registration marks of a shape lies at the core of most computer implementations of shape grammars (e.g., Hong and Economou 2022).

The construction lines ("carrier lines" or "line descriptors") function as an underlying organizational "grid" that contains the maximal line segments of the shape. The construction lines induce a natural partition of a shape into disjoint (nonempty) subshapes, so that each such subshape consists of one or more collinear line segments (perhaps separated by "gaps"); this partition is also called a diagonal decomposition (Krstic 2022). The reorganization of a shape's line segments provided by the construction lines has been used as a method of sorting the lines of a shape in order to store them in a computer's memory and to perform efficient comparisons and operations between shapes (Krishnamurti 1980). In general, any shape that is physically realizable with finitely many (maximal) line segments has a corresponding set of construction lines and a set of registration marks.

---

[1] The term *shape topology* is taken from Haridis (2020a; 2020b) and it refers to a topological structure induced on a shape on the basis of embedded parts (i.e., topology is not inherited from an ambient space).

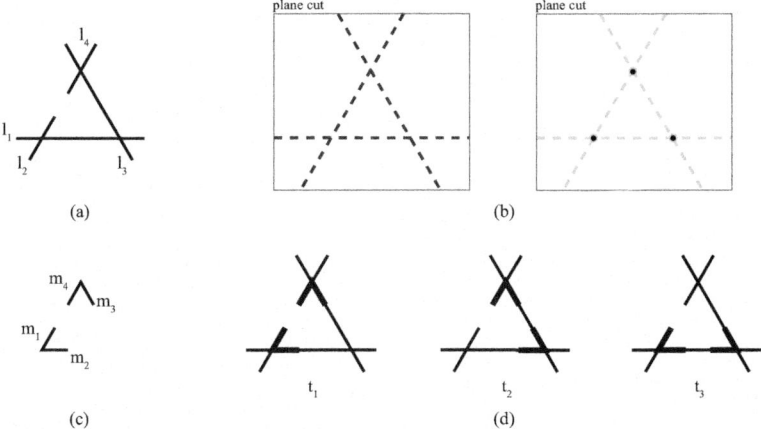

Fig. 1 A pictorial illustration of shape embedding. The shape in Fig. 1c can be matched or embedded into the shape in Fig. 1a under three distinct transformations, t_1, t_2, and t_3, illustrated in Fig. 1d. The plane cut denotes that the lines are infinitely extending, but it is mostly omitted from the figures in forthcoming sections

Point-Line Arrangements = Construction Lines + Registration Marks

In this paper, the two geometric concepts, "construction lines" and "registration marks," are unified as a single object called a *point-line arrangement* or simply an *arrangement* (Fig. 2). When the two are seen in this way, what is the geometric nature of the resulting point-line arrangement and what is its meaning and use in the context of visual calculating?

Point-line arrangements can be viewed as a kind of "local" geometric space, extracted directly from the basic elements of a given shape. In design and architecture, we are accustomed to using rules of geometry to describe or generate shapes. In this context, geometric notions are often used by designers in an informal sense as something that is interchangeable with "form" or "figure." But in the context of mathematics, geometry receives its formal meaning in terms of several viewpoints, characterized on the basis of the features that one chooses to pay attention to. There

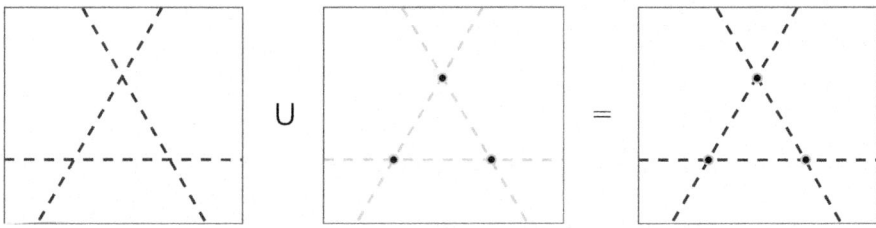

Fig. 2 A point-line arrangement is created from the union of a 1-dimensional object (collection of lines) with a 0-dimensional object (collection of points)

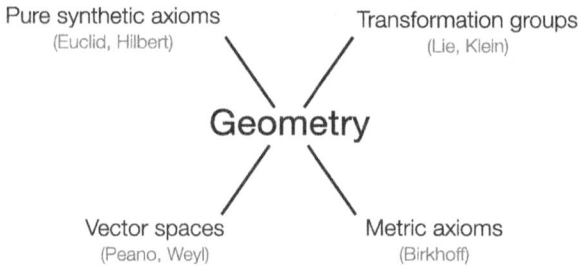

Fig. 3 Four different viewpoints of geometry in mathematics

is no single agreeable viewpoint for formally characterizing "a geometry" (Greenberg 1980; Stillwell 2005). Figure 3 illustrates diagrammatically a few of the most well-known viewpoints each one coming with its own assumptions of what geometry means: Euclid-style synthetic constructions, vector spaces in linear algebra, metric approaches through concepts of distance and angle measurement, transformation groups and the study of invariants (symmetries). New viewpoints continue to emerge to this day, such as the characterization of a geometry with its group of transformations using the tools of category theory (Marquis 2009).

With advancements in areas of technology such as digital manufacturing, digital scanning, social networks, computer vision and machine learning, we find a diversity of "shape" notions, including point-sets as "noisy" signals, networks of interconnected nodes (e.g., molecules in computational biology or nodes in social networks), unstructured point-clouds, discretized representations (e.g., triangle meshes), and other. Which viewpoint of geometry to use depends each time on the notion of shape adopted in an application area, and decisions of this kind have a bearing on how shapes are represented and on which operations are, or are not, possible or efficient (Arslan et al. 2022).

A point-line arrangement is studied in this paper in its own right as an independent object with special geometric features. I focus specifically on point-line arrangements that contain shapes of the algebra U_1, which covers the set of all two-dimensional shapes that can be made using straight line segments. The topics I develop range from the "algebra" of point-line arrangements and their formal definition to methods for the comparison and classification of shapes in the plane based on their underlying sets of registration marks and construction lines. In this way, I provide a more holistic view of construction lines and registration marks, beyond the usual roles they receive in computer implementations of shape grammars.

This paper's roadmap is given below for the reader's convenience:

Sections

1. Defining the point-line arrangements that contain shapes;
2. Arrangements are equivalence classes;
3. An algebra of point-line arrangements;
4. Comparison and classification in the plane and an empirical catalog;
5. Geometry of arrangements and a containment hierarchy;
6. References.

2 Defining the Point-Line Arrangements that Contain Shapes

In the following, S is a shape made with maximal line segments that can be (physically) realized in the two-dimensional plane (or in three-dimensional space). Given such a shape, the *point-line arrangement* that contains it consists of two things:

- A set H of straight lines called *construction lines*, and
- A set M of points called *registration marks*.

The set H of construction lines is a collection of infinitely extending lines while the set M of registration marks is a collection of points placed at the locations where these lines meet. The point-line arrangement underlying a shape S is a pair:

$$A_S = (H, M).$$

If the sets H and M are empty, then we have a special object called the *empty arrangement* which is denoted by $A_\varnothing$. The empty arrangement represents the arrangement associated with the empty shape.

The term "arrangement" is often used to refer to hyperplane arrangements in the real two-dimensional space—a subdivision of the real plane by a finite collection of lines (Stanley, 2007). Here, the term "arrangement" is used interchangeably with "point-line arrangement," or "arrangement that contains shapes," and refers to an arrangement of construction lines *and* registration marks, that is, it refers to both the hyperplanes and the points of their intersection.

Analytic Definition

The arrangement underlying a shape has a straightforward analytic definition that is constructive. A construction line is calculated from the line segments of the shape using the ordinary equation of lines in the Euclidean plane. A registration mark is defined with coordinates and represents a solution to (at least) two equations of lines. The same construction applies both in the two-dimensional plane and in the three-dimensional space. See, in particular, the following graphical illustration:

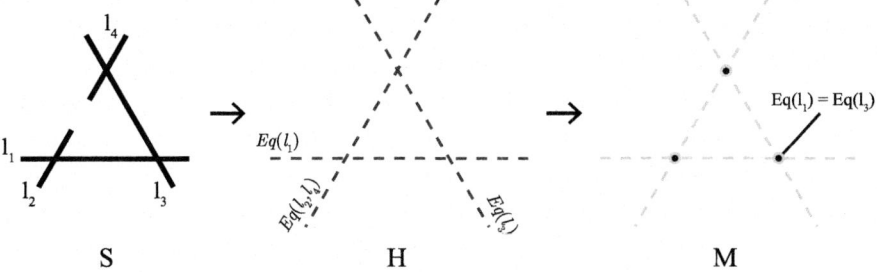

Rule-Based (Axiomatic) Definition

Given the line segments of a shape, it is a straightforward exercise to derive the point-line arrangement that contains them. But, in the opposite direction, given any set of straight lines and any set of points in the plane, how do we decide if the two sets form a valid arrangement of construction lines and registration marks? In other words, how do we decide if the two sets define an arrangement that contains shapes of the algebra U_1?

The following two conditions provide a decision method. A set H of straight lines and a set M of points in the Euclidean plane define an arrangement of construction lines and registration marks, if and only if:

(1) Any point in M is a point of intersection of at least two lines in H, and
(2) Any two non-parallel lines in H have a unique point of intersection in M.

The two conditions or rules (1) and (2) do not assume that there are any lines at all. In this case, there cannot be points and we get an empty arrangement. If there are lines, the rules do not assume there are points because the lines may be parallel. On the other hand, if there are points, then there must be lines that meet at those points.

The first rule is an incidence condition about the set M; it guarantees there are no "isolated" points (i.e., points not on any line) or points incident with one line only. The second rule is a parallelism condition for the set H; it guarantees that all intersection points of pairs of lines in the plane are in the set M. Both rules are needed to decide if the given sets M and H form a valid arrangement that contains shapes of the algebra U_1 and are used extensively in the forthcoming sections.

Some counterexamples are in Fig. 4a and b. The given sets of points and lines do not define arrangements that contain shapes because they fail to satisfy rule (1) or (2) or both. Many graphical depictions of configurations of "points" and "lines" in the plane that are familiar from geometry are, in fact, not arrangements that contain shapes; for example, those in Fig. 4b.

In three-dimensional space, *skewness* of lines must be considered in addition to parallelism, in which case the second condition, above, must be strengthened. Skew lines are neither parallel nor coplanar—for example, see Fig. 5. There are infinitely many different configurations of skew lines that do not have a point of intersection. In three dimensions, two nonparallel construction lines may still not intersect if they are skew. The registration marks in M are, therefore, created for all pairs of nonparallel lines in H that are also not skew.

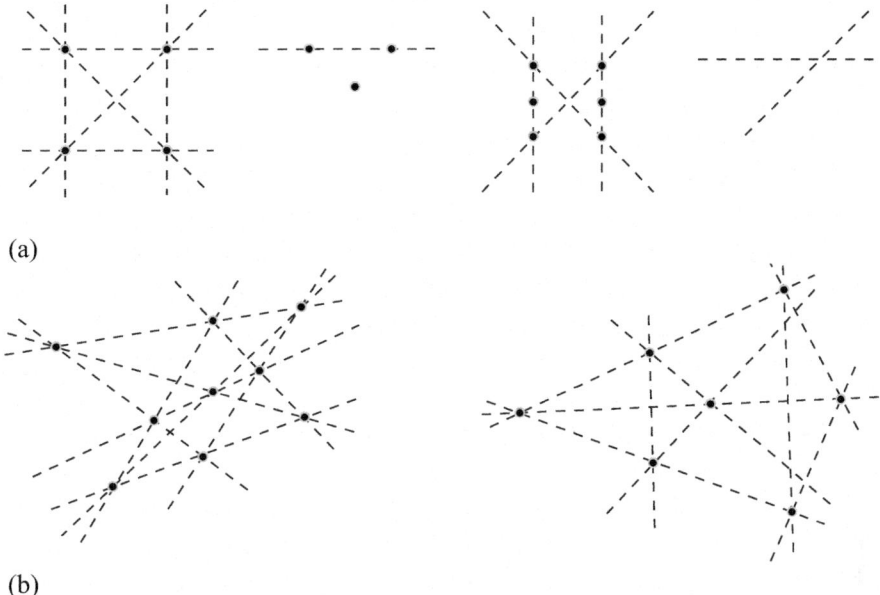

(a)

(b)

Fig. 4 a Counterexamples. b Familiar drawings from geometry depicting configurations of "points" and "lines" in the plane are not arrangements that contain shapes (here, the drawings depict the projective Pappus and the projective Desargues configurations)

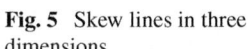

Fig. 5 Skew lines in three dimensions

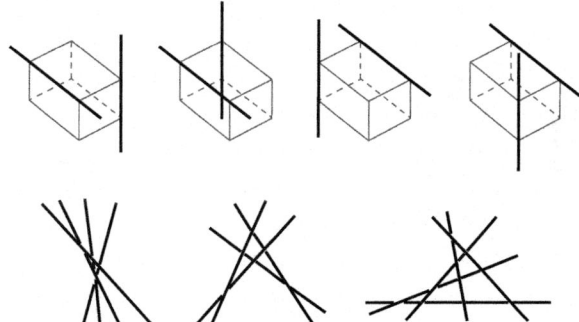

3 Arrangements Are Equivalence Classes

A point-line arrangement that is extracted from a shape contains the shape itself as well as infinitely many other shapes. For example, the arrangement that contains the shape in Fig. 1a also contains all the shapes in Fig. 6 because every shape instance in this collection determines the same construction lines and registration marks. Two shapes can determine the same point-line arrangement independently of whether one is a part of the other. In Fig. 6, only the four shapes marked with an asterisk are a part of the original shape in Fig. 1a.

An analogous example is in Fig. 7 where each shape instance is contained in the following point-line arrangement:

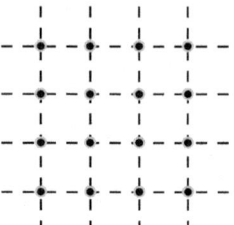

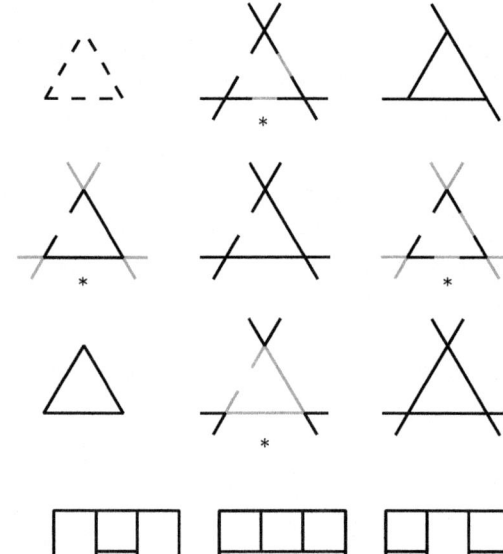

Fig. 6 An illustration of the concept of equivalence class that a point-line arrangement defines

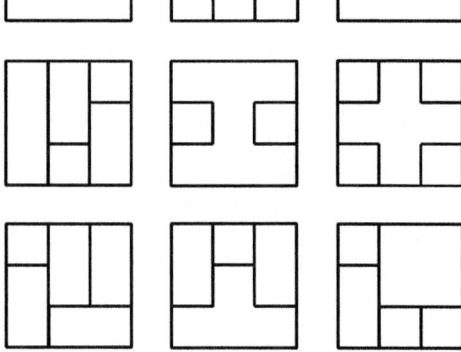

Fig. 7 An example of a class of shapes that are equivalent on the basis of their underlying point-line arrangement

More generally, a point-line arrangement satisfying the two conditions in Sect. 1 behaves as an *equivalence class* that contains infinitely many shapes, such that each shape yields the same construction lines and registration marks. This notion of equivalence can be formalized in a few different ways.

3.1 Equivalence and Part Equivalence

If S_1 and S_2 are two shapes and $A_{S_1} = (H_1, M_1)$ and $A_{S_2} = (H_2, M_2)$ are the point-line arrangements containing them, the shapes S_1 and S_2 are *equivalent*, or *point-line equivalent*, if there is a sequence of transformations t such that:

(i) $t(H_1) = H_2$ and
(ii) $t(M_1) = M_2$.

If both (i) and (ii) hold, then $t(A_{S_1}) = A_{S_2}$. Because this formulation of equivalence uses both the construction lines and the registration marks of a shape, certain transformations may lead to the introduction of new intersection points by forcing parallel construction lines to become intersecting. These are the transformations that do not preserve parallelism, namely, the transformations coming from the projective linear group; the underlying set of registration marks (and thus the point-line arrangement itself) is not invariant under transformations from this group as illustrated in Table 2. Parallelism between construction lines is preserved under at most the affine group in which case arrangements are said to be *affine equivalent*.

The notion of equivalence given in the conditions (i) and (ii) above has a generalization according to which equality (=) is replaced with subset ($\subseteq$) and "equivalence"

Table 2 An illustration of transformation groups acting on two separate point-line arrangements. The underlying set of registration marks is preserved up to affine transformations, which preserve the parallelism of construction lines

Identity	Isometry (Iso)	Similarity (Sym)	Affine (Aff)	Projective (P_{GL})

is interpreted as part equivalence. The two shapes S_1 and S_2 are called *part equivalent* if there is a sequence of transformations t such that:

(i)' $t(H_1) \subseteq H_2$ and.
(ii)' $t(M_1) \subseteq M_2$.

If both (i)' and (ii)' hold, then $t(A_{S_1}) \subseteq A_{S_2}$. When this relation holds, the arrangement of the first shape is a *subarrangement* of the arrangement of the second shape under the sequence of transformations t. It is important to note that "subarrangement" is not an ordinary subset. A subarrangement is itself an arrangement, that is to say, it satisfies the two rules (1) and (2) in Sect. 1—thus, not all subsets of an arrangement are subarrangements, only those that satisfy these two rules are.

3.2 Part Equivalence and Shape Embedding

If a shape is embedded in another shape, then the two shapes are part equivalent in the above sense. In particular, if S_1 is a part or subshape of S_2 under a sequence of transformations t, then $t(A_{S_1}) \subseteq A_{S_2}$ holds under that same sequence of transformations t. A graphical illustration of how shape embedding implies part equivalence is in Fig. 8. The shape x can be embedded in S under four different transformations. For each of these four ways of embedding, the underlying point-line arrangement of x becomes a subarrangement of the point-line arrangement of S under the same transformations—Fig. 8b illustrates this graphically for one of the four ways of embedding.

While shape embedding implies part equivalence on the basis of arrangements, part equivalence does not automatically imply shape embedding. That is to say, the relation $t(A_{S_1}) \subseteq A_{S_2}$ may be satisfied even if $t(S_1) \leq S_2$ is not (the point-line arrangement of S_1 can be equivalent to a subarrangement of S_2 even if the shape S_1 is not itself a part of S_2). For example, the following four shapes determine the exact same point-line arrangement as the triangle x in Fig. 8a. None of these four shapes, however, is a part of the shape S in that same figure.

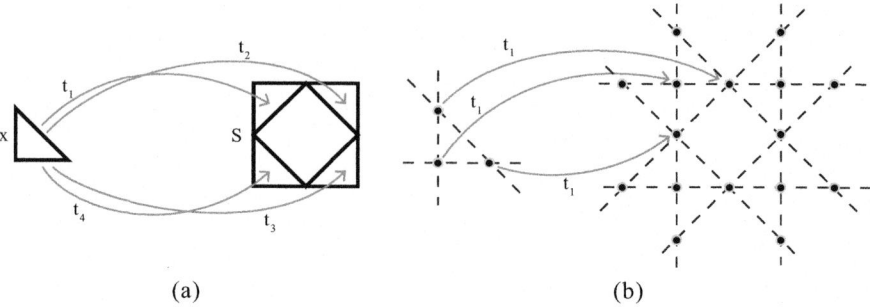

Fig. 8 An illustration of part-equivalence and subarrangement matching

It is possible to pursue a notion of equivalence between shapes on the basis of the registration marks only. This provides one method of classifying and comparing shapes according to how the set of registration marks belonging to one shape becomes a subset of those of another shape. Equivalence is established by condition (ii) and part equivalence by condition (ii)'. Here, however, the preferred method of establishing equivalence is through the combined use of registration marks and construction lines. The way in which the construction lines interact with (i.e., pass through) the registration marks has a geometric meaning, which is discussed in more detail in Sect. 5.

4 An Algebra of Point-Line Arrangements

The point-line arrangements that underlie the shapes in the algebra U_1 have their own algebra that specifies the allowable ways of combining them under various operations. Operations between arrangements can be viewed in two independent ways leading to different results.

4.1 Arrangement Sum, Product, and Difference

First, given two shapes, we can specify what happens to their corresponding point-line arrangements after we combine the two shapes in an algebraic operation—a sum, a product, or a difference. In particular, let the following three operations *arrangement sum* $(+^*)$, *arrangement product* $(\cdot^*)$, and *arrangement difference* $(-^*)$ correspond analogously to the operations of sum $(+)$, product $(\cdot)$, and difference $(-)$ for shapes in the algebra U_1. Then, the arrangement sum of A_{S_1} and A_{S_2} is the point-line arrangement of the shape $S_1 + S_2$, that is,

$$A_{S_1} +^* A_{S_2} = A_{S_1+S_2}.$$

Similarly, the arrangement product and the arrangement difference are the point-line arrangements of the shapes $S_1 \cdot S_2$ and $S_1 - S_2$ like so,

$$A_{S_1} \cdot^* A_{S_2} = A_{S_1 \cdot S_2} \text{ and } A_{S_1} -^* A_{S_2} = A_{S_1-S_2}.$$

All three operations are defined by closed-form expressions *relative to* the results obtained from the corresponding shape operations. This approach preserves the object-type under all three operations as desired (each operation results in a valid point-line arrangement).

4.2 Closure Under Set Union, Set Intersection, and Set Difference

Second, given any two point-line arrangements, we can directly work out the algebraic operations that combine them into a valid point-line arrangement. When two point-line arrangements are given without shape instances, the above expressions no longer apply. Instead, the operations must be defined directly in terms of the sets of construction lines and registration marks.

One way to approach this is to use the set-theoretic operations of set union ($\bigcup$), set intersection ($\bigcap$), and set difference ($\setminus$) and apply them component-wise for the set of construction lines and the set of registration marks. This approach quickly leads to various complications which are explained below.

Given two point-line arrangements in the plane $A = (H, M)$ and $A' = (H', M')$, their union $A \bigcup A'$ is the union of their respective sets of construction lines and registration marks, that is, $H \bigcup H'$ and $M \bigcup M'$. However, the pair $A \bigcup A' = (H \bigcup H', M \bigcup M')$ does not necessarily form a valid point-line arrangement—the resulting sets of points and lines may not satisfy the conditions (1) and (2) in Sect. 1. Figure 9 shows an example of a set union that results in a valid arrangement—the shape instances are given as visual helpers and as a way of comparing shape operations with corresponding operations between arrangements. On the other hand, each set union in Fig. 10 fails to produce a valid arrangement because the results are missing intersection points of newly intersecting lines.

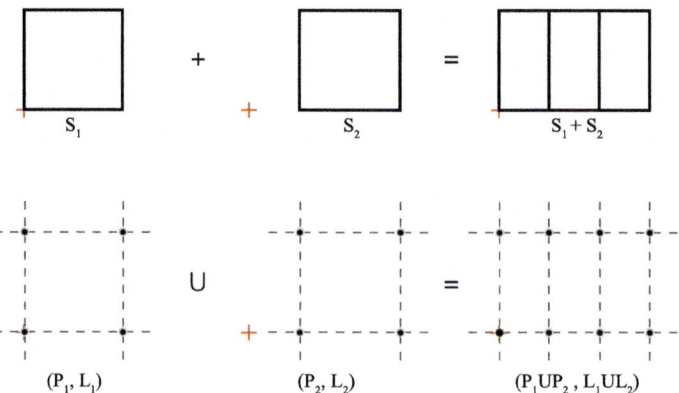

Fig. 9 Set union of two point-line arrangements that results in a valid point-line arrangement

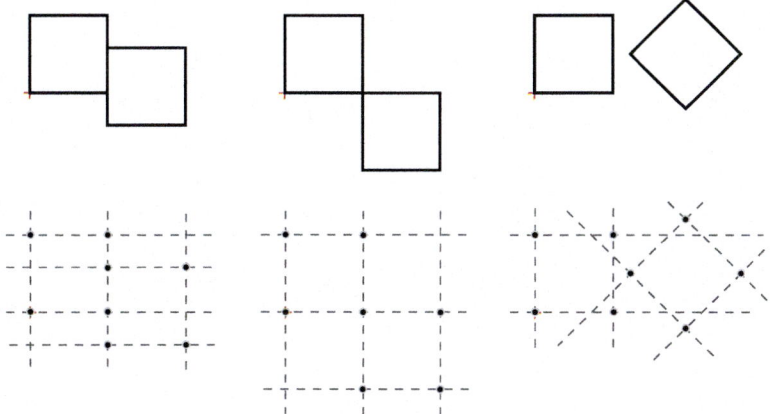

Fig. 10 Three set unions of point-line arrangements that result in an invalid arrangement

In general, the operation of set union produces an invalid point-line arrangement only by resulting in an object that does not have enough intersection points. The set union cannot result in an object that misses lines.

Unlike with set union, set intersection may result in an object that does not have enough lines while set difference may result in an object that does not have enough lines or enough intersection points. Examples of both are given in Figs. 11, 12, 13 and 14.

It is possible to consider "correcting operations" to fix the various cases of failure on a case-by-case basis. However, the sets of points and lines that intersection and difference produce cannot always be converted into point-line arrangements in a unique way (notice in the third intersection in Fig. 14 that an isolated point cannot be turned into a valid point-line arrangement uniquely). Even if there are cases where correcting operations are possible by adding missing lines or points (e.g., as

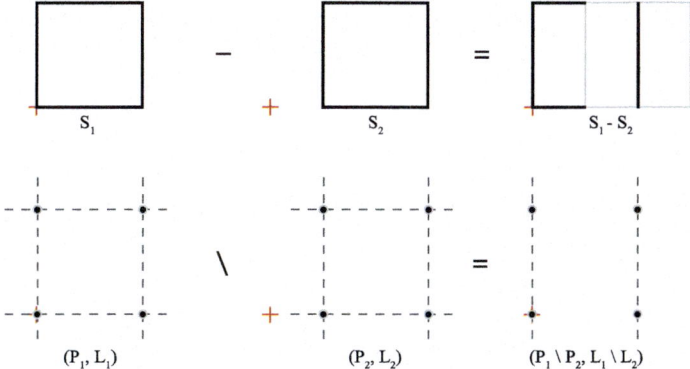

Fig. 11 Set difference between point-line arrangements that results in an invalid arrangement

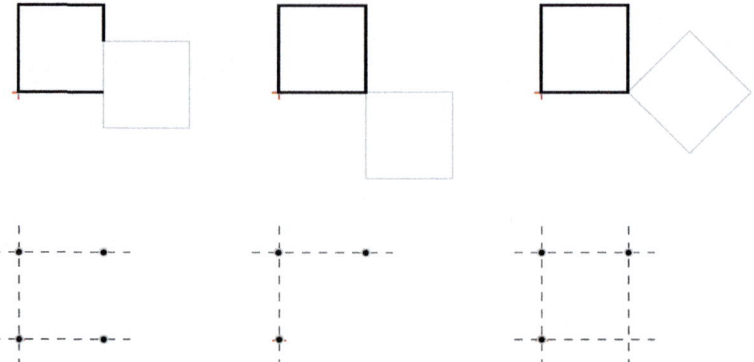

Fig. 12 Examples where set difference between point-line arrangements fails to produce a valid arrangement

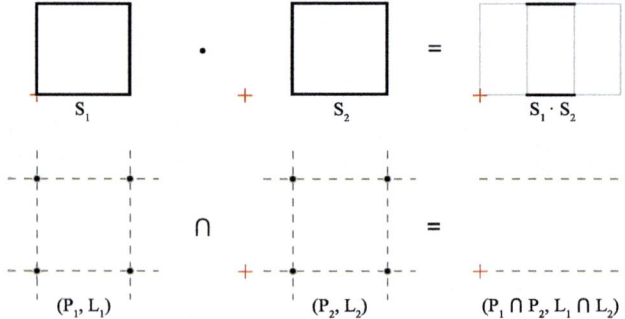

Fig. 13 Set intersection between point-line arrangements that produces a valid arrangement

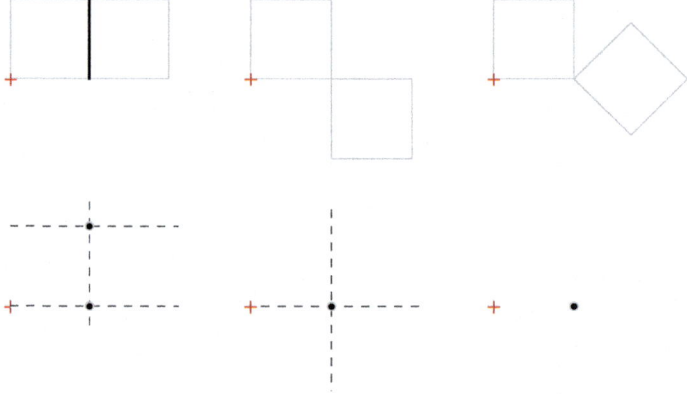

Fig. 14 Set intersections between point-line arrangements. The first two result in a valid arrangement, while the third one does not

in Fig. 12), a more general and systematic treatment is not possible to pursue because it would require us to take ad hoc decisions for every special case.

The set union is the only operation between point-line arrangements that has a natural and obvious correcting operation, that is, the insertion of registration marks for newly intersecting construction lines – for example, one may simply insert registration marks for all empty line intersections in Fig. 10.

From a physical standpoint, inserting missing registration marks is a way of "physicalizing" a collection of intersecting lines. If two seemingly intersecting lines are not equipped with a registration mark, then they essentially function as abstract connections with no physical meaning (this is a common approach, for example, in the study of abstract networks of nodes and edges in social science, network science, etc.). This act of signifying the physicality of intersecting lines has a direct and intuitive corollary in design and architecture. For example, in the representation of the structure of physical artifacts as line arrangements (e.g., trusses, building plans or structure systems) the insertion of registration marks signifies a physical connection between linear elements.

From a mathematical standpoint, inserting missing registration marks after taking the union of two point-line arrangements can be viewed as a *closure operation* (*cl*(·)). There is a direct correspondence between arrangement sum (+*) and closure of set union. In particular, if A_{S_1} and A_{S_2} are two point-line arrangements associated with the shapes S_1 and S_2, respectively, then the following holds:

$$cl(A_{S_1} \bigcup A_{S_2}) = A_{S_1} +^* A_{S_2} = A_{S_1+S_2}.$$

Thus, the universe (set) of all valid arrangements of registration marks and construction lines in the plane satisfying the conditions (1) and (2) in Sect. 1, closed under a single associative operation of arrangement sum, and equipped with the empty arrangement $A_\varnothing$ as the identity element, forms an algebra of arrangements with the structure of a monoid.

5 Comparison and Classification in the Plane and an Empirical Catalog

5.1 Comparison and Classification

When two point-line arrangements are affine equivalent one can be obtained as an affine image of the other. For example, in Fig. 15 (top) the three drawings represent the "same" point-line arrangement under affine equivalence. The physical realization of an arrangement in the plane (or in space) is therefore not unique. In general, the same arrangement of construction lines and registration marks can be realized with infinitely many equivalent drawings. When the drawings of point-line arrangements cannot be mapped onto each other by an affine transformation, the arrangements

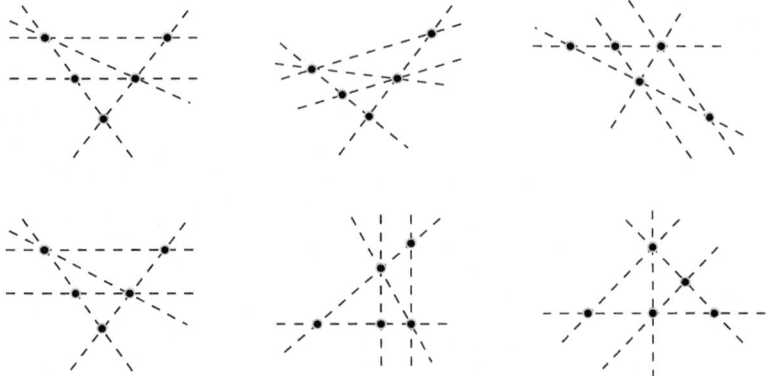

Fig. 15 (Top) affine equivalent arrangements. (Bottom) affine distinct arrangements but equivalent on the basis of incidence

are called *affine distinct*. Examples of affine distinct arrangements are in Fig. 15 (bottom). Affine distinct arrangements look visually dissimilar.

When two point-line arrangements are affine distinct in the above sense, they may still be equivalent in a combinatorial sense on the basis of information about incidences between their respective sets of registration marks and construction lines. In the following, a method for comparing and classifying affine distinct arrangements in the plane is summarized, and an empirical catalog that enumerates distinct arrangements in the plane is given. For a more detailed explanation, the reader may refer to the original study of point-line arrangements in Haridis (2020c).

5.2 Method for Comparing Affine-Distinct Arrangements

The method consists of the following three combinatorial features:

- The *triple of numbers* (k, n, d),
- The *incidence signature* $[2^{n_2}, 3^{n_3}, \ldots]$, and
- The *face structure*.

The triple of numbers (k, n, d) describes an arrangement in terms of the *number of registration marks* (k), the *number of construction lines* (n), and the *number of distinct slopes* or *directions* (d). The distinct slopes or directions are calculated from the construction lines by "removing" parallelism between the lines. While any given point-line arrangement can be described by the triple of numbers, these do not determine an arrangement uniquely. For example, the arrangements in Fig. 15 cannot be distinguished from each other on the basis of this triple of numbers (they result in exactly the same numbers).

When two arrangements have the same triple of numbers (k, n, d), we can separate them on the basis of information on how construction lines meet each other at every registration mark. This is captured by a particular construction that is associated with every arrangement called an *incidence signature*, or simply, *signature*. The signature of an arrangement is given by the vector:

$$[2^{n_2}, 3^{n_3} \ldots].$$

where n_m is the number of registration marks in which m construction lines intersect. The signature begins from $m = 2$ because every registration mark must be incident with at least two construction lines (condition (1), in Sect. 1). The length of the signature is always finite because we always deal with finitely many registration marks (here, the sum of the indices is expected to be equal to the number k of the registration marks, i.e., $n_2 + n_3 + \ldots = k$).

Some examples of point-line arrangements, their triple of numbers (k, n, d), and their incidence signatures are illustrated in Fig. 16.

If two arrangements are identical in terms of their incidence signatures, then these arrangements have the same *incidence relation* or "incidence structure" (see below). These arrangements are called *combinatorially equivalent*. An alternative but equivalent way of deciding if two arrangements have the same incidence relation is by labeling; see Fig. 17.

Because affine transformations preserve collinearity and incidence, they carry an arrangement to another with the same incidence relation. Thus, two arrangements that are affine equivalent are also combinatorially equivalent. Comparisons of point-line arrangements by incidence is, therefore, "coarser" than comparisons by affine equivalence.

Fig. 16 Four examples of point-line arrangements marked with the triple of numbers (k, n, d) and their incidence signature

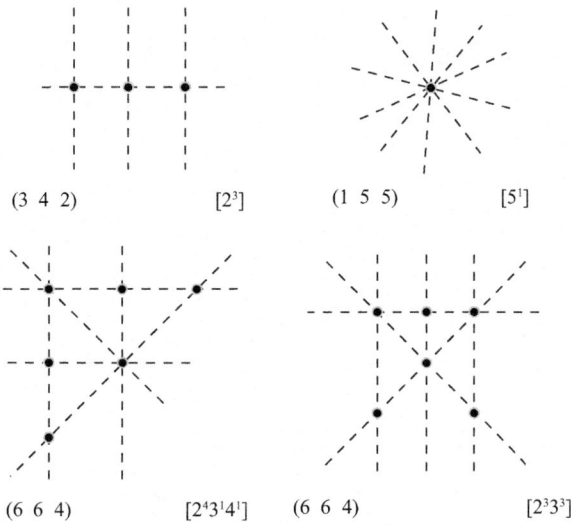

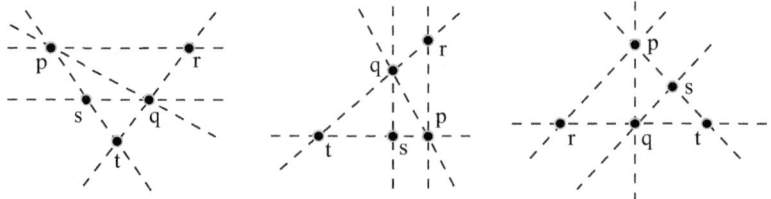

Fig. 17 Three arrangements with the same incidence relation. Their registration marks and construction lines can be given such labels that a registration mark and a construction line are incident in one of them if and only if they are incident in the other

When two arrangements are identical on the basis of the triple of numbers (k, n, d) as well as their incidence signatures, they can be further separated on the basis of a feature that the construction lines define called *face structure*.

The construction lines of any given point-line arrangement partition the two-dimensional plane into *bounded* and *semi-bounded regions*.[2] The following figure illustrates the two concepts for an arrangement of four construction lines:

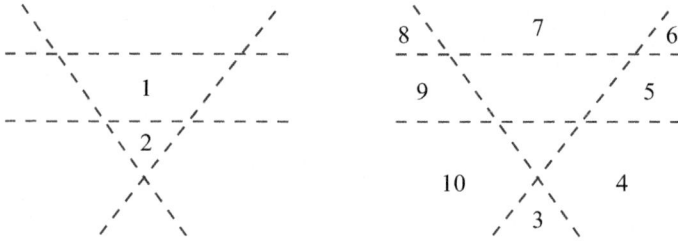

The regions labeled 1 and 2 are bounded, those labeled 3 through 10 are semi-bounded. The bounded regions are called *faces*. They are made by polygons with three, four, or more sides. These polygons are the face structure of an arrangement.

Two pairs of arrangements with identical triples (k, n, d) and incidence relations are shown in Fig. 18. The first pair can be separated on the basis of their face structures while in the second pair the face structures are identical. In the latter case, the arrangements cannot be separated by any of the aforementioned combinatorial tools, i.e., they represent identical arrangements.

---

[2] Though the exact formal terms somewhat vary in the literature (e.g., Stanley, 2007), the choice of these terms is suitable for our purposes here.

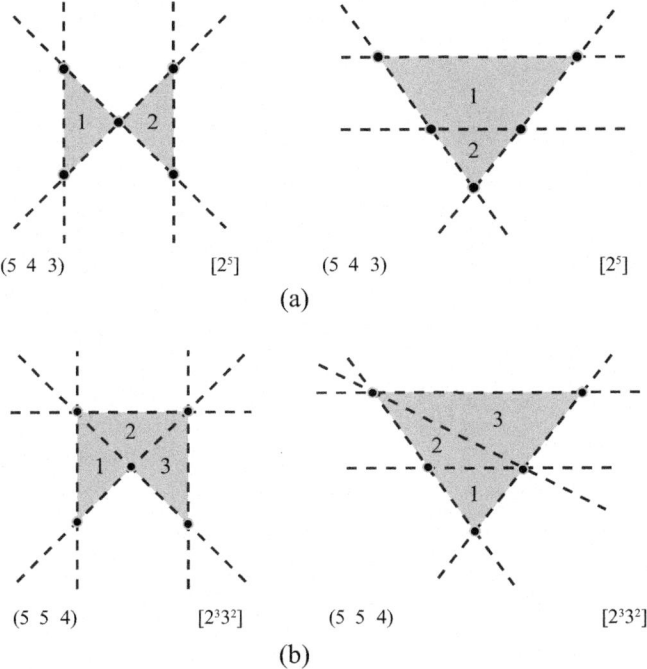

Fig. 18 Two pairs of point-line arrangements with identical triples of numbers (k, n, d) and incidence relations

5.3 An Empirical Catalog

With the tools introduced in the preceding sections, one can pursue an enumeration of the point-line arrangements in the plane for a given number of registration marks. Previous work in the literature has cataloged distinct (finite) arrangements by focusing on the lines, not the registration marks.[3] The empirical catalog below contains arrangements such that the lines meet at k points, for different values of k.

When $k = 0, 1, 2$ or 3, it is a straightforward exercise to describe and to visualize distinct point-line arrangements; see, in particular, Table 3. As the number k increases, it becomes harder to describe arrangements that meet at k points, let alone visualize them one by one. In Fig. 19, the empirical catalog continues with distinct point-line arrangements for $k = 4, 5$, and 6 registration marks. Each arrangement is marked by the triple (k, n, d) and its incidence signature. When two arrangements are identical in terms of these features, and they are both included in the catalog, then they have different face structures. When the face structures of two arrangements are identical, too, only one of them is included in the catalog.

---

[3] A first catalog can be found in Grünbaum (1972). See also Cai and Liao (2020).

Table 3 Empirical catalog of point-line arrangements for k = 0, 1, 2, and 3 registration marks

Registration	Construction Lines	Point-Line Arrangement
0	There are two cases: (i) The empty arrangement with no construction lines. (ii) One, two or more parallel construction lines.	(i) $A_\emptyset$ (ii)
1	Two or more construction lines intersecting at a common registration mark.	
2	Exactly three construction lines, two of which are parallel and a third one that intersects both.	
3	There are three cases: (i) Four construction lines, three of which are parallel and a fourth that intersects all three. (ii) Three pair-wise intersecting construction lines forming a triangle. (iii) Four construction lines, three of which are as in (ii) and a fourth one parallel to one (and only one) of them.	(i) (ii) (iii)

6 Geometry of Arrangements and a Containment Hierarchy

As explained in the introductory section, a point-line arrangement can best be understood as a local geometric space that contains (or "carries") shapes from the algebra U_1. Taking this approach further, is it possible to characterize this "local" space to understand how it relates to other types of spaces or geometries? This section characterizes the incidence relation that the construction lines and registration marks define through an application of the theory of *finite incidence geometry*. For lack of space, this section summarizes only the key points of this characterization, leaving the details to be found elsewhere (see, in particular, Haridis 2020c).

Incidence relations are described by "incidence structures." Incidence structures are at the core of the modern study of geometric spaces and their application extends to computer science, experiment design, machine learning, and other areas (Batten 1997; Dembowski 1997). An incidence structure is nothing more than the idea that two objects coming from different classes of things, e.g., "points" and "lines", can be "incident" with each other. In general, the influence of geometric thought in these expressions is misleading. The concepts "point" and "line" are meant to be undefined terms and to receive meaning in a particular application that may or may not be related to geometry. Here, the concepts "point," "line," and "incidence" receive

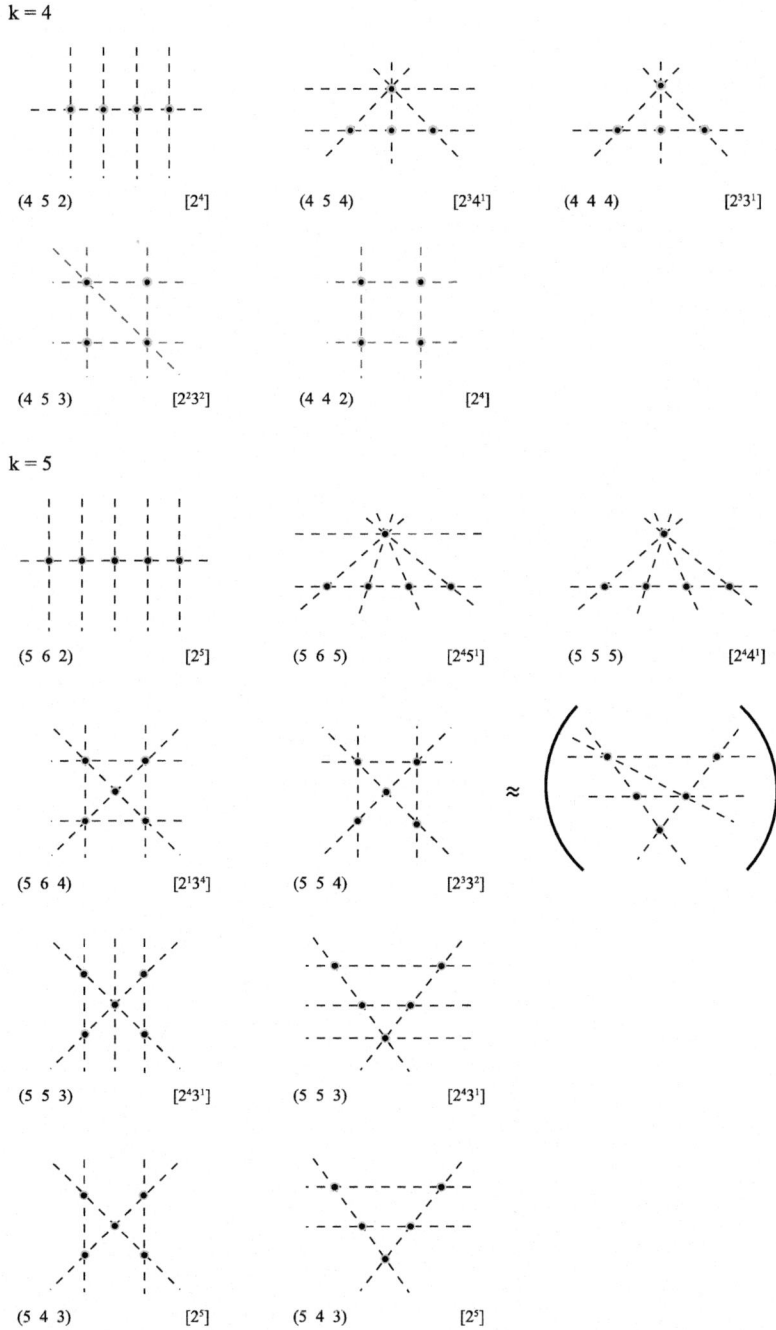

Fig. 19 Enumeration of distinct point-line arrangements in the plane for k = 4, 5, and 6 registration marks

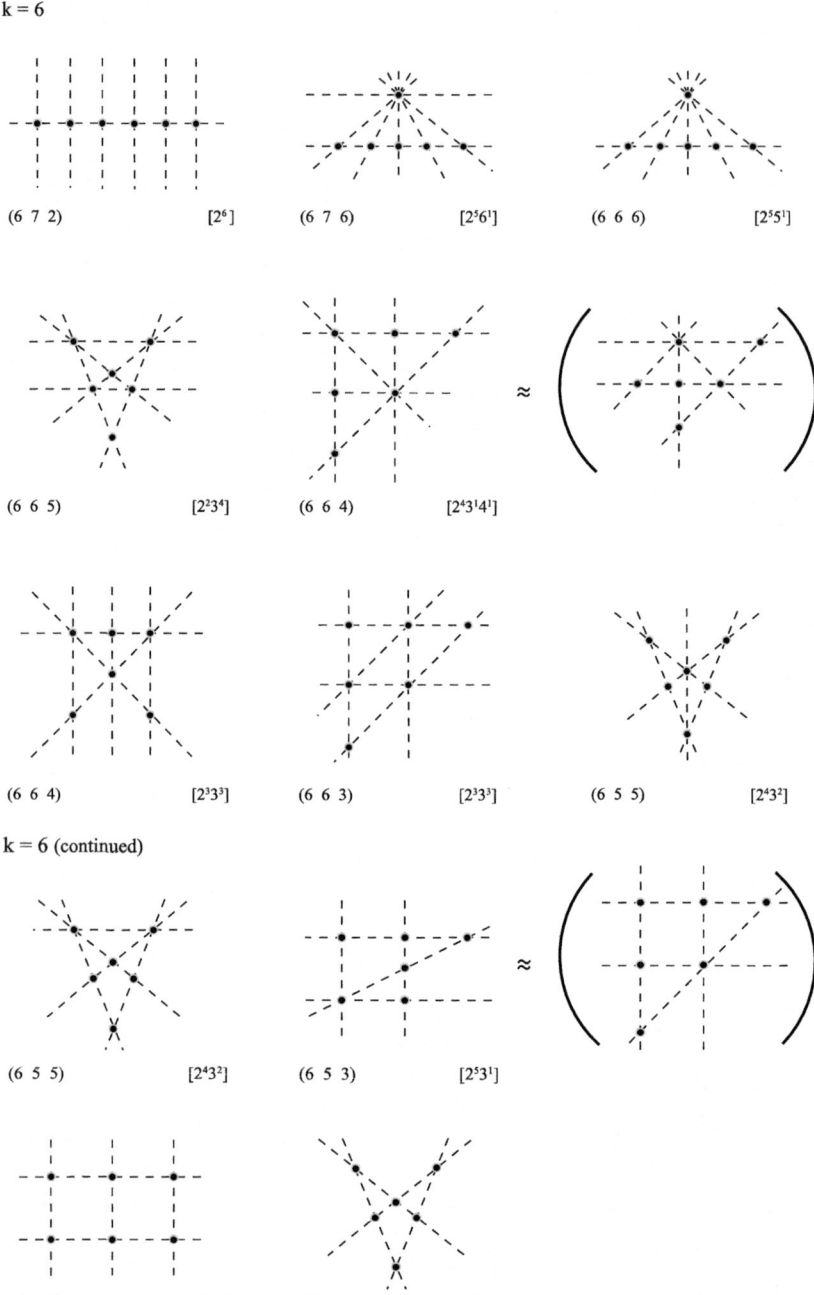

Fig. 19 (continued)

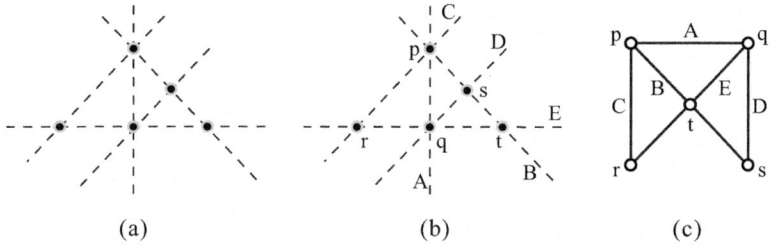

Fig. 20 a Point-line arrangement and its incidence structure. The registration marks and the construction lines in (**b**) are labeled with, respectively, small, and large letters, in no particular order. (**c**) The set of incidences is formed by marking for each registration mark the construction lines that pass through that registration mark, e.g., {(p, A), (p, B), (p, C), ..., (r, C), (r, E), ..., (t, B), (t, E)}

particular meanings: points represent registration marks, lines represent construction lines, and "incidence" represents our ordinary understanding of a registration mark being incident with a construction line, or that a construction line passes through a registration mark.

Formally, an *incidence structure* is a triple

$$(V, B, I)$$

where V and B are any two (finite) disjoint sets and I is a symmetric binary relation between V and B, $I \subseteq V \times B$. The elements of V are called "points," those of B "lines," and those of I "incidences" (or flags). Figure 20 illustrates how an incidence structure is derived from the drawing of a point-line arrangement. The elements of V are denoted by lower case letters (e.g., p, q, r), and those of B by upper case letters (e.g., A, B, C). Using this notation, the elements of I are pairs (p, A) consisting of a point p and a line A, indicating that "p is incident with A," or equivalently, that "A passes through p."

The drawing of an arrangement and the drawing of an incidence structure that represents this arrangement are not the same thing. Incidence structures cannot capture notions of distance and betweenness (i.e., order of points). As a result, the drawing of an incidence structure need not correspond visually to the drawing of the arrangement it represents as long as the two represent the same "incidences."

An incidence structure must satisfy some minimal rules that one assumes when speaking about "a geometry" of points and lines.[4] In particular, a *finite (point-line) geometry* is an incidence structure (V, B, I) subject to the following two rules:

(i) Every line in B is incident with at least two points of V, and
(ii) If two lines in B are distinct, then there is a point in V incident with one of the lines, which is not incident with the other.

---

[4] The minimal rules that specify "a geometry" vary from author to author. Here, I use the approach given in Shult (2011).

The first rule attributes to the lines the typical characteristic usually associated with them in geometry, that is, that a line is formed by "connecting" at least two points. In this case, a line is meant to be identified by the set of points it contains. The second rule certifies that there are no duplicate lines. Incidence structures that satisfy (i) and (ii) are called *geometric spaces*.

While all arrangements of construction lines and registration marks satisfy the second rule, there are arrangements that do not satisfy the first rule. In particular, (a) the lines do not necessarily pass through two points, or (b) all lines are parallel with each other and so there are no registration marks to begin with. The following arrangements are examples of (a) or (b):

In these instances, there is at least one construction line that cannot be identified uniquely by the set of registration marks it is incident with. In the case of the third arrangement above, the construction lines cannot be separated from each other on the basis of a single registration mark. Thus, the universe of arrangements that contains shapes of the algebra U_1 includes cases that do not form finite geometric spaces (in the above strict mathematical sense).

The point-line arrangements that *do* satisfy both rules (i) and (ii) above can be further characterized geometrically by introducing specialized rules of incidence. The smallest non-trivial geometric spaces are *near-linear* and *linear spaces*. They are the starting point for obtaining "higher-level" geometries; for example, the classical affine and projective spaces are linear spaces with extra specialization rules.

A finite geometry is a near-linear space if, and only if, in addition to (i) and (ii), any two distinct points in V are incident with *at most* one line in B. And if we strengthen this rule so that any two distinct points in V are incident with *exactly* one line in B, and further require that there at least three non-collinear points, the finite geometry becomes a linear space (the latter rule is a non-triviality condition, since without it, two points connected by a line form a linear space). Linear spaces are thus obtained from near-linear spaces by adding as lines all pairs of points not already on a line.

The majority of the point-line arrangements in the catalog in Sect. 4.3 form near-linear spaces. The smallest possible linear space is the triangular arrangement for $k = 3$ consisting of three construction lines and three registration marks. This arrangement is an instance of an infinite class called *near-pencils*, which can be defined for any $k \geq 3$, all of which are instances of linear spaces:

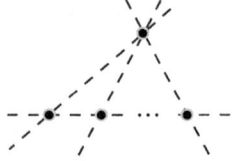

Fig. 21 The place of arrangements containing shapes relative to a hierarchy of finite geometries

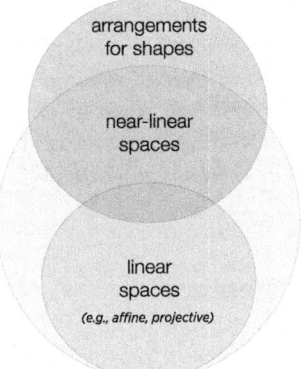

While near-linear spaces are the most common type of geometry that arrangements define, linear spaces are not. The main reason is the inability to arbitrarily draw construction lines between pairs of registration marks without altering the underlying set of registration marks, and without actually violating the rules in Sect. 1. One cannot pursue an approach in which the construction lines are treated as abstract lines (connections) passing through the desired points without physically intersecting; shapes in the algebra U_1 are spatial objects and so are the point-line arrangements that contain them. This distinguishes point-line arrangements that contain shapes from most abstract graph-theoretic constructions in the literature, including block designs and configurations (Grünbaum 2009; Shult 2011).

The containment hierarchy in Fig. 21 shows how arrangements that contain shapes are related to finite geometries. According to this diagram, there are arrangements that contain shapes that form near-linear and linear spaces and there are also arrangements that fall outside of the containment hierarchy of finite geometries. The latter are formed by shapes that we can draw in algebra U_1 to design or use them in shape computations but their underlying construction lines and registration marks do not yield "proper" forms of geometric spaces.

The arrangements that fall inside the hierarchy can be viewed as "weak" types of geometric planes that are equipped with finitely many points and lines. Unlike in a Euclidean plane (or space), where one is free to draw any line anywhere in it, the special plane that a point-line arrangement specifies restricts what can be drawn— one can "move/travel" in it only by following the available construction lines. Such planes exhibit almost no property that is conventionally studied for (finite) affine and projective planes. For example, a point-line arrangement that contains shapes satisfies only a weak form of parallelism; inserting special points or lines "at infinity" has no meaning in this context because such points or lines do not describe (finite) shapes; there are no arrangements that have an equal number of points on the lines as there are lines through each of those points (a requirement for finite projective planes); a "distance" cannot be defined for any two points but only for those for which there is a line crossing both, etc.

The fact that we can draw shapes in the algebra U_1 which do not lead to legitimate geometric spaces is another way of understanding why there can be indeterminate rules in shape grammars. The arrangements of these shapes do not have enough geometric features to determine definite transformations (isometries, similarities, and so forth), but we still use them in visual calculating because the shape instances they contain are perfectly legitimate and useful for developing design ideas.

References

Arslan, M.F., A. Haridis, P.L. Rosin, S. Tari, C. Brassey, J.D. Gardiner, A. Genctav, and M. Genctav. 2022. SHREC'21: Quantifying shape complexity. *Computers and Graphics* 102: 144–153.
Batten, L.M. 1997. *Combinatorics of finite geometries, Second Edition*. Cambridge: Cambridge University Press.
Cai, M., and N. Liao. 2020. On the enumeration of shapes. *Rose-Hulman Undergraduate Mathematics Journal* 21(1), Article 3.
Dembowski, P. 1997. *Finite geometries*. Berlin, Heidelberg: Springer.
Greenberg, M.J. 1980. *Euclidean and non-Euclidean geometries: development and history*. New York: W.H. Freeman & Co.
Grünbaum, B. 2009. *Configurations of points and lines*. American Mathematical Society.
Haridis, A. 2020a. Structure from appearance: Topology with shapes, without points. *Journal of Mathematics and the Arts* 14(3): 199–238.
Haridis, A. 2020b. The topology of shapes made with points. *Environment and Planning B: Urban Analytics and City Science* 47(7): 1279–1288.
Haridis, A. 2020c. *Some open problems regarding the number of lines and slopes in arrangements that contain shapes*. ArXiv:2011.10700.
Haridis, A., and G. Stiny. 2022. Analysis of shape grammars: Continuity of rules. *Environment and Planning B: Urban Analytics and City Science* 49(7): 1929–1948.
Hong, T.C., and A. Economou. 2022. What shape grammars do that CAD should: the 14 cases of shape embedding, *Artificial Intelligence for Engineering Design, Analysis and Manufacturing* 36(e4).
Krishnamurti, R. 1980. The arithmetic of shapes. *Environment and Planning B: Planning and Design* 7: 463–484.
Kristic, D. 1996. *Decompositions of shapes*. Ph.D. dissertation. Los Angeles: University of California, Los Angeles.
Kristic, D. 2022. Diagonal decompositions of shapes and their algebras. *Artificial Intelligence for Engineering Design, Analysis and Manufacturing* 36(e10).
Marquis, J.-P. 2009. Category theory and Klein's Erlangen program. In *From a geometrical point of view: Logic, epistemology, and the unity of science*, vol. 14, 9–40. Dordrecht: Springer.
Shult, E.E. 2011. *Points and lines: Characterizing the classical geometries*. Springer Berlin, Heidelberg: Springer.
Stanley, P. 2007. An introduction to hyperplane arrangements. In *Geometric combinatorics*, eds. E. Miller, V. Reiner, and B. Sturmfels, vol. 13 (IAS/Park City Mathematics Series).
Stillwell, J. 2005. *The four pillars of geometry*. New York: Springer.
Stiny, G. 1975 *Pictorial and formal aspects of shape and shape grammars*. Basel: Birkhäuser.
Stiny, G. 2022. *Shapes of imagination: Calculating in Coleridge's magical realm*. Cambridge: The MIT Press.

Open Access This chapter is licensed under the terms of the Creative Commons Attribution-NonCommercial-NoDerivatives 4.0 International License (http://creativecommons.org/licenses/by-nc-nd/4.0/), which permits any noncommercial use, sharing, distribution and reproduction in any medium or format, as long as you give appropriate credit to the original author(s) and the source, provide a link to the Creative Commons license and indicate if you modified the licensed material. You do not have permission under this license to share adapted material derived from this chapter or parts of it.

The images or other third party material in this chapter are included in the chapter's Creative Commons license, unless indicated otherwise in a credit line to the material. If material is not included in the chapter's Creative Commons license and your intended use is not permitted by statutory regulation or exceeds the permitted use, you will need to obtain permission directly from the copyright holder.

Shape Computation Research at the University of Sydney

John Gero

Abstract This chapter provides a brief overview of the shape computation research at the University of Sydney that was carried out over a 35-year period, between 1977 and 2012, referencing 17 Ph.D theses that discuss shape computation during that period. The chapter draws distinctions between shape computation using shape grammars and other approaches that parallel, supplement and diverge from shape grammars. It introduces a variety of methods for shape computation, including those based on infinite maximal lines, artificial intelligence, learning representations, learning generation rules, logic models, qualitative representation, qualitative reasoning, algebraic representation, shape semantics, situated learning, and shape interpretation.

1 Introduction

In 1966, I introduced computational modeling into the architecture teaching curriculum at the University of Sydney. Two years later, I developed and directed a research program that presented a continuing stream of shape computation, commencing in the mid-1970s, leading to the founding of the Key Centre of Design Computing and Cognition. This chapter provides a brief overview of this research. It draws distinctions between shape computation using shape grammars and other approaches that parallel, supplement and diverge from shape grammars. Each topic area is briefly introduced, with references to the primary papers (drawn from 17 separately listed PhD theses) so the reader can examine their approaches, methods and results in more detail. In 1975, Stiny (1975) at UCLA and Gips (1975) at Stanford published their two doctoral theses that together laid the foundation for a new approach to shape computation—shape grammars. They had published a paper earlier (Stiny and Gips 1972) in which shape grammars provide a formal algebra for shape computation (although initially it did not provide computable implementations). Prior to this work, shape computation was primarily focused on spatial layouts—the

---

J. Gero (✉)
University of North Carolina at Charlotte, Charlotte, NC 28223, USA
e-mail: john@johngero.com

synthesizing of plans given spaces and relational constraints between them. This was treated as part of a more general location-allocation problem, which became accessible to architects through the publication by Whitehead and Eldars in (1965) of easily implementable heuristics for the gravitational model approach. Mitchell et al. (1976) used a formal optimization approach for the same class of spatial synthesis problem. At the University of Sydney (1977), I demonstrated that this could be treated as an instance of a more general spatial optimization problem. The introduction of the concepts of shape grammars changed the focus of the field since they could be used directly to synthesize complex shapes. Additionally, shape grammars had the potential to operate with emergent shapes.

2 Extended and Infinite Maximal Lines

The concept of maximal lines was introduced by Stiny (1975). Maximal lines allow for line segment embedding. This eliminated the Euclidean constraint that lines had fixed end points and opened the possibility of generating segments from existing segments. However, two maximal lines that were collinear but separated remained two separate lines. Further, two maximal lines whose extensions intersected would not appear as related. For these reasons, two constructs were added to extend the definition of maximal lines. The first was the *extended maximal line* and the second was the *infinite maximal line* (Damski and Gero 1997; Gero and Yan 1994). An extended maximal line is made up of two collinear, non-touching, non-overlapping maximal lines. An infinite maximal line is an extended maximal line that has been extended to infinity in both directions. Examples of line segments, maximal lines, extended maximal lines and infinite maximal lines are given in the figure.

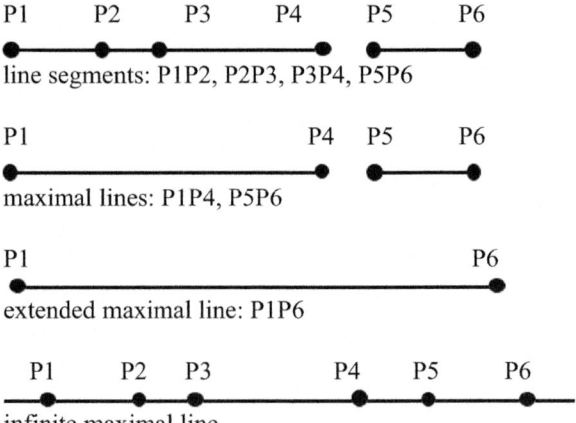

Extended maximal lines can be used to find emergent shapes beyond shapes that share lines with existing shapes—continuous illusory shapes. The dotted lines in the figure below are constructed using extended maximal lines.

Infinite maximal lines can be used to find emergent shapes that share some lines with existing lines but also lie outside any existing lines as extended illusory shapes. The dotted lines in the figure below are constructed using infinite maximal lines.

Extended maximal lines can be extended to be extended maximal planes. Infinite maximal lines can be extended to infinite maximal planes, as in the figure.

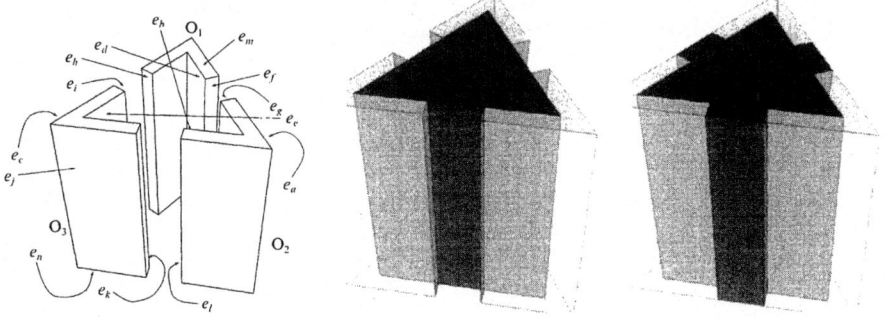

There is a third category of emergent or illusory shapes that neither maximal lines nor infinite maximal lines can find; these are called *corner illusory shapes* that share no lines with the shape from which they emerge. Corner illusory shapes, the dotted lines in the figure below, are emergent shapes constructed by the human vision system but, as yet, they have no direct computational model.

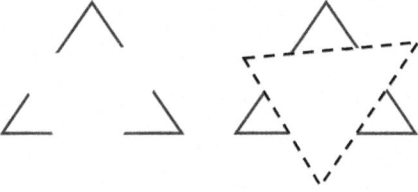

3 Non-Symbolic Systems for Emergent Shapes

Grossberg and Mingolla (1985) introduced a human perception neuronal approach to shape pattern recognition called the Boundary Contour system. The BC system does not use schemas (such as lines); rather it uses competitive neurons loosely based on human perception. Phil Tomlinson showed this approach is able to find both continuous and extended emergent or illusory shapes (Tomlinson and Gero 1997) as in the figure, with the potential to find corner illusory shapes but this has not been shown empirically.

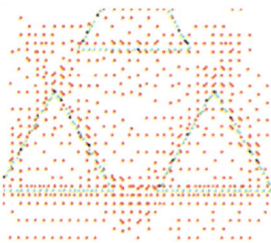

4 Models from Artificial Intelligence

Beginning in the late 1970s and early 1980s, ideas from artificial intelligence about symbolic reasoning began to be explored in representing and reasoning about shapes. The directions that were followed relate to symbolic representation and symbolic rules, although not necessarily algebraic rules. Using first-order predicate logic, objects and spaces could be represented symbolically such that "facts" about an object or space are separate from knowledge about it and that meaning, i.e., its semantics, could be attached or inferred (Akiner 1984; Ph.D. dissertation). Tuncer Akiner's research focused on a new form of representation that allowed reasoning about a range of spatial behaviors (Akiner 1986; Gero et al. 1983). So, it was possible to reason about topological spatial relationships, including the semantics of the spaces without recourse to geometry. Shape grammars were primarily being used to represent shapes and their relationships. Akiner showed that other representations could be used to reason about generated shapes in ways that shape grammars did not address and that these representations could be readily implemented computationally.

Richard Coyne took the artificial intelligence concepts of symbolic logic and symbolic planning and applied them to synthesizing spatial layouts where the basic unit was a space (Coyne 1986; Ph.D. dissertation). This logic-based planning model allows reasoning backwards from the goal state (what you want) to the initial state (where you have to start). Coyne developed this further into a hierarchy of languages that formed a rich design grammar (Coyne and Gero 1985, 1986). The three languages were: a language of form, a language of actions and a language of plans. These were implemented in Prolog, a logic programming language. These were not shape grammars in the sense of Stiny. What Coyne showed was that grammars could be used not as unconstrained generative tools but as the basis of generation with constraints. Further, they could be readily implemented, computationally, as demonstrated in the figure below, with the three languages represented on the left and a screen shot of their implementation on the right.

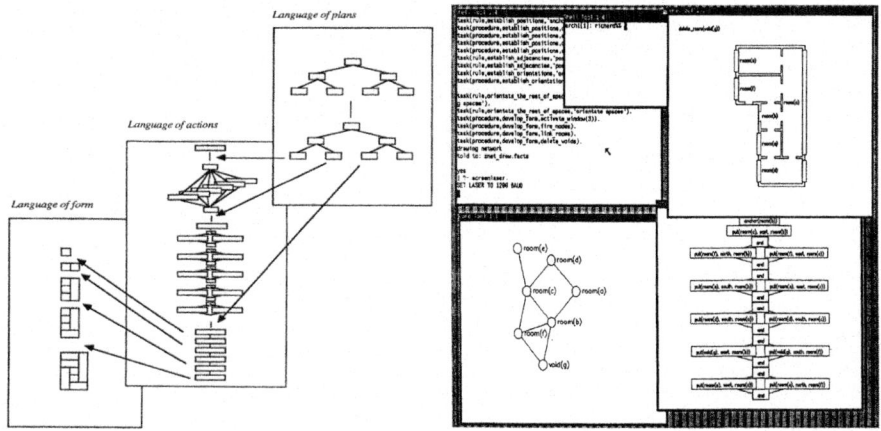

5 Genetic Algorithms

Genetic algorithms are computational constructs based on Darwin's theory of evolution, where genes of parents crossover to produce offspring. Jun Jo showed that this could be a model of generating spatial layouts, where the multiple genes represented spatial and relational elements (Jo 1993; Ph.D. dissertation). Representing shapes as genes and evolutionary processes as combination rules provided a readily implementable way to operationalize the execution of shape grammars where criteria needed to be satisfied. This approach was shown to match or improve on existing spatial layout algorithms (Jo and Gero 1995, 1998).

6 Learning Shape Representations and Shape Generation Rules

The development of grammars up to the mid-1980s was based on a hand-crafted approach. A large number of shape grammars were, and continue to be, developed for specific situations. However, such manual approaches are laborious and are not able to be generalized and made computable. Mackenzie (1990; Ph.D. dissertation) developed a machine learning approach that automatically generated the rules for a shape grammar (Mackenzie 1989). The automatic generation of shape representation and shape generation rules was a theme that was explored for many years from a variety of learning approaches.

Genetic engineering, the concept of intervening in evolution, was applied to design (Gero and Kazakov 1995, 2001a) to learn the genes that, when executed, generate shapes. The input is one, or a representative set of shapes, and what is learned is a set of genes that can be used to express shapes that belong to the same set as the input set but, as in natural evolution, new shapes can be generated that are grounded in the original shapes. Below on the left are genes, expressed as boundaries, for Frank Lloyd Wright Prairie house plans. On the right is a genetically engineered Frank Lloyd Wright Prairie house plan (Gero et al. 1997; Schnier 1999; Ph.D. dissertation).

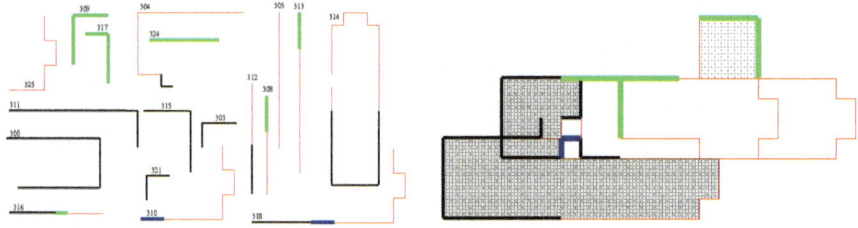

This approach had an interesting property that the genes for the different grammars are commensurate with each other and, therefore, can be mixed. It becomes possible to mix the genes of Frank Lloyd Wright's windows with those of Mondrian's paintings to produce "Flondrians" (Schnier and Gero 1998). Two Flondrians are shown.

This genetic engineering approach to learning grammar genes was applied to learning the control rules to apply grammar genes, thus learning design styles (Ding 1999; Ph.D. dissertation). Lan Ding applied this approach to learning the design style of traditional Chinese pagodas (Ding and Gero 2001; Gero and Ding 1997). The figure below shows two computer-generated Chinese pagodas with the same style.

Learning shape grammars is a way of capturing the past. What of producing novel shape grammars that extend existing grammars? Gero et al. (1994) showed how existing grammars could be used to produce novel grammars by taking genetic representations and using evolutionary crossover processes to generate novel design grammars. Novel grammars produced this way were applied to design optimization problems for cases where the Pareto designs were known. The novel grammars improved on prior Pareto optimal designs by generating a larger design space, as can be seen.

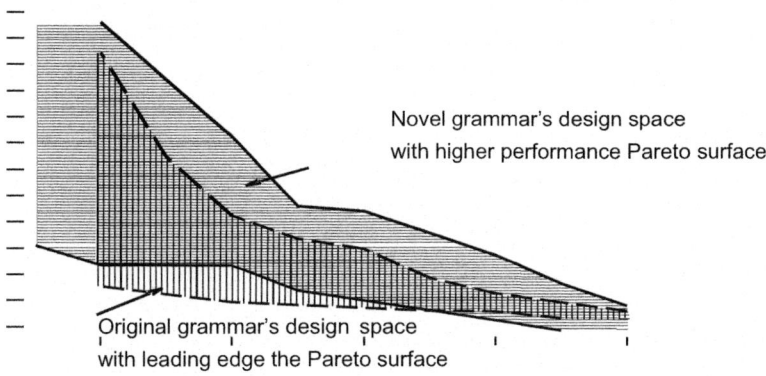

7 Logic Model of Shape Representation

While there is no inherent reason for representing shapes with straight lines, there is a basic distinction between straight-line bounded shapes and arbitrary-line bounded shapes. Damski (1996; Ph.D. dissertation) developed a formal logic-based representation of shapes and volumes that was unrelated to the form of boundaries. This was based on the axiom that any plane (or volume) could arbitrarily be divided into two half-planes (or volumes). From this axiom a number of theorems were developed that mapped onto topological operations on shapes (and volumes) (Damski and Gero 1996). For example, a subset of contiguous rooms in Botta's Family House in Stabio, Switzerland could be extracted using logic operations as topologically connected volumes (Damski and Gero 1998), as can be seen on the left below. Any spatially configured image could be modeled as a set of half-planes, irrespective of the shape of the boundaries, as in Miro's *Portrait of a Young Girl* (Damski 1996), on the right below. From this model, spatial operations can be carried out.

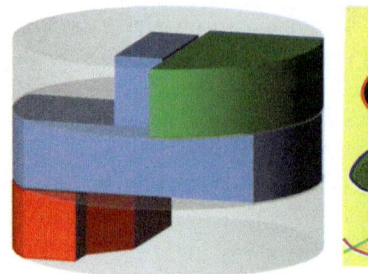
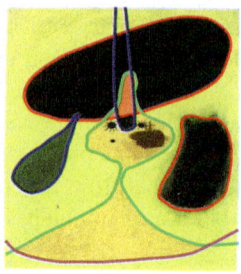

8 Qualitative Representation and Reasoning About Shapes

Studies of eye-tracking of humans indicate that they focus on landmarks rather than just edges when examining visual images. Park (1999; Ph.D. dissertation) developed a language for qualitative shape representation based on landmarks. Landmarks for shapes are any change in the boundary of the shape. Typical landmarks are corners or change of curvature. A simple symbolic language for such changes commences with: A- for a change in direction at a corner that is less than 180° and A+ for a change in direction at a corner that is greater than 180° (Gero and Park 1997). With only these two symbols it is possible to describe and categorize shapes as well as infer whether they are closed, they are convex or concave and other properties (Gero and Park 2000; Park and Gero 1999). A fuller set of qualitative codes for describing shapes is:

(i) angle measured at a node, A-code;
(ii) relative length of line segments, L-code;
(iii) angle measured at a node for two tangents, C-code;
(iv) relative curvature of a line segment, K-code.

It becomes possible to construct a hierarchy of code groupings that can be used to represent shapes at different levels of granularity. These can be described through a linguistic analogy. A word, as a minimum and discrete unit of information, can contain a basic shape feature. The words are then aggregated to construct more complicated expressions in the form of a phrase. A Q-phrase displays a certain syntactic structure of one or more Q-words by explicitly describing the structure with a set of syntactic operations such as "iteration", "alternation", and "symmetry". These smaller shape features are aggregated to form a complete and closed shape termed a Q-sentence as another level of shape features.

Q-code A basic symbol that refers to an atomic component of a shape attribute.

Q-word A sequence of Q-codes that refers to a shape pattern with distinctive design significance—a shape feature.

Q-phrase A sequence of Q-codes in which one or more Q-words show a distinctive pattern of structural arrangements.

Q-sentence An aggregation of Q-codes, Q-words, and Q-phrases so that it refers to a closed and complete contour of a shape.

Q-paragraph A group of Q-sentences where necessary spatial relationships are described with specific connectives.

Examples of each of these can be seen in the figure.

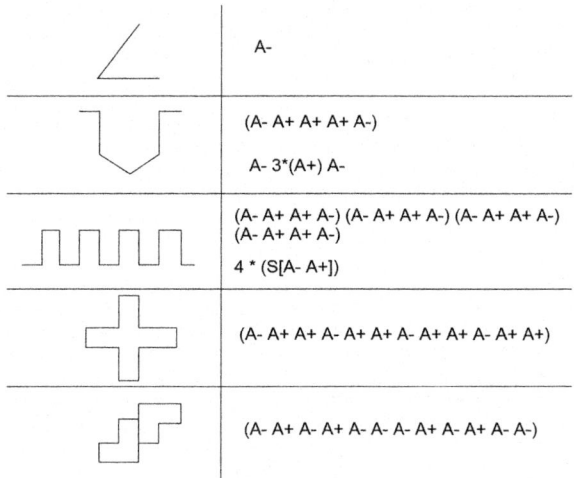

9 Algebraic Representation of Shape Patterns

Shape patterns capture structured relationships between shapes. Cha (1998; Ph.D. dissertation) produced an algebra of operations that could be used to represent patterns of shapes. It commences with the isometric transformations and proceeds to combine them such that any shape pattern can be represented. The four transformations of: translation, rotation, reflection and scale are shown below with their algebraic representations.

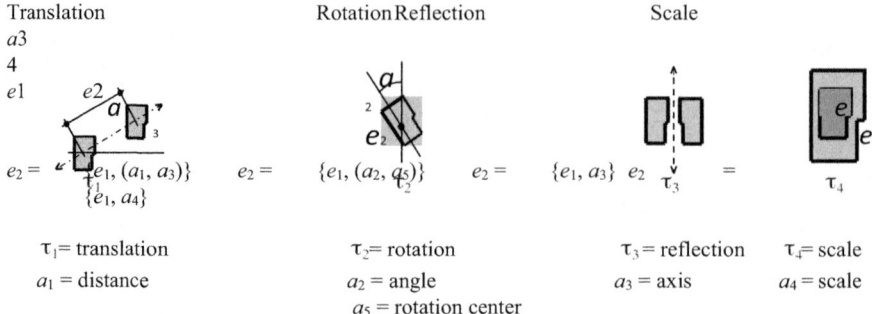

A shape recognition system was implemented where an image was scanned, converted to vectors and then into individual shapes. From the vectors of scanned shapes, algebraic expressions representing the patterns found in that image could be automatically generated (Cha 1998; Cha and Gero 1998). As an example, consider Michelangelo's Campidoglio and its two separate representations (Gero and Cha 1997), as in the figures.

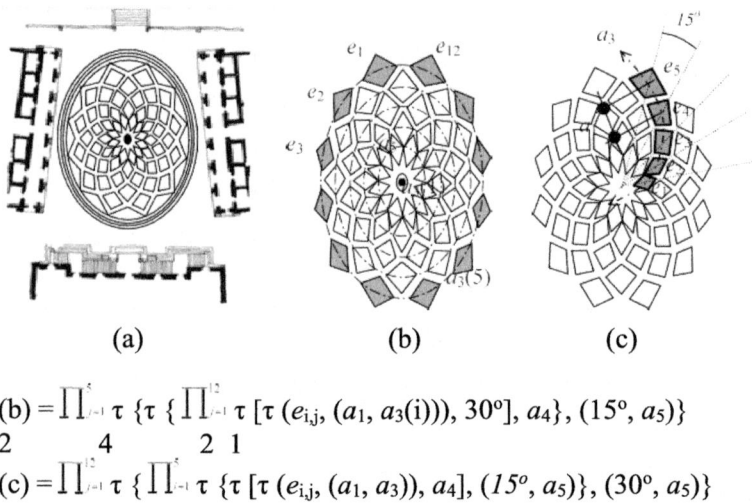

(a) (b) (c)

(b) = $\prod_{i=1}^{5} \tau \{\tau \{\prod_{i=1}^{12} \tau [\tau (e_{i,j}, (a_1, a_3(i))), 30°], a_4\}, (15°, a_5)\}$
    2   4   2 1

(c) = $\prod_{i=1}^{12} \tau \{\prod_{i=1}^{5} \tau \{\tau [\tau (e_{i,j}, (a_1, a_3)), a_4], (15°, a_5)\}, (30°, a_5)\}$

10 Shape Semantics

Configurations of shapes, such as those found in architectural floor plans and elevations, can be analyzed for emergent shape semantics. Jun (1997; Ph.D. dissertation) developed a system for emerging shape semantics, where the initial shapes were represented as lines. Four classes of emergent shape semantics were able to be found using a computational system: visual symmetry, visual rhythm, visual movement and visual balance (Gero and Jun 1995, 1998). The computational system used infinite

Shape Computation Research at the University of Sydney 327

maximal lines (Gero and Yan 1994), decomposition and Gestalt knowledge. As an example, a façade is represented as lines from which emergent visual rhythm is able to be found by the system, as in the right two images in the figure.

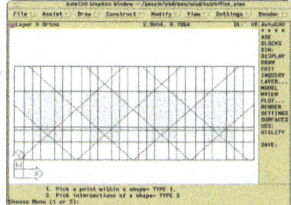 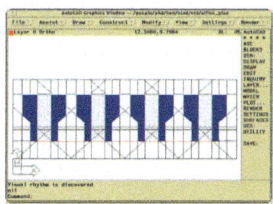 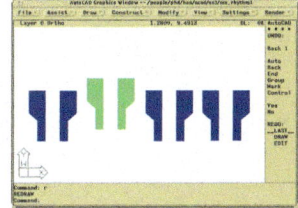

These can be used to produce new designs with an additional visual rhythm dimension, as.

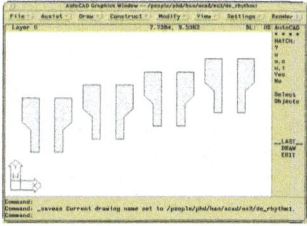

11 Situated Learning of Emergent Shapes and Patterns

Reffat (2000; Ph.D. dissertation) continued the work on emerging shapes by learning emergent shapes within a situation, still using infinite maximal lines (Gero 1998). This allowed a designer to guide the learning (Gero and Reffat 1997; Reffat and Gero 1998). A system was implemented that commenced with a raster scan of the plan of interest, converted into vectors and then situation-driven shapes and patterns emerged from that. In the example below the raster plan on the left is converted into vectors, then re-represented as infinite maximal lines. The designer explores rotation symmetry and a number of alternatives is generated from the original configuration of shapes, as in the images.

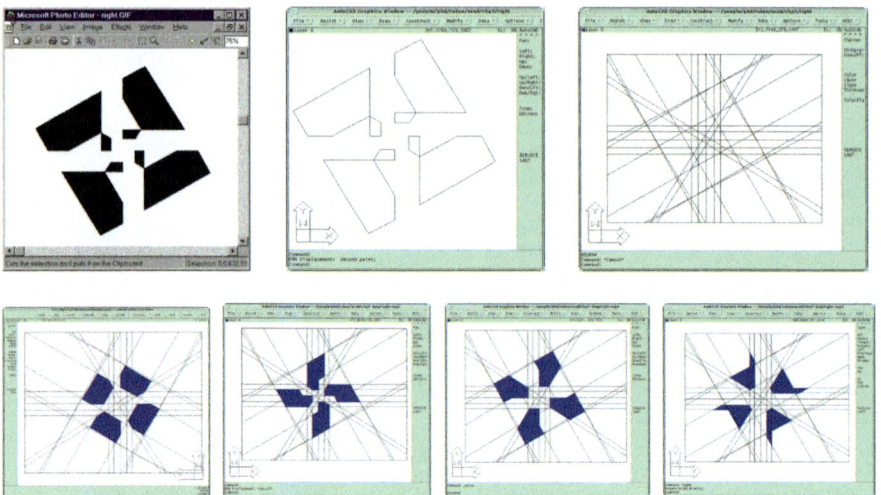

12 Visual Style

Julie Jupp combined qualitative representations of shapes with quantitative analysis and then with cognitive studies of visual style (Jupp 2005; Ph.D. dissertation). This research extended the qualitative shape representation of Park (1999) to include topology, shape and form categorization as the basis of visual style (Jupp and Gero 2006). Information theory is utilized to categorize style and neural networks are used to learn visual style. The research shows it is possible to recognize an architect's style from a corpus of their floor plans and to use the learned model to predict which architect designed any particular floor plan that was being viewed (Gero and Jupp 2002; Jupp and Gero 2004, 2010). Gero and Kazakov used entropy measures of qualitative shape representations to determine similarity of shapes (Gero and Kazakov 2001b). This was extended to 3D shapes (Gero and Kazakov 2003).

13 Shape Interpolation and Extrapolation

Typically, shape interpolation takes the form of homomorphism, i.e., the structure in one shape is preserved as the structure in the other shape. This is the result of a linear interpolation process. This process does not produce new shapes. However, if the interpolation is nonlinear the morphism is broken and novel shapes can be produced that have characteristics that are not derived from each of the two initial shapes. As a consequence, the space of possible shapes is expanded. Further, extrapolation is not possible with homomorphic processes, but can be executed non-homomorphically to generate an enhanced space of possible shapes. Some of the characteristics of the new shapes do not come from the initial two shapes (Gero and Kazakov 1999). Each non-linear interpolation function generates a new space of possible shapes. The two leftmost shapes in the figure below are the base shapes followed by a nonlinear interpolated shape between them, a shape from a left extrapolation and then a shape from a right extrapolation.

14 Shape Interpretation

Visual analogy, visual metaphor and visual allegory require associations between shapes. These associations are based on interpretations of the shapes that afford associations. Kazjon Grace investigated the role of interpretation in the generation of visual associations (Grace 2012; Ph.D. dissertation). Association can be decomposed into two processes: representation and mapping. In the representation process the agent builds representations of the objects. In the mapping process the agent searches those representations for the shared property or properties and relates them (Grace et al. 2012, 2015), as in the figure.

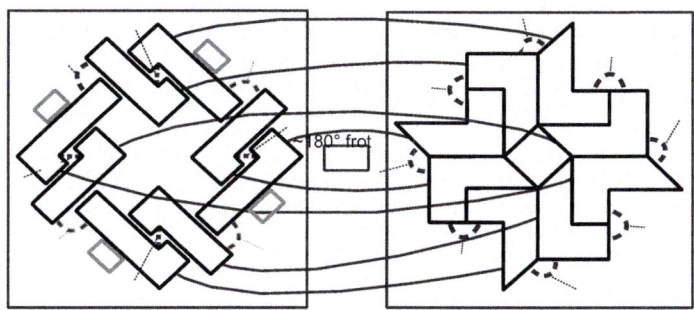

15 Situated Shape Interpretation

The interpretation of shapes from expectations grounded in their situation brings shape interpretation as a constructive act closer to human cognition. This approach takes shape construction beyond Phil Tomlinson's work on boundary contour systems

(Tomlinson and Gero 1997). Kelly (2011; Ph.D. dissertation) took this interpretation-driven approach where, for example, an agent learns characteristics of different architects' floor plans and uses these to construct an interpretation from some source data that need not be in a line representation. The representation can be as basic as pixels. The agent attempts to construct from the source representation a match with the characteristics of what is being looked for, i.e., it is expectation driven. This is the inverse of an approach where the shape representation is fixed and the agent looks for matching shapes. For example, if the source image, on the left in the figure below, is a set of pixels and the agent is attempting to interpret the image as a Frank Lloyd Wright plan, it determines whether it can construct a Wright plan from the source image, in the upper right image. Alternatively, with the same source image, the agent is attempting to interpret the image as a plan by Andrea Palladio to determine whether it can construct a Palladian plan from the source image, in the lower right image. This allows for multiple interpretations of the same image.

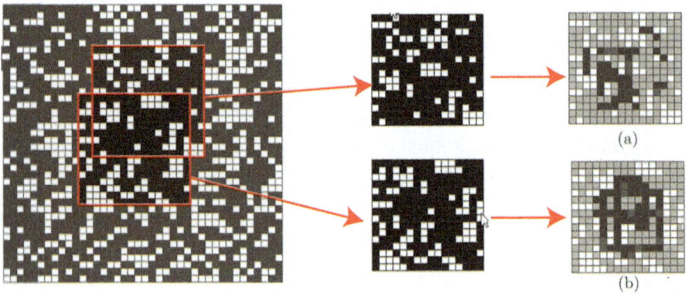

This chapter is not an exhaustive summary of the research and PhD theses on shape computation at the University of Sydney. Other research includes space planning (Yoon 1990; Ph.D. dissertation), grammars for the design of virtual worlds (Gu 2006; Ph.D. dissertation), and learning and constructing shapes using curious AI agents (Saunders 2002; Ph.D. dissertation).

Acknowledgements The research reported in this chapter was supported by multiple grants from the Australian Research Council, by University of Sydney PhD Scholarships, by University of Sydney Overseas Postgraduate Research Awards and by computing resources from University of Sydney's Key Centre of Design Computing and Cognition.

PhD Dissertations Referenced

Akiner, V.T. 1984. *Topology-1: A system that reasons about objects and spaces in buildings.* Ph.D. dissertation. Sydney: University of Sydney.

Cha, M.Y. 1998. *Architectural shape pattern representation and its application for design computation.* Ph.D. dissertation. Sydney: University of Sydney.

Coyne, R.D. 1986. *A logic model of design synthesis.* Ph.D. dissertation. Sydney: University of Sydney.

Damski, J. 1996. *Logic representation of shapes*. Ph.D. dissertation. Sydney: University of Sydney.
Ding, L. 1999. *An evolutionary model for style representation emergence in design*. Ph.D. dissertation. Sydney, Australia: University of Sydney.
Grace, K. 2012. *Interpretation-driven association in design*. Ph.D. dissertation. Sydney: University of Sydney.
Gu, N. 2006. *Dynamic designs of virtual worlds using generative design agents*. Ph.D. dissertation. Sydney: University of Sydney.
Jo, J.-H. 1993. *A computational process model using a genetic evolution approach*. Ph.D. dissertation. Sydney: University of Sydney.
Jun, H. 1997. *Emergence of shape semantics of architectural drawings in CAAD systems*. Ph.D. dissertation. Sydney: University of Sydney.
Jupp, J. 2005. *Diagrammatic reasoning: computational and cognitive studies in design similarity*. Ph.D. dissertation. Sydney: University of Sydney.
Kelly, N. 2011. *Constructive interpretation in design thinking*. Ph.D. dissertation. Sydney: University of Sydney.
Mackenzie, C. 1990. *Function and structure relationships and transformations in design processes*. Ph.D. dissertation. Sydney, Australia: University of Sydney.
Park, S.-H. 1999. *Qualitative representation and feature-based analysis of shape*. Ph.D. dissertation. Sydney: University of Sydney.
Reffat, R. 2000. *Computational situated learning in designing; application to architectural shape semantics*. Ph.D. dissertation. Sydney: University of Sydney.
Saunders, R. 2002. *Curious design agents and artificial creativity*. Ph.D. dissertation. Sydney: University of Sydney.
Schnier, T. 1999. *Evolved representations and their use in computational creativity*. Ph.D. dissertation. Sydney: University of Sydney.
Yoon, K.B. 1990. *A constraint model of space planning*. Ph.D. dissertation. Sydney: University of Sydney.

References

Akiner, V.T. 1986. Topology-1: A knowledge-based system for reasoning about objects and spaces. *Design Studies* 7 (2): 94–105. https://doi.org/10.1016/0142-694X(86)90022-0.
Cha, M.Y., and J.S. Gero. 1998. Shape pattern recognition using a computable shape pattern representation. In *Artificial intelligence in design '98*, ed. J.S. Gero and F. Sudweeks, 169–188. Dordrecht: Kluwer.
Coyne, R.D., and J.S. Gero. 1985. Design knowledge and sequential plans. *Environment and Planning B* 12: 401–418.
Coyne, R.D., and J.S. Gero. 1986. Semantics and the organization of knowledge in design. *Design Computing* 1 (1): 68–89.
Damski, J.C., and J.S. Gero. 1996. A logic-based framework for shape representation. *Computer-Aided Design* 28 (3): 169–181.
Damski, J., and J.S. Gero. 1998. Object representation and reasoning using halfplanes and logic. In *Artificial intelligence in design '98*, eds. J.S. Gero and F. Sudweeks, 107–126. Dordrecht: Kluwer.
Ding, L., and J.S. Gero. 2001. The emergence of the representation of style in design. *Environment and Planning B* 28 (5): 707–731.

Gero, J.S. 1977. Note on 'Synthesis and optimization of small rectangular floor plans' of Mitchell, Steadman and Liggett, *Environment and Planning B* 4: 81–88.

Gero, J.S. 1998. Conceptual designing as a sequence of situated acts. In *Artificial Intelligence in Structural Engineering*, ed. I. Smith, 165–177. Berlin: Springer.

Gero, J.S., V.T. Akiner, and A.D. Radford. 1983. What's what and what's where—knowledge engineering in the representation of buildings by computer. *PARC83*, Online, Middlesex, 205–215.

Gero, J.S., and C. Myung Yeol. 1997. Computer-based representation of patterns in architectural shapes. In *Proceedings of CAADRIA '97,* edited by Y.-T. Liu, J. -Y. Tsou, and J.-H. Hou, 377–388. Taipei: Hu's Publisher.

Gero, J.S., and J.C. Damski. 1997. A symbolic model for graphical emergence, *Environment and Planning B, 24*, 509–526. https://doi.org/10.1068/b240509.

Gero, J.S., and L. Ding. 1997. Exploring style emergence in architectural designs. In *Proceedings of CAADRIA '97,,* eds. Y.-T. Liu, J.-H. Tsou and J.-H. Hou, 287–296. Taipei: Hu's Publisher.

Gero, J.S., and H. Jun. 1995. Getting computers to read the architectural semantics of drawings. In *Proceedings of ACADIA '95: Computing in design : Enabling, capturing and sharing ideas*, eds. L. Kalisperis, B. Kolarevic, 97–112. Washington: ACADIA.

Gero, J.S., and H. Jun. 1998. Emergence of shape semantics of architectural shapes. *Environment and Planning B* 25 (4): 577–600.

Gero, J.S., and J. Jupp. 2002. Measuring the information content of architectural plans. In *Proceedings of SIGRaDi '02*, eds. P. L. Hippolyte and E. Miralles, 155–158. Caracas: Ediciones Universidad Central de Venezuela.

Gero, J.S., and V. Kazakov. 1995. Evolving building blocks for genetic algorithms using genetic engineering. In *1995 IEEE international conference on evolutionary computing*, 340–345.

Gero, J.S., and V. Kazakov. 1999. An interpolation/extrapolation process for creative designing. In *Computers in Building*, eds. G. Augenbroe and C. Eastman, 263–274. Boston: Kluwer.

Gero, J.S., and V. Kazakov. 2001a. A genetic engineering extension to genetic algorithms. *Evolutionary Computation*, 9 (1): 71–92.

Gero, J.S., and V. Kazakov. 2001b. Entropic similarity and complexity measures for architectural drawings. In *Visual and spatial reasoning in design II*, eds. J. S. Gero, B. Tversky, and T. Purcell, 147–161. Sydney: Key Centre of Design Computing and Cognition, University of Sydney.

Gero, J.S., and V. Kazakov. 2003. On measuring the visual complexity of 3D solid objects. In *E-Activities in design and design education*, eds. B. Tuncer, S. Ozsariyildiz, and S. Sariyilidiz, 147–156. Paris: Europia.

Gero, J.S., V. Kazakov, and T. Schnier. 1997. Genetic engineering and design problems. In *Evolutionary algorithms in engineering applications*, eds. D. Dasgupta and Z. Michalewicz, 47–68. Berlin: Springer.

Gero, J.S., S. Louis, and S. Kundu. 1994. Evolutionary learning of novel grammars for design improvement. *Artificial Intelligence for Engineering Design, Analysis and Manufacturing* 8 (2): 83–94.

Gero, J.S., and S.-H. Park. 1997. Computable feature-based qualitative modelling of shape and space. In *Proceedings of CAADFutures'97*, ed. R. Junge, 821–830. Dordrecht: Kluwer.

Gero, J.S., and S.-H. Park. 2000. Categorisation of shapes using shape features. In *Artificial intelligence in design '00*, edited by J.S. Gero, 203–223. Dordrecht: Kluwer.

Gero, J.S., and R. Reffat. 1997. Multiple representations for situated agent-based learning. In *Proceedings of ICCIMA '97*, eds. B. Varma and X. Yao, 81–85. Australia: Griffith University, Gold Coast, Queensland.

Gero, J.S., and M. Yan. 1994. Shape emergence using symbolic reasoning. *Environment and Planning B* 21: 191–218.

Gips, J. 1975. *Shape grammars and their uses*. Basel: Birkhauser.

Grace, K., J.S. Gero, and R. Saunders. 2015. Interpretation-driven mapping: A framework for parallel search and re-representation for computational analogy in design. *Artificial Intelligence for Engineering Design, Analysis and Manufacturing* 29 (02): 185–201.

Grace, K., R. Saunders, and J.S. Gero. 2012. Representational affordances and creativity in association-based systems. In *Proceedings of the third international conference on computational creativity*, eds. M.L. Maher, K. Hammond, A. Pease, R. Pérez y Pérez, D. Ventura, and G. Wiggins, 195–202.

Grossberg, S., and E. Mingolla. 1985. Neural dynamics of perceptual grouping. *Perception and Psychophysics* 38: 141–171.

Jo, J., and J.S. Gero. 1995. A genetic search approach to space layout planning. *Architectural Science Review* 38 (1): 37–46. https://doi.org/10.1080/00038628.1995.9696774.

Jo, J., and J.S. Gero. 1998. Space layout planning using an evolutionary approach. *Artificial Intelligence in Engineering* 12 (3): 149–162.

Jupp, J., and J.S. Gero. 2004. Qualitative representation and reasoning about shapes and spatial relationships. In *Visual and spatial reasoning in design III*, ed. J.S. Gero, B. Tversky, and T. Knight, 139–162. Key Centre of Design Computing and Cognition: University of Sydney.

Jupp, J., and J.S. Gero. 2006. Visual style: Qualitative and context dependent categorisation. *Artificial Intelligence for Engineering Design, Analysis and Manufacturing* 20 (3): 247–266.

Jupp, J., and J.S. Gero. 2010. Let's look at style: Visuo-spatial representation and reasoning in design. In *The structure of style: Algorithmic approaches to understanding manner and meaning*, eds. S. Argamon, K. Burns, and S. Dubnov, 159–195. Springer.

Mackenzie, C.A. 1989. Inferring relational design grammars. *Environment and Planning B* 16: 253–287.

Mitchell, W.J., J.P. Steadman, and R.S. Liggett. 1976. Synthesis and optimization of small rectangular floor plans. *Environment and Planning B* 3: 37–70.

Park, S.-H., and J.S. Gero. 1999. Qualitative representation and reasoning about shapes. In *Visual and spatial reasoning in design*, eds. J.S. Gero and B. Tversky, 55–68. Australia: Key Centre of Design Computing and Cognition, University of Sydney, Sydney.

Reffat, R., and J.S. Gero. 1998. Learning about shape semantics: A situated learning approach. In *Proceedings of CAADRIA '98*, eds. T. Sasada, S. Yamaguchi, M. Morozumi, A. Kaga and R. Homma, 375–384. Kumomoto: CAADRIA.

Schnier, T., and J.S. Gero. 1998. From Frank Lloyd wright to mondrian: transforming evolving representations. In *Adaptive computing in design and manufacture*, ed. I. Parmee, 207–219. London: Springer.

Stiny, G. 1975. *Pictorial and formal aspects of shape and shape grammars*. Basel: Birkhauser.

Stiny, G., and J. Gips. 1972. Shape grammars and the generative specification of painting and sculpture. In *Information processing, proceedings of IFIP congress 1971, Volume 2—Applications*, eds. C.V. Freiman, J.E. Griffith, and J.L. Rosenfeld, Amsterdam: North-Holland.

Tomlinson, P., and J.S. Gero. 1997. Emergent shape generation via the boundary contour system. In *Proceedings of CAAD futures '97*, ed. R. Junge, 865–874. Dordrecht: Kluwer.

Whitehead, B., and M.Z. Eldars. 1965. The planning of single-storey layouts. *Building Science* 1 (2): 127–139. https://doi.org/10.1016/0007-3628(65)90014-.

Open Access This chapter is licensed under the terms of the Creative Commons Attribution-NonCommercial-NoDerivatives 4.0 International License (http://creativecommons.org/licenses/by-nc-nd/4.0/), which permits any noncommercial use, sharing, distribution and reproduction in any medium or format, as long as you give appropriate credit to the original author(s) and the source, provide a link to the Creative Commons license and indicate if you modified the licensed material. You do not have permission under this license to share adapted material derived from this chapter or parts of it.

The images or other third party material in this chapter are included in the chapter's Creative Commons license, unless indicated otherwise in a credit line to the material. If material is not included in the chapter's Creative Commons license and your intended use is not permitted by statutory regulation or exceeds the permitted use, you will need to obtain permission directly from the copyright holder.

Computer Implementation of Shape Grammars

Shape Machine: Shape-Based Search and Replace in CAD

Athanassios Economou

Abstract Why haven't the Find and Replace operations, so essential in Word and Excel, been implemented yet in CAD? What would happen if we could seamlessly use vector-based shape rewrite operations with CAD operators in a logical processing framework to literally write programming code by drawing shapes? How would this affect our current view of computation, and what would it mean for design? The paper discusses the current state of Shape Machine, a shape-rewrite computational system that features shape-based operations and a logical processing framework developed at the Shape Computation Lab at the Georgia Institute of Technology, and currently integrated within Rhinoceros, a NURBS 2D/3D CAD software. Four applications across architecture and computer science are briefly discussed to showcase the potential impact of this new technology.

1 Introduction

Text searchable technologies have revolutionized the ways text is accessed, retrieved, and marked. One familiar example is given in Fig. 1a. The user types or highlights a letter, word, or a sentence in a document on their desktop, or the World Wide Web for that matter, and the instances of the word are immediately retrieved and highlighted on the screen. The reliance on a handful of fixed characters (letters, numbers, symbols) that, in combination, can make up all possible texts makes this proposition real and feasible. But what about drawing searchable technologies? One (un)familiar example is given in Fig. 1b. Here the user draws or highlights a particular spatial relation between a column and a window in a plan—presumably to test the visibility occlusion in a room—and two instances are highlighted on the plan. The scenario seems hypothetical (it is not!) and possibilities are immediately apparent. What would happen if such vector shape-based search operations were feasible? A designer could redline a vector drawing or a CAD model highlighting an aspect of the model that needs to be reworked or edited somehow and similar instances of

---

A. Economou (✉)
Georgia Institute of Technology, Atlanta, Georgia
e-mail: economou@gatech.edu

this redlined part would be highlighted on the fly. She could jot down the necessary operations to be made specifying the specific changes she wishes to see, and she could even do that by drawing and not by writing or typing. A myriad of problems and opportunities arise in any inspection of a drawing. And yet, the answer to this second speculation on drawing searchable technologies is… not yet. Still, the ubiquity of drawings, diagrams, illustrations, blueprints, 2D and 3D models and everything made up of vector lines clearly suggest the possibilities that such vector shape-based searchable technologies could support. Any vector files routinely used in mechanical engineering, electrical engineering, chemical engineering, biomedical engineering, aerospace engineering, Computer Aided Manufacturing (CAM), and more broadly, in science and art, could be searched and rewritten.

Why is it, then, that the (seemingly visual but not really) symbol-based Find and Replace operations, so essential in Word or Excel, have yet to be implemented in CAD or in vector graphics? What would happen if a truly visual implementation of *evanescent* spatial characters, akin to the fixed characters of Word but otherwise instantaneously defined and fused after their application, could be deployed to construct geometrical models that could be searched and edited like text documents in any way possible? And what would happen if we could seamlessly use these vector-based, shape-based operations with CAD operators and control flow to literally write programming code by drawing shapes? How would this affect our current view of computation and what would it mean for design?

The seed for these questions (and many more) was set up fifty years in the pioneering work of George Stiny and James Gips and their formulation of the shape grammar formalism—a unique formalism that advocated a completely new way to calculate with shapes (Stiny and Gips 1972; Gips 1975; Stiny 1975, 2006, 2022). Interestingly, shape grammars have always attracted the interest, and often the admiration of, scholars and researchers in CAD, and design research at large, but, except for the few researchers dedicated in the area, this interest (and admiration)

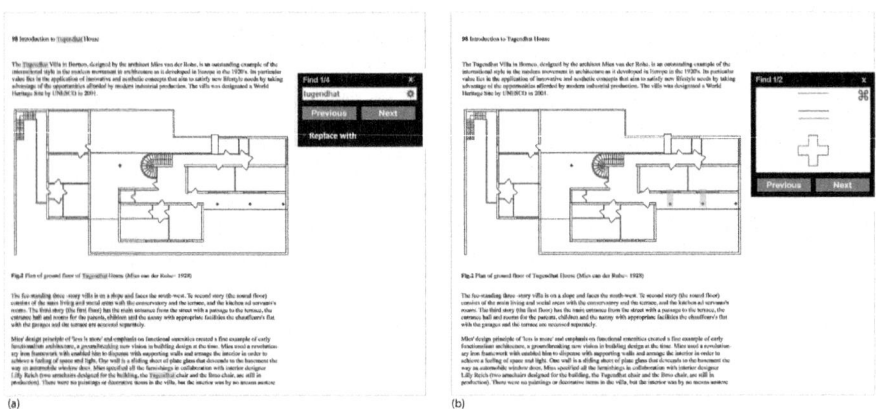

Fig. 1 A sketch of a comparison of a Find operation (⌘F) in text processing with a shape-based operation (⌘F) in vector processing

have remained idle and disengaged. Several expert testimonies on the formalism by researchers who otherwise did not pursue active research in the area can be easily found in CAD journals and design research journals over the fifty years of the formalism, at different times and with different agendas in mind. A representative assessment by McCullough, chief director at Autodesk in the early aughts, and published in Architectural Design (AD) taking on at the time the rising of coding practice in architecture and design, is as good as it can be:

"The universities were aiming much higher, of course. Also coming of age in the 1980s was a knowledge representation that has long remained an academic focus in architectural computing: shape grammars. (At Autodesk we read all the papers of the theory's inventor, George Stiny, with whom I had just studied briefly at UCLA.) Everyone knew that the discrete entity-and-layer structure of early CAD systems was not enough. Theoretically, grammars were a powerful way beyond this towards substitutions and subdivisions in articulating form. It was expected that these could become very practical expert assistants on well-formed classes of design problems. Unfortunately, the level of codification was high. Early applications operated within very specific architectural motifs, the most famous example of which was Wrightian Prairie-style houses. Even for applications in rote architectural production—say, laying out a set of hotel rooms—these better formal structures had less ready application to everyday architectural problem-solving" (McCullough 2006).

The pattern is easily discerned: researchers and engineers in software industry knew that CAD had not been set up properly; they were reading the papers of shape grammars' inventor; they recognized that shape grammars were a powerful mechanism to articulate form; they recognized that the level of codification for implementing shape grammars was high (too much work for the effort); they were unmoved by the early applications on style and/or franchise design as asymptote with the design workflows in architectural practice. The final verdict was, invariably, the kiss of (an admirable) death, and the bypassing of the formalism for the selection of some other current and useful form of design computation (coding, scripting, visual scripting, parametrics, AI, etc.)

The past future was set. Several shape grammar interpreters built upon the pioneering work of Ramesh Krishnamurti from the eighties onwards (Krishnamurti 1982; Krishnamurti and Giraud 1986; Chase 1989; Tapia 1999; Trescak et al. 2009; Jowers and Earl 2011; Grasl and Economou 2013a; Ruiz-Montiel et al. 2014; Stouffs 2022) verified the possibility of performing the miraculous calculations of the shape grammar formalism (or at least some of them)—under constrained conditions, but otherwise failed to compete with the seemingly unstoppable progress of mainstream CAD expanding to NURBS modeling, parametric modeling, building information modeling, digital fabrication, AI data models, and so forth.

It was clear that if shape grammars were to compete with this paradigm and become as useful in design discourse as McCullough had thought, they had to advance on three different fronts: (a) A software engine that would be able to showcase shape embedding from its simplest cases in shapes consisting of lines and arcs in two dimensional space to more complex curves to two- and three-dimensions, and to alternative CAD representations including b-rep, CSG, and parametric modeling; (b) interfaces that would put in practice the radical idea of shape embedding including seamless extraction of shape rules from actual workflows, exchange of rules across multiple machine and users, integration of shape rules with CAD operators and control flow frameworks to allow sequences, selections and loops in user-guided or stochastic automated workflows; and (c) applications that would show that these "better formal structures" are able to take on the full gamut of possibilities including, of course, rote production in architecture, design, engineering, science and art as well as integration with current state-of-the-art AI applications.

These three fronts provide the scaffold for the research program that underlies the design and engineering of the Shape Machine, a new shape-rewrite computational system developed at the Shape Computation Lab at Georgia Institute of Technology—and the subject matter of the work here (Shape Machine 2020). Shape Machine is new shape-rewrite computational system implemented in Rhino, a NURBS 2D/3D CAD software, featuring shape-based search and replace (⌘F/⌘–R) operations and a logical processing framework to allow programming code by drawing shapes (Economou et al. 2021, Economou and Hong 2023). A detailed account on the engineering of the Shape Machine can be found in Hong (2021); revised and sharpened versions of the indeterminate embedding for lines in planes and the point signature in shape embedding for lines in Hong and Economou (2022a, b, 2023); two paradigmatic cases of testing the shape embedding in design research in Ligler (2021) and Park (2023); and two case studies in new domains in Yu et al. (2021) and Okhoya et al. (2022), respectively.

The technical details underlying the implementation of shape embedding in Shape Machine and its integration with mainstream CAD, parametrics, BIM, control flow, and AI are staggering. The work here gives a sketch of the challenges underlying the development of the engine and its interface in the form of brief answers to three questions to provide some background for the motivation underlying the work, and a framework to contextualize the applications and the remarkable insights they have provided on the design of the engine and the interface themselves.

1.1 Three Questions

Question 1: Why is that the Find and Replace (⌘F/⌘R) operations, so essential in word processing, have yet to be implemented in CAD?

Vector shape-based Find and Replace (⌘F/⌘R) operations have not been implemented in CAD, and cannot be implemented in CAD as is, because CAD shapes have no unique representation. This certainly sounds wrong. How could that be? And yet, every practicing designer or engineer who uses CAD shapes for her work is very familiar with this situation. A line or an arc that may look like a well-modeled line or an arc might consist of several segments that overlap one another in no meaningful order or reason. The clean-up of the geometry of a 2D CAD drawing or a 3D model and its reduction to a set of well-crafted geometrical elements is part of the daily routine in an architectural office. No one is sure whom to blame. Is it the modeling skills of a novice? Is it the translational hick-ups between software schemes? Is it the differences in the version controls of a particular CAD software? It is reassuring (or not—depending on how one feels about it) to know that the problem is not related to any of these problems above but to something more profound and pertinent to the symbolic function of shapes themselves (Stiny 1975, 2006, 2022). Lines and arcs (and many other types of shapes) fuse with one another of their same kind to produce new shapes that were not instantiated by the designer of the model. For example, in CAD, if two squares combine one next to each other sharing one edge, the shared edge will consist of two lines, not one. A unique representation of shape should accommodate the ever-present possibility that shapes are constantly embedded upon one another. The powerful idea that underlines the need for radical new representation of shape is that a shape should not consist of the maximum number of smallest elements (points, lines, planes, or solids) but instead of the minimum number of maximal elements (points, lines, planes, or solids) to each other (Stiny 1980, 1991).

Once the maximal representation of shape is settled, the implementation of a vector-based, shape-based (⌘F/⌘R) Find and Replace operation is within reach. A successful shape grammar interpreter applies a replacement shape rule of the form $u \rightarrow v$, for u and v shapes, in an embed-fuse cycle that encompasses determining a transformation t that embeds u into the design w and calculating $w - t(u) + t(v)$ (Stiny 1980; Krishnamurti 1982). This process ends with the reduction of the modified design to its maximal representation, the smallest number of maximal elements to effectively describe the entire design.

An example of a recursive shape calculation in Shape Machine is given in Fig. 2. The shape generated, and subsequently transformed, pays homage to the familiar series of nested squares (or triangles, pentagons, hexagons, and so forth) that have characterized the field of shape grammars since its first appearance in Stiny (1975) and Gips (1975) (see, for example, Stiny 1980; Knight and Stiny 2001; Mitchell 2001; Jowers and Earl 2014, Stouffs and Krishnamurti 2019; Hong and Economou 2022a, b). All shapes, including the shape rules and the design, are modeled in the same working space (canvas) in Shape Machine for Rhino. The degree of perceptual change—between the initial series of the nested squares and the polygonal tessellation and the crenelated profile produced by the seamless application of shape rules—is stunning. The left column in Fig. 2 sums up the recursive redescription of the shape, while the rest of the columns showcase the individual steps of each rule application. The first two rows show the application of a shape rule that instantiates a square within a square, while changing the property (label) of the square it applies to. This shape rule applies twice under a similarity transformation to produce three squares (or four triangles and a square, four hexagons, 24 pentagons, and so forth). The third shape rule replaces all green labeled lines with black lines. The last two shape rules, one replacing an ad hoc k-shape defined at the intersections of the three squares with a bracket-shape that connects the endpoints of the k-shape, and a second one replacing the bracket-shape, above, with a pair of lines, produce two stunningly diverse results defying any expectation prior to the visual calculation. Note that the third shape rule that replaces all green labeled lines with black lines, and the fourth shape rule that replaces the ad hoc k-shapes with the bracket-shapes, are both indeterminate rules when applied under similarity transformations—and here they illustrate Shape Machine's ability to resolve specific classes of indeterminate matchings of shapes into sequences of steps of determinate ones.

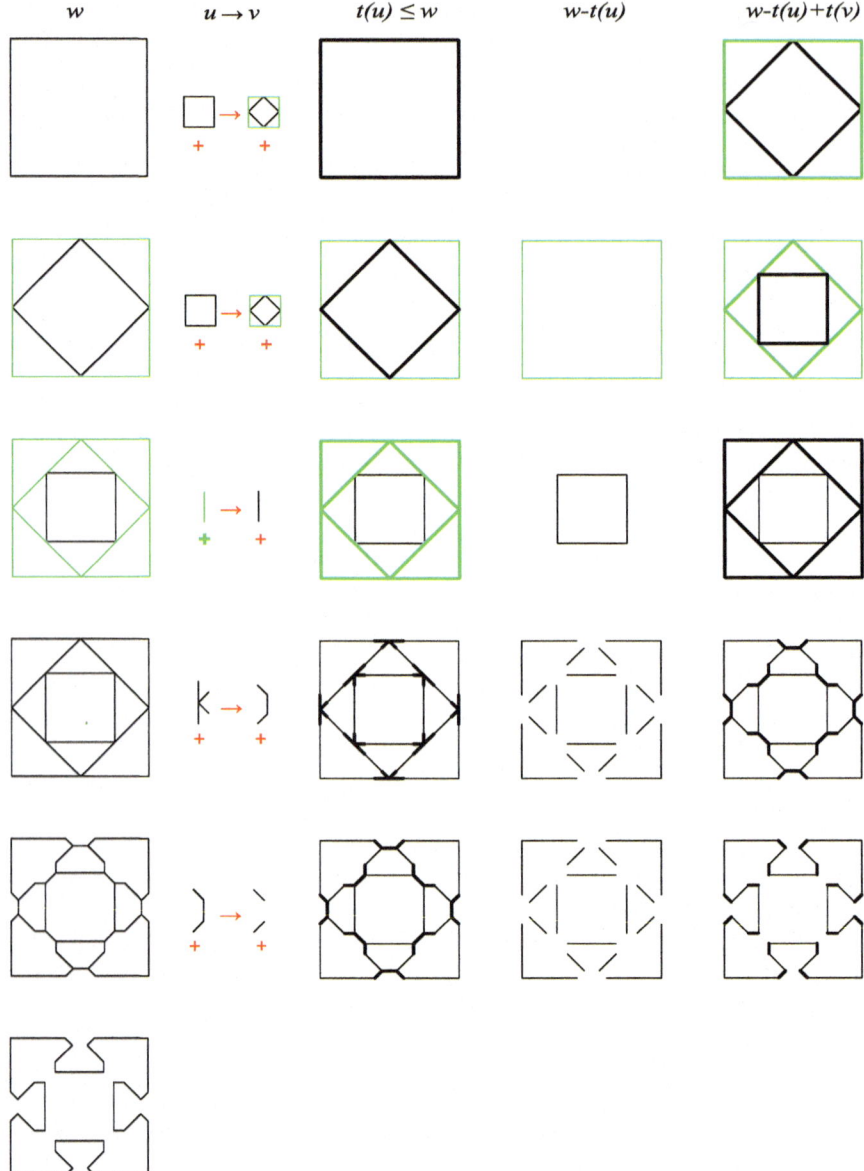

Fig. 2 A recursive calculation with labeled shapes in Shape Machine

Question 2. Can Find and Replace (⌘F/⌘R) operations alone simulate parametric design?

Parametric shape grammars introduce parametric composition and rewrites to the base formalism, enabling fully parametric design. Shape Machine and many of its predecessors, however, cannot interpret parametric grammars. Because of this, some of the most fundamental visual constructions in rule-based modeling, namely, any of Euclid's constructions with a straightedge and compass (Heath 1956; Martin 1998) appear awkward to simulate within the embed-fuse cycle of a shape grammar interpreter. There are, of course, good grounds for this apparent deficiency: the shape grammar formalism is a unique formalism constructed from scratch and entirely dispensing with the set theoretic foundations of CAD (Stiny 2022; Earl 1997; March 1996a, b, c): lines do not consist of an infinite set of points, endpoints of lines are not parts of lines, intersections of two lines do not make a point. Instead, shapes consist of four basic elements: points, lines, planes and solids and their combinations. Each basic element (except the points) consists of an unlimited number of parts that have the same dimension with the element itself; and each basic element (except the points) has boundaries that have one lesser dimension than the element itself (Stiny 2006, 2022). A generous framework indeed intended to capture the seamless redescription of shapes in creative design and reflection alike, and yet, somehow asymptote to the straight-edge and divider constructions characterizing Euclidean geometry and prevalent in the set-theoretic CAD geometry. Still, if transformational geometry and group theory managed to rework Euclid by implementing matrices and transformations (see, for example, Stewart 1995; Martin 1998) shape grammars ought to do so too.

The key idea for the solution of the riddle lies in the representation of shape in terms of basic elements (points, lines, planes, and solids) and their carriers (also known in the literature as equations, descriptors, co-descriptors, and/or hyperplanes) that the basic elements are defined in and classified by (Stiny 1975; Krishnamurti 1982; Krishnamurti and Earl 1997, more). The usage of the carriers for the registration of shape embedding has been pursued in most of the shape grammar interpreters that tackle emergence (Krishnamurti and Giraud 1986; Chase 1989; Tapia 1999; Stouffs 2022; Hong and Economou 2023). Shape Machine couples its native shape rewrite rules defined over basic elements of shapes with a new intersection operation defined upon the carriers of basic elements (point carriers, line carriers and arc carriers) to produce a highly expressive computational framework that seamlessly integrates calculations on basic elements with calculations on their carriers. The initial three intersection operations implemented in Shape Machine compute the intersections between line carriers, between arc carriers, and between line carriers and arc carriers; create point carriers at the identified intersections; and finally, instantiate basic points on the point carriers to be used for calculations with shapes that would be otherwise difficult or impossible to perform (Economou et al, 2024). By expanding Shape Machine with new operations (like the intersection operations) that are not element rewrite operations, Shape Machine can begin to encompass many of the missing aspects of parametric design.

An example of a shape calculation in Shape Machine, using shape embedding rules and intersection operations, is given in Fig. 3. Pedagogically, the calculation produces the nested square shape that was featured in the previous example. The production showcases the derivation of the nested square, using a shape grammar consisting of shape rules that operate under shape embedding and intersection operations that pick up the intersections of two arc carriers, and an arc carrier and a line carrier. The automated mechanical construction of the initial square is based on Euclid's construction of an inscribed square within a circle. Note that the two intersection operations are encoded with different colors to simulate the labeled shape calculations of the shape rules of the grammar.

Question 3. What would happen if we could seamlessly use vector-based, shape-based Find and Replace (⌘F/⌘R) operations in a logical processing framework using states, loops, jumps, and conditionals to literally write programming code by drawing shapes?

The spatial and the symbolic calculations in Shape Machine provide an expressive environment to perform elaborate and complex calculations. Every calculation requires the explicit specification of the shape rule and the vector drawing or the CAD model that the rule will apply on. The application of a shape rule in Shape Machine requires the manual selection of pairs of shapes—the shape rule and the model. Clearly, automating the execution of shape rules presents a completely new framework for the introduction of shape grammars in design. A first specification of the architecture of the environment to allow this programming with shapes is given

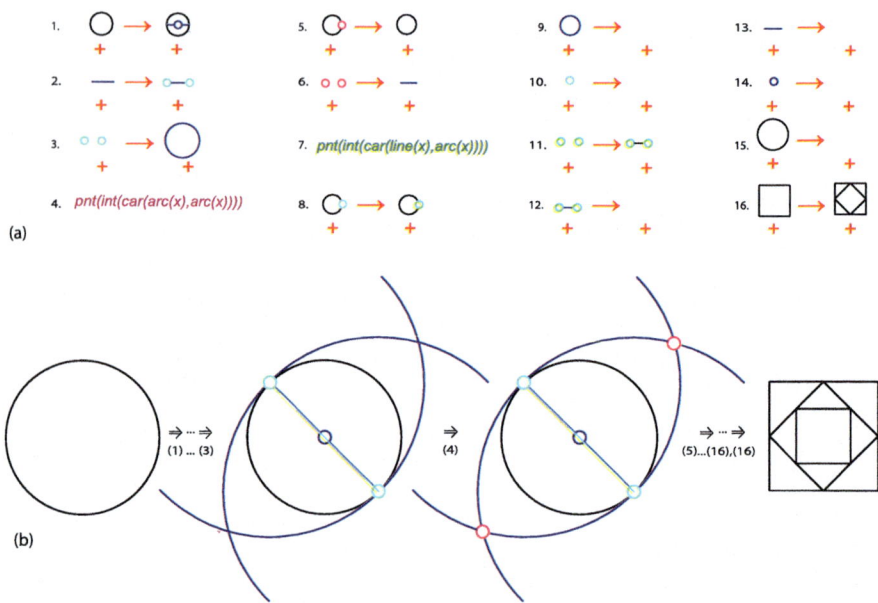

Fig. 3 An automated shape calculation in shape machine using shape embedding rules and intersection operations to produce the nested square of Fig. 2

by Hong (2021) and reworked in an object-oriented format by Newton (2025). Shape Machine tackles the automation problem with its imperative, interpreted programming language, DrawScript. In DrawScript, shape rules become the code for Shape Machine to interpret, the design becomes the program state. By sequencing rules together, users can create named routines that allow for clear organizational structure. Jump instructions, taken unconditionally or conditionally based on the presence or absence of shapes within a design, allow users to dynamically execute and reuse named routines to create powerful, complex programs, and establish DrawScript as a Turing Complete programming language for the application of shape grammars. Significantly, every shape rule in each process can be freely edited to determine the transformation it applies under, the number of loops it will execute before the next one in the sequence, and its mode of serial or parallel application for each embedding.

An example of a series of shape calculations in Shape Machine using DrawScript is given in Fig. 4. Pedagogically, the calculations produce one of the two variations upon the nested squares featured in Fig. 2. The first program showcases the usage of a for-loop. The second uses recursion to loop until a condition is met—in this case the embedding to a triangle consisting of a pair of red edges and one black edge.

The parameters of the execution of each set of rules and each individual shape rule are inserted as symbols on the interface of DrawScript including the specification of the ruleset itself through a number or a string of characters; the transformations the shape rules apply under (isometries, similarities, and affinities); the number of loops of the execution of the shape rules; and the manner of application of the shape rules (serial and parallel). Finer control allows for the specification of direct or indirect (handed) transformations that the rules apply under as well as the sequence of steps required for the translation of indeterminate shape rule applications to determinate ones.

2 The Way Ahead

The three sketches, above, provide some grounding on what has been done so far on the Shape Machine front, but more importantly, they are posed to prompt new questions that might have not been straightforward had this technology not been available. These questions span the whole gamut of the research program, outlined above, ranging from engineering issues related to the design of the engine and its algorithms, to the interface of the engine and its integration with current AI models and a variety of different workflows and contexts, and the fields of applications themselves. The range is impressive.

Several questions on the engine front are readily available: Is it possible to have robust shape embedding algorithms for conics, Bezier, and NURBS in 2D CAD? Is it possible to have robust shape embedding algorithms for wireframes in 3D CAD? How would these algorithms be expanded to include shape embeddings for B-rep representations for surfaces and solids? Is it possible to have shape embedding under projective transformations without significant cumulative rounding errors? Is

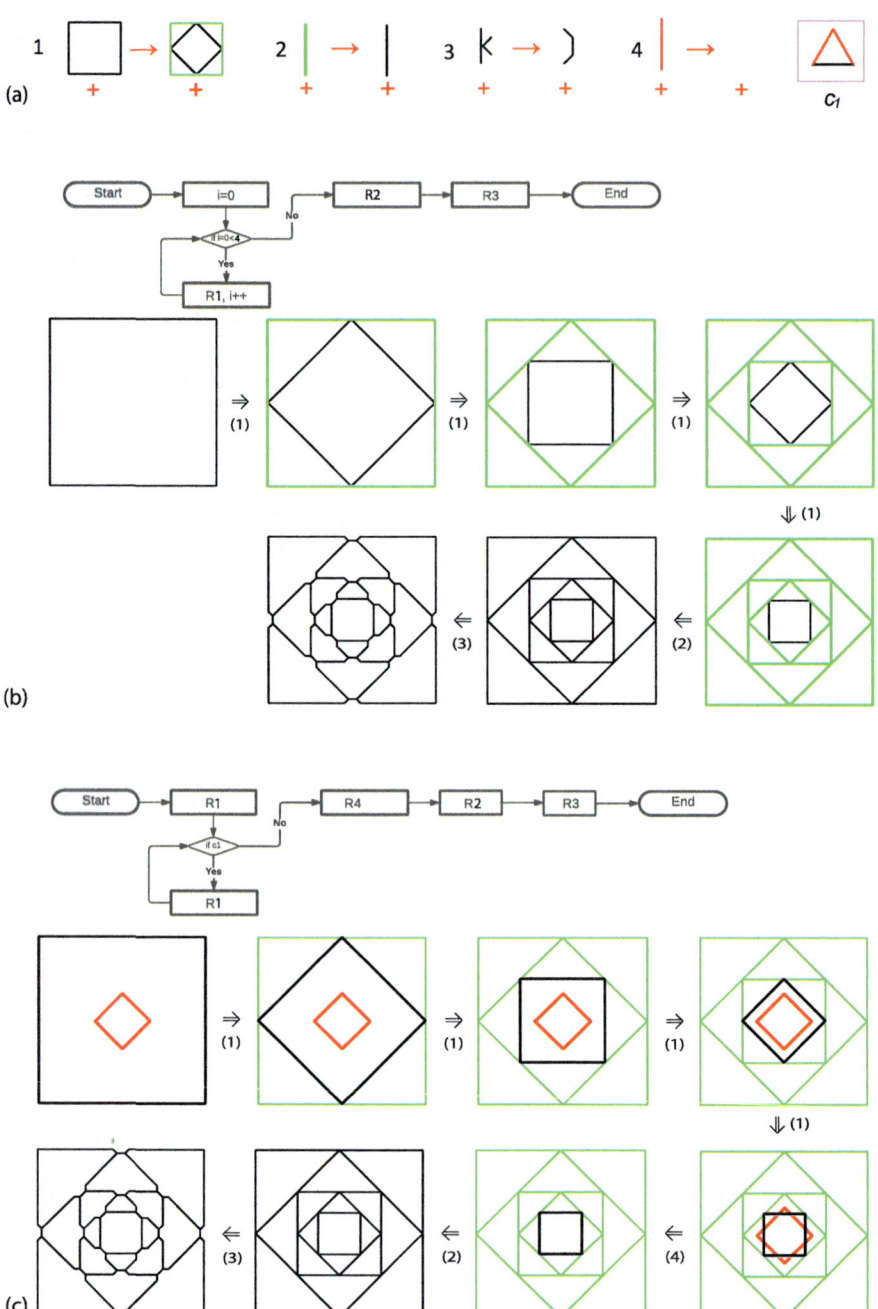

Fig. 4 Integration of control flow in shape machine **a** Four shape rules and one conditional statement **b** DrawScript featuring loops; **c** DrawScript featuring conditionals

it possible to have specification of indeterminate shape embeddings in stepped-wise rules for all shapes in 2D and 3D CAD? How will this work interface parametric modeling? How will parametric rule schemas be included in the architecture of the CAD engine? How much can the runtime be improved?

And even more questions arise on the interface front: What is the simplest interface to work with shape rules? Are there easier ways to specify shape rules? How does symbolic information pertaining to the definition of shape rules –e.g., the transformation under which the rules apply, the number of loops of rule applications, the serial or parallel execution of the rules, and the degree of randomness of an execution of a shape rule or rule set—can all be recorded without sacrificing the visual integrity of the process? Could the shape rules be captured in a temporal fashion by recording design actions within the design model? How would shape rules and sets of shape rules be shared and reused within design communities? Should all the different types of rules—shape rules, carrier schemas, and control flow—be streamlined in graphical interfaces or otherwise be presented in mixed visual and symbolic media?

However, as profound as these above questions may be, the list of questions pertaining to the application front is by far more open-ended, unsettled, and unruly! Not unlike McCullough's claim, quoted above, on the potential value of the formalism in applications in rote architectural production– reasoning and working with shapes in Shape Machine can indeed find uses in tasks and workflows previously unimagined, from the most mundane to the most exotic ones, including applications in computer-aided manufacturing, archaeology, digital heritage, real estate, kerfing, origami, electric circuitry, graphic music notation, knot theory, cryptography, modeling infectious disease spreads, game design, turn-based computer vs. player programs, and literally, any domain relying on shapes, diagrams, drawings, and models encoded in some form of vector graphics!

But first, where it all started, in architectural design research and design workflows: How can shape rules be used effectively for space planning and automated layout generation? How can shape rules be used to explore test fits? Can shape rules be formulated to simulate an expert's rule checking? Can shape rules provide a lean tool for Quantity Takeoff (QTO)? How would sequences of rules provide an automated means of filling in a level of detail (LOD) in CAD models? How can shape rules be used for shape mining to extract metadata from CAD models and drawings? How would sequences of rules capture expertise and signature in a design office? Clearly, the novelty of these questions has to do with the particularities of the subject matter—after all, for a long time the application of shape grammars had to do with issues of style and language, and this remains a field embedded in the heart of design discourse. How do we perceive a design as a Mies design? Or a SANAA design? How does the mechanical execution of shape rules Shape Machine show our understanding of the designs themselves? How does the usage of conditionals help explore design language and style? How does the usage of randomness and stochastic techniques help validate assumptions about design intentions? The applications of shape grammars are mind boggling and defy classification because like shapes themselves they can apply to many things and domains.

An initial set of applications in Shape Machine that show some of the possibilities this shape embedding technology enables—in architectural scholarly research (Ligler 2020, 2021; Ligler and Economou 2019a, b; Park 2023; Hong and Economou 2023; Economou and Hong 2023), in architectural practice (Okhoya et al. 2022) and in origami design (Yu et al. 2021)—is readily available.

2.1 Educational Paradigm Shift

Shape Machine provides a generous framework for taking on some of the questions suggested above. And yet, the road ahead is by no means royal, to paraphrase Proclus' response to the young Ptolemy Soter and his grudge on the complexity of Euclid's treatise (Heath, 1956). Shape grammars are intended to form a basis for purely visual computation (Gips 1999; Knight 2015; Stiny 2022), and, in that sense, any simulation of a shape embedding calculation in a digital computing medium is but a crude approximation (Krishnamurti 2015). And yet, the very attempt to put shape embedding next to symbolic procedures allows for a completely new approach to working with computers. This new synthesis, between open-ended seeing and problem solving, has been the underlying motivation for the design of a new curriculum at the School of Architecture and the School of Computing at the Georgia Institute of Technology, with its introduction of two joint classes ARCH 6508/CS 6492 Shape Grammars, and ARCH 8803/CS8803 Shape Machine Applications, open to students from both schools and other schools and programs too, including Industrial Design, Psychology, Engineering, Mathematics, Human Computer Interaction, and so forth. The curriculum is designed to explore the ways the shape grammar formalism and its implementation and augmentation, with additional programming rules in Shape Machine, engage a diverse body of students with different backgrounds and motivations, and has been set in motion since 2022 with the official inclusion of the CS 6492 Shape Grammars course in the graduate program of the School of Interactive Computing as a core elective.

A selected group of recent student work at Georgia Tech, exploring Shape Machine in speculative studies, is given below. Shape Machine may not be ready yet for prime time in a professional setting, but its impact in education is demonstrable. The list and the account are necessarily impressionistic to give a sense and range of the work, rather than to provide a comprehensive account of each project. All entries are given in the form of a sketch to denote the possibilities, rather than an in-depth presentation. The list includes two speculations in architectural analysis and design exploring issues of rule checking and automated floor plan generation, as well as a couple of exotic applications in other domains including modeling infectious disease spread and turn-based computer vs. player programs. The initial list intended to be shown in this work was much larger, but the length of the text would have become unwieldy; they should wait to be shown in independent scholarly publications.

2.2 Rule Checking

Designing a grammar is not different from designing any other artifact (Knight 2003; Economou and Grasl 2018). Shape rules may be imported ready-made, edited from existing ones, or created from scratch, and their value is evaluated by observing the production. Some rules that have been very useful in the past may be useless in a new setting and some rules that look opaque or insignificant when observed in isolation may be liberating in a contextual setting. In more ways than one, it is the when that matters, rather than the what or how (Goodman 1978).

A particularly interesting feature of Shape Machine is its support of identity rules (Stiny 1996) or alternatively, the visual review of the shape embeddings of the LHS of the shape rule. Any rules can be fed in these calculations. They might be found in the red lining of the print of a CAD file when an expert highlights an error, a code issue, or a change that occurred during the construction phase of a project. They might be found in the rule conflicts between different representations, e.g. an architectural, structural, mechanical, electrical, or HVAC plan of a project. They might be found in the need for rule adherence and validation that accompanies any building permit. In all cases, Shape Machine inspects visually all the instances of these shapes or parts of shapes in the drawing, or all the drawings of the project, for that matter, and the issue is foregrounded and reflected upon at once. Clearly, if the rules are immune (for example, building code) and there is a conflict, then the model must change to comply with them. Reversely, if the model is taken for granted (for example, a masterpiece or an architectural vernacular) and the inferred rules apply in undesired ways, then the rules should change to capture the intended process.

A project that nicely demonstrates (among other things) the usage of Shape Machine as a useful agent in the creative aspect of making a shape grammar, is the design of Entelechy grammar (Ligler 2020). The Entelechy I, designed by the architect John Portman as his own home in 1964, showcases a radical disposition in plan making taking on several aspects of mainstream modernism in architecture and transforming them through various formal ways to produce a unique architectural language that purportedly informed all his subsequent work—including of course, his signature atria that have since characterized the hospitality design all over the world (Portman and Barnett 1976).

An example of this creative usage of rule checking (and editing) in the design of the grammar is shown in the analysis of the distribution of the double-height volumes in the plan of the Entelechy I residence in Fig. 5. The distribution of the double-height volumes in the original plan can happen anywhere along the central circulation hall which is indicated by the smaller circular labels. In this example, Shape Machine finds six cases of an exterior major module at the perimeter connected directly to the central hall foregrounding possibilities for further testing.

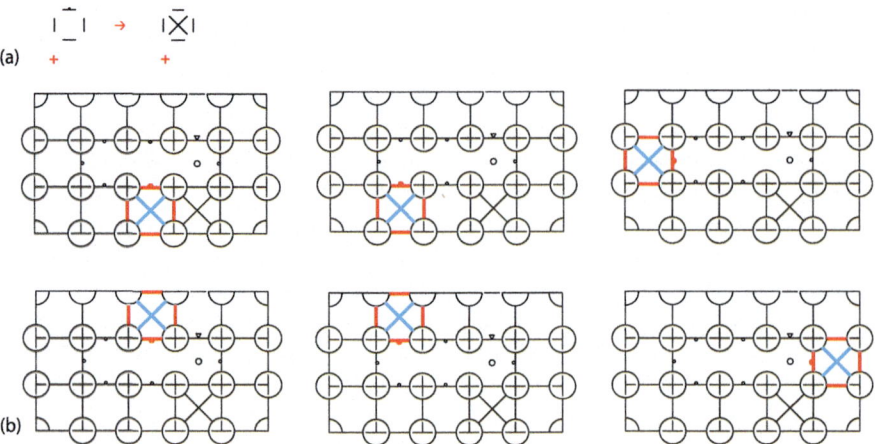

Fig. 5 Rule checking in Shape Machine. **a** Shape rule marking the double story volume; **b** Six possible location of double-height volumes in the plan (Ligler 2020, 2021)

2.3 Automated Floor Plan Generation

Automated floor plan generation has been a regular specimen in all futuristic encounters of architectural design with computing from the late 60s and 70s all the way to nowadays (Weber et al. 2022). A particularly interesting domain among these approaches is the automated completion of diagrammatic plans and sketches produced during the early phases of a design workflow (Eisenstadt et al. 2022). Is it possible to complete each of the design variations suggested by the preliminary representations during the early phases of architectural design? Can they be completed given the conventions of the practice? If not, is it because they are not productive or they are just not compliant with the final scheme? Shape Machine ought to be ideal for such tasks as it can perform routine calculations with shapes that denote architectural conventions of representation as well as bring new insights in the design development by suggesting possible shape embeddings and opportunities to carry forward the design process.

The project here takes on these questions (and many more) through the study of the archive of Mies Van Der Rohe's design process for the Dirksen United States Courthouse and its implementation in Shape Machine (Park 2023). The Dirksen federal courthouse, built in Chicago in 1964, is arguably one of the most significant buildings in the history of judicial architecture in the United States and abroad because of its transformative role in the formulation of the conventions underlying contemporary courthouse design (Blaser 1997). The archive of the design process consists of a wealth of sketches, diagrams, and drawings on alternative solutions, variational schemes, and sectional innovations that allow a very different look at the expressive range of the architectural language and the technical innovations proposed by one of the most influential architects of the twentieth century (Schulze 1992).

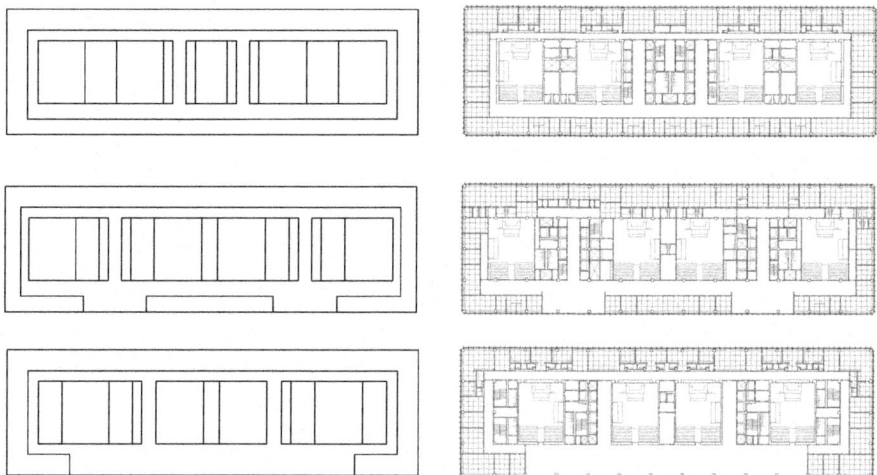

Fig. 6 From sketches to drawings. Automated completion of Mies Van Der Rohe's archival sketches in Shape Machine (Park 2023)

The research fleshes out the implicit design possibilities that the preliminary representations from the design process of the Dirksen Courthouse present through the proposition of an open-ended set of shape rules that can be readily expanded to complete the design variations documented in the archive and to generate hypothetical ones that can be in principle generated by this evolving grammar. Significantly, at any moment, new shape rules can be introduced seamlessly, as an intrinsic part of the design process of the grammar, without requiring reformatting of existing rules or advocating the design of a singular Miesian grammar. The shape rules introduced in the evolving grammar are derived from the study of the final design of the Dirksen courthouse as well as stylistic details from other works of Mies van der Rohe. Figure 6 illustrates three automated drawings produced in Shape Machine based on sketches archived at the Mies van der Rohe Archive in the Museum of Modern Art (MOMA). None of them shows the final state of the plan as built. All three productions have been made in the manual mode of software using automated shape embeddings and replacements. A significant novelty of the approach is its extensive usage of a particular kind of indeterminate rule applications under affine transformations that is useful to produce offsets of grids and walls based on a given set of skeletal axes.

2.4 Modeling Infectious Disease Spread

Modeling infectious disease spreading can be approached in a variety of ways (McLean 2013). Among them, the SIR model, for susceptible (S), infected (I), or recovered (R), continues to provide the most useful framework for the modeling of the spread (O'Leary 2009). There are three common conventional models to approach various types of the SIR models including Markov chains, continuum models, and cellular automata. A fourth method, particle modeling, takes inspiration from the cellular automata approach contextualizing the effects of infection in a real-world situation (De-Leon and Pederiva 2020).

The work here uses the particle approach for modeling infectious disease spread and reworks the basic rules in terms of spatial relations between people (Goode and Deshpande 2022). Compared to the dense math of the existing formal models, the SIR shape grammar may seem surprisingly elegant. Essentially, susceptible people have a chance to be infected if they are close enough to an infected person. Infected people have a chance to recover from the disease if they are lucky or die if they are unlucky. If an infected person and a susceptible person are separated by a wall, the disease cannot spread between them. The shape rules to determine the possibility of infection between two persons are straightforward: (1) Draw a circle around the infected person; (2) Draw a line connecting the infected and the susceptible; (3) Match the intersection of the line with the circumference of the circle; (4) Remove the circle; (5) Remove the marked lines; (6) If a line remains, the susceptible person was inside the radius.

A simple application of the SIR shape grammar simulating the spread of a disease through a floor of a hospital is given in Fig. 7. In timestep 0 one patient in the top middle room is infected. Two patients in isolation are infected. One patient in the bottom left room is infected. Everyone else on the floor, including doctors and nurses, are susceptible. After one timestep, the disease has spread to all four beds in the top middle room. One of the infected patients in isolation has died from the disease, although the disease has not spread to the two susceptible people in isolation.

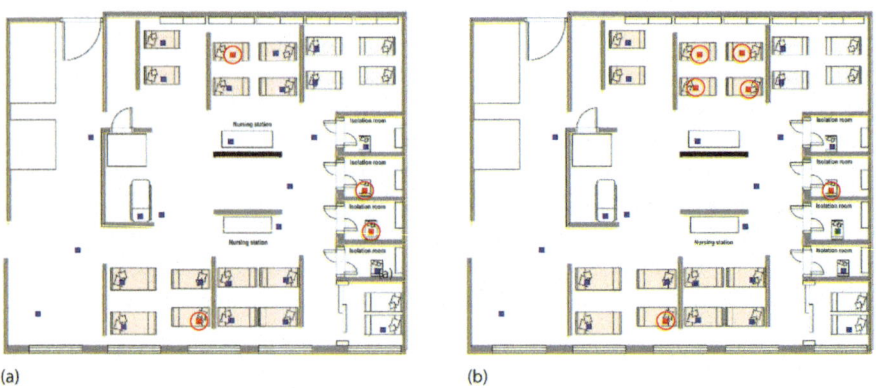

Fig. 7 An application of the SIR grammar for a given hospital plan in shape machine: **a** Timestep 0 (initial state); **b** Timestep 1 (Goode and Deshpande 2022)

2.5 Turn-Based Player Versus Shape Machine

Turn-based computer vs. player programs offer an impressive range of challenges to any player wanting to play against a computerized opponent. From their first encounter in Nimrod in 1951, the first computer game that formalized the game Nim (Bouton 1902) that featured the ability to play against a computer that made specific moves in response to differing scenarios that changed with user play, to the Checkers-playing Program by Artur Samuel—incidentally, the popularizer of the term "machine learning" in 1959—and his implementation of the Alpha–Beta Pruning algorithm (Knuth and Moore 1975)—to all recent AI applications, all utilize complex data structures, brute force and ML search methods and standard non-visual coding languages.

This project looks at the concept of playing against a machine through the visual computation tools provided via the Shape Machine (Hollosi 2020). The proposed visual algorithm chooses the best move based on a list of possible move scenarios, picking the first that it finds available. As a result, the program exhibits no foresight and no recognition of an entire board state, but instead recognizes smaller partitions of the entire board that it has pre-judged as generally more effective or less effective when compared to other moves. Note that while the program does play checkers against a user, it is not the typical version of checkers. This version is a simplified version played in a 6×6 board state and is played without the ability to king one's pieces, instead treating getting to the end of the other player's board as a win condition rather than getting a king. However, other than that, it is very similar to the standard game of Checkers. The player controls blue circle objects (O) at the bottom of the board whereas the Shape Machine controls orange "X"s located at the top.

The checkers shape rules and a complete game between a player and Shape Machine are given in Fig. 8. The first eight rules showcase the possible moves, ranked, hierarchically, based on game experimentation. Before the program makes a move based on the move list, the program finds all possible moves available to the computer and places a purple circle object as a "move marker" to show that the move is available to be made. This serves the purpose of letting the program know where it can move, but, more importantly, it ensures that the program does not make more than one move by deleting these markers after it has made its move. The second set of shape rules (9–12) implements a standard *if-else* statement so that multiple moves won't be illegally made by the computer. To prevent this, after every move, the series of the four shape rules extend a black circle or "removal marker" across the board and, subsequently, deletes all the pointers, including those that mark possible moves. Note that the rules are actual visual instructions, not diagrammatic representations of the permissible moves in the game.

It is worth reviewing the complete game between a player and Shape Machine shown in Fig. 8. In this game, after the player has made an initial move, the program employs Rule 5, pushing one piece up on the left and threatening to take the player's piece. The player responds by supporting the attacked piece with another piece, and Shape Machine responds by using Rule 4, supporting its own piece. The player moves

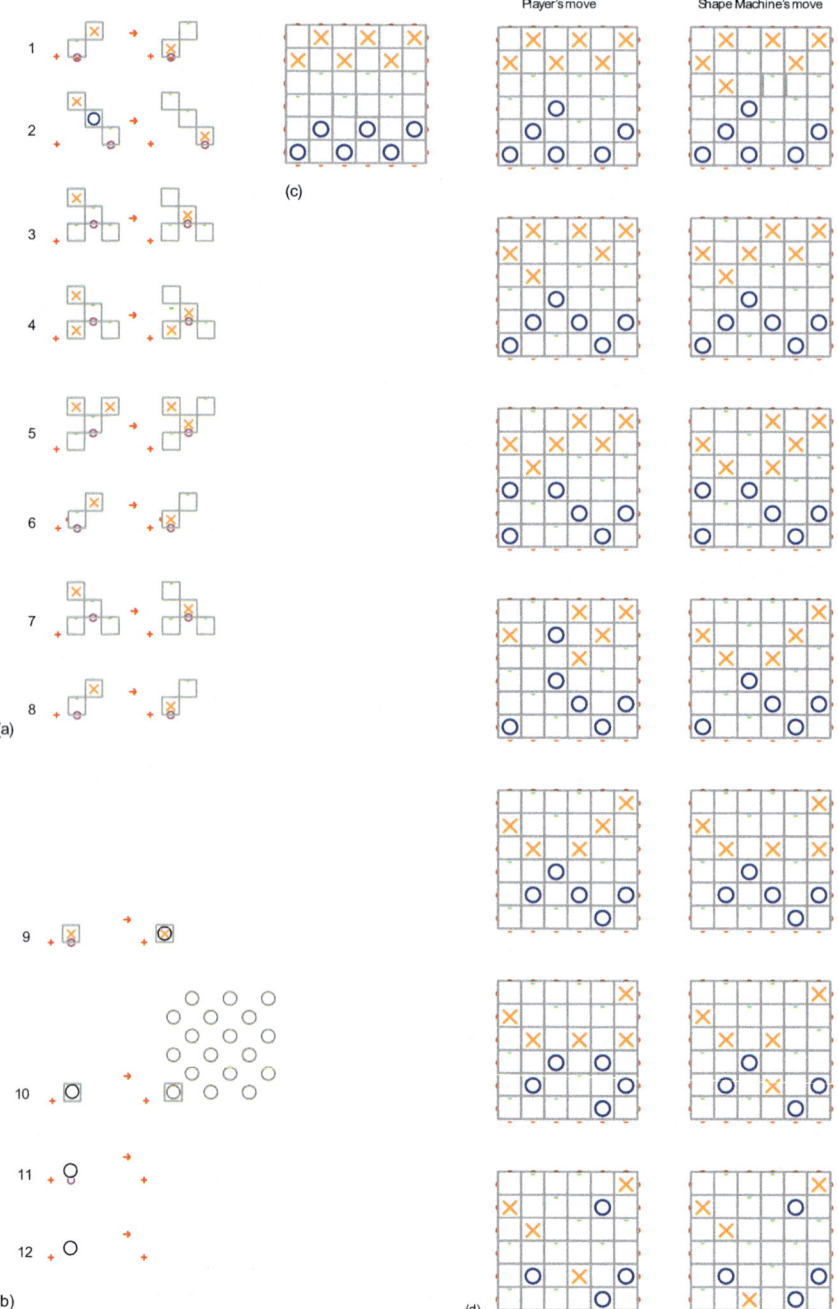

Fig. 8 Shape Machine checkers: **a** Checkers move shape rules; **b** *if-else* shape rules **c** Initial 6×6 board; **d** A complete game between a player (O) versus Shape Machine (X). Shape Machine wins (Hollosi 2020)

a piece to the left edge of the board, and Shape Machine moves its piece up once more, attacking the player's piece in the middle. The player takes Shape Machine's now undefended piece, and Shape Machine takes the piece back. The player supports the undefended piece, and Shape Machine moves a piece to the edge of the board. The player moves a piece to the center, Shape Machine takes this piece, and the player takes Shape Machine's piece too. The program gets a piece to the back edge and wins the game.

3 Discussion

It is difficult to pinpoint when the idea of the Shape Machine started, and even harder to say where it will go—the last examples in the previous section testify to this. And yet, if there is a zero-milestone that could stand for the origins of Shape Machine, then it is surely the body of weekly lectures by George Stiny and Lionel March at UCLA during the nineties. Both Stiny's lectures on Shape grammars and their algebras, rules, and applications—and March's lectures on Architectonics and its tripartite division on symmetry, proportion, and arrangement, provided the inspiration and the foundations for Shape Machine. Both series of lectures shared a relentless commitment in fusing design with math while embracing ambiguity and uncertainty, and yet both appeared asymptote to one another. It was not that Stiny and March had not tried to put together a shared corpus. In fact, the two papers they co-authored, one on Design Machines (Stiny and March 1981) and another on the history and logic of spatial systems in design (March and Stiny 1985) are still key foundational readings in the discourse and they were promptly followed by several more by March, who continued this very attempt (March 1996a, b, c, 1998). In the end, March decided to support shape grammars by deploying his uncanny facility with numbers and calculation in formal analyses of renaissance architecture (March 1997, 1999); it made sense, given the emergent need back then for parametric work in design.

And yet, I always believed there was a better way to combine the shape grammar formalism with March's work in graphs, sets, lattices, combinatorics, and the like; the scientific method and technical rigor in design research these formalisms aspire to did not necessarily have to lead to sterile, fixed, skeletal designs, like the ones produced during the Seventies and Eighties—for an excellent and synoptic criticism, see Stiny's dissections dissected (1979). March's later work stood aside from these; see, for example, March (1981).

It is suggested here that Shape Machine's current implementation of shapes as maximal points, lines and arcs defined over carrier points, lines, and arcs, respectively, appears to rework the two worldviews, above, in an uncanny blend. The implementation of the maximal points, lines, and arcs—the hallmark of the shape grammar formalism and a prerequisite for any calculation with shape rules—is clearly the key feature of the Shape Machine. The implementation of the carrier points, lines, and arcs is as important for the registration of shapes, the queries of shape embedding, and the constructions they enable with the maximal shapes. It would not be farfetched to claim that this dual representation of shapes in terms of basic elements and their carriers, revisits, yet again, the conventional dualism of drawings and grids, of models and guides, and of designs and regulating lines.

The implementation of existing shape grammar work drawn with pencil and eraser provided the initial insights. First, the simple pedagogical studies. Stiny has always enjoyed this part of the formalism and continues to cook impossible computations that pedagogically show the innate ability of shape to engage in embed and fuse computational cycles, from his early work with the rotation of triangles (Stiny 1989) to the translation of chevrons in his stars (Stiny 1994) and, most recently, to his translations of the two L-shapes (Stiny 2022). Many other examples of such impossible calculations can be found in the literature (see for example, Knight 2003; Kotsopoulos 2021). On the other extreme are the complex design studies. Several paper-based shape grammars have been proposed to capture complexity and hierarchy in design, using the resources of a pencil and an eraser, for example, the Palladian grammar (Stiny and Mitchell 1978), the Moghul Garden grammar (Stiny and Mitchell 1980), the Hepplewhite back-chair grammar (Knight 1980), the Japanese tea-houses (Knight 1981), the meander motif in Greek pottery (Knight 1986), and many more. And yet, this intuitive, tactile authoring of grammars (theories) has its limitation: the more complex and richer they become, the more unwieldy they are for others to evaluate and use. Are the authors correct? Can the rules execute what they purport? Would a manual check provide a definitive answer about whether the author or the evaluator is right (Economou and Grasl 2018)?

Clearly, only a mechanical, automated execution of the rules could provide the answer. The design of the underlying engine of Shape Machine is based on Krishnamurti's foundational work at CMU (1982, 1992; Krishnamurti and Giraud 1986). Gips' work in the Seventies won't count, as it did not accommodate shape embedding—an operation that several shape interpreters do not do anyway (Hong and Economou 2022b). A preliminary version of the software, proposed as a 2D/3D graph-based shape grammar interpreter and code-named GRAPE < GR-aph sh-APE (Grasl and Economou 2013a), successfully modeled some of the elusive shape calculations within the shape grammar discourse, while taking on a more ambitious constructive program in formal composition (Economou and Grasl 2018). The resulting software included a web application in HTML5, an AutoCAD plugin and a REVIT plugin, all showing the possibility that Stiny's calculations were indeed feasible, but with severe limitations because of the imported underlying graph engine: it was hard to integrate, cumbersome to troubleshoot, and impossible to develop. In retrospect, though, it is easy to explain the graph engine. The very same year that the work started in the Shape Computation Lab at Georgia Tech, it was asserted by the community that emergence was a very difficult task to pursue, and that spatial grammars interpreters—a particular version of set grammars (Stiny 1982)—had more possibility for limited but successful applications (McKay et al. 2012). In all cases, GRAPE provided the confidence that a shape embedding engine could be reworked from scratch, and this instilled enough confidence to take on the design of Shape Machine.

Questions pertaining to the design interface of the Shape Machine are open-ended. The preliminary work has been built upon a seamless graphical space to jot down the shape rules and apply to a 2D or 3D model within the same space. A second interface in Shape Machine allows the derivation of the shape rules as recordings of design actions. In this mode, the visual inspection of a shape rule can only be done in the context of a 2D, or 3D model; the seemingly gestalt fragments of shapes that are typically presented in a shape rule are unwieldy when they are shown on their own but fully meaningful when they are embedded in the geometrical model. A third interface in Shape Machine allows for the specification of the various parameters that specify the application of a shape rule or a set of shape rules: transformation that the shape rule applies under; serial or parallel execution of the shape rule; number of consecutive loops of the rule; additional information regarding the application of an indeterminate rule; and so forth. Different domains for shape queries (and possible shape replacements) suggest radically different environments for such operations.

4 Conclusions

Shape Machine provides a robust framework for performing vector-based shape-based search and replace (⌘ + F/⌘ + R) operations in 2D CAD, seamlessly integrated with control flow, to allow programming by drawing shapes consisting of lines, arcs, and their combinations. The challenges and opportunities that the technology suggests are staggering. The research program enabled by the technology is currently developed around three domains, namely, the engine and its algorithms, the interface of rule specification, and the fields of applications themselves. Work on the first domain includes the extension of shape embedding to additional types of shapes, including conics, Bezier curves, B-splines, NURBS, 3D geometries, B-reps, parametric representations, and so forth; additional types of transformations for the embedding of these shapes; and a significant emphasis on effective runtime calculation. Work on the second domain includes the development of interfaces to specify shape rewrite operations in 2D and 3D CAD models; integration of shape rewrite rules with symbolic procedural rules and control flow; design of interactive drawings and models to keep track of design actions as rules applied on them; and integration with ML/AI modules for usage of shape rules, among others. Work in the third domain is perhaps the most critical and the most urgent because it will demonstrate the value of the formalism and will help propel even more work on the engine and the interface. It is this last goal that provided the motive for the work presented here, and it is certainly hoped that the variety of the applications presented point to this future.

Acknowledgements I would like to thank George Stiny, Chris Earl, Ramesh Krishnamurti, and Rob Woodbury for their continuous support during the development of the Shape Machine: private discussions, manuscript reviews, and especially constructive feedback to the students at Georgia Tech have all profoundly shaped the research program underlying the development of Shape Machine. The work would have never happened without the commitment and enthusiasm of undergraduate, graduate and doctorate students, and especially, Kurt Hong, Heather Ligler, and James Park, now colleagues at various universities in the US.

References

Blaser, W. 1997. *Mies van der Rohe*. Basel: Birkhäuser. ISBN-13:978–3764356194.
Bouton, C.L. 1902. Nim, a game with a complete mathematical theory. *Annals of Mathematics, Second Series* 3 (14): 35–39. https://doi.org/10.2307/1967631.
Chase, S.C. 1989. Shapes and shape grammars: From mathematical model to computer implementation. *Environment and Planning B: Planning and Design* 16 (2): 215–242. https://doi.org/10.1068/b160215.
De-Leon, H., and F. Pederiva. 2020. Particle modeling of the spreading of coronavirus disease (COVID-19). *Physics of Fluids* 32 (8): 087113. https://doi.org/10.1063/5.0020565.
Earl, C.F. 1997. Shape boundaries. *Environment and Planning B: Planning and Design* 24 (5): 669–687. https://doi.org/10.1068/b240669.
Economou A., and T. Grasl. 2018. Paperless grammars. In *Computational studies on cultural variation and heredity. KAIST research series*, ed. J.H. Lee, 139–160. Singapore: Springer. https://doi.org/10.1007/978-981-10-8189-7_12
Economou, A., and T.K. Hong. 2023. Back to the drawing board: Shape calculations in shape machine. In *Design computing and cognition'22*, ed. J.S. Gero, 549–567. Springer, Cham. https://doi.org/10.1007/978-3-031-20418-0_33
Economou, A., T.K. Hong and J. Park. (forthcoming). *Shape signature* (unpublished manuscript).
Economou, A, T.K. Hong, H. Ligler, and J. Park. 2021. Shape machine: A primer for visual computation. In *A new perspective of cultural DNA. KAIST research series*, ed. J.H. Lee. Singapore:Springer. https://doi.org/10.1007/978-981-15-7707-9_6
Economou, A, T. K. Hong and R. Newton. 2024. Shape meets Euclid: Integrating shape computation with ruler and compass procedures. *Automation in Construction* 165, 105562. https://doi.org/10.1016/j.autcon.2024.105562
Eisenstadt, V., J. Bielski, B. Mete, C. Langenhan, K.D. Althoff, and A. Dengel. 2022. Autocompletion of floor plans for the early design phases in architecture: foundations, existing methods, and research outlook. In *Proceedings of the 27th international conference of the association for Computer-Aided Architectural Design Research in Asia (CAADRIA) 2022*, vol. 1, 323–332. Hong Kong. https://doi.org/10.52842/conf.caadria.2022.1.323
Gips, J. 1975. *Shape grammars and their uses: Artificial perception, shape generation and computer aesthetics (Interdisciplinary systems research series 10)*. Basel and Stuttgart: Birkhäuser Verlag. ISBN-13:978-3764307943.
Gips, J. 1999. *Computer implementation of shape grammars, workshop on shape computation*. MIT. https://www.academia.edu/3089939/Computer_Implementation_of_Shape_Grammars
Goode, E. and R. Deshpande. 2022. *Modeling infectious disease spread with shape grammars*. Atlanta: School of Interactive Computing, Georgia Institute of Technology.
Goodman, N. 1978. *Ways of worldmaking*. Indianapolis: Hackett Publishing Company, Inc. ISBN-13:978-0915144518.
Grasl, T., and A. Economou. 2013a. From topologies to shapes: Parametric shape grammars implemented by graphs. *Environment and Planning B: Planning and Design* 40 (5): 905–922. https://doi.org/10.1068/b38156.
Heath, T.L. 1956. *Euclid: The thirteen books of the elements*. New York: Dover Publications. ISBN 13:9780486600888

Hollosi, G. 2020. *Checkers in shape machine and the evolution of turn-based player vs. computer algorithms*. Atlanta: School of Interactive Computing, Georgia Institute of Technology.

Hong, T.K and A. Economou. 2022b. Five criteria for shape grammar interpreters. In *Design computing and cognition'20*, 189–208, ed. J.S. Gero. Springer. https://doi.org/10.1007/978-3-030-90625-2_11

Hong, T.K, and A. Economou. 2023. Implementation of shape embedding in 2D CAD systems. *Automation in Construction*, 146: 1–15, 104640. ISSN 0926–5805. https://doi.org/10.1016/j.autcon.2022.104640

Hong, T.C.K., and A. Economou. 2022a. What shape grammars do that CAD should: The 14 cases of shape embedding. *Artificial Intelligence for Engineering Design, Analysis and Manufacturing* 36 (E4): 1–20. https://doi.org/10.1017/S0890060421000263.

Hong, T.K. 2021. *Shape machine: Shape embedding and rewriting in visual design*. Ph.D. dissertation. Atlanta: Georgia Institute of Technology. http://hdl.handle.net/1853/67155

Hong, T.K. 2022. *Personal website*. Accessed 29 Dec 2023. https://k9krnd.net/.

Jowers, I., and C.F. Earl. 2011. Implementation of curved shape grammars. *Environment and Planning B: Planning and Design* 38 (4):616–635. https://doi.org/10.1068/b36162.

Jowers, I., and C.F. Earl. 2014. Shape interpretation with design computing. In *Design computing and cognition' 12*, ed. J.S. Gero, 343–360. Dordrecht: Springer. https://doi.org/10.1007/978-94-017-9112-0_19.

Knight, T. 1980. The generation of Hepplewhite-style chair-back designs. *Environment and Planning B* 7: 227–238. https://doi.org/10.1068/b070227.

Knight, T. 1981. The forty-one steps. *Environment and Planning B: Planning and Design* 8 (1): 97–114. https://doi.org/10.1068/b080097.

Knight, T. 1986. Transformations of the meander motif on Greek geometric pottery. *Design Computing* 1: 29–67.

Knight, T. 2003. Computing with emergence. *Environment and Planning B: Planning and Design* 30 (1): 125–155. https://doi.org/10.1068/b12914.

Knight, T. 2015. Shapes and other things. *Nexus Network Journal* 17: 963–980. https://doi.org/10.1007/s00004-015-0267-3.

Knight, T., and G. Stiny. 2001. Classical and non-classical computation. *Architectural Research Quarterly*. 5 (4): 355–372. https://doi.org/10.1017/S1359135502001410.

Knuth, D.E., and R.W. Moore. 1975. An analysis of alpha-beta pruning. *Artificial Intelligence* 6(4): 293–326. https://doi.org/10.1016/0004-3702(75)90019-3.

Kotsopoulos, S.D. 2021. Design without rigid rules. In *Design computing and cognition'20*, ed. J. S. Gero. Springer Nature Switzerland AG. https://doi.org/10.1007/978-3-030-90625-2_7.

Krishnamurti, R. 1992. The maximal representation of a shape. *Environment and Planning B: Planning and Design* 19 (3): 267–288. https://doi.org/10.1068/b190267.

Krishnamurti, R. 2015. Mulling over shapes, rules and numbers. *Nexus Network Journal* 17: 927–945. https://doi.org/10.1007/s00004-015-0269-1.

Krishnamurti, R., and C. Giraud. 1986. Towards a shape editor: The implementation of a shape generation system. *Environment and Planning B: Planning and Design* 13 (4): 391–404. https://doi.org/10.1068/b130391.

Krishnamurti, R. 1982. SGI: A shape grammar interpreter. In *Centre for configurational studies, design discipline*. The Open University, 1982. Accessed 29 Dec 2023. https://www.andrew.cmu.edu/user/ramesh/pub/distribution/technical/SGI.pdf.

Ligler, H. 2021. Reconfiguring atrium hotels: Generating hybrid designs with visual computations in shape machine. *Automation in Construction* 132: 103923. https://doi.org/10.1016/j.autcon.2021.103923.

Ligler, H., and A. Economou. 2019a. From drawing shapes to scripting shapes: Architectural theory mediated by shape machine. In *Conference proceedings of the symposium on simulation for architecture and urban design*. https://doi.org/10.5555/3390098.3390137.

Ligler, H., and A. Economou. 2019b. On John Portman's atria: Two exercises in hotel composition. In *Design computing cognition '18*, ed. J. S. Gero, 401–420. Dordrecht: Springer. https://doi.org/10.1007/978-3-030-05363-5_22.

Ligler, H. 2020. *The portman variations: A critical approach to entelechy i mediated by shape machine*. Ph.D. dissertation. Atlanta: Georgia Institute of Technology. http://hdl.handle.net/1853/64116.

March, L. 1981. A class of grids. *Environment and Planning B: Planning and Design* 8 (3): 325–332. https://doi.org/10.1068/b080325.

March, L. 1996a. The smallest interesting world? *Environment and Planning B: Planning and Design* 23: 133–142. https://doi.org/10.1068/b230133.

March, L. 1996b. Babbage's miraculous computation revisited. *Environment and Planning B: Planning and Design* 23: 369–376. https://doi.org/10.1068/b230369.

March, L. 1996c. Rulebound unruliness. *Environment and Planning B: Planning and Design* 23 (4): 391–399. https://doi.org/10.1068/b230391.

March, L. 1999. Architectonics of proportion: A shape grammatical depiction of classical theory. *Environment and Planning B: Planning and Design* 26 (1): 91–100. https://doi.org/10.1068/b260091.

March, L., and G. Stiny. 1985. Spatial systems in architecture and design: Some history and logic. *Environment and Planning B: Planning and Design* 12:31–53. https://doi.org/10.1068/b120031.

March, L. 1997. *Architectonics of humanism: Essays on number in architecture*. Wiley, SBN-13:978–0471977544

March, L. 1998. [8 + (6) + 11] = 25+x. *Environment and Planning B: Planning and Design Anniversary Issue*, 10–19. https://doi.org/10.1177/239980839802500702

Martin, G.E. 1998. *Geometric constructions*. New York: Springer. ISBN-13:9781461206293

McCullough, M. 2006. 20 years of scripted space. *Programming Cultures, Architectural Design* 76 (4): 12–15. https://doi.org/10.1002/ad.288.

McKay, A., S. Chase, K. Shea, and H.H. Chau. 2012. Spatial grammar implementation: From theory to useable software. *Artificial Intelligence for Engineering Design, Analysis and Manufacturing* 26: 143–159. https://doi.org/10.1017/S0890060412000042.

McLean, A.R. 2013. Infectious disease modeling. In *Infectious diseases* eds. P. Kanki and D. Grimes. New York: Springer. https://doi.org/10.1007/978-1-4614-5719-0_5.

Mitchell, W.J. 2001. Vitruvius redux. In *Formal engineering design synthesis*, 1–19. ed. E.K. Antonsson and J. Cagan. Cambridge: Cambridge University Press. ISBN 13:9780521792479.

Newton, R. 2025. *Manual of shape machine*. Atlanta: School of Interactive Computing, Georgia Institute of Technology.

Okhoya, V.W., M. Bernal, A. Economou, N. Saha, R. Vaivodiss, T.C.K. Hong, and J. Haymaker. 2022. Generative workplace and space planning in architectural practice. *International Journal of Architectural Computing* 20(3): 645–672. https://doi.org/10.1177/14780771221120580

O'Leary, D.P. 2009. Models of infection: Person to person. *Scientific Computing with Case Studies*, 213–219. https://doi.org/10.1137/9780898717723.ch19.

Park, J. 2023. *Sketches count: The Mies van der Rohe's dirksen courthouse archive redrawn*. Ph.D. dissertation. Atlanta: Georgia Institute of Technology. https://hdl.handle.net/1853/73159.

Portman, J., and J. Barnett. 1976. *The architect as developer*. New York: McGraw-Hill. ISBN-13,978–0070505360.

Ruiz-Montiel, M., M.V. Belmonte, J. Boned, L. Mandow, E. Millan, A.R. Badillo, and J.L. Perez. 2014. Layered shape grammars. *Computer-Aided Design* 56: 114–119. https://doi.org/10.1016/j.cad.2014.06.012.

Schulze, F., (ed). 1992. *The Mies van der Rohe Archive: An illustrated catalogue of the mies van der rohe drawings in the museum of modern art. Part II: 1938–1967, The American Work*. New York: Garland Publishing, Inc. ISBN:0–8240–5997–2.

Shape Machine OTL. 2020. Accessed 29 Dec 2023. https://licensing.research.gatech.edu/technology/shape-machine

Stewart, I. 1995. *Concepts of modern mathematics*. New York: Dover Publications. ISBN-13:9780486284248.

Stiny, G. 1975. *Pictorial and formal aspects of shape and shape-grammars: On computer generation of aesthetic objects (Interdisciplinary systems research series)*. Basel and Stuttgart: Birkhäuser. ISBN-13: 978–3764308032.

Stiny, G. 1979. Dissections dissected. *Environment and Planning B: Planning and Design* 6 (4): 469–470. https://doi.org/10.1068/b060469.

Stiny, G. 1980. Introduction to shape and shape grammars. *Environment and Planning B: Planning and Design* 7 (3): 343–351. https://doi.org/10.1068/b070343.

Stiny, G. 1982. Spatial relations and grammars. *Environment and Planning B* 9 (1): 113–114. https://doi.org/10.1068/b090113.

Stiny, G. 1989. Formal devices for design. In: *Design theory 88*, eds. S.L. Newsome, W.R. Spillers, and S. Finger, 173–188. New York: Springer. https://doi.org/10.1007/978-1-4612-3646-7_16.

Stiny, G. 1991. The algebras of design. *Research in Engineering Design* 2: 171–181. https://doi.org/10.1007/BF01578998.

Stiny, G. 1994. Shape rules: Closure, continuity and emergence. *Environment and Planning B: Planning and Design* 21: 49–78. https://doi.org/10.1068/b21S049.

Stiny, G. 1996. Useless rules. *Environment and Planning B: Planning and Design*, 23(2): 235–237. https://doi.org/10.1068/b230235

Stiny, G. 2006. *Shape: Talking about seeing and doing*. Cambridge: The MIT Press.

Stiny, G. 2022. *Shapes of imagination: Calculating in Coleridge's magical realm*. Cambridge: The MIT Press.

Stiny, G., and J. Gips. 1972. Shape grammars and the generative specification of painting and sculpture. In *Information processing*, vol. 71, 1460–1465. North-Holland Publishing Company.

Stiny, G., and L. March. 1981. Design Machines. *Environment and Planning B* 8: 245–255. https://doi.org/10.1068/b080245.

Stiny, G., and W.J. Mitchell. 1978. The Palladian grammar. *Environment and Planning B* 5: 5–18. https://doi.org/10.1068/b050005.

Stiny, G., and W.J. Mitchell. 1980. The grammar of paradise: On the generation of Moghul gardens. *Environment and Planning B* 7: 209–226. https://doi.org/10.1068/b070209.

Stouffs, R., and R. Krishnamurti. 2019. A uniform characterization of augmented shapes. *Computer-Aided Design*, 110:37–49. ISSN 0010–4485. https://doi.org/10.1016/j.cad.2018.12.004.

Stouffs, R. 2022. A multi-formalism shape grammar interpreter. In *Computer-aided architectural design. Design imperatives: The future is now. CAAD Futures 2021. Communications in computer and information science*, vol. 1465, eds. D. Gerber, E. Pantazis, B. Bogosian, A. Nahmad, and C. Miltiadis. Singapore: Springer. https://doi.org/10.1007/978-981-19-1280-1_17.

Tapia, M. 1999. A visual implementation of a shape grammar system. *Environment and Planning B: Planning and Design* 26 (1): 59–73. https://doi.org/10.1068/b260059.

Trescak, T., M. Esteva, and I. Rodriguez. 2009. General shape grammar interpreter forintelligent designs generations. In *Proceedings of the 6th international conference on computer graphics, imaging and visualization*, 235–240. https://doi.org/10.1109/CGIV.2009.74.

Weber, R.E., C. Mueller, and C. Reinhart. 2022. Automated floorplan generation in architectural design: A review of methods and applications. *Automation in Construction* 140: 104385, ISSN 0926-5805. https://doi.org/10.1016/j.autcon.2022.104385.

Woodbury, R. 2010. *Elements of parametric modeling*. Routledge. ISBN-13: 978-0415779876.

Yu, Y., K. Hong, A. Economou, and G.H. Paulino. 2021. Rethinking origami: A generative specification of origami patterns with shape grammars. *Computer-Aided Design*, 137: 103029, 1–14. ISSN 0010-4485. https://doi.org/10.1016/j.cad.2021.103029

Open Access This chapter is licensed under the terms of the Creative Commons Attribution-NonCommercial-NoDerivatives 4.0 International License (http://creativecommons.org/licenses/by-nc-nd/4.0/), which permits any noncommercial use, sharing, distribution and reproduction in any medium or format, as long as you give appropriate credit to the original author(s) and the source, provide a link to the Creative Commons license and indicate if you modified the licensed material. You do not have permission under this license to share adapted material derived from this chapter or parts of it.

The images or other third party material in this chapter are included in the chapter's Creative Commons license, unless indicated otherwise in a credit line to the material. If material is not included in the chapter's Creative Commons license and your intended use is not permitted by statutory regulation or exceeds the permitted use, you will need to obtain permission directly from the copyright holder.

Design Space Exploration in the Shape Machine Interface

Robert Woodbury

Abstract Shape grammars enable a tight loop comprising seeing and doing in design. Designers rapidly make many design alternatives as they work: they explore design spaces. The Shape Machine interface addresses the former, but not the latter. This paper presents lessons from an existing design space explorer and proposes extensions to the Shape Machine interface to better enable design space exploration.

1 Introduction

Designers cycle through seeing and doing—Stiny's (2006) maxim generalises far beyond shape grammars. Over time, designers make and examine many options as they work—they explore a design space. Together, seeing and doing plus exploration arguably comprise the core of Goldschmidt's (1991) *dialectics of sketching* and Schön's (1983) *conversation with the situation*. Currently, no human-computer interface for designers properly supports these two central acts.

But... two interface designs stand out.

The Shape Machine (Economou et al. 2021) makes visual almost all aspects of composing and invoking shape rules. Figure 1 shows a simple grammar in the Shape Machine interface. Designers can make a rule by seeing its LHS and RHS in a shape and then selecting and copying these to form the rule. By its design, the Shape Machine supports seeing and doing. Though seeing and doing has its origins in the shape grammar literature, any serious design instructor would recognise it as concisely stating perhaps the most important thing taught in early design education: draw, look and learn... again and again and again. Seeing and doing stands independent of the representation from which it originated.

D.Star (Mohiuddin and Woodbury 2022) supports collections of alternative parametric models and provides tools for making persistent collections and operating on the alternatives they contain. Figure 2 shows several design alternatives for a mixed-use development in both spatial views and in a *parallel coordinates* display showing

R. Woodbury (✉)
Simon Fraser University Surrey, 250-13450 102nd Avenue, Surrey, BC V3T 0A3, Canada
e-mail: robw@sfu.ca

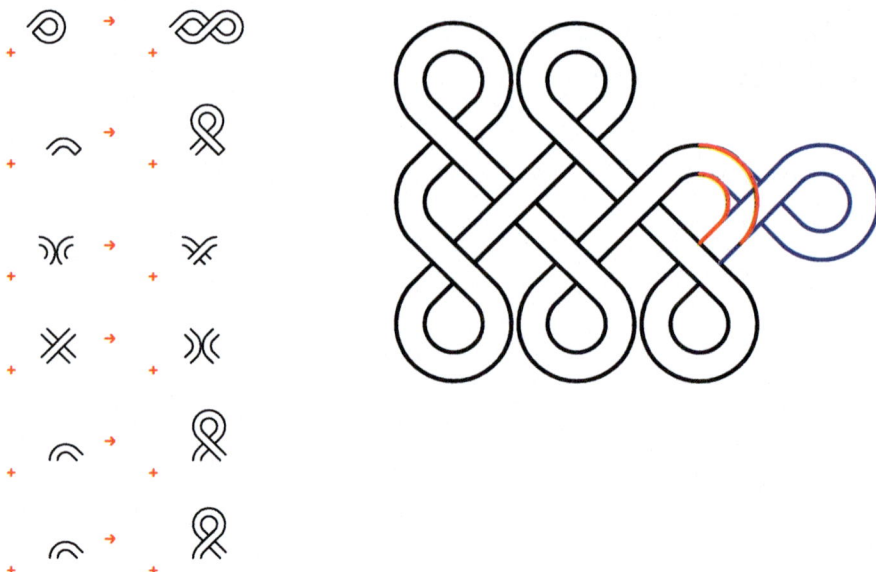

Fig. 1 The shape machine interface showing a grammar that derives Celtic knots. Both rules and shapes exist on the same infinite canvas. The Shape Machine interface imposes no location constraints: designers control the placement of shapes and rules. In the figure, red indicates a rule match (LHS) and blue a rule application (RHS). The same editing commands apply to rules and shapes. With the Shape Machine interface, much rule development occurs through copying and pasting parts of other rules and shapes

their parameters and performance. D.Star's collections support multiple episodes of design work over time. Though D.Star was designed specifically for parametric modelling, its central concept of collections of alternatives may well generalise to any digital representation.

Here I argue for a new shape grammar interface design that combines concepts from the Shape Machine and D.Star. This new interface builds on the minimalist design of the Shape Machine interface, adding a few interactions to support collections of shapes. I will argue that such an interface would enable a higher (that is, at a larger scale, not better) form of seeing and doing that operates over design spaces.

2 The Shape Machine Interface

The Shape Machine interface presents as an infinite canvas. Shapes and rules can occur anywhere on the canvas. Typically a designer makes rules by copying and editing parts of other shapes on the canvas, which may themselves be either target shapes or rule sides. Once seen, the interface's logic and interaction appear obvious.

Design Space Exploration in the Shape Machine Interface

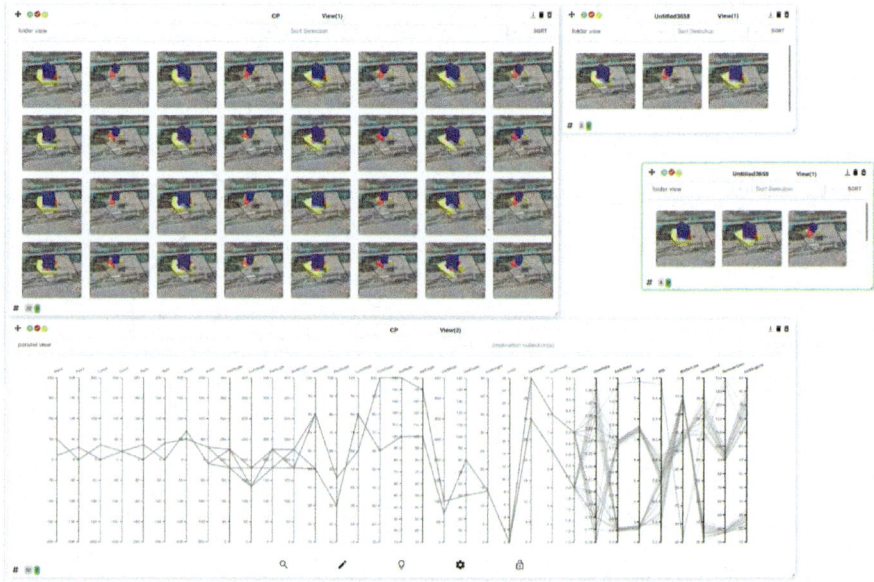

Fig. 2 D.Star enables designers to make ad-hoc collections of design alternatives. This figure shows three such collections. One, named *CP*, presents through two views, one showing thumbnails of 32 design alternatives and the other showing each alternative as a path through a parallel coordinates chart whose dimensions are selected inputs to the parametric model as well as selected outputs (performance variables). D.Star is independent from, but communicates with, its host parametric modeller. It provides commands not present in the modeller and relies on the modeller to provide all of the standard parametric modelling functionality

However, history shows it was not obvious in prospect—the Shape Machine interface is the first of its kind.

What makes the Shape Machine interface so good? Here I appeal to a well-known tool for heuristic analysis of interfaces: the *Cognitive Dimensions of Notations* (Blackwell et al. 2001). Comprising an open-ended set of questions and directives (the original version had 14 entries; the version I use here has 14 but includes *provisionality* and excludes *juxtaposability*), cognitive dimensions provide insight, both analytic (usability) and synthetic (heuristic design moves). The Shape Machine interface presents insights on all dimensions. I note that this is unusual. Table 1 shows a summary of the Shape Machine interface for each dimension.

While I am not aware of any usability studies of the Shape Machine interface, as an expert CAD user and designer I am struck by its directness and conciseness. It seems to me that using the Shape Machine interface closely approaches seeing and doing, closer in some ways than paper and pencil. But there remains a problem. The 36 published grammars on the Shape Machine website (Economou 2019) all show one target shape at a time. The Shape Machine interface falls into a nearly universal pitfall in interactive system design: the *single-state document model*.

Table 1 Summary of cognitive dimensions in the shape machine interface

Cognitive dimension	Shape machine interface
Abstraction gradient	Shape and shape rules comprise the two main supported abstractions. Selection provides a transient abstraction for limiting rule application scope
Closeness of mapping	The shape machine interface maps very closely to shape grammars. Two features stand out. First, shapes in the Shape Machine interface directly model shape grammar shapes. Second, shapes can be freely copied across the two main abstractions, enabling directness of seeing and doing
Consistency	The shape machine interface presents one syntactic form: shapes
Diffuseness/ terseness	A shape is always just that, a shape. No more concise form exists
Error-proneness	While shape rules are notoriously difficult to write, the shape machine interface provides accepted drawing tools (as implemented in an underlying CAD system) to enable geometric construction and control errors. Copying shapes reduces transcription errors
Hard mental operations	The shape notation introduces no hard mental operations in and of itself. Of course, seeing subshapes and composing shape rules may well be some of the hardest of mental operations, but these lie with shape itself, not the shape machine interface notation
Hidden dependencies	Shape rule execution and derivation sequences remain largely hidden in the shape machine interface. In shape grammars, it often occurs that a rule's apparent effect changes along a branch in a derivation sequence
Provisionality	Shape copying and selection make recording provisional ideas very easy
Premature commitment	Copying shapes (and thus parts of shape rules) almost eliminates constraints on the order of doing things. Shape editing, rule composition and rule application can occur in any order
Progressive evaluation	The shape interpreter permits work to be checked at almost any time
Role-expressiveness	The single base representation (that is, shape) greatly reduces role-expressiveness issues. Rules differ from simple shapes only in the origin and inference symbols—these cues may sometimes be too subtle
Secondary notation	The Shape Machine interface has minimal secondary notation, with three notable exceptions. That rules and shapes can be located anywhere on the canvas introduces a secondary notation: location can carry meaning for a designer. A designer can add text and graphical marks that do not participate in a grammar, enabling epistemic actions that may assist the work being done. A designer can select a subpart of a shape and this limits the scope of rule matching and application. All of the secondary notations (location, graphical marks and selection) exist directly in the infinite canvas and follow widely accepted editing practices. Designers do not have to leave the main shape machine interface to use them
Viscosity	Shape rules can be viscous: changing a label structure may require changing many rules. The shape machine interface provides little help with inherent shape grammar viscosity
Visibility	The shape machine interface makes visible all definitional elements of a shape grammar (shapes and rules). However, derivation remains less well-visualised

Almost universally, CAD system interfaces have been constructed around the single-state document model (Terry and Mynatt 2002), which '... requires a document to be in one, and only one, state at any particular time, thereby imposing a serial, linear progression through a task that is at odds with the "messy", highly iterative creative process.' A series of publications in human-computer interaction (HCI) and computer graphics report results on various techniques and tools for 'parallel setup, viewing and control of scenarios' (Lunzer and Hornbæk 2008), labelled as subjunctive interfaces, arguably the most well-known descriptor from HCI. None of these ideas is easy to realise in systems designed around the single-state document model, that is, virtually every extant CAD system, including the Shape Machine interface. In contrast, D.Star explicitly implements a multiple-state document model, precisely to enable design space exploration. I use its two key ideas (alternatives and collections) here as, at the time of writing, D.Star is unique.

Both the research literature and professional practice show that designers, especially experts, make sets of designs repeatedly as they work. They do so whatever their design medium: manual (e.g., a sketchbook) or digital (e.g., a CAD system), and, in each medium, they use different techniques to represent and examine such sets. In other words, they explore design spaces, and any medium suitable for design must support such exploration in some way as a necessary condition for being useful to designers. 'Necessary' does not imply 'sufficient': other things are needed in a design medium, else all design media would have strong core functionality for design exploration. The single-state document model fails to provide this needed functionality. Thus the Shape Machine interface also fails here.

In the absence of a proper tool, people invent hacks and workarounds. Expert users of conventional CAD systems often store alternatives in separate layers. Less expert users employ separate files. Novice users seldom store alternatives at all. We should expect people using the Shape Machine interface to do similar things. And they do. Several of the Shape Machine studies (Economou, 2019) show multiple alternatives as a necessary part of their argument. For example, see The Atrium Hotel Grammar, Mughal Gardens Redreamed, Palladio Computas and Port-ino Automated. These figures were prepared external to the Shape Machine interface and present prima facie evidence that the Shape Machine interface misses something important.

3 Design Space Exploration

But... the Shape Machine interface design anticipates multiple alternatives. In theory, nothing prevents a designer from copying target shapes to other parts of the canvas and using selection to direct the Shape Machine interface to apply rules to only desired targets. Coupled with the secondary notation of notes and graphical marks, this mimics a well-known manual strategy of making multiple alternative sketches on a single sheet of paper. This strategy founders on two rocks. The first is computational scaling—it would quickly overwhelm system capacity unless anticipated and addressed in the system architecture. The second is cognitive scaling—a single infinite canvas with many alternatives would quickly overwhelm the senses.

More fundamentally, using the Shape Machine interface to store and access multiple alternatives introduces a new task: design space exploration. To explain this, I return to Stiny's seeing and doing, in which he uses shape embedding as a necessary and central component. Seeing some of the infinite subshapes of a shape and acting on them makes shape both different from and more interesting than more atomic CAD representations. The main parts of the Shape Machine interface specifically support such seeing and doing. Design space exploration in the current Shape Machine interface must rely on its limited secondary notation to organise shapes and limit rule application.

Kirsh and Maglio (1994) distinguish *epistemic action* (taken to uncover or organise information) from *pragmatic action* (taken to move towards a problem-solving goal). I will argue two things here. First, the Shape Machine interface's secondary notation supports epistemic action in Stiny's seeing and doing. It helps designers organise their rules and applications to make more sense in context. Second, in the current Shape Machine interface, all design space exploration occurs through its secondary notation.

And now I make a claim: Stiny's seeing and doing, and design space exploration differ. Shape embedding dominates the former, whereas the latter comprises comparing entire alternatives (in this case, these are still shapes). Further, design space exploration sits on top of Stiny's seeing and doing: its elements are shapes and it uses shape, but the alternatives that comprise it have an identity of their own in addition to being shapes. Of course, the two unify if one looks at alternatives as zero-dimensional shapes that store higher-dimensional shapes inside. This claim has a point: design space exploration introduces new tasks not well supported in the current Shape Machine interface.

I argue that the two can be combined by observing that some epistemic actions in the Shape Machine interface become pragmatic actions in design space exploration and that these can be supported by modest extensions to the Shape Machine interface that largely maintain its minimalist design. Further, preserving the Shape Machine interface's simplicity and proximity to the shape domain is a necessary condition for success in extending the Shape Machine interface to design space exploration—Stiny's seeing and doing must remain paramount.

4 Pragmatic Action in Design Space Exploration

I present an initial set of pragmatic actions that would enable design space exploration in the Shape Machine interface. I describe these functionally here, that is, without reference to how they might be designed and implemented in an augmented Shape Machine interface. That is the topic of the next section.

Identify a shape as an alternative. An alternative *names* a shape—it provides a separate identity for the shape. A name may be a textual string or simply a graphical mark on a shape showing it to comprise an alternative. The existing Shape Machine interface mechanism of defining the scope of rule application by selection would remain: a selected alternative would be included in rule application. Selection would also be a part of creating alternatives: an additional action on a selection would turn the selection into an alternative.

Alternatives proliferate fast. Collections name sets of alternatives. They provide a way to organise parts of design space during exploration. A collection can include both a shape and one or more alternatives. The name 'collection' originates in D.Star in which it identifies a set of alternatives. In the Shape Machine interface, the term 'subcanvas' may be better—the existing Shape Machine interface comprises a single canvas; an augmented Shape Machine interface would just make canvasses hierarchical and not introduce an entirely new interaction object. I replace 'collection' with 'subcanvas' in the sequel.

Design space exploration requires focusing attention on the alternatives of interest at any particular time. Some form of zoom interface on alternatives and subcanvasses could provide such a focus device.

Rules apply to alternatives and subcanvasses. Of course, rules may vary across a design space and selecting which rules apply determines how an explicit design space develops. Which rules apply when remains an open question and one to be resolved interactively when working with a grammar. As an initial proposal, consider that rules apply by default in the canvas in which they originate and in any subcanvasses of that canvas (in object-oriented terms, subcanvasses inherit rules from their parents). This decision would introduce a significant abstraction mechanism and possibly hidden dependencies in cognitive dimension terms, requiring an abstraction management tool such as a checklist of which inherited rules apply.

5 Augmenting the Shape Machine Interface

Remember that the Shape Machine interface provides nearly direct interaction with the elements of a shape grammar: shape and shape rules. It imposes minimal additional interaction mechanisms, rather providing a small set of well-known secondary notations (selection, labels and graphical marks, and location control) to support needed epistemic action. In making the initial proposals below, I take the design stance of making the minimal possible additions to the Shape Machine interface, particularly preserving its infinite canvas, minimal command interface and positive aspects revealed by the cognitive dimensions analysis to the greatest degree possible. An initial set of augmentations follows.

Figure 3 shows that naming a selection makes a shape into an alternative. I have used the term 'name' loosely here—a name is simply some identifying mark. It may be an actual textual name or a graphical mark that indicates that a shape has become an alternative. Any selection (which feature already exists in the Shape Machine interface) can be made into an alternative. Note well that this policy means that any subshape can be an alternative, as can any arrangement of spatially separate

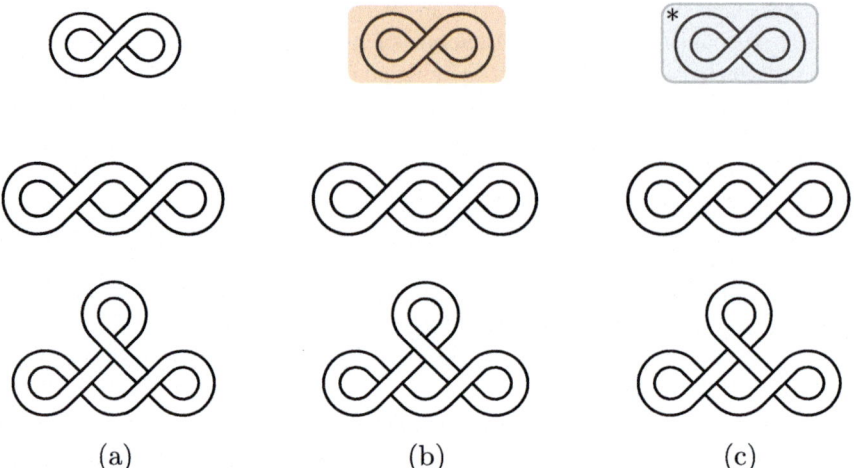

Fig. 3 a A celtic knot shape. **b** Selecting a subshape makes it available for other operations. **c** Distinguishing a selection makes it an alternative—a persistent object. The ∗ denotes that the alternative has a name, that is, a unique and persistent identifier. A specific naming convention awaits further interface design work

shapes. Thus, the augmented Shape Machine interface would set no policy on what can comprise an alternative, maintaining the Shape Machine interface's closeness of mapping to the shape domain. Alternatives can be freely located on their canvas, preserving the secondary notation of location for epistemic action in both shape and design space exploration tasks.

Figure 4 shows that naming a selection of alternatives makes a subcanvas that contains all of the selected alternatives. This interaction would be nearly identical to naming a shape to become an alternative.

Figure 5 shows that iconifying an alternative, or subcanvas, provides information zooming. The icon for a subcanvas provides a counter of contained alternatives.

Grammars have many rules. Controlling which applies, where and when, presents an important (and challenging) aspect of writing and running grammars. As an initial design proposition, I propose perhaps the most ancient scope control mechanism: declaring variables local to a subroutine (in object-oriented terms, if a sub-canvas is an object, then it inherits the rule property of its parent). In the present context, this means that rules defined in a subcanvas apply in that subcanvas, as do any rules defined in containing subcanvasses. Of course, controlling these scope rules introduces a new abstraction (in cognitive dimension terms) and will likely require an abstraction manager, which would add to the complexity of the Shape Machine interface. Figure 6 demonstrates this basic local variable scoping mechanism.

D.Star has a type of display object called a *view*, which applies to both alternatives and collections. Views support multiple and variant displays of the same information, in shapes for instance, shape, list and editing-time order views. These may be needed, but introduce a new level of abstraction that would likely obscure both the

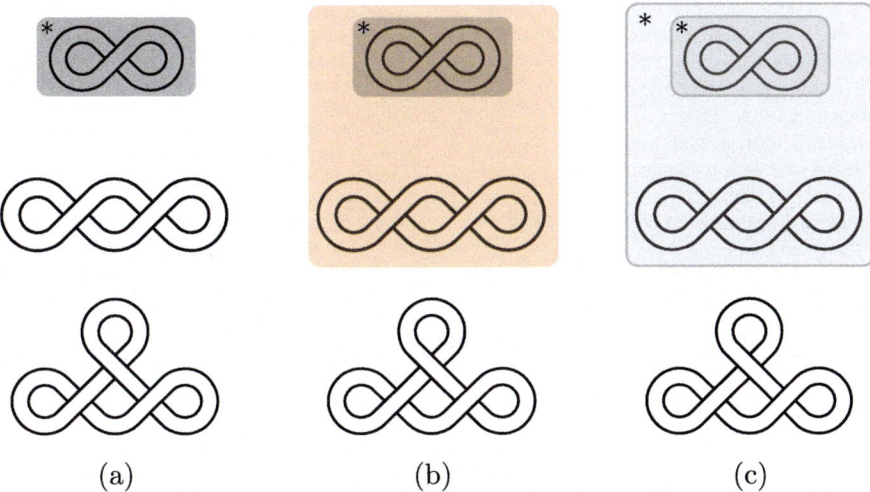

Fig. 4 **a** A celtic knot shape with a subshape denoted as an alternative. **b** Selecting a subshape makes it available for other operations. **c** Distinguishing a selection makes it a subcanvas—a persistent object. The ∗ denotes that the subcanvas has a name, that is, a unique and persistent identifier. A subcanvas is any distinguished selection that contains an alternative. A specific naming convention awaits further interface design work

Fig. 5 **a** A celtic knot shape with a subshape denoted as a subcanvas. **b** Iconifying a subcanvas hides it from view. The ∗ denotes that the subcanvas has a name, that is, a unique and persistent identifier. The :n gives the number of contained subcanvasses, in this case, $n = 1$

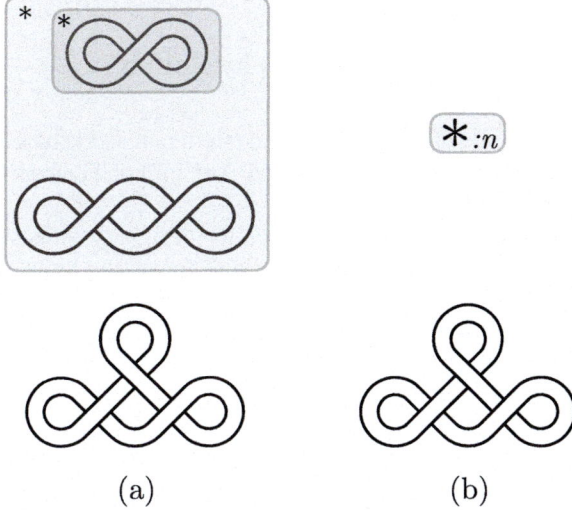

infinite canvas and initial efforts to augment the Shape Machine interface with design space exploration. Views introduce new interactions to handle cross-view effects, so implementing them would be a significant project.

Note well that these proposals will impose at least two new abstractions on the Shape Machine interface. The first would support rule scope: which rules apply to

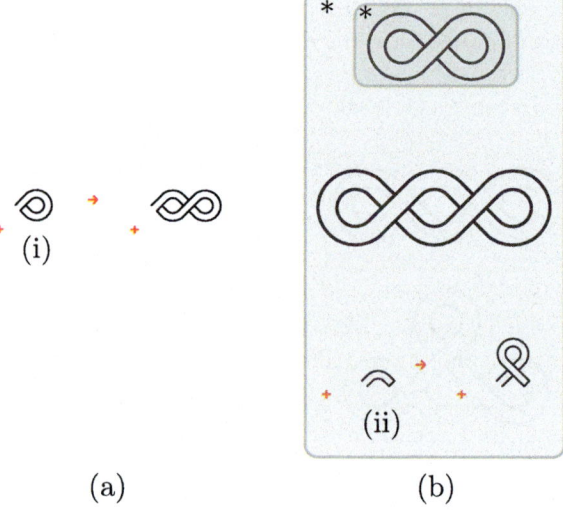

Fig. 6 Preliminary proposal for rule scope. Subcanvasses inherit rules from their enclosing canvas. Any rule defined in a subcanvas is local to it. There would need to be an abstraction manager to deactivate rules in a subcanvas (unspecified here). **a** The main shape machine interface canvas. Rule (i) lies in this canvas. **b** A subcanvas containing rule (ii). Both rules (i) and (ii) apply in subcanvas (**b**)

which alternatives. The second would control semantic zooming and arrangement of subcanvasses. I expect that these new abstractions will take many design iterations to become stable.

6 In the 'Age of AI' Why Bother?

At the time of writing, artificial intelligence (AI) based on neural nets dominates the computational design literature. And for good reason: such AI promises to automate many design tasks.

Human-computer interfaces for design systems stand apart from AI, the most frequent approach in the literature for using digital alternatives in design. Papers taking the AI stance (and these have a history nearly as long as computational design itself) employ a heuristic search algorithm to sample a typically informally specified design space and argue that the outputs of such a process have utility as designs. There exist many such papers, using, for example, evolutionary algorithms, simulated annealing, Pareto optimisation, tabu search, dispersion sampling, and (more recently) generative adversarial networks. I label research of this type as *appealing to an oracle*. In computer science, an oracle machine is a Turing machine augmented by a black box, an *oracle,* able to solve a decision problem in a single operation. In mythology, an *oracle* can divine the future, typically by appeal to a higher power. To appeal to an oracle means accepting what the oracle produces as useful and not inquiring about how it does its work. From a research perspective oracles are attractive as low-hanging fruit: undertaking such a project follows an established path, can be done with modest effort and yields a demonstrable result. As research contributions though, three problems emerge. The first is that such works are incremental: they elaborate one particular approach to using alternatives in design, whereas the domain may need new approaches. The second lies in the difficulty of demonstrating progress: oracles produce results and that is what most papers conclude, leading to circular argumentation. The third problem is that the approach itself has a fatal flaw: it does not support design tasks as done in the wild.

Why not? I offer two arguments. The first is that designers design—they devise things to achieve goals. In the task of designing, a design plays roles in critique and subsequent re-work. Schön's (1983) account of the reflective practitioner argues this invariant thoroughly. Thus, almost inevitably, an oracular result becomes the seed for yet more design work. Bradner et al. (2014) state this well in their study of designers using optimisation systems.

> Professionals reported that the computed optimum was often used as the starting point for design exploration, not the end product. (Bradner et al. 2014).

The second argument is that designers explain their work. In such explanations agency is important. *'We thought this through.'* presents a more credible argument than *'The algorithm told us to use this'*. Although facetiously expressed, the way designers explain and understand (i.e., explain to themselves) designs involves having worked with and through the design's ideas over time. Both of these arguments imply a strong need for basic and practical knowledge of designing interactions and systems supporting design alternatives.

Designers will appeal to oracles. We have done so throughout the history of computer-aided design: a BIM component is an oracle, as is a structural analysis. What differs here is only scale—oracles over design spaces operate at large scale. If the research literature mirrors practice, oracles must be widely accepted indeed–a large portion of the literature reports oracles. But there is evidence from practice as well. For example, Peters (2007) reports on developing an oracle (albeit an explainable one) for the roof of a Smithsonian building. As I have argued above, oracles produce new designs by sampling a design space. Against Dennett's (1995) Vastness of design space, oracles must sample. Against the ill-defined nature of design problems, designers will use oracular results as input to the next phase of design dialogue. Almost universally though, papers in the oracular literature end with the design space sampled, and that is exactly where directly interactive design exploration starts. Oracles produce collections: interactive design explorers enable rich interaction with such collections. The proper place for oracles in design is inside a directly interactive loop of a design medium. The proposal described here aims at a simple structure for such integration in one particular representation: shapes.

We know designers work with alternatives for which current systems provide impoverished support. We should expect such support to be specific to the design media (e.g., shape grammars or parametric models) being supported, and to be used by designers in ways both specific to the media and unexpected. We should not directly adapt existing work on design space explorers if it imposes design space structure on an interface. Here I have attempted to translate some of the lessons learned from my work in design space exploration for parametric models into shape notation.

7 Summary

Of course, I make a design proposition. Reifying it will require multiple design iterations and significant implementation work. I argue though that augmenting the Shape Machine interface with design space exploration would situate it in a larger task context and make it intrinsically more useful to both researchers and designers. This paper simply presents one approach to new and more useful shape grammar interfaces.

Competing Interests The author has no conflicts of interest to declare that are relevant to the content of this chapter.

References

Blackwell, A.F., C. Britton, A. Cox, T.R.G. Green, C. Gurr, G. Kadoda, M.S. Kutar, M. Loomes, and C.L. Nehaniv. 2001. Cognitive dimensions of notations: Design tools for cognitive technology. In *Cognitive technology: Instruments of mind*, vol. 2117, Heidelberg Berlin: Springer.

Bradner, E., F. Iorio, and M. Davis. 2014. Parameters tell the design story: Ideation and abstraction in design optimization. In *Proceedings of the symposium on simulation for architecture & urbam design, SimAUD '14*, vol. 26, 1–26. San Diego: Society for Computer Simulation International.

Dennett, D.C. 1995. *Darwin's dangerous idea: Evolution and the meanings of life*. Simon & Schuster.

Economou, A., T.-C. Hong, H. Ligler, and J. Park. 2021. Shape machine: A primer for visual computation. *A new perspective of cultural DNA*, 65–92. KAIST research series. Singapore: Springer.

Economou, A. 2019. *Shape computation lab research*. Accessed 25 Jan 2024 at https://shape.gatech.edu/Research

Goldschmidt, G. 1991. The dialectics of sketching. *Creativity Research Journal* 4 (2): 123–143.

Kirsh, D., and P. Maglio. 1994. On distinguishing epistemic from pragmatic action. *Cognitive Science* 18 (4): 513–549.

Lunzer, A., and K. Hornbæk. 2008. Subjunctive interfaces: Extending applications to support parallel setup, viewing and control of alternative scenarios. *ACM Transactions on Computer-Human Interaction (TOCHI)* 14 (4): 17.

Mohiuddin, A., and R. Woodbury. 2022. Interactive visualization for design dialog. In *Design, computing and cognition '20*, ed. J.S. Gero, 491–508. Cham: Springer International Publishing.

Peters, B. 2007. The Smithsonian courtyard enclosure: A case-study of digital design processes. In *Expanding bodies: Art, cities, environment: Proceedings of the 27th annual conference of the association for computer aided design in architecture*, 74–83. Halifax (Nova Scotia): Riverside Architectural Press and TUNS Press

Schön, D. 1983. *The reflective practitioner: How professionals think in action*. Basic Books.

Stiny, G. 2006. *Shape: Talking about seeing and doing*. Cambridge: The MIT Press.

Terry, M., and E.D. Mynatt. 2002. Recognizing creative needs in user interface design. In *Proceedings of the 4th conference on creativity and cognition, C&C '02*, 38–44. New York: ACM.

Open Access This chapter is licensed under the terms of the Creative Commons Attribution-NonCommercial-NoDerivatives 4.0 International License (http://creativecommons.org/licenses/by-nc-nd/4.0/), which permits any noncommercial use, sharing, distribution and reproduction in any medium or format, as long as you give appropriate credit to the original author(s) and the source, provide a link to the Creative Commons license and indicate if you modified the licensed material. You do not have permission under this license to share adapted material derived from this chapter or parts of it.

The images or other third party material in this chapter are included in the chapter's Creative Commons license, unless indicated otherwise in a credit line to the material. If material is not included in the chapter's Creative Commons license and your intended use is not permitted by statutory regulation or exceeds the permitted use, you will need to obtain permission directly from the copyright holder.

From Lines to Curves: Implementation of Shape Embedding for Circular Arcs and Circles in 2D CAD Systems

Tzu-Chieh Kurt Hong

Abstract It is not doubtful on the success of shape grammar formalism, especially the calculation of straight lines, which is implemented by multiple robust interpreters such as SGI (Krishnamurti in SGI: a shape grammar interpreter. Centre for configurational studies, design discipline, 1982), GRAPE (Grasl and Economou in Environ Planning B: Planning Des 40:905–922, 2013), SortAl GI (Dy and Stouffs in Combining geometries and descriptions a shape grammar plug-in for grasshopper, 2018), Shape Machine (Hong and Economou in Autom Constr (AiC) 146, 2023) and sort forth. Calculation of straight lines is adopted in multiple fields such architecture, industrial design, mechanical engineering and sort forth (McKay et al. in Artif Intell Eng Des Anal Manuf 26:143–159, 2012), however, calculation of curves, for example, arcs, circles, parabola, hyperbola, ellipse, elliptical arcs, piece-wise curves, NURB curves, is still a task that remains unresolved. Countable applications that tackled this task can be found in the literature, for instance, Curve-based SGI (McCormack and Cagan in Environ Planning B: Planning Des 33:523–540, 2006), QI (Jowers and Earl in Environ Planning B: Planning Des 38:616–635, 2011), and GRAPE (Grasl and Economou in Environ Planning B: Planning Des 40:905–922, 2013; Grasl and Economou in Artif Intell Eng Des Anal Manuf 32(2):208–224, 2018) But still, it is not ready to adopt curve calculation in the practical projects since curves, especially arbitrary curves, are hardly represented with point boundaries since the boundary representation of curves is not unique (Jowers and Earl in Environ Planning B: Planning Des 38:616–635, 2011). This study proposed an algorithm which is currently implemented in the Shape Machine, for maximization of circular arcs and circles in computer-aided design systems, with a calibration that can mitigate the precision errors in geometry modeler (Hong and Economou in Autom Constr (AiC) 146, 2023), as well as an algorithm for shape embedding of circular arcs and circles. By calculating circular arcs, and circles in shape grammar interpreter, great number of applications are possible. In the second section of this paper, one example is introduced to demonstrate the applications of curve calculation. In the end, a discussion of visual equivalency is given to provide an alternative perspective about the standard for shape calculation.

T.-C. K. Hong (✉)
University of Kansas, Lawrence, KS, USA
e-mail: kurt.hong@ku.edu

1 Introduction

The implementation of shape grammar interpreters in digital environment have been studies for decades, it can be found in the literatures that showed several purpose-built shape grammar interpreters were adopted in multiple domains resolving problems in design and engineering (Chau 2004; Eloy et al. 2018; Gips 1999; McKay et al. 2012). Moreover, a further challenge implementing general-purposed shape grammar interpreters as general visual computing systems had been taken by precedents (Chase 1989; Dy and Stouffs 2018; Grasl and Economou 2013, 2018; Hong and Economou 2023; Jowers and Earl 2011; Krishnamurti 1982; Krishnamurti and Giraud 1986; Ruiz-Montiel et al. 2014; Stouffs 2022; Stouffs and Li 2020; Tapia 1999; Trescak et al. 2012). A comprehensive literature review of shape grammar interpreters can be found in the work by Hong and Economou (Hong and Economou 2022a), where 73 shape grammar interpreters are reviewed and 24 of them are general-purpose shape grammar interpreters. Among the 24 applications, the Shape Machine is one of very few works that successfully manage to resolve the critical issues in shape grammar interpreters (Hong and Economou 2023) and are currently and adopted in education, research, and practices (Hong and Economou 2023; Okhoya et al. 2022). However, despite the success of the Shape Machine, the geometry type that most interpreters support is straight lines, and the absence of curves highly confines the possibility of design and further applications. Few accounts in the literatures tackled the challenge to implement curve-based shape grammar interpreters (Grasl and Economou 2013, 2018; McCormack and Cagan 2006; Jowers and Earl 2011), but yet, most of them are struggling in the similar issues that occur in the straight-line-based shape grammar interpreters, that is, the precision errors in modeling and rounding mechanism, rising algorithmic complexity of transformation derivation, and insufficient support to indeterminate shape embedding (Hong and Economou 2022b). Among these three issues, precision errors can be viewed as the most critical one since it causes a gap between visual perception and the data presentation. For example, a shape u is visually considered as embedded in another shape w, however, a small number of numerical differences in the actual data representation of the shapes might fail the embedding relation $u \leq w$ (Stiny 1975, 1980). More specifically, precision errors occur in the following conditions: (1) examining if a shape u is embedded into another shape w, $u \leq w$; (2) the derivation of the transformations f_s that embed shape u into $w, f(u) \leq w$; (3) examining the conditions if two shapes shall be maximized; (4) Boolean addition of two shapes; and (5) Boolean subtraction of two shapes. Since shape grammar is a visual computation system, it is critical to achieve visual equivalency, that is, matching visual perception and the data representation. To address this issue, a mechanism is proposed in the literature (Hong and Economou 2023) which calibrates the shapes in 2D computer-aided design systems for the shapes consisting of lines. However, there is no such a mechanism to calibrate the shapes consisting of curves, more specifically, arcs and circles. The applications of shape grammar interpreters are highly confined due to the limited shape types. This study proposed a calibration system for the shapes consisting of arcs and circles to address the issue of precision error, so that the shape embedding for shapes consisting of arcs and circles can be achieved. The proposed calibration system is implemented in the Shape Machine to evaluate this system.

2 Representation of Circles and Arcs: Center, Radius, Endpoints, and Radian

In this study, a circular arc is viewed as a segment embedded in a circle descriptor $x^2 + y^2 + axy + bx + cy + d = 0$, which can be treated as the underlying structure of shape. This notion is similar to the hyperplanes or line descriptors (Hong and Economou 2023; Krishnamurti 1992) where lines segments are embedded. And note that, a circle is an enclosed arc fully embedded in a circle descriptor, hence a circle does not have endpoints. Here, a circle C is represented with three items: (1) the coordinates of the center which is denoted O_C, (x_{O_C}, y_{O_C}) consisting of two real numbers and (2) the radius which is real number denoted as r_C and (3) its circle descriptor. Figure 1 shows the diagram of the representation of circular arcs. A circular arc A is represented with five items: (1) the coordinates of the center which is denoted O_A, (x_{O_A}, y_{O_A}) consisting of two real numbers; (2) the radius which is real number denoted as r_A; (3) the coordinates of the two endpoints H_A and T_A which are denoted as (x_{H_A}, y_{H_A}) and (x_{T_A}, y_{T_A}) consisting of four real numbers; and (4) the radian of the arc which is a region Θ_A consisting of two real numbers. It is denoted as $[\theta_{A_H} : \theta_{A_T}]$ where $-\pi \leq \theta_{A_H} \leq \pi$ and $-\pi \leq \theta_{A_T} \leq \pi$; and (5) its circle descriptor. In this study, the radian of an arc is derived with a datum which represents the point of 0 radian, and the opposite point to it is a marker that represents the point of π and $-\pi$ radian. The direction from center point to the datum is identical for any arc, that is, the radian region of any arc is calculated with the same convention. For some cases, there might be multiple regions of radian of an arc when the $\pi/-\pi$ point is located between the two endpoints. For instance, the radian of the arc in Fig. 1b has two regions $\Theta_{A_1} = [-\pi : \frac{\pi}{4}]$ and $\Theta_{A_2} = [\frac{\pi}{2} : \pi]$. Note that the circle descriptors of arcs and circles are mainly used while deriving registration points for calculating transformation in shape embedding process (see chapter "Reflections on Interpreting Shape Grammars").

It helps refine the argument if the spatial relations between two arcs and the spatial relations between two circles are defined. In this study, three spatial relations are defined: (1) two arcs (circles) are considered as concentric if they have identical center points, which is processed with a numerical comparison: $x_{O_A} = x_{O_B}$ and $y_{O_A} = y_{O_B}$; ; (2) two arcs (circles) are considered as co-circular if they have identical center points and identical radii, which is processed with a numerical comparison: $x_{O_A} = x_{O_B}, y_{O_A} = y_{O_B}$, and $r_A = r_B$; ; (3) two arcs are considered as connected if they have identical center points, identical radii and the identical boundaries of radian regions, which is processed with the numerical comparison: $x_{O_A} = x_{O_B}, y_{O_A} = y_{O_B}$,

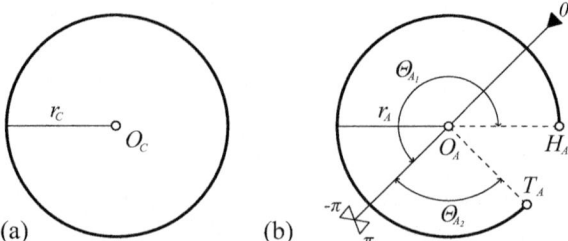

Fig. 1 Representation of circles and arcs. **a** The representation of a circle including the coordinates of the center consisting of two real numbers (x_c, y_c), a real number representing radius r_c; **b** The representation of an arc including the coordinates of the center point consisting of two real numbers (x_c, y_c), coordinates of two endpoint (x_H, y_H) and (x_T, y_T), and the radian of the arc θ_A

$r_A = r_B$, and $\theta_{A_H} = \theta_{B_T}$ (or, $\theta_{A_T} = \theta_{B_H}$). Figure 2 shows the three spatial relationships of two arcs.

The numerical comparison might fail due to a small amount of precision errors, which cannot be visually perceived. Therefore, the gap between visual perception and the actual numerical comparison is violating the fundamental principles of shape grammars, that is, numerical calculation is overriding the visual perception, which should be the only standard of the shape calculation. It is critical to introduce a calibration mechanism to eliminate the gap between numerical comparison and visual perception. This is a known and on-going issue that geometry modelers are discrete machines where geometries can hardly be represented perfectly, especially then the geometries are involved in infinite numbers and irrational numbers, and the imperfections, that is, precision errors, will cause failures of shape embedding (Hong and Economou 2023; Yue and Krishnamurti 2013, 2014). In the following chapter, an algorithm to calibrate the shapes consisting of arcs and circles in 2D CAD systems is proposed.

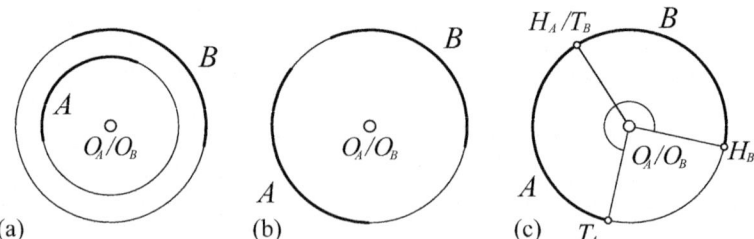

Fig. 2 Three spatial relationships of two arcs. **a** Concentric arcs; **b** co-circular arcs; **c** connected arcs

3 Calibration for Circles and Arcs

A calibration mechanism for shapes consisting of lines is proposed by Hong and Economou (Hong and Economou 2023), which is adopted in the Shape Machine to address the issues while calculating with lines. Here, a calibration mechanism proposed for shapes consisting of arcs and circles is processed for three cases: (1) calibration of two center points when two arcs (circles) that are visually considered as concentric; (2) calibration of two radii when two concentric arcs (circles) that are visually considered as co-circular; and (3) calibration of two radians when two co-circular arcs (circles) that are visually considered to be maximized.

3.1 Calibration of Two Center Points for Two Concentric Arcs

Figure 3 shows an example, where two arcs are visually considered as concentric, but actual numerical values of the center points are not equal. This condition can be described as that there is an actual distance ε_O between the two center points O_A and O_B.

The calibration of the center points for concentric arcs (circles) is processed in three steps: (1) examining if the distance ε_O between O_A and O_B is smaller than a tolerance d_O; if yes, (2) removing both arcs A and B; (3) remodeling both arcs $A\prime$ and $B\prime$ with a new center point $O_{AB\prime}$, which is derived by interpolating the coordinates O_A and O_B, that is, $(\frac{x_{O_A}+x_{O_B}}{2}, \frac{y_{O_A}+y_{O_B}}{2})$. Figure 4 shows the process of the calibration of two center points.

Fig. 3 An example of the precision error between two center points when two arcs are visually considered as concentric

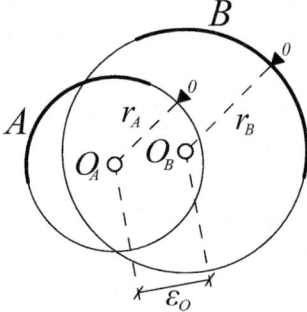

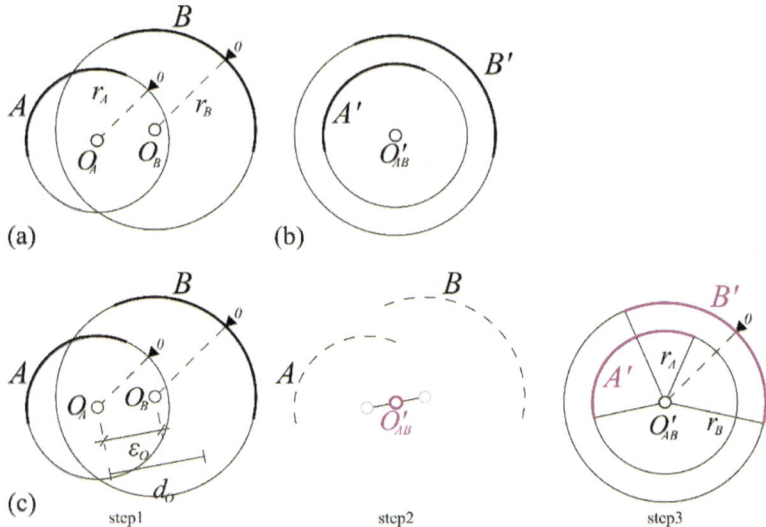

Fig. 4 An example of the calibration for two concentric arcs processed in three steps. **a** Two arcs A and B before calibration; **b** two concentric arcs A' and B' after calibration; **c** three steps of calibration

3.2 Calibration of Two Radii for Two Co-Circular Arcs

Figure 5 shows an example, where two concentric arcs are visually considered as co-circular, but actual numerical values of the center points and the radii are not equal. This condition can be described as that there is an actual distance ε_O between the two center points O_A and O_B, and there is a length difference ε_r between the two radii r_A and r_B.

The calibration of the radii for two arcs, that are visually considered as co-circular, is processed in three steps: (1) examining if the difference ε_r between r_A and r_B is smaller than a tolerance d_r, if yes, (2) removing both arcs A and B; (3) remodeling both arcs A' and B' with a new radius r'_{AB}, which is derived by interpolating the radii r_A and r_B, that is, $\frac{r_A+r_B}{2}$. Figure 6 shows the process of the calibration of two radii.

Note that calibration of radii for two arcs is processed only when they are concentric, thusly the calibration of radius is always after the calibration of center points, which is the calibration for the concentric relation.

Fig. 5 An example of the precision error between two radii that are visually considered as co-circular

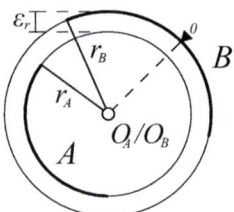

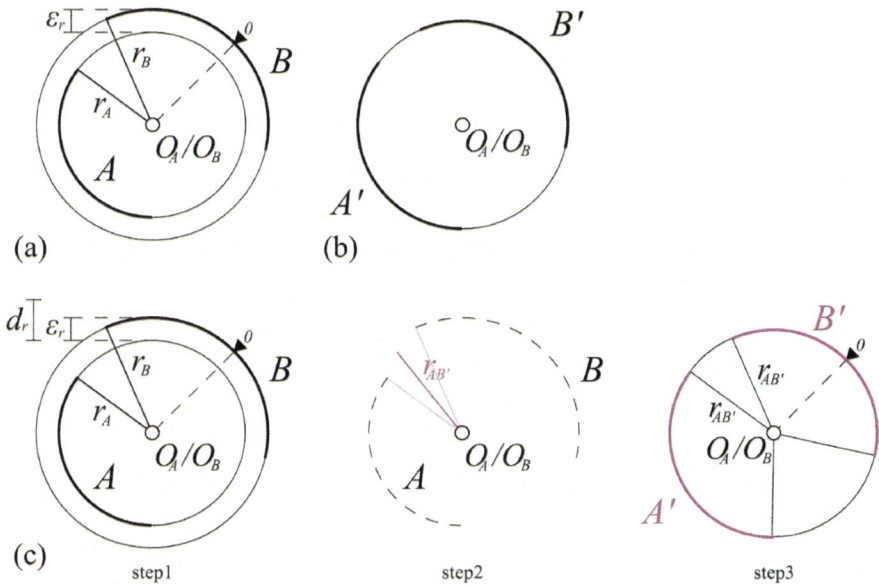

Fig. 6 An example of the calibration for two co-circular arcs processed in three steps. **a** Two arcs A and B before calibration; **b** two co-circular arcs A' and B' after calibration; **c** three steps of calibration

3.3 Calibration of Two Radians Regions for Two Connected Arcs

Figure 7 shows an example, where two arcs are visually considered to be maximized, but actual numerical values of the boundaries of the radian regions are not equal. This condition can be described as that there is a difference ε_θ between the upper bound θ_{A_H} of the radian region $[\theta_{A_H} : \theta_{A_T}]$ and the lower bound θ_{B_T} of the radian region $[\theta_{B_H} : \theta_{B_T}]$.

The calibration of the radian regions for two arcs, that are visually considered to be maximized, is processed in three steps: (1) examining if the difference ε_θ between θ_{A_H} and θ_{B_T} is smaller than a tolerance d_θ, if yes, (2) removing both arcs A and B;

Fig. 7 An example of the precision error between two endpoints two arcs that are visually considered to be maximized

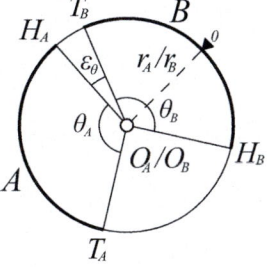

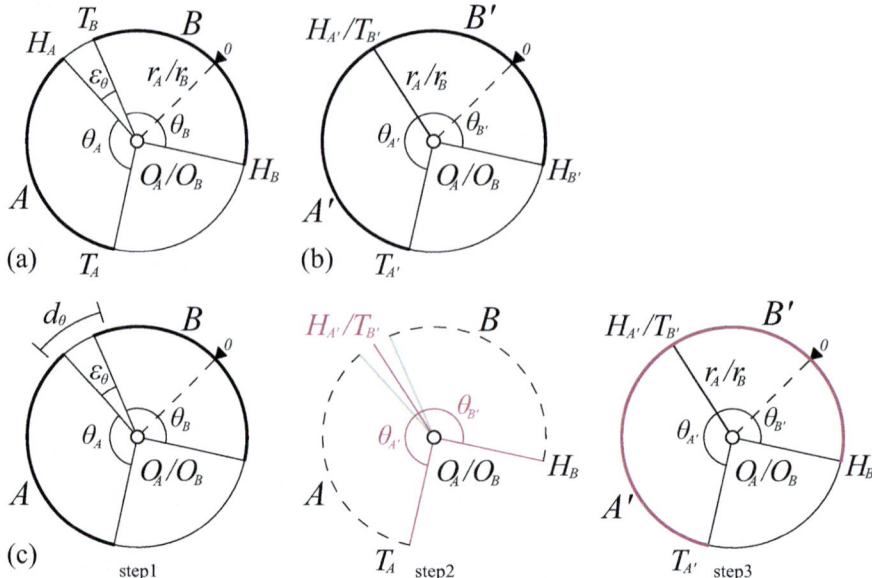

Fig. 8 An example of the calibration for maximization of arcs is processed in three steps. **a** Two arcs A and B before calibration; **b** a maximized arc C; **c** three steps of calibration

(3) remodeling both arcs A′ with a new radian region $[\theta_{A_H}' : \theta_{A_T}]$ and B′ with a new radian region $[\theta_{B_H} : \theta_{B_T}']$, where $\theta_{A_H}' = \theta_{B_T}'$ and it is derived by interpolating the θ_{A_H} and θ_{B_T}, that is, $\frac{\theta_{A_H}+\theta_{B_T}}{2}$. Figure 8 shows the process of the calibration of the radian regions.

Note that calibration of radian regions for two arcs is processed only when they are co-circular, thus, the calibration of radian regions is after the calibration of radii, which is the calibration for the co-circular relation.

3.4 Hierarchical Examination and Interpolation Algorithm

The three calibrations are in a hierarchical order where center points are calibrated firstly, the radii are calibrated secondly, and radian regions are calibrated thirdly. That is, maximization is processed if two arcs are co-circular, and two arcs are co-circular if two arcs are concentric. Hence, the first condition to examine is $\varepsilon_O \leq d_O$, the second condition to examine is $\varepsilon_r \leq d_r$, and the third condition to examine is $\varepsilon_\theta \leq d_\theta$. Table 1 shows all the conditions when the corresponding calibrations are applied.

From Lines to Curves: Implementation of Shape Embedding ...

Table 1 Conditions for calibration

$\varepsilon_O \leq d_O$	$\varepsilon_r \leq d_r$	$\varepsilon_\theta \leq d_\theta$	Center	Radius	Radian Region
False	False	False	No	No	No
False	False	True	No	No	No
False	True	False	No	No	No
False	True	True	No	No	No
True	False	False	**Applied**	No	No
True	False	True	**Applied**	No	No
True	True	False	**Applied**	**Applied**	No
True	True	True	**Applied**	**Applied**	**Applied**

The new numerical value is derived based on averaging the original values of the two arcs. This method is adopted to reduce the chance that the accumulated errors in the worst cases exceed the tolerance, thereby failing the system. An example is given in Fig. 9, where a point A is given and it is assumed that no precision error occurs on it, that is, it is a correct point. Another point B is modeled to be equal to point A. A precision error, ε, occurs in the modeling process, so that B is an incorrect point. The best case is to correct point B by calibrating B to A, that is, moving B to A. Nevertheless, it is clueless to determine which point is correct. If A is calibrated, that is, moving A to B, a precision error ε is added to the calibrated point A'. It is assumed that the chance to pick up the correct point is 50% without further information, hence, the average precision error is $\frac{\varepsilon}{2}$. However, precision errors are accumulated in the on-going modeling process. Assuming another point C is modeled based on the calibrated points A' (note that the original correct point A is calibrated to point B), and a precision error ε occurs again. The worst case is to calibrate the points A' to point C, that is, another precision error ε is added of the newly calibrated point A'', and the accumulated error is 2ε. Even though the chance to keep accumulating precision errors to the level that exceeds the tolerance is very low, still, interpolation is adopted to mitigate the accumulated errors. Figure 9b shows that, after two iterations of worst cases, the precision error of interpolation is ε, which is half of 2ε.

4 Maximization for Shapes Consisting of Arcs and Circles

After calibration, maximization for arcs and circle is processed in three steps: (1) examining the spatial relations between two arcs; (2) removing both arcs; (3) remodeling the maximized arc. In the maximization process, no calibration is required since all arcs are calibrated. However, after maximization, the calibration is applied again on all arcs to assure the data representation is synchronized with visual perception. For maximization for two arcs, two conditions are examined: (A) two co-circular arcs are maximized if they are co-circular and the intersection is not empty, $A \cap B \neq \varnothing$;

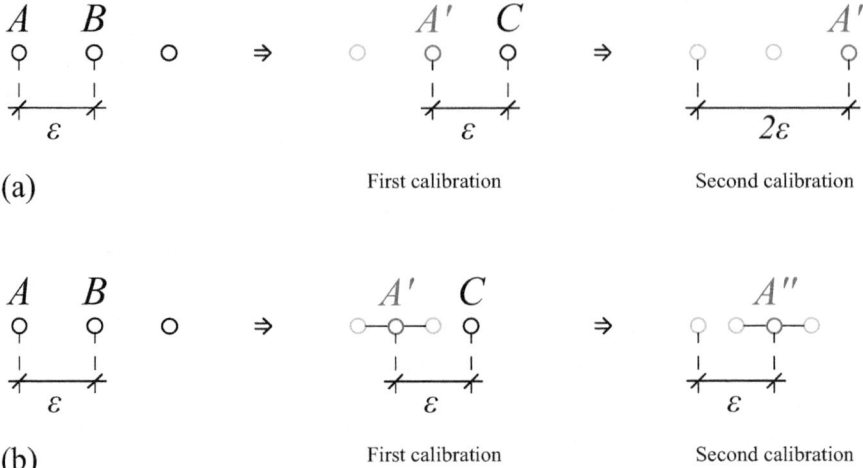

Fig. 9 An example of calibration with/without interpolation. **a** The calibration without interpolation; **b** The calibration with interpolation

and (B) two co-circular arcs are maximized if they are connected at one or two endpoints, even though the intersection of two arcs is empty, $A \cap B = \varnothing$.

The examination is processed by comparing the radian regions of two arcs, however, the radian regions are cyclic and defined with a datum, so that it is not straightforward to make comparisons in the circular domain. Hence, in this study, the radian regions are mapped from circular domain to linear domain with the datum, so that the comparison can be easily made. Note that, even though the examination is in linear domain, the maximization is still processed in the circular domain. Figure 11 shows two examples of mapping radian regions to linear domain for two arcs.

More specifically, the mapping process is to parameterize the radian to the length with a datum point, which is the zero point with a black triangle mark shown in Fig. 11. The point $-\pi/\pi$ represents the cutting point of the circular domain with two white triangle marks, also, it represents the boundaries of the domain which has lower bound $-\pi$ and upper bound π. The total the range (length) of the domain is 2π. For some arcs, there are multiple radian regions when the $-\pi/\pi$ point is located on the arc. Hence, for the arc that has multiple radian regions, the linear mapping will lead to multiple lines in the linear domain (see Fig. 11b). Figure 12 shows the 29 cases of radian regions mapping of the first condition, where two the intersection of the co-circular arcs is not empty. Note that the 29 cases are more than the eight cases shown in Fig. 10, because the mapping might be different if the location of the datum is different.

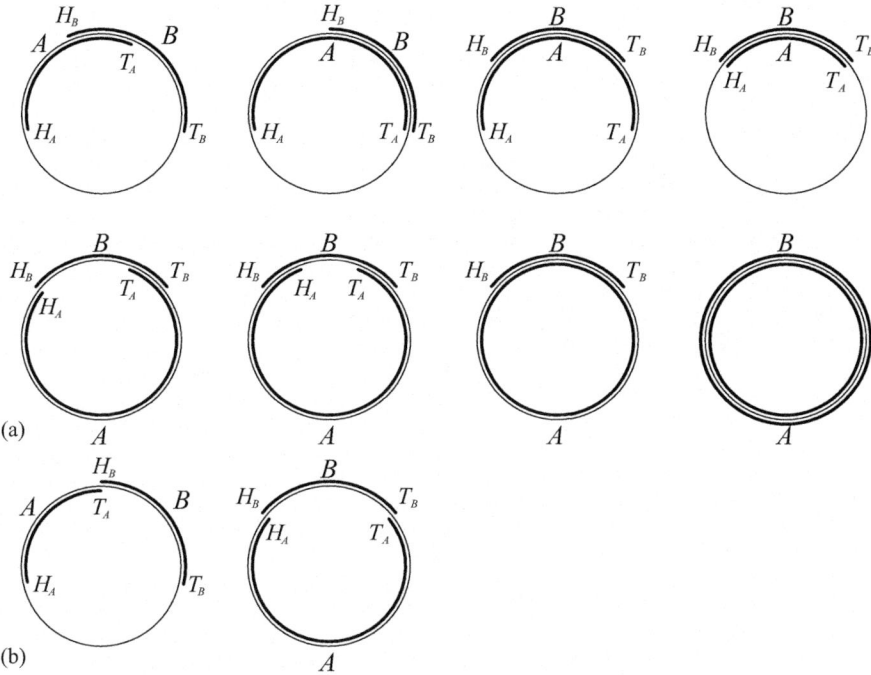

Fig. 10 Ten cases of two co-circular arcs or circles that need to be maximized. **a** The condition where the intersection of two co-circular arcs is not empty; **b** the condition where the intersection of two co-circular arcs is empty but they are connected

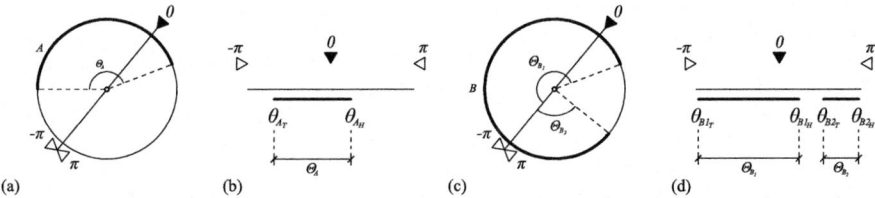

Fig. 11 Two examples of mapping radian regions to linear domain for two arcs. **a** Arc A with radian region $\Theta_A = [\theta_{A_T} : \theta_{A_H}]$; **b** linear mapping of Θ_A; **c** Arc B with radian regions $\Theta_{B1} = [\theta_{B1_T} : \theta_{B1_H}]$ and $\Theta_{B2} = [\theta_{B2_T} : \theta_{B2_H}]$; **d** linear mapping of Θ_{B1} and Θ_{B2}

After mapping radian regions to linear domain, the examination can be processed by comparing the numerical values of the boundaries of the radian regions. For the first condition, all 29 cases are listed in Table 2, and the examination is a table-looking-up process.

Figure 13 shows the six cases of the second condition that the two co-circular arcs are connected, and the intersection is empty.

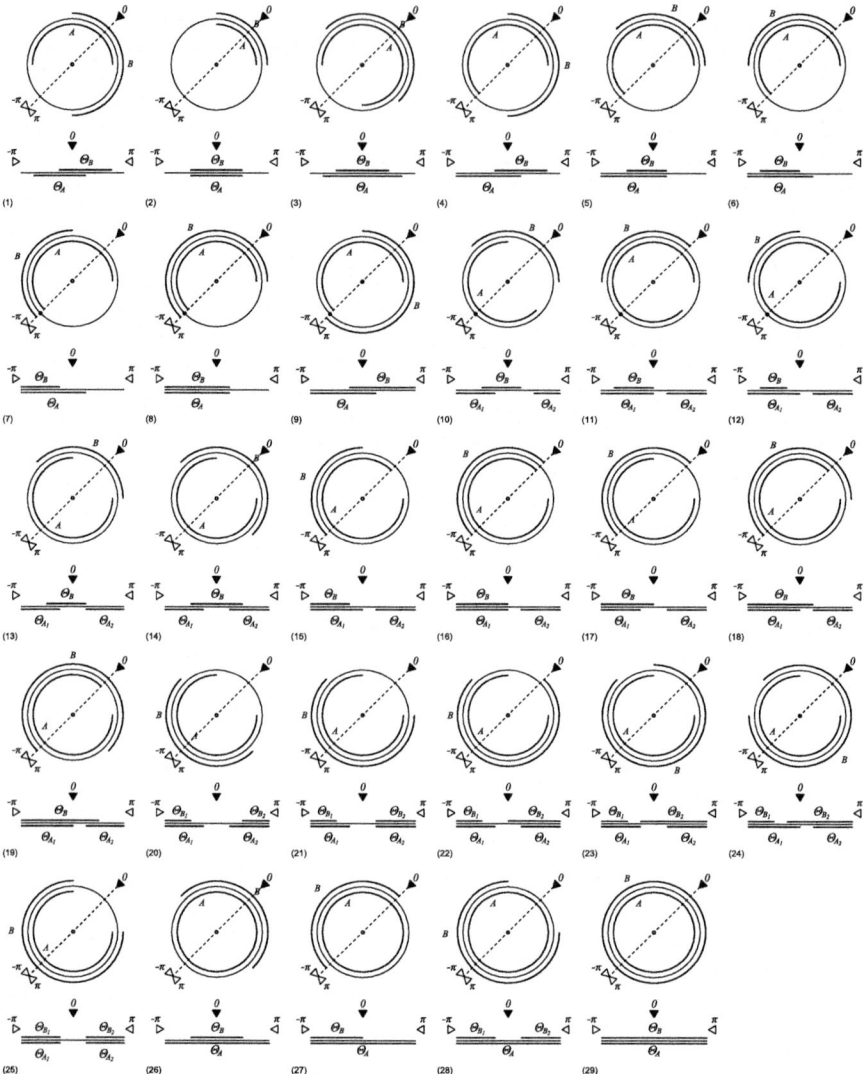

Fig. 12 29 cases of mapping the radian regions to linear domain for examining the first condition that the intersection of two co-circular arcs or circles is not empty

The examination of the second condition is processed with the comparison of the radian regions, which is listed out in Table 3. Hence, the examination is a table-looking-up process.

After the examination of the conditions, both arcs are removed, and a maximal arc is remodeled with the new radian region derived from the linear domain. An example of remodeling a maximal arc under the first condition, that is, two arcs A and B are co-circular, and the intersection of A and B is not empty, is shown in Fig. 14. The

Table 2 29 cases of the first condition

Case	Condition
1	$-\pi < \theta_{A_T} < \theta_{B_T} < \theta_{A_H} < \theta_{B_H} < \pi$
2	$-\pi < \theta_{A_T} = \theta_{B_T} < \theta_{A_H} = \theta_{B_H} < \pi$
3	$-\pi < \theta_{A_T} < \theta_{B_T} < \theta_{B_H} < \theta_{A_H} < \pi$
3	$-\pi = \theta_{A_T} < \theta_{B_T} < \theta_{A_H} < \theta_{B_H} < \pi$
5	$-\pi = \theta_{A_T} < \theta_{B_T} < \theta_{A_H} = \theta_{B_H} < \pi$
6	$-\pi = \theta_{A_T} < \theta_{B_T} < \theta_{B_H} < \theta_{A_H} < \pi$
7	$-\pi = \theta_{A_T} = \theta_{B_T} < \theta_{B_H} < \theta_{A_H} < \pi$
8	$-\pi = \theta_{A_T} = \theta_{B_T} < \theta_{A_H} = \theta_{B_H} < \pi$
9	$-\pi = \theta_{A_T} < \theta_{B_T} < \theta_{A_H} < \theta_{B_H} = \pi$
10	$-\pi = \theta_{A1_T} < \theta_{B_T} < \theta_{A1_H} < \theta_{B_H} < \theta_{A2_T} < \theta_{A2_H} = \pi$
11	$-\pi = \theta_{A1_T} < \theta_{B_T} < \theta_{A1_H} = \theta_{B_H} < \theta_{A2_T} < \theta_{A2_H} = \pi$
12	$-\pi = \theta_{A1_T} < \theta_{B_T} < \theta_{B_H} < \theta_{A1_H} < \theta_{A2_T} < \theta_{A2_H} = \pi$
13	$-\pi = \theta_{A1_T} < \theta_{B_T} < \theta_{A1_H} < \theta_{B_H} = \theta_{A2_T} < \theta_{A2_H} = \pi$
14	$-\pi = \theta_{A1_T} < \theta_{B_T} < \theta_{A1_H} < \theta_{A2_T} < \theta_{B_H} < \theta_{A2_H} = \pi$
15	$-\pi = \theta_{A1_T} = \theta_{B_T} < \theta_{B_H} < \theta_{A1_H} < \theta_{A2_T} < \theta_{A2_H} = \pi$
16	$-\pi = \theta_{A1_T} = \theta_{B_T} < \theta_{B_H} = \theta_{A1_H} < \theta_{A2_T} < \theta_{A2_H} = \pi$
17	$-\pi = \theta_{A1_T} = \theta_{B_T} < \theta_{A1_H} < \theta_{B_H} < \theta_{A2_T} < \theta_{A2_H} = \pi$
18	$-\pi = \theta_{A1_T} = \theta_{B_T} < \theta_{A1_H} < \theta_{B_H} = \theta_{A2_T} < \theta_{A2_H} = \pi$
19	$-\pi = \theta_{A1_T} = \theta_{B_T} < \theta_{A1_H} < \theta_{A2_T} < \theta_{B_H} < \theta_{A2_H} = \pi$
20	$-\pi = \theta_{A1_T} = \theta_{B1_T} < \theta_{B1_H} < \theta_{A1_H} < \theta_{A2_T} < \theta_{B2_T} < \theta_{A2_H} = \theta_{B2_H} = \pi$
21	$-\pi = \theta_{A1_T} = \theta_{B1_T} < \theta_{B1_H} < \theta_{A1_H} < \theta_{A2_T} = \theta_{B2_T} < \theta_{A2_H} = \theta_{B2_H} = \pi$
22	$-\pi = \theta_{A1_T} = \theta_{B1_T} < \theta_{B1_H} < \theta_{A1_H} < \theta_{B2_T} < \theta_{A2_T} < \theta_{A2_H} = \theta_{B2_H} = \pi$
23	$-\pi = \theta_{A1_T} = \theta_{B1_T} < \theta_{B1_H} < \theta_{A1_H} = \theta_{B2_T} < \theta_{A2_T} < \theta_{A2_H} = \theta_{B2_H} = \pi$
24	$-\pi = \theta_{A1_T} = \theta_{B1_T} < \theta_{B1_H} < \theta_{B2_T} < \theta_{A1_H} < \theta_{A2_T} < \theta_{A2_H} = \theta_{B2_H} = \pi$
25	$-\pi = \theta_{A1_T} = \theta_{B1_T} < \theta_{A1_H} = \theta_{B1_H} < \theta_{A2_T} = \theta_{B2_T} < \theta_{A2_H} = \theta_{B2_H} = \pi$
26	$-\pi = \theta_{A_T} < \theta_{B_T} < \theta_{B_H} < \theta_{A_H} = \pi$
27	$-\pi = \theta_{A_T} = \theta_{B_T} < \theta_{B_H} < \theta_{A_H} = \pi$
28	$-\pi = \theta_{A_T} = \theta_{B1_T} < \theta_{B1_H} < \theta_{B2_T} < \theta_{A_H} = \theta_{B2_H} = \pi$
29	$-\pi = \theta_{A_T} = \theta_{B_T} < \theta_{A_H} = \theta_{B_H} = \pi$

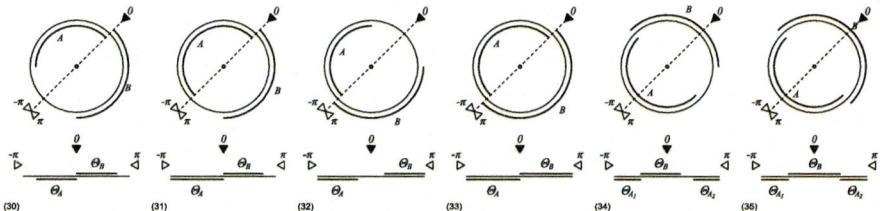

Fig. 13 Six cases of mapping the radian regions to the linear domain for examining the second condition that two co-circular arcs are connected, and the intersection is empty

Table 3 Six cases of the first condition

Case	Condition
30	$-\pi < \theta_{A_T} < \theta_{A_H} = \theta_{B_T} < \theta_{B_H} < \pi$
31	$-\pi = \theta_{A_T} < \theta_{A_H} = \theta_{B_T} < \theta_{B_H} < \pi$
32	$-\pi = \theta_{A_T} < \theta_{A_H} < \theta_{B_T} < \theta_{B_H} = \pi$
33	$-\pi = \theta_{A_T} < \theta_{A_H} = \theta_{B_T} < \theta_{B_H} = \pi$
34	$-\pi = \theta_{A1_T} < \theta_{A1_H} = \theta_{B_T} < \theta_{B_H} < \theta_{A2_T} < \theta_{A2_H} = \pi$
35	$-\pi = \theta_{A1_T} < \theta_{A1_H} = \theta_{B_T} < \theta_{B_H} = \theta_{A2_T} < \theta_{A2_H} = \pi$

remodeling process is consisting of three steps: (1) removing both arcs A and B; (2) deriving new radian region $\Theta_C = [\theta_{C_T} : \theta_{C_H}]$; and (3) instantiating the maximal arc C.

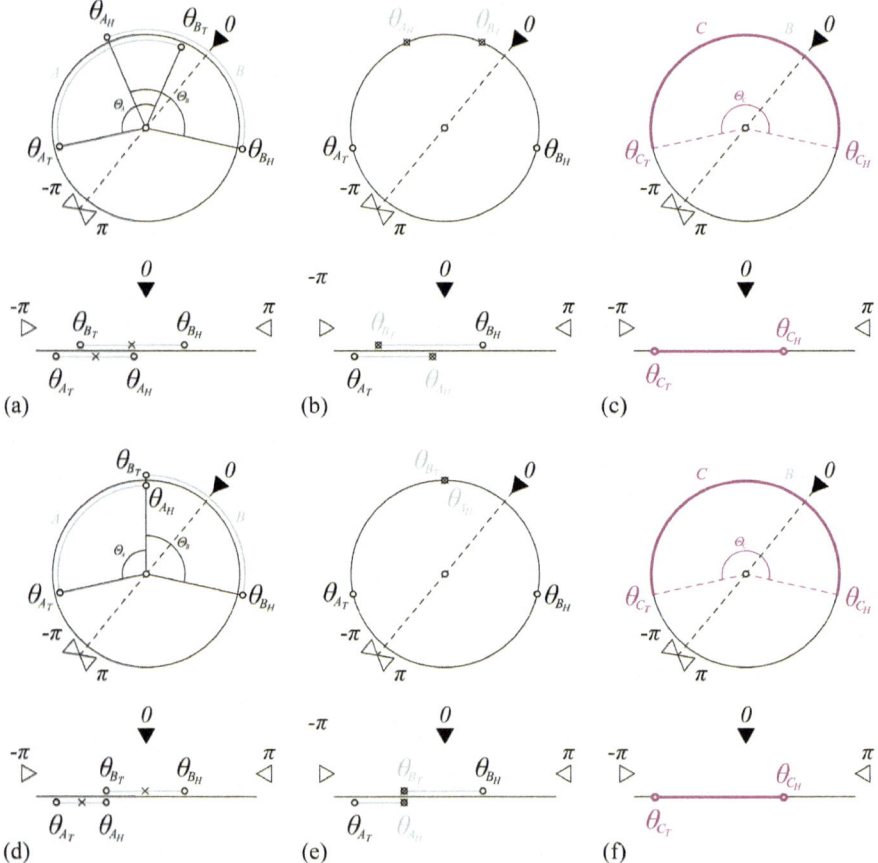

Fig. 14 Two cases of maximization for arcs by mapping the radian regions to the linear domain and examining the conditions that two co-circular arcs are connected

In the example shown in Fig. 14, the new radian region is derived by removing the boundary θ_{A_H} of radian region Θ_A, and the boundary θ_{B_T} of radian region Θ_B. The boundary θ_{A_T} is preserved as the boundary θ_{C_T} of new radian region Θ_C, and the boundary θ_{B_H} is preserved as the boundary θ_{C_H} of new radian region Θ_C. Tables 4 and 5 show respectively the derivation of the new radian regions for the 29 cases under the first condition and the six cases under the second condition.

After derivation of the new radian region, a new arc C is instantiated according to the radian region. If there are two radian regions, an arc C is instantiated by combining the two regions at the boundaries $-\pi$ and π. And, if the radian region is $[-\pi : \pi]$, a circle C is instantiated. Figure 15 and Fig. 16 show respectively the 29 cases of the maximization for two arcs under the first condition and the six cases of the maximization for two arcs under the second condition.

A test was run in the Shape Machine to test the performance of the maximization of arcs and circles. When the precision errors are larger than the tolerances, which are set as 0.1% unit in the CAD system, the calibration is not applied on the arcs. Hence, the two arcs are not maximized. This test is set up by creating a number of

Table 4 29 cases of the maximal radian regions under the first condition

Case	Radian region(s)	Case	Radian region(s)
1	$[\theta_{A_T} : \theta_{B_H}]$	2	$[\theta_{A_T} : \theta_{A_H}]$
3	$[\theta_{A_T} : \theta_{A_H}]$	4	$[\theta_{A_T} : \theta_{B_H}]$
5	$[-\pi : \theta_{A_H}]$	6	$[-\pi : \theta_{A_H}]$
7	$[-\pi : \theta_{A_H}]$	8	$[-\pi : \theta_{A_H}]$
9	$[-\pi : \pi]$	10	$[-\pi : \theta_{B_H}]$ and $[\theta_{A2_T} : \pi]$
11	$[-\pi : \theta_{A1_H}]$ and $[\theta_{A2_T} : \pi]$	12	$[-\pi : \theta_{A1_H}]$ and $[\theta_{A2_T} : \pi]$
13	$[-\pi : \pi]$	14	$[-\pi : \pi]$
15	$[-\pi : \theta_{A1_H}]$ and $[\theta_{A2_T} : \pi]$	16	$[-\pi : \theta_{A1_H}]$ and $[\theta_{A2_T} : \pi]$
17	$[-\pi : \theta_{B_H}]$ and $[\theta_{A2_T} : \pi]$	18	$[-\pi : \pi]$
19	$[-\pi : \pi]$	20	$[-\pi : \theta_{A1_H}]$ and $[\theta_{A2_T} : \pi]$
21	$[-\pi : \theta_{A1_H}]$ and $[\theta_{A2_T} : \pi]$	22	$[-\pi : \theta_{A1_H}]$ and $[\theta_{B2_T} : \pi]$
23	$[-\pi : \pi]$	24	$[-\pi : \pi]$
25	$[-\pi : \theta_{A1_H}]$ and $[\theta_{A2_T} : \pi]$	26	$[-\pi : \pi]$
27	$[-\pi : \pi]$	28	$[-\pi : \pi]$
29	$[-\pi : \pi]$		

Table 5 Six cases of the maximal radian regions under the second condition

Case	Radian region(s)	Case	Radian region(s)
30	$[\theta_{A_T} : \theta_{B_H}]$	31	$[-\pi : \theta_{B_H}]$
32	$[-\pi : \theta_{A_H}]$ and $[\theta_{B_T} : \pi]$	33	$[-\pi : \pi]$
34	$[-\pi : \theta_{B_H}]$ and $[\theta_{A2_T} : \pi]$	35	$[-\pi : \pi]$

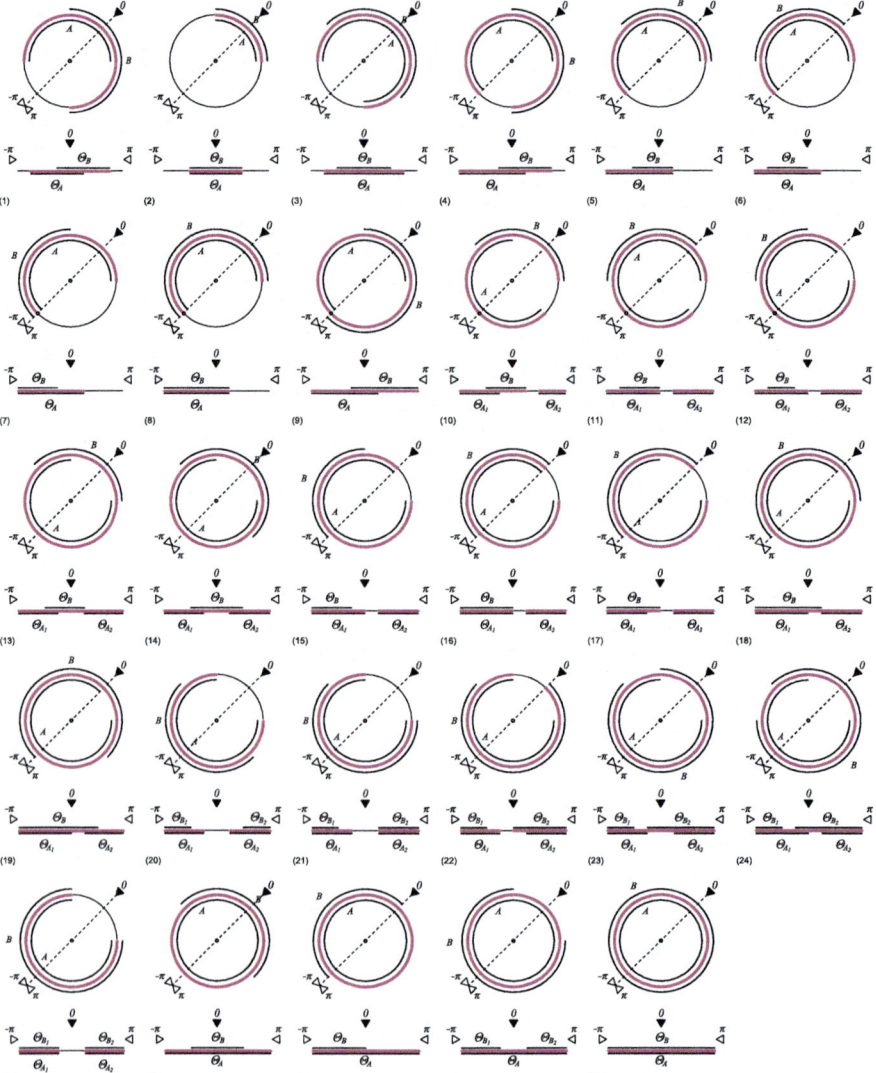

Fig. 15 29 cases of maximization for two arcs under the first condition that the intersection of two co-circular arcs is not empty

concentric arcs and the radius of every arc is slightly longer than the previous arc with 0.01% unit. Calibration and maximization are applied on these arcs and the results are shown in Table 6.

The results show that the calibration fails when the accumulated precision error is larger than the tolerance 0.1% unit. For example, 50 arcs are instantiated with incremental precision error 0.01% unit, therefore, the accumulated precision error

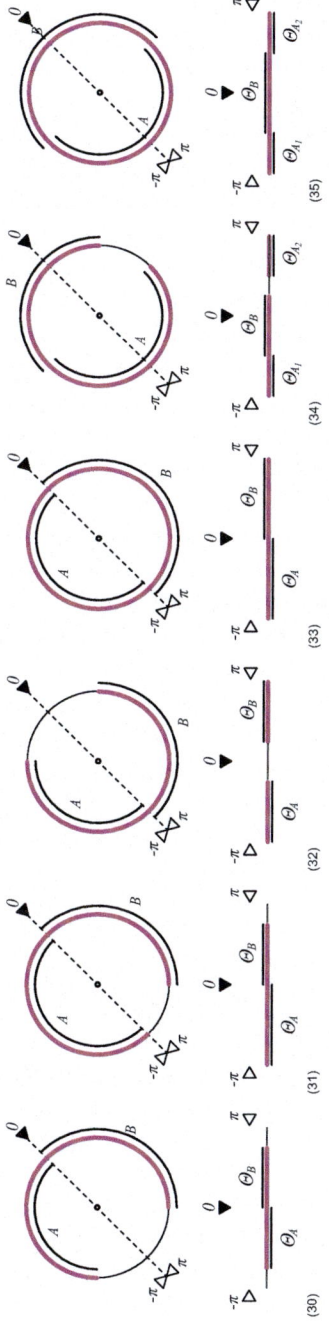

Fig. 16 Six cases of maximization for two arcs under the second condition that two co-circular arcs are connected the intersection is empty

Table 6 Results of the test of calibration and maximization of arcs

# of input arcs	# of maximal arcs derived	Process time (seconds)
10	1	1.775
50	5	3.677
100	10	3.720
200	20	4.013
300	29	4.577
400	39	5.182
500	48	5.816
600	58	7.129
700	67	7.901
800	77	9.121
900	86	10.531
1000	96	12.066

reaches the tolerance 0.1% for every ten arcs instantiated. Hence, these arcs are not maximized into one single maximal arc, instead, there are five maximal arcs after the procedure and there is an accumulated precision error 0.1% unit between every two maximal arcs. However, the cases in this test are extreme since it is rare to have incremental precision errors.

5 Transformation Derivation of Shape Embedding for Shape Consisting of Lines, Arcs, and Circles

The algorithm used in this study to achieve shape embedding for the shapes consisting of lines, arcs and circles is transformation derivation with registration points, which is commonly adopted in the literatures, such as: SGI (Krishnamurti 1982; Krishnamurti and Giraud 1986); SGS (Chase 1989); GRAIL (Krishnamurti 1992); GEdit (Tapia 1999); Shape Machine (Hong and Economou 2023) and so forth. A brief description to this approach is: given a shape u and a shape w, find all transformations fs that make shape u embedded into another shape w. The fs, that are represented as 3×3 matrices, can be reversely calculated by sampling registration points from both shape u and shape w.

5.1 Registration Points of Shapes Consisting of Lines, Arcs, and Circles

Taking the shapes consisting of lines as an example, the registration points of shapes consist of lines are defined as the intersection points of hyperplanes of lines (Hong and Economou 2023; Krishnamurti and Earl 1992). Hyperplane is also well-known as line descriptor and line equation, which is the underlying structure of a line. For the shapes consisting of lines, arcs and circles, the underlying structure of an arc or a circle, circle descriptor (circle equation) is introduced to define the registration points. The registration points here are defined in seven types: (1) a center point of a circle descriptor; (2) intersection points of two circle descriptors; (3) a tangent point of two circle descriptors; (4) an intersection point of two hyperplanes; (5) intersection points of a hyperplane and a circle descriptor; (6) a tangent of a hyperplane and a circle descriptor; (7) a projected point from the center point of a circle descriptor perpendicularly to a hyperplane. Note that the notion of the projected point can be found in the literature (Krishnamurti and Earl 1992). Figure 17 shows the examples of the seven types of registration points of a shape consisting of lines and arcs.

The derivation of registration points for lines and arcs is processed by simply calculating the solutions of line equations and circle equations. For instance, the general form of a line equation of a hyperplane can be described as $ax + by + c = 0$; the general form of a circle equation of a circle descriptor can be described as $x^2 + y^2 + dx + ey + f = 0$, and the coordinates of the intersection points can be derived as (x_1, y_1) and (x_2, y_2) where the solution of x_1 is $\frac{-B+\sqrt{B^2-4AC}}{2A}$; the solution of x_2 is

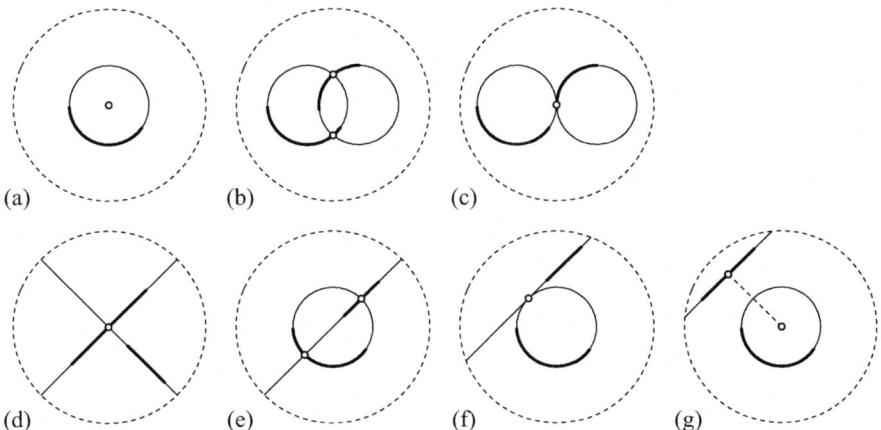

Fig. 17 An example of the seven types of registration points of a shape consisting of lines and arcs. **a** Center point of an arc or a circle; **b** Intersection points of two circle descriptors; **c** Tangent point of two circle descriptors; **d** Intersection of two hyperplanes; **e** Intersection points of a circle descriptor and a hyperplane; **f** Tangent point of a circle descriptor and a hyperplane; **g** Intersection point of the hyperplane and a perpendicular line through the center of an arc

$\frac{-B-\sqrt{B^2-4AC}}{2A}$, and $y_1 = \frac{-(ax_1+c)}{b}$; $y_2 = \frac{-(ax_2+c)}{b}$, where $A = 1 + \frac{a^2}{b^2}$, $B = d + \frac{2ac}{b^2} - \frac{a}{b}$, $C = f + \frac{c^2}{b^a} - \frac{ce}{b}$. Even though seven types of registration points are defined in this study, the shape embedding is processed mainly by sampling the registration points of type (1), (2) and (7).

5.2 Transformation Derivation for Shape Embedding

Like the shape embedding for the shape consisting of lines (Hong and Economou 2023), shape embedding for arcs and circles are processed by deriving the transformation matrices with the registration points. For example, derivation of similar transformation including translation, rotation, reflection and scaling, two registration points from shape u and two registration point from shape w are required to resolve three variables of a similar transformation matrix. However, different from the registration points of the shapes consisting of lines, that are the incident points instantiated by the intersection of hyperplanes (Hong and Economou 2023), the registration points of the shape consisting of arcs and circles are the centers of arcs and circles. Figure 18 shows an example of the shape embedding which is processed under similar transformation derived by two registration points (center points).

The process of shape embedding for the shape consisting of lines, arcs and circles are using both types of registration points, that is, the registration points instantiated by the sections of hyperplanes, and the registration points instantiated by the center or arcs and circles. Note that, the intersections of hyperplanes and circle descriptors do not instantiate registration points in this study, however, the registration points instantiated by the intersection of hyperplanes and circle descriptors can be used for more applications (Hong and Economou, forthcoming). For shape embedding, both types of registration points are adopted to derive the transformations that embed shape u into shape w. Figure 19 shows an example of shape embedding under similar transformation for the shapes consisting of lines, arcs, and circles.

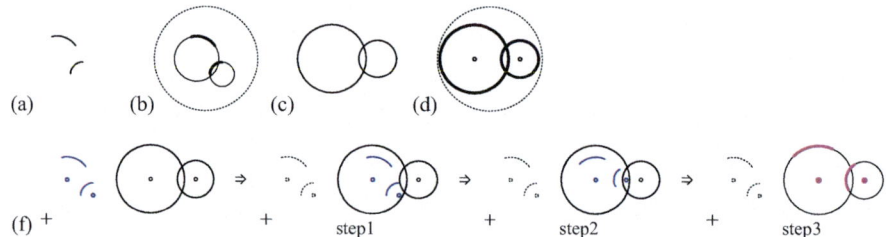

Fig. 18 An example of shape embedding for the shapes consisting of arcs and circles. **a** Shape u; **b** Registration points and circle descriptors of shape u; **c** Shape w; **d** Registration points and circular descriptors of shape w; **e** Process of shape embedding consisting of three steps (step 1, translation; step 2, rotation; step 3, scaling)

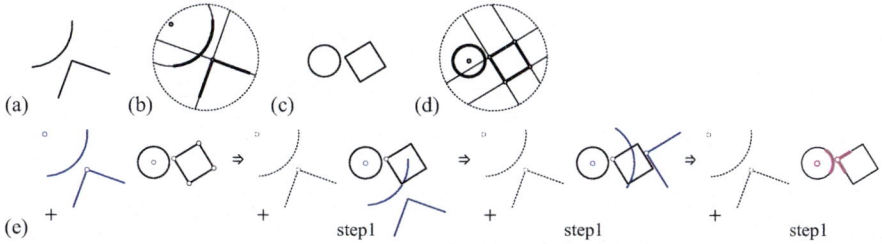

Fig. 19 An example of shape embedding for the shapes consisting of lines, arcs and circles. **a** Shape u; **b** Registration points, hyperplanes and circle descriptors of shape u; **c** Shape w; **d** Registration points, hyperplanes and circle descriptors of shape w; **e** Process of shape embedding consisting of three steps (step 1, translation; step 2, rotation; step 3, scaling)

6 Applications

Here a well-known example, automating Celtic Knots, is given to show that shape embedding for shapes consisting of lines, arcs and circles can be adopted to in design automation. The grammar of Celtic knot is proposed in the literature (Jowers and Earl 2011), which shows that the design language of Celtic knot can be captured by the shape rules, so that the variations of Celtic knots that are generated by the grammars are following the logic. Figure 20 shows an example of automating Celtic knot.

In the example above, there are four shape rules that are applied to automate Celtic knot. The rule production is: (1) rule 1 is applied to grow the knot to the right; (2) rule 2 is applied to grow the knot upward; (3) rule 3 is applied to grow the knot downward; (4) rule 4 is applied to grow the knot on the right-bottom corner; (5) rule 4 is applied to grow the knot on the left-top corner; (6) rule 4 is applied to grow the knot on the left-bottom corner; and (7) rule 4 is applied to grow the knot on the right-up corner. Note that any thread in a Celtic knot follows the logic that is going over and under another thread repeatedly. Figure 21 shows ten variations with three additional shape rules.

In addition to design automation, implementation of arcs and circle in Shape Machine also supports those architectural studies that require curves. For example, a study on the sub-symmetry of a well-known design, Midnight Sun cocktail bar by John Portman (Ligler 2022) demonstrates the potential of the implementation of shape embedding and rewriting for arcs and circles in architecture research, suggesting that arcs and circles can be used as basic elements to rationalize complex curves for further applications. Moreover, by introducing more well-defined curves such as ellipses, elliptical arcs, parabolas, hyperbolas, and so forth, more complex curves might be implemented and used in shape grammar interpreters.

Fig. 20 An example of Celtic knot automation. **a** Initial shape; **b** Final design; **c–f** Shape rule 1–4; **g** Production of the grammar

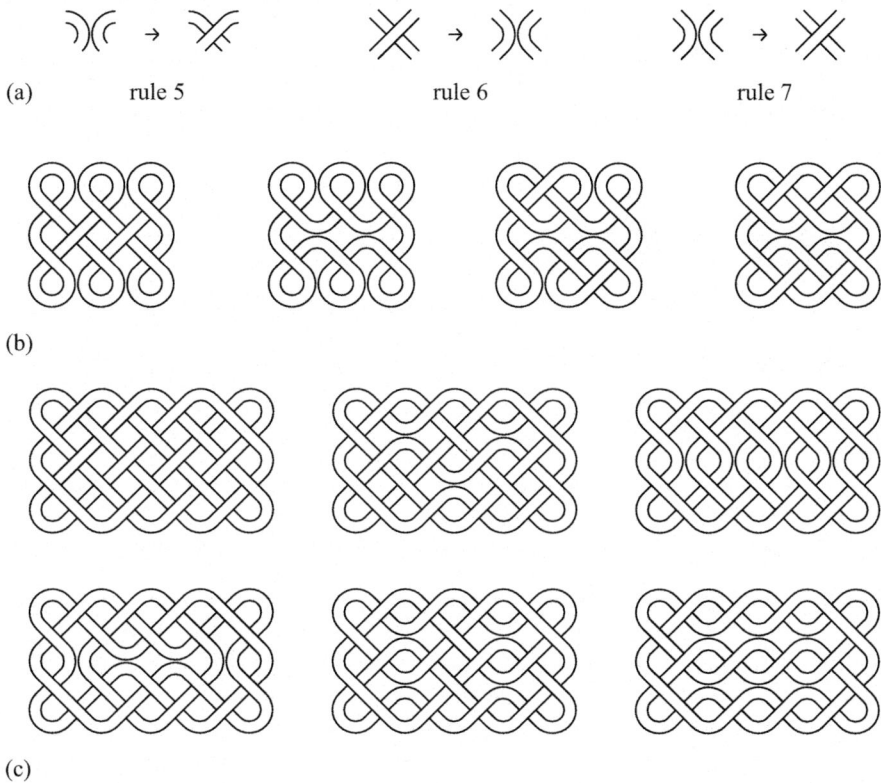

Fig. 21 Three additional shape rules and ten variations of Celtic knots. **a** Shape rules 5, 6 and 7; **b** Four variations of the 3 × 3 Celtic knot; **c** Six variations of the 5 × 3 Celtic knot

7 Discussion

In this study, the algorithm of shape embedding for shapes consisting of lines, arcs and circles is implemented in the Shape Machine to expand the horizon of the geometry types of shape grammar interpreters. The works implemented in Shape Machine can be found in the literatures (Hong and Economou 2023; Ligler 2021, 2022; Okhoya et al. 2022) and demonstrate the advantages of arcs and circles. However, some challenges still need to be resolved for further development.

Firstly, to enable the computation for shapes, it is required to derive the registration points, the boundaries, and the descriptors (hyperplanes) of shapes. However, the derivation process is becoming more complicated while introducing more shape types. For the shapes with higher degree such as planes, the representation of the shapes in CAD system is more complex. For instance, a line is represented in CAD

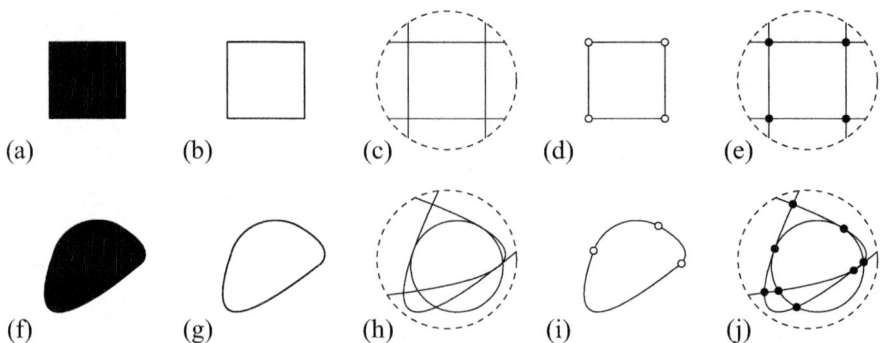

Fig. 22 Two planes with different boundaries. **a** A rectangular plane; **b** four boundary lines of the rectangular plane; **c** four hyperplanes of the four boundary lines; **d** four boundary points of the boundary lines; **e** four registration points of the hyperplanes; **f** a plane with complex boundary; **g** three boundary curves of the complex plane; **h** three hyperplanes (a circle equation, two parabola equations) of the three boundary curves; **g** three boundary points of the three boundary curves; **h** Eight registration points of the three hyperplanes

system with its boundaries, that is, two endpoints. A rectangular plane is represented in CAD system with its boundaries, that is, four lines. And the boundaries are represented with their boundaries, that is, eight endpoints (Earl 1997; Stiny 2006). Thus, shapes with higher degree are represented with a multi-layered structure and that causes more complexity on deriving the registration points, the boundaries and the descriptors. Furthermore, the derivation process is even more complicated when the profile of a plane is not only consisting of lines but also curves. Figure 22 shows two planes with different boundaries.

The plane with complex boundary in Fig. 22 shows that the boundary of the plane is consisting of three curves: an arc and two parabola curves. The hyperplanes of the three boundary curves are: a circle equation with a general representation as $(x-a)^2 + (y-b)^2 = r^2$, two parabola equations with general representation as $ax^2 + bxy + cy^2 + dx + ey + f = 0$. If the plane is in 3D space, the plane equation $ax+by+cz+d = 0$ is involved too. The complexity of the boundary curves and their hyperplanes might cause severe precision errors because of the heavy computation load.

Secondly, the implementation of shape grammar interpreter usually supports one specific shape type. For instance, QI (Jowers and Earl 2011) is an implementation for quadratic Bezier curves; SGIRF (Trescak et al. 2012) supports rectilinear lines; SGI for Curves (McCormack and Cagan 2006) is for piecewise curves, and so forth. However, many shape types are commonly used in CAD systems, thusly each shape type requires a specific implementation to process the computation. The shape types commonly used in the CAD systems are listed in the Table 7 including the shapes in U_{i0}, U_{i1}, U_{i2} and U_{i3}.

Table 7 Shape types commonly used in CAD systems

U_{i0}	U_{i1}	U_{i2}	U_{i3}
Point	Point	Point	Point
	Straight line	Straight line	Straight line
		Arc/Circle	Arc/Circle
		Elliptical arc/Ellipse	Elliptical arc/Ellipse
		Parabola	Parabola curve
		Hyperbola	Hyperbola curve
		Bezier curve	Bezier curve
		Plane	Flat plane
			Spherical surface
			Cylindrical surface
			Conical surface
			Elliptical surface
			Hyperbolical surface
			Gaussian surface
			Spline surface

For every shape type listed in Table 6, the implementation includes maximization, shape embedding, Boolean addition and Boolean subtraction for the specific shape type. That is, 15 different implementations of maximization, shape embedding, Boolean addition and Boolean subtraction for 15 shape types are needed to fully support the applications in design. Moreover, each component requires a calibration system. Hence, the scale of the implementation will be large, and the complexity of the work will be very high as well if each shape type needs to be implemented individually.

Thirdly, the performance of shape embedding might be lower when the complexity of the design increases. The complexity shows up not only in the derivation of the registration points, boundaries and descriptors, but also in the derivation of transformation and the embedding process. Figure 23 shows an example of shape embedding where a shape u consisting of a curve is embedded in the shape W consisting of an enclosed curve under isometric transformation.

The derivation of the transformation requires much complexity; as well, the examination of shape embedding for curves involves much computation since the boundaries of shapes are complicated. It is imaginable that the complexity of the calculation will be much higher in 3-dimensional space.

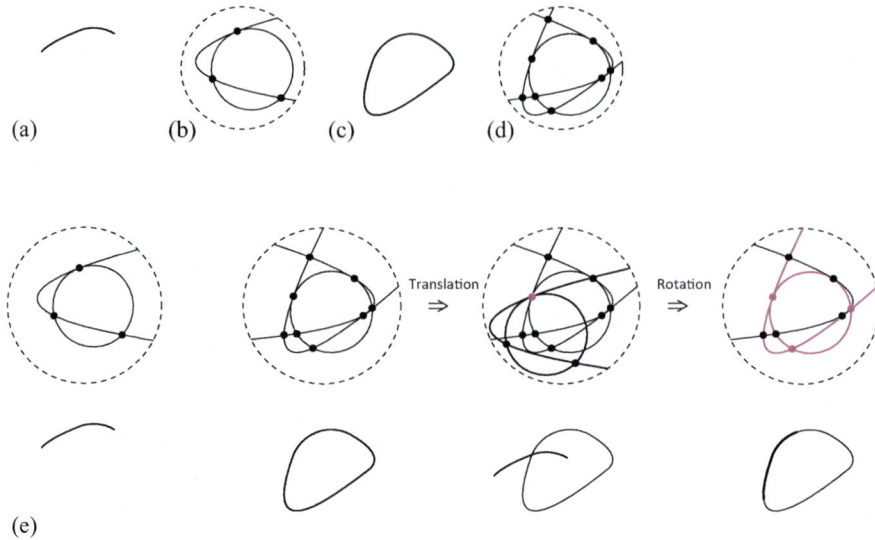

Fig. 23 An example of shape embedding under isometric transformation. **a** Shape u; **b** hyperplanes of u; **c** shape W; **d** hyperplanes of shape W; **e** the embedding process including translation and rotation

8 Conclusion

The major contribution of this study is the calibration mechanism that mitigates the gap between data representation of arcs and circles in CAD systems and the visual perception. The mismatching between data and visual perception is caused by the imperfection of the discrete representation, that is, the limited number of digits a computer represents a real number. A promising method can be found in the literature (Yue and Krishnamurti 2013, 2014) which is adopting tractable representations to avoid real numerical representations. For instance, a number $\sqrt{2}$ is represented with two symbols $\sqrt{}$ and 2, so that the numerical comparison can be altered to the comparison of symbols, thereby avoiding the rounding errors. This method is to eliminate imperfections to obtain the purely perfect data representation. Nevertheless, not all the numbers, note that there are infinite numbers, can be represented with tractable format (Krishnamurti 2015), therefore, the set of tractable numbers is a subset, and it might be short for some applications.

The calibration system, instead of seeking for a perfect standard (error-free data representation) of calculation, is aiming for an alternative standard of calculation, that is, visual perception. Hence, for some cases, the calibration mechanism even adds more errors just to achieve visual equivalency. Hence, the idea of the calibration mechanism proposed here, as well as the calibration for lines implemented in the Shape Machine, is to introduce visual equivalency as the new standard for the shape calculation. From this perspective, the precision error here is not an error from the absolute correctness, instead, it is treated as an error between two things that are supposed to be one thing. That is, neither the thing is absolutely correct, as long as the two things are brought together so that the error is removed, the correctness is achieved.

By bringing these two things together as one, visual equivalency is achieved because the small gap between two things is not perceivable. On the other hand, if the calibration removes the gap which is perceivable, even though the gap is truly an error, it is viewed as a failure. Thus, the setting of the tolerance is critical because visual perception is subjective. For example, if the tolerance is too small, the calibration fails when the precision errors are out of the tolerance but still cannot be perceived. In this case, it fails because the calibration should be applied but it is not. Vice versa, if the tolerance is too big, the calibration fails when the precision errors are within the tolerance, but they can be perceived. In this case, it fails because the calibration should not be applied but it is.

The numerical representation of shapes has been a hurdle since the algebraic structure was imposed on the shapes. This hurdle is even enlarged while hosting the numerical representation of shapes in computer. It is acceptable in most contexts since the resolution is high enough for handling most of the use cases. For example, most CAD models are used for communicating ideas and manufacturing products. However, both scenarios do not require high resolutions of CAD models, even the most precise machineries do not reach the limitation of the resolution. However, calculation relies on a standardized environment where no errors are allowed. The comparison needs to be unambiguous, thus, shape calculation fails when imperfection occurs. In the end, standard is used to determine what the correctness is, and this study is trying to introduce an alternative perspective of view on the standard in visual computation.

References

Chase, S.C. 1989. Shapes and shape grammars: From mathematical model to computer implementation. *Environment and Planning B: Planning and Design* 16 (2): 215–242. https://doi.org/10.1068/b160215.

Chau, H.H. 2004. Evaluation of a 3d shape grammar implementation. In *Design computing and cognition'04*, ed. J.S. Gero, 357–376. Dordrecht: Springer. https://doi.org/10.1007/978-1-4020-2393-4_19.

Dy, B., and R. Stouffs. 2018. *Combining geometries and descriptions a shape grammar plug-in for grasshopper*. https://doi.org/10.52842/conf.ecaade.2018.2.499

Earl, C.F. 1997. Shape boundaries. *Environment and Planning B: Planning and Design* 24 (5): 669–687. https://doi.org/10.1068/b240669.

Economou, A., and T.K. Hong. 2022. Back to the drawing board: Shape calculations in the shape machine. In *Design computing and cognition '21*, ed. J.S. Gero, 549–567. Dordrecht: Springer. https://doi.org/10.1007/978-3-031-20418-0_33

Eloy, S., P. Pauwels, and A. Economou. 2018. Advances in implemented shape grammars: solutions and applications. *AI EDAM Special Issue: Artificial Intelligence for Engineering Design, Analysis and Manufacturing* 32 (2): 131–137. https://doi.org/10.1017/S0890060417000634

Gips, J. 1999. Computer implementation of shape grammars. In *Proceedings of the workshop on shape computation*. Cambridge: Massachusetts Institute of Technology. Accessed 11 Oct 2022. https://www.academia.edu/3089939/Computer_Implementation_of_Shape_Grammars

Grasl, T., and A. Economou. 2018. From shapes to topologies and back: an introduction to a general parametric shape grammar interpreter. *Artificial Intelligence for Engineering Design, Analysis and Manufacturing* 32 (2): 208–224. https://doi.org/10.1017/S0890060417000506

Grasl, T., and A. Economou. 2013. From topologies to shapes: Parametric shape grammars implemented by graphs. *Environment and Planning B: Planning and Design* 40 (5): 905–922. https://doi.org/10.1068/b38156.

Hong, T.K., and A. Economou. 2022a. Five criteria for shape grammar interpreters. In *Design computing and cognition'20*, ed. J.S. Gero, 189–208. Cham: Springer. https://doi.org/10.1007/978-3-030-90625-2_11

Hong, T.K., and A. Economou. 2022b. What shape grammars do that CAD should: The 14 cases of shape embedding. *Artificial Intelligence for Engineering Design, Analysis and Manufacturing* 36: E4. https://doi.org/10.1017/S0890060421000263

Hong, T.K., and A. Economou. 2023. Implementation of shape embedding in CAD systems. *Automation in Construction (AiC)* 146. ISSN 0926–5805. https://doi.org/10.1016/j.autcon.2022.104640

Jowers, I., and C.F. Earl. 2011. Implementation of curved shape grammars. *Environment and Planning B: Planning and Design* 38 (4): 616–635. https://doi.org/10.1068/b36162.

Krishnamurti, R. 1982. SGI: A shape grammar interpreter. In *Centre for configurational studies, design discipline*. The Open University. Accessed 11 Oct 2022. https://www.andrew.cmu.edu/user/ramesh/pub/distribution/technical/SGI.pdf

Krishnamurti, R. 1992. The arithmetic of maximal planes. *Environment and Planning B: Planning and Design* 19: 431–464. https://doi.org/10.1068/b190431.

Krishnamurti, R. 2015. Mulling over shapes, rules and numbers. *Nexus Network Journal* 17 (3): 927–945. https://doi.org/10.1007/s00004-015-0269-1.

Krishnamurti, R., and C.F. Earl. 1992. Shape recognition in three dimensions. *Environment and Planning B: Planning and Design* 19 (5): 585–603. https://doi.org/10.1068/b190585.

Krishnamurti, R., and C. Giraud. 1986. Towards a shape editor: The implementation of a shape generation system. *Environment and Planning B: Planning and Design* 13 (4): 391–404. https://doi.org/10.1068/b130391.

Ligler, H. 2021. Reconfiguring atrium hotels: Generating hybrid designs with visual computations in Shape Machine. *Automation in Construction (AiC)* 132. ISSN 0926–5805. https://doi.org/10.1016/j.autcon.2021.103923

Ligler, H. 2022. The subsymmetry analysis and rule-based synthesis of John Portman's midnight sun. *Nexus Network Journal* 25: 409–437. https://doi.org/10.1007/s00004-022-00644-6

McCormack, J.P., and J. Cagan. 2006. Curve-based shape matching: Supporting designers' hierarchies through parametric shape recognition of arbitrary geometry. *Environment and Planning B: Planning and Design* 33 (4): 523–540. https://doi.org/10.1068/b31150.

McKay, A., S. Chase, K. Shea, and H.H. Chau. 2012. Spatial grammar implementation: From theory to usable software. *Artificial Intelligence for Engineering Design, Analysis and Manufacturing* 26 (2): 143–159. https://doi.org/10.1017/S0890060412000042.

Okhoya, V., M. Bernal, A. Economou, N. Saha, R. Vaivodiss, T.K. Hong, and J. Haymaker. 2022. Generative workplace and space planning in architectural practice. *International Journal of Architectural Computing* 20 (3): 645–672. https://doi.org/10.1177/14780771221120580

Ruiz-Montiel, M., M.V. Belmonte, J. Boned, L. Mandow, E. Millán, A.R. Badillo, and J.L. Pérez. 2014. Layered shape grammars. *Computer-Aided Design* 56: 114–119. https://doi.org/10.1016/j.cad.2014.06.012.

Stiny, G. 1975. *Pictorial and formal aspects of shape and shape grammars*. Birkhäuser Basel: Interdisciplinary Systems Research. ISBN 978-3-7643-0803-2

Stiny, G. 1980. Introduction to shape and shape grammars. *Environment and Planning B, Planning and Design* 7 (3): 343–351. https://doi.org/10.1068/b070343.

Stiny, G. 2006. *Shape: Talking about seeing and doing*. Cambridge: The MIT Press. ISBN 0-262-19531-3
Stouffs, R. 2018 Where associative and rule-based approaches meet. In *Proceedings of computer-aided architectural design research in Asia 2018*, vol. 2, eds. T. Fukuda, W. Huang, P. Janssen, K. Crolla, and S. Alhadidi, 453–462. https://doi.org/10.52842/conf.caadria.2018.2.453
Stouffs, R. 2022. A multi-formalism shape grammar interpreter. In *Computer-aided architectural design. Design imperatives: The future is now. computer-aided architectural design futures 2021. Communications in computer and information science*, vol. 1465, eds. D. Gerber, E. Pantazis, B. Bogosian, A. Nahmad, and C. Miltiadis. Singapore: Springer. https://doi.org/10.1007/978-981-19-1280-1_17
Stouffs, R., and A. Li. 2020. Learning from users and their interaction with a dual-interface shape-grammar implementation. In *Proceedings of computer-aided architectural design research in Asia 2020*. https://doi.org/10.52842/conf.caadria.2020.2.153
Tapia, M. 1999. A visual implementation of a shape grammar system. *Environment and Planning B: Planning and Design* 26 (1): 59–73. https://doi.org/10.1068/b260059.
Trescak, T., M. Esteva, and I. Rodriguez. 2012. A shape grammar interpreter for rectilinear forms. *Computer-Aided Design* 44 (7): 657–670. https://doi.org/10.1016/j.cad.2012.02.009.
Yue, K., and R. Krishnamurti. 2013. Tractable shape grammars. *Environment and Planning B: Planning and Design* 40 (4): 576–594. https://doi.org/10.1068/b38227.
Yue, K., and R. Krishnamurti. 2014. A paradigm for interpreting tractable shape grammars. *Environment and Planning B: Planning and Design* 41 (1): 110–137. https://doi.org/10.1068/b39107.

Open Access This chapter is licensed under the terms of the Creative Commons Attribution-NonCommercial-NoDerivatives 4.0 International License (http://creativecommons.org/licenses/by-nc-nd/4.0/), which permits any noncommercial use, sharing, distribution and reproduction in any medium or format, as long as you give appropriate credit to the original author(s) and the source, provide a link to the Creative Commons license and indicate if you modified the licensed material. You do not have permission under this license to share adapted material derived from this chapter or parts of it.

The images or other third party material in this chapter are included in the chapter's Creative Commons license, unless indicated otherwise in a credit line to the material. If material is not included in the chapter's Creative Commons license and your intended use is not permitted by statutory regulation or exceeds the permitted use, you will need to obtain permission directly from the copyright holder.

Rewriting Shape Rules

Developing Implemented Shape Grammars to See Anew

Heather Ligler

Abstract Over the last half-century, shape grammar formalism has equipped designers with a visual framework for interpreting architectural languages through computation. The canonical shape grammars are valued primarily for their manual encoding of shape rules and subsequent design generation through analog drawing. This approach has captured the intuitive and immediate power of visual computing for design. Still, it has yet to focus equally on the processes of designing and implementing shape rules digitally to further the discourse. This chapter presents insights from designing shape rules and programming with them to initiate discussions on the value of encoding shape grammars to rewrite design narratives that impact our reading of architectural histories and theories over time. The process of rule implementation utilized here, to develop from analog to animate, is achieved by the computer-aided drawing of rules, closing the distance between shape grammar theory and digital mediums for computer-aided grammars. The research engages with the possibilities and current limitations of animating shape grammars with machine feedback to reflect on how an evolving, automated take on designing shape rules can enliven our approach to visual computing in architectural theory, history, and future practices in architecture and design.

1 Introduction

Shape grammars are defined as a type of "non-classical/classical" computation because the formalism relies on visual representation and visual process to carry out a generative design production in a fully visual computation. This is unique compared to prevalent computational practices in design, which are characterized by "classical" symbolic representations increasingly processed in a "non-classical" way that evades clear communication of the computations performed on the backend, behind the user interface we interact with (Knight and Stiny 2001). For example, a state-of-the-art "classical-non/classical" computation that has been increasingly utilized by

---

H. Ligler (✉)
Florida Atlantic University, Ft. Lauderdale, USA
e-mail: hligler@fau.edu

designers since its debut in 2022 is Midjourney, which generates images from natural language descriptions, or prompts, without communicating the process from text to visual outcome. The codependence in shape grammars on both visual representation and visual process offers a complement and alternative to these mainstream tendencies that foregrounds the exposure of visual computations that are valued for their ability to deliver understanding, step-by-step, and to make use of emergent potentials discovered in the design and generation processes. However, how shape grammars are designed or developed, along with their inherent suggestion of constant evolution and reflection, are less studied topics that would benefit from visualization to advance understanding and development of the formalism. This is especially true for grammars that evolve from analog to animate versions, codified under the conventions of available technologies.

The research presented in this chapter aims to advance discussion on the processes of designing and implementing shape grammars (Stiny 2006, 2022). In doing so, it emphasizes the fluidity of the shape grammar formalism and its ability to support constant reinterpretation (Stiny 2015). Knight (1998) emphasizes this quality of shape grammars—that they "allow for multiple and unlimited interpretations of designs and design styles by allowing for unlimited decompositions of designs into shapes and spatial relations between them." In the most straightforward way, this ongoing facility for interpretation and decomposition highlights the notion that shape grammars and their "rules" are not forever fixed, but rather endlessly supportive of multiple, varied, and changing points of view. In this context, each shape rule communicates a particular and momentary perspective—one representative of the designer(s) who created it, as well as their intuitions and/or objectives in designing a grammar. A shape grammar formalizes and extends these perspectives to exercise its range and expressiveness in generating design outcomes, while the sequential nature of shape rule application leaves the process open for pause, surprises, modification, reflection, and/or extension at any interval. The visual basis of the formalism, and this open-ended flexibility, are two essential characteristics of shape grammars that contribute to their unique explanatory and generative power.

In this chapter, these capacities are engaged to discuss how shape grammars enable the animate and alternate representation of design languages as well as their histories, theories, and narratives. As the basis for this discussion, the development of one shape grammar over time and a variety of media—from an analog, manually drawn grammar to a digital, computer-aided grammar—demonstrates the constructive, creative, and critical power of redrawing shape rules and redefining shape grammars in a discontinuous, adaptable process with evolving technologies.

2 Catching Up

The notion of visual algorithms specified with shape rules for the description, interpretation, and evaluation of designs emerged with the invention of the shape grammar formalism. Stiny and Gips introduced shape grammars for generating two-dimensional paintings and three-dimensional sculptures (1972), a project that grew to address how computer models could enliven criticism and design across the arts in a theory of algorithmic aesthetics (1978). The first expansion of shape grammars to address architectural design focused on Andrea Palladio's villas in the Veneto region of Italy, resulting in the Palladian grammar (Stiny and Mitchell 1978). Palladio wrote about the concepts guiding these designs and illustrated them in idealized drawings. The shape rules of the grammar reinterpret Palladio's theory and practice with a visual generation process for producing villa designs using analog shape computations directly drawn with paper and pencil. The designs produced with the grammar include those originally created by Palladio as well as new prototypes, demonstrating how a shape grammar expands our understanding of the architecture of the past through rule-based interpretation and generation. This algorithmic approach to architectural history, theory, and practice inspired another project in the 1990s to produce custom software so that possible Palladian villas could be generated in plans and elevations. The project was inspired by shape grammar theory, and especially the Palladian grammar, yet it was committed to a digital interpretation of Palladian algorithms to test the usefulness of computer programs for architectural study. The details are articulated by the authors, but what is interesting to consider here is how the research motivated Hersey and Freedman to argue for an algorithmic approach to architectural theory that would partner with imperfect implementations to strengthen architectural reasoning in history and theory coursework. More precisely, they argued for the value of studying possible and impossible villas together to sharpen the critical reading of architecture by teaching how to distinguish "good" and "bad" plans or those that are reasonable fakes and those that are unrecognizable as Palladian. This required familiarity with Palladian design principles and the ability to assess them given a generated design. These interactions and lessons were mediated by two computer programs—Planmaker and Façademaker, each of which could generate a range of designs, "good" and "bad," based on an interpretation of Palladian logic (Hersey and Freedman 1992).

These initial strands of computational studies on Palladio share an ambition to reinterpret design narratives, particularly those related to architectural history and theory, with algorithms. The project described here revisits this aim in the context of recent developments in shape computation. More specifically, it focuses on translating from an analog grammar to a digital grammar implemented in Shape Machine, a shape grammar plug-in for Rhinoceros that had an initial prototype available in 2018 (Economou et al. 2020). The Entelechy I grammar (Ligler and Economou 2018) forms the basis for three proto-implementations that will be discussed here as process work, as well as a first implementation that successfully generates families of two-dimensional plans with the Shape Machine software. Uniquely, the evolution of the implementations corresponds with the development of the technology, to reiterate Mitchell's point that the languages of our design tools are not neutral: their data structures limit or expand our design languages too (Mitchell 1973, 1990). The work here aligns with previous efforts to incorporate shape grammar implementation as part of the analysis and creation of architectural designs (Flemming 1987; Duarte 2005). In this way, it continues the argument that rewriting shape rules is yet another way of looking, testing intuitions, and furthering visual algorithms in creative, critical, and constructive ways. Even more, rewriting shape rules provides novel workflows for collaboration to progress the project of shape computation in architectural theory, history, and future practices in architecture and design. In other words, this is also a response to Knight's roughly twenty-five-year-old criticism that we need much "catch-up work in programming" with shape grammars (1998).

3 The Entelechy I Grammar

The analog shape grammar that forms the basis for the work here is the Entelechy I grammar (Ligler and Economou 2018), a shape grammar developed to study John Portman's domestic language, as expressed in his 1964 house, Entelechy I. Portman described the house as a generator for architectural principles that can be adapted to a variety of scales and conditions (Portman and Barnett 1976). This theory is tested in the grammar to evaluate how a single house can be interpreted as a system for designing a family of residential designs with varying parameters. Beyond this initial work, the Entelechy I grammar structures a larger project looking at Portman's myth—the design narrative that relates the house to work at other scales. The broader project employs multiple transformation grammars (Knight 1994) implemented in Shape Machine to relate principles and shape rules from the original house to a selection of Portman's other interior, hospitality, urban, and residential designs (Ligler 2020, 2021, 2022).

A few challenges have been continuously present in the work: first, how to represent the grammar; and second, how to generate designs with the shape rules. The first conference paper on the Entelechy I grammar includes two-dimensional rules that generate designs with analog computations (Ligler and Economou 2015a). The extensive development of that grammar resulted in a parallel two-dimensional and three-dimensional ruleset that also generated designs manually by drafting or modeling the derivations and design outcomes. Due to limitations of space and format in the publication, the three-dimensional rules are included in a supplemental appendix (Ligler and Economou 2018). Concurrent with the development of the Entelechy I grammar, efforts to implement the shape rules in GRAPE (Grasl and Economou 2013) were attempted in both two-dimensional and three-dimensional programs—both of which were partial (Ligler and Economou 2015b). In the end, for clarity in the dissemination of this initial work, these attempts to automate the grammar in a digital environment were paused in favor of analog drawing and modeling. This resulted in a first set of

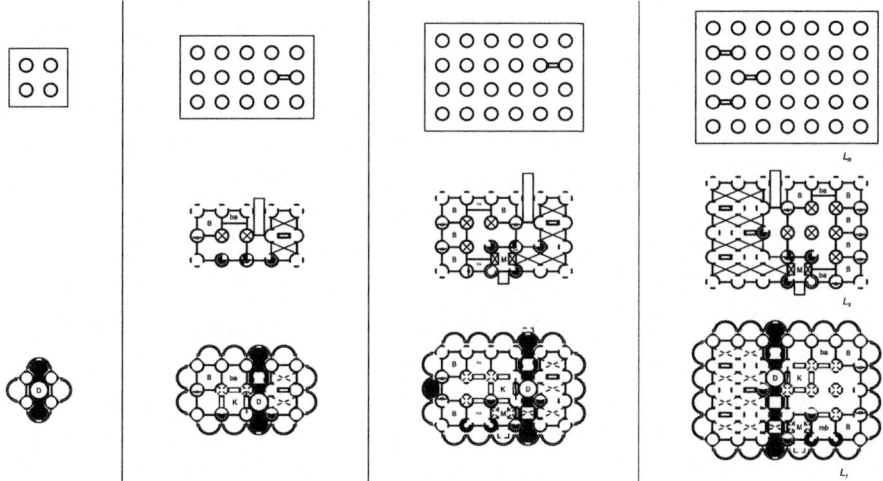

Fig. 1 Four designs produced with the Entelechy I grammar

designs generated with the Entelechy I grammar, including the original design plus three variations that demonstrate the minimum to communicate the viability of the grammar (Fig. 1).

Each of these designs illustrate the efforts of specifying parallel shape rules and productions to coordinate between multiple plans. Looking from left to right, the first design, Entelechy S, the simplest configuration, consists of lower level (L_1) and roof level (L_R) plans to capture the essence of the composition. This design also foregrounds the base unit of the Entelechy system, which combines two spatial modules: (a) a major space to house primary rooms; and (b) four minor spaces structured by hollow or exploded columns that house support spaces at the corners of major spaces. The remaining designs produced with the grammar consist of three plans: lower level (L_1), upper level (L_2), and roof level (L_R). Each configuration increases in size to illustrate incremental variations based on a 2 × 4 grid (Entelechy M), a 3 × 5 grid (Entelechy L, which matches Portman's original design), and a 4 × 6 grid (Entelechy XL). All these designs maintain strict divisions of public and private space defined by the axis of the entry bridge and the subsequent sequence of foyer and double-height spaces on the upper level, which is reiterated on the lower level by the continuous indoor/outdoor water channel shown with a black fill. These correspondences are achieved with shape rules that interpret this division in relation to cross axes of entry and circulation that limit where certain types of spaces and volumes can happen in a configuration.

Upon reflection, this limited interpretation as well as the challenges of working with a complex rule-based design system in a manual way constrained the potential of the grammar. In addition, this self-criticism sympathizes with McCullough's reflection that shape grammars were originally seen as powerful generative systems in computer-aided design research in the 1980s but that their "level of codification was too high" for the architectural and design fields (McCullough, 2006). Fortunately, the initial analog Entelechy I grammar was published just as a working prototype of the Shape Machine plug-in for Rhino (Economou et al., 2020) was available for testing the rules, developing the shape grammar, and advancing the productions in a digital environment.

4 Proto-Implementations

A first proto-implementation of the Entelechy I grammar was completed by programming seven two-dimensional shape rules redefined from the analog ruleset in Shape Machine. Shape Machine is a plug-in for Rhinoceros that enables shape rules to be programmed by drafting two-dimensional rules directly in Rhino. This is achieved using a rule template that can be copied anywhere within the default canvas of the software. Rule application happens using the buttons of the plug-in, which guide the selection of the individual rules and the design(s) to which they will be applied. The shape rules and designs can be arranged freely within the Rhino workspace, providing a flexible design environment. This fluid canvas can be configured to produce multiple design states at once as rules are applied in an interactive mode, allowing users to interact with design alternatives (Woodbury 2021).

At first, the working version of the Shape Machine plug-in was limited by a vocabulary of two-dimensional lines under Euclidean transformations. This resulted in what was colloquially referred to as the "8-bit" implementation of the Entelechy I grammar, since the curvilinear geometries of the plan were simulated with 8-gon, octagonal figures. Figure 2 illustrates the analog shape rules of the Entelechy grammar (left), the implemented "8-bit" shape rules (center), and the derivation of the first production achieved in this proto-implementation (right). The numbers of each shape rule correspond to the shape rules of the original Entelechy I grammar for ease of comparison to the ruleset (Ligler and Economou 2018).

This first draft implementation was focused on proof-of-concept, where the goal was to achieve the lower-level plan of the Entelechy S design produced with the analog shape rules (Fig. 1). Seven shape rules are required to generate this configuration. Rule 1 is the initializing rule, which starts the generation of a design by producing a single major space. After Rule 1 starts the process, Rule 7 adds the geometry of a minor space to the corners of the initial shape. These two rules allow for the generation of the base unit of the Entelechy system to foreground the importance of this underlying compound module throughout the design process. Once this is in place, Rule 18 modifies the geometry of each minor space to include the four parts that define the structural components of a hollow or exploded column. Rule 31 adds an island slab to the central major space of the design. Rule 33 extends pools on two opposite sides of the central island. Rule 52 adds perimeter half-walls around the pools. Finally, Rule 53 adds perimeter half-walls around the other two sides to enclose patios without pools. As illustrated in the derivation on the right in Fig. 2, these steps each contribute to the generation of the Entelechy S variation.

Rewriting Shape Rules

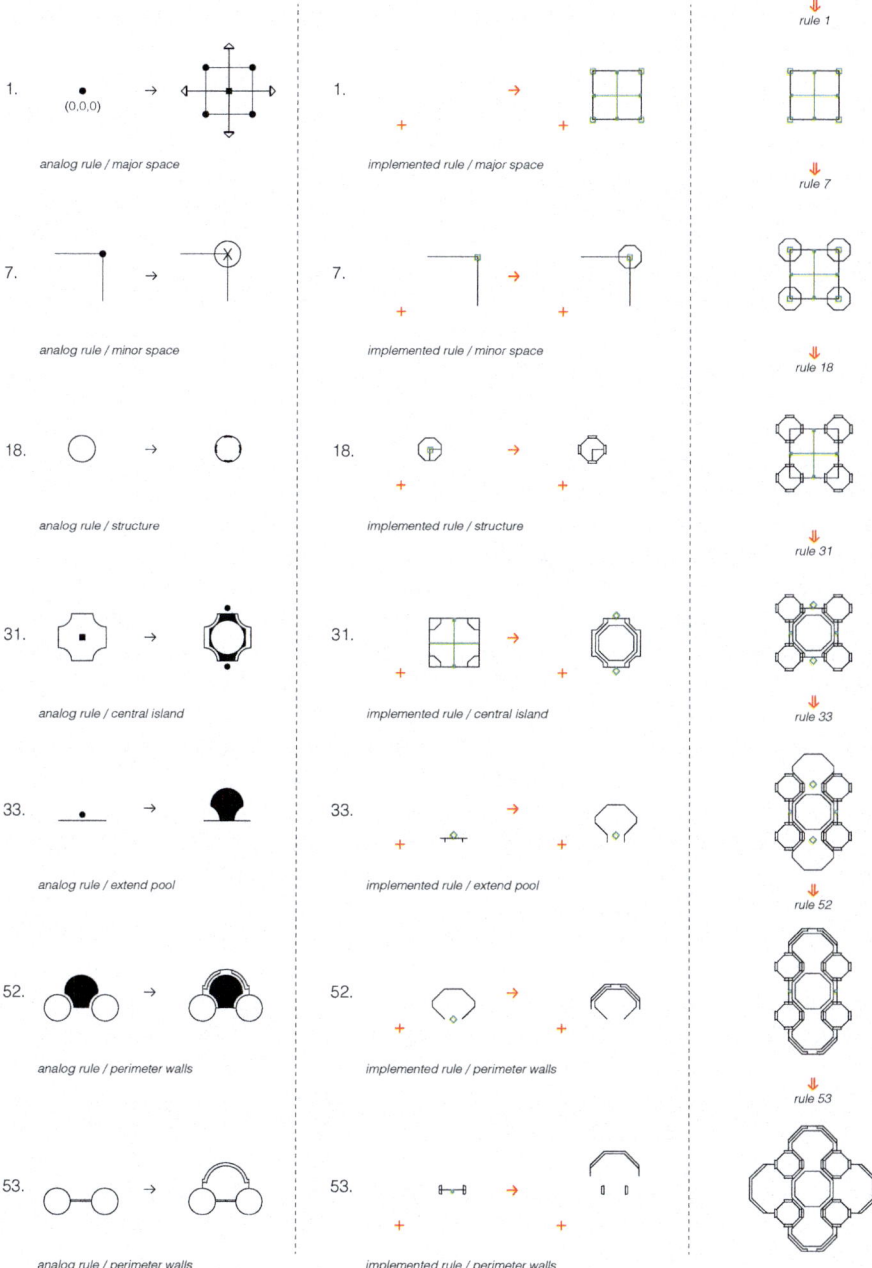

Fig. 2 Proto-implementation 1: analog rules (left), implemented rules (center) and derivation for Entelechy S (right)

In all, the implemented shape rules are quite similar to the original analog ones in this proto-implementation. Besides the geometric conversions, the conventions and details of each rule are modified only minimally, demonstrating how conducive the software is for translating two-dimensional shape rules from analog versions to digital ones. It is worth mentioning here that, although the development of Shape Machine is limited by specific geometries (lines only in the initial version), the software's recognition and transformation algorithms operate on the maximal representation of a given shape vocabulary, as defined in the shape grammar formalism (Stiny 1986; Krishnamurti 1981, 1992). Therefore, the shape rules can be specified with embedded and emergent subshapes and their definitions are non-indexical. In this way, the software allows for the application of rules to match with parts of shapes and to modify those parts, not only their original wholes. This allows for "non-classical/classical" computations in Rhino, characterized by the visual representation of shape rules with geometry and the visual processing of those rules to generate geometric designs (Knight and Stiny 2001). These computations align with the theoretical foundations of shape grammars (Stiny 2006) and allow for translation from analog to digital explorations of the formalism (Ligler and Economou 2019).

A closer look at each rule of the first proto-implementation demonstrates these partial or emergent definitions in the left-hand-side (LHS) of the shape rules given in the central column of Fig. 2 as well as other unique conditions. Specifically, Rule 1 can be applied anywhere in the Rhino workspace to start the design generation process because the LHS of the rule template is left as a blank canvas (or, more precisely, an empty shape). When this rule is applied, the initial major space is generated as a closed square. Rule 7 is subsequently used to operate successfully on the two embedded lines that define each corner of this square to add the minor spaces. Rule 18 then depends on the part of these embedded corner lines that is within the octagon of the minor space to generate the structural components of the hollow columns. The LHS of Rule 31 reverses this to utilize the parts of the octagonal minor spaces that overlap with the square defining the major space to replace these parts with the central island. Rule 33 depends on the two remaining partial sides of the initial square that are marked with the larger labels of Rule 31 to extend the pools. Then, the LHS of Rule 52 specifies part of each extended pool to define the perimeter walls around it. Lastly, Rule 53 looks for the last remaining labeled partial sides of the initial square, as denoted on the LHS of the rule, to complete the definition of the perimeter walls. Additional refinements to the implemented shape rules were made incrementally to speed up the algorithms. In particular, labels drawn with shapes are strategically used to expedite the matching algorithms as the geometry gets more complex in its quantity and arrangement. The extensive use of partial definitions of subshapes in the LHS of the shape rules were also developed to accelerate the processing of each step.

The Entelechy S variation is the first architectural design generated in Shape Machine. This partial implementation of the Entelechy I grammar demonstrates how the drafting of shape rules that depend on embedding and emergence can be directly produced in a computer-aided design environment. The success of this small proof-of-concept led to the implementation of the complete set of rules needed to generate the lower-level plan of the original house. The outcome of this second proto-implementation (the "8-bit extended") is given in Figs. 3 and 4 as proof-of-concept for a more complex design output based on the logic of the original grammar. A selection of these rules is shown in the right column of Fig. 3 alongside their analog versions in the left column. Conceptually, the shape rules in this exploration closely follow the logic of the original grammar. However, these rules also demonstrate where some of the original shape rules required more specificity to work properly in the implementation. For example, the initial shape of the original grammar sets up a pair of compositional axes (see Rule 1 in Fig. 2). Rules 8 and 9 are used to give labels to these axes which specify zones of a house layout based on two key areas: (1) a public entertaining zone, E; and (2) a private family zone, F. In the implemented version, these labels were completely rethought to work in an incremental and graphic way (at the time, the software did not support the use of text or dot labels). The implemented versions of rules 8 and 9, illustrated in Fig. 3, show two different geometric labels that were used instead of the letter labels to achieve this. Following this setup, another example of rule development required for the implementation is found in rule 11. Originally, it allowed for the placement of a zone label within every major space of a design based on the compositional axes. Two rules were developed to achieve this in the implementation, 11a and 11b. These rules populate the graphic labels in a stepwise manner throughout the plan. A third set of rules offer another example for comparison. Rules 20–22 define the generation of three stairs corresponding to the E or F spatial zones. These were also refined for the implementation to follow the graphic labeling conventions developed in this version. Each of these modifications can be observed in more detail through the derivations given in Fig. 4.

Despite these adjustments, this partial implementation is similar to the first proto-implementation in its adherence to the original grammar and can likewise be categorized as an illustration of the existing shape rules edited to work within the constraints of the software. Nevertheless, the ability to generate a more extensive layout in a straightforward way began to inspire more questioning as to how the shape rules could be developed to yield further possibilities. By simultaneously gaining confidence in the logic of the original shape grammar and the logic of the software as a medium for design exploration, the inquiry began to shift towards how to enlarge the interpretation of the original house through more expressive and productive interpretations.

A third proto-implementation began to take this on with more design intention. This implementation is characterized by the "24-bit" configurations produced with 24-gon or icositetragon resolution to substitute for curvilinear geometries in the shape rules. It was in the process of working on editing and refining this version of the implemented grammar that it became evident that it would be beneficial to develop shape rules alongside the development of design variations. Figure 5 gives an example of a landscape of design variations organized and generated within the same workspace in Rhino. With each shape rule application, these outcomes could be studied in a number of design contexts at once to aid in rewriting the rules with additional precision and insight.

This proto-implementation also included experimentation with using parallel shape rules to generate the lower and upper-level plans of the Entelechy I grammar in a

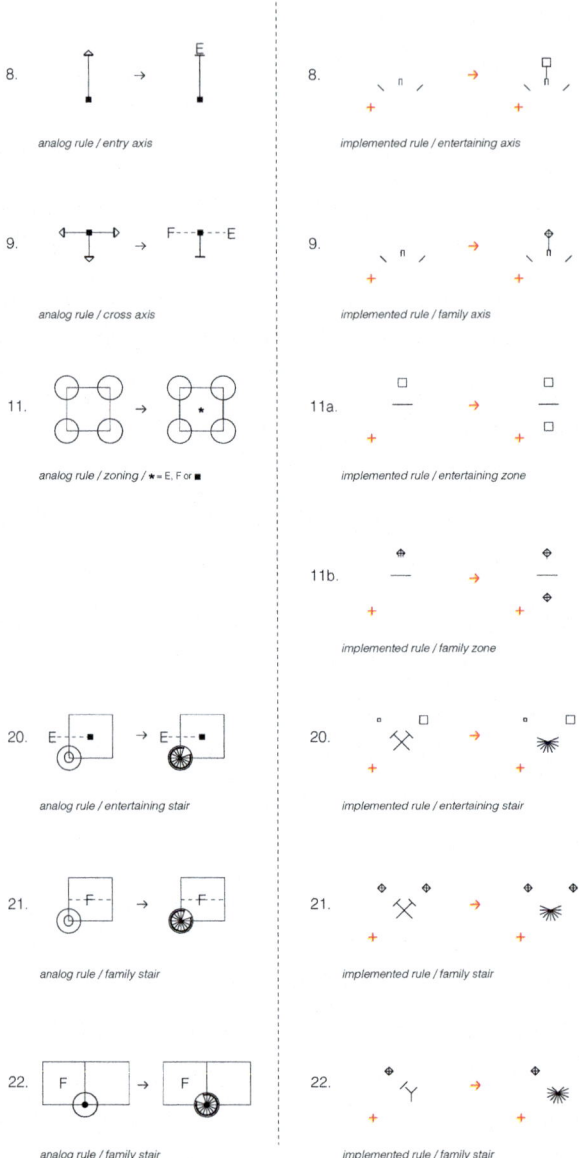

Fig. 3 Proto-implementation 2: analog rules (left) and implemented rules (right)

coordinated sequence. This development of a compound or parallel shape grammar is characterized by parallel rules, where rules operate on more than one view of a design to generate more than one coordinated representation of the design. For example, plans, sections, elevations, and three-dimensional views can be defined in parallel with rules that apply in direct product algebras, like in the shape rules of

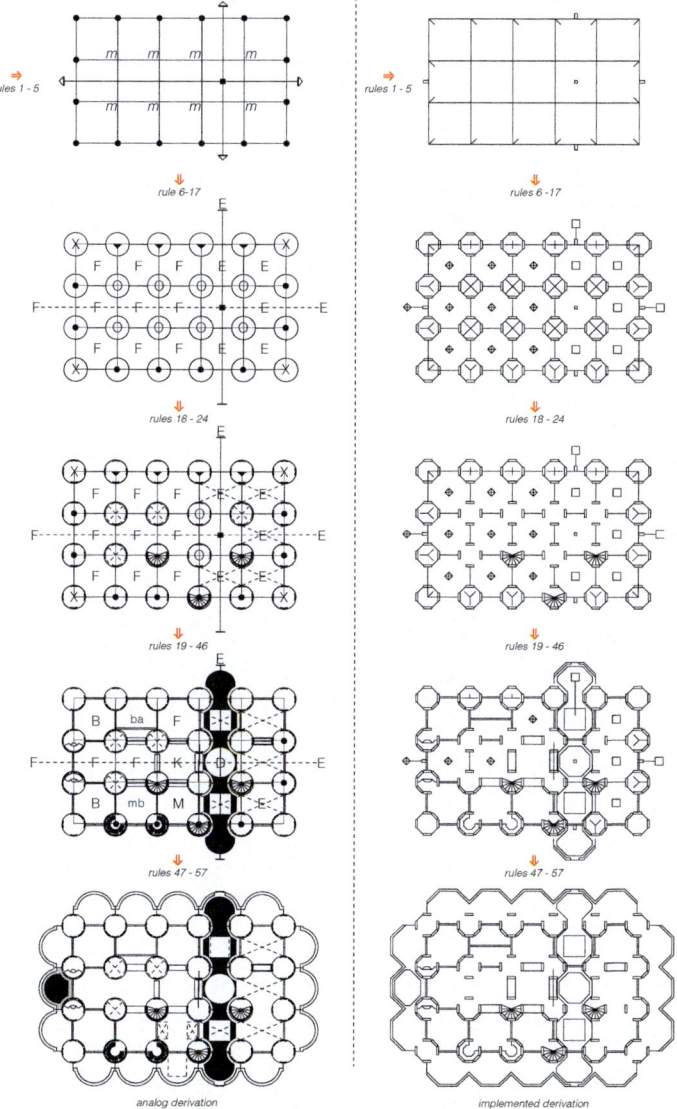

Fig. 4 Proto-implementation 2: analog derivation (left) and implemented derivation (right)

Duarte's Malagueira grammar (2005) and those of the original Entelechy I grammar that produced multiple plans and three-dimensional representations (Ligler and Economou 2018). In this "24-bit" parallel implementation, two floor plans, upper and lower, are defined in the Cartesian product of the algebras U_{12} and V_{02}. Together, they provide a pair of two-dimensional representations of the three-dimensional spatial relations of the house designs. The application of these rules entails defining

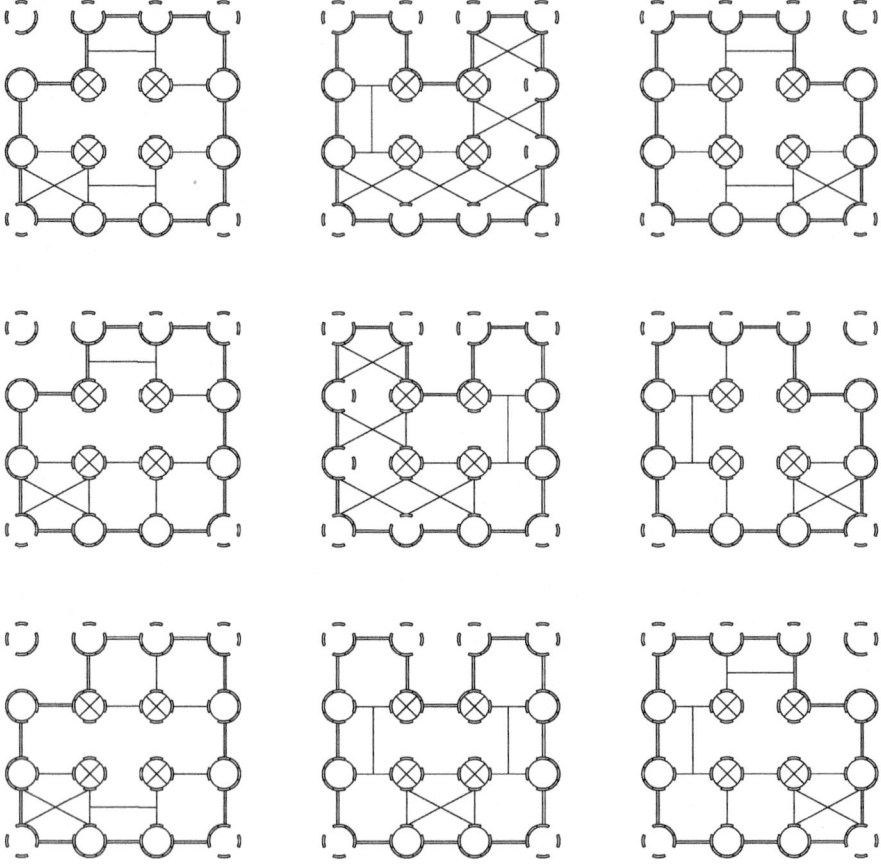

Fig. 5 Proto-implementation 3: a sample workspace for testing rules

a workspace in Rhino that includes a view for each representation of the compound grammar. To apply each parallel rule, partial selection was used to select each part of the rule, followed by its corresponding plan representation to apply it in the proper context.

Figure 6 illustrates five compound shape rules (right) compared to their basis in the analog rules of the original grammar (left). These rules are selected to visualize both their parallel representation and their evolution in comparison to the "8-bit" and "8-bit extended" implementations. For example, the evolution of Rule 7 highlights changes in the rule sequence as well as revisions that locate minor spaces and label them graphically, rather than symbolically. More specifically, the upper-level part of Rule 7 finds a minor space at a cross-intersection of the grid and labels it as an interior space by rotating that cross-intersection forty-five degrees to achieve a graphic commonly used to denote a double-height space opening (an open-to-below condition) on an architectural floor plan. Then, the lower-level part of Rule 7 removes

Rewriting Shape Rules

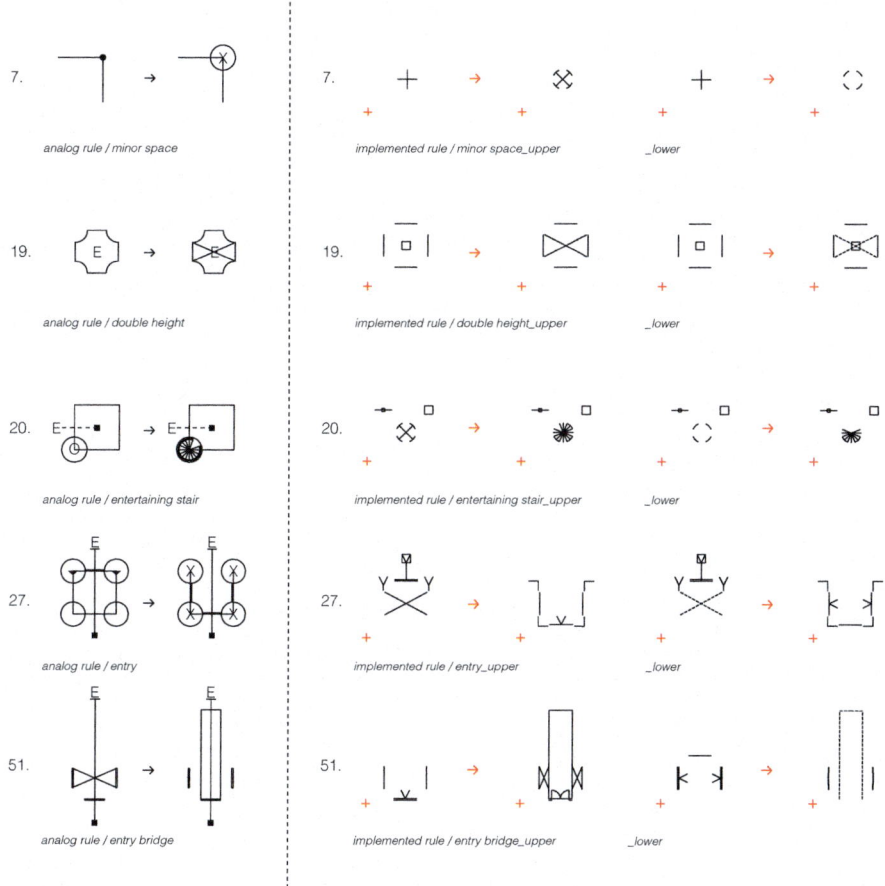

Fig. 6 Proto-implementation 3: analog rules (left) and implemented rules (right)

the cross-intersection and replaces it only with the four arc parts needed to complete the definition of an open circular space in the floor plan. Rules 19 and 20 are given here to build on their previous development in their "8-bit extended" versions, shown in Figs. 3 and 4. In the parallel implementation, the placement of double-height spaces and staircases is defined by specifying graphic conventions that distinguish how they work for two plans, upper and lower. Lastly, rules 27 and 51 are shown to capture the articulation of the recessed entry and entry bridge in parallel rules that build on the graphic labels for the zoning features developed in the "8-bit extended" version. This entry sequence became more and more important in the grammar as it emerged as a new starting point redefining how circulation and spatial zoning worked with the shape rules. The shift in this logic is based on incremental movement from one major space to another rather than the initial zoning structured by the compositional axes of the analog grammar. To see these revisions more clearly, Fig. 7 provides four steps

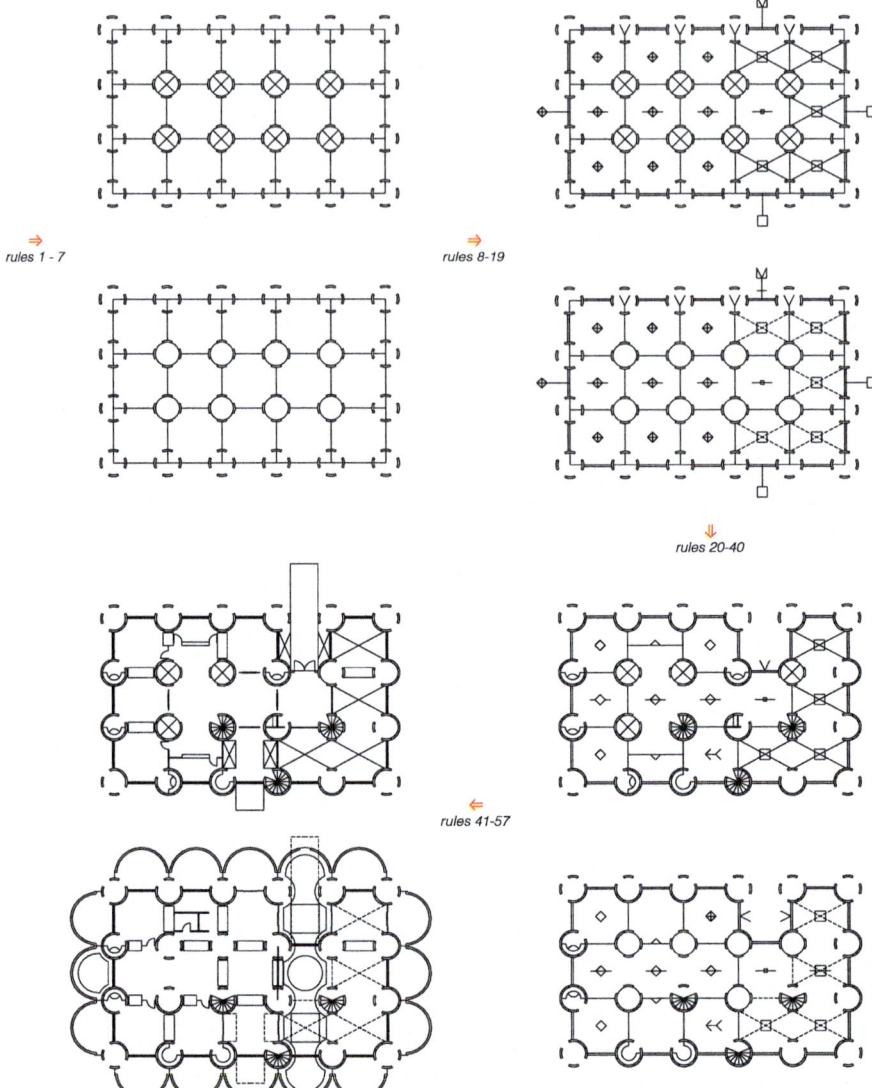

Fig. 7 Proto-implementation 3: a derivation illustrating the generation of both floor plans of Entelechy I in parallel

of the derivation to generate the "24-bit" upper and lower-level plans that match the original configuration of Entelechy I. In these steps, the rule modifications given as examples are shown in their context of application alongside the overall development of each floor plan. The first step corresponds to the development of rules 1–7, the second step to rules 8–19, the third step to rules 20–40, and the fourth step to rules 41–57.

As expected, the icositetragon resolution is deceptive visually so that the absence of the characteristic curves of the original design is less recognizable in the final design outcomes. However, according to Rhino, the lower-level plan alone is composed of 2,282 geometric parts. This complexity slowed down the production process in Shape Machine as each progressive design state would add to the sheer number of geometries that needed to be matched and/or replaced when applying rules with the plug-in. A strategic workaround used in these implementations to speed up the design generation process is partial selection in the design space to effectively limit the scope of rule application, meaning that when a rule is applied the designer selects the part of the design where they want to apply the rule, instead of the entire geometry of a given design state. This is an especially resourceful move for testing rules as they are edited and refined.

The three proto-implementations are proof-of-concept for the logic of the Entelechy I grammar and demonstrate how the concurrent refinements of the software impact the development of the implemented shape grammar and vice versa. However, these three versions are characterized as proto-implementations due to their partial, experimental nature as well as the technical challenges of simulating the curvilinear geometry, which significantly slowed down the process of design generation and stepwise refinement. Still, some highlights in these design processes developed the grammar in ways that began to enlarge its potential. First, the development of visual labels gave more elegance, specificity, and incremental adjustment to the grammar. This is primarily seen in how the labeling of the E and F zones is translated in Figs. 3 and 4 as well as how this impacts the definition of the entry sequence as shown in Figs. 6 and 7. Second, the development of the workspace as a place for testing rules incrementally and in comparable variations as illustrated in Fig. 5 began to provide new ways of studying the impact of particular rules and rule sequences, which also led to experiments implementing parallel rules to coordinate multiple plans. While the individual rules provide hints of these changes, comparing the derivations of Figs. 4 and 7 captures their overall impact more effectively. In all, these developments provided a context for seeing more potential in each move, spatial relationship, and specification of the grammar. The software provided a medium for testing, recording, editing, and enlivening these findings. This foundational process work paves the way for a further evolution of the Entelechy I grammar, where the goal was not only implementation but to thoroughly improve the range and quality of design variations by rewriting shape rules to expand the narrative on the house as a potential design system.

5 Computer-Aided Grammars

A first implementation that extended the expressiveness of the Entelechy I grammar was completed by programming two-dimensional shape rules developed and revised from the proto-implementation rulesets. These shape rules were rewritten in an updated and extended version of the Shape Machine for Rhino plug-in to accommodate computations with both lines and arcs (Hong 2021; Hong and Economou 2023). This extension of the technology alone simplified the complexity of the computations. For example, the number of geometric elements in the lower-level plan was reduced by over fifty percent (from 2,282 parts in the "24-bit" proto-implementation to 1,124 parts in the implementation using lines and arcs). The process of implementation was the same as before, involving translation from the previous rulesets, however, the technical improvements in the plug-in benefitted the process tremendously and allowed for extended experimentation with individual rules in an incremental feedback loop. The quick interactions with the interpreter were also more strategically applied to larger families of designs. Figure 8 illustrates a workspace setup in the Rhino canvas to test shape rules for a range of fifteen underlying grid sizes. The grid sizes vary incrementally from a 2 × 2 to a 4 × 6, and each size is duplicated so that parallel plans are studied for each spatial constraint. The resulting implementation achieved a variety of design configurations represented in pairs of two-dimensional plans. The ability to expand the outcomes of the grammar so dramatically was a direct result of having a technology for implementation that aligns with shape grammar theory. This allowed for the programming of each shape rule to proceed as an intuitive and reflective process of design and development. The ease of programming by drafting rules also significantly enhanced the ability to be critical and imaginative in advancing the grammar.

Specifically, the previous developments in the proto-implementations, especially those of using visual zoning labels and the initial interest in developing the entry sequence, gained further definition so that new configurations were achieved. The major modifications of the shape rules that evolved in this process were primarily related to reconceiving the primary circulation paths to open more potentials in the language. For example, a recessed entry could be located in any major space on the entry side of the major grid, including corner conditions. In addition, constraints to the arrangement of double-height spaces were loosened to allow for volumes to be carved adjacent to any major circulation hall in a design. These revisions will be discussed in detail in this section to elaborate on how they work in the implementation.

The back-and-forth process of editing and revising the rules also enlarged the narrative on how the original Entelechy I could be interpreted as a generator of broad design principles too as the dual framework of minor and major spaces provides a flexibility for volumetric play that aligns with the notorious features of Portman's broader corpus. In turn, this led to the redescription of the design variations as Portm-Inoes to characterize Entelechy I as a postmodern reinvention of Le Corbusier's Dom-Ino housing system (Ligler 2020; Le Corbusier 1970). The Dom-Ino system materializes Corbusier's five points of architecture—the pilotis, the free ground plan, the free façade, the horizontal window, and the roof garden—as the locus of domestic (*dom*) innovation (*ino*). The generation of Portm-Inoes likewise articulates how these elements can be reinterpreted in a postmodern context as the exploded columns, the porous floor plan, the undulating façade, the skylight matrix, and the indoor/outdoor water garden, respectively. First, the exploded, hollow columns of the Portm-Ino celebrate structure as space where the pilotis of the Dom-Ino minimized structure.

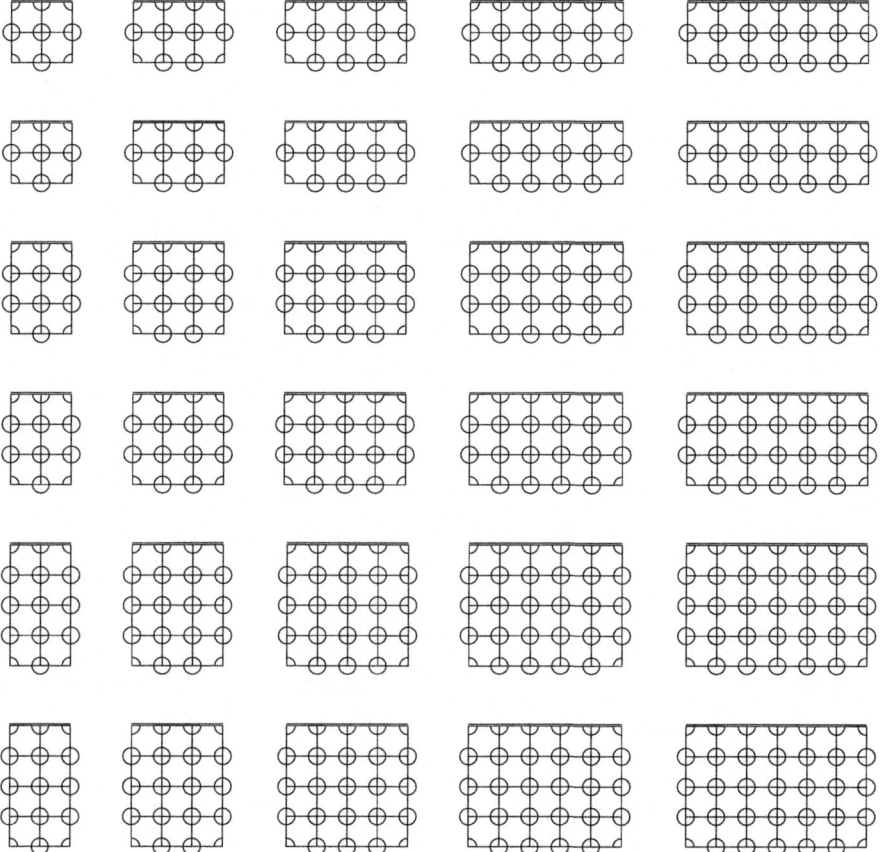

Fig. 8 Portm-ino implementation: a sample workspace for testing rules

Second, the porous floor plan of the Portm-Ino reinterprets the free ground plan of the Dom-Ino as a punctured surface that can connect spaces horizontally and vertically through a number of means. Third, the undulating façade of the Portm-Ino integrates the façade and exploded column to hybridize structure, space, and enclosure—a move that challenges the definition of interior and exterior to contrast the free façade of the Dom-Ino. Fourth, the skylight matrix infuses natural light into the depths of Portm-Ino interiors at repeated intervals customized to the human scale in counter distinction to the ubiquitous horizontal window of the Dom-Ino. Fifth, and lastly, the indoor/outdoor water garden blurs the boundaries between the architecture and the site, promoting the merging of environments as opposed to the separation and elevation of the roof garden in the Dom-Ino.

Together, these five points and their related principles aid in reconceptualizing Entelechy I, systematically, to expand the narrative on this design and its value in Portman's corpus, while the generation of families of Portm-Inoes give flesh and bones to these points and their potential. Figure 9 provides a sample of four Portm-Inoes generated with parallel rules. Each design includes both upper and lower-level plans. The variation in the lower right corner corresponds to the original design of the house. The other three variations illustrate how the development of the grammar led to more productive results that engage a wider range of design potentials including corner entries, mixed zones of use, and the dispersion of double-height spaces to engage volumetric play. It was through the generation of this set of parallel plans that it became evident that there were some drawbacks to the parallel productions. Mainly, explaining the parallel process of design generation with implemented, parallel rules presented new challenges in representing the logic of the grammar clearly and concisely. At the same time, in the process of developing the grammar to focus on the entry sequence and circulation, the upper-level plan emerged as the most important one that controlled the overall logic of each variation. This realization of the significance of the upper-level as a piano nobile motivated a simplification of the work for the dissemination and presentation of the implemented grammar to focus on the generation of the upper-level exclusively.

Figure 10 presents five shape rules reworked for the Portm-Ino Implementation (right), compared to their basis in two of the analog rules of the original grammar (left). These rules are selected to relate to the previous discussions of rule modifications in the proto-implementations, especially those in Fig. 6. The reordering and expansion of the logic of these rules is reflected in their renumbering, which corresponds to their sequencing in the implementation. Implemented rules 12–16 were rewritten based on the concepts behind analog rule 27 to define the recessed entry. Rule 12 specifies the position of the entry within any major space on the front side of a composition (always towards the top of the workspace or page in the implementation) rather than in relation to an axis predetermined by the initial shape. Some consequences of this revision can be seen in the first row of Fig. 11. This row illustrates two of the five potential locations for the application of this rule. The red color represents the matching of the LHS of the rule and the blue color represents the preview of the RHS application of that rule. These calculations and their visualizations in the figure simulate the interactive feedback of Shape Machine.

After Rule 12, rules 13 and 15 define the entry foyer and a double-height space related to it, respectively. These three rules characterize a complete entry sequence in the Portm-Ino Implementation. Rule 16 is then applied recursively to open up a circulation hall off of the entry. Rule 20 is also emphasized here to demonstrate how the exploration of volumes is more playfully achieved in the implementation as compared to its original definition as Rule 19 in the analog grammar. The complete set of potential applications of this rule in the context of the 3×5 grid of Entelechy I is given in the second, third, and fourth rows of Fig. 11. To further express these revisions in this context, Fig. 12 provides the derivation of the upper-level plan of Entelechy I based on eight key steps in the Portm-Ino Implementation. The derivation expresses more drastically the contrast in this implementation compared to those shown in Figs. 4 and 7. In particular, the labeling of the spatial zones has been eliminated in favor of subtle graphic labels and the compositional development of spaces is based on their relation to the entry and circulation paths.

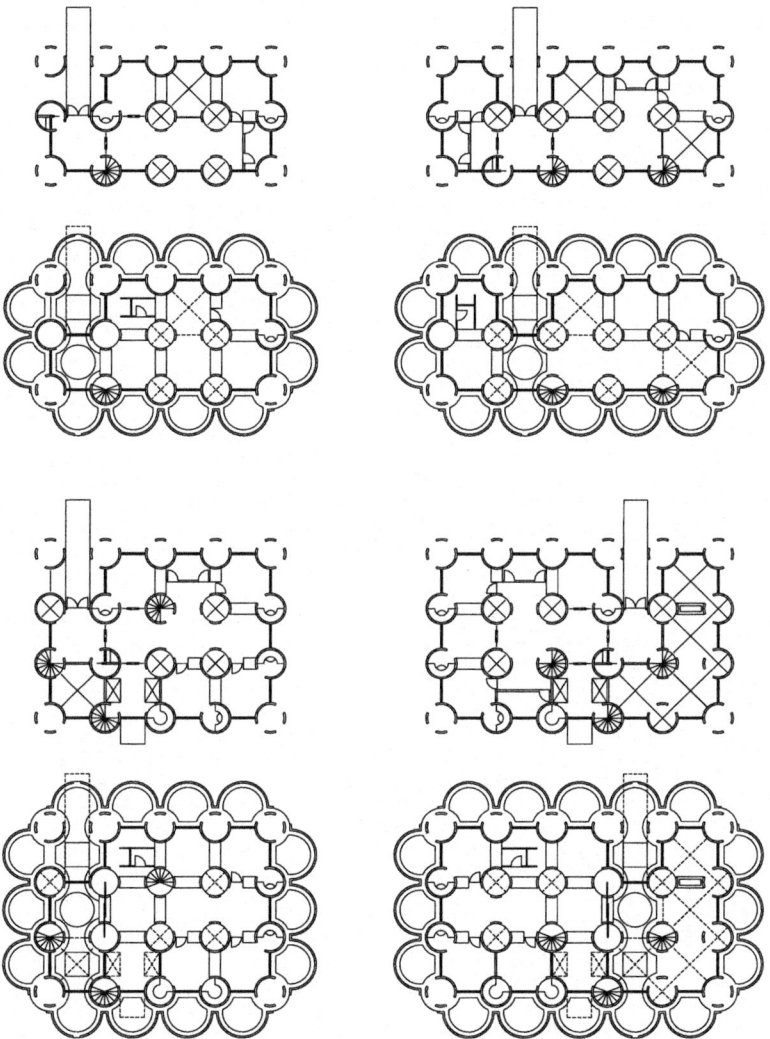

Fig. 9 Portm-ino implementation: four designs generated with parallel rules

To elaborate on these refinements and their larger consequences with more immediacy, a sample of twelve Portm-Ino design variations produced with the complete set of rewritten, implemented rules are given in Fig. 13. These variations include the four shown with parallel plans in Fig. 9 for ease of comparison. The designs materialize the potential of the Portm-Ino as a design system that yields house variations ranging from small to large, from closed to open, from symmetrical to asymmetrical, and more. More specifically, these Portm-Inoes feature circulation halls that are crafted as bars, elongated T-shapes, squat T-shapes, U-shapes, L-shapes, and cross-shapes. The spaces around these halls also have varying levels of porosity characterized by

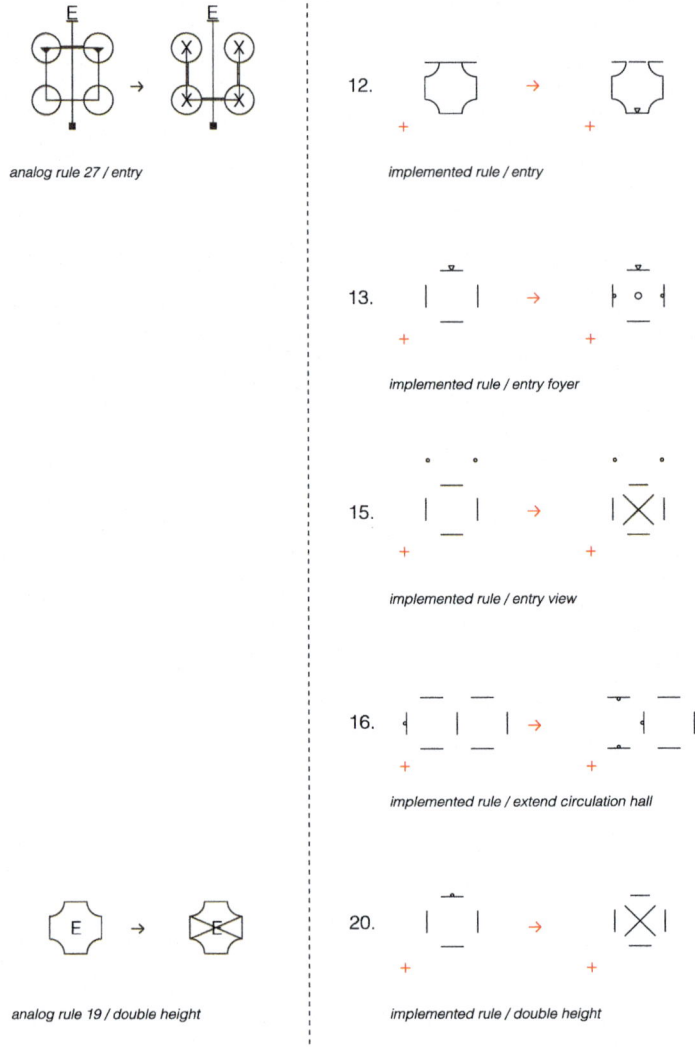

Fig. 10 Portm-ino implementation: analog rules (left) and implemented rules (right)

the number of double-height spaces developed in each design that provide spatial, visual, and site-specific connections within a house layout.

In the context of these variations, the potential of computer-aided grammars comes forward as a critical, creative, and constructive method for redescribing and expanding research in shape grammars. Computer-aided grammars are digitally implemented shape grammars written in environments that support embedding and emergence. Shape Machine for Rhino offers one such environment for designers to evaluate these possibilities. The interpreter allows a shape grammar to be represented

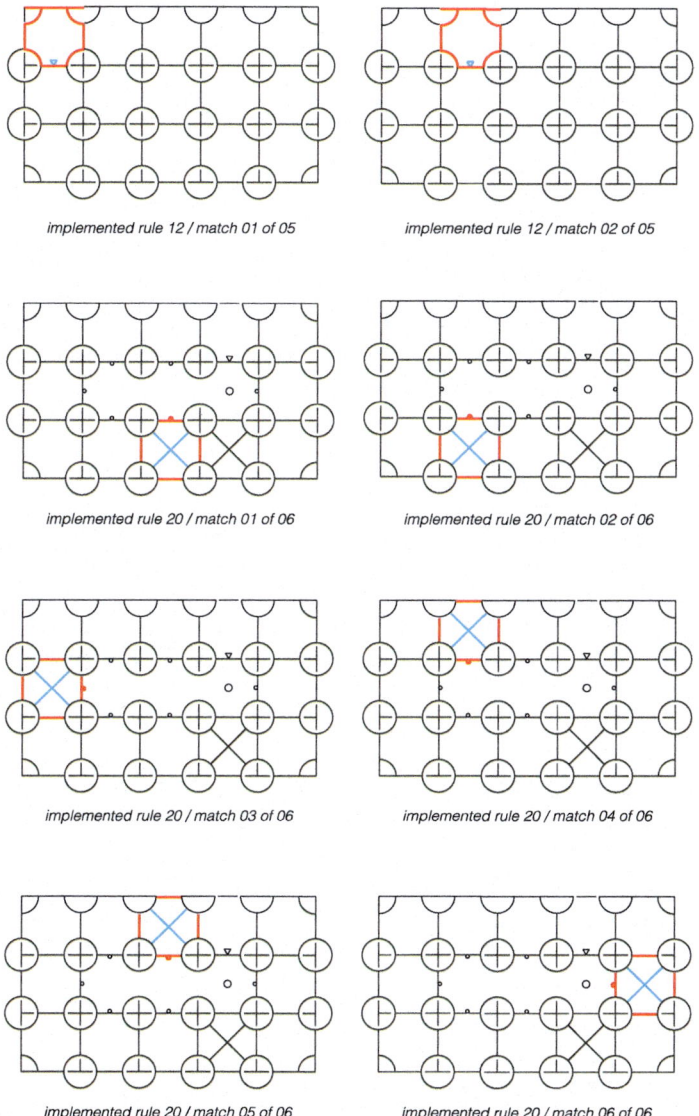

Fig. 11 Portm-ino implementation: sample rule matches in shape machine

by drawing shapes in visual rules to specify formal relationships and to visually automate production processes to generate designs without requiring users to interface directly with the background script. The only programming required of the grammar designer is authored in a visual language by drawing shapes directly in the Rhino workspace. This flexibility permits the development of novel and intuitive workflows

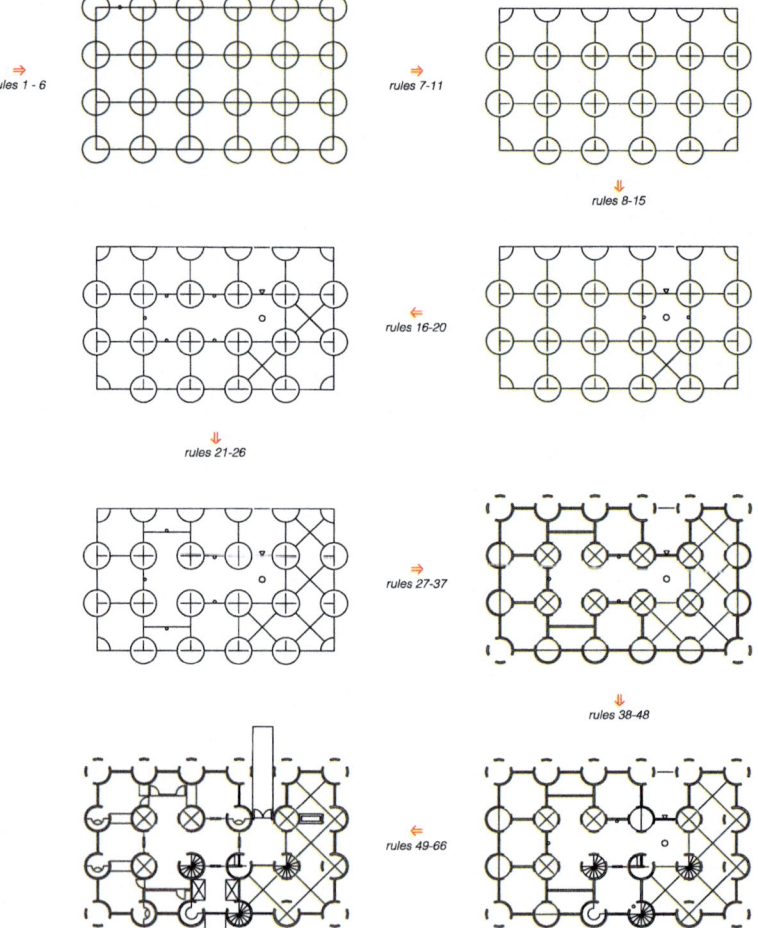

Fig. 12 Portm-ino implementation: a derivation illustrating the generation of the upper-level floor plan of Entelechy I

for designing and developing shape rules. In turn, this methodology revisits the ambition to reinterpret design narratives pertinent to architectural history and theory with digital algorithms.

6 Expanding Narratives with Shape Rules

Shape grammars empower the ongoing interpretation of languages of designs. Rewriting shape rules in analog and digital media allows for a range of feedback mechanisms that expand possibilities for engaging shape rules in design research,

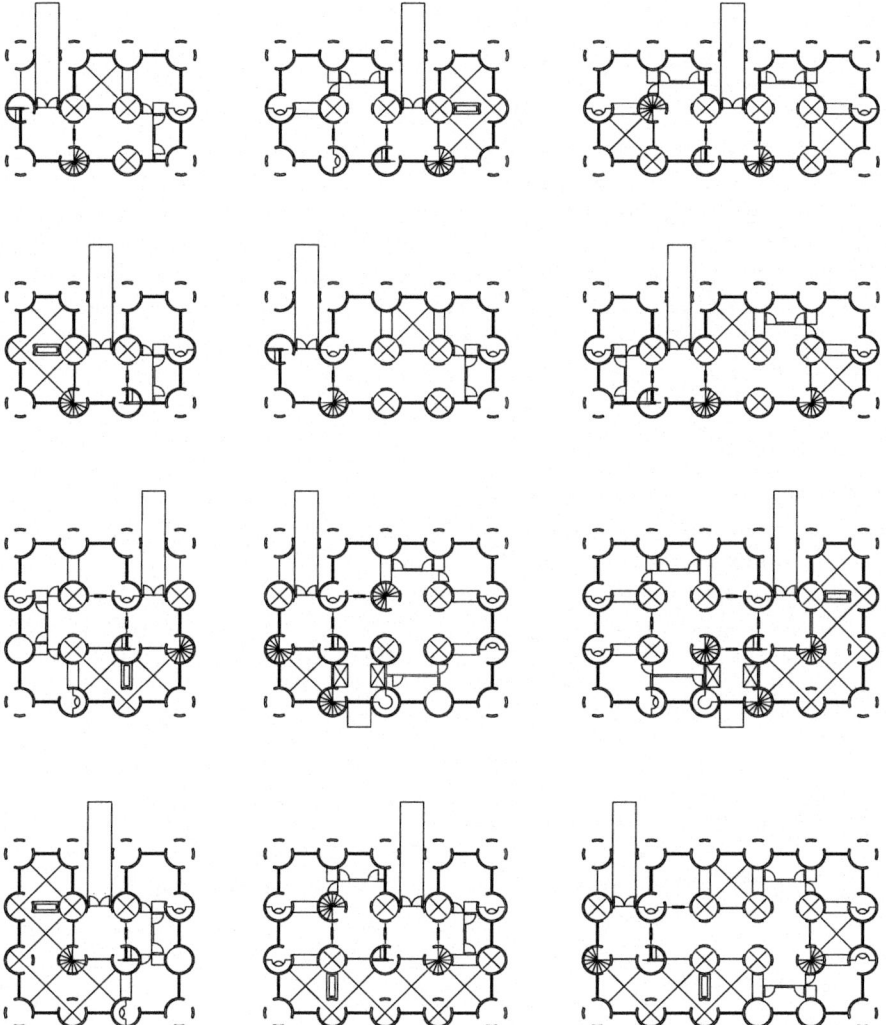

Fig. 13 Portm-ino implementation: a sample of twelve Portm-Inoes

education, and practice. These mediums enliven and animate shape computations interactively toward further understanding of the formalism. The iterative process of implementation presented here suggests how this work can continue.

A simple example is offered to inspire future research and design. It addresses a constant constraint in the Entelechy I grammar: a regular grid. This condition in Portman's original house, especially the precise utilization of a 3×5 grid, aligns Entelechy I with the "ideal villas" by Palladio and Le Corbusier, which share this geometric correspondence (Rowe 1976). Nevertheless, irregularities are the reality of many site conditions, contexts, and configurational approaches in architectural design. To test a potential expansion of the narrative on Entelechy I and to evaluate how the house-as-system is interpreted in the implemented shape grammar, a modified Portm-Ino or "Portm-Ino extended" language is proposed that can address domestic constraints with irregular grids and boundaries. A design outcome from this approach is given in Fig. 14 to round out the work presented here. This design was generated in Shape Machine using the shape rules of the Portm-Ino Implementation, plus four additional shape rules to deal with new spatial conditions related to the irregularities of the grid. The simplicity of this approach, and the ability to expand a shape grammar with incremental tests and edits, suggest new and open workflows for collaborating on shape rules and grammars that challenge the assumptions of our descriptions, interpretations, and evaluations of design languages. By engaging with machine feedback and file systems that can store shape rules ready for working and reworking, rewriting shape rules provides a framework for discontinuous reuse and transformation in shape computation. While, for now, this work is limited by two-dimensional geometries, the ambition is to extend this vocabulary to encompass a complete range of shape vocabularies and mechanisms for architectural design. Even so, current progress suggests the possibility for networks and interactions of design machines (Stiny and March 1981) formalized with implemented shape rules to materialize and prompt novel forms of algorithmic sharing and exchange.

This automated and animated approach provides ways to see anew through an ongoing process of designing and developing shape rules. It incorporates the reflection-in-action that Schön captured as characteristic of design activity (1983),

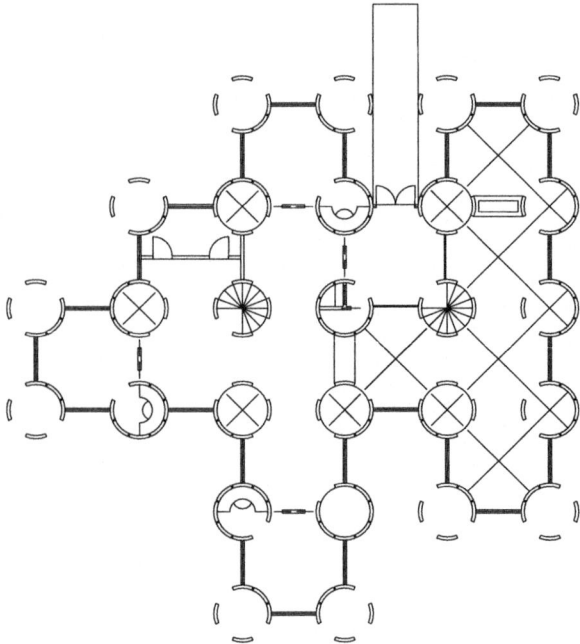

Fig. 14 Portm-ino implementation extended: a design variation generated with an irregular underlying grid and four additional shape rules

with a workflow that depends on the rewriting of rules. The examples discussed here, to revisit Portman's architectural language, align this method with previous efforts to animate architectural histories and theories about Palladian villas in analog and digital formats (Stiny and Mitchell 1978; Hersey and Freedman 1992). In addition, it offers a path towards adopting these workflows in future synthetic processes in architectural practices to rewrite design narratives with shape rules and their implementations.

References

Duarte, J.P. 2005. Towards the mass customization of housing: The grammar of Siza's houses at Malagueira. *Environment and Planning B: Planning and Design* 32 (3): 347–380.
Economou, A., T.C.K. Hong, H. Ligler, and J. Park. 2020. Shape machine: A primer for visual computation. *A New Perspective of Cultural DNA*, 65–92. KAIST research series. Singapore: Springer.
Flemming, U. 1987. The role of shape grammars in the analysis and creation of designs. In *Computability of designs*, ed. Y.E. Kalay. New York: John Wiley.
Grasl, T., and A. Economou. 2013. From topologies to shapes: parametric shape grammars implemented by graphs. *Environment and Planning b: Planning and Design* 40 (5): 905–922.
Hersey, G., and R. Freedman. 1992. *Possible Palladian Villas*. Cambridge: The MIT Press.
Hong, T.C.K. 2021. *Shape machine: Shape embedding and rewriting in visual design*. Ph.D. dissertation. Atlanta: Georgia Institute of Technology.
Hong, T.C.K., and A. Economou. 2023. Implementation of shape embedding in 2D CAD systems. *Automation in Construction*, 146
Knight, T. 1994. *Transformations in design: A formal approach to innovation in the visual arts*. Cambridge: Cambridge University Press.
Knight, T., and G. Stiny. 2001. Classical and non-classical computation. *ARQ: Architectural Research Quarterly* 5: 355–372.
Knight, T.W. 1998. Shape grammars. *Environment and Planning B: Planning and Design*, Anniversary Issue, 86–91.
Krishnamurti, R. 1981. The construction of shapes. *Environment and Planning b: Planning and Design* 8 (1): 5–40.
Krishnamurti, R. 1992. The maximal representation of a shape. *Environment and Planning B: Planning and Design* 19 (3): 267–288.
Le Corbusier. 1970. Five points toward a new architecture. In *Programs and manifestos in twentieth century architecture*. Cambridge: The MIT Press.
Ligler, H. 2020. *The portman variations: A critical approach to Entelechy I mediated by Shape Machine*. Ph. D. dissertation. Atlanta: Georgia Institute of Technology.
Ligler, H. 2021. Reconfiguring atrium hotels: Generating hybrid designs with visual computations in Shape Machine. *Automation in Construction*, 132.
Ligler, H. 2022. The subsymmetry analysis and rule-based synthesis of John Portman's Midnight Sun. *Nexus Network Journal* 25: 409–437.
Ligler, H., and A. Economou. 2015a. Entelechy I: Towards a formal specification of John Portman's domestic architecture. In *Proceedings of the 33rd eCAADe conference*. Vienna, Austria: eCAADe.
Ligler, H., and A. Economou. 2015b. Lost in translation: Towards an automated description of John Portman's domestic architecture. *Blucher Design Proceedings* 2 (3): 657–661.
Ligler, H., and A. Economou. 2018. Entelechy revisited: On the generative specification of John Portman's architectural language. *Environment and Planning B: Urban Analytics and City Science* 45 (4): 623–648.

Ligler, H., and A. Economou. 2019. From drawing shapes to scripting shapes: Architectural theory mediated by shape machine. In *Proceedings of the Symposium on Simulation for Architecture and Urban Design' 19*, Article 39, 1–8. San Diego: SimAUD.

McCullough, M. 2006. 20 years of scripted space. *Architectural Design* 76: 12–15.

Mitchell, W.J. 1990. *The logic of architecture: Design, computation, and cognition*. Cambridge: The MIT Press.

Mitchell, W.J. 1973. Vitruvius computatus. In *The proceedings of the fourth environmental design research conference, EDRA 4*, 384–386. Blacksburg: Virginia Polytechnic Institute and State University.

Portman, J.C., and J. Barnett. 1976. *The architect as developer*. New York: McGraw-Hill.

Rowe, C. 1976. The mathematics of the ideal villa. In *The mathematics of the ideal villa, and other essays*. Cambridge: The MIT Press.

Schön, D.A. 1983. *The reflective practitioner: How professionals think in action*. New York: Basic Books.

Stiny, G. 1986. A new line on drafting systems. *Design Computing*. 1: 5–19.

Stiny, G. 2006. *Shape: Talking about seeing and doing*. Cambridge: The MIT Press.

Stiny, G. 2015. The critic as artist: Oscar Wilde's prolegomena to shape grammars. *Nexus Network Journal* 17 (3): 723–758.

Stiny, G. 2022. *Shapes of imagination: Calculating in Coleridge's magical realm*. Cambridge: The MIT Press.

Stiny, G., and J. Gips. 1972. Shape grammars and the generative specification of painting and sculpture. In *Information processing*, vol. 71, ed. C.V. Frieman.

Stiny, G., and J. Gips. 1978. *Algorithmic aesthetics: Computer models for criticism and design in the arts*. Berkeley: University of California Press.

Stiny, G., and L. March. 1981. Design machines. *Environment and Planning B: Planning and Design* 8 (3): 245–255.

Stiny, G., and W.J. Mitchell. 1978. The Palladian grammar. *Environment and Planning B: Planning and Design* 5 (1): 5–18.

Woodbury, R. 2021. Simon's ant: Towards new task environments for design alternatives. *A new perspective of cultural DNA*, 1–15. Singapore: Springer.

Open Access This chapter is licensed under the terms of the Creative Commons Attribution-NonCommercial-NoDerivatives 4.0 International License (http://creativecommons.org/licenses/by-nc-nd/4.0/), which permits any noncommercial use, sharing, distribution and reproduction in any medium or format, as long as you give appropriate credit to the original author(s) and the source, provide a link to the Creative Commons license and indicate if you modified the licensed material. You do not have permission under this license to share adapted material derived from this chapter or parts of it.

The images or other third party material in this chapter are included in the chapter's Creative Commons license, unless indicated otherwise in a credit line to the material. If material is not included in the chapter's Creative Commons license and your intended use is not permitted by statutory regulation or exceeds the permitted use, you will need to obtain permission directly from the copyright holder.

Automating the Archaeological Reconstruction of Classical Greek and Roman Architecture in Shape Machine

Myrsini Mamoli, Yichao Shi, and Violet Cerbone

Abstract This paper demonstrates how shape grammars can be applied in archaeological research to deduce design principles and apply them to reconstruct fragmentary artifacts. The paper presents two examples, two archaeological shape grammars implemented in Shape Machine that showcase the range of possibilities: a grammar analyzing a single, well-documented building, the Ionic Porch at the Sanctuary of the Greater Gods in Samothrace, to generate variant possible reconstructions effectively; and a grammar analyzing the Roman aedicular façade to assist in the reconstruction of a fragmentarily surviving one, the main hall of Hadrian's library and help visualize and understand new hypotheses about its use. These two examples manage to (a) visually document the reconstruction process an archaeologist follows in recreating the building and defining the relationships of its architectural components; (b) generate variant possible reconstruction drawings in parallel, in plan and elevation, quickly; (c) comply with the principles of archaeological reconstruction to showcase variation and probability, and finally; (d) automate the reconstruction process that can be expanded to other buildings as well.

1 Introduction

Reconstruction of fragmentary archaeological evidence almost always entails speculation based on other parallel and precedent comparable designs (comparanda). Often, variation in reconstruction is needed to convey the ambiguity of the data and the probability with which they may have occurred. With the rise of digital technologies used in archaeological reconstruction and heritage visualization, the London Charter established principles and objectives to ensure that digital heritage visualization would be as intellectually and technically rigorous as longer established research and communication methods (Denard 2012), with evidence validated in similar ways as scholarly publications. The point for the digital visualization is to

---

M. Mamoli (✉) · Y. Shi · V. Cerbone
School of Architecture, Georgia Institute of Technology, Atlanta, Georgia
e-mail: Myrsini@gatech.edu

distinguish between hypotheses and the evidence uncovered so that the proposed reconstruction is not perceived to be more predictive than it actually is.

Parametric Shape Grammars can be used in this way, to capture the design and reconstruction process, step by step, with visual rules that document evidence in metadata. Variant alternative possibilities can be visualized and incorporated into rules that can be applied, or not applied, in each iteration. In a combinatorial way, these iterations present a series of possible reconstructions of the same artifact, ranked according to the occurence of each individual feature in the archaeological record. An excellent example of archaeological parametric grammars is the library grammar (Mamoli 2014, 2015, 2020), an analogue shape grammar that analyzes the design principles of ancient Greek and Roman libraries and encodes them in visual rules. Each rule corresponds to a building component and its possible architectural form. The grammar can generate all 17 known cases in the record, and through statistical analysis of the strings of rules that apply in each generation, it can establish a probabilistic model for the building type of the ancient library. The higher the frequency with which a rule applies in the grammar-generated reconstructions of all 17 libraries, the more important the feature that the rule generates is, and the more probable for it to have occurred in ambiguous cases (Mamoli 2018; 2020). When a rule applies without secure archaeological evidence, it is denoted with an asterisk, and it is not included in the frequency analysis. Additionally, the asterisks point to hypothetical parts of the reconstruction, and thus distinguish between evidence and hypothesis.

Archaeologists have identified the applicability of grammars in the archaeological reconstruction and classification since long ago (Chippindale 1992; Hodder 1982; Mamoli and Knight 2013; Mamoli 2021). Most recently, Bonna Wescoat (2019), in her paper "Architectural Documentation and Visual Evocation: Choices, Iterations, and Virtual Representation in the Sanctuary of the Great Gods on Samothrake," stated that "Greek architecture favors such an approach because its orders are composed of defined elements of certain shape, proportion, and relationship," while she expressed the fear that the computational investment required to generate the fully diverse range of possible reconstructions for a single, partially preserved, and fairly idiosyncratic structure may not currently be time effective. It is true that analyzing a building type, and authoring a shape grammar to generate one case of it or the whole class, is a very rigorous and time-consuming process. However, as in any investment, after the initial set up of the implementation of the grammar, we can manipulate the parametric grammar to generate a whole range of variations quickly, almost miraculously! Further, as Terry Knight has demonstrated (Knight 1994), we can transform the grammar to account for stylistic or typological change and to reconstruct other related buildings or even building types.

Here, we suggest a methodology of two parallel grammars that generate reconstruction drawings of Greek architecture. Parallel grammars for the analysis of an architectural style were introduced in the work of Li for the analysis of Chinese Architecture in the *Yingzao Fashi* architectural manual (Li 2005). The *Yingzao Fashi* grammar, developed in successive stages, generates Chinese buildings successively in plan, elevation and section. Here, we rework the concept of parallel grammars to work, literally, in parallel and to concurrently generate the floorplan and the front and side elevations of the building. Users see the building being reconstructed in front of their eyes, and in a very didactic way, can grasp how the generation of each architectural component is represented in plan and elevation at the same time, and at the same scale. The first part of the grammar generates the basic subdivisions in the drawings, and the architectural components in basic, schematic forms, while at the same time, places them in different layers. The detailed part of the grammar replaces the basic schemata with detailed architectural components.

The two parts of the grammar — the schematic and the detailed—are authored in stages in Shape Machine (Economou et al. 2024; Hong and Economou 2022, 2023), a shape grammar interpreter, a plug-in, in Rhino, that has been developed at the Shape Computation Lab at Georgia Tech. Within each stage, rules are structured vertically. Each rule, in addition to the visual rule with the shapes in the left- and right-hand side, includes a header, with values for Transformation, Selection, and Loop among others. Transformation defines under which transformation the rule applies: the value 10 stands for identity transformations (translation, rotation, reflection), 20 for similarity (scale), and 30 for affinity (stretch and smear). Selection defines whether the rule will be applied to one (value = 1) or all embedded instances in the drawing (value = 2) or to a manual selection by the user (value = 3). The Loop count, like state labels, as defined by Knight (1994; see, for example, the grammar for the Greek Meander) in analogue grammars, is defined by the user and refers to how many times the rule will be applied before the computation moves to the next rule. The rules are structured in layers. Each shape in each rule is placed on a certain layer in Rhino that gives that shape a distinct color and line weight (in Rhino 8 only). This layer becomes part of the rule itself, in a similar way that shape labels specify to which part of the design the rule can be applied (Stiny 2006, 2022; Stiny and Gips 1972). The advantage of the layering is, first, that the shape rules can be simplified, as there is no need for other labels. In fact, rules turn out to be very schematic, as the information can be encoded in very simple shapes. Additionally, the computation works faster because the machine embeds each rule in specific layers only, and finally, shapes can be assigned line weights to follow the conventions of architectural representation.

Two case studies are illustrated here to map the possibilities that shape grammars and Shape Machine open to archaeological research, in visualizing already formulated or new hypotheses. Firstly, we present a shape grammar implementation of the reconstruction of the Ionic Porch, a well-studied Hellenistic building in Samothrace, Greece, to demonstrate how a shape grammar can visually document the reconstruction process of an archaeologist in recreating the building and defining the relationships of its architectural components. We show how a visual script enables the archaeologist to extrapolate with little effort various possible reconstruction drawings in plan and elevation. Secondly, we implement a shape grammar that generates Imperial Roman aedicular facades in the case study of the interior of the main hall (oikos) of the Library of Hadrian in Athens to show how shape grammars can help test a new hypothesis. The possibility of an interior façade with "aedicule", temple-like formations framing every other niche and creating an angulating entablature, for the interior of a library room has not been explored in traditional archaeological research due to now-debunked theories related to its functionality. On the other hand, the building remains of the Library of Hadrian that showcase large beam sockets for perpendicular entablatures open the possibility of an aedicular façade and have weighed in the debate on its function—whether a library or something else. The most recent architectural reconstruction contemplates this new interpretation (Kanellopoulos 2020), but to showcase the partiality of the evidence, stops short of completing the crowning of the façade, and it does not visualize the whole range of possibilities. We show how shape grammars can complement traditional archaeological research and generate multiple possible reconstructions that can help us speculate about its use.

2 Reconstructing the Ionic Porch in Samothrace

In her monograph, *Monuments of The Eastern Hill* (Wescoat 2017) at the Sanctuary of the Greater Gods in Samothrace, Bonna Wescoat painstakingly presented each building, the archaeological evidence of each individual component, proposed possible reconstructions, and drew conclusions about the design of the building and a possible module of measure. Among all the monuments, the Ionic Porch, a small-scale structure built in the late 3rd or second century BCE against the back wall of an earlier monument, presented the challenge and opportunity for multiple possible reconstructions. The reason for this is that it was preserved only at its foundation, and not all of its architectural components were identified on site, which left several parameters open to speculation. First, regarding the exact floor plan of the building, the dimensions of the foundations allowed the archaeologists to identify it as a small building with four supports in its façade. However, several aspects of it are open to speculation, because the final course of cut stone where the columns stepped (stylobate) did not survive to show the exact position of the columns.

First of all, it is not clear whether the building had four columns, and therefore, was a tetrastyle building, or it had just two columns embedded into antae (piers formed by thickening the ends of the lateral walls) and thus was a "distyle in antis" building. Second, the exact dimensions of the stylobate could be restored either to the full extent of the foundation (krepis), or one step inwards. Third, the spacing of the columns (intercolumniations) could be restored at either three equal spaces or two smaller spaces on the sides, and a widened one in the center. Additionally, the front and side elevations can also be restored in variant ways: while archaeological fragments of the shaft and the capitals secure the order as Ionic, the height of the columns is not known and it can be restored as a variant parametric ratio to the Lower Column Diameter, measured either at the shaft or the column base (at the apophyge); moreover, evidence of a frieze course has not been found, and therefore its restoration over the architrave can only be speculated, in which case its possible height can have a range of acceptable values. Finally, the evidence of the roof includes fragments of only roof tiles and a gutter (sima) and based on the context, its slope has been restored at between 11 and 12°. However, the lack of other evidence is not exclusive of certain roof types, so three different types of roofs can be considered: a ridge roof

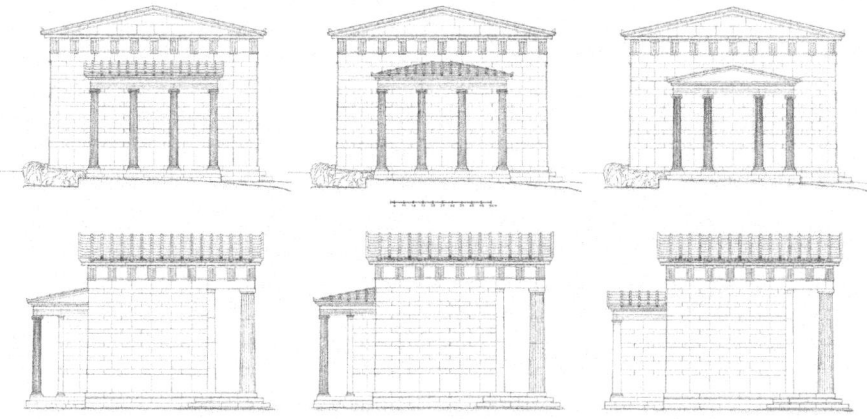

Fig. 1 Three possible reconstructions of the Ionic Porch in front and side elevation: tetrastyle with shed roof; a tetrastyle with hip roof; distyle in antis with ridge roof (American Excavations Samothrace)

with a ranking sima and a pediment, a shed roof with sima in the front and ranking on the sides, or a hip roof with the same sima on all sides and no pediment.

In proposing a reconstruction for the building, the archaeological team's main concern was to showcase the partiality of the evidence, and to illuminate how different scenarios in plan and elevation would foreground a different function of the building. These functions could be either a stoa-like place of gathering in the case of a shed roof, or a naiskos-like display place in the case of a pedimented roof. Thus, to communicate the uncertainty and the possible range of variations, Wescoat chose to illustrate three indicative combinatorial scenarios, two tetrastyle plans, one with a shed roof and one with a hip roof, and one distyle in antis with a pediment (Fig. 1).

Wescoat (2019) recognized that twenty years ago, the production of these three variations, by hand, was a time and energy consuming project, and that with digital tools, the same outcome could be produced with fewer resources and in less time. Further, Wescoat challenged us to automate the reconstruction process with shape grammars to demonstrate whether this is an appropriate methodology for archaeological reconstruction, since it simulates the design process and captures the design space of variations within the given parameters. Another challenge was to identify whether, in visualizing multiple possible variations of a building, shape grammars can make the process faster and more effective and thus offer insights that might not have been obvious to the archaeologists (Wescoat 2019, 309).

2.1 A Shape Grammar for the Automation of the Reconstruction Process of the Ionic Porch

Following Wescoat's lead, we went back to the archaeological record as it was documented in the text (Wescoat et al. 2017) and translated the rules she followed in reconstructing the building into visual rules. We drew the rules in Shape Machine, and we authored a grammar in stages (Fig. 2). Each stage generates one of the architectural components Wescoat had analyzed in her documentation of the restoration and more or less follow the same sequence: (1) restored krepis (foundation, step or euthynteria, stylobate), (2) restored plan (centerlines, interaxials, plinths, column bases, anta bases), (3) restored columns (shaft height, capitals), (4) restored entablature (epistyle, possible frieze, dentil-geison course), (5) restored roof (sima with lion waterspouts, possible ranking sima, antefixes, roof tiles).

The grammar has been directly drawn into Rhino with the Shape Machine plug-in, developed at the Shape Computation Lab at Georgia Tech. The grammar is structured into two parts, one that generates the overall schema with the basic subdivisions and all the architectural components with the correct proportions and dimensions and a second one that generates the details of each one of these components. All rules simultaneously generate each component in plan, front and side elevation. Within each stage, there are alternative rules that generate alternative possible architectural forms, for example in the stage of krepis, there are rules that may or may not generate an additional step. In the case of the roof type, there are three different blocks of

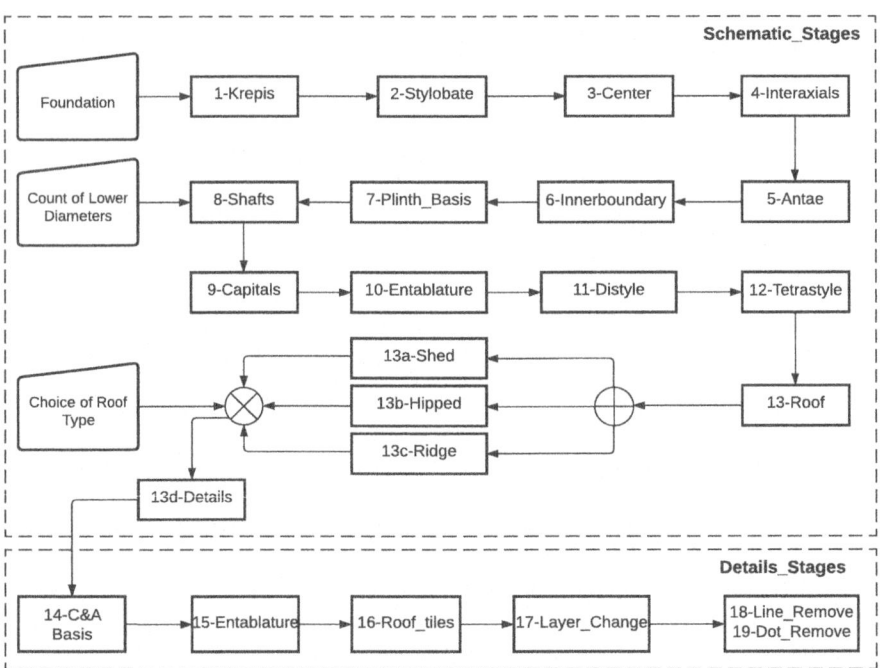

Fig. 2 The two-phase Ionic Porch grammar shows the sequence of the different stages, corresponding to the different components and the different options within each that generate the schematic and detailed drawings in plan and elevation

rules that correspond to the three different roof types, to which the production jumps based on the user's choice.

2.2 Initial Shape and Grammar Rules

The production of the reconstruction of the Ionic Porch starts with the identification of the initial shape, which is the outline of the foundations (krepis) of the porch in plan, a simple rectangle. The rules of the schematic grammar are given in Fig. 3.

The first stage of rules generates the axes of symmetry of the krepis in plan and generates the krepis in front and side elevation, so that the basic views of representation in 2D representation are generated in parallel throughout the grammar. The dimensions of the krepis, the width, depth and height are given in the archaeological record. Stage 2 rules generate the stylobate. The first three rules add an additional

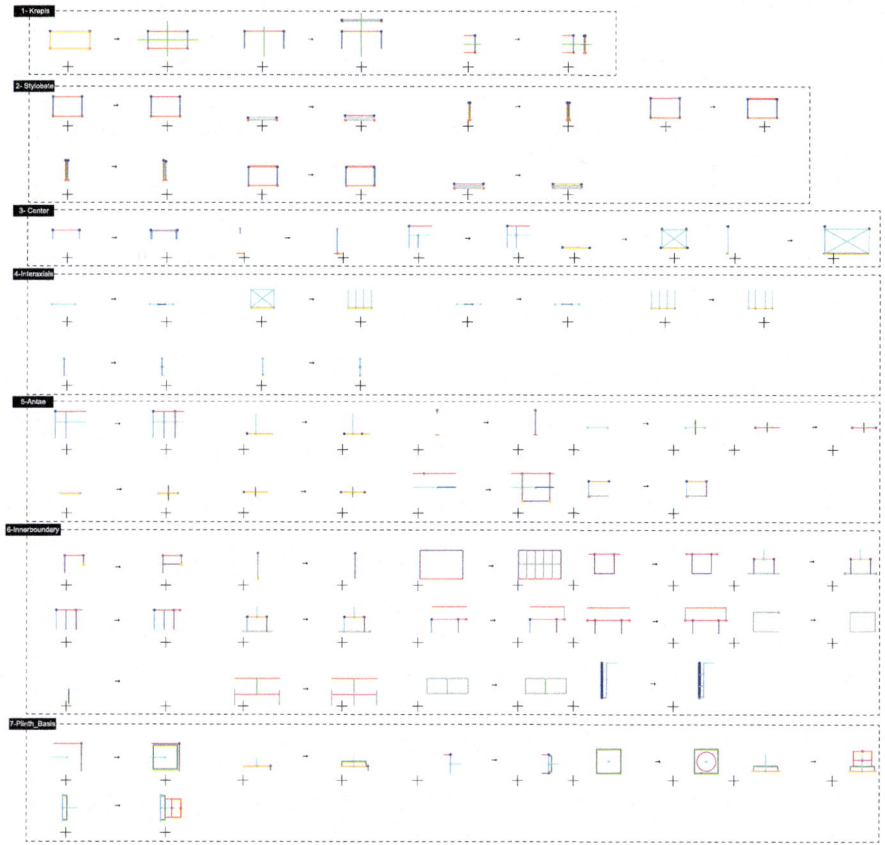

Fig. 3 Stages 1–7: rules generating the schematic Ionic porch in plan, front and side elevation

course of cut stone (euthynteria), flush with the profile of the upper course of the foundations (stereobate), whereas the following three rules offer the option to set it back to convert it into a step. Stage 3 rules generate the centerlines of the walls of the front and side stylobate, where the axes of the columns and walls are going to be placed and extend them to the back wall. All these rules work under similarity to account for various wall lengths, and they place the centerline exactly half lower diameter (LCD) of the column so that the columns are flush with the edge of the stylobate. Subsequent rules add vertical lines to indicate the axes of the corner supports (columns or antae) bounding with diagonal lines the space where the elevation will be developed in subsequent stages.

Stage 4 rules generate the interaxials of the façade. Two alternative sets of rules subdivide the front façade into three interaxials, either of equal length or with a wider central interaxial. During the computation, the user can change the loop in the header and activate either of the two options by setting the loop of the corresponding rule into 1. Further rule change can be achieved by spreading the interaxials further manually to create a wider intercolumniation. The following two stages place the antae, the rectilinear supports at the extension of the side walls. The antae are instantiated with the first two rules that place their centerpoint in the center of the side wall. Then these points are moved, either flush with the front façade to restore a distyle in antis plan or at any point within the side walls all the way to the back wall to restore a tetrastyle prostyle plan. The plan is complete with the subsequent stages that add the inner boundary of the stylobate, defined as another half diameter from the centerlines; subdivide the courses of the krepis into blocks so that each plinth fits within one block of the krepis, and finally add the plinths, the column bases, and the anta bases within the boundaries of the stylobate. In all these rules, in the header the selection is set to 2, which means that the rule applies in parallel at once to all instances identified.

The rule in stage 8 extends the column drum by half diameter at a time until they reach their possible full height. The user can set the loop into any number, so that the height is restored as a ratio of up to 1:10 Lower Column Diameters, as given by Vitruvius and/or identified in the archaeological record. The rules in stage 9 add capitals to the columns with dimensions based on the archaeological record. Subsequent stages apply under similarity to add an entablature (stage 10) consisting of an epistyle on top of the capitals, a possible frieze (that does not survive in the record but is considered necessary for structural and aesthetic reasons by archaeologists), and a final course with a series of cubical projections (dentils) and a molding (geison). The following two stages modify the side walls to create two different floor plans: either a distyle in antis or a tetrastyle building. Finally, stage 13 adds a schematic roof that jumps to one of the three blocks of rules that generate a shed, a hip, and a ridge roof (Fig. 4).

After the construction of the overall schema of the reconstruction the second part of the grammar generates the details of each component, e.g. the column bases, the flutes on the column shafts, volutes and other details on the schematic column capital; the three-stepped epistyle, the details of the dentil-geison course, the gutters (sima and the ranking sima), the roof tiles and the antefixes, the decorative terminating

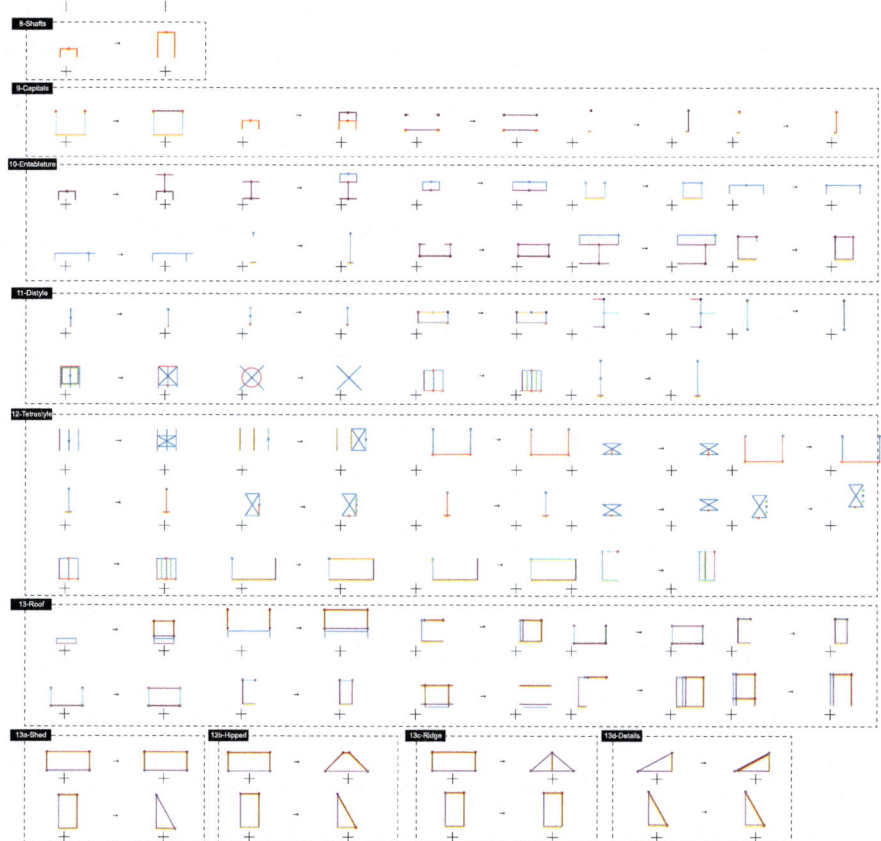

Fig. 4 Stages 8-13 with rules that complete the generation of the schematic porch

blocks of the cover tiles. Also, there are rules that redraft all layers with black color to match the conventions of architectural representation and clean up the lines and points of the grammar. A sample of these rules is given in Fig. 5.

The grammars are still work in progress, as more rules are needed to add a higher level of detail in the entablatures and roofs, work out certain refinements, such as the entasis of the columns, and clean up some construction lines. Also, the interface is not very straightforward. The rules are based on simple shapes in layers, and since these shapes are defined by points, and the rules are embedded under affinity transformation, the shapes have unintuitive proportions and seem simplistic. They do not immediately convey their computational power, and they do not look like an intuitive tool that archaeologists can use.

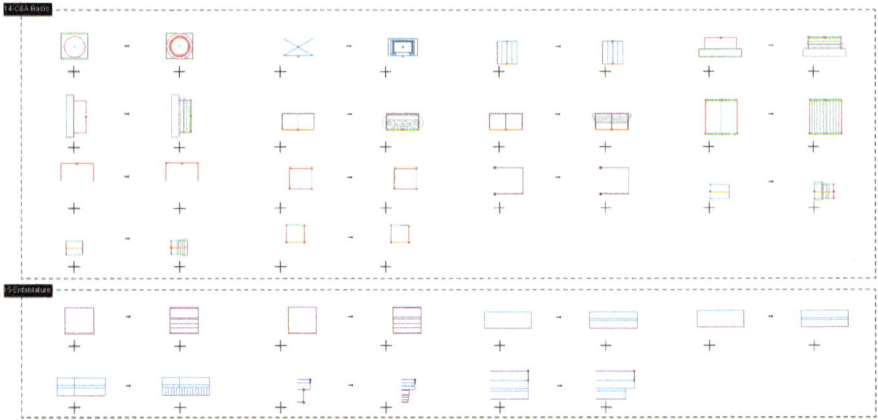

Fig. 5 Rules that add details to each architectural component, column bases, anta bases, column shafts, capitals, etc.

2.3 Computation

A possible stage-by-stage computation of the schematic and detailed grammars is given in Fig. 6. The computation starts with the input by the user of the initial shape, which is the outline of the krepis. From there the front and side elevations are generated. Then the computation generates the plan as a tetrastyle prostyle building with an additional step, four columns placed at equal interaxials, the anta placed very close to the column and thus creating a very deep shelter. The height of the column shafts is defined at 9 lower shaft diameters, and the roof type as a hip roof.

Eight variations upon this reconstruction are shown in Fig. 7, six for each of the types that are proposed by Wescoat, as the matrix of possibilities of two possible floor types and the possible types of roofs, and two more that show further parametric variations that expand the range of possibilities.

2.4 Conclusions

The advantage of these eight sets of reconstruction drawings over the six sets drafted by hand 20 years ago and published by Wescoat (2017) is minimal. However, these drawings, despite their incompleteness, exemplify the potential of a user being able to illustrate any parametric variation and any combination of components that seems appropriate for any hypothesis. If a new component is discovered on site, the evidence can easily be incorporated in the grammar and new drawings can be generated easily. Any member that joins the archaeological team any time can generate any variation and can utilize Shape Machine as a tool to visualize their own ideas without being a programmer or an architect.

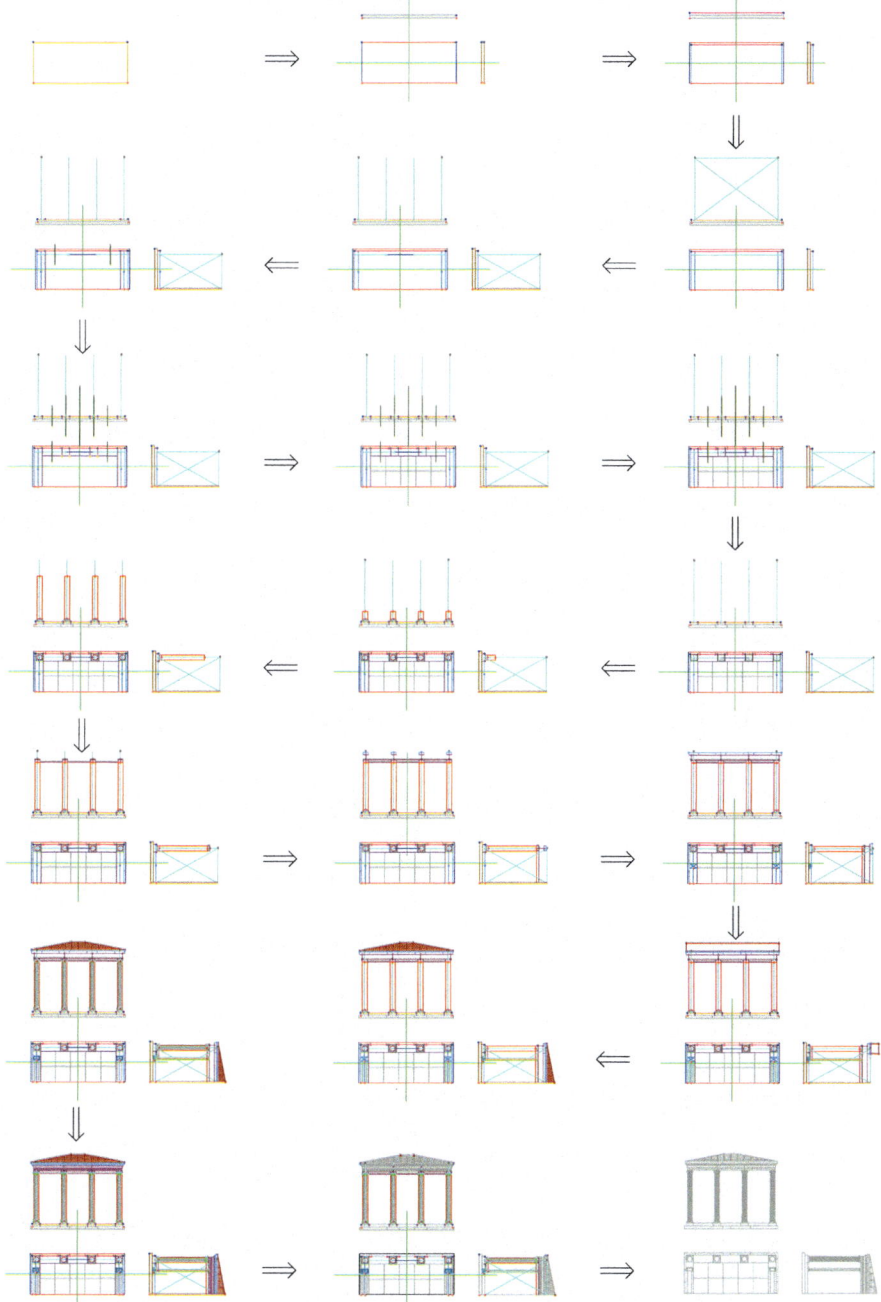

Fig. 6 A stage-by-stage computer-generated computation of the reconstruction drawings of the Ionic Porch as a tetrastyle prostyle building with a hip roof

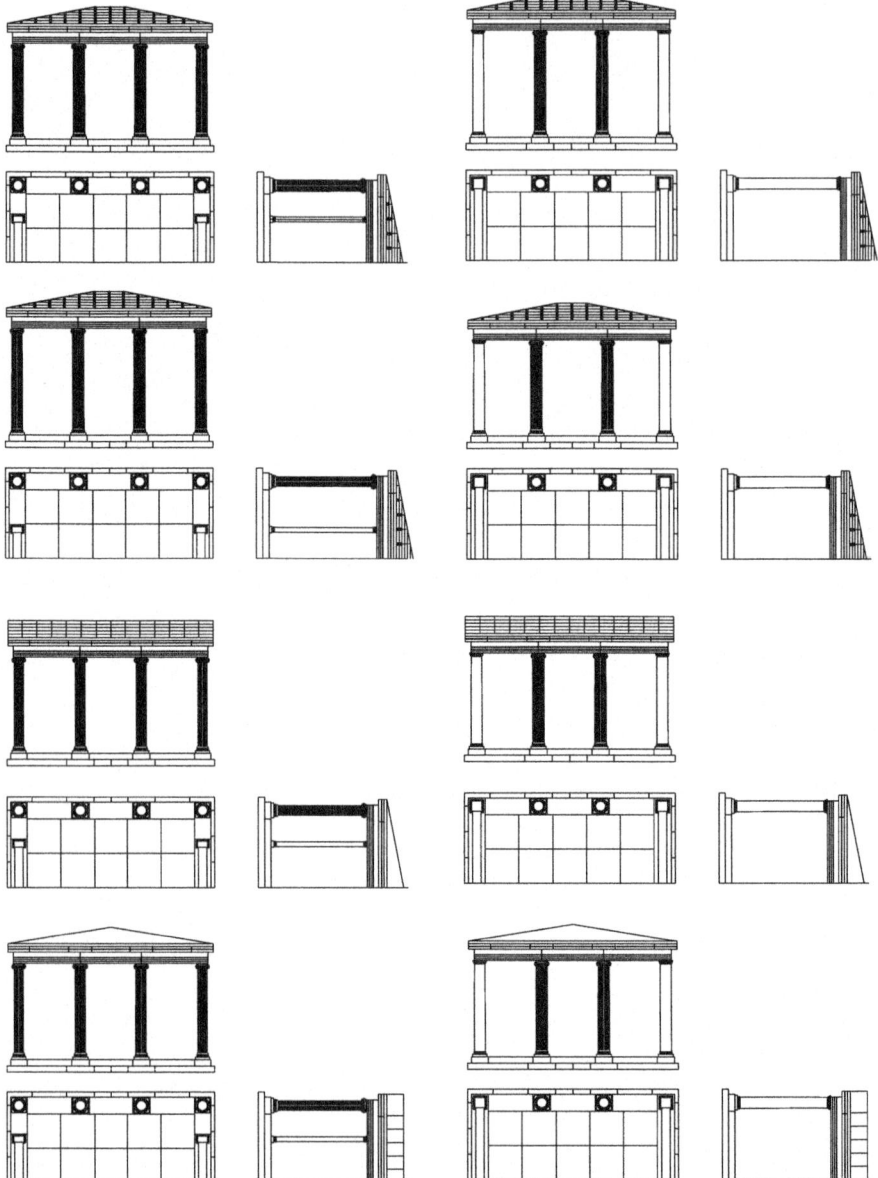

Fig. 7 Eight grammar-generated reconstructions in plan, front and side elevation of the Ionic Porch in Samothrace showcasing different floor plan types, roof types and different instantiations of their different architectural components

By authoring a parametric shape grammar with which we can visualize the design principles of buildings of this scale and form, we can have a system with which we can automate the visualization of any instance within the same language of designs. Further, we can expand the grammar to the overall corpus of 84 tetrastyle and distyle in antis buildings of Greek architecture (Kylindreas 2015) and with rule change (in proportions, dimensions, loops), we can generate reconstruction drawings for each of them.

3 The Aedicular Façade Grammar and the Reconstructions of the Main Hall of the Library of Hadrian in Athens

The aedicular façade grammar illustrates the opposite problem. Instead of starting from a well-documented case and extrapolating principles to apply to the whole type, we start from the analysis of a type to extrapolate the design principles that we can then apply to fragmentary cases.

The Library of Hadrian in Athens, built by emperor Hadrian around 130 CE, was one of the most monumental building complexes of antiquity. Its main hall, located on the longitudinal axis of the complex at the end of a vast landscaped colonnaded courtyard is referenced in literary sources as a luxurious room made of polychromatic marble. However, few fragments survive today and the evidence comes mainly from the back wall of the room with the niches and the podium in front of it, which give us the interaxials of the columns framing bays of varying widths with niches on the back wall of the room, where books and other artifacts were placed. The architectural reconstruction of the room is at the core of the argument about its interpretation: scholars who interpret the building as a cultural center or a university with a library, argue for a continuous colonnade supporting a gallery that facilitated access to the niches of the upper floor. Scholars who emphasize the evidence of the large beam sockets on the back wall, argue that these sockets supported the perpendicular blocks of an angulating entablature of aedicule, temple-like projections from the back wall with two columns, and argue that this aedicular room was a hall for the imperial cult. The aedicular facade was indeed the staple of Imperial Roman Architecture in the High Empire and it was utilized for a variety of buildings (Burrell 2006; Jones 2000). Most recently, Kanellopoulos (2020) re-confirmed the reconstruction of the aedicular facade and proposed three different options to allow for circulation on the upper level and maintain its interpretation as a library room.

Our work expands upon Kanellopoulos' reconstruction and interpretation as we explore the "design space" within this theory, the range of possible solutions that an aedicular facade could have had within the envelope of Hadrian's Library in Athens. In doing so, we first eliminate the requirement to provide circulation on the upper level of a library, as our previous work has demonstrated that on one hand, there is no evidence of stairs, and on the other, there is ample material and literary evidence that a library hall in antiquity had books in conjunction with artworks and sculpture

(Mamoli 2014, 185). Secondly, we revisit the evidence of the varying interaxials from the center towards the sides to explore the different forms of focal points. The main question is in how many ways can we reconstruct the entablatures and the pediments given the dimensions of the room and the known schema of vertical supports framing the bays? Can we identify configurations that are more probable than others?

To answer these questions, we turn to the other parallel buildings with aedicular facades, such as main halls in gymnasia, scaenae frons of ancient theaters, gates, and nymphaea, identified by Burrell (2006) and Kanellopoulos (2020) and we focus on 11 examples that are appropriate for the scale and interiority of a room with one focal point. These facades were articulated with great variability based on different criteria: (a) the bay rhythm A-A-C-A-A, A-B-C-B-A or A-A-A-A-A as; (b) the focal point as single bay or tripartite; (c) the superimposition of the upper level as aligned or offset in relationship to the lower level; (d) the focal point pediment as triangular, or triangular nested in split pediments, or triangular interrupted by an arcuated lintel; and (e) the side bays carrying triangular or segmental only pediments, or an alternation of the two, or split pediments framing the focal point, or no pediment at all.

3.1 The Grammar

The aedicular façade grammar encodes these design principles in visual rules. In a similar way to the previous case of the Ionic Porch, it consists of two parts, the schematic and the detailed. The schematic grammar is structured in 11 stages and the final stage replaces the schemata with detailed drawings of the different components with the appropriate line weights. The structure of the grammar is given in the Table 1.

The initial shape of the grammar is the floorplan of the main hall, unfolded, so that the elevation is seen as a continuous façade with the corners indicated by dashed lines (Fig. 8). This representation does not depict the corner columns as well as the inner corner of the room and the two corner niches appear almost continuous. We accepted the loss of information of the corner column because in all instances in the corpus the corner entablatures on the side elevation are the reflected version of the front elevation. The corner condition is usually treated differently either with single free-standing columns supporting a section of an entablature, the ressaults, or with columns supporting on one side the mirrored version of the other, so that they can be inferred by the rest. On the other hand, this representation is the only way that the dimensions of the podium and the main hall are preserved without distortion. An additional advantage is that this "unfolded" representation works with curvilinear main halls as well. The floorplan is layered so that each component is in a different layer and has its own color.

The rules in the first stage identify the edge of the podium, the folding axes, the edges of the wall, the centerline of the podium, and the edges of the column bases and extend them to generate the vertical grid on which the façade will be developed; the second stage generates the horizontal grid of the lower and upper edges of the podium and the entablatures. These rules basically copy the groundline vertically and assign them to appropriate layers. The subsequent four stages generate the focal

Table 1 Aedicular façade grammar structure

Stage 1: Vertical grid
Stage 2: Horizontal grid
Stage 3: Lower level of central bay: (1) protruding or (2) recessed
Stage 4: Upper level of central bay: (1) protruding or (2) recessed
Stage 5*: Tripartite focal point: (1) no or (2) yes
Stage 6: Focal point completion
Stage 7: Lower-level side bays: (1) alternating, (2) protruding and (3) recessed
Stage 8: Upper-level side bays alignment: (1) aligned or (2) offset
Stage 9: Pediments for the focal point: for (1) single bay (1.1) triangular or (1.2) single bay triangular framed by split pediment; (2) tripartite (2.1) split pediment, (2.2) triangular with a nested arch
Stage 10: Pediments for side bays: (1) none, (2) all triangular, (3) alternating triangular and segmental, (4) split
Stage 11: Niches
Stage 12*: Details

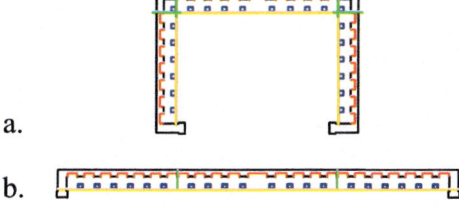

Fig. 8 The initial shape of the aedicular façade grammar as it applies at the instance of the Hadrian's Library in Athens: **a** the restored floorplan of the main Hall, **b** the unfolded version of the restored floor plan

point. Stage 3 gives the option to apply two alternative blocks of rules, one that generates a protruding entablature to the lower lever of the central bay, and one that generates a recessed entablature, supported on the back wall of the hall instead. Stage 4 does the same at the upper level of the central bay. Stage 5 is optional- it combines the central bay with the lateral one on either side to create a tripartite focal point that accounts for the cases with rhythm A-B-C-B-A. Finally, stage 6 completes the assignment of any remaining horizontal entablatures of the tripartite focal point-if any- and generates the entablatures of the lateral bays to the focal point at the lower level. All these rules work under isometry transformation.

The subsequent two stages generate the aedicule on the lower level and a second row of aedicule on the upper level, either aligned with the ones below, or offset by one bay, so that they create an alternation of protruding and recessed entablatures. The following two stages generate the pediments at the focal point and the sides. At the focal point, there are the options of (a) a single-bay focal point with triangular pediment, (b) a single-bay focal point nested within a split pediment, (c) a tripartite focal point with split pediments, and (d) a tripartite focal point with a triangular pediment and an arcuated lintel. For the sides, the grammar generates (a) no pediments on entablatures, (b) all triangular pediments, (c) alternating triangular and segmental pediments, and (d) split pediments. The next stage generates the niches on the walls, one in each bay, framed by the columns, half of them within an aedicula and the other half framed by an aedicula on either side.

Finally, the final stage generates the details of the different architectural components by finding and replacing the schematic parts with more detailed weighted line drawings of seven different combinations of line weights and colors. The thickest line is the ground line (1.5 pt.), followed by the cut line (1 pt.) in the section of the wall of the main hall and the folding axes (1pt dashed). Finally, there is a thicker line (0.6 pt) for the outline of the architectural members, and a finer line (0.25 pt.) for their details, black for the protruding components of the façade and gray for the recessed. The level of detail at this point is not very high, and some details such as the acanthus leaves in the capitals and the flutes on the column shafts are omitted.

Depending on the user's selections when applying the rules, a different schematic reconstruction is being generated. The string of numbers that refers to the different selections within each stage gives the "DNA" of each reconstruction. The stage-by-stage computation of the possible reconstruction with DNA 1.2.2.2.2.3 is illustrated in Fig. 9. It includes a protruding central bay at the lower level with a recessed above, a tripartite focal point, offset aedicule on the upper level of the side bays, and a triangular pediment, with a nested arcuated lintel crowning the focal point.

3.2 Enumeration of Computations

The grammar generates 128 possible reconstructions, equal to the product of the different options within each stage. Among them, the ones that reflect instances in the corpus are considered more probable than the others, and some of them are unlikely if they do not occur at all in the corpus. Figure 10 shows ten of these 128 that demonstrate a range of what the grammar can generate.

3.3 Conclusions

The computation of these possible facades illustrates several breakthroughs. Firstly, the degree of detail of these fully automated and computer-generated drawings, in plan and elevation (Fig. 10), is unprecedented. There is no doubt that the grammar can be enriched, but even at its current level, it can automatically generate drawings that follow the conventions of architectural drawing representation.

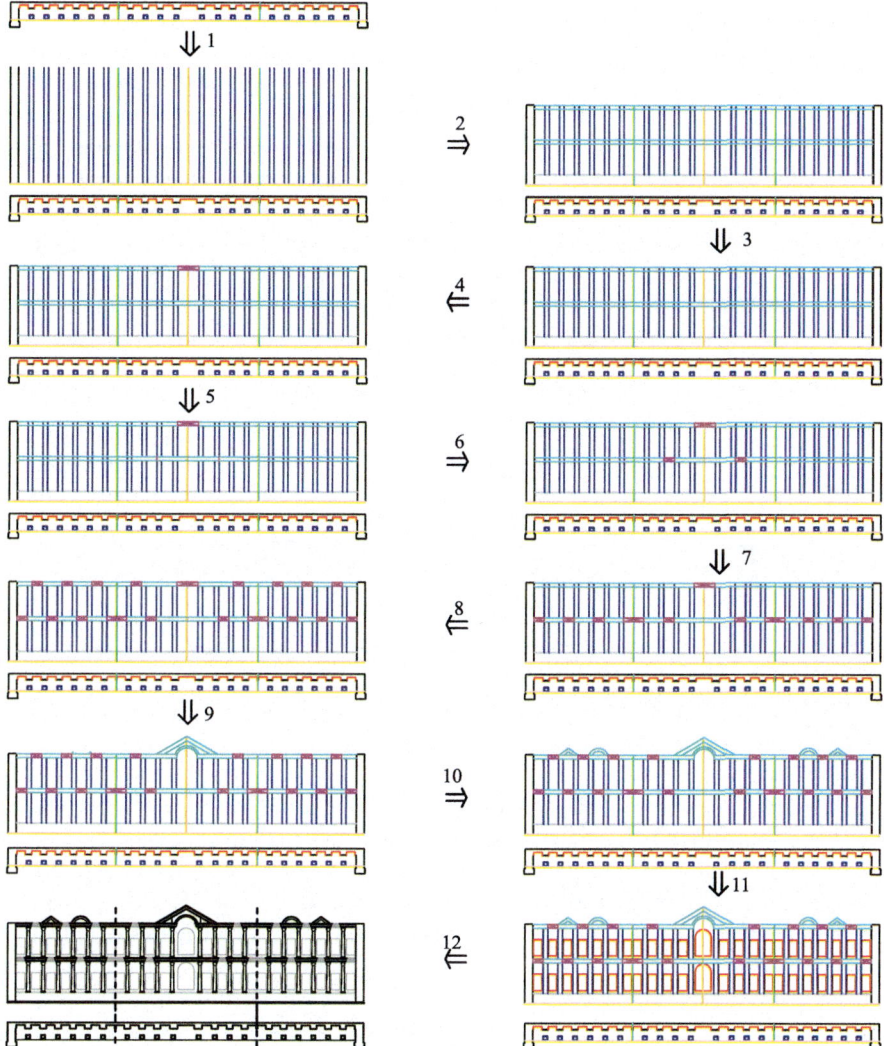

Fig. 9 Stage-by-stage computation of the (1.2.2.2.2.3) possible reconstruction of Hadrian's Library as an aedicular room with a protruding lower-level and a recessed upper-level central bay, and, a tripartite focal point crowned by a triangular pediment with an arcuated lintel at the central bay, within split pediments

Moreover, this example demonstrates how, by following a systematic combinatorial approach in the derivation, we can systematically generate the widest range of iterations, including variations that we might not have been predisposed to consider, and thus provide new insights into the architectural reconstruction and interpretation of the building, as Wescoat (2019, 309) had predicted. Particularly in the case

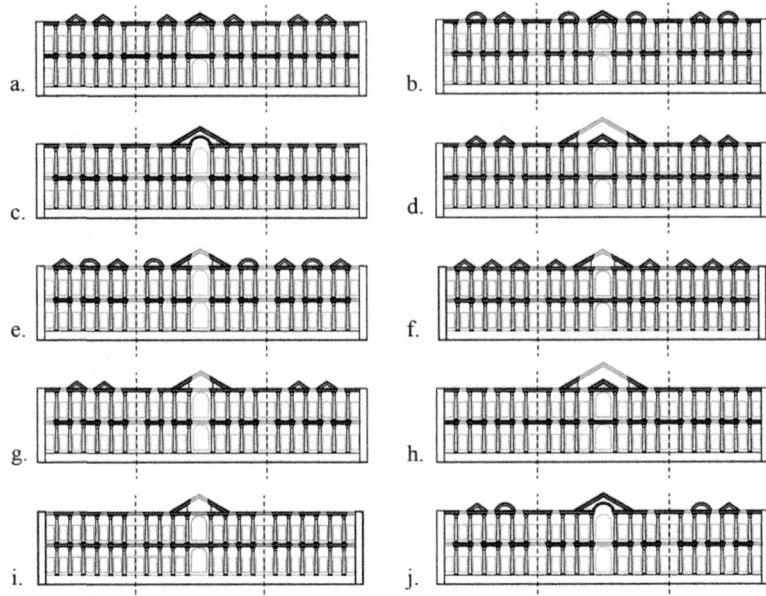

Fig. 10 Ten possible grammar-generated reconstructions of the unfolded interior façade of the main hall of the library of Hadrian in Athens with design principles after: **a** Building M in Side, **b** the Library of Celsus, Ephesus, **c** the Marble-Hall of Sardis, **d** the Hydrekdochein in Ephesus, **e** the theater of Aspendos, **f–j** hypothetical variations

of the interior elevation of the main hall of Hadrian's library, the evidence shows that there were three types of intercolumniations: the central one, the ones immediately after, and everything else. Even though this evidence was documented from the beginning (Tigginaga 1999), scholars did not discuss it or justify it as part of the design and bypassed it as they were predisposed to the long established theories about the design of libraries with a continuous colonnade. With shape grammars, we can include possible scenarios, even when they don't seem possible at first, so that we can use the visual evidence as an additional tool to argue for their potential efficacy.

Additionally, shape grammars help us overcome the problem of visualizing uncertainty when there is lack of evidence. Archaeologists avoid visualizing parts when there is no material evidence to rely on. In his most recent reconstruction of Hadrian's library, Kanellopoulos avoids resolving the architectural form of the crowning of the aedicule and visualizes a plain horizontal entablature. While this is a viable scenario, there are more scenarios that are equally viable. Shape grammars presents a more inclusive approach, one that visualizes uncertainty through the visualization of multiples.

Finally, this methodology allows us to explore a hypothesis that was never formulated: that the Roman imperial library like other imperial buildings could have had a monumental, lavish interior, which would have done better justice to the mixed function of a "bibliotheca" in antiquity, exhibiting manuscripts and other works of art within the same space. The lower niches, easily accessible from the podium, could have been dedicated to the storage of books and possibly sculptures, while the upper niches could have been used solely to exhibit sculptures. Shape grammars allowed us to explore the variation with which this hypothesis can be visualized and argue about its viability or not.

4 Conclusions and Future Directions

These two grammars showcase two opposite approaches in archaeological shape grammar applications. The first one analyzed a single building and encoded in visual rules what the archaeologists had already established for its reconstruction. The goal of this grammar was to provide a tool that can visualize a possible reconstruction in plan and elevation effectively in a small amount of time. The second grammar analyzed a corpus of 11 facades sharing the same language of design, with a goal to visualize possible hypothetical reconstructions of a building whose façade did not survive, yet stylistically and typologically shares the same language of design.

Within a language of design, the first grammar can be expanded to account, parametrically, for other buildings of the same type, whereas the second grammar expands the design principles of known cases in a corpus to apply to a particular building, aiding speculation about possible architectural forms it may have had. Ultimately, both grammars generate classical buildings, and both grammars can be perceived as variations of a proto-grammar of classical architecture. In both cases, key for the applicability of the grammar in more than one case is the parameterization of the rules, so that they can apply in an initial shape of different dimensions and proportions. At this point, the parameterization of the rules has been achieved with different techniques. First, it has been achieved with the help of labels in the header. For some rules, the transformation under which the rule is allowed to apply is defined as a similarity transformation (scale) and for some others as an affinity transformation (stretch and sear). Another technique is the loop count. The user can change the number of loops a rule can apply and generate components of different proportions, as, for example, in the generation of the column shaft height as a ratio of the lower column diameter. Finally, the user of the grammar can modify the shape rules, e.g. the distance of the antae from the back wall in a tetrastyle prostyle plan, and the widened central interaxial in a distyle in antis plan can be manipulated to generate different variations that can test different hypotheses; but this requires a manual rule change that may be challenging for some users.

The parameterization of the grammar so that it can be reused in more than one building raises the question of the interface. Researchers who need grammars are those who need guidance for reconstructing a fragmentary design. A grammar should make it easy for a user to understand the range of possibilities and either select the manual generation of any possibility one wants to visualize or select an automated generation of all possibilities. Moreover, grammars need to respond to the issue of probability. When more than one possibility is present, which is more probable? The authors of grammars frequently analyze the evidence, sort through it, and rank it. This knowledge can be included as a comment next to each rule to indicate its probability. A more impactful way to incorporate probability would be if we could incorporate it into the rules, so that in each computation we could decide which block of rules to apply, when more than one choices are available. Further, in an automated computation in Drawscript, the machine could make that decision, and in the end, it could generate its own corpus of randomized versions, ranked by their degree of probability.

Finally, a future goal should be to bridge the gap between the state of preservation and the possible reconstruction or reconstructions, the part and the whole. This requires the recognition of fragmentary lines and the replacement of potential lines of maximum length that might correspond to their reconstruction. We are hopeful that Shape Machine and Drawscript can accomplish a part-recognition and thus, complete automation of archaeological reconstruction, from the state of preservation to its restored state. It's important to keep in mind that the goal of archaeological reconstruction is not to present a definitive proposition, but rather to convey the intellectual process that an archaeologist follows in creating a reconstruction from a fragmentary artifact, and to invite the readers and users to participate by developing their own hypotheses and testing them. At its core, shape grammars is a methodology that responds to the principles of archaeological research, which is to discover impartial evidence and the possibilities that arise from it, and demonstrate probabilities as they arise.

References

Burrell, B. 2006. False-frons: Separating the aedicular façade from the imperial cult in Roman Asia minor. *American Journal of Archaeology* 110 (3): 437–469.

Chippindale, C. 1992. Grammars of archaeological design. A generative and geometrical approach to the form of artifacts. In *Representations in archaeology*, eds. C. Peebles and J.-C. Gardin, 251–276. Bloomington: Indiana University Press.

Denard, H. 2012. A new introduction to the London charter. In *Paradata and transparency in virtual heritage*, eds. A. Bentkowska-Kafel, D. Baker, and H. Denard, 57–71. Farnham: Ashgate.

Economou, A., T.C.K. Hong, and R. Newton. 2024. Shape meets Euclid: Integrating shape computation with ruler and compass procedures. *Automation in Construction*, 165: 105562, ISSN 0926-5805. https://doi.org/10.1016/j.autcon.2024.105562

Hodder, I. 1982. *Symbols in action. Ethnoarchaeological studies of material culture*. Cambridge: Cambridge University Press.

Hong, T.C.K., and A. Economou. 2023. Implementation of shape embedding in 2D CAD systems. *Automation in Construction*, 146: 1–15, 104640, ISSN 0926-5805. https://doi.org/10.1016/j.autcon.2022.104640

Hong, T.C.K., and A. Economou. 2022. What shape grammars do that CAD should: The 14 cases of shape embedding. *Artificial Intelligence for Engineering Design, Analysis and Manufacturing*, 36 (E4): 1–20. https://doi.org/10.1017/S0890060421000263

Kanellopoulos, C. 2020. The lost skin of the library of Hadrian in athens. *Athens University Review of Archeology* (3): 121–149

Knight, T. 1994. *Transformations in design: A formal approach to stylistic change and innovation in the visual arts*. Cambridge: Cambridge University Press.

Kylindreas, M. 2015. *The tetrastyle prostyle temple in relation with the distyle in antis oikos during the Archaic: Classical and Hellenistic Periods*. Master's Thesis. Athens: University of Athens.

Li, A.I. 2005. *A shape grammar for teaching the architectural style of the Yingzao Fashi*. Ph.D. dissertation. Cambridge: Massachusetts Institute of Technology.

Mamoli, M. 2020. A shape grammar for the building-type definition of the ancient Greek and Roman library and the evaluation of library plans. *Artificial Intelligence for Engineering Design, Analysis and Manufacturing*, 34: 191–206. https://doi.org/10.1017/S0890060420000189.

Mamoli, M. 2021. Shape grammars as the decoder of cultural DNA. In *A new perspective of cultural DNA*, ed. J. Lee, 93–110. KAIST research series. Singapore: Springer.

Mamoli, M. 2014. *Towards of a theory of reconstructing ancient libraries*. Ph.D. dissertation. Atlanta: Georgia Institute of Technology.

Mamoli, M. 2015. Library grammar: A shape grammar for the reconstruction of fragmentary ancient Greek and Roman libraries. In *Proceedings of the 33rd eCAADe conference: Real time*, eds. B. Martens, G. Wurzer, T. Grasl, W.E. Lorenz, and R. Schaffranek, vol. 1, 463–470. Vienna: Vienna University of Technology. https://doi.org/10.52842/conf.ecaade.2015.1.463

Mamoli, M. 2018. Shape grammars as a probabilistic model for building type definition and computation of possible instances: The case study of ancient Greek and Roman libraries. In *Design computing and cognition'*, ed. J.S. Gero, vol. 18, 499–518.

Mamoli, M., and T. Knight. 2013. Reconstructing fragments: Shape grammars and archaeological research. In *Proceedings of the 40th Annual conference of computer applications and quantitative methods in archaeology: Archaeology in the digital era*, eds. A. Chrysanthi, K. Papadopoulos, P. Murrieta-Flores, T. Sly, E. Earl, D. Wheatley, I. Romanowska, P. Verhagen, 888–896. Amsterdam: Amsterdam University Press.

Stiny, G. 2006. *Shape: Talking about seeing and doing*. Cambridge: The MIT Press. https://doi.org/10.7551/mitpress/6201.001.0001

Stiny, G. 2022. *Shapes of imagination: Calculating in Coleridge's magical realm*. Cambridge: The MIT Press. http://mitpress.mit.edu/9780262544139

Stiny, G., and J. Gips. 1972. Shape grammars and the generative specification of painting and sculpture. In *Information processing*, vol. 71, ed. C.V. Frieman. Amsterdam: North-Holland.

Tigginaga, I. 1999. Η μεγάλη ανατολική αίθουσα της βιβλιοθήκης του Αδριανού (Βιβλιοστάσιο). In *Archaeologiko Deltion*, vol. 54, 42.

Wescoat, D. 2019. Architectural documentation and visual evocation: Choices, iterations, and virtual representation in the sanctuary of the Great Gods on Samothrake, In *New directions and paradigms for the study of Greek architecture: Interdisciplinary dialogues in the field*. eds. Sapirstein and Scahill. Leiden: Brill.

Wescoat, D., et al. 2017. *The monuments of the Eastern Hill. Samothrace 9*. Princeton: The American School of Classical Studies at Athens.

Wilson Jones, M. 2000. *Principles of Roman architecture*. New Haven: Yale University Press.

Open Access This chapter is licensed under the terms of the Creative Commons Attribution-NonCommercial-NoDerivatives 4.0 International License (http://creativecommons.org/licenses/by-nc-nd/4.0/), which permits any noncommercial use, sharing, distribution and reproduction in any medium or format, as long as you give appropriate credit to the original author(s) and the source, provide a link to the Creative Commons license and indicate if you modified the licensed material. You do not have permission under this license to share adapted material derived from this chapter or parts of it.

The images or other third party material in this chapter are included in the chapter's Creative Commons license, unless indicated otherwise in a credit line to the material. If material is not included in the chapter's Creative Commons license and your intended use is not permitted by statutory regulation or exceeds the permitted use, you will need to obtain permission directly from the copyright holder.

Shape Grammars and Artificial Intelligence

A Generative Grammar for the Design of Adaptive Intelligent Virtual Worlds

Ning Gu and Mary Lou Maher

Abstract Inspired by the notions of shape grammars, a generative design grammar is a set of rules that describe a design style. This chapter describes and demonstrates how such a grammar can be developed to design places in 3D virtual worlds, using the scenario of a virtual gallery. By encoding the design style as a grammar, the rules can be executed to generate many different individual designs sharing the style, in contrast to the current practice of handcrafting each individual design. The intelligent design of places in 3D virtual worlds is then realized through a Generative Design Agent (GDA) by at first applying the grammar to generate the initial design, and then adapting the design during real-time interactions in virtual worlds. As a result, this design grammar is integral to the 3D model, providing the algorithmic process through GDAs to modify the virtual world in response to the needs of its inhabitants in real time. This represents a new role for design grammars in which the grammar enables real-time design and modification of a virtual world.

1 Introduction

Virtual worlds are places that exist entirely in networked environments in which people co-exist, communicate and interact through their avatars. These worlds are dynamic and interactive environments that support a broad range of social, entertainment, educational, and productive activities that are loosely based on activities in the physical world. A common metaphor for the design of these worlds is the concept of place. Inspired by the notions of shape grammars, design grammars provide a formalism for generating virtual worlds and, due to their inherent ability to capture transformation through a rule-based formalism, these design grammars provide a

N. Gu (✉)
UniSA Creative, University of South Australia, Adelaide, SA 5000, Australia
e-mail: ning.gu@unisa.edu.au

M. L. Maher
College of Computing and Informatics, University of North Carolina Charlotte, Charlotte, NC 28223, USA
e-mail: m.maher@uncc.edu

representation for virtual worlds that can automatically adapt to the changing needs of their users. In this chapter, we describe a design grammar that is integral to the 3D model, providing the algorithmic process of agents that modify the virtual world in response to the needs of its inhabitants in real time. This is a new role for design grammars in which the grammar enables real time design and modification of a virtual world. We show how this can be applied to an intelligent virtual gallery.

The design of places in virtual worlds draws on our experience and knowledge of architectural design in the physical world. The metaphor of place and reference to concepts from architectural design provides a consistent and familiar base for people in the virtual world and also for designers of virtual worlds. This metaphor and reference facilitate the interaction of virtual world users with the designed environments and with each other. Effectively, the physical world provides the inspiration for the design of virtual worlds. The underlying rationale for using the place metaphor is that, because to a large extent our social and cultural behaviors are organized around the physical world's spatial elements, we can carry over these patterns of behaviors to virtual worlds by designing them to have the same potentials for conception and interaction as do the physical world exhibits (Kalay and Marx 2001; Champion and Dave 2002; Kalay 2004). The patterns of behavior we learn in the physical world become useful in virtual worlds. By structuring virtual worlds in a way that allows us to apply these learned traits from the physical world, we can reduce the cognitive efforts needed to inhabit the worlds. Adopting the metaphor of place, designing virtual worlds as a relatively new area is able to make analogical references to place design which has been developed for centuries, rich with its own theories and practice. The analogy provides a base to understand and further extend the use of these networked environments.

Early developments in the design of virtual worlds took two major paths: (1) The conceptual development of the design and purpose of virtual worlds; and (2) the technical development of the design and implementation of virtual worlds. In the conceptual development of virtual worlds, design practitioners and researchers explore the possibilities of virtual worlds, illustrate the future of virtual worlds, and study the impact of virtual worlds on existing design theories and practices. For example, Benedikt (1991) collects a series of influential writings by designers, artists, novelists, engineers, businessmen and academics to predict and illustrate the future of cyberspace from different perspectives, where everything seems possible. These writings also outline the dramatic changes in the physical world, and our future daily life due to the influence of cyberspace. Woolley (1993) discusses the emergence of virtual worlds, which changes public reality through virtual reality and artificial reality. Anders (1998) presents theories and examples that use space as a cognitive tool for managing our daily activities in the physical world, showing how these concepts may be extended to cyberspace. Wertheim (2000) follows the history of the Western conception of space from the Middle Age to the information age, critically accessing cyberspace and cyber culture. Currently, in the 21st Century, as we look back and re-examine some of these concepts and predictions, virtual worlds are indeed challenging, and are gradually changing traditional forms of communication, education, entertainment, business and so on. However, as some have predicted, the physical world has not been radically eclipsed by the emergence of virtual worlds, which coexist with, and supplement, the physical world. The concept of ubiquitous computing, initiated by Weiser (2002) and researched by many others (Abowd and Mynatt 2000; Dourish 2001), is one example of the seamless blend of the virtual and the physical). This concept has also been applied on an urban scale to demonstrate the development of ubiquitous cities in a real context (Lee et al. 2008). Another early development in the design of virtual worlds emerged from advances in design and implementation, in which design practitioners and researchers in the 1990's worked on technically realizing and building virtual worlds, when the internet had become more accessible. This served two purposes: (1) to simulate battlefields for military training by; for example, SIMNET (simulator networking) developed by the US Department of Defense that simulates battlefields for military training purposes; and (2) to become the basis for networked games; for example, Doom and Quake, 3D networked games first released in the 1990s, downloaded and shared by millions, even today. Beyond military simulation and networked games, virtual worlds have been designed and implemented for numerous purposes. The internet has accommodated many different technologies that support early text- based virtual worlds, graphic virtual worlds, and 3D virtual worlds. This has expanded to include social communication, education, design collaboration, e-commerce, and many other areas.

Key literature on virtual world design and implementation, as suggested by Maher (1999), can address one or more of the following issues: (1) implementation—technologies for realizing virtual worlds; (2) representation—a consideration of digital representation and management of various components of virtual worlds; (3) interface—the types of interface that allow access to virtual worlds for interacting with different environments, and for interacting with each other.

At the implementation level, 3D models have become the dominant form of visualizing virtual worlds beyond text-based and graphic representations. At the representation level, the use of metaphors is used, for example, in the study of how text-based virtual worlds, mainly MOOs, are represented and designed (Cicognani and Maher 1998); and its extensions from the linguistic characterizations to include graphical and spatial characterizations (Maher et al. 2000a, b, 2001) using a place metaphor as an analogy to the built environment, and to create a coherent hierarchy of architectural elements, such as buildings, rooms and objects representing object-oriented virtual worlds. More recently, computational models have been used to represent virtual worlds as a way to integrate artificial intelligence and virtual worlds. This includes cognitive agent models (Maher and Gero 2002; Smith et al. 2003; Maher et al. 2005) and generative design algorithms (Gu and Maher 2003; Müller et al. 2006). At the interface level, the input and output devices for virtual worlds include desktop and mobile screen-based interaction, augmented and mixed reality, and immersive reality headsets.

Emerging in the 1990s, virtual worlds are an online environment that allows multiple users to have access to a single application in real time on the internet. Singhal and Zyda (1999) describe virtual worlds as software systems where multiple users connect from different geographical locations and interact with each other in real time. They also characterize virtual worlds with five common features: (1)

Fig. 1 Left: a virtual cinema designed in the Active Worlds University of Sydney Universe, by the students; right: a public place in Second Life

A shared sense of location; (2) a shared sense of presence; (3) a shared sense of time; (4) online communication; and (5) interaction with the virtual environments. Highlighted by these features, virtual worlds are capable of providing multiple users with the ability to interact with each other, to share information, and to interact with the virtual environment by manipulating virtual world objects in the environment through immersive 3D models. The following provides an overview of some typical virtual world examples.

Active Worlds (AW) is one of the earliest platforms for designing and operating 3D virtual worlds. Using AW, virtual worlds are designed and implemented based on the AW object library. The object library provides a list of 3D models that simulate place elements. Users can build virtual places using these models. The object library can be expanded by using external 3D modeling and translation tools. Virtual world objects can use AW triggers and commands, a simplified scripting language. Figure 1 shows a virtual cinema designed and implemented using AW.

Second Life (SL), as one of the most popular 3D virtual world platforms, has a strong focus on virtual communities that support a diverse range of online activities. Besides gaming, social communication and e-learning activities, SL is particularly known for its e-commerce activities, including trading for virtual estates and properties. SL provides basic in-world modeling tools for designing and implementing virtual worlds through the direct manipulation of geometric primitives. Behaviors of virtual world objects and avatars can be controlled and customized using Linden Scripting Language (LSL). Figure 1 shows a public place in SL.

Having introduced a formal computational approach to designing places in virtual worlds based on design grammars and computational design agents, we now describe a framework for a generative grammar for virtual worlds (Sect. 2); demonstrate how this grammar is used for designing an adaptive virtual gallery (Sect. 3); and provide a summary and key conclusions (Sect. 4).

2 A Generative Design Grammar Framework

A generative design grammar is a set of rules that describe a design style. By encoding the design style as a grammar, the rules can be executed to generate many different individual designs sharing the style, in contrast to handcrafting each individual virtual world. This section presents a framework that provides organizing principles for developing a grammar that addresses the essential considerations in designing virtual worlds. The framework specifies the general structure of the design grammar and its basic components: the design rules. Designers can develop their own grammars for designing virtual worlds by using the framework to develop rules that describe their style. A Generative Design Agent (GDA) applies the grammar to first generate the initial design, and then adapt it during real time interactions in virtual worlds. The agent model is wrapped around the design grammar, providing the mechanisms to sense and change the virtual world. Each GDA can be associated with a user of 3D virtual worlds, serving as their personal design agent to intelligently design virtual worlds, satisfying the person's changing needs and capturing her style. For demonstration purposes, we present a generative design grammar example for a virtual gallery that adapts in eight stages. These stages present various changes in a virtual gallery during its use, for example, changes of activities, exhibition requirements, visitors, and so on. The demonstration shows that the artist's GDA analyzes these changes and responds by dynamically generating and modifying the design of the virtual gallery.

2.1 Generative Design Grammars for Place Design in 3D Virtual Worlds

Our concept and development of generative design grammars is inspired by shape grammars (Stiny 2006, Stiny and Gips 1972), stemming from shape grammars' suitability as a formalism for both describing and generating designs. A generative design grammar describes place designs in 3D virtual worlds where the basic components of its design rules are virtual world objects.

2.1.1 A Place Design in 3D Virtual Worlds as "Objects in Relations"

A design generated by a shape grammar is viewed as "elements in relations" (Stiny 1990, 1999). To apply a shape grammar for design generation is basically to identify the "elements" of design, and define and alter (add, subtract or replace) the "relations" among the "elements" via shape rule applications. In this manner, shape grammars can generate complex designs based on simple design elements. This view of design is consistent with the object-oriented nature of virtual worlds. A place design in 3D virtual worlds can be viewed as "objects in relations".

Three-dimensional virtual worlds are object-oriented systems in which a design is constructed through the placement and configuration of virtual world objects. The virtual world objects can then be configured via scripts or codes to have certain behaviors that enable people to interact with the place and with each other. Therefore, a place design in 3D virtual worlds basically comprises various virtual world objects that visually and functionally support intended human activities (Gu and Maher 2003) in the following ways:

1. Visually/spatially, via the use of the place metaphor, the 3D models are composed to form an ambient environment where people can inhabit and the intended activities can take place.
2. Functionally, selected virtual world objects are ascribed with behaviors to support the intended activities of people in the virtual world.

When designing places in 3D virtual worlds, generative design grammars describe virtual places in terms of virtual world objects and their relations, in the forms of design rules, applied to generate different designs.

2.1.2 Design Phases of 3D Virtual Worlds

Our approach considers virtual worlds as functional places that support an extended range of activities online. Based on this understanding, designing virtual worlds can be divided into four phases:

1. Layout: the layout of the virtual place defines how areas are related to each other spatially, in a way that defines and accommodates intended activities in the 3D virtual world.
2. Objects: each place is then configured with a number of virtual world objects, such as walls and floors that provide visual boundaries of the different areas in the place and more generally, objects such as the information desk or paintings that provide visual cues for supporting the intended activities.
3. Navigation: navigation in virtual worlds can be facilitated byway finding aids and hyperlinks for assisting users' movements from one place to another.
4. Interactions: this phase ascribes scripted behaviors to selected virtual world objects so that the users can interact with the objects and with each other by triggering those behaviors.

Places can then be generated in 3D virtual worlds in terms of visualization design (place layout and object design), navigation design and interaction design. These four inseparable design phases create the integral structure of a virtual place.

2.1.3 Addressing Design Requirements of 3D Virtual Worlds

Generative design grammars are able to address both visual and spatial requirements as well as non-visual, non-spatial requirements for designing 3D virtual worlds. The design requirements that are related to the functional aspect of virtual worlds are often non-visual/spatial requirements. They ascribe scripted behaviors to selected virtual world objects to support the intended activities. Similar functional problems have been addressed using shape grammars in design. Although shape grammars are spatial, Stiny (1981) demonstrates the use of description functions to address the composition of designs in other terms that are non-spatial, for example, those related to the functions, purposes, uses and meanings of the designs. Knight (1999) further suggests the linkage of an original shape grammar with a parallel description grammar to generate designs and their non-spatial descriptions in parallel. Subsequent developments such as discursive grammar (Duarte 2005) and transformation grammar (Eloy and Duarte 2011) also aim to address the semantic and syntactic issues in design.

2.1.4 Capturing Stylistic Characterizations of Virtual Places

A specific style is exemplified when several designs "each create a similar impression" (Stiny and Mitchell, 1978). 3D virtual worlds can be considered in terms of visualization design (place layout and object design), navigation design and interaction design. Compared to many novice designs, virtual worlds designed with a specific style in mind will achieve better consistency, making it easier for users to get oriented and interact in the virtual world. A shape grammar is capable of generating design instances that belong to an existing language of design as well as defining a new language of design. March and Stiny (1985) use "syntax" and "semantics" as the two major factors to distinguish designs from one another. "Syntax" determines how the shapes are composed to represent the designs. "Semantics" describe the designs in terms of functions, purposes, uses or meanings other than shapes. Generative design grammars describe and generate languages of design for 3D virtual worlds, through categories of design rules to specify place design in 3D virtual worlds in terms of syntax or visualization: place layout and object design, and semantics: navigation and interaction.

2.2 The Generative Design Grammar Framework

The generative design grammar framework outlines the structure of a generative design grammar and its basic components: design rules. Following the structures suggested in the framework, and integrating with different design and domain knowledge, designers will be able to develop their own grammars for designing places in 3D virtual worlds. A generative design grammar G is comprised of design rules R, an initial design Di, and a final state of the design Df. The basic components of the grammar are design rules R. The general structure of our grammar comprises four sets of design rules applied in order: layout rules Ra, object design rules Rb, navigation rules Rc, and interaction rules Rd as illustrated in Fig. 2.

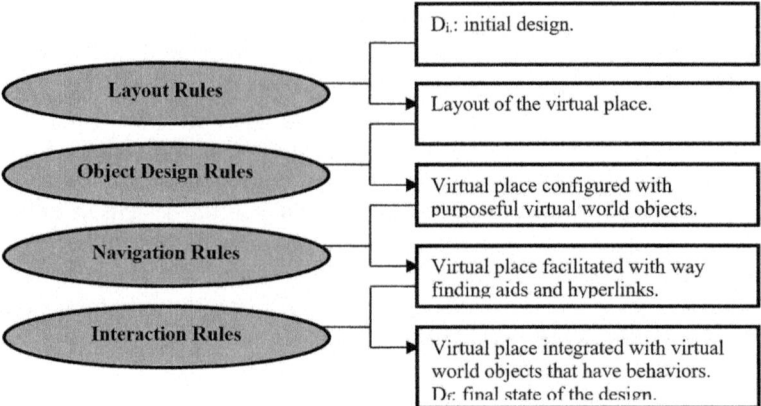

Fig. 2 The generative design grammar framework

2.3 Design Rules

The basic components of a generative design grammar are design rules. The general structure of design rules is similar to that of shape rules. In shape grammars, a shape rule can be defined as: LHS → RHS, specifying that when the left-hand-side shape (LHS) is recognized in the design, it will be replaced by the right-hand-side shape (RHS). The replacement of shapes is usually applied under a set of shape operations or spatial transformations. The shapes are labeled (using spatial labels and state labels) for controlling the shape rule applications. A design rule of a generative design grammar is defined as: LHO + sL → RHO, which specifies that when the left-hand-side object (LHO) is recognized in the 3D virtual world, and the state labels sL are matched, the LHO will be replaced by the right-hand-side object (RHO). The term "object" used here can refer to a virtual world object, a set of virtual world objects or virtual world object properties.

In the design rules, state labels are expressed explicitly as sL. The use of state labels is essential to the generative design grammar application as they direct the application to ensure that the generated virtual world design satisfies the requirements. Each design rule is associated with certain state labels representing specific design contexts. In order for a design rule to be applied, virtual world objects need to be recognized in the 3D virtual world that match the LHO of the design rule, and the design context as represented by the sL of the design rule needs to be relevant to the current design needs. The basic components of design rules are virtual world objects, not shapes. Therefore, they are not entirely visual or spatial. As a result, for the interaction rules and parts of the navigation rules, the replacement of LHO with RHO may not change the way a virtual world object looks or is placed but may add scripted behaviors to an existing object.

Layout rules are visual/spatial rules that generate the layout of the place according to the kinds of intended activities to be supported in the 3D virtual world. The use of divided virtual areas for different activities provides a way of organizing and allocating activities, and creates a sense of movement for people when changing from one activity to another. Because of the use of the place metaphor, layout problems in 3D virtual worlds can have similar solutions to the ones in the physical world. However, designers do not have to strictly obey physical constraints, since virtual places can also be hyper-linked.

Object design rules are applied after the layout rules, and they are also visual/spatial rules. After a layout of the virtual place is produced, object design rules further configure the place to provide visual boundaries and visual cues for supporting the intended activities through virtual world object design and placement. Object design rules can also refer to principles and examples in built environments. 3D virtual worlds designed with a consistent use of forms, colors and other visual/spatial elements are more effective in assisting users' orientation and interaction.

Next, navigation rules are applied to provide way finding aids and hyperlinks in the generated virtual place to assist users' navigation. Way finding aids in virtual worlds have been studied with direct references to those in the built environment (Vinson 1999; Darken and Sibert 1993, 1996). Unlike their physical counterparts, virtual places can also be hyper-linked, allowing users' avatars to move directly between any two locations. Navigation rules are not entirely visual/spatial. Their application involves virtual world object design and placement for defining way finding aids and hyperlinks in the generated place. However, before these object design and placement are conducted, navigation rules are mainly about recognizing the connectivity within the generated place and with other places as well as finding appropriate navigation methods for users.

Interaction rules are the final set to be applied. The application of interaction rules ascribes scripted behaviors to selected virtual world objects in the generated place. People can interact with the virtual place and with each other by triggering these behaviors. Interaction rules are non-visual, non-spatial rules that recognize selected virtual world objects and ascribe appropriate behaviors to these objects. There can be at least two different types of interaction rules. One supplements object design rules and the other supplements navigation rules. The first type of the interaction rule ascribes scripted behaviors to relevant virtual world objects in order to support the intended activities in the virtual place. The other type of interaction rule looks for way finding aids and hyperlinks generated by navigation rules and ascribe scripted behaviors to activate them.

2.4 The Generative Design Agent Model

The Generative Design Agent (GDA) model is the reasoning mechanism of the adaptive virtual world, based on five computational processes of sensation, interpretation, hypothesizing, designing and action (Fig. 3), where sensors and effectors act as the interface between a GDA and the virtual world. In this chapter we focus on the

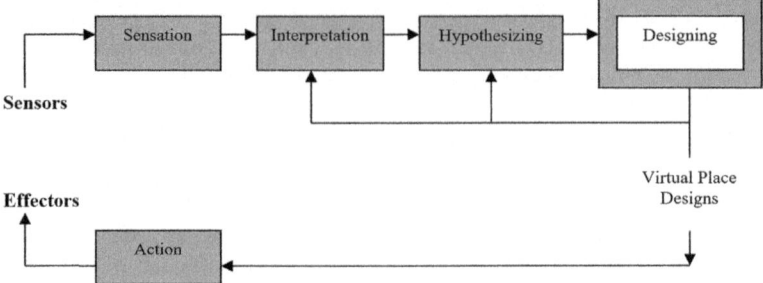

Fig. 3 Five computational processes of the GDA model

designing component of a GDA which is enabled by the application of a generative design grammar.

3 Demonstrating the Grammar for a Virtual Gallery that Adapts

We present a demonstration of the design grammar for designing a virtual gallery that adapts to changing needs, such as types of activities, exhibition requirements, and number of visitors. This demonstration is a virtual gallery designed for an aspiring artist as a personal studio and exhibition space. The purpose is to provide online places to display artists' work and to serve as the main portal to access their online presence, similar to that of social media or websites. It also provides alternative places supporting other key activities of the artist; for example, art creation, collaboration, meeting, public lecture and other functions. The GDA assists the artist by reasoning, designing and acting on their behalf in the virtual world. The design component of the GDA is realized by the application of this example grammar. In our virtual gallery, different designs of the gallery are dynamically generated, and different arrangements of the exhibitions are provided accordingly, adapted for changes during its use, to better serve the needs and to optimize the individual experiences of different parties. The visual forms of the gallery provide a spatial awareness by defining a 3D ambient environment. The virtual gallery displays both digital replicas of the physical artworks, and digital or network-based exhibitions that cannot be displayed in physical galleries. The exhibitions are arranged in different gallery areas for viewing and interaction, rather than being integrated into an electronic database or web pages for browsing. Visitors are represented as avatars, providing an awareness of self and others. Through these avatars, visitors are able to explore the gallery and move from area to area for different activities online.

3.1 A Generative Design Grammar for Intelligent Virtual Gallery Design

The generative design grammar framework has layout rules to provide gallery layouts by allocating different areas; object design rules that configure each area with certain purposeful objects which provide visual boundaries and visual cues for supporting the intended activities; navigation rules that specify navigation methods in the gallery by using way finding aids and hyperlinks for assisting the visitors' visits; and interaction rules that ascribe scripted behaviors to selected objects so that exhibitions and other activities can function. Here, we include selected rules for demonstration and a complete list of design rules can be found in Gu and Maher (2014). We group each set of design rules into additive and subtractive rules to show how the rules enable the virtual world design to adapt to real time needs of users.

3.1.1 Layout Rules

The virtual gallery grammar has 15 additive and 17 subtractive layout rules. The virtual gallery can have five areas: the reception, galleries 1 and 2, artist's personal studio, and multi-function areas.

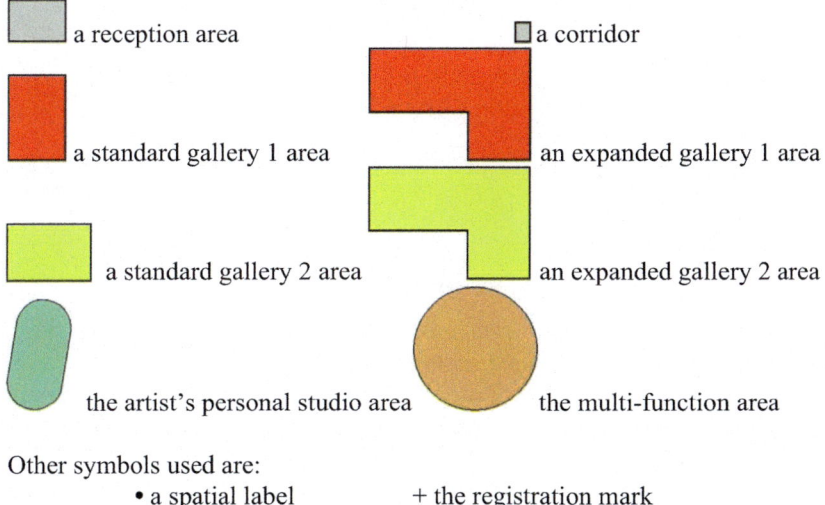

Other symbols used are:
• a spatial label   + the registration mark

The initial design is:

A reception area provides visitor information, and access to other areas. The grammar allows for a reception area to provide visitor information and access to other areas, 2 gallery areas, a personal studio area is private for the artist and their guests, and a multi-function area can serve as a conference venue for exhibition opening and public lectures.

The spatial labels are used to control the layout rule application. For example, ■ is the layout of a standard gallery 1 area, and is marked with a spatial label on its left edge. This indicates that a layout rule can be applied once to this symbol to compose the gallery layout from this left edge. The state labels can be roughly divided into three groups: state label 1, state labels used in additive, and subtractive rules. $sL = 1$ is used in all layout rules being the first set of design rules to be applied. Other state labels are matched if the design context they represent is related to the GDA's current design goals. For example, to match $sL = S$ (the personal studio area for the artist is needed), firstly in the process of interpretation, the artist's GDA interprets that: The artist is present in the virtual gallery. There is no studio space available for the artist. The GDA hypothesizes a design goal based on these interpretations—a personal studio area for the artist is needed. As a result, $sL = S$ is matched. A design rule is selected for application only if the LHO of the rule is found, and the state labels of the rule are matched.

Additive Layout Rules add different areas for generating the gallery layout. For example, rule 1 shows that the artist's personal studio area can be added spatially adjacent to a reception area if: The reception area is found, and $sL = S$ is matched.

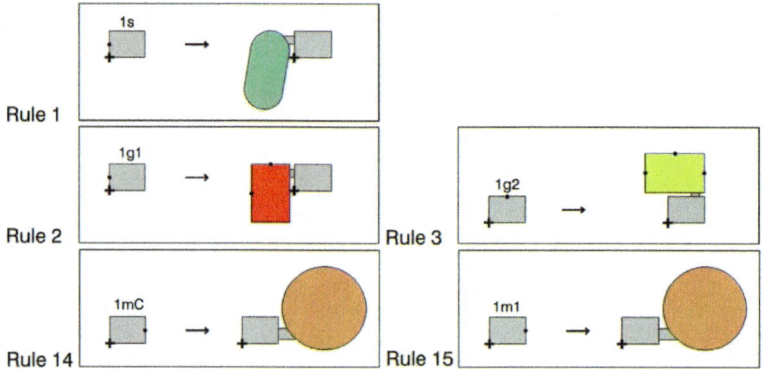

Through these 15 rules, the artist's personal studio, initial standard gallery and multi-function areas can be added accordingly, adjacent to a reception area. Additional gallery areas can be added to accommodate more avatars once the visitor limits are reached. The standard gallery area can also be expanded to accommodate more exhibition items.

Subtractive Layout Rules remove or reduce different areas from a gallery layout. For example, Rule 16 shows that the artist's personal studio area can be removed if the personal studio area is adjacent to a reception area, and $sL = cS$ is matched.

A Generative Grammar for the Design of Adaptive Intelligent Virtual ... 469

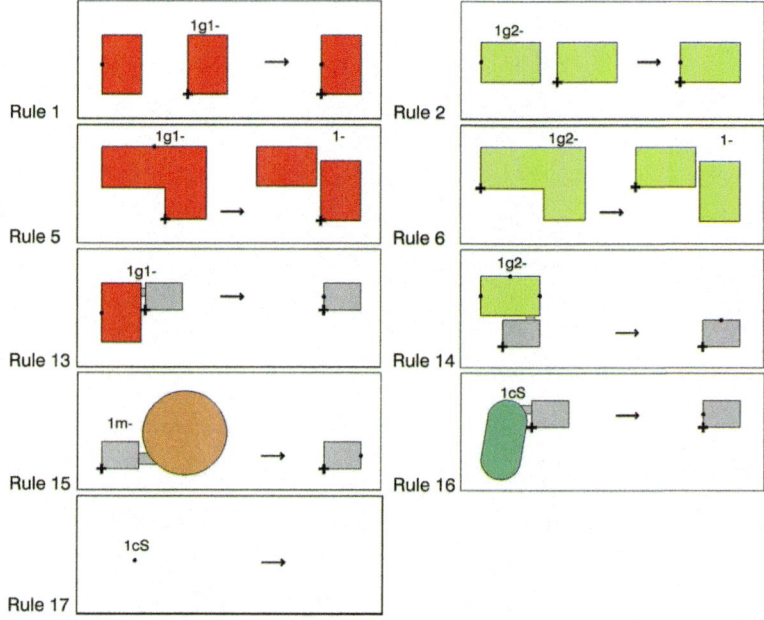

These 17 rules can be applied to remove the initial standard gallery, multi-function, and personal studio areas accordingly to remove redundant gallery areas to accommodate smaller visitor numbers or fewer exhibition items, and to remove spatial labels so that the grammar application is terminated.

3.1.2 Object Design Rules

There are 24 additive and 9 subtractive object design rules. The objects appear in the virtual gallery as various 3D models. Figure 4 illustrates the visual boundaries of a reception area, and the artist prefers a cold-color scheme for the design.

Various purposeful objects are designed for the gallery, for example, a helpdesk is placed in a reception area to provide visitor information (Fig. 5). The information, together with hyperlinks for assisting navigation and for providing access to other media and networked environments, are visualized by the warm-color cubes to provide a contrast to the ambient environment.

The state labels can be divided into two groups: state Label 2, and state labels used in additive object design rules. Subtractive object design rules share the same state labels with subtractive layout rules. $sL = 2$ is used in all object design rules that are the second set of rules to be applied. Additive Object Design Rules are developed for transforming a 2D layout into a 3D environment and arranging 3D objects for specific purposes. In the 2D to 3D transformation rules the LHO represents the layout

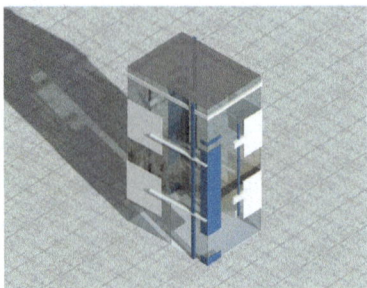

Fig. 4 The visual boundaries of a reception area

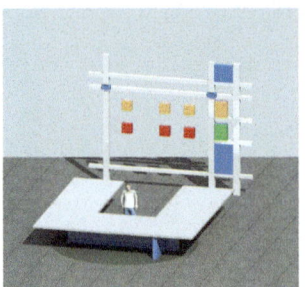 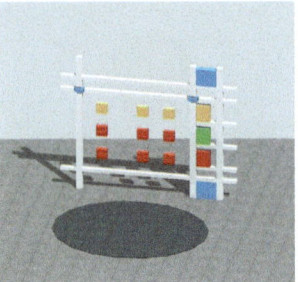

Fig. 5 A helpdesk (left) and a frame-like partition with hyperlinks (right)

of an area to be replaced by the RHO representing relevant 3D objects. For example, Rule 3 shows that a standard gallery 1 area can be provided with relevant 3D objects, to define its visual boundaries, and specify purposeful objects if: The layout of a standard gallery 1 area has been generated, and sL = cC is matched.

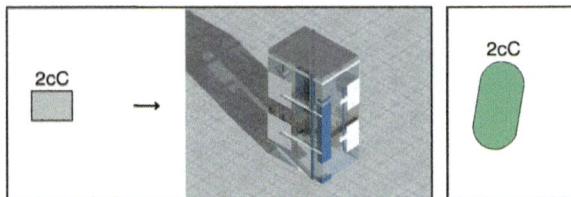

Additive object design rules 1 (left) and 2 (right).

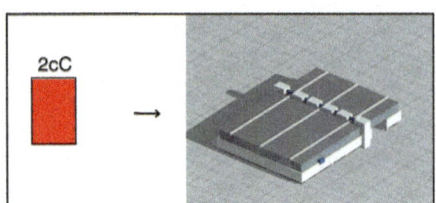 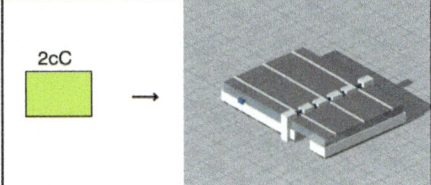

Additive object design rules 3 (left) and 4 (right).

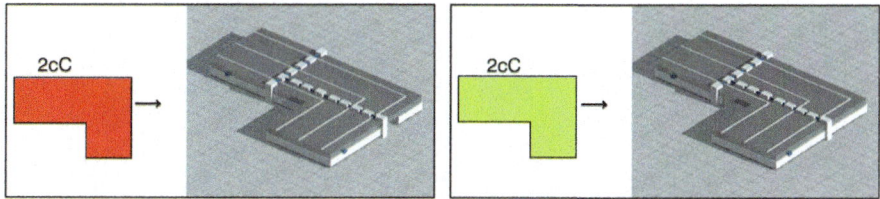

Additive object design rules 5 (left) and 6 (right).

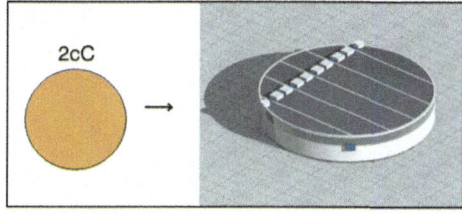

Additive object design rule 7.

Other additive object design rules, as illustrated below, further arrange each area for different purposes. The ceiling of each area is removed to show the interior. For example, Rule 8 shows that a standard gallery 1 area can be arranged for displaying digital images using configuration 1 if: A standard gallery 1 area is found, and sL = gIM1 is matched.

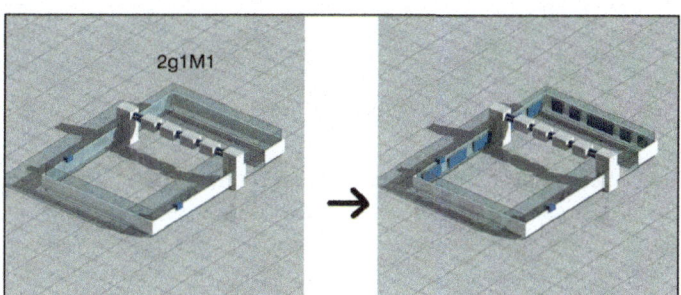

Rule 22 arranges a meeting area inside the artist's personal studio area. Rules 23–24 arrange the multi-function area as a conference venue for various public events serving a larger crowd, and as an additional gallery area for exhibiting large-scale installations.

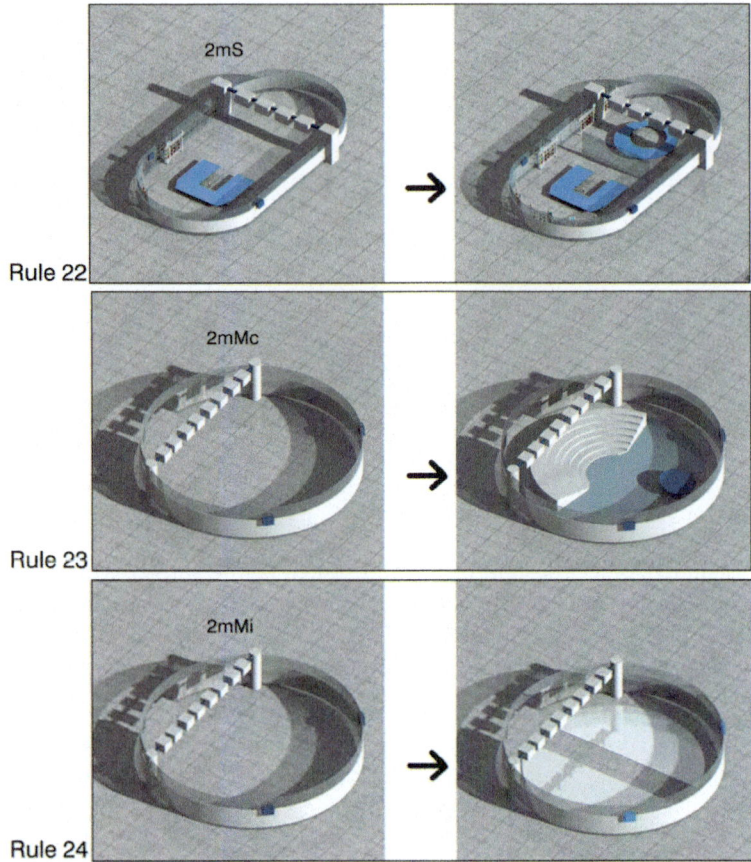

Subtractive Object Design Rules are closely related to the subtractive layout rules, directed by the same group of state labels. They remove visual boundaries and purposeful objects from gallery areas. For example, Rule 8 shows that the artist's personal studio area can be removed if the artist's personal studio area is adjacent to a reception area, and sL = cS is matched.

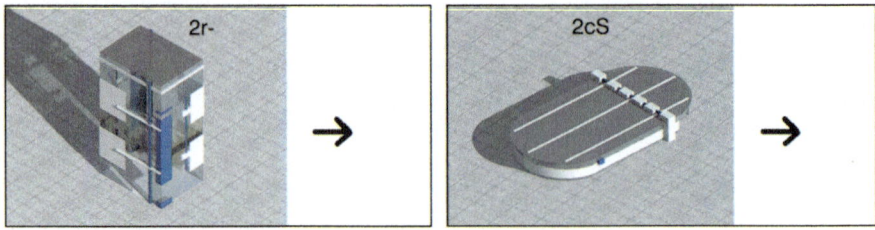

Subtractive object design rules 7 (left) and 8 (right).

3.1.3 Navigation Rules

Navigation rules have 23 additive and 6 subtractive rules. In the example grammar, the main navigation method for short-distance travel is through "walking" by the avatars, and the main navigation method for long-distance travel is through the use of hyperlinks. They are enriched to include a set of guidelines reflecting the preferences of the artist. Navigation rules share some of the state labels, defined earlier for object design rules. The only states label specific for navigation rules is sL = 3, which is used in all rules indicating they are the third set of design rules to be applied. Additive Navigation Rules add way finding aids and hyperlinks to the generated virtual gallery design. Way finding aids and hyperlinks appear as 3D models. However, for the ease of presentation, navigation rules are illustrated in 2D plan view in this chapter. They are applied to add hyperlinks to reception areas (e.g. rules 1–3); to lay paths in gallery areas to guide visitors through exhibitions (e.g. rules 4, 17); and to lay paths between different areas for directing visitors when connecting via reception areas (e.g. rules 18, 21–22). For example, rule 4 shows that paths can be laid for a standard gallery 1 area to guide visitors through the exhibition if: The standard gallery 1 area is found, and sL = gIM1 is matched.

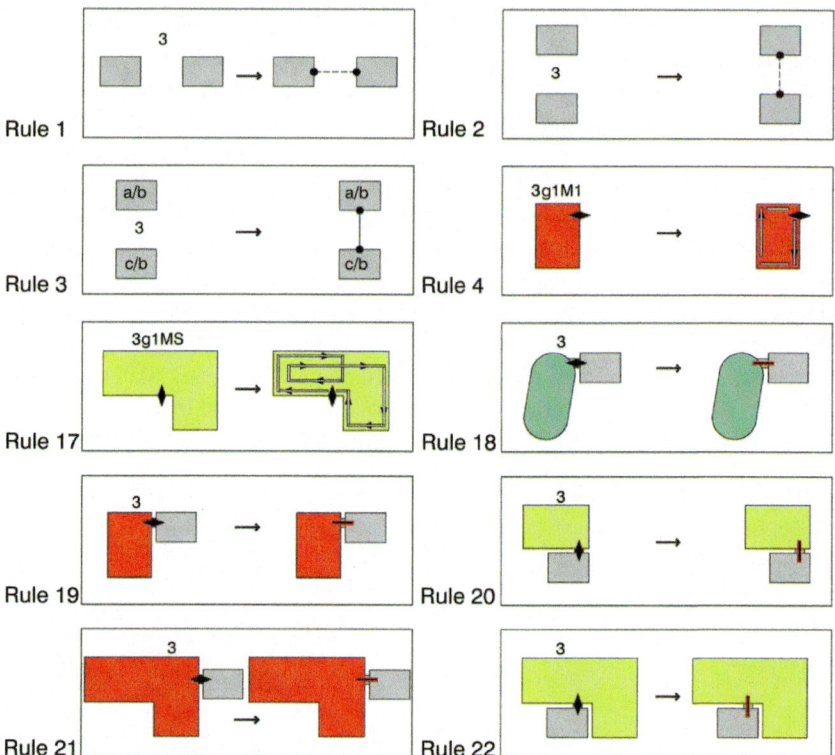

Subtractive Navigation Rules are not visual/spatial. They are mainly about recognizing the connections among different virtual gallery areas and removing different types of way finding aids and hyperlinks once the connections are lost or changed.

3.1.4 Interaction Rules

The final set of design rules have 6 additive interaction rules and no subtractive rules. Interaction rules do not operate at a visual/spatial level. $sL = 4$ is the only state label used in interaction rules indicating they are the final set of rules to be applied. Additive Interaction Rules has two different kinds in this grammar: rules that ascribe scripted behaviors to selected objects so that exhibitions and other intended activities can function and rules that look for way finding aids and hyperlinks and ascribe appropriate behaviors to activate them.

3.2 An Eight-Stage Design Scenario

To demonstrate the application of the GDA and the example grammar for the design of an intelligent virtual gallery, our design scenario shows how a GDA reasons, designs and acts in the virtual gallery for some typical situations where changes occur in the gallery during its use. The application of the grammar is directed by a set of special state labels in order to generate designs that meet the GDA's current design goals. In cases where there is more than one design rule that meets the conditions a control mechanism is needed to resolve the conflict. In general, there are three main methods for such control. They are random selection, human designer or user intervention and agent learning mechanism. The scenario applies the human intervention method.

The scenario starts with the login of the artist to the gallery, being represented by a GDA in the virtual world. This marks the beginning of the intelligent design of the gallery, which is dynamically designed, implemented and manipulated as needed, adapting to its use. At the start of the scenario, a static design of the virtual gallery is the result of the artist's previous visit (Fig. 6). This static design comprises one gallery area for displaying one of the exhibitions. The gallery area connects to a reception area from Floor 1.

Fig. 6 The visualization of the static virtual gallery design used at the start of the design scenario

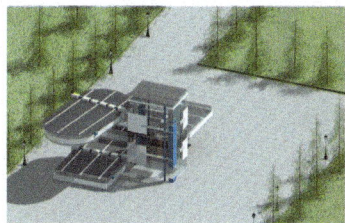

Fig. 7 Left: the visualization of the virtual gallery design for Stage 1; right: the visualization of the virtual gallery design for Stage 4

At Stage 1, the artist connects to their virtual gallery, with the intention to display two new Exhibitions 1 and 2 and the GDA senses a number of visitors. The GDA demolishes the initial gallery and at the same places the initial design of the example grammar: the layout of a reception area. After applying the design rules, the gallery has four areas: the reception area, personal studio for the artist, and two gallery areas that are configured for displaying the two exhibitions. The visualization of the generated virtual gallery design is shown in Fig. 7.

At Stage 2, more visitors visit the virtual gallery, soon reaching the maximum capacity of a standard Gallery Area (for this grammar, a gallery area has two different sizes and each gallery is designed for hosting a certain number of avatars for viewing comfort and the ease of avatars' movements in the area). The artist's GDA senses this change and applies the grammar to add an additional gallery area for displaying exhibition 1. Any future visitors to Exhibition 1 will be automatically transported to this new gallery, until the visitor number in the original decreases.

At Stage 3, the artist is browsing some digital information in their personal studio, and instructs the GDA to organize a meeting with Guest B, who will be arriving soon. The GDA applies the grammar to arrange a meeting area inside the personal studio. When user B connects to the virtual world the GDA transports B's avatar to the generated meeting area.

At Stage 4, the artist decides to give an opening address for the new exhibitions after the meeting. The artist instructs the GDA to organize this public event. The GDA applies the grammar to generate the multi-function area as a conference venue. After the artist finishes the preparation the GDA sends an invitation to the visitors and transports those who accept the invitation to the multi-function area. Compared to the previous stage, the current design (Fig. 7) has a multi-function area for the artist's opening address.

At Stage 5, after the talk, visitors return to the exhibitions. At one point, the visitor number in Standard Gallery 2 also reaches the maximum capacity. The artist' GDA senses this change and applies the grammar to add an additional gallery for displaying exhibition 2. Any future visitors to Exhibition 2 will be automatically transported to this newly generated gallery until the visitor number in the original decreases.

At Stage 6, the artist returns to the personal studio and decides to add more items to Exhibition 2. The artist instructs the GDA to accommodate these changes. The GDA informs the visitors about the changes and applies the grammar to expand and

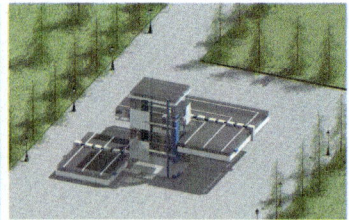

Fig. 8 Left: the visualization of the virtual gallery design for Stage 6; right: the visualization of the static virtual gallery design generated in the end of the design scenario

rearrange relevant gallery areas. The new design (Fig. 8) has therefore expanded the two Gallery 2 areas.

At Stage 7, increasingly more visitors disconnect from the virtual world. Eventually, one of the Gallery 1 areas has no visitors. The GDA senses this change and applies the grammar to remove this area. Similar situations soon occur in one of the Gallery 2 areas and in the multi-function area. As a result, two gallery areas, the multi-function area, and a reception area to which they connect are removed.

At the final Stage 8, the artist disconnects from the virtual world. The GDA records the current gallery design for future references. The GDA then applies the grammar to remove the artist' personal studio and removes all spatial labels of the design to terminate the grammar application, before terminating its own agent program. This generates a static design of the virtual gallery (Fig. 8), comprising a standard gallery 1 area, an expanded gallery 2 area, and a reception area that connects to the two galleries. Visitors can continue their visits to the exhibitions. The intelligent gallery design will restart when the artist returns.

The artist's GDA has generated an intelligent virtual gallery with its design evolving over eight stages of the design scenario, adapting to its use (Table 1). The gallery designs generated for the scenario are visualized in sequence from the top to the bottom in the left-hand-side column. The layouts of these designs are illustrated in the right-hand-side column.

At both Stages 2 and 5, alternative designs could have been generated if the artist had preferred to expand the virtual gallery with a different orientation. This would have changed the subsequent designs in the scenario. These alternative designs have been added to the corpus to enrich the design language of the grammar. They are also included in the right-hand-side column of Table 1 in their layout form. They are marked with a darker background color to distinguish from the ones selected for the scenario.

3.3 Reflections on the Intelligent Virtual Gallery Design

The design scenario demonstrates the effective use of generative design grammars and GDAs both for intelligent place design in 3D virtual worlds. As a design

A Generative Grammar for the Design of Adaptive Intelligent Virtual ... 477

Table 1 The corpus of virtual gallery designs generated for the design scenario

formalism for virtual worlds, a generative design grammar is able to describe and generate a design language for virtual worlds that captures stylistic characterizations. The corpus of gallery designs in the scenario presents a small set of samples from the design language defined by the grammar. Although these designs are generated for different purposes, at different moments during the use of the gallery, they provide a similar impression by sharing a sense of design coherency. Their stylistic characterizations can be described as follows.

Visualization: the gallery designs generated by the grammar are coherent in terms of the spatial relations defined for different areas of the gallery, and the use of forms and color schemes for visualizing the virtual world objects.

Navigation: the generated gallery designs follow the same guidelines for providing way finding aids and hyperlinks to assist the visitors' navigation.

Interaction: the generated gallery designs also have similarities in activating object behaviors in virtual worlds. Therefore, similar experiences can be gained and applied to assist the visitors participating in activities in the virtual gallery.

The virtual gallery grammar presented is only a simplified example for demonstration purposes and has limitations. These limitations do not necessarily apply to other generative design grammars and can be optimized. Each rule set in the example grammar is grouped into additive and subtractive rules. This method of design generation is simple and clear for demonstration. However, this is not the only way to arrive at design generation; for example, using a parametric grammar will enable the design to accommodate more complex and dynamic forms and spatial relations. Object design rules in the example are rigid, and they can be enriched to enable wider varieties of design considerations and styles. Currently, designers strictly control the gallery layouts and configurations. Nevertheless, design rules can be developed to allow suitable levels of ambiguity or randomness, to enable the generation of novel designs that are less predictable.

4 Conclusions

Generative design grammars adopt the descriptive and generative nature of shape grammars, but modify some of the original properties to suit the purpose of designing 3D virtual worlds. The characteristics of generative design grammars can be highlighted based on a comparison to shape grammars in terms of grammar components, meeting design requirements, and computation, as follows.

Grammar components: in shape grammars, a set of shape rules, an initial shape and a final state constitute a shape grammar (Knight 1994). Shape rules are applied recursively in a step-by-step manner to generate designs. Spatial and state labels are used to control the shape rule application. A generative design grammar also has the same components. More specifically, it has four sets of design rules tailored for designing in 3D virtual worlds.

Generative design grammars: also use spatial labels and state labels to control the rule application. The original use of state labels is to control the sequence of shape rule application. In our generative design grammar, this original purpose is maintained. In addition, a special set of state labels are developed to represent a set of design contexts that relate the current design needs.

Meeting design requirements: there are generally two different ways to connect shape grammars with design goals (Knight 1998). The first is to integrate the constraints into the shape rules so that the generated designs will meet the given design requirements. The second is to develop the original, unrestricted grammars but apply a manual or automated search and test strategy to the generated designs in the end to select the designs that satisfy the requirements.

Generative design grammars are restricted. They are similar to set grammars (Stiny 1982). The main difference between set grammars and the original shape grammars is the way designs are decomposed (Knight 1998). In a set grammar, each generated design can only be viewed and decomposed in one definite way, while in a shape grammar, it can be viewed and decomposed in unlimited ways. This liberty of the original shape grammars supports two very important aspects of shape grammars' generative power: emergence and ambiguity. However, it also makes it very difficult to predict the behaviors and design outcomes of grammar applications. Similar to set grammars, a place in the 3D virtual world, designed and composed using generative design grammars, can be decomposed uniquely into a set of purposeful virtual world objects and their properties. The application of a generative design grammar also needs to be controllable and predictable to a certain extent, as functional networked places. Using a special set of state labels, the application of the generative design grammar is directed to meet the given requirements.

Generative design grammar applications can be carried out manually by designers or be automated by computational agents, during real-time interactions in a 3D virtual world. These require each generated design to meet the current needs, and with efficiency in order to keep up with the fast pace of real-time interactions in the virtual world. The search and test approach is therefore impractical.

Computation: initial shape grammar studies were performed by hand and time-consuming. Automating this process has many advantages, and the most common purpose of automation is to assist the generation of designs (Gips 1999). Gips further points out that the challenge in automation lies in the tension between the spatial nature of shape grammars and the symbolic nature of the underlying computer representations and processing. To implement shape grammars as a computational process or system, it requires a different kind of thinking and representation (Stiny 2006). Generative design grammars are intended for computer implementations.

The GDA approach to intelligently designing 3D virtual worlds is distinctive in terms of design automation and the role of the human designer. Different from the conventional approaches, the GDA approach automates the design and implementation processes. Place design in 3D virtual worlds becomes an internal process of the virtual worlds and adapts to their uses. Previously, these processes were mainly controlled by human designers. In contrast to other approaches where the agency is provided to existing virtual world components, the GDA approach gives agency to the users of the virtual world. Shifting the agency from existing virtual world components to people frees virtual worlds from being pre-defined. Rational design agents are applied to intelligently design virtual worlds during their uses, rather than being limited to reasoning about and modifying existing designs. In terms of the roles of human designers: Using the GDA model, human designers define generative design grammars that produce different design languages for the virtual worlds, rather than pre-define every virtual place. The actual design tasks are carried out by the GDAs during real-time interactions in the virtual worlds, by applying different grammars. In summary, following the general framework illustrated above, different gram- mars can be developed for designing specific places in 3D virtual worlds with different stylistic characterizations. Designers and users may adopt them to reflect their own identities and preferences in virtual worlds. A GDA is a computational agent, which has five computational processes defined specifically for reasoning and designing in 3D virtual worlds. The designing process of a GDA is supported by the application of a generative design grammar. Through this approach, rule-based place design in 3D virtual worlds is achieved, as virtual places can be intelligently designed, implemented and manipulated as needed, adapting to their uses.

Acknowledgements The authors would like to acknowledge the support of Nova Ebrahimi for her assistance in formatting this chapter.

References

Abowd, G.D., and E.D. Mynatt. 2000. Charting past, present, and future research in ubiquitous computing. *ACM Transactions on Computer-Human Interaction (TOCHI)* 7 (1): 29–58.
Anders, P. 1998. *Envisioning cyberspace: Designing 3D electronic spaces*. New York: McGraw-Hill Professional.
Benedikt, M. 1991. *Cyberspace: First steps*. Cambridge: The MIT Press.
Champion, E., and B. Dave. 2002. Where is this place. In *Proceedings of the 2002 Annual conference of the association for computer aided design In architecture ACADIA 02: Thresholds*, 87–97. Pomona: ACADIA.
Cicognani, A., and M.L. Maher. 1998. Two approaches to designing virtual worlds. In *Proceedings of design computing on the net*, vol. 98. Sydney: University of Sydney.
Darken, R.P., and J.L. Sibert. 1993. A toolset for navigation in virtual environments. In *Proceedings of the 6th annual ACM symposium on user interface software and technology*, 157–165. New York: Association for Computing Machinery.
Darken, R.P., and J.L. Sibert. 1996. Wayfinding strategies and behaviors in large virtual worlds. In *Proceedings of the SIGCHI conference on human factors in computing systems*, 142–149. 157-165. New York: Association for Computing Machinery.
Dourish, P. 2001. *Where the action is: The foundations of embodied interaction*. Cambridge: The MIT Press.
Duarte, J.P. 2005. A discursive grammar for customizing mass housing: The case of Siza's houses at Malagueira. *Automation in Construction* 14 (2): 265–275.
Eloy, S., and J.P. Duarte. 2011. A transformation grammar for housing rehabilitation. *Nexus Network Journal* 13: 49–71.
Gips, J. (1999). Computer implementation of shape grammars. In *NSF/MIT workshop on shape computation*, vol. 55, 56. Cambridge: Massachusetts Institute of Technology.
Gu, N., and M.L. Maher. 2003. A grammar for the dynamic design of virtual architecture using rational agents. *International Journal of Architectural Computing* 1 (4): 489–501.
Gu, N., and M.L. Maher. 2014. *Designing adaptive virtual worlds*. Warsaw: De Gruyter Open.
Kalay, Y.E. 2004. *Architecture's new media: Principles, theories, and methods of computer-aided design*. Cambridge: The MIT Press.
Kalay, Y.E., and J. Marx. 2001. Architecture and the internet: Designing places in cyberspace. In *Proceedings of the 21st annual conference of the association of computer aided design in architecture ACADIA '01*, ed. J. Wassim, 230–240. Buffalo, NY: ACADIA.

Knight, T.W. 1994. *Transformations in design: A formal approach to stylistic change and innovation in the visual arts*. Cambridge: Cambridge University Press.
Knight, T.W. 1998. Shape grammars. *Environment and Planning B: Planning and Design* 25 (7): 86–91.
Knight, T.W. 1999. Applications in architectural design, and education and practice. In *NSF/MIT workshop on shape computation*, vol. 67. Cambridge: Massachusetts Institute of Technology.
Lee, S.H., J.H. Han, Y.T. Leem, and T. Yigitcanlar. 2008. Towards ubiquitous city: Concept, planning, and experiences in the Republic of Korea. In *Knowledge-based urban development: planning and applications in the information era*, 148–170. United Kingdom: Information Science Reference.
Maher, M.L. 1999. Designing the virtual campus as a virtual world. In *Proceedings of computer supported collaborative learning, CSCL 99, Stanford University*, eds. C. Hoadley, J. Roschelle, 376–382. Palo Alto: SRI International.
Maher, M.L., and J. Gero. 2002. Agent models of virtual worlds. In *Proceedings of the 2002 Annual conference of the association for computer aided design In architecture ACADIA 02: Thresholds*, 127-138. Pomona: ACADIA.
Maher, M.L., N. Gu, and F. Li. 2001. Visualisation and object design in virtual architecture. In *Proceedings of the sixth conference on computer aided architectural design research in Asia, CAADRIA '01*, 39-50. Sydney: CAADRIA.
Maher, M.L., S. Simoff, N. Gu, and K, Lau. 2000a. Two approaches to a virtual design office. *International Journal of Design Computing* 2.
Maher, M.L., S. Simoff, N. Gu, and K. H. Lau. 2000b. Designing virtual architecture. In *Proceedings of the fifth Conference on computer aided architectural design research in Asia, CAADRIA '00*, 481-490. Singapore: CAADRIA.
Maher, M.L., P.S. Liew, N. Gu, and L. Ding. 2005. An agent approach to supporting collaborative design in 3D virtual worlds. *Automation in Construction* 14 (2): 189–195.
March, L., and G. Stiny. 1985. Spatial systems in architecture and design: Some history and logic. *Environment and Planning B: Planning and Design* 12 (1): 31–53.
Müller, P., P. Wonka, S. Haegler, A. Ulmer, and L. Van Gool. 2006. Procedural modeling of buildings. In *ACM SIGGRAPH 2006 papers*, 614–623. New York: Association for Computing Machinery.
Singhal, S., and M. Zyda. 1999. *Networked virtual environments: Design and implementation*. New York: ACM Press/Addison-Wesley Publishing Co.
Smith, G.J., M.L. Maher, and J.S. Gero. 2003. Designing 3D virtual worlds as a society of agents. In *Proceedings of the 10th international conference on computer aided architectural design futures: Digital design, research and practice*, eds. M.L. Chiu, J.Y. Tsou, T. Kvan, M. Morozumi, T.S. Jeng, 105-114. Tainan: Kluwer Academic Publishers.
Stiny, G. 1981. A note on the description of designs. *Environment and Planning B: Planning and Design* 8 (3): 257–267.
Stiny, G. 1982. Spatial relations and grammars. *Environment and Planning B: Planning and Design* 9 (1): 113–114.
Stiny, G. 1990. What is a design? *Environment and Planning B: Planning and Design* 17 (1): 97–103.
Stiny, G. 1999. Shape. *Environment and Planning B: Planning and Design* 26 (1): 7–14.
Stiny, G. 2006. *Shape: Talking about seeing and doing*. Cambridge: The MIT Press.
Stiny, G., and J. Gips. 1972. Shape grammars and the generative specification of painting and sculpture. In *Information processing*, vol. 71, 1460–1465. Amsterdam: North Holland.
Stiny, G., and W.J. Mitchell. 1978. The palladian grammar. *Environment and Planning B: Planning and Design* 5: 5–18.
Vinson, N.G. 1999. Design guidelines for landmarks to support navigation in virtual environments. In *Proceedings of the SIGCHI conference on human factors in computing systems '99*, 278–285. New York: Association for Computing Machinery.
Weiser, M. 2002. The computer for the 21st century. *IEEE Pervasive Computing* 1 (1): 19–25.

Wertheim, M. 2000. *The pearly gates of cyberspace: A history of space from dante to the internet.* New York: WW Norton & Company.

Woolley, B. 1993. *Virtual worlds: A journey in hype and hyperreality.* London: Penguin Books.

Open Access This chapter is licensed under the terms of the Creative Commons Attribution-NonCommercial-NoDerivatives 4.0 International License (http://creativecommons.org/licenses/by-nc-nd/4.0/), which permits any noncommercial use, sharing, distribution and reproduction in any medium or format, as long as you give appropriate credit to the original author(s) and the source, provide a link to the Creative Commons license and indicate if you modified the licensed material. You do not have permission under this license to share adapted material derived from this chapter or parts of it.

The images or other third party material in this chapter are included in the chapter's Creative Commons license, unless indicated otherwise in a credit line to the material. If material is not included in the chapter's Creative Commons license and your intended use is not permitted by statutory regulation or exceeds the permitted use, you will need to obtain permission directly from the copyright holder.

What Does Shape Grammar Implementation Tell Us About Artificial Intelligence in Architectural Design?

Thomas Wortmann

Abstract This chapter, which uses the mirror of shape grammar implementation to reflect on why shape grammars, or, more precisely, visual calculating, isn't more popular among computational designers, proposes that visual calculating might be better understood as a tool for thinking about design, rather than as a method for designers. The chapter compares visual, symbolic and subsymbolic calculating in architectural design, concluding that, while specific shape grammars resemble symbolic artificial intelligence and foreshadowed today's use of computational design for design automation, visual calculating appears to evade the distinction between symbolic and subsymbolic artificial intelligence methods as a less popular, third alternative. The chapter then briefly summarizes the current state of shape grammar implementation, concluding that visual computing's lack of popularity can no longer be blamed only on an absence of powerful, usable, and faithful enough shape grammar interpreters. The chapter discusses subshape recognition as a fundamental barrier to an ideal shape grammar interpreter that allows designers to use any rules they want, whenever they want. This barrier is not so much that subshape recognition is NP-hard—which could potentially be overcome by advances in computing—but that the number of subshapes is unbounded in the most general case. As such, shape grammar implementation becomes tractable only through human judgments that limit the exponential increase of possible rule applications. This principled resistance of visual computing to computer implementation —which arises from the requirement that designers should be able to do anything they want to—demonstrates that visual computing more convincingly serves as a striking reminder of the limits of artificial intelligence than as a design method. This conclusion leads to a remarkable convergence with Rittel, who identified "Sollsetzung," i.e., arbitrary judgment, as the limit of artificial intelligence in design. Finally, the chapter proposes that, over the next fifty years, shape computation could ask what more we can learn from visual calculating about design, and how we might use contemporary artificial intelligence methods to support further advancements in shape grammar implementation.

T. Wortmann (✉)
Department for Computing in Architecture, Institute for Computational Design and Construction (ICD/CA), University of Stuttgart, Keplerstr. 11, 70174 Stuttgart, Germany
e-mail: thomas.wortmann@icd.uni-stuttgart.de

© The Author(s) 2025
S. Kotsopoulos (ed.), *Shape Computation*, Mathematics and the Built Environment 9,
https://doi.org/10.1007/978-3-031-81623-9_22

1 Introduction

The "fifty years" in this volume's title are counted from the publication of "Shape Grammars and the Generative Specification of Painting and Sculpture," widely regarded as the starting point of the field that is variously called shape grammars (SG) and visual computing (VC) (Stiny and Gips 1972; Stiny 2006). The mention of "generative specification" reminds readers that generative design approaches have been studied far longer than the more recent popularity of computational design in architecture and related fields indicates. Caetano et al. (2020) locate the emergence of this popularity in the past two decades, while acknowledging that computational design "emerged in the 60s".

Caetano et al. also provide a useful taxonomy that defines generative design as the subset of computational design approaches, where "after starting the generative process, the system executes encoded instructions until the stop criterion is satisfied." This definition includes SG in the sense that "a shape is generated from a shape grammar by beginning with the initial shape and recursively applying the shape rules" (Stiny and Gips 1972). In their current incarnation, generative design methods draw on recent advances in artificial intelligence (AI), and, specifically, learning with deep artificial neural nets, to generate two-dimensional images and, in some cases, three-dimensional shapes (del Campo and Leach 2022). A related, accompanying trend is the encoding of practical architectural design knowledge, on, for example, parking layout, in rule-based systems by start-ups such as Archistar Property Insights, Digital Blue Foam, Finch3D and Giraffe (Wortmann and Gänshirt 2022).

But, despite the success of SG and VC in establishing the long-lasting research field this volume celebrates, mentions of SG and VC are largely absent in discussions about generative architectural design, and especially generative AI methods. For example, Caetano et al. (2020) barely mention SG in their analysis of computational design papers from 1978 to 2017, and they are completely absent in the AD issue on "Architecture and Artificial Intelligence" (del Campo and Leach 2022).

One reason for this absence might be that, although it is relatively easy to create simple designs with VC, it is more difficult to create more complex designs, analogous to what Leitão et al. (2012) have observed in a comparison of visual and textual programming. Pauwels et al. (2015) present "a system that is capable of reframing its grammar," in other words, a system that facilitates a transition from fixed SG to more flexible VC, noting that "the main difficulty in using this kind of shape grammar system … is that the grammar can very quickly become inconsistent and thus unusable."

Another, often-cited reason for this absence is that, despite efforts dating back to 1974 (Gips 1999), there is still a perceived lack of computer implementations that are faithful to the "very core values" of SG (Hong and Economou 2022). Recently, there has been notable progress in shape grammar implementations (SGI) (Economou et al. 2021; Stouffs 2022) that approximate the "Big Enchilada" that Gips (1999) memorably has called for—"the shape grammar interpreter program with Everything—a slick interface, parametric application, 2D and 3D, curves, extensions, all kinds of tools and goodies for the user." But the wider adoption of SG by computational designers that one might have hoped for hasn't manifested itself yet. Perhaps uniquely among generative design approaches, the computer implementation of SG is thus regarded as a topic that remains (1) challenging and (2) somewhat separate from the field of SG, as such. The remainder of this chapter reflects on this absence by discussing some relationships between SG and generative AI for architectural design, from the perspective of shape grammar implementation.

2 Visual, Symbolic, and Subsymbolic Calculating in Architectural Design

An important aspect of SG and VS is that they are visual, as opposed to symbolic, in contrast with grammars that govern natural or mathematical languages. In contrast with such grammars and other structured systems, the operands—the maximal elements—in SG are not discrete, but continuous. Maximal elements are created by a geometrical union that is called the embedding relation. This embedding relation is related to how humans perceive shapes (as undivided) and thus is what makes calculations with SG "visual" (Stiny 2006). Embedding ensures that geometrical elements are always represented maximally, i.e., to their largest continuous extents.

The embedding relation implies that SG operates on shapes in ways that are far more flexible than the CAD programs, typically used by architectural designers (Stiny 2006, p. 134). For example, embedding allows the visual identification of an indefinite number of line segments from a single line segment. According to Jowers et al., this visual character of SG also accounts for the difficulty of implementing them on digital computers (2019). Gips identifies a tension "between the visual nature of shape grammars and the people who want to use them, and the inherently symbolic nature of the underlying computer representations" (1999).

The AI community commonly distinguishes symbolic and subsymbolic, or neural methods (Sarker et al. 2021). Symbolic methods explicitly operate on symbols. These symbols represent concepts that can be understood by humans. Symbolic methods are thus explainable (at least in principle) and predictable in the sense that conclusions are drawn with formal logic and related approaches. Primary examples of subsymbolic methods are those using artificial neural nets. In neural nets, representations are distributed across a network's weights, in a manner not easily understood by humans, but one that is highly flexible in terms of possible representations because appropriately sized neural nets can represent any continuous function (Kratsios and Papon 2022). Advantages of neural nets include their scalability in terms of data processing, and their flexibility for different kinds of representations and applications, including generative design (Goodfellow et al. 2016). Other kinds of machine learning methods can be classified as subsymbolic as well. An important aspect of the symbolic/subsymbolic dichotomy is that, in symbolic AI, the representations and operations are defined by humans, whereas in subsymbolic AI, they are learned from data. Recently, there has been a lot of interest in neuro-symbolic AI methods that combine the flexibility and scalability of artificial neural nets with the explainability and rigor of symbolic reasoning (Sarker et al. 2021).

Specific SG, such as the Palladian grammar (Stiny and Mitchell 1978) and the Hepplewhite-style chair-back grammar (Knight 1980), resemble symbolic AI in the sense that they constitute rule sets defined by humans, representing human (design) knowledge, in an explicit, albeit not symbolic form. Some canonical SG have been successfully implemented on computers (Gips 1999; Hong and Economou 2022), for example, the grammar of Queen Anne Houses (Flemming 1987). Not coincidentally, the latter also appears in a book on "expert systems," i.e., on applications of symbolic AI (Flemming 1988). VC, on the other hand, appears to be a third, in-between category. As discussed above, while the operands in VC are not symbolic but visual, the operations, i.e., the shape rules, are clearly intended to be defined by humans (Stiny 2011).

Undeniably, "AI and evolutionary algorithms are key in many computer applications in design" (Stiny 2006, p. 51). As mentioned above, architectural designers use symbolic methods to automate the design of standard design tasks, such as housing, offices, and parking (Zwierzycki 2020). This design automation typically uses symbolic methods instead of VC, but it achieves results that conceptually resemble specific SG: it represents specific kinds of design tasks in explicit rule sets. More recently, architectural designers also use subsymbolic methods to generate images and three-dimensional shapes (del Campo and Leach 2022). While, similar to BIM (Scheer 2014), design automation with symbolic methods goes hand in hand with standardization, and the results that are generated with subsymbolic methods can be considered creative, but they often lack architectural constraints, such as a recognizable spatial order or structure (Wortmann and Gänshirt 2022).

However, in contrast to symbolic and subsymbolic AI methods, architectural designers apparently do not use VC much, or at least are hesitant to talk publicly about this use (Stiny 2022, p. 52). In short, while specific SG and their implementations have foreshadowed today's use of computational design for design automation, VC represents a third, little-used alternative to the current use of symbolic and subsymbolic AI methods by architectural designers.

3 Limits of Shape Grammar Implementation

As discussed in the previous section, the visual character of SG is what distinguishes them from other computational design approaches, and from symbolic and subsymbolic AI methods. This visual character makes shape grammar implementation challenging (Jowers et al. 2019; Hong and Economou 2022). This challenge ultimately arises from the fact that digital computers operate on binary symbols (IEEE Standards Board 1985). This discreteness holds, even in the case of subsymbolic AI, where the weights of a neural net are represented as floating point numbers. This discreteness (1) makes it difficult to represent shapes while preserving the embedding relation and (2) can hinder rule application due to tolerances that are caused more directly by floating point precision (Wortmann 2013; Hong and Economou 2022).

But the third, and most fundamental, challenge to shape grammar implementation is the problem of subshape recognition, i.e., the problem of finding shapes that are candidates for rule application (Wortmann 2013; Hong and Economou 2022; Stouffs 2022). Yue and Krishnamurti (2013) have shown that subshape recognition is NP-hard for parametric transformations in a graph-based shape representation, and Wortmann and Stouffs (2018) have generalized this proof to all representations. Specifically, the computational complexity of subshape recognition increases exponentially relative to the input shape, which implies that, in practice, only rules with very simple shapes (e.g., triangles or quadrilaterals) are possible. Jowers et al. (2019) have shown that the alternative of using predefined shape representations does not work even for some comparatively simple cases.

The NP-hardness of subshape recognition is also unlikely to be overcome by advances in computing. Of various avenues for improvement outlined by Shalf (2020), quantum computing appears the most promising due to its "ability to solve combinatorially complex problems in polynomial time". But—unlike classic combinatorial optimization problems such as Traveling Salesman (Korte and Vygen 2018, p. 413)—subspace recognition lacks a single optimal solution and instead requires enumeration of all possible solutions. This enumeration takes exponential time, irrespective of the employed computing architecture or programming model. These results imply that an SGI that can do everything a computational designer might want it to do—as Stiny (2011) writes, "use any rule(s) you want, whenever you want to"—will remain out of reach for the foreseeable future. Conversely, SGI becomes tractable only through human judgments that limit the exponential increase of possible rule applications with suitable constraints.

Importantly, these challenges are relevant mostly for so-called general shape grammar interpreters (SGI). Implementations of specific grammars suffer from these challenges to a much smaller extent, since the possible operations and the shapes that arise from them are determined and known. But the implementation of specific grammars is also far less relevant when one intends VC as a computational design method, since architectural designers clearly should be free to define und refine operations and shapes during their design processes (Stiny 2011; Pauwels et al. 2015; Jowers et al. 2019; Hong and Economou 2022; Stouffs 2022).

There currently are two ambitious SGI that aim to address these challenges: Shape Machine and Sortal GI. Shape Machine supports two-dimensional line segments and arcs as operands and isometric, similarity, and affine geometric transformations (Economou et al. 2021). Sortal GI goes beyond this and supports "a broad variety of spatial and non-spatial elements" as operands and "two alternative matching mechanisms for spatial elements, a non-parametric mechanism matching shapes under similarity transformations … and a parametric-associative mechanism under some topological constraints as well as associations of perpendicularity and parallelism" (Stouffs 2022).

As such, Sortal GI seems to offer most of the big enchilada's features (cf. Gips 1999). Both Shape Machine and Sortal GI are implemented as "plug-in[s] for a traditional computer aided design program" (ibid.), specifically in Rhinoceros 3D and its visual programming platform, Grasshopper, respectively. Shape Machine hasn't been publicly released yet and, so far, Sortal GI hasn't been widely adopted.

At the time of writing, food4rhino, Grasshopper's "app store," counts almost 4.000 downloads of Sortal GI. But this respectable number is dwarfed by the most popular plug-ins such as Lunchbox, Ladybug Tools, and Kangaroo Physics, each of which count hundreds of thousands of downloads. This relatively limited adoption of a powerful interpreter, integrated in the most popular platform among computational designers (Wortmann et al. 2022), shows that SG and VC haven't been held back only by a lack of faithful SGI. In short, it appears that the implementation of VC on digital computers is limited, both fundamentally and in terms of its adoption, with the latter being due to fundamental usability issues like those mentioned by Pauwels et al. (2015).

4 Visual Calculating as a Critique of Symbolic Reasoning

Ultimately, the NP-hardness of parametric subshape recognition (Wortmann and Stouffs 2018) shows that VC, in its most general form, is something that, in the words of Dreyfus (1992), "computers still can't do." This result aligns with the contention that "insight and imagination subsume and exceed fancy in visual analogies (descriptions)—in visual calculating in order to bring in art and design. Computer descriptions in words, in numbers, in data and statistics, etc. limit what they describe, so that experience is final, and things (objects) are fixed and dead" (Stiny 2022, p. 58). In other words, VC exhibits a richness that symbolic reasoning does not, due its inflexible representations.

The relationship between VC and the critique of computers and, especially symbolic AI, is not coincidental. Stiny repeatedly references a debate at MIT in 1966 between Marvin Minsky, as the champion of symbolic reasoning, and the opposing Hubert Dreyfus, who objected to the idea of simulating human thought with symbolic reasoning from a philosophical, phenomenological perspective (2006, pp. 17, 36, 51, 2022, p. 159).

But considering contemporary advances in mostly subsymbolic AI, Dreyfuss' argument appears problematic: "These procedures, however, when used by human beings, depend upon one or more of the specifically human forms of 'information processing.'" For human beings, at least, the use of chess heuristics presupposes fringe consciousness of a field of strength and weakness; the introduction of means-ends analysis eventually requires planning, and thus a distinction between essential and inessential operations; semantic considerations require a sense of the context" (Dreyfus 1992, p. 295). Arguably, AlphaZero has a subsymbolic representation of a field of strength and weakness (Silver et al. 2018); RT-2 can distinguish between essential and inessential operations (Brohan et al. 2023); and GPT-4 exhibits a sense of context (Bubeck et al. 2023). Such advances are thus not only technical, but also change the definition of what "specifically human forms of information processing" are. Accordingly, one may conclude that the critique of AI from a VC perspective has held up better than the one from a phenomenological perspective, at least insofar as VC has resisted computer implementations.

In his prescient discussion of "the influence of the computer on the future role and the profession of architects," which appeared in a book on automated architecture in 1986, Rittel (1986) formulates a third critique, where the human core of architectural design is not the inability of using computers for VC but for "Sollsetzung," i.e., arbitrary judgment. According to Rittel, the construction of Mies-machines and Palladio-generators shows that "when there's decent rules, the computer works!" (ibid.). Conversely, computers cannot make decisions that aren't grounded in other decisions. Although Rittel specifically argues against VC, or at least SG, as a fundamental limit on architectural design with computers, his argument intersects with Stiny (2011): as discussed above, what makes VC hard to implement on digital computers ultimately is not just their visual character, but also the potential arbitrariness of architectural designers, which VC accommodates with the embedding relation.

5 Conclusion

In short, (1) in architectural practice, computational designers appear to embrace symbolic and subsymbolic AI, more readily than VC; (2) the computer implementation of VC faces a fundamental limit; (3) the freedom of VC represents a fundamental challenge to these AI methods, which (4) is ultimately rooted in the potential arbitrariness of designers. Numbers (2) and (3) are in conflict in the sense that the more flexible, powerful, and faithful SGI is, the weaker the challenge to AI methods, in the sense that VC turns out to be implementable on digital computers, after all. Accordingly, VC may be better understood as a tool for thinking about design, rather than a method for designing.

From this perspective, a critical engagement with the question of what contemporary AI methods can mean for computational designers seems an important and productive program for the next fifty years of shape computation. This program would consider (1) what more we can learn from VC about design, (2) what this understanding of design implies in terms of requirements for generative AI methods, and (3) how we can leverage contemporary AI methods for SGI.

A promising direction for the last point is neuro-symbolic AI methods, with artificial neural nets offering the representational flexibility required for SGI, and symbolic AI methods as the mechanism for defining and applying shape rules. This approach would offer a highly flexible shape representation, but would still be limited by the exponential increase of possibilities that needs to be constrained by human judgment. Ultimately, the enduring value of VC for the next fifty years may lie less in the answers it provides, but in the challenging questions it raises.

Acknowledgements Supported by the German Research Foundation (DFG) under Germany's Excellence Strategy—EXC 2120/1—390831618.

References

Brohan, A., N. Brown, J. Carbajal, et al. 2023. RT-2: Vision-language-action models transfer web knowledge to robotic control. arXiv:2307.15818.
Bubeck, S., V. Chandrasekaran, R. Eldan, et al. 2023. Sparks of artificial general intelligence: Early experiments with GPT-4. arXiv:2303.12712.
Caetano, I., L. Santos, and A. Leitão. 2020. Computational design in architecture: Defining parametric, generative, and algorithmic design. *Frontiers of Architectural Research* 9: 287–300.
del Campo, M., and N. Leach (eds.). 2022. *Machine hallucinations: Architecture and artificial intelligence*. John Wiley & Sons.
Dreyfus, H.L. 1992. *What computers still can't do: A critique of artificial reason, revised edition of: What computers can't do, 1979*. Cambridge: The MIT Press.
Economou, A., T.-C. Hong (Kurt), H. Ligler, and J. Park. 2021. Shape machine: A primer for visual computation. In *A new perspective of cultural DNA*, ed. J.-H. Lee, 65–92. Singapore: Springer.
Flemming, U. 1987. More than the sum of parts: The grammar of Queen Anne houses. *Environment and Planning B* 14: 323–350.
Flemming, U. 1988. Rule-based systems in computer-aided architectural design. In *Expert systems for engineering design*, 93–112. New York: Academic Press.
Gips, J. 1999. Computer implementation of shape grammars. In *NSF/MIT workshop on shape computation*, 56. Cambridge: Massachusetts Institute of Technology.
Goodfellow, I., Y. Bengio, and A. Courville. 2016. *Deep learning*. Cambridge: The MIT Press
Hong, T.-C.K., and A. Economou. 2022. Five criteria for shape grammar interpreters. In *Design computing and cognition'20*, ed. J.S. Gero, 191–207. Cham: Springer International Publishing.
IEEE Standards Board. 1985. IEEE Standard for binary floating-point arithmetic.
Jowers, I., C.F. Earl and G. Stiny. 2019. Shapes, structures and shape grammar implementation. *Computer-Aided Design* 111: 80–92.
Knight, T.W. 1980. The generation of Hepplewhite-style chair-back designs. *Environment and Planning B* 7: 227–238.
Korte, B., and J. Vygen. 2018. *Combinatorial optimization: Theory and algorithms*. Berlin, Heidelberg: Springer.
Kratsios, A., and L. Papon. 2022. Universal approximation theorems for differentiable geometric deep learning. *Journal of Machine Learning Research* 23: 1–73.
Leitão, A., L. Santos, and J. Lopes. 2012. Programming languages for generative design: A comparative study. *International Journal of Architectural Computing* 10: 139–162.
Pauwels, P., T. Strobbe, S. Eloy, and R.D. Meyer. 2015. Shape grammars for architectural design: The need for reframing. In *Computer-aided architectural design futures*, eds. G. Celani, D.M. Sperling, J.M.S. Franco, 507–526. The next city—new technologies and the future of the built environment. Berlin: Springer.
Rittel, H.W. 1986. Über den Einfluss des Computers auf die zukünftige Rolle und das Berufsbild von Architekten. In *CAD: Architektur automatisch? Texte zur Diskussion*, eds. W. Ehlers, G. Feldhusen, C. Steckeweh, 205–230. Braunschweig: Friedr. Vieweg & Sohn.
Sarker, M.K., L. Zhou, A. Eberhart, and P. Hitzler. 2021. Neuro-symbolic artificial intelligence: current trends. arXiv:2105.05330
Scheer, D.R. 2014. *The death of drawing: Architecture in the age of simulation*. London: Routledge.
Shalf, J. 2020. The future of computing beyond Moore's Law. *Philosophical Transactions of the Royal Society a, Mathematics, Physics, and Engineering Science* 378: 20190061. https://doi.org/10.1098/rsta.2019.0061.

Silver, D., T. Hubert, J. Schrittwieser, et al. 2018. A general reinforcement learning algorithm that masters chess, shogi, and go through self-play. *Science* 362: 1140–1144. https://doi.org/10.1126/science.aar6404.
Stiny, G. 2006. *Shape: Talking about seeing and doing*. Cambridge: The MIT Press.
Stiny, G. 2011. What rule(s) should i use? *Nexus Network Journal* 13: 15–47.
Stiny, G. 2022. *Shapes of imagination: Calculating in Coleridge's magical realm*. Cambridge: The MIT Press.
Stiny, G., and J. Gips. 1972. Shape grammars and the generative specification of painting and sculpture. In *Information processing*, vol. 71, ed. C. Freiman, 1460–1465. Amsterdam: North Holland.
Stiny, G., and W.J. Mitchell. 1978. The Palladian grammar. *Environment and Planning B: Planning and Design* 5: 5–18.
Stouffs, R. 2022. A multi-formalism shape grammar interpreter. In *Computer-aided architectural design*, eds. D. Gerber, E. Pantazis, B. Bogosian, et al., 268–287. Design imperatives: The future is now. Singapore: Springer
Wortmann, T. 2013. *Representing shapes as graphs*. SMArchS Thesis. Cambridge: Massachusetts Institute of Technology.
Wortmann, T., J. Cichocka, and C. Waibel. 2022. Simulation-based optimization in architecture and building engineering—results from an international user survey in practice and research. *Energy and Buildings* 259: 111863.
Wortmann, T., and C. Gänshirt. 2022. Entwerfen im Zeitalter der künstlichen Intelligenz. *Manege Für Architektur* 2: 28–29.
Wortmann, T., and R. Stouffs. 2018. Algorithmic complexity of shape grammar implementation. *Artificial Intelligence for Engineering Design, Analysis and Manufacturing* 32: 138–146.
Yue, K., and R. Krishnamurti. 2013. Tractable shape grammars. *Environment and Planning B: Planning and Design* 40: 576–594.
Zwierzycki, M. 2020. On AI adoption issues in architectural design. In *Anthropologic architecture and fabrication in the cognitive age—Proceedings of the 38th eCAADe conference*, 515–524. Berlin: eCAADe.

Open Access This chapter is licensed under the terms of the Creative Commons Attribution-NonCommercial-NoDerivatives 4.0 International License (http://creativecommons.org/licenses/by-nc-nd/4.0/), which permits any noncommercial use, sharing, distribution and reproduction in any medium or format, as long as you give appropriate credit to the original author(s) and the source, provide a link to the Creative Commons license and indicate if you modified the licensed material. You do not have permission under this license to share adapted material derived from this chapter or parts of it.

The images or other third party material in this chapter are included in the chapter's Creative Commons license, unless indicated otherwise in a credit line to the material. If material is not included in the chapter's Creative Commons license and your intended use is not permitted by statutory regulation or exceeds the permitted use, you will need to obtain permission directly from the copyright holder.

The *Shape* of Generative AI

Onur Yüce Gün

Abstract Over the past fifty years, shape grammars helped develop a critical outlook in the development of computational design by emphasizing the complexity and interplay between algorithmic processes and artistic intuition. Having previously shown how computation could be contextualized without representations or electronic digital computers, shape grammars hold the potential to provide a critical framework for generative artificial intelligence (AI) development. To this end, this paper outlines shape grammars' historical evolution at the intersection of visual creativity and computational thinking, with a focus on the concepts of "seeing" and "doing." The identified gaps between human vision and perception, language, and the inner workings of artificial intelligence systems are further studied through three interwoven questions that relate to visual ambiguity, processes of meaning-making, and mereology. The presented inquiry aims at fostering a deeper understanding of human creativity and computational design, which are much-needed ingredients that can help shape the future of generative AI.

1 Introduction

Shape grammars provide rule-based design systems for creating visual arrangements and designs, using shapes and shape rules, and often intentional ambiguities. More than just a tool for design, shape grammars also serve as a tool of analysis that unravels inherent design patterns present in visual compositions and architectural designs. Applications encompass a wide variety of visual meaning-making, from the generation of classical architectural plan layouts (Stiny and Mitchell 1978) to knots (Knight and Stiny 2015), to watercolor paintings (Gün 2017), and a generalized understanding of making (Gürsoy and Özkar 2015).

The way in which shape grammars work is equally, if not more important than what it does, to build a strong insight into computational design. The first part of this

---

O. Y. Gün (✉)
New Balance Athletics, Inc., 100 Guest St, Boston, MA 02135, USA
e-mail: onuryucegun@alum.mit.edu

investigation takes a deep look into the phenomenon of *seeing and doing* (visual-making), in a historical context, observing how efforts began on the side of formalization (computation), yet shifted towards the human component. Electronic digital computation operates on symbols—that is, through structures representing other structures—in shape grammars, computation runs directly with shapes.

This contrast between the counter-representational foundation of shape grammars and symbolically and combinatorically-driven electronic digital computation gets echoed more and more through Stiny's latest books on shapes (Stiny 2006, 2022), consequently helping shape grammars become a tool for studying the potentials and limitations of electronic digital computation in visual meaning-making—shape grammars reminds us "what computers (still) cannot do" (Dreyfus 1992).

This research examines the foundational principles of shape grammars to offer a critique and a way to navigate the recent, unprecedented propagation of AI-driven image generation tools, likened to "rocks [falling] from the sky" (Steinfeld 2023). To this end, while acknowledging the seven decades of artificial intelligence research, the paper focuses on the technological leap taken over the last decade in terms of synthetic image generation, following the introduction of Generative Adversarial Networks (GANs) (Goodfellow et al. 2014). The goal here is to clarify that the fundamental dynamics in visual meaning-making remain the same over the past decades, purely dependent on human intention, perception, and intuition, by strategically employing the shape grammarian perspective of computational progress in art and design.

The struggle of defining aesthetics through numbers, a problem stated by shape grammars numerous times, can be observed in computational research that aims to devise purely computational solutions for problems relating to visual creativity and aesthetics. A recent paper sets their AI image generator that is developed on Generative Adversarial Networks (GANs) to "generate novel works" but that "the novel work should not [be] too novel" and "should increase the stylistic ambiguity." These inherently quality-focused goals are then quantified by using Wundt curves, the psychological concept that illustrates the relationship between arousal and pleasure (Elgammal et al. 2017). The research struggles to define what is novel, not living up to the promise as it fails to measure creativity and originality. On the flip side, Stiny affirms that "creativity [remains] forever out of reach, as long as everything is kept only to a number of data" (Stiny 2022) and puts the act of *seeing* front and center in creative visual endeavors.

The following sections historically study shape grammars—with its foreground emphasis on 'seeing' and 'doing' to provide insight into the divide that resides between the embodied ways of visual meaning-making and making images through descriptions that help run artificially intelligent algorithms. It thus highlights how much more work needs to be done to explicate visual ambiguity, meaning-making, and mereology in the context of digital computing.

2 Seeing

2.1 Shape Grammars and AI: Algorithmic Aesthetics (1975+)

Inspired by the "creation" and "valuation" concepts of the time in artificial intelligence, in 1975, Stiny and Gips framed a system concerned with aesthetics that consists of "design" and "criticism" algorithms. Detailed further in 1978, in their book *Algorithmic Aesthetics* (Stiny and Gips 1978), their framework introduced "receptors" and "effectors", reminiscent of the discriminative and generative AI systems of modern days.

Stiny and Gips describe receptors as sensory linkages between the outside world and the algorithm of a computing machine, they are input devices. The receptors consist of a transducer, and a linked algorithm. The transducer portion of the receptor, consisting of a "color television camera or a microphone, or a combination of these" (Stiny and Gips 1978), is sensitive to electromagnetic or mechanical energies. What the transducer senses is expressed by a sequence of symbols produced by the linked algorithm, and this sequence of symbols is the description of an object to be recognized and judged. The output of the receptor is the specification of the initial conditions for making an object.

Effectors, in turn, are output mechanisms for criticism or design algorithms. They act on the analysis outcome—encompassing description, interpretation, and evaluation of an object—to generate a tangible response in the external world. This response could manifest in diverse forms, including light, sound, or pressure. Integral to psychology and computer science, effectors are pivotal in translating algorithmic analysis into actionable outcomes or objects, serving as a foundational element in related investigations.

Strikingly, these descriptions are closer to the mechanics of machine learning as we know it today than those of shape grammars in its fiftieth anniversary. In their paper from 1975, Stiny and Gips conclude that "the ability to specify a criticism algorithm or a design algorithm may well presuppose the ability to formalize a wide range of perceptual and cognitive skills and a wide range of knowledge … [and thus] … constructing particular criticism algorithms and design algorithms can be extremely difficult … at the present time." After around half a century, although we have much more advanced tools, a working practice, and a clear strategy for developing artistically quality-aware algorithms remain elusive.

2.2 Seeing as Shape Recognition: Shape Interpreters (1975–2020)

Can machines see shapes? Shape interpreters are developed towards this very goal through algorithms that aim to identify, construct, and manipulate shapes.

A general-purpose shape grammar interpreter (SGI) aims to work with all kinds of shapes, from triangles to circles to sophisticated structures, using pre-compiled shape and rule descriptions. A specific-purpose SGI extracts the shapes within specific geometry types, and is thereby more efficient at performing the designated function, albeit with a more limited application area (Eloy et al. 2018).

These interpreters use geometric algorithms, with the more advanced ones utilizing reinforcement learning to recognize shapes and apply grammar (Ruiz-Montiel et al. 2013). The shape computation lab at the Georgia Institute of Technology lists 73 computational implementations of SGIs developed between 1975 and 2020 (Hong and Economou 2022).

A correlation between shape interpreters and the "synthesis algorithm" linked to "receptors" mentioned in *Algorithmic Aesthetics* can be drawn (Stiny and Gips 1978). They are both tasked with interpreting and evaluating shapes. Similarly, the challenges mentioned by Stiny and Gips (1975) in developing intelligent systems for shape calculations apply to the development of shape interpreters (Wortmann and Stouffs 2018).

The main challenge is emulating what seeing–human perception–does without effort. For shape grammars, seeing is the source of open-ended visual emergence, which happens naturally. Effective implementation of shape grammars must support design emergence without relying on pre-defined geometrical parts, allowing for greater creative freedom—yet without pre-defining parts, a computational shape grammar interpreter cannot be built. Moreover, sub-shape detection, a core task for SGIs, becomes a computational challenge, considering shape grammars constantly demonstrate how a shape can be visually dissected into infinitely many parts in infinitely many ways. The algorithmic complexity of these systems adds further challenges to the development of general-purpose shape interpreters.

Stiny speaks of shape interpreters in a limited fashion, investing in the open-endedness of shape interpretation and calculating, and assigning the role of interpretation to artists and designers. This way, he builds the discussion around imagination and creativity (Stiny 2022) instead of algorithmic implementation, emphasizing the endless possibilities for constructing and describing shapes differently.

One may bring forth the capacity of advanced contemporary vision systems in powering self-driving cars or bi-pedal robots that demonstrate human-like, if not more efficient and precise locomotion (Kuindersma et al. 2015). Once deemed an unattainable feat, such tasks have become applicable with the technological advancements in hardware development, such as graphical processing units (GPUs) (Pandey et al. 2022), and through socio-contextual and infrastructural developments (Steinfeld 2023)—but with a caveat: in a precisely defined world, where every environment, object, and shape is precisely and, more importantly, immutably defined, the task gets reduced to collecting, sorting, and utilizing massive amounts of data. Yet, shape grammars' main concerns revolve around the ever-changing dynamics of visual inspiration: what shapes look like and what they can look like next, including "see[ing] things as in themselves they really are not" (Stiny 2022).

2.3 Seeing as Data Averaging (2014–2024)

"Computers deserve our lasting trust—[...] the day isn't too far off when AI will do what we do better than we can—it's so long to appetite, mood, taste, and whim. Of course, AI extends only to useful things; it goes without saying, which is why I'm saying it, that AI won't do for calculating in art and design, where ambiguity—seeing things as in themselves they really are not—is the very heart and soul of insight and imagination in the embed-fuse cycle" (Stiny 2022).

The endeavor to create intelligent machines, a cornerstone of artificial intelligence (AI), can be traced back to the mid-twentieth century. This pursuit began following the seminal work of Alan Turing, particularly his 1950 paper 'Computing Machinery and Intelligence,' in which he proposed what is now known as the *Turing Test*, a criterion for machine intelligence.

The term "Artificial Intelligence" was first coined by John McCarthy during the *Dartmouth Summer Research Project on Artificial Intelligence* conference in 1956. One of the other conference organizers, Marvin Minsky, published *Steps Toward Artificial Intelligence* later in 1961 to "organize and survey the field and suggest how research ought to proceed" (Winston 2012).

In the context of synthetic image generation, specifically in reference to today's text-to-image tools, Steinfeld points out the "Perceptron," the classification algorithm conceptualized in 1943 by McCulloch and Pitts (2023). Much of the history of artificial intelligence, especially in connection to arts and design, has been written about numerous times (Vardouli 2012).

However, the last decade of AI research has produced unprecedented results. The recent advancements in AI image-making tools can be primarily attributed to critical technological and methodological developments within the last decade. The exponential increase in computational power, facilitated by GPUs and TPUs, has vastly improved the ability to process more complex and extensive algorithms (Goodfellow et al. 2014).

The unprecedented progress in deep learning, particularly with improved convolutional neural networks (CNNs), has been essential for processing images and sequences (Krizhevsky et al. 2012).

The development of Generative Adversarial Networks (GANs) and the integration of advanced natural language processing with image processing technologies have enabled the generation of high-quality images from textual descriptions (Goodfellow et al. 2014; Ramesh et al. 2022). The development of CLIP-guided diffusion models, which combine the reasoning systems of CLIP (contrastive Language-Image Pre-training) regarding text-image perception with the generative capacities of the diffusion models, made the creation of detailed, context-sensitive images from text more feasible (Rombach et al. 2021).

Alongside these advances, the availability of large quantities of images and text suitable for training algorithms and networks, such as open-source LAION-400M and LAION-5B data sets, was essential (Steinfeld 2023).

All this development has led to a burst in image-making tools, including Stable Diffusion, Midjourney and DALL-E, and subsequently, a flood of AI-generated images, concept art, and architectural visualizations emerged. Midjourney, which launched its services in February 2022 for a limited audience, reached 16 million users by November 2023 (2024). This acceleration, though, included both exponential growth and a subsequent slowdown in user count as more tools became available and as potential users needed help to assess the value of having the power to generate hundreds, if not thousands, of images per day.

The immense power of AI image generation tools potentially made it harder for creative ones to reflect on their work as the creative impulse to keep producing the next image made it harder to slow down and ask: so, what?

AI image generation entails a negotiated semiosis built into the generation process. However, the meaning of these images is not limited to the semantic content of the input text.

In simple terms, AI-generated imagery would be linear extensions of the input text. Between the potential of language to generate meaning and the unfathomable and often unforeseen consequences of those meanings in the AI frame, we find a gap shaped by experience and analysis. This phenomenon asks us to rethink the process of text-in to image-out and—perhaps the most crucial reconstruction of all—the complex meaning-making process (Larsson 2021).

This framing resonates more broadly with discussions of AI and art, where emergent visual output is not only understood as a consequence of algorithmic function but also of generative, creative practice that combines the literal and the figurative, and the concrete and the imaginative. Hence, although text-to-image tools mark a significant technological innovation, the way they produce visual meaning is characterized by interpretive multivalences that complicate and challenge the longstanding structures of meaning-making in art (Mazzone and Elgammal 2019).

On the flip side, AI image generation tools leave creators, although massive, in the ocean of an averaged-out dataset full of the best photography, compositions, and former artwork. Nevertheless, all these systems operate without a detailed structure of insight for criticism, evaluation, and nuanced understandings of aesthetics.

2.4 Seeing as Everything (∞)

While the shape grammarian train of thought evolves from *Algorithmic Aesthetics* (1975) to Shape (2006), "receptors and effectors," Stiny replaces these concepts and linking algorithms with "eyes and hands." This preference for descriptions is manifested on Stiny's book cover: *Shape: Talking about Seeing and Doing*.

Seeing, fundamentally a sophisticated physiological phenomenon (Metzger 2009), is for shape grammars simple, as in "what you see is what you get" (Stiny 2006). Seeing is about perceiving, recognizing, and interpreting. The receptor, formulated and described as a digital, mechanical, or photo-receptive device in 1975, leaves its place to the "fabulous winged eye on the reverse side of Alberti's portrait medal" as it symbolizes "vigorous creative activity that illuminates in a flash of insight and a surge of imagination" in Stiny's terms (2022). For Stiny, the lightning bolts and the flash depicted on this medal represent the moment of discovery, precious epiphanies. In simpler terms, seeing identifies elements in a design or pattern that can be manipulated or understood based on shape grammar principles. It is a (the) way of calculating.

Doing involves applying or manipulating shapes and patterns per established shape grammar rules. It includes engaging in the design process—crafting, adjusting, and generating forms. Doing is simply putting insights and structures perceived through seeing into practice to make things.

Shape grammars purposefully incline towards the terms "seeing and doing" for the sake of preserving human intuition, intention, and interpretation in the context of shapes. By doing so, it does something very crucial: it purposefully "decouples" (Llach 2021) itself from electronic digital computation. Stiny and Gips take on the challenge of computing, or rather calculating without "physical computers"—in Cardoso's terms—to demonstrate that calculation with shapes in a formal way, and without electronic digital computers, is possible. This helps develop an expansive rule-based, but not representation-based, formalism and offers profound perceptions about how computations can be performed and understood.

In other words, what, on the surface, appears to be visual meanderings in the two-dimensional world of shapes extends to the very roots of meaning-making.

3 Doing

3.1 Visual-Making

> There are many cases in art and design that have nothing to do with making—that are described in terms of seeing. Duchamp's *Fountain* is a nice example. He didn't make anything, and it's one of the most famous works of art in the twentieth century! (Stiny in Gün et al. 2012)

> The gold standard for creativity assessment is the consensual assessment technique, in which human raters meticulously evaluate ideas and products on the basis of their own judgment and expertise (Elgammal et al. 2017)

In a debate based on the two above contrasting claims, one would perform better explaining why *Comedian,* which "appears as a fresh banana affixed to a wall with duct tape" has been acknowledged as a work of art—and it is not the consensual assessment of people. The art world keeps creating shocks through artwork, seen and done with free rules and shapes. Potentially similar motivations that led to Duchamp's *Fountain* led to the emergence of *Comedian,* also known as the *Art Basel Banana* (2019) (Fig. 1).

Shape grammars explain such creations that are sometimes created "without [even] lifting a finger" (Stiny 2011) through using identity operations, which is

Fig. 1 Comedian, reproduced by the author in 2024

about mapping a shape to an identical output. The trick is that a change or switch in a viewpoint or perceptual change in a context of appreciation can lead to a new perceptual or appreciative discovery of new parts and relations of the shapes in the same construction.

For *Fountain*, the "x goes to x" schema applies:

$$\text{fountain} \rightarrow \text{fountain}$$

While the assessment of creativity, or the quality generated through this artwork, would be hardly handled in computational means – *Comedian* comes with "detailed diagrams and instructions for its proper display" (Wikipedia 2022). For *Comedian*, we can talk about potential translation rules in the making as the two objects moved in space and were carefully placed on the wall. The calculation looks simple; however, it is not straightforward. Nevertheless, a calculation is possible.

$$< \text{banana, tape} > \rightarrow < t(\text{banana}') + t(\text{tape}) >$$

Other weight rules can be written, especially considering the banana's changing skin color, or deformations, that take place while it is being placed, hence the apostrophe ('). Considering the banana was eaten after being sold for $120,000, we can confidently use the schema:

$$x \rightarrow 0$$

Or as a shape rule,

$$\text{banana} \rightarrow \text{nothing}$$

Or

$$\$120,000 \rightarrow \$0$$

Within this pure investment into seeing with zero returns, the fate of the duct tape remains unknown.

As speculative as it may be, this is how humans take on risky visual endeavors to create meaning and something genuinely new through art: we cannot always explain phenomena with pure analytical reasoning. Shape grammars aim to capture such ambivalent moments, to establish a frame of reasoning no matter how fuzzy the action of visual-making might be.

3.2 Meaning-Making

In an averaged-out world, we experience the averages—*the Fountain* and *the Comedian* would not come to life if they were intended to fit into a bell curve, simply because they are singular outliers. In a large dataset (of artwork), they would vanish.

Taleb recurrently explains the shortcomings of conventional statistical methods, particularly their failure to anticipate or explain the "impact of improbable events," building up on the black swan theory (Taleb 2010). These "highly improbable" events are the focal subjects of shape grammars in the visual realm. The motivation is that what can be found through the anomalies that reside outside the known, the obvious and statistically available (Kozyrkov 2023), hold the potential to create unprecedented quality—which is "not a thing, it's a way of looking at things; not a static concept, it's a dynamic and evolving one; not something that can be defined or measured objectively, it's a subjective experience" and it "is the source of all values and morals" (Pirsig 2005).

Stiny's two squares or four triangles are the crux of this phenomenon, as the same shapes keep giving birth to new shapes that were not thought of previously. The breadth of the spectrum of emergent shapes is hard to pre-conceive as "…the shape can be divided arbitrarily, in any way whatsoever […] divided by drawing a crazy curve through it so that pieces on one side are one part, and pieces on the other side are another part" (Knight and Stiny 2001) (Fig. 2).

While yielding unexpected shape formations, these shapes again visualize what is so challenging in developing shape interpreters that rely on formerly defined shape structures. Visual meaning-making is about searching for meaning beyond the parts and consistencies of a visual representation. The artist's intention and presumptions about the visual gathering can influence the emergence of meaning, but the actual meaning may not be known until the process of making is complete. Moreover, interestingly, such making, as one keeps seeing, is never complete (Gün 2016).

Fig. 2 Reproduced drawing of the two parts shown by Knight and Stiny (2001)

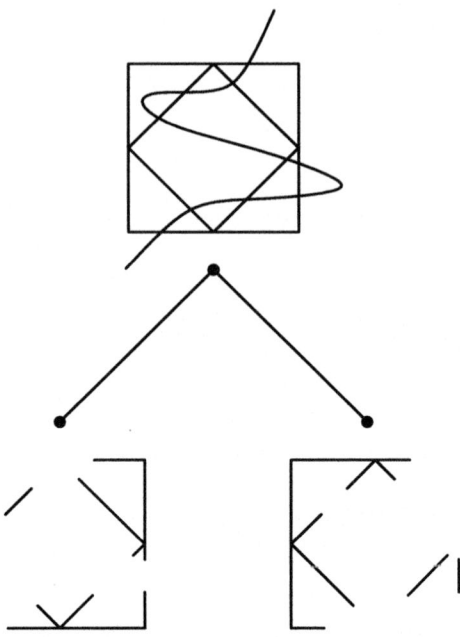

If we take drawing as a mode of visual-making, one does not draw what one (already) knows, or, more importantly, one does not draw to know in the first place. One draws because drawing helps one learn, know, but also forget, which in turn constitutes one's becoming. The same goes for seeing. This suggests that drawing is not just a means of producing a visual representation but also a means of exploring and understanding the world and oneself. Visual meaning-making is a dynamic and iterative process involving the artist's intention, the act of drawing, and the emergence of meaning through the process of becoming. For shape grammars the becoming process happens through talking about seeing and doing.

3.3 "Talking About Seeing and Doing"

The historical overview of shape grammars, concentrating on "seeing" and "doing," sheds light on the disparities between visual meaning creation and the mechanisms of artificial intelligence systems. This realization emphasizes the need to gain a deeper understanding of visual ambiguity, meaning-making, and mereology, especially in the context of electronic digital computers. To facilitate this, three interrelated questions can help structure how we approach the core concerns of shape grammars concerning human perception and creativity.

First, *duck or rabbit?*—relating to the wicked problem of ambiguity; second, *inside or outside?*— relating to the wicked problem of meaning-making at the

The *Shape* of Generative AI   503

nexus of human vision and perception; third, *part or whole?*—relating to the wicked problem of mereology, concerning the intricacies of living, being, and the engineering of human life in its entirety. Answering these three questions from a shape grammarian perspective helps build insights for navigating the less human parts of generative AI systems in a more human manner.

Question 1. duck or rabbit?—the wicked problem of ambiguity;

Question 2. inside or out-side?—the wicked problem of meaning-making at the nexus of human vision and perception;

Question 3. part or whole?—the wicked problem of mereology, concerning the intricacies of living, being, and the engineering of human life in its entirety.

A concise way to encapsulate the subject matter of shape grammars is to say that the world is in perpetual motion. Within this endless dance, phenomena emerge and fade, and we work with what we see and what we do—and we talk about what we see and do in the process (Gün and Stiny 2012).

In this section, what Stiny calls talking will be handled in terms of descriptions. Descriptions will help us understand how we handle shape calculations, communicate visual ideas, and, in a broader perspective, understand our surroundings. As we delve into the differences between how humans work with descriptions and how we implement descriptions in artificial intelligence technologies, we will start seeing the contrasts between the dynamics of making sense of images in the human and machine realms.

For shape grammars, visual creation is a free journey that runs with unbounded perception. Perception grapples with change while we seek moments of stasis for comprehension— when the (linguistic) descriptions crystallize. In shape grammars, the moments of stasis serve as checkpoints for describing what looks like what, making calculating possible. The moment we apply a shape rule, the perception starts running wild again—descriptions can disappear, reappear, or be changed entirely throughout seeing and calculating (Stiny 2006).

In electronic digital computing, descriptions work differently. Regardless of the computational system's sophistication or algorithm, every bit must be described precisely and explicitly. Even ambiguities need to be classified and described, which, in essence, is a fundamental contradiction in logic.

The moment you explicitly describe the ambiguous—it is no more.

AI diffusion models are trained by using datasets consisting of images and accompanying text descriptions (Steinfeld 2023). By adding and removing noise, starting from a base bitmap and a text prompt, these tools generate images that try to match the description (Ramesh et al. 2022). This approach emphasizes "prompt engineering," a method dedicated to crafting optimal descriptions to produce desired visual results. Whether or not visual phenomenon can be described in its entirety using natural language remains as one of the biggest challenges for generative AI technologies.

3.4 Duck or Rabbit?

We will talk about ducks and rabbits, but first, let us answer one of the simple but also most investigated questions in shape grammar about the infamous shape—the two squares (or four triangles). These shapes have been dissected in virtually unlimited ways before, starting with Stiny own (Fig. 3).

Stiny and others showed there are numerous other ways of employing the

$$x \rightarrow \text{prt}(x)$$

Schema using the same shape (Stiny 2016, Hong and Economou 2022). Various calculations yield unforeseen results, and each result produces families of further interpretations depending on mirrored applications or by altering split locations (Fig. 4).

The crucial lesson here is that "in fact, no finite description says everything there is to say about the shape" (Knight and Stiny 2001). Knight and Stiny demonstrate how "envelopes, blunt arrows, pencil points, overlapping Us" can be created. Finally, they use a "crazy curve" to cut the shape into two indescribable parts, as formerly shown.

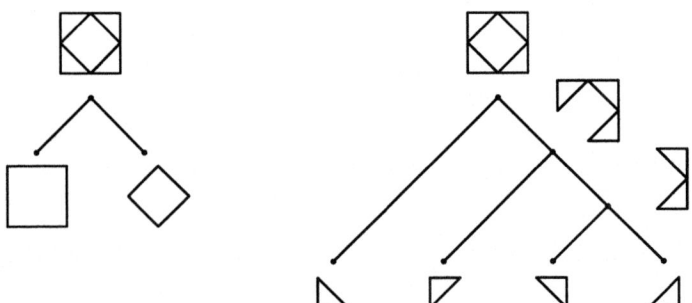

Fig. 3 Two different segmentations of a shape into its constituting parts, two squares on the left, four triangles on the right

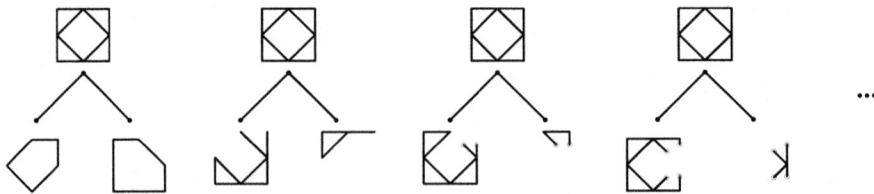

Fig. 4 Part schema applied in various fashions to obtain parts. The first calculation can yield 4, the second one 8, and the latter two can yield infinitely many configurations through application of the same schema

The *Shape* of Generative AI 505

While generative AI has made remarkable strides in responding to real-world problems, it remains a product of its training data and the algorithmic and computational structures that define it. Shape grammars behave as a loose cannon in a sense, letting the beholder change the rules and come up with new shapes anytime (Kotsopoulos 2022).

The disparity between how dynamic human perception deals with ambiguity and how electronic digital struggles with it set the stage for this first question.

3.4.1 The Wicked Problem of Ambiguity

Visual perception not only interprets and broadens our understanding of visual narratives but also thrives amid the nuances of visual ambiguities. This raises the questions:

How many different meanings can an image hold? Can the creator of an image dictate what the viewer should perceive or interpret?

In essence, a viewer peering into any drawing can derive countless meanings, often deviating from the creator's original intent. Consider visual ambiguity through gestalt switch drawings, epitomized by Jastrow's iconic duck-rabbit illustration. Such depictions still bind us to limited interpretations as they often oscillate between two definitive states, the *duck* and the *rabbit*. Yet, as Stiny proposes, the expanse of what can be perceived transcends the finite labels attached to these images. In a novel interpretation of Jastrow's drawing, one may discern another creature, perhaps a concealed *crocodile* nestled in the rabbit's ears or camouflaged within the duck's beak (Gün 2016). Taking the rabbit as x, we can take the part, the ear, and then add the body of the crocodile, applying the rule:

$$x \rightarrow prt(x) + y$$

In preconceived rules of electronic digital computing, the crocodile remains unseen. Interpretations are far from exhausted in this third perception (*crocodile*); the visual journey can widen indefinitely. A viewer's gaze can transcend the artist's attributed descriptions, extracting an array of meanings from a single image. One may even dare to claim that virtually anything can be perceived in any shape. Stiny's two squares, or four triangles, or the intermediary shapes in between illustrate this phenomenon. Stiny starts with such shape descriptions and leaves the rest to the intentions and interpretations of the gazer (Fig. 5).

Stiny notes, "A given vocabulary may limit the rules I try. Designers, especially architects, are invariably looking for one to inform their work" (Stiny 2016). In other words, the connotations extracted from visuals are primarily a product of the viewer's perceptions, influenced by inherent perceptual abilities and individual intentions. This dominance of open-ended perception over specific intent serves as the nucleus of creativity in visual endeavors. A viewer continually reimagines through observation, recreating, and revisiting visuals.

Fig. 5 Duck, rabbit, or crocodile?—line art by the author

3.4.2 Dragons as Soft Shapes

The *Broadened Drawing-Scape* is a painting apparatus that harmonizes hardware and software components for artistic innovation. It features a transparent drawing surface, a computational camera, and a digital projector, all integrated through *Processing* software (Gün 2016). This setup aims to bind the open-ended gestural painting process with rule-based design operations. Artists merge hand-drawn elements with digital projections during the visual-making process while augmenting their creative journey. A unique aspect of this apparatus is its ability to handle unlabeled or non-described shapes. While the shape rules and transformations are defined computationally, the watercolor-painted shapes retain their original, undefined form. Instead of labeling a shape—say as two squares or four triangles—the shape gets recorded as a variable only for the sake of computation. However, the working bit is the shape projection, which does not involve a description or label. The handling of unlabeled shapes can best be illustrated when computing with bits that inherently possess ambiguity and cannot be perfectly duplicated. Using watercolor shapes for computation serves as an apt example of this (Gün 2017). In the following example, a simple shape, x, is captured and then multiplied to generate a digital pattern (template) to work with (Figs. 6 and 7).

The painting starts with a motif, a product of a gentle gesture. Then, in the digital environment, the shape is transformed several times using the transformation rule:

$$x \rightarrow x + t(x)$$

Following three consecutive steps of transformations, the entire calculation becomes:

$$x \rightarrow x + t(x) + t'(x) + t''(x)$$

The resultant digital image is then projected underneath the analog drawing surface, for the painter to trace over the shapes in red.

$$< x, 0 > \rightarrow < x, x' >$$

The *Shape* of Generative AI   507

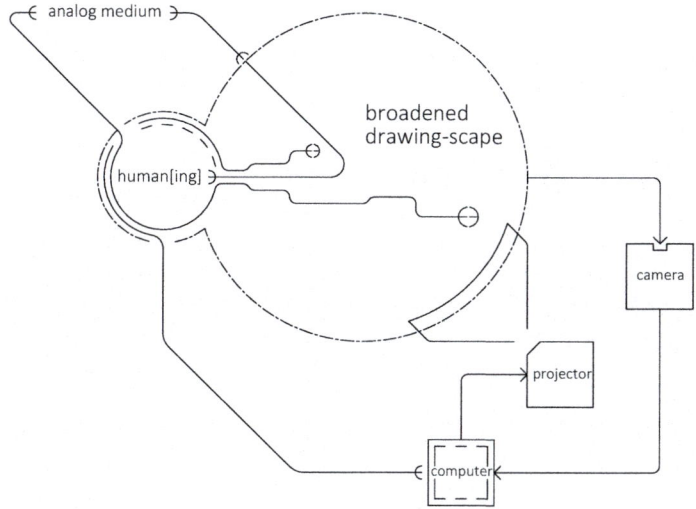

Fig. 6 The "un-diagram" showing the working principles of the *broadened drawing-scape*—(Gün 2016)

Fig. 7 Spatio-temporal representation of development of a watercolor painting process—(Gün 2017)

…to generate the watercolor painting. Finally, the artist sees dragons, birds and fish in boundaries of watercolor shapes:

$$< x, x', 0 > \rightarrow < x, x', b(x) + b(x') >$$

The artist evaluates the painting in retrospect to find out the repeating features and she recalls the actions that she has taken over the course of painting.

#1 shows the initial shape (the very first brush stroke, the *motif* in blue) and the stroke that is created by tracing over the translated shape. The painter uses the schema:

$$x \rightarrow x + t(x)$$

#2 comes to life as an "explosion" of the copy of the first stylistic curve, so that the artist uses the part schema:

$$x \rightarrow prt(x)$$

The artist paints a similar shape after shifting the pattern. #3 is painted after a watercolor wash appears in the projection and gets duplicated after the former shift. #4 is generated when the artist does a computational operation (image difference) using two of the latest captured patterns before projecting it underneath the painting. #5 utilizes the same technique as #4 and finally rotates the final projection to paint an upside-down and traces over a less precise duplicate.

The process demonstrated here is a non-symbolic, un-labeled, open-ended, and trace-based way painting process. No ducks, no rabbits, only traces. This method allows for the computation of watercolor paintings using soft, imperfectly reproduced, unlabeled shapes, presenting a novel way to integrate traditional artistic techniques with digital computation, enriching the artist's palette with endless possibilities.

The *Broadened Drawing-Scape* shifts the focus back to the importance of tactile and visual elements in the artistic process, challenging the nature of digital tools. It offers a more immersive and interactive experience, rekindling the artists' intimate connection with their artwork in an increasingly digital world (Fig. 8).

Fig. 8 Analysis of the developed painting, isolating and marking the imperfectly repeated patterns, applying the $x \rightarrow t(x')$ rule —(Gün 2017)

Fig. 9 The inside-outside problem, as set by Ullman, drawing reproduced by the author

3.5 Inside or Outside?

3.5.1 The Wicked Problem of Perception

Determining whether a point lies inside or outside a closed curve (a polygon) is instinctual for humans. However, beneath this apparent simplicity lies a series of intricate perceptual and cognitive processes (Ullman 1984) (Fig. 9).

Human perception utilizes specialized mechanisms in the visual cortex, adapted to line contrasts and orientations, for edge and boundary identification, aiding form recognition. This process is aided by figure-ground differentiation, enabling the distinction of individual points from their background. Evolutionarily established spatial understanding aids in instinctive recognition of concepts such as *inside* and *outside*, with prior experiences strengthening interpretations of visual patterns (Metzger 2009). In electronic digital computing, algorithms like ray-casting determine spatial relationships. These computational methods operate differently from the intricate processes of human perception.

3.5.2 The Ray Casting Method

To enable a computer to determine whether a point is inside a given shape, we must first translate this inherently visual and intuitive concept into a language the machine can interpret. One established technique to achieve this is the ray intersection method or ray-casting algorithm.

This method involves shooting rays from a designated point within or around the shape and counting the number of times each ray intersects with the shape; a point within the shape will have rays that intersect the shape an odd number of times, while a point outside will intersect an even number of times. To illustrate, using Ullman's ray intersection visualization, the rays intersect the shape 1 and 3 times (odd numbers) on the left side. In contrast, on the right side, they intersect 0, 2, or 4 times (even numbers) in the other. Thus, the task transforms from discerning inside or outside to simply counting intersections and determining evenness or oddness (Fig. 10).

While humans innately process such visual information in a parallel and holistic manner, algorithms function sequentially, adhering to predetermined steps, and rely on precise mathematical and geometric calculations to derive conclusive results. Much of what we visually discern and draw on computers is quantified and expressed

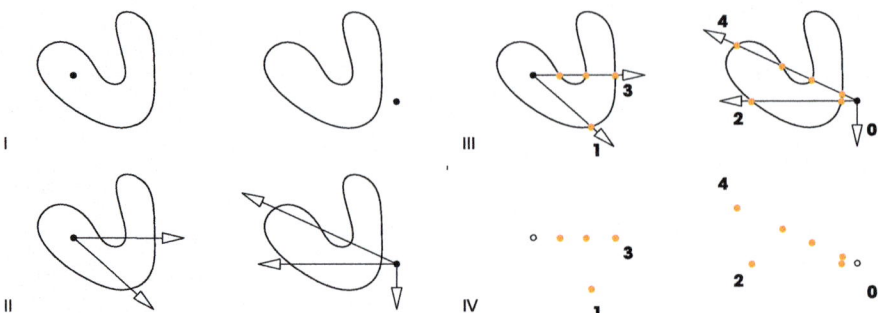

Fig. 10 Ullman's ray casting method, drawing reproduced by the author. Eventually, the closed curve is unseen for the sake of computation

numerically. Notably, computation often comes at the cost of a genuine visual understanding. In employing algorithms, we essentially "unsee" the shape to generate the desired output. In this fundamental problem, the polygon (closed curve) goes unseen. In other words, a major challenge in electronic digital computation is the abstraction of complex, multifaceted concepts into computable entities. Unlike human cognition, which can fluidly interact with ideas without needing to formalize them, computation demands the conversion of these concepts into concrete, defined forms. This often means simplifying, categorizing, or stripping nuances from a concept to make it digestible to a machine.

This requirement to convert concepts for digital computation applies to any visual act. While visual tasks are naturally processed by human perception, computational implementations require abstraction and reduction. While abstraction facilitates computation, it inherently discards the richness and granularity of the real-world phenomena it represents, potentially misrepresenting or oversimplifying them. This constant need to reduce and categorize contrasts shape grammars, which puts human perception in the front and center.

In essence, while Ullman's inside-outside problem underscores the challenges faced by artificial visual systems, Metzger's insights into the deep-seated connection between visual perception and cognition in humans illuminate the sophisticated mechanisms that allow us to navigate and interpret our visual environment seamlessly. These perspectives collectively emphasize the profound intricacies of visual perception and the significant strides yet to be made in replicating such capabilities in artificial systems (Ullman 1984; Metzger 2009). Although Wei et al. conclude that "most of the current visual perception computational models oriented artificial intelligence are helpful to human life and production," they highlight the challenge of working around human intuition and imagination–the key points in creating intelligent models capable of complex planning, creativity, and problem-solving–in the fast-evolving domain of AI (2021).

Yet the problem does not end there since the actuality of a specific condition does not necessarily mean that it would be perceived in that exact actuality. The actuality

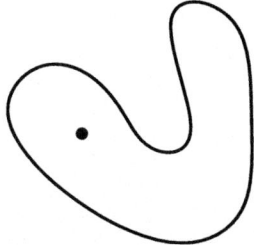 

Fig. 11 The ground truth is that the point computationally resides outside the curve on the right side. But environmentally and sensibly, does it?

of a specific condition may not align with the perception and interpretation of the same condition (Fig. 11).

Creativity takes place over extended periods of exploration and uncertainty with fuzzy dynamics, a far cry from the data-driven absolutes of AI, typically draped in hard metrics and ground truths. Once again, this contest points to a deeper problem in AI research—how to achieve the creative capability humans seem to crave, a capability that frequently utilizes ambiguity. While AI seeks to solve problems by relying on certainty, granularity, and the hard metrics of ground truths, humans use soft and ephemeral, unquantifiable acts to be creative. Creative processes require an uncertain amount of time over which humans deal with a roller coaster of emotions, both positive (of satisfaction) and negative (of failure).

3.6 Part or Whole?

3.6.1 The Wicked Problem of Mereology

The third and last question challenges the procreative capacity of generative AI, a question that has occupied the minds of philosophers for eons. At what point does a finger cease to be and the palm begin? Where does the boundary between the palm and the forearm exist? No intricate plane or meticulously crafted point cloud can pinpoint the precise threshold between the finger and palm— it is an elusive threshold.

This desire to distinguish and segment is fundamental to any *scientific* study. The crystallization of such notions, as observed by Descartes' dualism, had impacts that echoed for generations: Descartes' framework separated the ethereal *mind* (res cogitans) from the tangible *body* (res extensa) (Damasio 2000). In the problem of duality, one can discern a latent interplay of part and whole, envisioning humans as blends of two distinct components. Bohm studies the phenomenon of duality by discretizing the term *psychosomatic* into its building blocks, whereby *psyche* means mind or soul, and *soma* means body. In this context, these two aspects of being are handled separately for analytical reasons and for the purposes of analysis, but in reality, they exist inseparably infused (Bohm 1989). Consequently, wholes resist using dismantling as a method to be fully comprehended by their so-called *parts*.

3.6.2 A Handful of Sand

For the sake of analysis and computation, such discretization and dismantling of the world can best be read through Pirsig's *analytic knife* (Pirsig 2005). In *Zen and the Art of Motorcycle Maintenance*, the metaphor of Phædrus' knife is presented as an insightful lens into mereology, explicitly addressing the interplay between selective attention, categorization, and discrimination.

In Pirsig's masterwork, an individual tasked with sorting grains of sand into categories based on their physical and visual properties (such as color, size, overall shape, translucency, sharpness, and like) walks to a beach and tries to devise a categorization method. Soon, he realizes that an imminent sand piece would not fit into his previously conceived categories. What comes next?

The individual standing on a beach immediately becomes confronted with the vastness of sand— a symbol of the overwhelming totality of our awareness. From this overwhelming landscape, one consciously can select a mere handful, representing only a subset of reality one chooses to engage with. This selection is not just a passive act but a crucial process, like wielding a knife that discriminates, divides, and categorizes. The sorting process hit a wall imminently as the individual cannot put the sand grain in hand into a formerly defined category.

Pointing to the insufficiency of preconceived categories, Pirsig advocates for an integrative worldview that neither rejects meticulous categorization nor holistic contemplation.

The metaphor of Phædrus' knife and sorting sand grains are tools to think about the challenges in AI training where, just as in sorting grains of sand from massive mountains of data, humans decide what information to include—overtime, selecting what we can sort to build reductive AI systems (Goodfellow et al. 2016).

The essence captured is a call to recognize and appreciate the vast, endless landscape of awareness from which our selective consciousness emerges, and within this vastness, to acknowledge the ever-present figure— akin to shape grammars, which insists on keeping the perception and intention of the maker at the center of shape calculations. So, instead of keeping human in the loop (Doan 2022) for AI development, we can claim:

Human IS the loop.

4 Talking Prompts

4.1 Seeing as (Merely) Describing

Diffusion models are generative models used in AI to create images. These models begin with a distribution of random noise and then gradually transform one distribution of random noise into another distribution of data., through a reversible process of adding and removing noise (Ho et al. 2020).

Specifically, diffusion models are trained on a large corpus of text and corresponding images, making them proficient at generating an image that fits the input text when provided a text prompt, which is done by iteratively refining a random noise into a picture with a word prompt (Ho et al. 2020). This is complemented by the model's good grasp of image space and language, enabling a large enough range of imaginative choices.

CLIP-guided diffusion involves coupling a diffusion model with a CLIP (Contrastive Language–Image Pretraining) model trained to associate text and images, guiding the diffusion process to generate images that ensemble with the descriptive text (Steinfeld 2023). By iteratively adjusting the image toward a more minimal discrepancy between the text and the image as perceived by the CLIP model, CLIP-guided diffusion tries to generate outputs that represent the textual description. The intricacies and challenges recently yielded a new sort of (controversial) occupation called prompt engineering, essentially the practice of carefully crafting text inputs (prompts) that guide the AI text-to-image system.

The core challenge here is capturing real-world and human experiences in natural language, especially in visual creativity (Zhou et al. 2018). Same applies to producing or refining visual compositions through the complexities of natural language (Goodfellow et al. 2016).

AI's grasp of meaning is constrained by the language guiding it. Stiny often highlights a paradox within the very tools—words—that diffusion models rely upon (2022). Words, by their nature, strip away the silent depth of visual experience, flattening the broader meanings of images when transformed into a structure generative AI systems can interpret. In essence, while prompts are crafted to inspire creativity through descriptions shaped by language, these very definitions raise the question of whether AI can genuinely capture the indescribable essence of the visual through the very same language. Stiny's observation serves as a reminder that beyond the structured framework of prompt engineering lies an entire realm of nuance that words alone may never fully express, challenging the true potential of such models in capturing the intricacies of human experience and emotion.

4.2 Many Dragons or Lernaean Hydra?

Consider the Lernaean Hydra, the multi-headed creature from Greek mythology, as a compelling example. Since the Hydra exists only in stories, our understanding of its form is inferred from evolving descriptions over time. Each retelling adds layers to its visual complexity, making it challenging to capture the Hydra's essence in a single description using language. Language offers a spectrum of interpretations rather than providing a definitive image, each contributing to a multifaceted conceptualization that aligns with human interpretive processes and our diverse perceptions of the Hydra's appearance.

This challenge becomes pronounced with CLIP interrogation in Stable Diffusion Models and the */describe* function exhibited by Midjourney. When leveraging diffusion models for image generation, the early versions of these systems falter. Uncertain about the Lernaean Hydra's visual structure, it describes it as "many dragons" instead of depicting it as a creature with multiple heads. The descriptions remain bounded by limitations of natural language—two squares or four triangles, duck or rabbit, many dragons or Lernaean Hydra?

Similarly, CLIP guided diffusion models struggle to accurately translate the image content in stable diffusion, failing to capture the Hydra's unique visual characteristics. This gap may be bridged in future diffusion models or subsequent AI systems once datasets incorporate accurate descriptions for this specific instance. However, this solution raises another concern: as the visual realm remains open-ended, there will always be untold concepts and ambiguities not accounted for in datasets (as in seeing the unfathomed *crocodile* in the *duck or rabbit* equation). For example, while Midjourney v3 failed to render the Lernaean Hydra in 2022, their v6 release in 2024 handled the task with significantly improved fidelity. Yet, a prompt such as "a geometric line drawing of a mythological five-headed Lernaean Hydra, composed of squares and triangles, in the style of George Stiny's shape grammars drawings" remains an extraordinarily complex challenge for diffusion models to resolve.

Furthermore, once an image or concept, like the Lernaean Hydra, is encapsulated in linguistic terms, the diffusion model embarks on interpreting that language. The realm of natural language is vast and complex, packed with potential ambiguities and shades of meaning. For instance, if someone describes the Lernaean Hydra's heads as ever-growing, does that signify their regenerative capabilities or overall scale? The AI model must have the intricate task of discerning context, sentiment, literal implications, and even potential metaphors or symbolic meanings to generate or modify images accurately.

These layers of abstraction in describing and interpreting offer potential and pose challenges. Diffusion models utilize these layers of abstraction to interact with users and deliver the anticipated outputs proficiently. However, this also emphasizes the importance of honing our linguistic directives and being aware of the models' capabilities and constraints (Fig. 12).

Image-to-image, image blending, and image referencing techniques offer more versatile, imaginative, and visually stimulating solutions than text-to-image methods.

Fig. 12 Lernaean Hydra generated in Midjourney v6.1 by the author. The intent of creating *"a geometric line drawing ... in the style of George Stiny's shape grammars drawings"* falters

Often, the system generates unanticipated outcomes rather than aiming to align with a preconceived mental image shaped by language precisely. Running counter to the precision that prompt engineering strives for, these more ambiguous methods frequently yield more dynamic and intriguing results.

The intricate endeavor of refining prompts to yield specific images highlights a fundamental challenge: the vast spectrum of meanings an image can encompass beyond mere words' descriptive capacity.

5 Contributions

In 1972, Stiny and Gips planted a foundational seed for contextualizing computation within human creativity, a vision deliberately positioned outside the bounds of traditional electronic digital computation at a time when the field moved aggressively in the opposite direction. Through a critical examination of generative AI via the lens of shape grammars, the approach advocated here establishes a discourse grounded in the inherently ambiguous and interpretive processes of human perception and intention. This repositioning underscores shape grammars' value as a framework for understanding AI's potential and boundaries in art and design.

The presented inquiry:

Establishes a theoretical lens for AI critique by proposing a distinct theoretical outlook on generative AI that departs from conventional models rooted in digital electronic computation. Addressing the limitations of symbolic and language-based systems in capturing visual experience highlights shape grammars as a critical foundation for assessing AI's reach and boundaries in artistic production.

Introduces a framework for navigating visual ambiguity using three guiding questions—'duck or rabbit?', 'inside or outside?', and 'part or whole?'—These are structured as conceptual pillars, separated for analytical clarity yet inseparable in intent, to deepen our understanding of generative AI's inherent limitations and possibilities without directly conflating or contrasting these systems with human creative capacities.

Advances a multidisciplinary perspective by drawing on the interdisciplinary nature of shape grammars, emphasizing the importance of integrating scientific rigor and humanistic insight in AI critique. This approach encourages a comprehensive understanding of generative AI technologies in art and design, positioning these systems as more than mere tools but as participants in a broader discourse on meaning-making and perception.

Examines prompt engineering and the role of ambiguity by questioning the emphasis on linguistic precision within text-to-image systems. To this end, linguistically precision-driven approaches can be compared with image-based techniques, such as image-to-image translation and blending. Such systems allow for unanticipated outcomes that better capture the essence of visual creativity than those relying solely on linguistic guidance.

Projects future directions for AI and creativity by underscoring ambiguity as an essential driver of creativity.

These progressive suggestions are designed to be open-ended, aiming to inspire the development of AI systems that are more adept at handling ambiguity, thus increasing their effectiveness for creators as they swing between moments of discovery and disaffection, both of which are required for visual discovery.

AI technologies will keep developing, but significant divides will remain as we discuss ambiguities in their connection to visual creativity. Generative AI remains fundamentally a human creation, constructed on principles that, while advanced, are bound to human-defined parameters. Overstated comparisons or anthropomorphic interpretations of these technologies risk obscuring the unique dynamics that shape both AI and human creativity.

AI, looking at what it has tried to do since its inception—solving world problems through labeling and precision—will try to resolve those divides to offer a more stable world. However, ambiguities will need to be sacrificed in the process. In turn, ambiguity will remain the source of visual creativity and design thinking. Shape grammars will keep empowering us in uniquely mysterious ways, encouraging us to keep "talking about seeing and doing"—ambiguities included.

References

Bohm, D. 1989. Meaning and information. In *The search for meaning: The new spirit in science and philosophy*, ed. P. Pylkkanen, 43–66. Wellingborough: Thorsons Publishing Group.
Comedian (Artwork). 2024. In Wikipedia. https://en.wikipedia.org/w/index.php?title=Comedian_(artwork)&oldid=1194838354.
Damasio, A.R. 2000. *The feeling of what happens: Body and emotion in the making of consciousness*. New York: Harcourt Inc.
Doan, D. 2022. 2023 Guide to effective human-in-the-loop automation. Klippa. https://www.klippa.com/en/blog/information/human-in-the-loop/.
Dreyfus, H.L. 1992. *What computers still can't do: A critique of artificial reason*, 1st ed. Cambridge: The MIT Press.
Elgammal, A., B. Liu, M. Elhoseiny, and M. Mazzone. 2017. CAN: Creative adversarial networks, generating 'art' by learning about styles and deviating from style norms. Preprint at arXiv: https://doi.org/10.48550/arXiv.1706.07068.
Eloy, S., P. Pauwels, and A. Economou. 2018. AI EDAM special issue: Advances in implemented shape grammars: solutions and applications. *Artificial Intelligence for Engineering Design, Analysis and Manufacturing*. 32 (2): 131–37. https://doi.org/10.1017/S0890060417000634.
Goodfellow, I.J., J. Pouget-Abadie, M. Mirza, B. Xu, D. Warde-Farley, S. Ozair, A. Courville, and Y. Bengio. 2014. Generative adversarial networks. Preprint at arXiv: https://doi.org/10.48550/arXiv.1406.2661.
Goodfellow, I.J., Y. Bengio, and A. Courville. 2016. *Deep learning (Illustrated edition)*. Cambridge: The MIT Press.
Gips, J., and G. Stiny. 1975. Artificial intelligence and aesthetics. In *Proceedings of international joint conference on artificial intelligence*. https://www.semanticscholar.org/paper/Artificial-Intelligence-And-Aesthetics-Gips-Stiny/cfae035e124de9557783891ccf4c93025cb0106e.
Gün, O.Y. 2016. A place for computing visual meaning: The broadened drawing-scape. Ph.D. dissertation. Cambridge: Massachusetts Institute of Technology. https://dspace.mit.edu/handle/1721.1/106364.
Gün, O.Y. 2017. Computing with watercolor shapes: Developing and analyzing visual styles. In *Proceedings of Computer-aided architectural design: Future trajectories. CAAD futures 2017*, vol 724, eds. G. Çağdaş, G., M. Özkar, L. Gül, E. Gürer. Singapore: Springer. https://www.academia.edu/34122178/Computing_with_Watercolor_Shapes_Developing_and_Analyzing_Visual_Styles.
Gün, O.Y., and G. Stiny. 2012. An open conversation with George Stiny about calculating and design. *Dosya 29: Computational Design* 29: 6–11.
Gürsoy, B., and M. Özkar. 2015. Visualizing making: Shapes, materials, and actions. *Design Studies, special issue: Computational making* 41 (November): 29–50. https://doi.org/10.1016/j.destud.2015.08.007.
Ho, J., A. Jain, and P. Abbeel. 2020. Denoising diffusion probabilistic models. Preprint at arXiv. https://doi.org/10.48550/arXiv.2006.11239.
Hong, T.C.K., and A. Economou. 2022. Five criteria for shape grammar interpreters. In *Design computing and cognition'20*, ed. by J. S. Gero, 191–207. Cham: Springer. https://doi.org/10.1007/978-3-030-90625-2_11.
Knight, T., and G. Stiny. 2001. Classical and non-classical computation. *Arq: Architectural Research Quarterly* 5 (04): 355–372. https://doi.org/10.1017/S1359135502001410.
Knight, T., and G. Stiny. 2015. Making grammars: From computing with shapes to computing with things. Prof. Knight via Anna Boutin. https://dspace.mit.edu/handle/1721.1/111993.
Kotsopoulos, S.D. 2022. Design without rigid rules. In *Design computing and cognition '20*, ed. J. S. Gero, 107–128. Cham: Springer. https://doi.org/10.1007/978-3-030-90625-2_7.
Kozyrkov, C. 2023. Fooled by statistical significance. Medium. https://towardsdatascience.com/fooled-by-statistical-significance-7fed1bc2caf9.

Kuindersma, S., R. Deits, M. Fallon, A. Valenzuela, H. Dai, F. Permenter, T. Koolen, P. Marion, and R. Tedrake. 2015. Optimization-based locomotion planning, estimation, and control design for the atlas humanoid robot. *MIT Web Domain.* https://dspace.mit.edu/handle/1721.1/110533.

Krizhevsky, A., I. Sutskever, and G. Hinton. 2012. ImageNet classification with deep convolutional neural networks. *Neural Information Processing Systems* 25. https://doi.org/10.1145/3065386.

Larson, E.J. 2021. *The myth of artificial intelligence: Why computers can't think the way we do.* Cambridge: Belknap Press, An Imprint of Harvard University Press.

Cardoso Llach, D. 2021. Sculpting spaces of possibility: Brief history and prospects of artificial intelligence in design. In *The Routledge companion to artificial intelligence in architecture*, 13–28. London: Routledge. https://www.taylorfrancis.com/chapters/edit/10.4324/9780367824259-3/sculpting-spaces-possibility-daniel-cardoso-llach.

Mazzone, M., and A. Elgammal. 2019. Art, creativity, and the potential of artificial intelligence. *Arts* 8 (1): Article 1. https://doi.org/10.3390/arts8010026.

Metzger, W. 2009. *Laws of seeing.* Cambridge: The MIT Press.

Midjourney Statistics (Updated February 2024). 2023. April 25, 2023. https://photutorial.com/midjourney-statistics/.

Pandey, M., M. Fernandez, F. Gentile, O. Isayev, A. Tropsha, A.C. Stern, and A. Cherkasov. 2022. The transformational role of GPU computing and deep learning in drug discovery. *Nature Machine Intelligence* 4 (3): 211–221. https://doi.org/10.1038/s42256-022-00463-x.

Pirsig, R.M. 2005. *Zen and the art of motorcycle maintenance: An inquiry into values.* New York: William Morrow Paperbacks.

Ramesh, A., P. Dhariwal, A. Nichol, C. Chu, and M. Chen. 2022. Hierarchical text-conditional image generation with CLIP latents. Preprint at arXiv: https://doi.org/10.48550/arXiv.2204.06125.

Rombach, R., A. Blattmann, D. Lorenz, P. Esser, and B. Ommer. 2021. High-resolution image synthesis with latent diffusion models. Preprint at arXiv: https://arxiv.org/abs/2112.10752v2.

Ruiz-Montiel, M., J. Boned, J. Gavilanes, E. Jiménez, L. Mandow, and J.L. Pérez-de-la-Cruz. 2013. Design with shape grammars and reinforcement learning. *Advanced Engineering Informatics* 27 (2): 230–245. https://doi.org/10.1016/j.aei.2012.12.004.

Steinfeld, K. 2023. Clever little tricks: A socio-technical history of text-to-image generative models. *International Journal of Architectural Computing* 21 (2): 211–241. https://doi.org/10.1177/14780771231168230.

Stiny, G. 2006. *Shape: Talking about seeing and doing.* Cambridge: The MIT Press.

Stiny, G. 2011. What rule(s) should I use? *Nexus Network Journal* 13 (1): 15–47. https://doi.org/10.1007/s00004-011-0056-6.

Stiny, G. 2016. *The critic as artist: Oscar Wilde's prolegomena to shape grammars.* MIT Open Access Articles. https://dspace.mit.edu/handle/1721.1/103155.

Stiny, G. 2022. *Shapes of imagination: Calculating in Coleridge's magical realm.* Cambridge: The MIT Press.

Stiny, G., and J. Gips. 1978. *Algorithmic aesthetics: Computer models for criticism and design in the arts.* Berkeley: University of California Press.

Stiny, G., and W.J. Mitchell. 1978. The Palladian grammar. *Environment and Planning B: Planning and Design* 5 (1): 5–18. https://doi.org/10.1068/b050005.

Stiny, G., and J. Gips. Artificial intelligence and aesthetics. The Encyclopedia of Aesthetics. https://www.academia.edu/44344316/Artificial_intelligence_and_aesthetics. Accessed 1 Feb 2024.

Taleb, N.N. 2010. *The black swan: The impact of the highly improbable: With a new section: On robustness and fragility*, 2nd edn. New York: Random House Trade Paperbacks.

Ullman, S. 1984. Visual routines. *Cognition* 18 (1–3): 97–159. https://doi.org/10.1016/0010-0277(84)90023-4.

Vardouli, T. 2012. Computer of a thousand faces: Anthropomorphizations of the computer in design (1965–1975). *Dosya 29: Computational Design*: 24–31.

Wei, B., Y. Zhao, K. Hao, and L. Gao. 2021. Visual Sensation and Perception Computational Models for Deep Learning: State of the Art, Challenges and Prospects. CoRR, abs/2109.03391. https://doi.org/10.48550/arXiv.2109.03391, https://arxiv.org/abs/2109.03391

Winston, P.H. 2012. The next 50 years: A personal view. In Biologically Inspired Cognitive Architectures, Vol. 1, 92–99. Elsevier. https://doi.org/10.1016/j.bica.2012.03.002.

Wortmann, T., and R. Stouffs. 2018. Algorithmic complexity of shape grammar implementation. In *Artificial intelligence for engineering design, analysis and manufacturing*.

Zhou, B., Y. Sun, D. Bau, and A. Torralba. 2018. Interpretable basis decompositionmfor visual explanation. In *Proceedings of the European conference on computer vision, ECCV '18*, eds. V. Ferrari, M. Hebert, C. Sminchisescu, Y. Weiss, 119–134. Cham: Springer. https://openaccess.thecvf.com/content_ECCV_2018/html/Antonio_Torralba_Interpretable_Basis_Decomposition_ECCV_2018_paper.html.

Open Access This chapter is licensed under the terms of the Creative Commons Attribution-NonCommercial-NoDerivatives 4.0 International License (http://creativecommons.org/licenses/by-nc-nd/4.0/), which permits any noncommercial use, sharing, distribution and reproduction in any medium or format, as long as you give appropriate credit to the original author(s) and the source, provide a link to the Creative Commons license and indicate if you modified the licensed material. You do not have permission under this license to share adapted material derived from this chapter or parts of it.

The images or other third party material in this chapter are included in the chapter's Creative Commons license, unless indicated otherwise in a credit line to the material. If material is not included in the chapter's Creative Commons license and your intended use is not permitted by statutory regulation or exceeds the permitted use, you will need to obtain permission directly from the copyright holder.

Shape Grammars in Design and Education

Computing Chinese Architecture

Andrew I-kang Li

Abstract The fundamental text on Chinese architecture is the building manual *Yingzao fashi* (1103). The first—and still the most important—scholar of that text was Liang Sicheng (1901–72), who called it a "grammar book of Chinese architecture," citing its use of "formulas based on principles and proportions" (Liang 1984b, 358). In other words, as the essence of the text, he recognized what we would now call its algorithmic nature. His insight suggests that if we investigate this quality, we will better understand the text and, indeed, Chinese architecture generally. Accordingly, to this text we apply computational tools and methods that were not available to Liang. This enables us to identify and formally restate the algorithms it contains, both individually and as a collection. In addition, we consider explicitly the kind and conduct of this computational investigation. Finally, we suggest how to extend this approach from the *Yingzao fashi* not only to extant buildings of the Tang through Yuan dynasties (618–1368), but also to those of the Ming and Qing dynasties (1368–1911) and the comparable building manual of that era, the *Gongcheng zuofa zeli* (1734).

1 Before We Begin

Readers of this essay may wonder what is new. The short answer is that it summarizes in one place my work in computational design and explains what it can do for non-grammarists. The long answer, for computational insiders, is that it depends on who is asking. I will explain.

I have worked mainly on two projects in computational design. In the first, I used shape grammars to understand what the *Yingzao fashi* had to say about architectural design. I developed a grammar that characterized the schematic design of a single building type (Li 2001). This work was related to Chinese architectural history, but its contributions were to computation. The first contribution was using a novel approach to a plausible, if simple, design question. In this approach, I treated designs as n-tuples

A. I. Li (✉)
Kyoto Institute of Technology (retired), Tokyo, Japan
e-mail: i@andrew.li

of drawings, descriptions and other devices. The other contribution was clarifying Stiny and Mitchell's (1978) explication of a computational understanding of style. They said that to have such an understanding was to have three capabilities: synthetic, analytic, and descriptive.[1] Generative algorithms, such as shape grammars, have all three capabilities. We develop these algorithms—and thus our understanding—by repeatedly generating and evaluating the predicted designs.

In practice, this means we need to be able to do two things. The first is to generate designs quickly, reliably, and with minimum distraction from the domain-level tasks of developing algorithms and evaluating designs. That is, we need a tool for subdomain design-crunching. The other thing we need is to understand how to evaluate the designs so crunched. This is a less obvious task than it might seem, as it is not objective, but subjective (Li 2011). It deserves the user's attention more than design-crunching does. As a result of this work, I perceived how to generalize from schematic designs in a text belonging to a specific time and place to actual extant buildings over a span of centuries and an entire country. All these insights were possible because the visuality and formality of shape grammars provided a congenial entry into design computation. The richness of the insights suggested that shape grammars would remain a suitable tool for my work in Chinese architectural history.

The next step was to develop a design cruncher. At the time, developers of shape grammar interpreters (e.g., Krishnamurti 1982; Tapia 1999) were concerned with the fundamental technical issues of implementing subshape detection and rule application. These interpreters were non-parametric, did not support advanced devices like descriptions, were not primarily concerned with usability, and so did not support design-crunching. I had crunched my designs by hand, but it was not practical to continue doing so.

Thus, my second project in computation was to develop a suitable sub-domain interpreter. It seemed to me that such a tool would support both designers and analysts (Li 2005). Doubling the audience doubled the motivation. I worked on this project first with Hau Hing Chau, then with Rudi Stouffs, each of whom had built an interpreter engine with a minimal interface (Chau et al. 2004; Stouffs 2018). Ultimately, I produced for Stouffs's engine an intuitive interface in a geometric modeling environment, a non-parametric design cruncher (Li 2016; Li and Stouffs 2021). It supported an iterative digital workflow. Several generations of students, all grammar novices, have enjoyed using it and been doing impressive work with it. This tool was important progress, but it did not support the advanced technical devices—descriptions, parameters, *n*-ary designs, etc.—in my original complex grammar. A grammatical contribution beyond the domain of design computation was still out of reach.

This has brought me to my current project: to rework the algorithms from visual to symbolic form, by writing Python programs that crunch geometric models in Rhinoceros3D. It is true that we will lose the immediacy of visual algorithms, but we will gain the speed and invisibility necessary to support serious analytical work in Chinese architectural history. The essay that follows is an explanation for those in that domain.

---

[1] More on this below.

I see my current task as "just doing the work,"[2] both in computation, where programming is routine, and in Chinese architectural history, where it is too early for meaningful results. Having subjected readers to a long explanation, I now suggest this answer to the original question. For computationalists, the new is past; for historians, the new is yet to be.

2 The *Yingzao Fashi* and Grammar

In the history of Chinese architecture, one artifact stands out above the others: the *Yingzao fashi* [Building Standards], a book about building, first published in 1103, but still well known today to anyone who has studied architecture in a Chinese-speaking country. It is one of only two complete surviving books in Chinese history dealing with building, a subject which was traditionally thought to be undeserving of written attention. The other is the *Gongcheng zuofa zeli* [Code of Building Practice], published six centuries later in 1734 (Qing dynasty). We will encounter this second book again shortly.

The *Yingzao fashi* was written by the court architect of the Huizong emperor (r. 1101–26) of the Song dynasty (960–1127), Li Jie (d. 1110), to facilitate communication between two groups of people: government officials, who commissioned buildings, and craftsmen, who constructed the buildings. Li did this by systematizing and recording existing building practices. With this basis for a common understanding, he aimed to reduce corruption and increase efficiency.

The earliest research into the *Yingzao fashi* was by Liang Sicheng (1901–72) and his associates, Lin Huiyin (1904–55) and Liu Dunzhen (1897–1968). It is fair to say that Liang's whole career was shaped by the *Yingzao fashi;* his comprehensive study, the *Yingzao fashi zhushi* [The Annotated *Yingzao fashi*] (Liang 1984a), is an essential study.[3] Liang considered both building manuals to be "grammar books of Chinese architecture," with the *Yingzao fashi* as the more commendable. "The *Gongcheng zuofa zeli* ... lists dimensions in a mechanical fashion. [By contrast] the *Yingzao fashi* in all cases gives formulas based on principles and proportions" (Liang 1984b, 358). In other words, he recognized as the essence of the text what we would now call its algorithmic nature. His insight suggests that if we investigate this aspect, we will better understand the text and, indeed, Chinese architecture in general.

Accordingly, to the *Yingzao fashi* we apply computational tools and methods that were not available to Liang. This enables us to identify and formally restate the algorithms it contains, both individually and as a collection. In addition, we explicitly consider the kind and conduct of this computational investigation. Finally, as noted earlier, we suggest how to extend this approach from the *Yingzao fashi* not only

---

[2] Stiny's advice to me when I left MIT to return home and work on my dissertation: "Don't try to be brilliant. Just do the work."

[3] For Liang's life and career, see Fairbank (1995). Other useful studies of the *Yingzao fashi* are by Takeshima (1970–72), Glahn (1984), Guo (1999), and Feng (2012).

to extant buildings of the Tang through Yuan dynasties (618–1368), but also to the *Gongcheng zuofa zeli* (1734) and to extant buildings of the Ming and Qing dynasties (1368–1911).

3 An Algorithm in the *Yingzao Fashi*

An excellent example of a "formula based on principles and proportions" is the method for determining the curved roof section. It is known as *juzhe*,[4] which means "raise and depress" (we'll see why shortly). Liang's drawing of *juzhe* (Fig. 1) shows half of a roof section, consisting of five purlins: the ridge purlin at the upper left, the eaves purlin at the lower right, and three purlins in between. The total horizontal distance between the ridge purlin and the eaves purlin is B.

To calculate a roof section, we use *juzhe* to find the height of each purlin relative to the eave purlin. The first step is to find the height of the ridge purlin. We "raise" the ridge purlin to a height $R = B/4$ or $R = B/3$, depending on the building type. We draw a line from the newly raised ridge purlin to the eave purlin; this is the working line.

The second step is to find the height of the first mid purlin below the ridge purlin. Starting from the working line, we "depress" a distance of R/10, find the height of the first mid purlin, and draw a new working line to the eave purlin. We repeat this step for the second mid purlin (depressing a distance of R/20), and again for the third mid purlin (R/40). Each pair of neighboring purlins supports a rafter; the four abutting rafters form the curved roof.

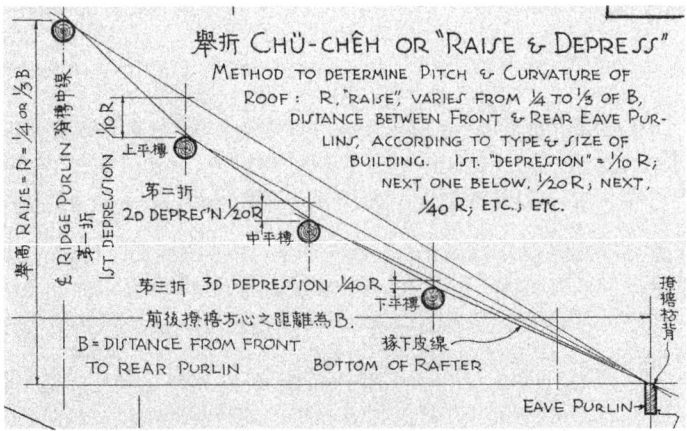

Fig. 1 The algorithm *juzhe*, as drawn by Liang (1984c, Fig. 7)

---

[4] Pronounced something like *jyew juh*.

What makes *juzhe* a "formula based on principles and proportions" to Liang is that it is a finite set of unambiguous steps that solve a problem; in other words, it is an algorithm. And the fact that *juzhe* is an algorithm is what helps us understand what makes some roof sections acceptable and others not.

Stiny and Mitchell (1978) explain why. They define a style as a set of buildings[5] that "create a similar impression," and then propose that understanding a style means having three capabilities: synthetic, analytic, and descriptive. An algorithm, such as *juzhe,* supplies all three. The first capability is to create a new roof section. You may be a rookie builder, but if you apply *juzhe* correctly, you are assured of producing an acceptable roof section.

The second capability is to verify whether a given roof section is permissible. Suppose now that you are the building inspector, and you need to approve a roof under construction. You count and measure the rafters and apply *juzhe.* If the roof in front of you matches your result, then it is acceptable; if not, then not. You have a means of verification, whether or not you have ever encountered that particular case.

The third capability is to explain why all the roof curves appear similar: not too high, not too low, not too curved, not too straight. Here, *juzhe* is itself the explanation, and it is so straightforward that you will probably still remember it tomorrow.

4 A List in the *Yingzao Fashi*

As Liang implied, the alternative to the formula is a mechanical account: a straightforward enumeration or list. We have an example of this as well, in the *Yingzao fashi:* a set of 18 *ting* hall sections with varying arrangements of interior columns (Fig. 2). Each purlin is supported either by a column standing on the floor or by a post supported in turn by a beam spanning between two columns. Put another way, a column is always located below a purlin, but below a purlin there is not always a column.

Clearly, the 18 sections do not show all the possible ways to insert columns. A 6-rafter building, for example, has 32 possibilities,[6] but only three are included. It is safe to assume that these three are acceptable, but what of the other 29? The list does not tell us. This is unlike *juzhe,* which, for any permissible starting condition, delivers one permissible solution. How does the list perform with respect to the three capabilities? First, what the list explains—if one can call it explaining—is only that the sections on the list are on the list. It does not explain whether the list is exhaustive, random, or representative (much less in what way). As for creating an (acceptable) section, it follows that the only reliable way is to select one from the list. An unlisted section, even if it is acceptable (i.e., one of the 29 above), is simply not available.

---

[5] This can be generalized to designs, including roof sections.

[6] A 6-rafter building has seven purlins and seven column positions. Assuming that the front and rear columns are always present, this gives five possible column positions; $2^5 = 32$.

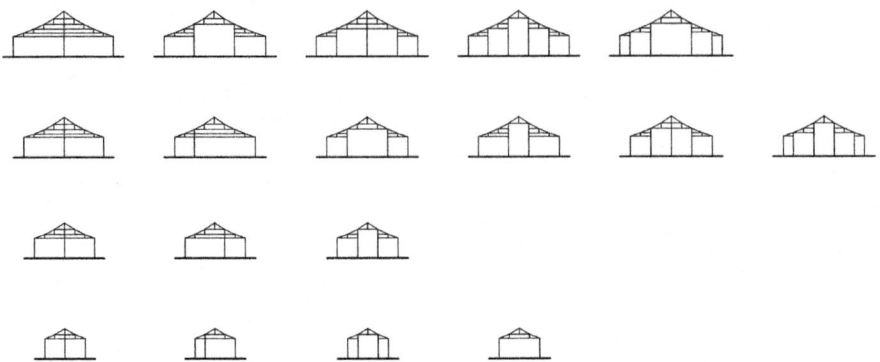

Fig. 2 Sections of 18 *ting* halls (10, 8, 6, and 4 rafters deep) in the *Yingzao fashi*

Finally, the only sections that can be verified are those that are already on the list. On all three counts, the list disappoints.

5 Formulating a Hypothesis

Now, imagine that the builder and the inspector disagree. The builder argues: "Your list has a 4-rafter clear-span *ting* hall. My *ting* hall is a 6-rafter clear-span. Why can't you approve it?" The inspector, of course, replies simply that not on the list means not on the list (Fig. 3).

The builder is pointing out that his section makes a similar impression to one on the list. This suggests that we can propose an algorithm that creates all similar sections, whether they are on the list or not. Such an algorithm might look like this.

- Step 1 (required). Begin with a building section of $2n$ rafters and $n + 1$ purlins (column positions).
- Step 2 (required). Insert the front column and the back column (eaves columns).
- Step 3 (optional). Insert a (central) column under the ridge purlin.
- Step 4 (optional). Do nothing. We decide we are done.
- Step 5 (optional). Insert a column and a beam. The column goes in an available position; the beam spans from the last added column on the same side of the building to the new column. If the beam passes below a purlin, then the column

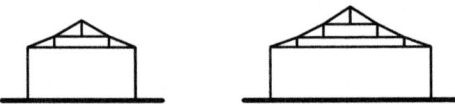

Fig. 3 A 4-rafter clear-span *ting* hall on the list (left). A 6-rafter clear-span *ting* hall not on the list (right)

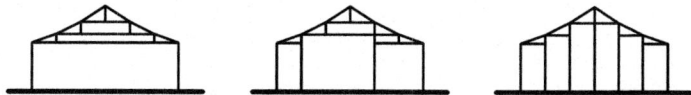

Fig. 4 Three 6-rafter *ting* hall sections created by the algorithm. The builder's disputed section is on the left

position below is no longer available. Repeat as desired or until there are no more available column positions.
- Step 6 (required). If there remains an available column position without a column, insert a beam. We are done.

This algorithm is more complicated than *juzhe,* but it is still a finite set of unambiguous steps.[7] By applying the steps differently, it creates all the sections on the list (Fig. 2) and some that are not, including our builder's disputed section (Fig. 4).

6 Evaluating Predictions

The disagreement between the builder and the inspector leads to a question. The builder's section above, created by our algorithm, is not on the list, but is it acceptable? Unlike the inspector, we assume that the possibility exists. Put another way, the algorithm is a hypothesis that the section is acceptable; how do we evaluate whether the prediction is true? If this were science, we would do an experiment. That is, we would appeal to the ultimate independent authority: nature.

But this is not science, and there is no independent authority. Li Jie has been dead for nine centuries. It is up to us, not only to evaluate these predictions, but also to consider how to evaluate them. For instance, when considering our builder's disputed section, we might think that a 6-rafter beam is too long and unsafe. Then we might notice that there is an 8-rafter section on the list, and that it contains a 6-rafter beam. Further, there are no 8-rafter beams in any of the sections on the list (Fig. 5). From this we might conclude that 6-rafter beams are acceptable, that the disputed 6-rafter section is acceptable, but that sections with longer beams are *not* acceptable. We can incorporate this conclusion into the algorithm by modifying step 5.

Or we may feel we need to know more to make an informed evaluation. Is a 6-rafter beam inside an 8-rafter building structurally different than one inside a 6-rafter building? A structural analysis might provide an answer. Are 8-rafter beams too scarce? An historical investigation might show whether and where trees of the right size and species were available. Whatever we decide, confronting the fact and method of evaluation contributes to our understanding.[8] Now that we know how to proceed with both principled and mechanical descriptions in the *Yingzao fashi,* we

---

[7] There are more steps to complete the roof structure. For the complete version, see Li (2001).
[8] For more on evaluation, see Li (2011).

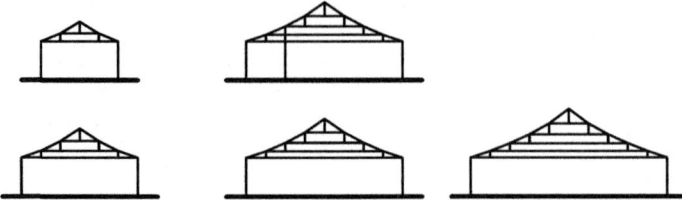

Fig. 5 Evaluating the disputed section (bottom left) by examining other sections: two on the list (top); three not on the list (bottom)

can handle complete massing diagrams of *ting* halls and other building types. How does our approach fit in with mainstream scholarship?

7 Mainstream Scholarship

One of the most frequent comments about Chinese architecture is that it is remarkably consistent over time and space. Steinhardt (2022, 6) puts it like this:

> Is it really possible that so much Chinese architecture looks like so much other Chinese architecture, and that it appears so similar to so much architecture in Korea, Japan, and Mongolia for two thousand years? ... One can call the question why Chinese architecture changes so little over such a long period a field-defining question, for only when one accepts that sameness exists for so long can she begin to answer it.

Nevertheless, even within this field-defining sameness, Steinhardt does allow for evolution. Indeed, Liang highlights it as the subtitle of his book, *A pictorial history of Chinese architecture: a study of the development of its structural system and the evolution of its types* (Liang 1984c). There he distinguishes three periods: the period of vigor (c. 600–c. 1050), the period of elegance (c. 1000–c. 1400), and the period of rigidity (c. 1400–c. 1900) (Fig. 6). The first two periods evolve continuously, whereas the break between the second and third periods, in the early years of the Ming dynasty, is particularly pronounced.

> The change is very abrupt, as if some overwhelming force had turned the minds of the builders toward an entirely new sense of proportion (Liang 1984c, 103).

This makes two major divisions, in the middle of each of which was published one of our two building manuals, first the *Yingzao fashi* and then the *Gongcheng zuofa zeli*. As a result, researchers discuss extant buildings with reference to the corresponding text.

> We owe to [the two grammar books] all the technical terms that we know and all the criteria that we employ today for the comparative study of the architecture of different periods (Liang 1984c, 14).

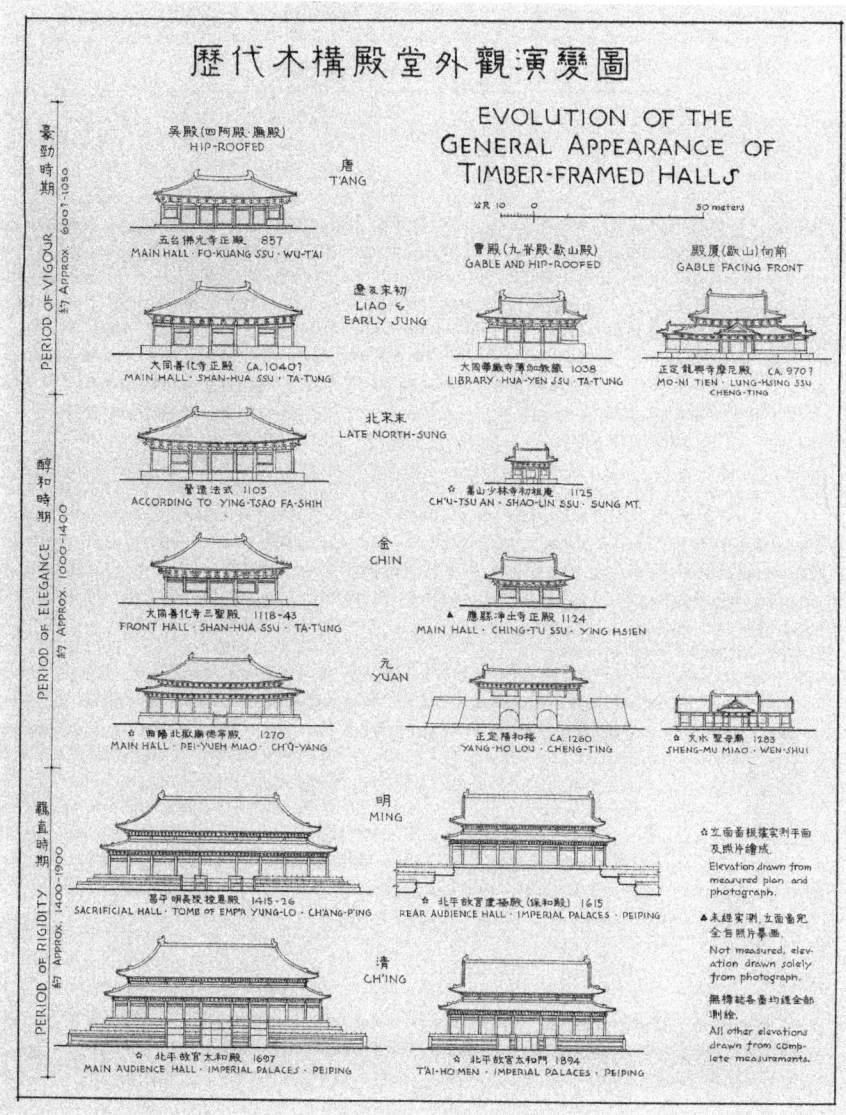

Fig. 6 Evolution of wood-frame buildings. (Liang 1984c, Fig. 20)

During the first major division, extant buildings evolve, so it is no surprise that most do not follow the *Yingzao fashi* exactly. In fact, only one—Chuzu an of Shaolin si (1125)—is reasonably faithful. To put it computationally, the set of buildings that can be created by our *Yingzao fashi* algorithms include Chuzu an, but not other extant buildings. Yet those extant buildings are traditionally considered to be related to the *Yingzao fashi*. How do we account for this computationally?

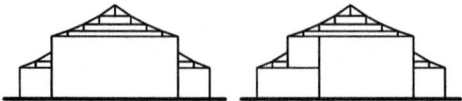

Fig. 7 A *dian* hall with a structurally dependent porch (left); Shengmu miao of Jin ci (1023–31) (right), with a structurally integrated porch and a truncated eaves column

8 Evolution

Knight (1994) provides an elegant method. Simply put, as the buildings change, so change the algorithms that create them: by steps added, steps deleted, or steps changed. As an example, notice that over time roofs start low and become higher (Fig. 6). Remember further that the height of the ridge purlin is set by the "raise" step of the *juzhe* algorithm. In the *Yingzao fashi*, the "raise" is B/4 or B/3, but it is a simple matter to produce a different height by adjusting this value to, say, B/2 or B/5. In this case, the number and locations of components remain fixed; only their dimensions change. These changing proportions underlie mainstream approaches to the dating of buildings (see Qi 1981, for example).

But there are other variations as well, with differences not of proportion but of structure. For example, a hall sometimes has a covered porch wrapped around it. Structurally, the two parts are clearly differentiated: the hall proper is independent, but the porch is dependent. The differentiation is spatial as well: the indoor space is contained within the hall proper, and the outdoor space is defined by the porch (Fig. 7).

From the outside, the Shengmu dian of Jin ci (1023–31) (Fig. 7) appears to be such a case. But in fact, its two structures have merged, and the two spaces no longer align with the roofs. The outdoor space takes over part of the hall proper, and the columns below the upper eaves do not extend to the ground.

However, the evidence in the *Yingzao fashi* is that in all building types the outermost columns—those under the eaves—stand on the ground. Indeed, this is the assumption behind the first step of our algorithm.[9] To alter the algorithm to account for the truncated eaves column in Shengmu dian, we need either to insert a column and then truncate it, or to insert a truncated column in the first instance. In algorithmic terms, we either add a step for truncation, or we make step 1 optional and add a step for inserting a truncated column. Through the changing composition of its steps, our algorithm can evolve to explain how buildings evolve over both time and space, as Liang (1984c, 37) points out:

> [A] building of an early date may be found heralding a new style of feature, or in regions far from the cultural and political centers, a late structure may tenaciously cling to a bygone tradition.

---

[9] Shengmu dian is a *dian* hall, which has a slightly different structure from a *ting* hall. But it is the same with respect to the outermost columns, so we can refer to our *ting* hall section algorithm.

Similarly, we can propose an algorithm of the *Gongcheng zuofa zeli* as a variation of that of the *Yingzao fashi,* and Ming–Qing buildings as variations of the *Gongcheng zuofa zeli* algorithm. For instance, in the *Yingzao fashi,* the roof section calculates the "raise" first, which means that the overall proportion is consistent (B/4 in our example), regardless of the building's depth. But in the *Gongcheng zuofa zeli,* we start at the bottom and "raise" each purlin (*jujia*). This means that the height of the ridge purlin, as a proportion of the building depth, can vary from 50% (2-rafter building) to 100% (8-rafter building). Buildings of Liang's third period (the period of rigidity) vary hardly at all, other than in overall massing. For example, in terms of sections,

[a]fter about 1400 a column is seldom omitted to make room for a utilitarian requirement (Liang 1984c, 103).

Step 5 in our algorithm above would insert a column at every available position and so become simpler, and this would remain consistent throughout the third period. There are no structural adventures like Shengmu dian to accommodate.

9 Computing Chinese Architecture

The process we have described has three steps: formulating hypotheses, generating predictions from those hypotheses, and evaluating the predictions. The evaluation may in turn prompt us to reformulate the hypotheses and repeat the cycle. Of the three steps, the second—generating predictions—is the mechanical one, and for this reason we have postponed discussion until now. It should be quick, easy, and accurate—in a word, transparent—and the best way to do this is automatically. We used plain English to express our algorithm, but natural language is not formal and so generation cannot be reliably automated. Better than plain English are shape grammars, which can express algorithms visually. In addition, they are formal, which makes it easier to generate designs accurately. In fact, our plain English algorithms above are translations of shape grammars used in Li (2001).

General shape grammar interpreters are being developed (e.g., Economou et al. 2021; Stouffs and Li 2020), but do not yet support all the technical capabilities that make shape grammars theoretically powerful. A passable workaround is to translate the shape grammars into computer programs and run these in a modeling application to get drawings and models. We could, for example, implement a shape grammar as a Python program and run the program in Rhinceros3D. Our real work is the first and third tasks—hypothesizing and evaluating—and it is noteworthy that we, the investigators, do both, because this is not always the case. For example, in scientific experiments the hypothesis is formulated by the investigator, while the evaluation is performed by that independent authority, nature. Similarly, with evolutionary algorithms, "evolution" is independent, while the fitness function is determined by the investigator. In other words, our work is entirely subjective. But it always was; the difference now is that it is more apparent.

Style is not "out there," waiting to be discovered. Rather, it is constructed, both through the grammar and through the mechanism by which we evaluate our grammar's predictions (Li 2011, 192).

We took as our point of departure Liang's interpretation of the *Yingzao fashi* as a grammar book of architecture. We then developed this idea with a computational understanding of grammar and its relation to style and evolution. This framework allows us to examine traditional texts, extant buildings, and mainstream research in a unified, transparent, and rigorous way. Thus capacitated, we compute Chinese architecture.

References

Chau, H.H., X. Chen, A. McKay, and E. de Pennington. 2004. Evaluation of a 3D shape grammar implementation. *Design Computing and Cognition* 4: 357–376.
Economou, A., T.C. Hong, H. Ligler, and J. Park. 2021. Shape machine: A primer for visual computation. In *A new perspective of cultural DNA*, ed. J. H. Lee. Singapore: Springer.
Feng, J. 2012. *Chinese architecture and metaphor: Song culture in the Yingzao fashi building manual*. Honolulu: University of Hawai'i Press.
Fairbank, W. 1995. *Liang and Lin: Partners in exploring China's architectural past*. Philadelphia: University of Pennsylvania Press.
Glahn, E. 1984. Unfolding the Chinese building standards: Research on the *Yingzao fashi*. In *Chinese traditional architecture*, ed. N.S. Steinhardt, 47–57. New York: China Institute in America.
Guo, Q. 1999. *The structure of Chinese timber architecture*. London: Minerva.
Knight, T. W. 1994. *Transformations in design: A formal approach to stylistic change and innovation in the visual arts*. Cambridge: Cambridge University Press.
Krishnamurti, R. 1982. SGI: A shape grammar interpreter. N.P.
Li, A.I. 2001. *A shape grammar for teaching the architectural style of the Yingzao fashi*. Ph.D. dissertation. Cambridge: Massachusetts Institute of Technology.
Li, A.I. 2005. Thoughts on a designer-friendly shape grammar interpreter. In *Proceedings of 23nd eCAADe conference, Digital design: The quest for new paradigms*, eds. J.P. Duarte, G. Ducla-Soares, A.Z. Sampaio, 523–528. Lisbon: eCAADe.
Li, A.I. 2011. Computing style. *Nexus Network Journal* 13 (1): 183–193.
Li, A.I. 2016. The interpreter project. http://andrew.li/interpreter/index.html. Accessed 5 Feb 2024
Li, A.I., and R. Stouffs. 2021. Towards a useful grammar implementation: Beginning to learn what designers want. In *A new perspective of cultural DNA*, ed. J.-H. Lee, 55–64. Singapore: Springer.
Liang, S.C. 梁思成. 1984a. Yingzao fashi zhushi 營造法式註釋 [The annotated *Yingzao fashi*] . Beijing: Zhongguo jianzhu gongye.
Liang, S.C. 梁思成. 1984b. Zhongguo jianzhu zhi liangbu "wenfa keben" 中國建築之兩部「文法課本」 [The two "grammar books" of Chinese architecture]. In Liang Sicheng wenji 梁思成文集 [The collected works of Liang Sicheng], 357–363. Beijing: Zhongguo jianzhu gongye.
Liang, S.C. 1984c. *A pictorial history of Chinese architecture: a study of the development of its structural system and the evolution of its types*, ed. W. Fairbank. Cambridge: The MIT Press.
Qi, Y.T. 祈英濤. 1981. Zenyang jianding gu jianzhu 怎樣鑒定古建築 [How to appraise premodern architecture]. Beijing: Wenwu.
Steinhardt, N.S. 2022. *The borders of Chinese architecture*. Cambridge: Harvard University Press.
Stiny, G., and W.J. Mitchell. 1978. The Palladian grammar. *Environment and Planning B: Planning and Design* 5: 5–18.
Stouffs, R. 2018. Sortals.org. http://www.sortal.org/downloads/index.html. Accessed 5 Feb 2024.

Stouffs, R., and A.I. Li. 2020. Learning from users and their interaction with a dual-interface shape grammar implementation. In *Proceedings of the 25th international conference on computer aided architectural design research in Asia CAADRIA 2020: Anthropocene, design in the age of humans*, eds. D. Holzer, W. Nakapan, A. Globa, I. Koh, 153–162. Hong-Kong: CAADRIA.

Takeshima, T. 竹島卓一. 1970–72. *Eizō hōshiki no kenkyū* 営造法式の研究 [A study of the *Yingzao fashi*]. Chūō kōron bijutsu shuppan.

Tapia, M. 1999. A visual implementation of a shape grammar system. *Environment and Planning B: Planning and Design* 26: 59–73.

Open Access This chapter is licensed under the terms of the Creative Commons Attribution-NonCommercial-NoDerivatives 4.0 International License (http://creativecommons.org/licenses/by-nc-nd/4.0/), which permits any noncommercial use, sharing, distribution and reproduction in any medium or format, as long as you give appropriate credit to the original author(s) and the source, provide a link to the Creative Commons license and indicate if you modified the licensed material. You do not have permission under this license to share adapted material derived from this chapter or parts of it.

The images or other third party material in this chapter are included in the chapter's Creative Commons license, unless indicated otherwise in a credit line to the material. If material is not included in the chapter's Creative Commons license and your intended use is not permitted by statutory regulation or exceeds the permitted use, you will need to obtain permission directly from the copyright holder.

Using Shape Grammar as an Analytical Tool for Shelters in Protracted Refugee Camps: The Zaatari Camp Grammar

Dima Abu-Aridah, José P. Duarte, and Rebecca L. Henn

Abstract Refugee camps around the world emerge as aid landscapes intended to provide temporary emergency settlements and other basic needs to people who are fleeing conflict in their home region. However, despite the term "temporary," many camps endure for decades where inhabitants find themselves facing a protracted state of 'bare life', experiencing isolation from the rest of society, and are often treated as humanitarian subjects and aid recipients for generations. While refugee camps and settlements are created according to site planning guidelines outlined in emergency response handbooks and manuals, refugees face the challenge of recreating their lives socially and spatially, actively reforming their identity. Based on the narratives and experiences of refugees in the Zaatari camp of Syrians in Jordan, this chapter presents a novel approach to studying the informal self-led and needs-based re-creation of refugee shelters using an analytical shape grammar. The grammar offers a systematic framework to analyze and generate spatial configurations of shelters based on identified rules and constraints. Through field observations, in-depth interviews with camp residents, and documentation of shelter layouts and conditions, the research examines a corpus of twenty-two shelters that have undergone gradual transformations by their inhabitants over ten years. A comprehensive analytical shape grammar identifies recurring patterns, evaluates functionality, and uses human-centered patterns for generating spatial structures to enhance the living conditions of refugees. The grammar encodes the conventions of existing modes of spatial generation and introduces new conventions that can be used to explore novel spatial arrangements that meet resident needs.

D. Abu-Aridah · J. P. Duarte (✉) · R. L. Henn
The Pennsylvania State University, 151 Stuckeman Family Building, University Park, PA 16802, USA
e-mail: jxp400@psu.edu

D. Abu-Aridah
e-mail: dabuaridah@outlook.com

R. L. Henn
e-mail: rhenn@psu.edu

1 Introduction

Protracted refugee camps are initially established to offer temporary aid for displaced individuals escaping conflicts, persecution, or natural disasters (Etzold et al. 2019; Lambert 2017; Latka 2017). The transition from makeshift shelters to dignified living environments faces various complex challenges stemming from political, social, and economic factors (Ansar and Md Khaled 2021; Papatzani et al. 2022). Therefore, an analysis of refugee shelters' initial plan and reconfiguration helps researchers and aid workers better understand the specific humanitarian challenges posed by refugee camps, thereby improving both policy decisions and the living conditions of displaced populations. Studying the design and functionality of refugee shelters enables design and planning researchers to identify gaps in current planning approaches in order to propose innovative solutions for current design challenges. Evaluating current shelter models can address evolving displaced population needs, foster respectful, culturally sensitive designs, and inform inclusive socio-cultural spaces (Betts et al. 2017; Elorduy 2020; Hart et al. 2022; Sanyal 2017). In protracted camps, the analysis of refugee shelters is crucial for understanding the challenges faced by displaced populations and identifying opportunities for improvement of the built environment. It enables the development of innovative solutions, enhances livelihood opportunities, promotes dignity and well-being, and informs policy and planning decisions (Carter-White and Minca 2020; Elorduy 2020).

This chapter uses shape grammar to analyze informal spatial patterns in the evolution of self-made refugee shelters in prolonged refugee camps. Following an analytical shape grammar approach, this chapter investigates shelter layout, typology, and adaptability. Shape grammar helps to identify recurring patterns, evaluate their functionality and usability, and use those human-centered patterns to enhance the living conditions of refugees. This study introduces a comprehensive grammar that encompasses both existing and novel conventions to generate spatial structures, emphasizing the exploration of variations that enhance resident needs. The research is based on twenty-two shelters transformed by inhabitants over a decade in Zaatari camp (2012–2022). Data was collected in August 2022 through field observations, interviews, and shelter documentation. Employing analytical shape grammars, this study evaluates diverse design possibilities, considering different parameters' impact on shelter suitability and livability. The grammar also explores adaptable shelter expansions for evolving socio-spatial needs in protracted camps' dynamic environments. The grammar was developed in 2D, but it can be extended to 3D in the future.

2 Literature Review: Protracted Refugee Camps and Shape Grammar

It is essential to address and understand the challenges encountered by refugees in protracted settlements. This section highlights the significance of spatial dynamics within protracted camps, examining how refugees shape their built environment and negotiating their experiences. Furthermore, it highlights the importance of conducting further investigations into refugee housing within similar protracted conditions. This section investigates the existing techniques in studying the built environment in refugee camps, and how shape grammar can enhance our understanding of the physical space in these camps.

2.1 Protracted Refugee Camps and Refugee Shelters

Protracted refugee camps, intended as temporary short-term solutions, often transform into long-term or semi-permanent settlements due to the enduring and protracted nature of displacement. These camps endure for extended periods, often spanning decades, as the challenging circumstances in refugees' home countries prevent a swift resolution to their displacement. Factors such as ongoing conflicts, political complexities, and limited opportunities for rebuilding elsewhere contribute to the lack of viable options for refugees, prolonging their stay in these temporary camps beyond what was originally anticipated (Conti et al. 2020; Tobin et al. 2022; Turner 2015).

Dantas et al. (2021) study on refugee settlement creation emphasizes the lack of analysis in emergency settlement planning. Their research highlights the need for tailored planning and management strategies to address the unique characteristics and challenges of emergency settlements, including protracted camps (Dantas et al. 2021). This need for tailored planning underscores the complexity of providing suitable solutions for displaced populations, due to the fact that many refugee camps are spatially created using a grid-inflexible layout that assumes that human needs are similar all around the world (Alnsour and Meaton 2014; Turner 2015). This grid planning pattern made of identical housing units can be clearly seen in the three settlement examples shown in Fig. 1.

While camps are created following the grid-based model, the spaces in refugee camps typically change gradually as they undergo processes of morphological change informally led by residents (Turner 2015). Spatial dynamics strongly influence shaping the experiences of refugees in protracted camps. Researchers, like Al-Harithy et al. (2021) and Hanna et al. (2022) examine camp architecture and spatial practices, highlighting the transformation of temporary emergency shelters into complex-built structures. They argue that the built environment—though often perceived as chaotic and improvised—carries cultural significance and reflects displaced communities' socio-spatial requirements. This perspective challenges the notion that protracted

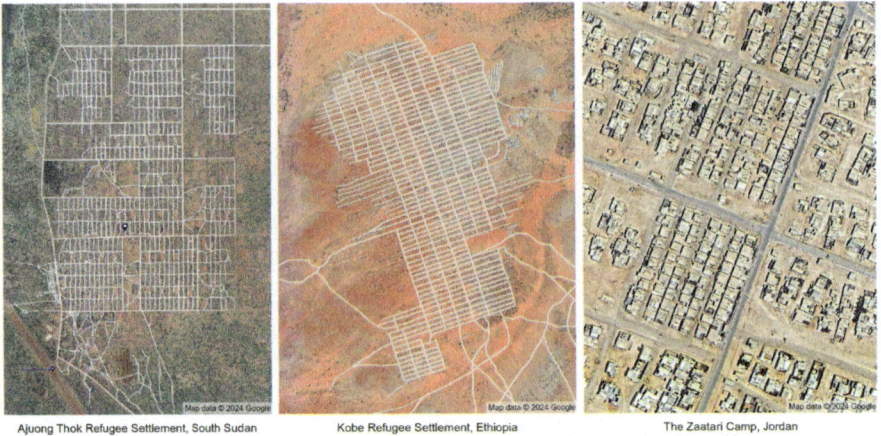

Fig. 1 Examples of the spatial organization of planned refugee camps [UNHCR 2016, Google Earth 2023]

camps are solely characterized by disorder and sheds light on the ways in which refugees shape their built environment to create a sense of home and belonging (Hanna et al. 2022).

Current planning approaches and consequent spatial shortcomings in creating proper refugee settlements lead to challenges in living conditions for refugees. Overcrowding is a common issue, with densely populated camps leading to cramped living spaces that challenge residents' growing needs. Insufficient resources compound the difficulties related to shelter and access to services, intensifying the hardships faced by those residing in the camps. Additionally, freedom of movement is often restricted for refugees within protracted camps due to security concerns or governmental policies. These restrictions limit refugees' ability to explore alternative living arrangements or seek better opportunities outside the camp and urge them to cope with the living conditions and challenges in the camps (Conti et al. 2020). In an equivalent manner, Papatzani et al. (2022) underscores the disruptive potential of limited mobility in challenging and reshaping dominant socio-spatial structures associated with displacement. The study highlights the agency of refugees in shaping their experiences by examining how displaced individuals negotiate and resist these figurations through their mobility practices.

At the socio-economic level, self-reliance emerges as a critical issue for refugees in protracted camps. Shalan (2019) emphasizes the importance of self-reliance as the social and economic ability of individuals, households, or communities to meet their essential needs in a sustainable manner. Refugees must navigate the challenges of ensuring the long-term viability of their livelihoods within the constraints of the camp environment (Shalan 2019). Similarly, Bulley (2014) and Carter-White and Minca (2020) urge researchers to give greater attention to the role of the community in the governance of refugee camps and emphasize its resistance to formal types of governance. They highlight the significance of understanding the community's meaning within refugee camps, arguing that camps serve as spaces of empowerment and self-agency for their residents, including the refugees' desire to recreate their physical spaces.

The reviewed studies enhance our understanding of the complexities of protracted refugee camps, offering insights into the spatial dynamics and transformations of those camps. This area demands closer examination for innovative approaches and design strategies that can enhance the living conditions and overall well-being of refugees in protracted camps. The examined studies acknowledge the shortcomings of the grid-based and conventional planning approaches, along with the consequent spatial challenges. Thus, there is a need to investigate alternative planning and design methodologies that consider the diverse needs and experiences of displaced populations, promote flexibility, adaptability, and human-centered design principles, and foster community engagement in shaping their living spaces. These combined efforts promote creating more effective and sustainable refugee settlements.

2.2 Shape Grammar and the Analysis of the Built Environment in Refugee Camps

Employing shape-grammars-based design approaches can enhance flexibility in the design process and assist designers in tackling complex design problems and aid in identifying spatial patterns in existing spatial configurations (Duarte and Beirão 2011; Lambe and Dongre 2019). Shape grammars, introduced by Stiny and Gips in 1972, are rule-based formalisms that generate a design language through step-by-step processes, employing visual computations of shape production and transformation (Gips 1999; Stiny and Gips 1972). Shape grammars fall into categories: analytical grammars used for analyzing existing designs, and original grammars focused on creating new designs (Colakoglu 2005). The step-by-step nature of shape grammars makes them a valuable tool for testing and evaluating different design alternatives (Schirmer and Kawagishi 2011). Shape grammars can also serve as stand-alone design systems in architecture, simplifying the design process for professionals (Eloy et al. 2017).

Shape grammars have found application in understanding and generating spatial configurations at the urban and architectural scales. Duarte et al. (2007), for instance, employed parametric shape grammar to study the urban form of the Medina of Marrakech, capturing the features of the existing urban fabric. Similarly, shape grammars have been applied to examine informal settlements in Brazil by Brazil by Dias (2014) and Verniz and Duarte (2020). Dias (2014) used shape grammars to generalize the rules for describing the morphological features of favelas in Rio de Janeiro, which capture the interplay among diverse spatial components constituting the informal architecture, resulting in a distinctive urban development pattern. The developed grammar enables new housing initiatives to explore integrating buildings and architectural elements, creating fresh avenues for interaction with the built environment's spatial arrangements (Dias 2014). Verniz and Duarte (2020) developed an analytical shape grammar for an informal settlement in Brazil to recreate the urban fabric of the settlement. The grammar illustrates how the favela's organic growth relies on implicit rules guided by its residents, leading to resource-efficient construction. This grammar was one step forward to gather insights from informal settlements, fostering an innovative design approach for low-income urban areas (Verniz and Duarte 2020).

Previous studies underscore the significance of shape grammars in architectural and urban design, emphasizing its flexibility and ability to address complex design problems, its potential to be used for design exploration, variability of design solutions and adaptability to different contexts, and customization of user-oriented designs. Shape grammars have been used in analyzing existing designs, generating new designs, and understanding spatial patterns. They have proven valuable in studying both urban environments, architectural forms, and informal settlements, offering insights into the morphological features of various urban contexts.

However, despite its effectiveness as a computational tool for investigating the design process and capturing incremental and spontaneous spatial changes (Dias 2014; Duarte et al. 2007; Eloy et al. 2017; Lambe and Dongre 2019; Schirmer and Kawagishi 2011; Verniz and Duarte 2020), it has not been used in studying protracted refugee contexts. This is also evident when looking at the common spatial analysis tools used to study refugee settlements. Existing studies focused on specific spatial analysis approaches to decipher physical space, such as descriptive qualitative approaches and spatial mapping using GIS tools (Elorduy 2021; Herz 2013). While these methods have been useful in exploring and understanding the evolution of the built environment at the urban level, they have limited ability to address the architectural scale of settlements.

Therefore, exploring shape grammars to examine the built environment of refugee camps holds promise to benefit residents. Such an investigation could contribute to understanding the spatial morphology of camps considering their purported temporariness. Moreover, shape grammars could enable the derivation of design guidelines and best practices for shelter interventions, resulting in context-specific innovative solutions that can be combined with cultural, social, and environmental aspects to inform the development of designs that can expand or transform gradually to accommodate changing needs and population dynamics. The study described in this chapter constitutes a first step in this direction.

3 Methodology: Research Design and Approach

This study explores the Zaatari camp, a settlement for Syrian refugees in Jordan, as a case study. The research employs the shape grammar formalism as an analytical tool to interpret the spatial configurations of shelters within the camp. The study lays the groundwork for designing new strategies centered around the concepts of flexibility and adaptation, aiming to anticipate and accommodate self-build processes. Through this approach, the study illustrates the ways refugees shape their living spaces and develops innovative solutions that can enhance the well-being of the residents.

3.1 The Case Study: The Zaatari Camp

The Zaatari camp is one of the largest refugee camps in the world. It is a home for more than 80,000 refugees and it has more than 26,000 occupied shelters. It was established in 2012 close to Jordan's northern border with Syria. It is under the administration of the Syrian Refugee Affairs Directorate, a governmental Jordanian entity, and the United Nations Higher Commission for Refugees (UNHCR 2021). Since its establishment, Zaatari Camp has evolved from a small collection of tents into an urban settlement (UNHCR 2021). The Zaatari camp primarily accommodates refugees originating from the southern part of Syria, particularly from *Reef Daraa* (Ledwith 2014).

Occupying an area of 530 hectares correspond to 530 hectares on land owned by the Jordanian armed forces (UNHCR 2021) the Zaatari Camp was designed following a grid system based on the guidelines outlined in the UNHCR Handbook for Emergencies (UNHCR 2007) and the Sphere Handbook (SphereAssociation 2018). At the camp's establishment, tents were used to shelter the refugees. Later, prefabricated units locally referred to as *"Caravans"* were arranged in rows within the grid system. However, significant changes have been made by users to the original arrangement of shelters.

The camp is divided into 12 districts (Fig. 2). Those districts are subdivided into blocks and sub-blocks using streets and circulation paths. The sub-blocks represent the smallest units within the layout of the camp, with each sub-block containing a certain number of shelters or housing units.

3.2 Data Collection

The data used for this study were obtained from field trips and in-depth interviews that were conducted in August 2022. During the field trips, each shelter was documented using photographs and on-site sketches, and measurements were taken using a laser measurement tool. These shelters were randomly selected from households in District No. 1 of the camp (Fig. 3).

Each of the households interviewed resided in a shelter that initially started as a tent but was eventually replaced with Caravans. While tents were initially provided during the early years of the camp, Caravans have become the predominant form of shelter offered by the UNHCR and other donors. Figure 4 depicts the tents and Caravans that are found in the camp. The initial shelters made of Caravan shelters were further expanded using locally claimed temporary construction materials.

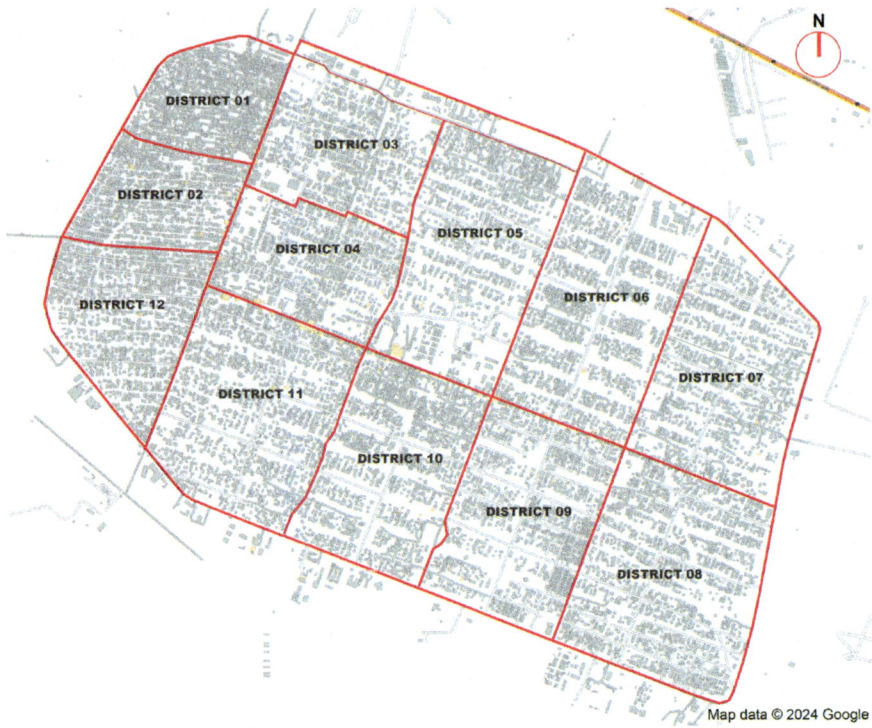

Fig. 2 The Zaatari Camp's districts

Fig. 3 Layout of the studied area in district 1, where 22 shelters were selected for analysis

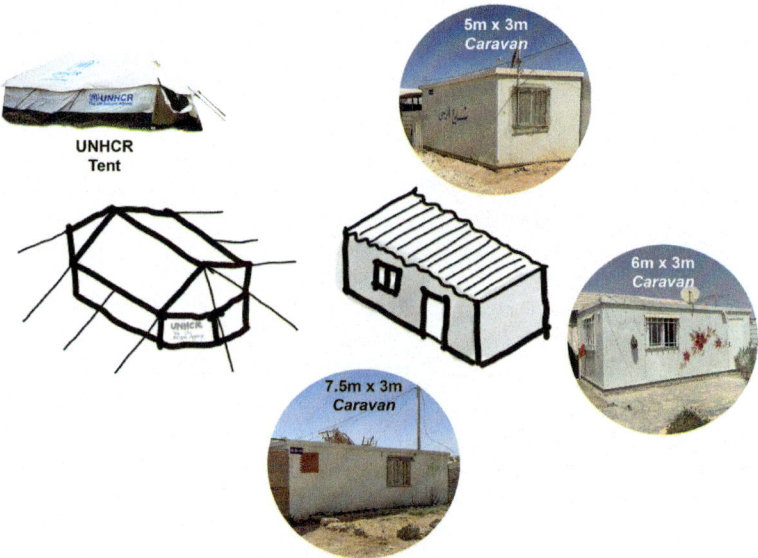

Fig. 4 Images and depictions of tents and prefabricated shelters found in the Zaatari Camp

Through field observations and interviews, it became evident that the refugees exhibited remarkable resourcefulness in transforming their shelters into more livable spaces, using available materials such as metal sheets, sandwich panels from dismantled Caravans, and leftover tent fabrics, despite the challenging circumstances they faced such as shortage of financial sources. Gradually, the participants experienced incremental improvements in their physical environment and living spaces. The shelter layouts presented in this chapter represent the conditions of the selected shelters as of August 2022 as illustrated in Fig. 5.

3.3 Main Spatial Features of the Zaatari Camp Shelters

As mentioned above, currently, the shelters in the Zaatari Camp primarily consist of prefabricated Caravans. These Caravans, with sizes ranging from 15 to 20 m^2, are rectangular in shape with dimensions of approximately 5 to 7.5 m in length and 3 m in width. The available sizes vary due to different donors, including relief organizations from Saudi Arabia, Kuwait, Qatar, and Oman, depending on the location within the camp and the timing of donations.

Caravans serve as the core units of each shelter, as illustrated in Fig. 6, upon which later expansions are dependent. Residents reported that the ones they own were even provided by managing agencies within the camp or informally purchased from households who have left the camp. During fieldwork, Caravans were noted to be constructed from sandwich panels consisting of two thin corrugated metal sheets

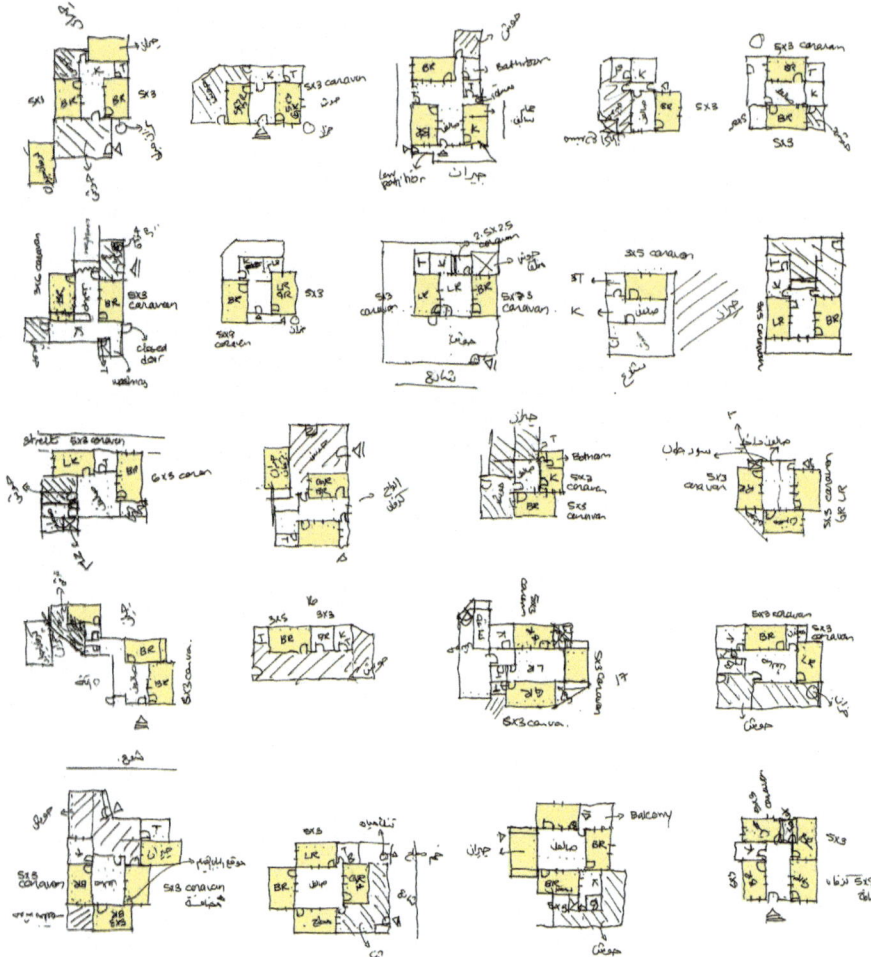

Fig. 5 Layouts of the 22 selected shelters manually documented during the fieldwork

with an insulation layer. Given the extreme climate of the Zaatari camp, characterized by intense heat in summer and cold and wind in winter, Caravans offered a degree of protection against heat, heavy seasonal rains, and occasional snow, exceeding the protection provided by structures made from single metal sheets and fabrics. Refugees also favored Caravans for the sense of security and privacy they offered. An interviewee expressed this sentiment, stating, *"We are grateful for having doors that can be locked with a key and a secure space for sleeping."*

Caravans serve predominantly as sleeping, living, and guest areas. It is common to use Caravans as multipurpose spaces accommodating daily household activities during the day and providing rest and sleep during the night.

Using Shape Grammar as an Analytical Tool for Shelters in Protracted ... 547

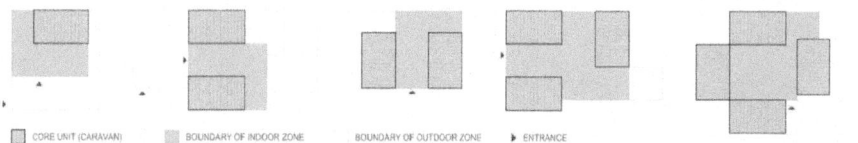

Fig. 6 Illustration of core and non-core structures in shelters

The number of Caravans in each shelter varies, based on household size and its financial capacity. The minimum number is one, while the maximum ranges from three to four housing up to nine household members. But some households were able to acquire additional ones to enhance the quality of their shelters. Expansions to the main core Caravan were achieved using temporary construction materials. A typical shelter layout includes a main living space that serves as a multipurpose area used for activities such as eating, gathering, and socializing. Designated spaces are provided within the shelters to accommodate essential facilities such as a kitchen, a toilet, and a bathroom. An area for hosting guests, known as *Madafa*, is often found as a separate space or integrated into the common living area. Most shelters also feature a front patio called *Sibat*, which can be shaded or unshaded, serving as a transitional space between the outdoors and indoors. The *Sibat* is commonly used as an outdoor seating and gathering area, particularly during late cool afternoons, and after sunset time. It also serves as an extension of the living space where residents can socialize with family, neighbors, and guests. Furthermore, many shelters have a front yard, backyard, or both, known as *Hawsh* or *Hakoura*. These spaces are created by constructing high fences, providing an outdoor private area for families. This design reflects the ' pre-displacement rural lifestyle of the camp's majority of refugees who come from the southern rural areas of Syria, as inferred from in-depth interviews.

Residents' resourcefulness observed in the camp led to adaptations of the shelters to better suit their needs. Temporary partitions made of curtains are used to create separate spaces within the limited area, ensuring privacy and better use of the space. The adaptability of the shelters allowed families to personalize their living spaces, fostering a sense of home within the challenging circumstances of the camp. Figure 7 illustrates the most prevalent features and elements observed in most of the shelters across the camp, offering an overview of the typical shelter configuration there.

3.4 Corpus of Shelter Designs

The corpus of the grammar comprises 22 shelters that exhibit a rectilinear composition, reflecting the informal approach developed by refugees to expand their shelters and accommodate their spatial needs. The corpus emphasizes the common spatial features found in refugee shelters in the camp. As the core units of the shelter designs, Caravans play a pivotal role in determining the overall layout configurations. Figure 8 shows the detailed layouts of the selected shelters.

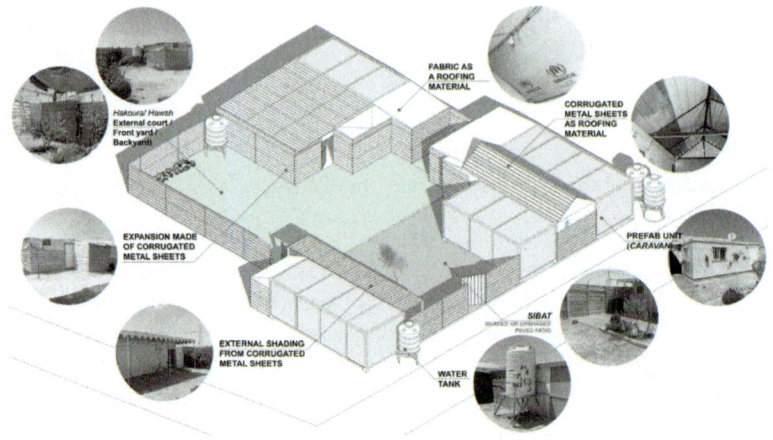

Fig. 7 The most common spatial features of the shelters in the Zaatari Camp

Fig. 8 Corpus of design: layouts of the 22 shelters used for this research The Caravans are shown in yellow

4 The Grammar

The Zaatari Camp Grammar is an analytical parametric grammar developed to analyze and explain the informal configurations of refugee shelters and to investigate the application of shape grammars in generating layout solutions for refugee shelters in protracted situations. The Zaatari Camp Grammar establishes a strategic approach by defining grammar rules that guide the initiation of the shelter, the allocation of further spaces, and the placement of basic architectural elements like window and door openings. These rules provide a structured framework to ensure efficient and thoughtful design decisions throughout the shelter construction process.

The development of the Zaatari Camp Grammar involved five steps, starting with an analysis of the spatial organization of shelters and the creation of the shape vocabulary. It then proceeded with a study of functions, followed by the development of rules, validation of the grammar, and the application of the grammar:

1. **Analysis**: Analysis of spatial layouts and functional relationships of selected shelters;
2. **Vocabulary**: Definition of functional uses and shape vocabulary for refugee shelters;
3. **Rules**: Design of parametric rules for current shelter conditions;
4. **Validation**: Application of rules to generate and validate existing shelter designs;
5. **Application**: Derivation of new shelter layouts using shape rules to address refugee needs.

4.1 Analysis: Analysis of Spatial Layouts and Functional Relationships of Selected Shelters

To comprehend the corpus of designs and explore spatial relationships, a study of the functional layouts of all shelters was conducted, as depicted in Fig. 9. The analysis examined and addressed the spatial characteristics of a sample of shelters from the larger set of 65 shelters surveyed during this study's fieldwork. The spatial characteristics of the shelters, including functional and formal aspects, have evolved over more than ten years through progressive spatial adaptations led by users. As such, the current state of the camp's shelters reflects the collective contributions of its residents. This analysis enables us to gain insights into the detailed configurations within each shelter, revealing patterns of space utilization and usage. This analysis is also important to identify commonalities and variations among the shelter designs, facilitating the development of a holistic understanding of the overall spatial organization of the camp and it also serves as a crucial foundation for subsequent phases of the research.

Based on the previous examination of the functional analysis of the layout, a partial tree diagram for a subset of selected shelters was developed to depict primary spatial relations and to facilitate the establishment of a shape vocabulary used in formulating shape rules, contributing to a deeper understanding of the overall design patterns observed in the camp. The tree diagram, shown in Fig. 10, illustrates how the accumulation of spaces follows an informal grid-like organization, with each new space relating to the core structure. This incremental logic inspired the creation of a flexible yet semi-modular parametric shape grammar to interpret the underlying design process. This grammar allows further interpretations of the underlying

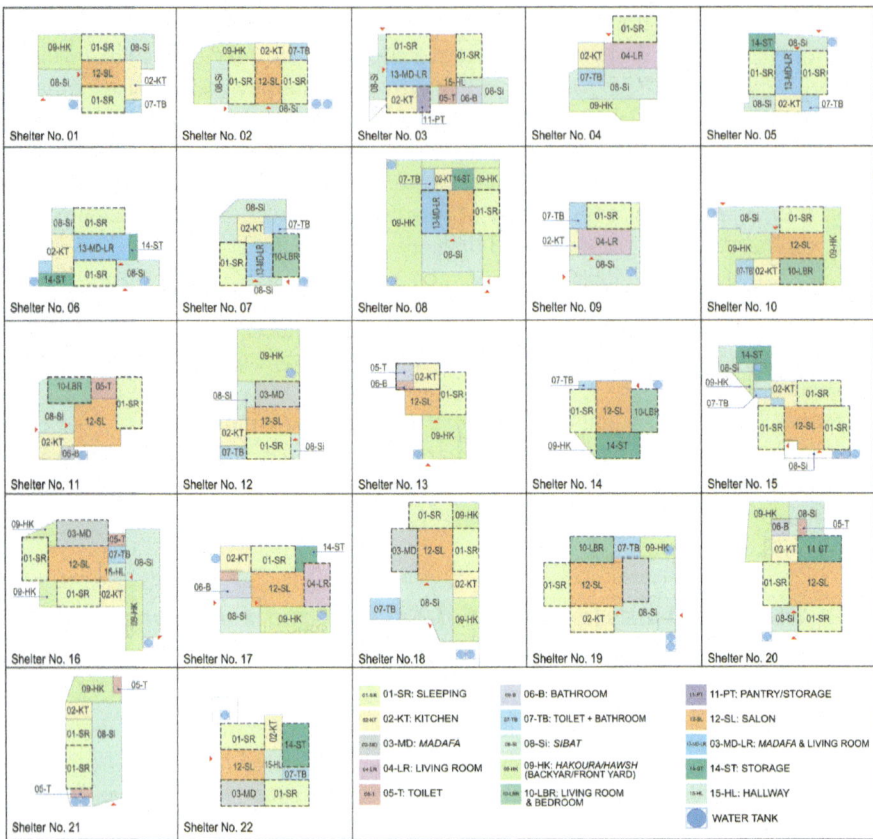

Fig. 9 Functional layouts of the analyzed shelters

informal refugee-led design process, providing insights into the adaptive nature of the shelter architecture in the camp.

4.2 Vocabulary: Definition of Functional Uses and Shape Vocabulary for Refugee Shelters

The functional shape vocabulary of the Zaatari Camp grammar, illustrated in Fig. 11, has been developed through the analysis of the spatial elements present in the shelter's layout. The shape vocabulary comprises polygons that represent the different spatial functions within the refugees' shelters, encompassing a total of 15 functions.

Table 1 provides an overview of the functions and spaces that constitute the shelters in the Zaatari camp. These spaces and functions serve as the basis for defining the vocabulary of the grammar. It is important to emphasize that all definitions of spaces

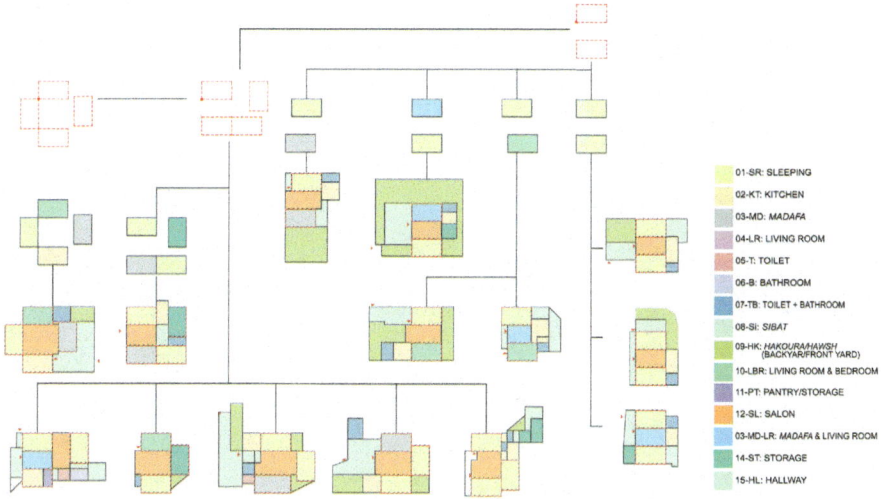

Fig. 10 Tree diagram unveiling the spatial relations of shelters in the Zaatari Camp

Fig. 11 Shape vocabulary of the Zaatari Camp grammar

and functions are derived directly from on-the-ground observations and interviews conducted with refugees, ensuring an authentic and firsthand perspective.

4.3 Rules for Current Shelter Conditions

The Zaatari Camp grammar provides a systematic approach based on a methodology that examines layout variations for refugee shelters to address households' spatial

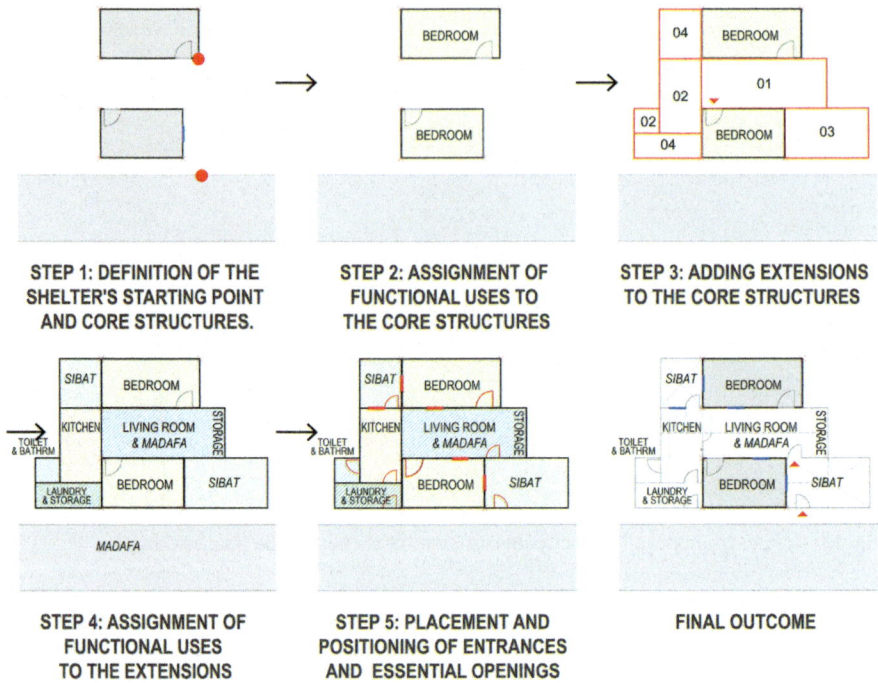

Fig. 12 The five steps in deriving a shelter following the grammar

needs. The result is a layout that maximizes the available resources while creating livable and dignified spaces within the constraints of the camp environment. The Zaatari Camp grammar consists of five sets of shape rules. These rules correspond to the five main steps involved in the creation of shelters, depicted in Fig. 12.

The first step initiates the creation of the shelter by defining the main Caravan structures (Fig. 13). The second step involves assigning functional uses to these core structures (Fig. 14). The third step adds extensions made of temporary construction materials to the core structure (Fig. 15). In the fourth step, functional uses are assigned to these added extensions (Fig. 16). Lastly, the fifth step defines the placement and positioning of essential openings such as doors and windows (Fig. 17).

Step 1 Definition of the Shelter's Starting Point and Core Structures (Caravans)

This step applies four main rules that comprise the initial stage of shelter creation. It begins with defining the entry point from the street, followed by positioning the first Caravan and the arrangement of subsequent Caravans. The first rule (1.1.) defines the entrance point to the shelter from the street, while the second rule (1.2.) includes a set of labels and determines the initial shape of the first core structure (Caravan). It is worth mentioning that the Zaatari camp does not have defined plots; instead, it consists of large blocks and sub-blocks defined by streets. Consequently, the shelters

Table 1 Definition of the functions found in shelters

Function	Definition and system of activities
Sleeping area[1] (**01-SR**)	A space designated for sleep and rest. It is also used for storing personal belongings and clothes. The provision of such a space within shelters is of utmost importance for households, often located within the structure of the *Caravan*
Kitchen[2] (**02-KT**)	A space designated for cooking and preparing food, washing, and cleaning dishes and cooking equipment, food storage, and occasionally for doing laundry
Madafa (guest room)[1] (**03-MD**)	*Madafa* literally means "the space of hospitality" (Dalal 2021). It is used for welcoming and receiving guests. It can be also used for holding public family events, such as family celebrations and occasions
Living room[1] (**04-LR**)	A space designated for a household's daily activities such as gathering, relaxation, dining, studying, watching TV, and similar activities
Toilet[2] (**05-T**)	A designated space used only for "using the toilet" and maintaining personal hygiene
Bathroom[2] (**06-B**)	A designated space used only for bathing and showering
Toilet/bathroom[2] (**07-TB**)	A double function space that combines both bathroom and toilet activities. It mostly exists when there is not enough space for building separate toilets and bathrooms
Sibat (shaded or unshaded Patio) (**08-Si**)	*Sibat* in the Syrian context has colloquially been used to refer to a 'covered passage' (Dalal 2021). However, in the Zaatari camp context, it is used to describe a paved front patio or terrace. It serves as a transitional area between the indoors and the outdoors. It can be shaded or unshaded and is typically paved with a concrete screed
Hakoura/Hawsh (**09-HK**)[3]	*Hawsh/Hakoura* is a term commonly used in rural regions of Syria, referring to farmland or a garden/farm attached to a house. It is more prevalent among villagers and farmers than city dwellers. This concept is significant in rural areas, where many dwellings integrate agricultural spaces into their living environment (Dalal 2021)

(continued)

[1] Those spaces are predominantly situated within the structure of the *Caravan* (prefabricated unit).

[2] Those spaces are typically located within temporary structures that are attached to the *Caravan* and constructed using metal sheets, sandwich panels, or both materials.

[3] In the Zaatari camp, the *Hawsh/Hakoura* refers to an external court or yard found in the shelters. It is commonly situated at the front, back, or even on both sides of the shelter. It serves as a special outdoor area, offering residents a private space where they can enjoy the outdoors while remaining within the boundaries of their shelter. The *Hawsh/Hakoura* is used for gardening, storing materials, and drying clothes. It is often connected to the *Sibat* area, creating an extended outdoor living space for the residents.

Table 1 (continued)

Function	Definition and system of activities
Living room/Bedroom[4] (10-LBR)	A multi-functional that is used for the household's daily activities during the day and for resting and sleeping during the night
Pantry/storage (11-PT)	A space designated for storing food items and other essential goods
Salon (12-SL)	It is an interior space that functions as an intermediate area, accommodating a range of activities such as activities from the *Madafa*, living room, relaxation, and occasionally even food preparation
Living room/*Madafa*[3] (13-MD-LR)	A multi-functional space that is alternately used for the household's daily activities and for receiving guests
Storage (14-ST)	A space designated for storing miscellaneous household tools and objects
Hallway (15-HL)	A corridor space that connects different rooms or areas within a shelter

[4] Those spaces are predominantly situated within the structure of the Caravan (prefabricated unit).

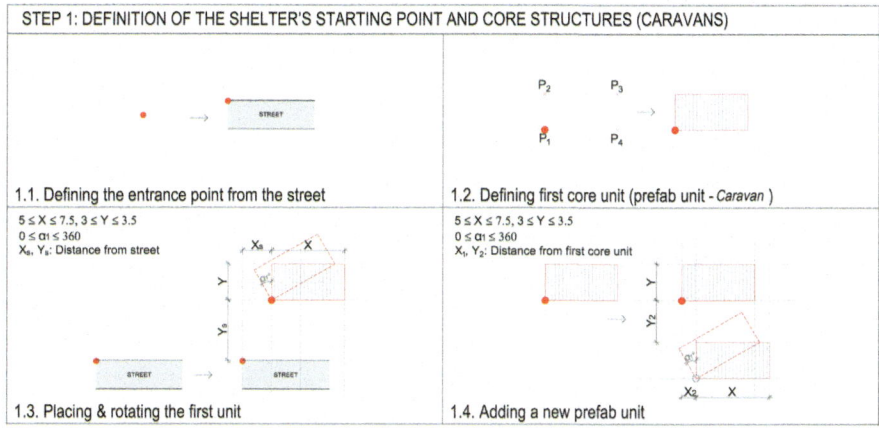

Fig. 13 Ruleset for the Zaatari Camp grammar, step 1

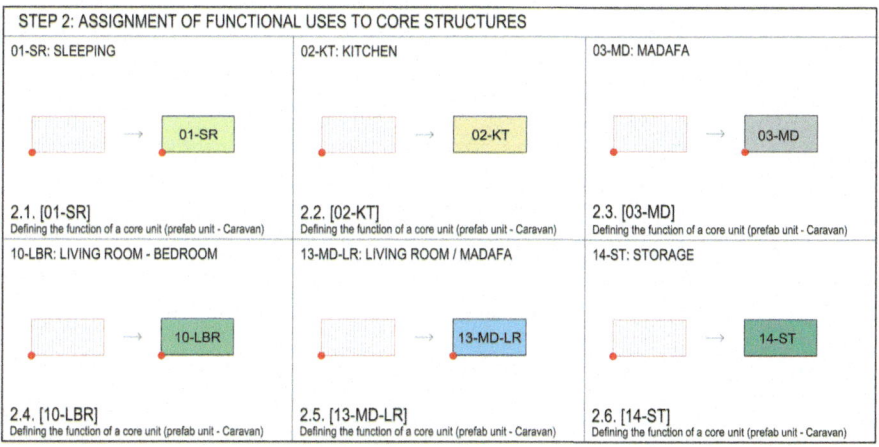

Fig. 14 Ruleset for the Zaatari Camp grammar, step 2

have flexible boundaries that are influenced by the boundaries of neighboring shelters. Moreover, the spacing between shelters in the camp is not consistent. In densely populated districts like one, two, and twelve, shelters are closely adjacent, leading to visual and acoustic privacy concerns among neighbors. In other districts, shelters are spaced around four to six meters apart, and in some areas, this distance can extend to 10 to 12 m.

The third rule (1.3.) positions the core structure in relation to the entrance point from the street, and specifies its rotation. This rule establishes a set of parameters, X_s and Y_s, which determine the distance between the label of the new core unit and the entrance point on the street. The parameters X and Y are used to determine

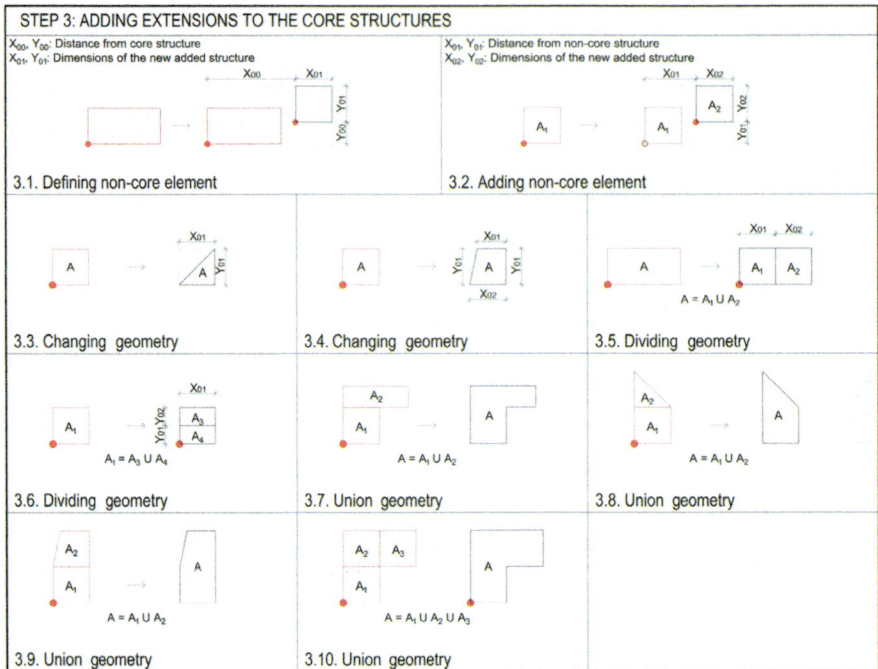

Fig. 15 Ruleset for the Zaatari Camp grammar, step 3

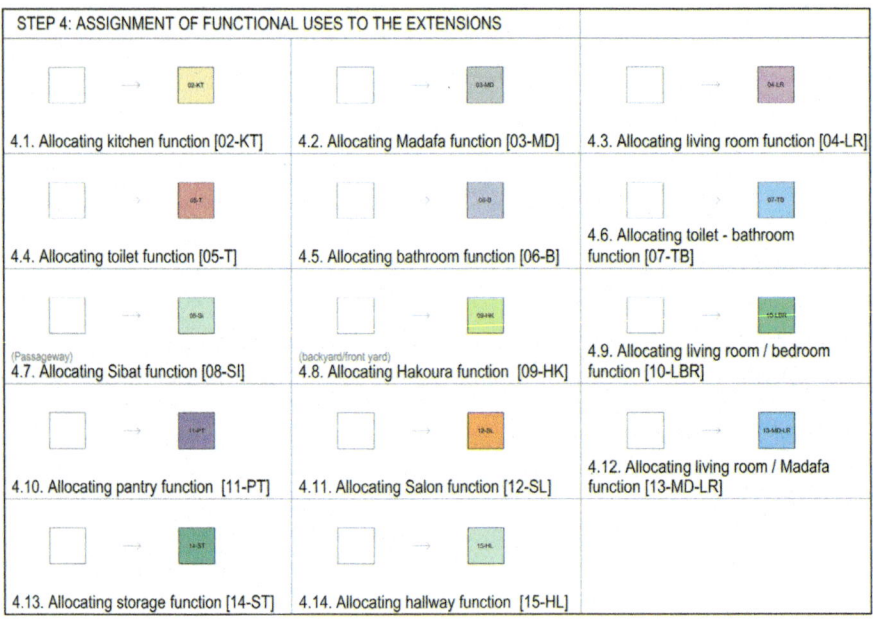

Fig. 16 Ruleset for the Zaatari Camp grammar, step 4

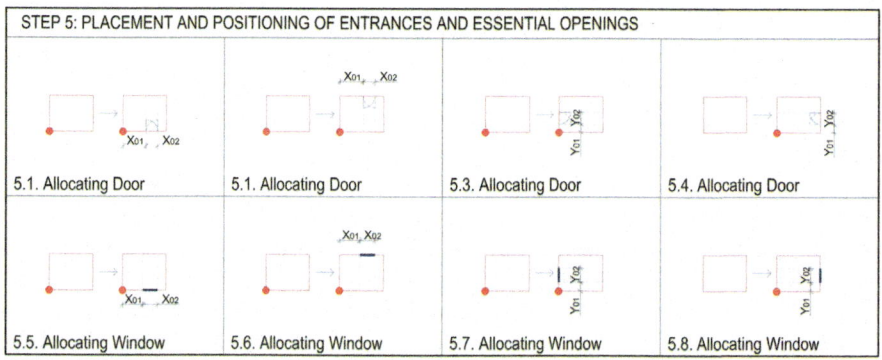

Fig. 17 Ruleset for the Zaatari Camp grammar, step 5

the dimensions of the new Caravan. The angle parameter α is used to determine the rotation of the added structure.

Lastly, the fourth rule in this group (1.4.) adds another core Caravan to the original structure. If more than one Caravan is required in the shelter, this rule can be applied multiple times. This rule also introduces a set of parameters to determine the distance between the new structure and the previously added one, as well as the dimensions and the rotation angle of the new structure.

Step 2 Assignment of Functional Uses to Core Structures

In this step, a set of six rules assigns functional uses to the core structures added in the first step. These functional uses encompass sleeping areas (2.1), kitchen (2.2), *Madafa* (2.3), a combined living room and sleeping area (2.4), a combined *Madafa* and living room (2.5), and storage (2.6). These functions were identified during fieldwork as the most common ways in which Caravans are utilized.

Step 3 Adding Extensions to the Core Structures

This step encompasses a set of rules that add extensions to the core structures (Caravans). These extensions predominantly have rectangular shapes, although some may possess more complex geometries that require additional geometric processing. Therefore, this set of rules incorporates various procedures to transform, dissect, and concatenate geometries, enabling the creation of flexible shelter forms (Rules 3.3 to 3.10). This flexibility arises from the absence of defined boundaries established by the managing agency in the camp.

Rule 3.1 is employed to establish and calculate the dimensions of the extensions added to the core unit. It utilizes a set of parameters, namely X_{00} and Y_{00}, which determine the distance between the labeled point of the new geometry and the labeled point of the original core unit. Additionally, X_{01} and Y_{01} are utilized to determine the dimensions of the new extension.

Rule 3.2 is used to add a new rectangular extension in reference to the previously added extensions. It utilizes parameters X_{01} and Y_{01} to determine the distance between the label of the new extension and the label of the previous extension. Additionally, parameters X_{01} and Y_{02} are introduced to determine the dimensions of the new rectangular extension.

Rules 3.3 and 3.4 transform rectangular geometries into triangular and quadrilateral shapes. For those rules, X_{01}, X_{02}, Y_{01}, and Y_{02} are defined to determine the dimensions of each side of the new geometry. Rules 3.5 and 3.6 are employed to dissect rectangular geometries. They encompass the necessary parameters to determine the dimensions of the dissected parts. Additionally, to concatenate two adjacent shapes, Rules 3.7 to 3.8 are utilized. These rules ensure that the two shapes seamlessly connect, creating a unified structure. Additionally, Rule 3.10 is employed to concatenate three adjacent shapes, allowing for the creation of more complex configurations.

Step 4 Assignment of Functional Uses to the Extensions

This step has 14 rules and is utilized to assign functional uses to the added extensions. These functions have been identified through interviews with refugees and shelter documentation during fieldwork. For instance, Rule 4.1 is defined to assign the kitchen function, Rule 4.2 for the *Madafa* function, Rule 4.3 for the living room function, Rule 4.4 for the toilet function, Rule 4.5 for the bathroom function, Rule 4.6 for the combined toilet and bathroom function, Rule 4.7 for the *Sibat* function, Rule 4.8 for the *Hakoura/Hawsh* function, Rule 4.9 for the combined living room and bedroom function, Rule 4.10 for the pantry/storage function, Rule 4.11 for the Salon function, Rule 4.12 for the combined living room/*Madafa* function, Rule 4.13 for the storage function, and finally, Rule 4.14 for the hallway function.

Step 5 Placement and Positioning of Entrances and Essential Openings

The fifth step applies eight specific rules designed to address the placement and allocation of essential openings within the shelter. In this step, four rules allocate the door openings (Rules 5.1 to 5.4), ensuring that access points are appropriately positioned to facilitate entry and exit. Additionally, another set of four rules (Rules 5.5 to 5.8) allocates window openings.

4.4 *Validation: Application of Rules to Generate and Validate Existing Shelter Designs*

To further articulate the Zaatari Camp grammar, this section provides a detailed derivation exemplifying the application of shape rules to generate one shelter. Additionally, it includes an explanation of the reasons behind the inclusion of each function, based on insights from an interview conducted with the household head for this particular shelter. Figure 18 visually presents a step-by-step illustration of the rule application process, facilitating a clear understanding of the design generation.

For the presented shelter, the household initially lived in a tent for three years, provided by the UNHCR. During this time, they utilized some leftover tent fabrics

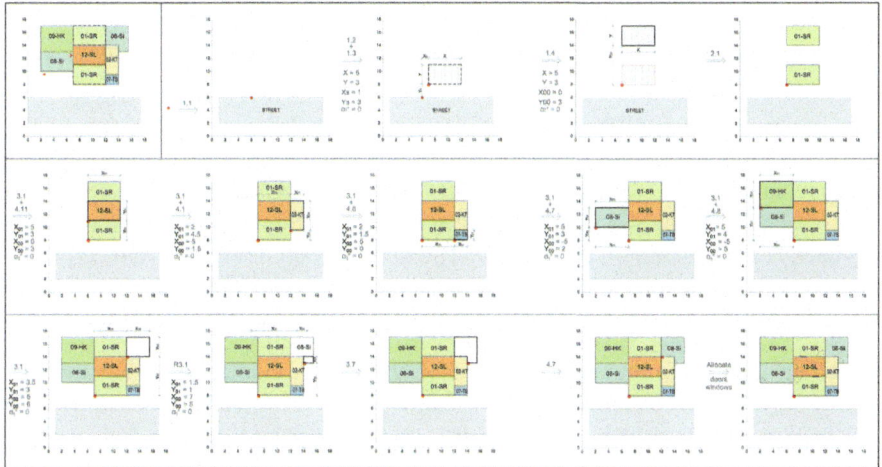

Fig. 18 Step-by-step existing shelter design generation

to create additional living space and protect against rain and occasional snow. After three years in the tent, the household received their first 3 m x 5 m Caravan from a donor and placed it parallel to the street. This process is implemented using two rules. Rules 1.1 is used to determine the dimensions of the Caravan, while Rule 1.2 is utilized to place the Caravan in its designated location.

The household then purchased a new 3 m × 5 m Caravan and positioned it facing their first Caravan, which is implemented using Rule 1.4 and the two Caravans were assigned to be sleeping areas by applying Rule 2.1 The two facing Caravans created an expanded living area in between, which is implemented in the grammar using Rules 3.1 and 4.11. However, due to the lack of access to construction materials and restrictions on importing such materials into the camp, the space between the Caravans remained covered using leftover tent fabric and without flooring.

Eventually, when construction materials became available, the household added concrete screed flooring to the salon space and constructed a private toilet and a kitchen. By applying Rules 3.1 and 4.1, a 2 m x 4 m kitchen space was added and by applying rules 3.1 and 4.6, a toilet and bathroom space of 2 m × 1.5 m was added. However, the *Hakoura* and *Sibat* spaces were added much later, several years after constructing the kitchen and toilet. This delay was due to their limited financial capacity and restricted access to construction materials at the time. The front *Sibat* space was added by applying Rules 3.1 and 4.7 and the 4 m x 4 m *Hakoura* space was added by applying Rules 3.1 and 4.8. The *Sibat* space, which is L-shaped, is adjacent to the kitchen. It was added by applying Rule 3.1 twice to create two spaces of 3.5 m x 3 m and 1.5 m x 1 m. The two shapes were concatenated using Rule 3.7 and the *Sibat* function was assigned using Rule 7.4. In the final step, doors and windows were added to form the final layout of the shelter.

4.5 Application: Derivation of New Shelter Layouts Using the Zaatari Camp Grammar to Address Refugee Needs

This section produces different layout variations for a shelter that has two Caravans as the core units and has the following functions: two sleeping areas, *Madafa*, a living area or a salon space, whereas the *Madafa* and living area could be combined into one space or with the sleeping space, one kitchen, one toilet, one bathroom, front and back *Hakoura/Hawsh*, and a *Sibat*. The tree diagram in Figs. 19 and 20 present 32 variations that can be generated using the rules, given that the rules can generate around 131,072 different layouts according to the following explanation:

- In Step 1, there are four core structures (Caravans) available. Therefore, there are four ways to choose the first core structure (Rule 1.3). For each subsequent core structure, we have three options (Rule 1.4), considering that two Caravans have already been used.
- In Step 2, each core structure can be assigned one of the four functional uses determined in the required scenario (Rules 2.1, 2.2, 2.4 and 2.5). As there are Caravans, there are $4 \times 4 = 14$ ways to assign functional uses to them.
- In Step 3, extensions are added four times. Rule 3.1 determines the dimensions of the extensions, and Rule 3.2 adds a new rectangular extension. The order of adding extensions can vary, so we need to consider the different combinations of applying Rule 3.1 and Rule 3.2 four times.
- In Step 4, Rule 4.1 assigns the kitchen function. Rule 4.2 assigns the *Madafa* function or Rule 4.12 assigns the combined living room and *Madafa* functions. Rule 4.3 assigns the living room function or Rule 4.12 assigns the combined living room and *Madafa* functions. Rule 4.4 assigns the toilet function. Rule 4.5 assigns the bathroom function. Rule 4.7 assigns the *Sibat* function. Rule 4.8 assigns the *Hakoura* function. Since there are multiple options for some rules, all possible combinations of functional uses for the extensions should be considered.
- In Step 5, for the sake of simplicity, it is assumed that there is only one way to assign the door and window openings.

Based on this explanation, we can outline the range of possible design solutions. Consider a shelter with two caravans as core units, incorporating the following spaces: two sleeping areas, a Madafa, a living area or salon, a kitchen, a toilet, a bathroom, front and back Hakoura/Hawsh spaces, and a Sibat. The total number of design solutions is calculated as follows. In Step 1, there are four possible ways to arrange the two caravans based on Rules 1.3 and 1.4. In Step 2, each of the two caravans can be assigned one of four functions, leading to $4 \times 4 = 16$ configurations. In step 4, considering that extensions can be added up to four times, with each addition having two options ("included" or "not included"), there are: $2^4 = 16$ configurations. In step 4, the programmatic layout must account for various functional rules: sleeping area (one rule), Madafa (one rule), living area (one rule) or salon (one rule), then Madafa and living area can be combined (one rule), living area can be combined with a sleeping space (one rule), and Madafa can be combined with a sleeping space

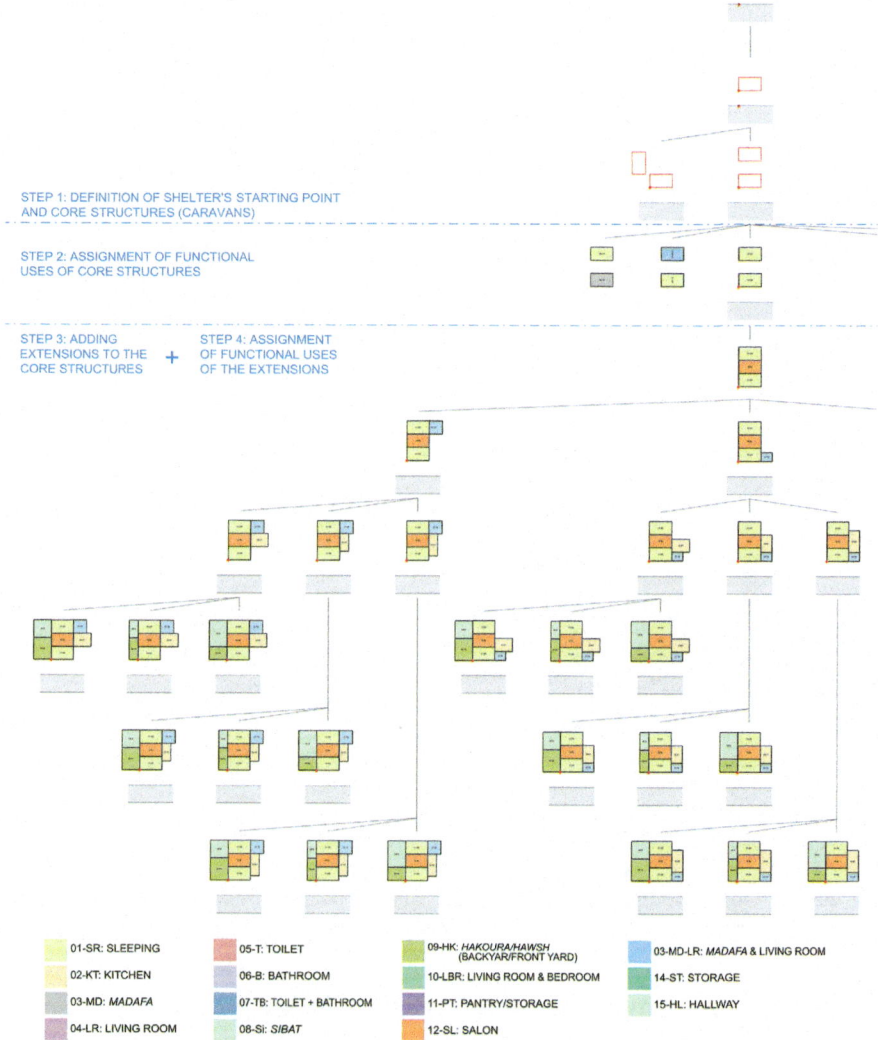

Fig. 19 Derivation tree (left half): some design variations obtained by applying the Zaatari Camp grammar

(one rule). To simplify calculations, we assume each rule has two options, leading to 2^7 = 128 configurations. Multiplying all factors together:

$$\text{Total Design Variations} = 4 \times 16 \times 2^4 \times 2^7 \times 1 = 131,072$$

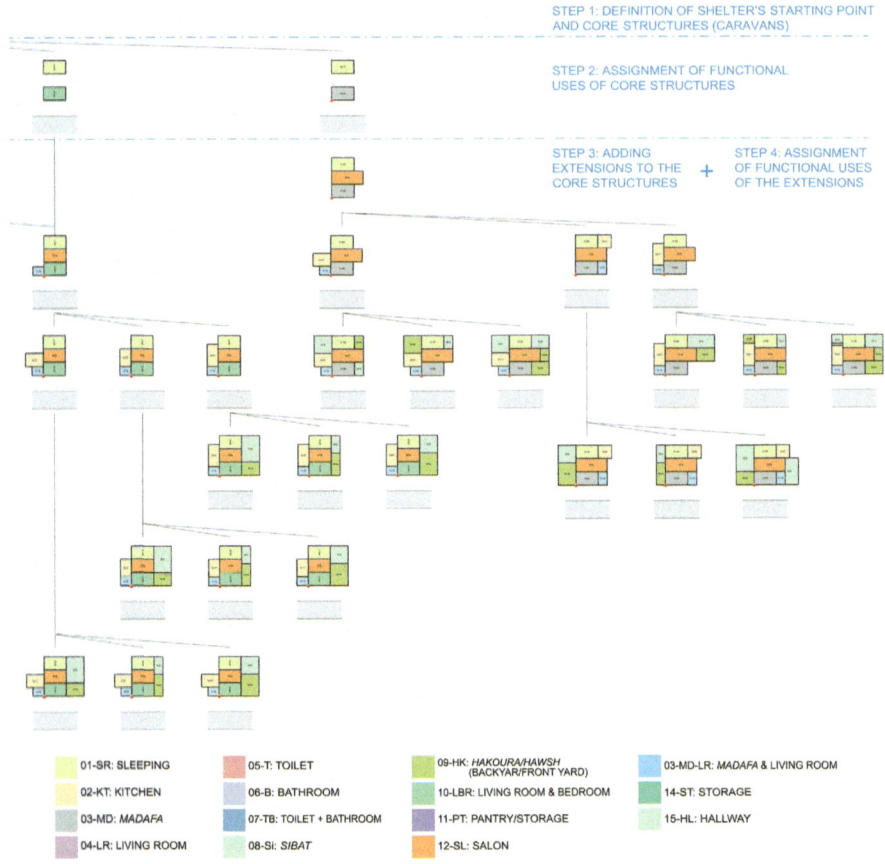

Fig. 20 Derivation tree (right half): some design variations obtained by applying the Zaatari Camp grammar

5 Discussion, Conclusions, and Future Work

This study examined the spatial configurations of shelters in the Zaatari Camp, an important settlement for Syrian refugees in Jordan. The Zaatari Camp, initially a collection of tents, evolved into an urban settlement accommodating more than 80,000 refugees in prefabricated units known as "*Caravans*". These Caravans serve as core structures, around which the refugees construct extensions and additions to fulfill their various functional needs.

The study highlights the remarkable resourcefulness and agency of refugees in shaping and transforming their living spaces within the constraints of the camp environment. Utilizing locally available materials and gradually expanding and modifying their shelters, refugees create environments that better meet their evolving needs and preferences. These findings challenge common narratives that portray refugees solely as passive recipients of aid, emphasizing their active role in defining their living conditions.

Through the application of shape grammar, the research provided a comprehensive analysis of how refugees adapt and shape their living spaces to better suit their needs and aspirations. This study revealed a degree of modularity and incremental expansion in shelter design, which allows for organic growth aligned with refugees' preferences and circumstances. The methodology involved five sequential steps: analysis, vocabulary definition, rules for shelter conditions, validation, and application. These steps facilitated a systematic exploration of spatial layouts and functional relationships within the corpus of designs provided by the camp. The study developed an iterative process that captured the incremental adaptations made by refugees, revealing the evolution from basic shelters to multifunctional shelters.

Findings show that the grammar can effectively capture the informal and adaptive nature of refugee-led design processes within the camp, demonstrating its potential to generate new shelter layouts that respond to the spatial needs of households. These findings hold significant implications for refugee camp planning and shelter design. Understanding the adaptive strategies employed by refugees can help humanitarian organizations and governments develop more effective and responsive shelter solutions that resonate with the cultural norms and aspirations of the affected population. The emphasis on flexibility and adaptation in design aligns with the dynamic nature of refugee communities, empowering residents to actively shape their living spaces and foster a sense of home amidst challenging circumstances.

This research contributes to the body of knowledge on refugee camps by providing insights into spatial configurations of shelters and adaptive processes of refugee communities. Future interventions can promote the well-being and resilience of displaced populations by acknowledging refugees' efforts to integrate their needs and preferences into shelter design. Future research suggests extending the grammar to 3D, conducting further investigations and applications of the Zaatari Camp grammar in other contexts, which is necessary to validate its effectiveness and adaptability beyond the Zaatari Camp. Further comparative analyses of other refugee camps and settlement contexts could provide a broader perspective on informal spatial practices and adaptive design strategies.

Acknowledgements We thank the Stuckeman Center for Design Computing (SCDC) at Penn State University and the Alma Heinz and August Louis Pohland Graduate Student Fellowship for supporting our research. Special thanks to SCDC for supporting this chapter's production. We also acknowledge the UNHCR-Jordan technical team for their invaluable assistance during data collection in the Zaatari camp.

References

Al-Harithy, H., A. Eltayeb, and A. Khodr. 2021. Hosting Syrian refugees in Saida (Lebanon) under protracted displacement: Unfolding spatial and social exclusion. *International Journal of Islamic Architecture* 10 (2): 413–439. https://doi.org/10.1386/ijia_00050_1.

Alnsour, J., and J. Meaton. 2014. Housing conditions in Palestinian refugee camps, Jordan. *Cities* 36: 65–73. https://doi.org/10.1016/j.cities.2013.10.002

Ansar, A., and A.F. Md Khaled. 2021. From solidarity to resistance: Host communities' evolving response to the Rohingya refugees in Bangladesh. *Journal of International Humanitarian Action* 6 (1): 1–14.

Betts, A., L. Bloom, J.D. Kaplan, and N. Omata. 2017. *Refugee economies: Forced displacement and development*. Oxford University Press.

Bulley, D. 2014. Inside the tent: Community and government in refugee camps. *Security Dialogue* 45 (1): 63–80. http://www.jstor.org/stable/26292275

Carter-White, R., and C. Minca. 2020. The camp and the question of community. *Political Geography* 81: 102222. https://doi.org/10.1016/j.polgeo.2020.102222

Colakoglu, B. 2005. Design by grammar: An interpretation and generation of vernacular Hayat houses in contemporary context. *Environment and Planning B, Planning and Design* 32 (1): 141–149. https://doi.org/10.1068/b3096

Conti, R.L., J. Dabaj, and E. Pascucci. 2020. Living through and living on? Participatory humanitarian architecture in the Jarahich refugee settlement, Lebanon. *Migration and Society* 3 (1): 213–221. https://doi.org/10.3167/arms.2020.030117.

Dalal, A. 2021. The Refugee camp as urban housing. *Housing Studies*, 1–23.

Dantas, A., D. Banh, P. Heywood, and M. Amado. 2021. Decoding emergency settlement through quantitative analysis. *Sustainability* 13 (24) (Article 13586). https://doi.org/10.3390/su132413586

Dias, M.A. 2014. Informal settlements: A shape grammar approach. *Journal of Civil Engineering and Architecture* 8 (11).

Duarte, J.P., and J. Beirão. 2011. Towards a methodology for flexible urban design: Designing with urban patterns and shape grammars. *Environment and Planning B: Planning and Design* 38 (5): 879–902. https://doi.org/10.1068/b37026.

Duarte, J.P., J.M. Rocha, and G.D. Soares. 2007. Unveiling the structure of the Marrakech Medina: A shape grammar and an interpreter for generating urban form. *Artificial Intelligence for Engineering Design, Analysis and Manufacturing* 21 (4): 317–349. https://doi.org/10.1017/S0890060407000315.

Elorduy, N.A. 2020. Learning in and through the long-term refugee camps in the East African Rift. In *Refuge in a moving world*, ed. E. Fiddian-Qasmiyeh, 362–381. UCL Press. https://doi.org/10.2307/j.ctv13xprtw.31

Elorduy, N.A. 2021. *Architecture as a way of seeing and learning: The built environment as an added educator in east African refugee camps*. UCL Press. https://doi.org/10.2307/j.ctv1gn3t8t.

Eloy, S., P.E. Vermaas, and M.A.P. Andrade. 2017. The quality of designs by shape grammar systems and architects: A comparative test on refurbishing Lisbon's Rabo-De-Bacalhau apartments. *Journal of Architectural and Planning Research*, 34(4), 271–294. http://www.jstor.org.ezaccess.libraries.psu.edu/stable/44987237

Etzold, B., M. Belloni, R. King, A. Kraler, F. Pastore. 2019. Transnational figurations of displacement: Conceptualising protracted displacement and translocal connectivity through a process-oriented perspective.

Gips, J. 1999. Computer implementation of shape grammars. In *NSF/MIT workshop on shape computation*. Cambridge: Massachusetts Institute of Technology.

Hanna, D.M., L. Buys, and A. Kumarasuriyar. 2022. Domestic refugee architecture in Jordan: A socio-spatial analysis of chaotic camps. *Journal of Architecture* 27 (1): 44–70. https://doi.org/10.1080/13602365.2022.2062034.

Hart, J., D. Albadra, N. Paszkiewicz, K. Adeyeye, and A. Copping. 2022. End user engagement in refugee shelter design: Contextualising participatory process. *Design Studies* 80: 101107.

Herz, M. 2013. Refugee camps of the Western Sahara. *Humanity: An International Journal of Human Rights, Humanitarianism, and Development*, *4*(3), 365–391.

Lambe, N.R., and A.R. Dongre. 2019. A shape grammar approach to contextual design: A case study of the Pol houses of Ahmedabad, India. *Environment and Planning B, Urban Analytics and City Science*, *46*(5), 845–861. https://doi.org/10.1177/2399808317734207

Lambert, H. 2017. Temporary refuge from war: Customary international law and the Syrian Conflict. *International and Comparative Law Quarterly* 66 (3): 723–745.

Latka, J.F. 2017. Emergency and relief architecture: Motivation and guidelines for temporary shelters. *A+ BE|Architecture and the Built Environment* (19), 267–330.

Ledwith, A. 2014. Zaatari: The instant city. *Affordable Housing Institute*.

Papatzani, E., P. Hatziprokopiou, F. Vlastou-Dimopoulou, and A. Siotou. 2022. On not staying put where they have put you: Mobilities disrupting the socio-spatial figurations of displacement in Greece. *Journal of Ethnic and Migration Studies* 48 (18): 4383–4401.

Sanyal, R. 2017. A no-camp policy: Interrogating informal settlements in Lebanon. *Geoforum* 84: 117–125.

Schirmer, P., and N. Kawagishi. 2011. Using shape grammars as a rule based approach in urban planning-a report on practice. In *Proceedings of the 29th conference on education in computer aided architectural design in Europe, eCAADe '11: Respecting fragile places*, 116-124. Ljubljana: eCAADe and University of Ljubljana.

Shalan, M. 2019. In pursuit of self-reliance—perspectives of refugees in Jordan. *Archnet-IJAR* 13 (3): 612–626. https://doi.org/10.1108/ARCH-04-2019-0085.

SphereAssociation. 2018. *The sphere hanbook: Humanitarian charter and minimum standards in humanitarian response*, 4th edn. Geneva, Switzerland.

Stiny, G., and J. Gips. 1972. Shape grammars and the generative specification of painting and sculpture. In *Information processing*, 71. Amsterdam: North-Holland.

Tobin, S.A., F. Momani, and T. Al Yakoub. 2022. 'The war has divided us more than ever': Syrian refugee family networks and social capital for mobility through protracted displacement in Jordan. *Journal of Ethnic and Migration Studies* 48 (18): 4365–4382. https://doi.org/10.1080/1369183X.2022.2090157.

Turner, S. 2015. What Is a refugee camp? Explorations of the limits and effects of the camp. *Journal of Refugee Studies* 29 (2): 139–148. https://doi.org/10.1093/jrs/fev024.

UNHCR. 2007. *Handbook for emergencies*. United Nations High Commissioner for Refugees.

UNHCR. 2016. *Settlement folio*. Planned settlement chapter. Geneva: UNHCR.

UNHCR. 2021. *Refugee statistics*. USA for UNHCR. https://www.unrefugees.org/refugee-facts/statistics/. Accessed 5 Jan 2022.

Verniz, D., and J.P. Duarte. 2020. Santa Marta urban grammar: Unraveling the spontaneous occupation of Brazilian informal settlements. *Environment and Planning B: Urban Analytics and City Science* 48 (4): 810–827. https://doi.org/10.1177/2399808319897625.

Open Access This chapter is licensed under the terms of the Creative Commons Attribution-NonCommercial-NoDerivatives 4.0 International License (http://creativecommons.org/licenses/by-nc-nd/4.0/), which permits any noncommercial use, sharing, distribution and reproduction in any medium or format, as long as you give appropriate credit to the original author(s) and the source, provide a link to the Creative Commons license and indicate if you modified the licensed material. You do not have permission under this license to share adapted material derived from this chapter or parts of it.

The images or other third party material in this chapter are included in the chapter's Creative Commons license, unless indicated otherwise in a credit line to the material. If material is not included in the chapter's Creative Commons license and your intended use is not permitted by statutory regulation or exceeds the permitted use, you will need to obtain permission directly from the copyright holder.

Designing with Visual Calculation is Child's Play

Derek Ham

In this essay, I argue for using shape grammars in teaching Beginning Design classes in a way that bridges the gap between STEM and STEAM education. The course description in this essay captures one instance of a course that has been taught several times over the last eight years. What I have found to be true is that shape grammars are the ultimate way to stimulate "play" in the design studio.

1 The Rules of Play

Good design requires creativity.
Good design is the manifestation of creativity.
Creativity requires play or playful thinking.
Every form of play has components that make it a game.
Every game has several components, one of which is play.
Design = creativity = play = game.
Therefore design = game
All games have rules.
Since design is a type of game, it must have rules.

D. Ham (✉)
College of Design, NC State University, 50 Pullen Road, Raleigh, NC 27695-7701, USA
e-mail: daham@ncsu.edu

2 Introduction

I spend a lot of time looking at and thinking about curricula in design schools. I look at what students do and learn at my institution and in many foundational design programs, where I've observed ongoing discussions about sustainability, social responsibility, design research methods, and design technology–and very few on how to stimulate the imagination and spur creativity. This paper argues that creativity can and should be at the core of design education. It demonstrates that students must learn to play again, and that one key to doing this is through computational thinking and shape grammars. As Thomas and Brown (2011, p.18) put it, "when play happens within a medium for learning like a culture in a petri dish—it creates a context in which information, ideas, and passions grow".

Fortunately, in the spring of 2017, faculty in the Graphic Design Department at NC State's College of Design were open to the idea of altering the curriculum to remove "D105" (a course focused on entry-level undergraduate graphic design topics) in favor of a new class, *GD 210 Image & Tech Tinkering*. I was tasked with setting up this new undergraduate course and creating the pathway for students to embrace a curriculum that bridged both analog and digital systems. My approach was to create three primary modules: Computational Systems, Spatial Systems, and Blended Systems. Each module was built on the next and would cover theoretical material through hands-on class exercises and design projects.

Following this pedagogy, I used shape grammars to get students to explore the design process as a designed system with rules. In doing so, through play, they learned about the structure of forms, the shaping of visual language, and the use of expression as a means toward these ends. I gave students a series of assignments and critiques in which they were encouraged to think deductively and intuitively, and investigate the creative potential of various materials. Although the teaching team (Instructor and TAs) offered critiques during the process and at the end of each project, the most critical dialogue took place between peers, as this proved vital in helping students reach their full potential.

3 How Shape Grammars Facilitate Play in the Design Process

In one of the first exercises, students were asked to explore a design process over three weeks driven by shape grammars. The first task was introducing the students to this methodology through a tabletop game called *"On the Line,"* which I created some years earlier (distributed by *Brainwright Games*). Students were asked to solve specific design compositions as a form of puzzle play by manipulating transparent shapes on a table. The game requires the player to perform several shape rules to solve each of the pattern cards. Each card has a unique set of rules that make up the

Designing with Visual Calculation is Child's Play

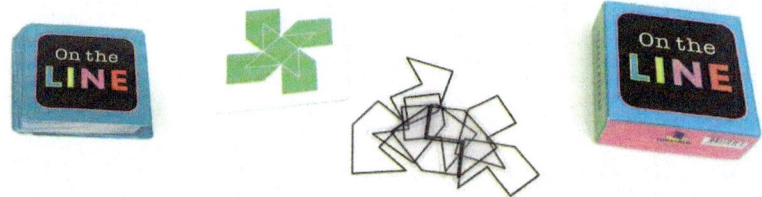

Fig. 1 "On the Line" shape grammar game

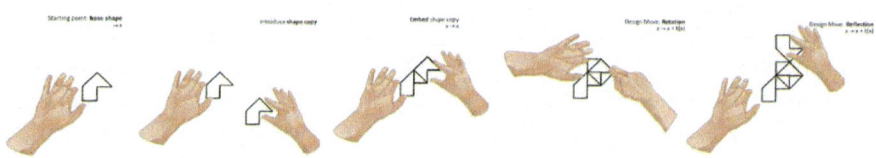

Fig. 2 "On the Line" schemas and rules

algorithm of its composition. Through the game, students began to reverse-engineer the designs by manually carrying out the rules with their hands (Fig. 1).

At this point, students were not required to be fully fluent in the algebraic representation of the rules found in shape grammars. Instead, they were asked to play with compositions and then find their own words to articulate what they had done. Once they began arranging the shapes onto one another, students would comment on the novelty of embedding and shape emergence through what Terry Knight calls "slow computing" (Knight 2012). Through each applied rule and each hand gesture using the rule, new patterns were discovered. From there, we began to name them, and at this point introduced the formal descriptions of the shape rules (Fig. 2).

After discussing the formal definition of rules, we returned to playing the shape game. Now, equipped with this new vocabulary, students could use the rules and schemas to explain the various approaches to solving the designs found on the pattern cards. The resulting conversations among the students from this exercise were full of insight. Students could now grasp how the simplicity of a visual system could open up a vast array of designs. They could communicate these concepts through formal descriptions in ways they had never done before. With this new knowledge, students were prompted to create their shapes, make their own rules, and generate their designs (Fig. 3).

The first order of business for any design instructor hoping to stir innovation and creativity in their students is teaching them how to play. The misconception is that "play-states" are intuitive and have no structure. Some would say that play is being ruined by teaching it, but those who think of "play" in this way regard play as something unruly, wild, and not having rules. Nevertheless, all forms of play have rules, and these rules fuel play, enabling the players to reach a higher intensity of play and creativity. Teaching students the "game of design" without

Fig. 3 Students making their own compositions through schemas and rules

explicitly teaching them the rules would be akin to dropping them in the middle of a rugby match, demanding that they "pick it up" through experimentation, an approach wrongly endorsed in design schools. This is where shape grammars come along. Shape grammars are, and have long been, a valuable strategy to get students to think differently about how rules are used in art and design, and to give structure to their intuition and design creativity.

4 How Shape Grammars Facilitate Playing with Computer Code

The *Image & Tech Tinkering* studio had a second agenda: to give students a foundational year with a pathway to embracing technology without fear. Surprisingly, many students who enroll in design schools come from educational experiences in the arts that have resisted the complementary use of digital tools and technology. In this course, coding became an object of study. Students learned to hack and play with computer code as a malleable component of the design process, designing objects and programs of their own throughout the semester.

To test their shapes and rules, and explore the affordances of playing with code, students were prompted to use the online tool "Scratch" as a prompt to continue shape play and deep visual exploration. In many ways, the play within Scratch is combinatorial. The interface not only borrows from Legos but, visually, has strong similarities. Combinatorial play is often a great starting point for young learners, but it should be seen as something other than the end-all goal; even within the realm of coding, several in-between moments can be wonderful and should be explored. If students or artists only limit themselves to seeing variables as zero-dimensional units, they will miss opportunities to see the emergence of new components. Nonetheless, in this class, this combinatorial exercise in visual calculation was the ideal prompt to get students to make the cognitive leap from "rules" to "scripts".

To further stress the relationship between shape rules and "computer commands," the coding blocks found in *Scratch* were translated into physical handheld coding blocks. This exercise enables students to play in the same manner as the shape

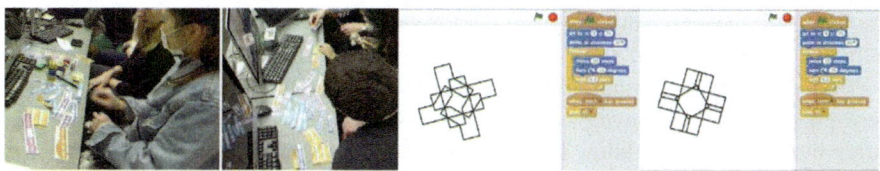

Fig. 4 Students playing with handheld Scratch cards and exploring rules through coded animation

game with their hands, using the physical table to think through the logic and their understanding of applied rules (Fig. 4).

Another benefit from the tangible play in this activity was its impact on collaborative group learning. Traditional computer programming courses are designed for students to learn independently, following instructions while staring at a syntax-based code editor, an approach that differs from how artists and designers interact and learn. While teaching programming principles through tangible manipulatives is useful in primary education (Patterson 2016), the critical element in these activities is play. In fact, students in art and design happily accepted this playful approach, affirming that it softened the subject matter and positioned them to learn concepts many said they had previously rejected.

When the students returned to the computer screen, their approach to the tools continued to be playful. In object-oriented programming, statements that have parameters attached to them are called conditional statements. The same occurs in mathematics and is fundamental to rule creation and algorithmic thinking. Students were able to make the connection here because of the exercises in playing with shape grammars. "If–then" and "if–then–else" statements became their new mechanisms of play. With them, they could increase the complexity of their compositions over time. Simple rules were now animated, making the designs much more visually robust. While the software could not articulate the embedded forms, the students themselves continued to use their eyes to find the emergent delight. Visual calculation constantly operates under this kind of playful logic. When a student observes the spatial relationship of shape variables, their configuration communicates the emerging shape. Clearly, the "if" statements were played out in the cause and effect observed by the eye during each frame of the animation.

5 Final Remarks

In design education, I have seen firsthand the benefits of using shape grammars to give students a deeper, more fundamental understanding of visual systems, styles, and design languages. The idea here is to gamify design elements, mechanics, and principles in non-gaming contexts to engage and motivate individuals to achieve specific goals or solve specific problems. Gamification is implemented to make activities more enjoyable, competitive, and interactive, but, as this paper has demonstrated, it can be much more; it can offer deep learning and a better understanding of complex subjects and systems.

Further, today's fascination with AI tools embedded in creative practices offers a new play method. I believe those who regard this as a threat to human creativity are often those who do not know how creativity works in the first place. Most contemporary AI tools are copy bots, with access to a vast library of designs for generating schemas and rules and applying them to new generative art. To quote Stiny (2006), "If you want to know how to be creative, then learn how to copy." Artists and designers have been doing this all along. AI gives us tools that enable us to play with schemas and rules faster One student's end-of-the-semester course evaluation captures the intended learning objective described in this paper, that rules are foundational to play and creativity:

> Having a recipe book of codes organized is so essential…because it allows the coder to understand what is happening and combine these tools in a creative and new way….Being able to reference early class notes constantly and being able to research different methods not used in class allowed me to think creatively and not stress about doing it the 'right way.'

Such is the power of play, the power of visual calculation permeating its way into design.

References

Brown, S. 2009. *Play: How it shapes the brain, opens the imagination, and invigorates the soul.* New York: Penguin Group.
Ham, D. 2015. Playful calculation. In *Revolutionizing arts education in K-12 classrooms through technological integration*, ed. N. Lemon, 125–144. Accessed 22 Oct 2015. https://doi.org/10.4018/978-1-4666-8271-9.ch006.
Huizinga, J. 1950. *Homo Ludens: A study of the play-element in culture.* Beacon Press.
Knight, T. 2012. Slow computing: Teaching generative design with shape grammars. In *Computational design methods and technologies: Applications in CAD, CAM and CAE education*, 34–55. IGI Global. Accessed 27 Aug 2023. https://doi.org/10.4018/978-1-61350-180-1.ch003.
Patterson, S. 2016. *Programming in the primary grades: Beyond the hour of code.* United Kingdom: Rowman & Littlefield Publishers.
Singer, I. 2011. *Modes of creativity.* Cambridge: The MIT Press.
Stiny, G. 2006. *Shape: Talking about seeing and doing.* Cambridge: The MIT Press.
Thomas, D., and J.S. Brown. 2011. *A new culture of learning: Cultivating the imagination for a world of constant change.* Lexington: Soulellis.

Open Access This chapter is licensed under the terms of the Creative Commons Attribution-NonCommercial-NoDerivatives 4.0 International License (http://creativecommons.org/licenses/by-nc-nd/4.0/), which permits any noncommercial use, sharing, distribution and reproduction in any medium or format, as long as you give appropriate credit to the original author(s) and the source, provide a link to the Creative Commons license and indicate if you modified the licensed material. You do not have permission under this license to share adapted material derived from this chapter or parts of it.

The images or other third party material in this chapter are included in the chapter's Creative Commons license, unless indicated otherwise in a credit line to the material. If material is not included in the chapter's Creative Commons license and your intended use is not permitted by statutory regulation or exceeds the permitted use, you will need to obtain permission directly from the copyright holder.

EthnoComputation: An Inductive-Deductive Shape Grammar on Toraja Glyph

Rizal Muslimin

Abstract This paper highlights the ways in which Shape Grammar inductive reasoning can analyze and represent design knowledge in a tacit environment. Deductive Shape Grammar has effectively examined designs from the past, where access to the artifacts' authors is not possible. However, in a condition where access to the craftsperson and the making process is possible, there is an opportunity to induce design grammar from the evidence on-site. Nevertheless, in such contexts, direct access to the craftsperson does not necessarily mean that access to their design knowledge is straightforward, as reflected in our case study, *Passura*: a Traditional Glyph in Toraja, Indonesia. In this article, the formulation of inductive Shape Grammar is provided, and applications on the tacit environment are discussed.

1 Introduction

Shape Grammar has been effectively used as a rewriting system to analyze design artifacts from the past, namely for Chinese Ice-Ray lattice windows, Hepplewhite Chairs, the Yingzao Fashi, Mayan ornaments, and the Mughal Gardens (Knight 1980; Li 2001; Müller et al. 2006; Stiny 1977; Stiny and Mitchell 1980). By analyzing a series of shapes as design evidence and rewriting them into a new set of design rules, the Shape Grammar method can reproduce similar as well as new emergent designs within the same design language. The deductive aspect of Shape Grammar is useful in situations where access to the artifacts' authors is not possible and needs to be substituted by a deductive approach, hence the adoption of a set of rules. With less attachment to the original context, the deducted rules provide new paths to deduce other artifacts. For instance, the Ice-Ray's subdivision rules have inspired many other designs, including Mughal-Garden, Hepplewhite and Mondrian grammars. Nevertheless, in a situation where the design practice remains active and access to the craftsperson and the design process are present, there is an opportunity to

R. Muslimin (✉)
School of Architecture, Design, and Planning, The University of Sydney, 148 City Road, Darlington, NSW 2008, Australia
e-mail: rizal.muslimin@sydney.edu.au

incorporate evidence not only from the artifact but also from the way in which the craftsperson perceives and produces the shapes. This type of study is imperative in a context where shape, meanings, and production systems simultaneously symbolize the multifaceted concepts of culture (e.g. indigenous culture).

However, in such contexts, direct access to the craftsperson does not necessarily mean that access to their design knowledge is straightforward, as the method might be passed on in a tacit environment. Pure inductive analysis, such as grounded theory or a phenomenological approach may be effective in capturing the anthropological reflection of the designer or an artifact's appearance (Glaser and Strauss 1999; Heidegger 2008; Husserl 2012). Yet, there are limits to representing a mode of design inquiry in textual narratives. Unless the authors articulate their design knowledge explicitly (e.g. via textual, verbal, or visual means), one would need to interpret and rewrite the knowledge comprehensively. This article highlights the ways in which the Shape Grammar inductive approach can analyze and represent design knowledge in a tacit environment, where a researcher has access to the designer and the process in situ.

Our case study is the '*Passura*' engraved ornamentation style, which remains actively practiced in Toraja, Indonesia. The *Passura* ornament serves as a visual narrative (glyph) to address the encoded messages of the house's owner (Fig. 1). While the ornaments have many variations (about 72 to 100 motifs), the style across some motifs is consistent and can be easily recognized in defining the Torajan style (Adams 2006; Kis-Jovak 1988; Waterson 1988).

Ornamentation is integral to the local religious principles (*Aluk Todolo*) which govern world-views and ways of life, including the way in which an ornament has to represent an appropriate story about a person's rite of passage (Duli 2003). For more references about the semantic and cultural symbolism of Passura see studies by Duli (2003), Pakan (1961), Thosibo (2005), Said (2004), and Waterson (1988). Existing studies on Passura have classified the ornament into four categories for such a passage: *Passura todolo* represents the values for a married person's life until his/her death, *Passura malollek* represents the values for a teenager's journey until his/her marriage, and *Passura pakbarean* symbolizes happiness (possibly representing the values for a person's childhood). The fourth category, *Garontok passura*, represents the main religious principles from which the three Passura categories should pertain.

Fig. 1 Passura engraved in Toraja's traditional house (Tongkonan)

These motifs are engraved in the houses, barns and coffins for the family member to relate to their elders' history and to communicate the ritual messages that are recorded via the Passura glyphs (Pakan 1961; Thosibo 2005). However, while the *Passura's* glyphs have been properly catalogued, the underlying production method is mostly passed on tacitly from generation to generation within a master-apprentice relationship. This study holds that in the absence of explicit documentation, Passura production knowledge can be interpreted through the making process on the site, in this case, from drawing activities prior to engraving and carving stages.

2 Methods

The data for this study is collected from observation of the crafting process in Kete Kesu village, Toraja, Indonesia and an interview with a local engraver to obtain more information about the design principles. Based on this, Shape Grammar embedding and generative analysis were applied to induce and deduce findings from the observation. The Shape Grammar representation helps to interpret the role of shapes when a maker engages with their object. Such that underlying an object transformation in the making process, the maker constantly performs visual calculation that drives their decision for subsequent iteration; and recurring attention and modification on a specific shape indicates that they are pivotal in the making process (Muslimin 2014). The following section discusses the way in which Shape Grammar principles could serve for inductive reasoning and whether such an inductive approach could work in tandem with the classical deductive method.

2.1 Framework for Observing Tacit Design/Making

In a traditional culture, most of which consists of tacitly passed-down traditions, access to an explicit representation of a rule is not always available. For instance, let us assume a hypothetical designer or craftsperson uses rule R1 in Fig. 2a to generate the artifact in Fig. 2b. Later on, if an observer, other than the craftsperson, finds the artifact, either directly or indirectly, the original rule is not always immediately obvious. Each sub-shape in the craft objects could lead to different rule deductions. For instance, based on different observations of the embedded shapes in Fig. 2b, different rules in Fig. 2c (i.e. R2, R3, and R4) can be deduced to interpret how a craftsperson generates the craft object.

As such, in deductive Shape Grammar, the correlation between the rule, initial shape, and the computation is less bounded to the original context. Different deduced rules and initial shapes in Fig. 2d are fine as long as they produce similar artifacts in Fig. 2b. In contrast, in in situ observation, where the making process is accessible, the bound between an initial shape and the making process is visible and may constraints the number of possible rules. By interacting with the maker, we may not only induce

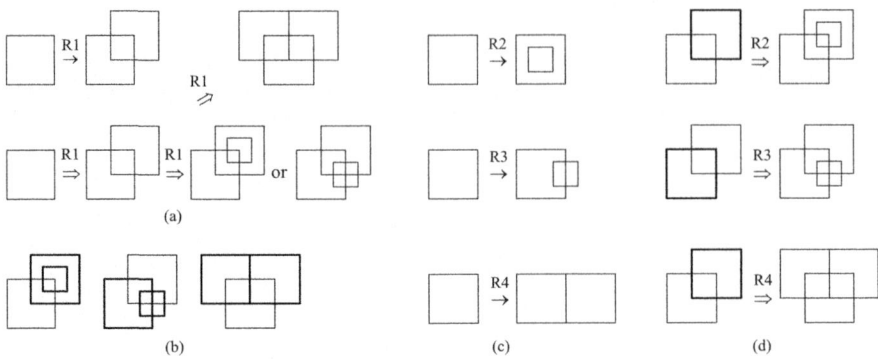

Fig. 2 Computation from original rules (R1) to deductive rules (R2, R3, R4)

the rules, as in Fig. 2a, but also the reason behind them, which could yield a different insight into such rules.

Yet, direct access to the craftsperson's making process could be problematic either because of issues in articulating the rules or because they might inherit different types of rules, as the craftsperson may not be the author of the design rule in Fig. 2a. When such issues occur, inductive reasoning needs to be carefully conducted in tandem with deductive reasoning. For this reason, it is necessary to reconfigure the correlation between rule, initial shape and the computation process for implementing deductive and inductive reasoning in Shape Grammar.

The correlation can be identified by the arrows that define the rules and the double-arrows that indicate steps in the making process. Such that, *if* the double arrows ($\Rightarrow$) in the rule iteration signify the process of applying a rule (r) for a shape (x) in computation (r, x), assuming that rule (r) and shape (x) exist, *then*, when a computation (r, x) exists, the observable shapes in the computation can be used to infer the rules (r) and the initial shape (x). This can be obtained by assuming the observable making process as the iterative computation to retrieve rule (r) and shape (x).

Figure 3 shows a diagram of inducing rules from a given sequence of making process. Through this diagram, we can record the observable making process to induce shape (x) and rule (r). In the making sequence, the earlier shape provides an initial shape for the left-hand side (LHS) of the rules while the subsequent shapes represent the transformed (or new) shape for the right-hand side of the rule (RHS). The inverse process from iterative computation to rule definition is notified by double-arrows that turn into single-arrows. The embedded shape, mentally applied in the rule, is represented with the schema $x \rightarrow x$ to identify the visual embedding process, while the shape transformation in the making process is represented with the schema $x \rightarrow t(x)$, $x \rightarrow y$ or $x \rightarrow x + t(x)$. As such, results from inductive reasoning will be constrained by the observer's sequencing strategy when splitting a continuous making process into discrete steps. Discretising a long process into one step will return a generic process while splitting a short sequence into several steps provides more detailed processes.

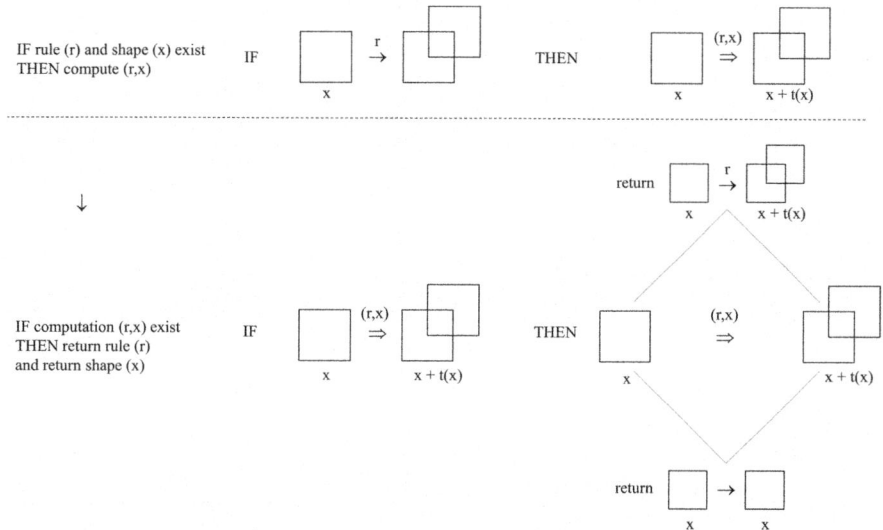

Fig. 3 Inverse-process from existing computation (top) into inductive rules (bottom): *If* a making process exists (marked with ⇒), *Then* induces the making rules from the visible transformation process x → x + t(x) (top of the lattice), and the implied visual matching process x → x, (bottom of the lattice)

Based on this framework, the following experiments are structured in two rounds based on the level of generality and specificity of the Passura design. In each round, the study will alternate between inductive shape grammar, where evidence of the making process is directly observable, and deductive grammar, where evidence is indirectly available.

3 Round One

3.1 Inductive Shape Grammar

We observed and interviewed the local engraver as he demonstrated the ornamentation process through drawing. To clarify the visual data, we simplify the engraver's drawing into line drawings. The drawing sequences are parsed by the difference between the spatial compositions in the drawing, notated by ⇒ in Fig. 4, based on the following routines.

For each shape i and shape i + n that appear in the making process, where 'n' indicates the degree of difference between an initial shape and its subsequent appearance during the making process:

1. If shape i is different from shape i + n, then return x → y, where x is shape i and y is shape i + 1

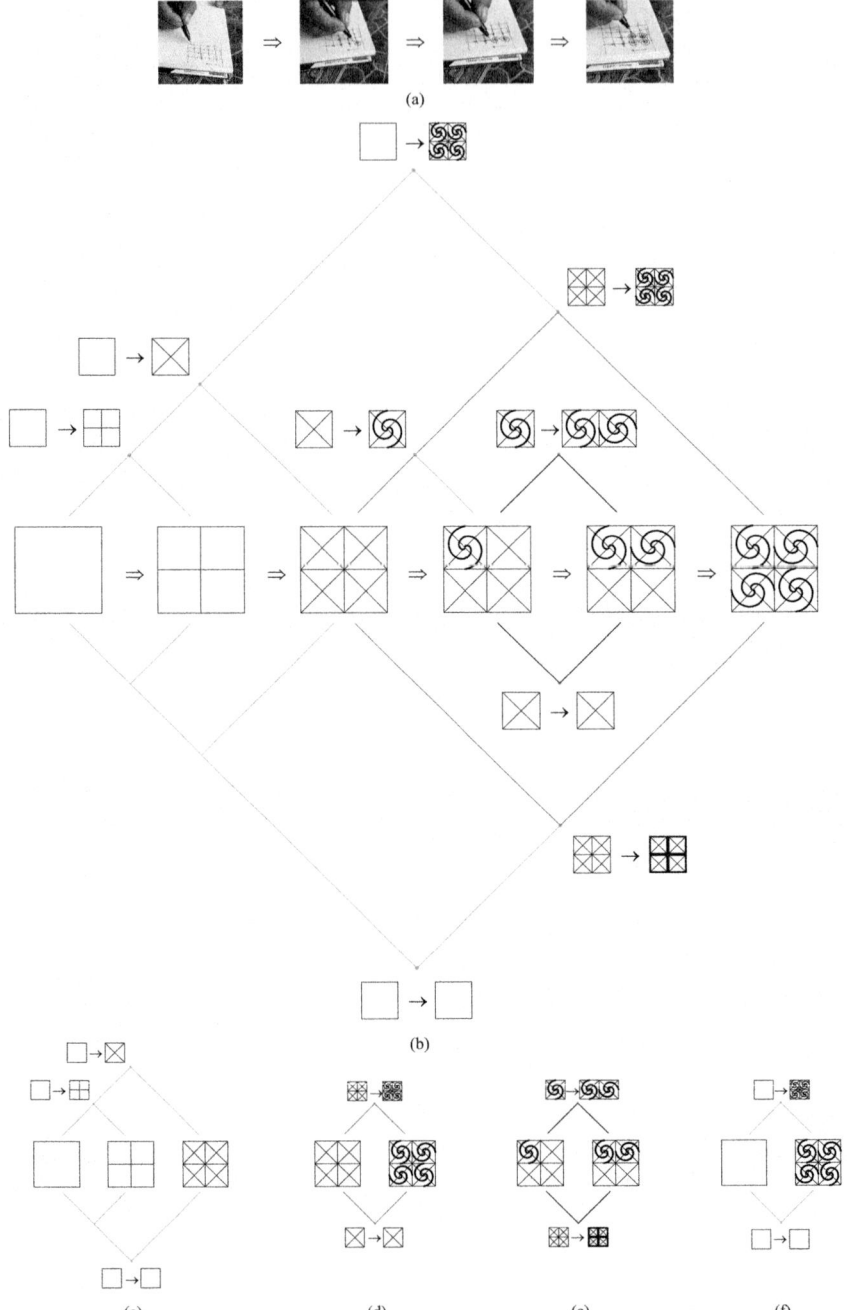

Fig. 4 Inductive Shape Grammar (**b**) generates rules from the making process (**a**). **c–f** Inductive rules presented separately based on the number of steps

2. If shape i is repeated in shape i + n, then return x → x + t(x), where x is shape i and t(x) is shape i's transformation
3. If shape i is identical to part of shape i + 1, then return x → x, where x is shape i

The degree of differences (n) is based on the observer's decision. It can be driven by the number of lines, subdivisions, and types of curves developing in a process, among many others. Lower n values will result in the subtle difference between the two designs, while higher n suggests more contrast between the two designs. For example, the n value in Fig. 4 is higher than in Fig. 7.

In this example, n signifies the differences in terms of types of subdivisions and curves. The sequencing process was started by subdividing a square, and then crosses were added over the square to create a rectilinear grid (the important role of the crossed shape in the engraving was emphasized by the local engraver). After the initial grid was finished, a new shape (a spiral) was drawn, and then he continued drawing another spiral next to the first one until about four spirals were composed. The drawing iteration in Fig. 4a serves as an initial dataset.

The lattice in Fig. 4b shows a set of nine rules induced from the datasets, where the rules below the making iteration signify the visual-embedding rules, and the ones above it shows the transformation rules. As can be seen, the induced rules appear in a different hierarchy based on how far the rules can cover the steps in the making process in the dataset. Low-level rules are more specific and located closer to the making steps iteration, while higher-level rules are further near the edge of the lattice to cover multiple steps at once. The latter is more detailed to highlight an intention for making a specific design, while the former is simpler to articulate steps for the making process.

The smaller diagrams in Fig. 4c–f magnify the different hierarchy groups. The local engravers stated that the crossed-linear grid is the main template to develop most of the Passura motifs (note: the cross square is also a motif in itself). Based on this, we induced the rules for making the grid from the first two steps: two subdivision rules and one embedding rule based on schema x → div(x) and x → x, respectively (Fig. 4c). The group starting at the second step represents an intention to engrave a chosen motif after the grid is drawn (Fig. 4d). The other two groups indicate the completion of the designs on two different levels. The third group, starting from step three, signifies the process of adding a transformed spiral inside the cross-grid modules, which induces an additive-transformative rule, *add a rotated spiral*, and another embedding rule *see a cross-grid* module as to where to add the new spiral (Fig. 4e). The last group covers the first and last step represents a decision to make a chosen motif, starting with a blank square (Fig. 4f).

3.2 Deductive Shape Grammar

The previous inductive process generates a preliminary Passura grammar and highlights the roles of their underlying schemas. For instance, the schema x → x to embed squares and crosses is prerequisite in transforming the designs. The schema $x \rightarrow x + y$ applies in drawing new shapes on the grid, and the schema $x \rightarrow x + t(x)$ encapsulates the way shapes are transformed to finish the motif.

The recurring appearance of these schemas and similarities between some Passura motifs suggest they might also be required for the other designs. In the following, the deductive Shape Grammar is exercised to speculate how far the schemas can generate the other motif designs. Starting with the schema $x \rightarrow x$ for focusing our visual attention on the square and the cross, we deduced several embedded shapes from the crossed-grid, as seen in Fig. 5.

There are at least ten squares and six crosses from the crossed-grid, which can serve as placeholders to insert a shape. Following our observation of the motifs surrounding the villages, we found several shapes, such as circles, spirals, and stars, to be paired with the embedding schema. Figure 6 shows our deductive iteration to reproduce several ornament types, namely Circular type (a), Stars Type (b) and Meander Type (c). Note that there are fewer double-arrows in Fig. 6, compared to Fig. 4, as they are not derived from a making process observation. The schema $x \rightarrow x$ allows different matches for adding new shapes to create the nine sample motifs in Fig. 6. The deduced types are correlated to about 60 ornament designs out of approximately 70 designs.

Within the circular and star types, we also deduce that some motifs could be a precursor to another one (marked with $\Rightarrow$ in the diagram) by applying the embedding rule twice for the same additive rule (further deduced association can be inferred between the circle and the star's motif as intersecting circles could eventually produce some of the star's motif). Interestingly, while the basic shape for circle and star has a minimum transformation in the addition process, this is not quite the case for the

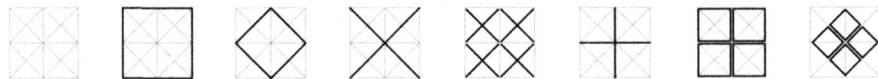

Fig. 5 Embedded shapes from the crossed-grid

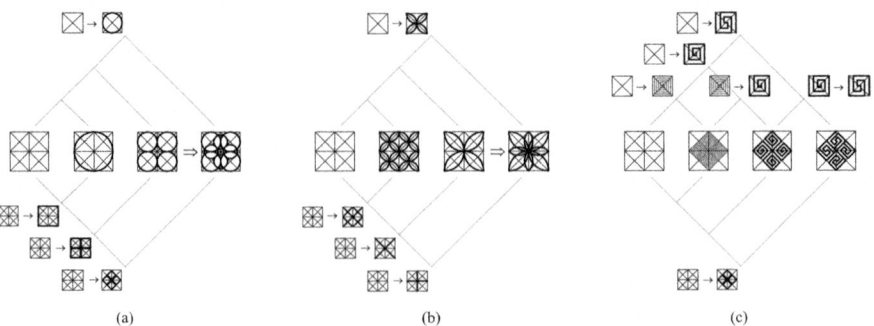

Fig. 6 Deductive rules on synthesizing ornamental types

meandering types (Fig. 6c). There are subtle differences between the basic meander module, in which we deduce that there should be low-level rules that vary the meander transformation with schema x → t(x). What follows in the next round is inductive and deductive Shape Grammar to investigate a more refined aspect of the Passura Meander and Spiral design.

4 Round Two

4.1 Inducing Passura-Meander Grammar

To investigate the meander transformation in Fig. 6c, we asked the engraver to draw one of the meander designs (*paqpapan kandaure*). The induction process is similar to the iteration in Sect. 3.1. However, instead of parsing the making process based on types of subdivisions and curves, this time, we examine the process by parsing the line development within one design, where n (degree of difference) refers to the completion of one maximal line after another.

The engraver first drew a rectilinear 6×6 grid, which was much denser than the grid drawn earlier for the spiral (Fig. 7top). He then bold the middle lines to subdivide the grid into four 3×3 quadrants. In one of the quadrants, he located the quadrant's mid-point, drew an x–y axial line from it, and then drew a square inside the quadrant and subdivided it into four more smaller squares. It was in this smaller squares that he developed the meander shape (see the diagram in Fig. 7). He started drawing a line from the edge of the square and went from outside to inside to turn the lines perpendicularly inward, and this went the same way until meeting the smallest cell in the grid. After the meander in one quadrant was created, he repeated the same process for the other quadrants.

Figure 7 shows the picture and the simplified drawing. The induction process started with a rectilinear grid as the initial shape and a square for the embedding rule. Each time a line was added, new transformation rules were induced at the top. The process generated several rules with the grid and the line as the left-hand shape. In order to obtain a recursive function, part of the right-hand shape of the rule has to be identical to the left-hand shape. Additionally, the rule needs to have constrained parameters in order to prevent it from becoming too loose and generating outputs that are either out of the corresponding style, or not specific enough so that they do not have the flexibility to cover various designs within a certain style. As seen in Fig. 7, Rules R5 and R6 are too generic as they can easily produce arbitrary L and U shape compositions, whereas Rules R8, R9, and R11 are too specific, as they can only generate an intricate spiraling style. Rule R7 is the first rule that indicates a meander pattern as the line turns inward, yet it is also too loose as the straight line in its LHS can be oriented in multiple ways. Rules R10 can generate the meander pattern by mirroring and scaling an L-shape that makes it recursive and specific enough to constrain the meander orientation.

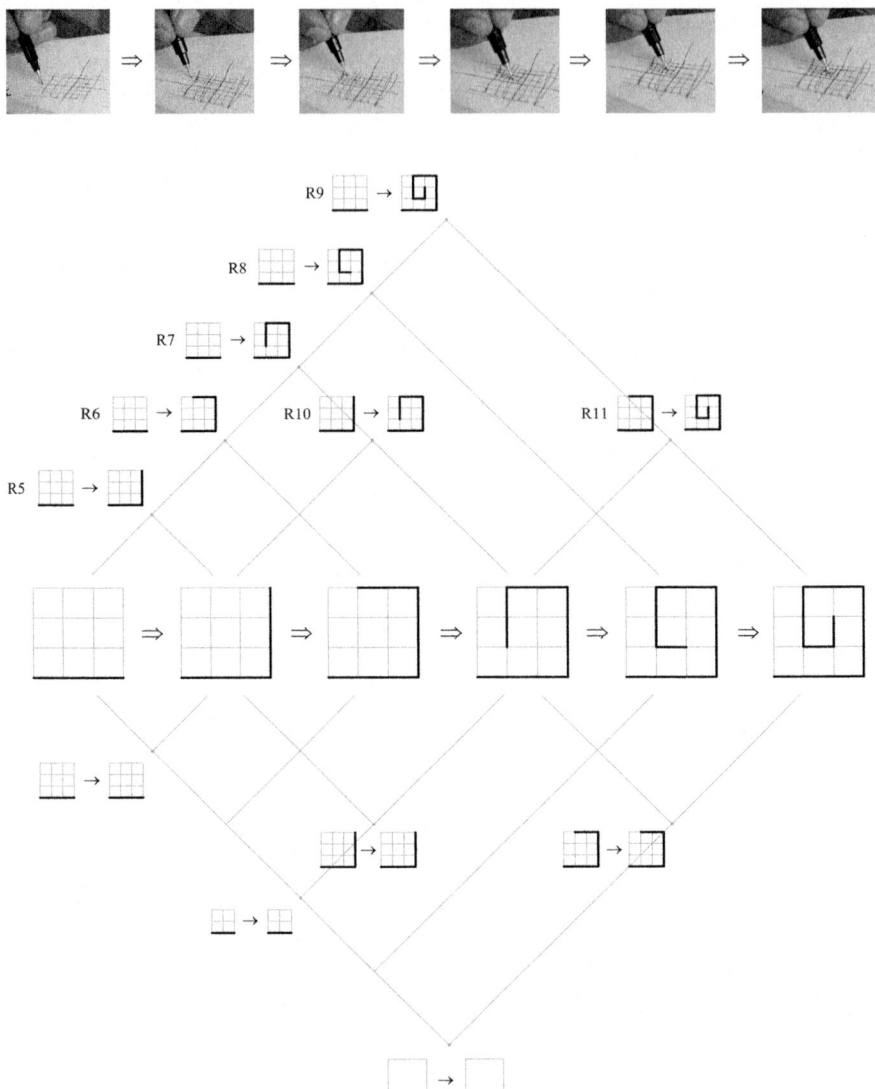

Fig. 7 Pictures of engraver's drawing process (top) and the meander pattern induction rules (bottom)

4.2 Deducing Passura Meander and Spiraling Grammar

We deduce a meander rule R12 using the embedding rule *see a square* and Rule R10 in Fig. 7. The rule mimics the way the engraver develops the meander outside-in from a square or the quadrants (Fig. 8a). It starts with a square, and then a new square is added inside with an opening on one corner. Mirroring labels by the square's diagonal axis helps to orient the square for recursive iteration to generate a smaller square. A counter (i) defines the maximum iteration to end the computation with rule R16 (e.g., when i = 5, it will stop after five steps). Additionally, to include the interlocked meander in the third step of Fig. 6c, a rotation rule R13 is added. Figure 8c shows the rule iteration and design results that resemble Passura meandering modules. Additive rules to populate the motif on a larger surface (R17, R17, R18, R19) are deduced using reflection, rotation and translation (Fig. 8d).

In the same manner, the meander pattern can be generated using translation and rotation, the spiral pattern can also be deduced with the law of symmetry. Evidence of symmetrical law has been indicated in a previous study (Muslimin 2013; Oliver 1998) with linear transformation and additive rules applied with labels to identify the variation (Knight 1995; Stiny 1980). Figure 9 exemplifies this with a spiral pattern carried inside a square module with a crossed grid that has eight orders of symmetry. When this module assembled into four shape rules (R21, R22, R23, R24) and computed in three steps, there are at least 64 designs can be generated from iterations R21 $\Rightarrow$ R21 $\Rightarrow$ R21, R21 $\Rightarrow$ R21 $\Rightarrow$ R22, and all the way to R24 $\Rightarrow$ R24 $\Rightarrow$ R24. However, the number of spiral tilings in Passura is only 14. Furthermore, we found that some passura' designs can be developed from the same spiral design by seeing the shape differently and detailing them to portray a particular object, for instance, by highlighting parts of the spiral's boundary, the areas in between, or their combination (see the three designs at the bottom of Fig. 9b). This non-zero-sum design embraces the role of embedding as functioned in Shape Grammar.

The new design (stem) characteristic that emerged from the spiral patterns becomes clearer after they are multiplied on a surface by using rotation along a two-dimensional grid (Fig. 9c). Results from this periodical tiling indicate that in resembling the actual object, the spiral tessellation serves as a guideline to develop the final shape.

4.3 Inducing (the Making of) Spiral Motifs

This last section verifies the deducted iteration in Fig. 9c with the actual Passura application in the village (Fig. 10c left), where the spiraling motifs are applied in a rectilinear grid on a wall. At each intersection point of this grid, an array of small oval dots appears, which, at first, appear randomly alternating around the intersection points. However, on closer inspection, the oval dots' position on the site coincides with the spirals' rotation point in Fig. 9 (see the dots in Fig. 10c).

To investigate these oval dots, we revisited the induction process in Fig. 4 and examined more steps in between. Figure 10a reveals that the engraver drew an array of oriented ovals (which he called 'the eye') around the grid intersection. After which, he extended the ovals' edges out to develop the spirals and connected them with the spirals from the neighboring ovals. This implies that the designs (including the spiral orientation) were mentally projected onto the surface, after which the ovals were then placed to aid the drawing process (Fig. 10b).

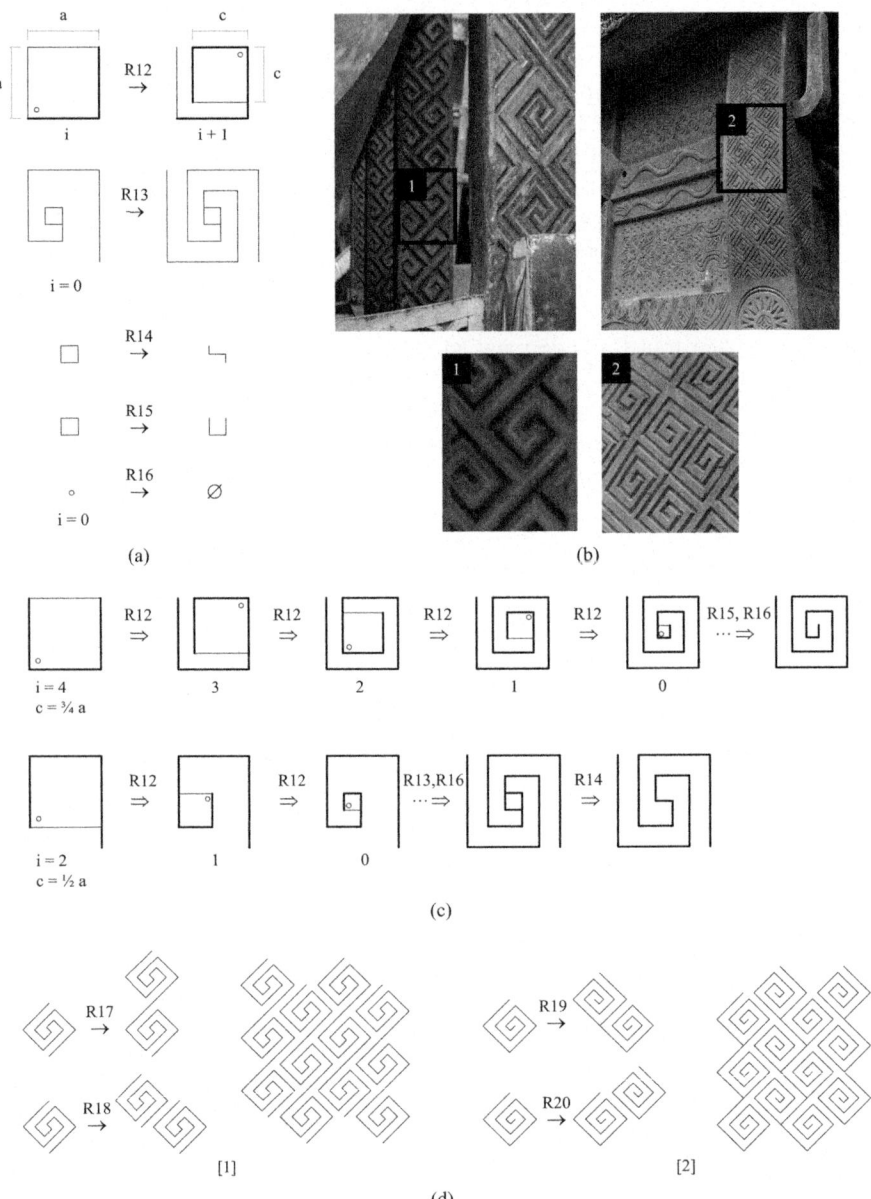

Fig. 8 Deducted rules to compute the Meander motif and its composition: **a** the deducted Meander rules: R12-R16; **b** the Meander motifs on the site; **c** the computation to generate the Meander motif's module; and **d** the modules assembled into a larger pattern with translation rules R17-R20

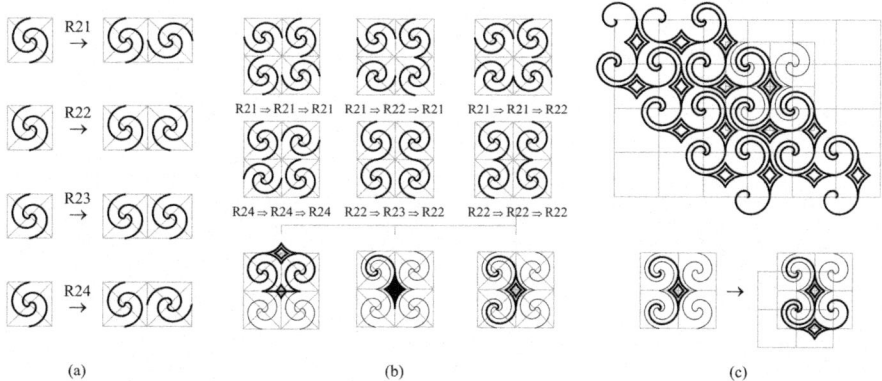

Fig. 9 **a** Deductive Rule of Passura Spiralling Motif with **b** linear transformation, embedding and **c** periodic tilings

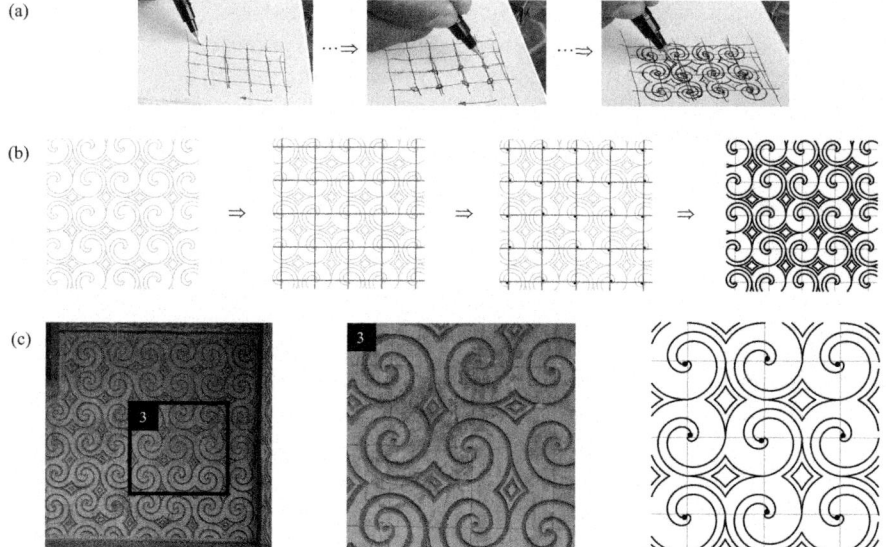

Fig. 10 **a** the engraver's drawing process as observed and **b** interpreted design-to-drawing process: (left) intended motif is mentally projected on a surface; (middle) grid is drawn and the motif's eye is placed along the grids intersection, and (right) the intended motifs are drawn; **c** Spiral orientation on the actual ornament (left) and the deducted version (right)

5 Conclusion

This article has formulated inductive Shape Grammar and exercised how it contribute to one another with the classical Shape-Grammar deductive approach in a tacit environment. On the one hand, the deductive rule is useful for representing the logic behind the mental design iteration, but it does not always represent the making process. For instance, the additive rules for the spiral and meander patterns may imply stamping, tiling, or drawing. On the other hand, the inductive rule is valuable in recording the making process. However, relying solely on the bottom-up drawing process might not be effective for synthesizing the design logic, especially for culturally restricted designs such as Passura motifs; the use of the induction process is to show more about how to make a design rather than the design process itself. As motif design and their making process have been well-established, today's ornamentation activities in Kete Kesu are essentially a reproduction of designs that have existed for hundreds of years. It is more analogous to writing than designing, in the sense that in writing, a set of well-known terms is assembled by following a certain consensus to address a message, whereas in designing, either a new term is introduced, or a set of well-known terms are creatively assembled to address a new message.

Nevertheless, both deductive and inductive Shape Grammar have shown promising results in interpreting knowledge from tacit design/making process more explicitly. As it is in the nature of Shape Grammar to embrace multiple interpretations, as can be seen in the case of the early grammar being rewritten (e.g., Palladian and Mughal-Garden grammar), we seek to find a better way to represent Passura grammar in further studies. This includes, but is not limited to, interpreting how actual objects are being distilled and translated into Passura designs and how Passura motifs are composed to address cultural/ritual messages.

Acknowledgements This research is supported by Sydney Southeast Asia Center. The author would like to thank Aleksander Panimba and Ting Ting for their support and contribution during the field study.

References

Adams, K.M. 2006. *Art as politics: Re-crafting identities, tourism, and power in Tana Toraja, Indonesia.* Honolulu: University of Hawaii Press.
Duli, A. 2003. *Toraja: Dulu Dan Kini.* Makassar: Pustaka Refleksi.
Glaser, B.G., and A.L. Strauss. 1999. *The discovery of grounded theory: Strategies for qualitative research.* London: Routledge.
Heidegger, M. 2008. *Being and time.* Reprint. New York: Harper Perennial Modern Classics.
Husserl, E. 2012. *Introduction to the logical investigations: A draft of a preface to the logical investigations (1913).* Springer Science & Business Media.
Kis-Jovak, J.I. 1988. *Banua Toraja: Changing patterns in architecture and symbolism among the Sa'dan Toraja, Sulawesi, Indonesia.* Amsterdam: Royal Tropical Institute.
Knight, T. 1980. The generation of Hepplewhite-style chair-back designs. *Environment and Planning B: Planning and Design* 7 (2): 227–238. https://doi.org/10.1068/b070227.
Knight, T. 1995. Constructive symmetry. *Environment and Planning B: Planning and Design* 22 (4): 419–450. https://doi.org/10.1068/b220419.
Li, A.I. 2001. *A shape grammar for teaching the architectural style of the Yingzao Fashi.* Ph.D. dissertation. Cambridge: Massachusetts Institute of Technology.
Müller, P., T. Vereenooghe, P. Wonka, I. Paap, and L.J. Van Gool. 2006. Procedural 3D reconstruction of Puuc buildings in Xkipché. In *VAST*, 139–146. Citeseer.

Muslimin, R. 2013. Decoding Passura–representing the indigenous visual messages underlying traditional icons with descriptive grammar. In *Proceedings of the 18th International Conference on Computer-Aided Architectural Design Research in Asia, CAADRIA 2013: Open Systems*, 781–790. Singapore: CAADRIA.

Muslimin, R. 2014. *EthnoComputation: On weaving grammars for architectural design*. Ph.D. dissertation. Cambridge: Massachusetts Institute of Technology.

Oliver, P. 1998. *Encyclopedia of vernacular architecture of the world*, vol. 2. Cambridge: Cambridge University Press.

Pakan, L. 1961. *The secret of typical Toraja's patterns*, 2nd, 1973rd ed. Ujung Pandang.

Said, A.A. 2004. *Simbolisme unsur visual rumah tradisional Toraja dan perubahan aplikasinya pada desain modern*. Yogyakarta: Ombak.

Stiny, G. 1977. Ice-ray: A note on the generation of Chinese lattice designs. *Environment and Planning B: Planning and Design* 4 (1): 89–98. https://doi.org/10.1068/b040089.

Stiny, G. 1980. Kindergarten grammars: Designing with Froebel's building gifts. *Environment and Planning B: Planning and Design* 7 (4): 409–462. https://doi.org/10.1068/b070409.

Stiny, G., and W.J. Mitchell. 1980. The grammar of paradise: On the generation of Mughul gardens. *Environment and Planning B: Planning and Design* 7 (2): 209–226. https://doi.org/10.1068/b070209.

Thosibo, A. 2005. *Mengungkap Makna Ornamen Gambar Passuraq Pada Arsitektur Vernakular Tongkonan Melalui Persepsi Indra Visual*. Bandung: ITB.

Waterson, R. 1988. The house and the world: The symbolism of Sa'Dan Toraja house carvings. *RES: Anthropology and Aesthetics* (15): 34–60.

Open Access This chapter is licensed under the terms of the Creative Commons Attribution-NonCommercial-NoDerivatives 4.0 International License (http://creativecommons.org/licenses/by-nc-nd/4.0/), which permits any noncommercial use, sharing, distribution and reproduction in any medium or format, as long as you give appropriate credit to the original author(s) and the source, provide a link to the Creative Commons license and indicate if you modified the licensed material. You do not have permission under this license to share adapted material derived from this chapter or parts of it.

The images or other third party material in this chapter are included in the chapter's Creative Commons license, unless indicated otherwise in a credit line to the material. If material is not included in the chapter's Creative Commons license and your intended use is not permitted by statutory regulation or exceeds the permitted use, you will need to obtain permission directly from the copyright holder.

Afterword

Computational design finds diverse applications in art, craft, architecture, product design, engineering, design computing, mathematics, education, and shape studies. The essays presented in this volume contribute to a growing body of interdisciplinary work that increasingly views design as a dynamic, open-ended computational process.

The 27 chapters of the book underscore references that elucidate key areas of study and suggest additional paths to pursue. This section outlines a selection of background readings in the broader field of computational design and shape computation that inspire in distinct ways and serve as valuable resources. Given the vastness of the present computational design literature, creating such a reading list can be daunting. The objective here is to propose a course reader for an imaginary introductory seminar on computational design. The proposed readings align with distinct areas of study and are presented chronologically to outline the historical development of the field.

Between 1974 and 2017, studies and views related to computational design and shape computation formalism were published in the journal *Environment and Planning B: Planning and Design*. Lionel March, the founding editor of *Environment and Planning B*, set the course for the journal from 1974 until 1982 when he joined the Royal College of Art as its new Rector. From the early days of computational design, designers, architects, and visual artists have sought to adapt to the new and often conflicting mathematics of the field. March (2002), in his retrospective account of the role of mathematics in architecture, confessed "the utter unmanageability of the topic". However, he made a valuable logical distinction: mathematics tends towards abstract, universal generalizations ($\forall$), whereas design is concretely particular ($\exists$). March believed that the architect starts from general typological notions about buildings and tries to devise a unique design to satisfy specific expectations. In this process, mathematics and computation play their part at three levels: generic knowledge, performance prediction, and specific design. Following the classification of form-making, made by the great German educator Friedrich Froebel (1782–1852), March associated the prediction of performance with quantitative computation and *forms of knowledge*; the configurational studies of geometrical arrangement and order with qualitative computation and *forms of beauty*; and last, the computation of relations with shape computation and *forms of life*. This three-part categorization is used here to help order the presentation of the computational design literature—but taxonomies are seldom perfect or complete, crisscrossing in unanticipated ways time and again.

A substantial area of computational design research has historically centered on quantification in architecture. Quantitative studies have been the traditional method for measuring and predicting physical performance across critical architectural domains, including structural integrity, acoustics, thermal behavior, and lighting. This process generates specific numerical data that architects use to optimize their designs for better performance.

The presumption is that key performance indicators and socio-economic development patterns can be effectively described, predicted, and controlled. Quantitative studies use deductive methods to transform mathematical concepts into concrete *forms of knowledge*. Models addressing building physics, energy performance, material usage, emissions, and socio-economic factors like occupancy, real estate, and economic impact all strive for optimization. These efforts are driven by pre-established objectives, assuming a collective agreement on the intended results.

At the far end of performance, George Birkhoff's *Aesthetic Measure* (1933) is a celebrated and seminal historical reference in quantitative design research, notable for introducing a framework to quantify beauty in art and design (Fig. 1). In this original work, Birkhoff suggests that an aesthetic measure M can be distilled into a mathematical formula, specifically in the relationship between order O and complexity C:

$$M = O/C$$

Leslie Martin and Lionel March's edited volume, *Urban Space and Structures* (1972), provides another early example of quantitative research. Originating at the Centre for Land Use and Built Form Studies in Cambridge, the book pinpoints measurable factors that guide decision-making for urban systems. Subsequently, March's edited collection, *The Architecture of Form* (1976), offers further essays exploring a quantitative approach to architectural design, accompanied by a

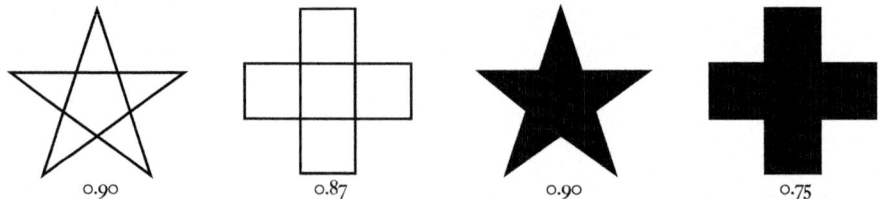

Fig. 1 Two polygons and their respective aesthetic measures as "ornaments" and "tiles" calculated by the formula $M = O/C$ in George Birkhoff's *Aesthetic Measure* (1933). (Figure adapted from Birkhoff's catalogue of 90 polygons by S. Kotsopoulos)

substantial editorial introduction examining the underlying logic of design and the challenging issue of value.

Expanding this discourse, Dean Hawkes's *The Environmental Tradition* (1996) presents a diverse collection of essays that trace the development of environmental awareness in architectural thought and execution, while also examining the wider theoretical and historical context of the discipline.

Nicholas Negroponte edited the volume, *Reflections on Computer Aids to Design and Architecture* (1975); it offers an early perspective on the shift from manual drafting to digital design tools, including initial thoughts on the potential of computer-aided design (CAD) to transform design practice. Roughly a quarter-century later, Charles Eastman's *Building Product Models* (1999) laid the groundwork for Building Information Modeling (BIM), outlining the core concepts, technologies, and methodologies associated with digital models for architecture, civil engineering, and building construction. Robert Woodbury's *Elements of Parametric Design* (2010) presents essential principles of parametric modeling relating to computer software and digital fabrication tools.

Quantitative approaches in contemporary research have undergone a remarkable transformation, with a clear movement away from the architectural scale towards the analysis of cities and urban dynamics. Michael Batty has been instrumental in defining and driving this trend, for example, in *The New Science of Cities* (2013) and *Inventing Future Cities* (2018).

In qualitative studies, abstract mathematical concepts are rendered into concrete designs to exemplify *forms of beauty*. Unlike quantitative studies, which usually depend on numerical equations, these mathematical concepts are not expressed in this way. Set theory, Boolean algebra, topology, connectivity graphs, trees, and planar maps are mathematical tools commonly used in constructing and deconstructing spatial configurations. Qualitative studies produce maps, graphs, and other configurations by assembling and disassembling discrete elements as in a Tinkertoy set.

In *The Geometry of Environment* (1971), Lionel March and Philip Steadman introduce fundamental analytical principles of spatial organization, making complex mathematical concepts accessible to designers through graphic architectural examples. The authors explain and illustrate various concepts of modern mathematics relevant to design, including relations and mappings, set theory and Boolean algebra, group theory and symmetry, spatial transformations and matrix representation, and graph theory.

Christopher Alexander's *Notes on the Synthesis of Form* (1964) is another notable early contribution to qualitative design research. Alexander argues for a rational and systematic approach to design, proposing that the correctness or incorrectness of a building should be treated as an objective matter of fact rather than a subjective judgment of value. He addresses the challenge of breaking down complex design programs into manageable sub-problems, tackling each by developing abstract diagrams that reconcile conflicting forces related to function. Alexander's set-theoretic methodology is further elaborated in his subsequent works: *A Pattern Language: Towns, Buildings, Construction* (1977) and *The Timeless Way of Building* (1979).

In *Architectural Morphology* (1983), Philip Steadman examines the combinatorial possibilities within specific architectural typologies. By establishing graph-theoretic definitions for particular classes of plans, he develops a systematic methodology enabling the enumeration of the complete range of possible plans within a given type. Steadman also explores how geometric relationships, spatial arrangements, and the assembly of components contribute to the diversity and evolution of building types. These plan-generating methods may have applications in architectural design and in the history and theory of architecture.

Bill Hillier and Julienne Hanson's *The Social Logic of Space* (1984) suggests that the initial act of design is space itself and its fundamental interrelation with patterns of human behavior. Hillier and Hanson show how analyzing spatial relationships like connectivity and integration can reveal an underlying social logic within architectural and urban forms. Their book establishes an analytical framework for spatial organization and planning in terms of "space syntax" for social and behavioral dynamics.

Lionel March's *Architectonics of Humanism: Essays on Number in Architecture* (1998) offers a fresh perspective on the use of symmetry and proportion in the architecture of Alberti and Palladio, as it introduces an innovative framework for architectural analysis and theory of architecture based on number and proportion. The book also re-examines the evolution of the Renaissance tradition, contrasting it with the views of Le Corbusier and the French school and highlighting its continuation and transformation in the Viennese and American practices of architects like Frank Lloyd Wright.

Quantitative and qualitative computation relies on counting and combining primitives. However, creative design involves a more dynamic and intuitive engagement (free play) with the basic elements of space: points, lines, planes, and solids to produce *forms of life*. Our natural capacity to divide and compose these elements into coherent forms through graphic representations suggests a computational framework that operates across spatial dimensions. The shape grammar formalism provides such a framework, enabling visual perception to directly inform computation with elements of any spatial dimension in art and design, thereby including counting and combining as a special case.

Allowing objects to exist without a strict representation is central to visual calculation in shape grammars, where ongoing perception often challenges prior knowledge to incite imagination. Engaging with shapes directly in calculations, rather than through symbolic descriptions, liberates us from the established norms of verbal thought and communication with symbols and numbers. By embracing the richness of experience and perception, shape grammars re-establish a more direct sense of human agency in the creative process.

In *Pictorial and Formal Aspects of Shape and Shape Grammars*, George Stiny (1975) explores the relationship between shape, form, and aesthetics to show how his shape formalism generates visually appealing objects. The pictorial properties of shapes, including color, texture, spatial relationships, and embedding, and their impact on perception and aesthetic value, are examined, concluding that shape and perception go beyond geometry to visual composition.

James Gips's *Shape Grammars and Their Uses* (1975) introduces shape grammars as a computational framework for interpreting and generating designs. Like Chomsky's phrase structure grammars, his shape grammars recursively specify forms by operating on shapes rather than symbols. Gips outlines a method for simulating Turing machines with shape grammars, demonstrating their computational power. The book also discusses computer implementation and aesthetics, highlighting the potential for automated art analysis and generation.

In *Algorithmic Aesthetics* Stiny and Gips (1978) go on to show how algorithms can be defined and implemented to generate, analyze, and evaluate aesthetic objects. They examine the formal structures that underlie aesthetic experience and the processes that capture the qualities contributing to our perception of beauty and artistic value. This includes the application of shape grammars and other generative systems to produce novel visual forms. Additionally, the book addresses fundamental questions about aesthetics within a computational framework, examining how concepts like style, variation, and originality can be algorithmically approached and understood.

The Logic of Architecture by William J. Mitchell (1990) examines principles of architectural design thinking. Drawing upon the theory of computation, Mitchell demonstrates how the languages of architectural form can be formally specified through shape grammars, and how these formal specifications can then be systematically interpreted verbally. This dual approach—developing verbal and formal descriptions of architecture—leads to a "critical language" for evaluating designs.

Terry Knight's *Transformations in Design* (1994) investigates the mechanisms by which styles evolve, and innovations arise within the visual arts. Using shape grammars for modeling and analysis, she examines the structure of visual styles, proposing that stylistic change can be understood as a process of modifying and transforming the rules within these formal systems. By explicitly defining the generative rules of a specific style, Knight demonstrates how new styles can emerge through the alteration, addition, or deletion of rules.

In *Shape: Talking about Seeing and Doing* (2006), George Stiny articulates a comprehensive theory of shape in which shapes are not static but dynamic entities actively participating in seeing and doing. Stiny develops shape algebras and shape grammars to calculate with shapes. In this context, seeing is a process of parsing and interpreting shapes according to rules and spatial relationships, while doing is the application of rules to create new shapes. This framework of seeing and doing offers a profound understanding of aesthetic experience, creativity, and our fundamental engagement with the visual world.

Finally, Stiny's *Shapes of Imagination* (2022) discusses the connection between shape computation and invention, drawing insightful parallels with Samuel Taylor Coleridge's distinction between "fancy" and "imagination." Stiny suggests that shape grammars intertwine with fancy and imagination. While fancy involves combining fixed and definite (0-dimensional) "counters,", imagination, with its vital power to "dissolve, diffuse, dissipate, in order to re-create," aligns with the dynamic processes of embedding and fusing shapes that are 0-dimensional and higher. By examining embedding and fusing in shape grammars alongside Coleridge's "esemplastic power of the imagination", Stiny offers a compelling view of the principles driving artistic invention for the artist and critic alike (see Figs. 3 and 4).

The graphic examples presented in Figs. 1, 2, 3 and 4 form the sequence Birkhoff, March, Stiny. They correspond to three distinct ways to calculate with shapes: quantitative, qualitative, and relational, respectively. This sequence establishes an ongoing and developing "aesthetic" tradition. Birkhoff's way of measuring O and C, and March's 0-dimensional Boolean descriptions are both subsumed in shape grammars.

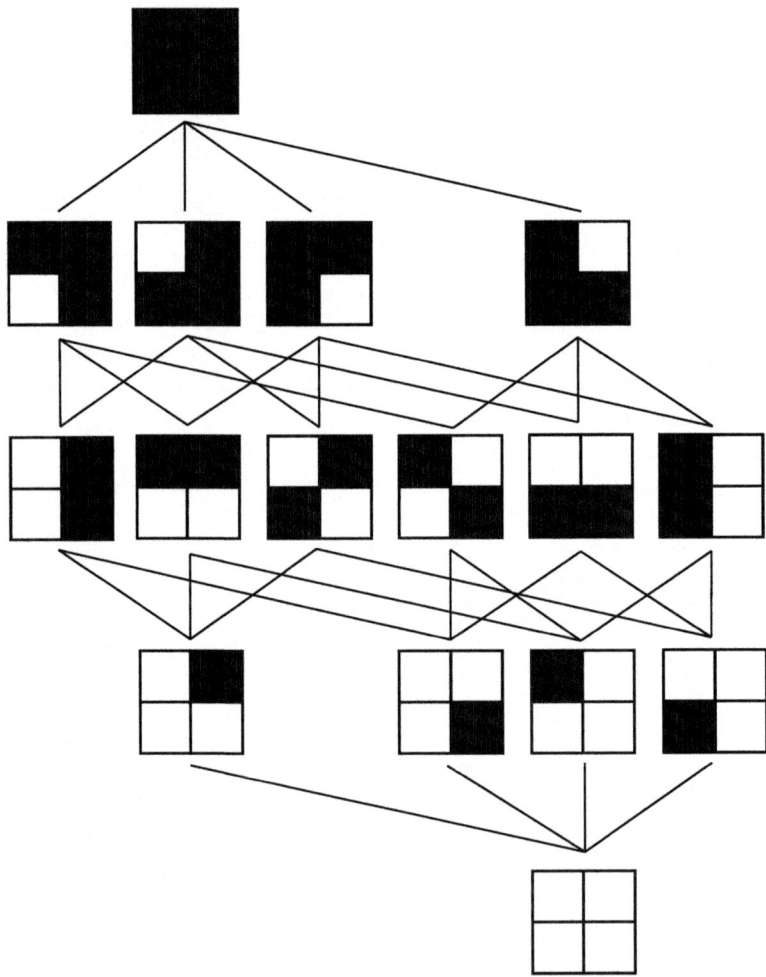

Fig. 2 The Boolean lattice (Hasse diagram) for the sixteen possible plan arrangements in a 2 × 2 grid. From *The Geometry of Environment* (1971), by Lionel March and Philip Steadman. (Figure adapted by S. Kotsopoulos)

Fig. 3 A pair of polygonal L's in two distinct spatial relations. But what does the right pair look like?

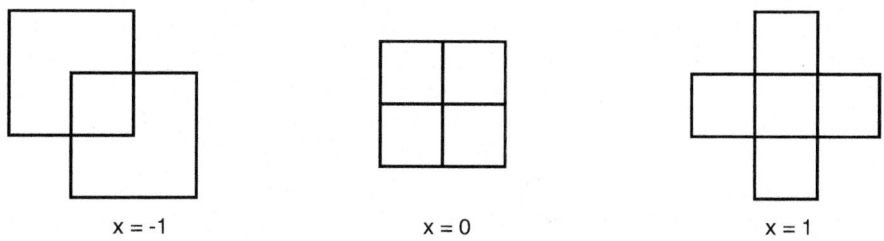

Fig. 4 Translating the L's in Fig. 3 back and forth on a diagonal axis with rules in a shape grammar can be nerve-wracking, not knowing what shapes are involved or what comes next. From George Stiny's *Shapes of Imagination* (2022). (Figs. 3 and 4 adapted by S. Kotsopoulos)

While not exhaustive, the background readings on computational design given here contextualize the contributions of this book within the historical trajectory and inbuilt tensions of the field. Despite the tensions and the continuous emergence of "big ideas" in contemporary computing, these readings underscore a remarkable consistency in the underlying research themes of computational design over more than 50 years, and they hint at a wealth of compelling concepts yet to be realized.

Turing machines founded on symbol manipulation, have long been the standard model of computation, but shape grammars present a remarkable extension. They satisfy the fundamental criteria of a Turing machine by substituting discrete 0-dimensional symbols with shapes, going from points, to lines, planes, and solids. This shift from symbolic to spatial calculation puts computation squarely in the visual realm.

The prevailing contemporary outlook of computational design as building science, incorporating quantitative and qualitative habits of thought, has also been typical since the beginning of the field. It is based on the acknowledgement that, much like in the natural sciences, designers must understand the causal mechanisms behind phenomena, seeking not just improved descriptions but the reasons why things appear as they do. This reflects the scientific drive for truth, often coupled with the conviction that true explanations will exhibit a compelling beauty by unveiling a more profound reality beyond superficial appearances.

Shape grammars transcend the limitations of this methodology by embracing constant redescription. In art and design, this prioritizes seeing and appearance over certainty and truth, recognizing that visual form is perpetually open to further interpretation. Solid proof does not work outside *narrow* fields such as logic and math. One can only examine points of view—aesthetic, performative, social, or other—that are never conclusive. Shape grammars acknowledge that ingenuity and delight in art and design arise from the very act of free formation. This unconstrained process holds intrinsic value, independent of its proximity to an ultimate explanation. Aesthetic meaning and value reside in seeing, preceding any scientific reduction to systems of atoms and units. Such descriptive conveniences are defined retrospectively in the embed-fuse cycle and are subject to revision with each new rule application.

Today, symbolic computation and artificial intelligence strive to show that symbolic programming and datasets supplant the nuances of human perception and imagination fully. The premise is that extensive data entails these faculties when massive processing power is available. However, shapes in art and design do significantly more than symbols and data, with less complexity and cost. The profound implication of calculating with shapes instead of symbols is a calculating framework that considers how the eye and mind synthesize existing knowledge and new perception in terms of embedding and fusing. This facilitates a shift from a computational paradigm centered on calculating with static objects to one centered on dynamic shapes. Calculating with shapes offers a basis for a comprehensive theory of creativity, knowledge, and imagination that is perceptual and cognitive at the same time. In this spirit, the 27 chapters of this anniversary publication cover a broad spectrum of practices, insightful reflections, critical assessments, and forward-looking speculations, to mark 50 years of progress that look with excitement to the future. The thrills and challenges of seeing in terms of shape computation are unending.

Sotirios Kotsopoulos

References

Alexander, C. 1964. *Notes on the synthesis of form*. Cambridge: Harvard University Press.
Alexander, C., S. Ishikawa, and M. Silverstein. 1977. *A pattern language: Towns, buildings, construction*. Oxford University Press.
Alexander, C. 1979. *The timeless way of building*. Oxford University Press.
Batty, M. 2013. *The new science of cities*. Cambridge: The MIT Press.
Batty, M. 2018. *Inventing future cities*. Cambridge: The MIT Press.
Birkhoff, G.D. 1933. *Aesthetic measure*. Cambridge: Harvard University Press.
Eastman, C. 1999. *Building product models: Computer environments, supporting design and construction*. CRC Press.
Environment and Planning B: Planning and Design (1974–2017).
Gips, J. 1975. *Shape grammars and their uses*. Basel and Stuttgart: Birkhauser.
Hawkes, D. 1996. *The environmental tradition*. E & FN Spon.
Hillier, B., and J. Hanson. 1984. *The social logic of space*. Cambridge: Cambridge University Press.

Knight, T. 1994. *Transformations in design: A formal approach to stylistic change and innovation in the visual arts.* Cambridge University Press.

March, L., ed. 1976. *The architecture of form.* Cambridge Urban and Architectural Studies, Series Number 4: Cambridge University Press.

March, L. 1998. *Architectonics of humanism: Essays on number in architecture.* New York: John Wiley.

March, L., and P. Steadman. 1971. *The geometry of environment: An introduction to spatial organization in design.* RIBA Publications.

Martin, L., and L. March, eds. 1972. *Urban space and structures.* Cambridge Urban and Architectural Studies: Cambridge University Press.

Mitchel, W. J. 1990. *The logic of architecture.* Cambridge: The MIT Press.

Negroponte, N., ed. 1975. *Reflections on computer aids to design and architecture.* New York: Petrocelli-Charter.

Steadman, P. 1983. *Architectural morphology: An introduction to the geometry of building plans.* London: Pion Limited.

Stiny, G. 1975. *Pictorial and formal aspects of shape and shape grammars: On computer generation of aesthetic objects.* Basel and Stuttgart: Birkhauser.

Stiny, G. 2006. *Shape: Talking about seeing and doing.* Cambridge: The MIT Press.

Stiny, G. 2022. *Shapes of imagination: Calculating in Coleridge's magical realm.* Cambridge: The MIT Press.

Stiny, G., and J. Gips. 1978. *Algorithmic aesthetics.* Berkeley, Los Angeles, London: University of California Press.

Woodbury, R. 2010. *Elements of parametric design.* London: Routledge.

Index

A
Aedicular façade grammar, 445–447
Aesthetic value, 222, 594
Affine transformations
  distinct, 304
  equivalent, 304, 305
Agarwal, M., 186
Alberti, L. B., 3, 15, 18, 19, 34, 42, 157, 498, 594
Alexander, C., 61, 151, 593
Algebraic representation, 317, 325, 569
Algebra of shapes, 196
Algorithmic
  aesthetics, 409, 495, 496, 498, 595
  methodology, xiii
AlphaZero, 488
Ambiguity, 6, 11, 130, 164, 173, 180, 195, 230, 355, 433, 478, 479, 493, 494, 497, 502, 503, 505, 506, 511, 514, 516
Analog drawing, 407, 410, 506
Analogy, 6, 7, 9–11, 15, 16, 18, 19, 27–29, 324, 329, 458, 459, 488
Analysis and synthesis, 195
Anders, P., 458
Anxiety, 9
Appearance, 19, 34, 38, 170, 171, 222, 289, 341, 514, 576, 579, 582, 597, 598
Archaeological reconstruction, 433, 434, 437, 452
Architectural language, 29, 349, 350, 407, 431
Architecture and language, 29
Argument lattice, 199–201, 210, 211, 216, 217
Aristotle, 157, 222

Artificial Intelligence (AI)
  assistants, 257
  image generation, 494, 498
Artist's formula, 18
Assignment, 26, 54, 55, 85, 277, 278, 280, 281, 447, 557, 558, 568
Atoms, 15, 17, 214, 223, 224, 243, 253, 255, 256, 598
Augmented shape, 78, 368–370
Axiom, 8, 11, 323

B
Backtracking, 107, 265, 268–275, 282, 283, 285
Bandur, M., 39
Basic elements, 39, 174, 175, 196, 224, 225, 256, 289–291, 343, 356, 397, 594
Batten, L. M., 308
Beautiful form, 7, 9, 11, 14, 16–18
Beauty, 8, 11, 15, 17, 19, 170, 592, 593, 595, 597
Berger, J., 22, 102
Bézier curve, 174, 175, 357, 400, 401
Birkhoff, G., 197, 200, 222, 592, 596
Blaser, W., 350
Bohm, D., 511, 512
Boolean operations, 37–39, 197, 199, 201, 204
Boundary contour system, 320, 329
Bouton, C. L., 353
Brand
  identity, 163, 164, 168, 170–173, 187
  style, 164, 170–173, 179, 181
Branded product, 163, 164, 170, 172, 173, 177, 178, 180, 181

Brown, S., 568
Browning, J., 258
Bryson, N., 102
Buick, 164, 165, 168, 172, 187
Building Information Modeling (BIM), 6, 7, 339, 340, 373, 486, 593
Built form, 33, 34, 39, 592

C

Cagan, J., 74, 165, 168, 185–188, 377, 378, 400
Calibration mechanism, 380, 381, 402
Calibration system, 378, 379, 401, 402
Caravan, 543, 545–548, 552–554, 557, 559, 560, 562
Cardoso, L. D., 157, 499
Carpo, M., 157
Carrier, 75, 76, 82, 88, 197, 198, 201, 290, 343, 344, 347, 355, 356
Cartesian product, 198, 201, 417
Causal history, 222
Cellular automata, 101, 352
Celtic knot, 364, 370, 371, 397–399
Characteristic shape, 170
Chase, S. C., 74, 339, 343, 378, 394
Chau, H. H., 74, 163, 166, 167, 169, 171–173, 175, 177, 378, 524
Chinese architecture, 434, 523, 525, 530, 533, 534
Chinese lattices, 57, 60, 86
Chippindale, C., 434
Chomsky, N., 22, 29, 595
Circular arc, 13, 175, 177, 240, 377, 379
Co-circular arcs, 380–383, 385–390, 392, 393
Cognitive dimension, 365, 366, 369, 370
Coleridge, S. T., 3, 5, 7, 8, 12, 13, 15, 17–20, 27, 163, 595
Color grammar, 128
Combinatorics, 35, 355
Communication and evaluation, 172
Complexity value, 110, 111
Compositional
　axes, 415, 419
　basis, 43
　development, 424
　terms, 56
　units, 39
Compound
　rules, 202
　shapes, 198, 201–205, 209, 211, 218, 418

Computational
　complexity, 81, 175, 487
　geometry, 172, 175, 177, 179, 180
Computer Aided Design (CAD), 7, 30, 39, 91, 175, 177, 180, 190, 257, 258, 337–341, 343–345, 347, 349, 357, 365–368, 373, 377, 378, 380, 391, 399–403, 411, 414, 485, 488, 593
Computer aided grammar, 407, 408, 421, 426
Computer aided manufacturing, 338
Computer implementation of shape grammars, 290, 292, 485
Concentric arcs, 240, 380–382, 392
Conjunction, 38, 283, 284, 445
Construction lines and registration marks, 289, 292–294, 296, 297, 300, 304, 308, 312, 313
Constructive
　act, 329
　exercise, ix, x
　method, 426
　process, 65
　program, 356
　representation, 90
　tradition, 34
Consumer
　products, 170, 171, 173, 177, 181, 189
　goods, 163, 170, 173, 181
Conversation with the situation, 363
Coons, S., 158, 257, 258
Copying, 223, 231, 363, 364, 366, 367
Counterplastic, 41
Counters, 17, 41, 116, 134, 159, 234–237, 370, 423, 494, 515, 585, 595
Coyne, R., 320, 321
Craft
　object, 577
　practice, 126, 127, 157, 158, 180
Creativity, surprise, and style, xi
Criticism algorithms, 55, 495
Cybernetics, 8
Cyberspace, 458
Cycles of design generation and computation, 180

D

Dali, S., 14
Darwin, C., 321
Data
　and statistics, 488
　driven, 511

processing, 485
representation, 378, 385, 402
sets, 258, 497
Debugging, 270
Decision time, 106, 107, 109, 110, 116, 117, 119–121
Decomposition
 diagonal, 224, 227, 228, 232, 248, 256, 290
 discrete, 211, 227, 242
 natural, 211–216, 219, 227, 242, 243, 253, 256
Deductive, 283, 575, 577–579, 581–583, 587, 588, 592
Design machine, 55, 221, 224, 256, 257, 355, 430
Delight, 4, 70, 571, 598
Dembowski, P., 308
Dennett, D. C., 373
Density, 33, 61, 107, 110, 116, 117, 128, 134
De Pennington, A., 172, 180
Descartes, R., 511
Design
 configuration, 163, 181, 186, 422
 creativity, 174, 179, 570
 education, 363, 568, 571
 space exploration, 363, 367–371, 374
 thinking, 130, 516, 595
 variations, 350, 351, 415, 421, 422, 425, 430, 561, 562
De Stijl, 27, 28, 40, 41, 104
Deterministic
 non-deterministic, 21, 188
Dewey, J., 18, 20
Dexter, E., 102
Diagonal
 decomposition, 224, 227, 228, 232, 248, 256, 290
 elements, 224
 set of atoms, 243, 253, 256
 shape, 221, 224, 227, 237, 242, 255, 256
 shape algebras, 256
 space, 227, 228, 237, 242
Dialectic, 41, 42, 363
Digital fabrication
 devices, 121
 machines, 157, 158
 methods, 127
Dimension i ≥ 0, 4
Directive, 262–265, 278, 282, 365, 514
Direct product, 198, 201, 202, 204, 209, 416

Discrete
 decomposition, 211, 227, 242
 elements, 237, 593
 lines, 228
 machines, 380
 mathematics, 33, 34
 objects, 258
 representation, 402
 space, 58
 unit, 324
Discrete infinity, 15, 16
Disjunction, 283, 284
Dispositio, ordinatio, distributio, 55
DNA, 190, 448
Dirty marks on paper, 25, 257
Dom-Ino, 422, 423
Dourish, P., 459
Dove soap bar, 166, 171
Downing, F., 22
Drafting
 conventions, 246
 hierarchy, 221, 237, 243, 248, 253–255
Drawing
 blind time, 102
 diversity, 110
 process, 106, 128, 584, 585, 587, 588
 shapes, 221, 247, 337, 338, 340, 344, 357, 427
 tools, 366
DrawScript, 345, 346, 452
Dreyfus, H., 488, 494
D.Star, 363–365, 367, 369, 370
Duarte, J., 58, 187, 410, 417, 463, 541, 542
Duchamp, M., 499
Duck/rabbit, 11, 12
Dukta, 126, 145

E
Earl, C. F., 62, 74, 75, 85, 87, 174, 175, 177, 187, 232, 240, 339, 341, 343, 357, 377, 378, 395, 397, 400
Economou, A., 54–56, 60, 64, 70, 74, 85, 86, 89, 172, 176, 180, 218, 263, 290, 339–341, 343, 348, 349, 356, 363, 365, 367, 377–381, 394–396, 399, 409, 410, 412, 414, 417, 422, 435, 484, 486, 487, 496, 504, 533
Egan, P., 189
Embedding, 7, 26, 73, 126, 174, 225, 240, 256, 258, 261, 264, 285, 289–291, 298, 318, 339, 340, 343–345, 347–351, 356, 357, 368, 377–380,

394, 396, 397, 399, 401, 402, 414, 426, 485, 486, 489, 569, 577, 578, 581–585, 587, 594, 595, 598
Embed-fuse cycle, 3, 4, 6–8, 13–21, 25, 27, 149, 224, 256, 257, 341, 343, 497, 598
Emergency shelters, 539
Emergent
parts, 53, 54
shapes, 67, 91, 125, 174, 318, 319, 320, 326, 327, 501
Entelechy I grammar, 409–412, 414, 415, 417, 421, 429
Enumeration, 39, 283, 284, 307, 309, 448, 487, 527, 594
Environmental aspects, 154, 542
Equivalence class, 82, 83, 225, 292, 296, 297
Esemplastic power, 15, 18, 20, 595
Ethnocomputation, 575
Euclidean, 88, 293, 294, 313, 318, 343, 412
Eurythmia, symmetria and décor, 55
Evaluation, 55, 56, 170, 172, 177, 186, 189, 366, 409, 430, 495, 498, 529, 533, 572
Evocation, 434
Execution order, 224, 237, 240
Expression and interpretation, 36, 38
Eye and hand, 18, 234
Eye tracking, 173, 324
Eysenck, H. J., 222
Eizenberg, J., 187

F
Face structure, 304, 306, 307
Fairbank, W., 525
Fancy, 8, 15, 17, 18, 20, 488, 595
Feedback, 132, 153, 157, 189, 358, 407, 422, 424, 428, 430
Feng, J., 525
Fibonacci, 69
Figurative construction, 223, 224, 231, 243, 246, 254–256
Figure-ground relations, 11
Find and replace, 337, 338, 340, 341, 343, 344
Finite decomposition, 199, 211
Firmness, commodity, and delight, xi
Fixities and definites, 17
Flemming, U., 21, 22, 24, 29, 30, 58, 90, 187, 410, 486
Flondrians, 322

Flow, 165, 261, 264–271, 273–276, 278, 281–286, 338, 340, 346, 347, 357
Formal
description, 126, 127, 569, 595
representation, 145, 146, 195, 223
Forms of life, knowledge, and beauty, 17
Freedman, R., 409, 431
Free play, 16, 594
Froebel, F., 17, 30, 66–68, 591
Fusing, 216, 255, 355, 595, 598

G
Galileo, 15
Generalized Boolean algebra, 197, 198
Generative
AI, 154–156, 257, 484, 485, 489, 493, 495, 503, 505, 511, 513, 515, 516
design agent, 457, 461, 465, 466, 468, 475–477, 479
design grammar, 457, 461–464, 466, 467, 476–480
explanation, 223
grammar, 15, 457, 460
modeling, 513
process, 229, 262, 484
specification, 484
Generative Adversarial Networks (GANs), 180, 372, 494, 497
Genesis, 18, 99, 100
Genetic algorithms, 188, 263, 321
Genius, 9, 25
Gestalt, 11, 327, 357, 505
Gesture time, 107, 108, 110, 117, 120, 121
Gips, J., 22, 29, 73–76, 78, 185, 190, 195, 217, 317, 338, 341, 348, 356, 378, 409, 435, 461, 479, 484–487, 495, 496, 499, 515, 541, 595
Glahn, E., 525
Glarner, F., 99, 100, 103–107, 109–111, 121
God, 3, 8, 10, 11, 13–16, 19, 433, 434, 436
Goethe, J. W. von, 17
Golden ratio, 34, 69
Goldschmidt, G., 363
Gongcheng zuofa zeli, 523, 525, 526, 530, 533
Goodman, N., 150, 349
Gopalan, V., 245, 246
Grasl, T., 86, 263, 339, 349, 356, 377, 378, 410
GRAPE, 356, 377, 410
Graphic composition, 246
Graphs, 8, 10, 18, 19, 90, 202, 313, 356, 487, 593, 594

Gray, J., 195
Greek spirit, 17
Greek architecture, 434, 445
Greek meander, 229, 435
Greenberg, M. J., 292
Grid as generator, 53, 56, 59, 60, 61, 63, 64, 70
Grid-like shapes, 54
Ground truth, 106, 511
Grünbaum, B., 307, 313
Gu, N., 330, 457, 459, 467
Guo, Q., 525
Gün, O., 128, 493, 499, 501, 503, 505–508

H
Habit, 16, 17, 597
Hadrian's library, 433, 445, 447, 449, 450
Ham, D., 567
Harley-Davidson, 164, 165, 168
Harmony, 10, 42, 46, 229, 506
Hand
　drawn, 128, 506
　crafted, 321
Hard coded, 187, 257
Hawsh, 547, 553, 558, 560
Hersey, G., 409, 431
Heidegger, M., 100, 576
Hepplewhite chair back, 167
Heuristics, 17, 189, 283, 318, 365, 372, 488
Hierarchical
　assembly, 132
　order, 384
Hierarchy, 7, 9, 11, 186, 205, 221, 223, 224, 226–229, 232, 233, 237, 242, 243, 248, 253–256, 292, 308, 313, 320, 324, 353, 356, 369, 384, 459, 581
Hixon Symposium, 6
Hodder, I., 434
Hogarth, W., 17
How question, 100
Human computer interaction, 151, 348, 367
Hume, D., 11
Husserl, E., 576

I
Ice-ray, 86, 87, 164, 165, 166, 174, 176, 575
Identity
　rule, 134, 349
　transformation, 435
Illusory shapes, 319, 320
Imagination
　primary, 13
　secondary, 7, 13, 17
　strong, 19
　tricks of, 16
Implemented shape grammars, 188, 407, 421, 426, 430
Incidence signature, 304–307
Indifference, 12
Inductive reasoning, 575, 577, 278
Infinite
　canvas, 364, 366, 368, 369, 371
　decomposition, 199
　line, 82
　maximal lines, 317–319, 327
　space, 186
　set, 197, 209, 343
　sum, 199, 255, 256
Infinite use of finite means, 14
Ingold, T., 100, 102, 132, 157
Initial shape, 56, 58–60, 62, 63, 89, 107, 108, 111, 112, 120, 139, 164, 165, 168, 176, 177, 219, 244, 253, 262, 274, 277, 278, 326, 328, 398, 412, 415, 424, 439, 442, 446, 447, 451, 478, 479, 484, 507, 552, 578, 579, 583
Insight, 4, 15, 20, 23, 57, 70, 126, 130, 149, 159, 171, 189, 190, 254, 258, 340, 350, 356, 365, 407, 415, 437, 449, 484, 488, 493, 494, 497–499, 503, 510, 512, 516, 523–525, 541, 542, 549, 550, 558, 563, 569, 578, 595, 598
Intention, 15, 36, 37, 54, 70, 150, 155, 156, 347, 415, 475, 493, 494, 499, 501, 502, 505, 512, 516, 581
Intelligent design, 457, 474
Interaction rules, 463–465, 467, 474
Intuition, 16, 21, 23, 25, 290, 408, 410, 493, 494, 499, 510, 570
Invention, 409, 595, 596
Ionic porch, 433, 435–439, 443, 444, 446

J
Jaffé, HLC, 27, 28
James, W., 18
Jowers, I., 53, 74, 87, 174, 175, 201, 232, 240, 339, 341, 377, 378, 397, 400, 485–487
Judgment, 4, 17, 483, 487, 489, 499, 593
Juzhe, 526, 527, 529, 532

K

Kanellopoulos, C., 436, 445, 446, 450
Kelly, E., 4, 330
Kindergarten, 17
Kit Kat, 178
Kit of parts, 29, 91, 224, 242, 247, 254–256
Klee, P., 17–19, 53, 63, 99–102
Knight, T. W., 78, 87, 99, 100, 103, 104, 106, 125, 127, 128, 131, 164, 165, 167, 218, 223, 229, 231, 246, 255, 263, 341, 348, 349, 356, 407, 408, 410, 414, 434, 435, 463, 478, 479, 486, 493, 501, 502, 504, 532, 569, 575, 585, 595
Knuth, D., 353
Koning, H., 187
Kotsopoulos, S., 19, 54, 56, 127, 218, 356, 505, 592, 596, 597, 598
Krishnamurti, R., 44, 62, 70, 73, 75, 78, 80, 81, 83–88, 90, 91, 175, 177, 187, 262, 290, 339, 341, 343, 348, 356, 357, 377, 378–380, 394, 395, 402, 414, 487, 524
Krstic, D., 75, 87, 195, 197, 199–202, 208–210, 212, 216, 219, 222, 226, 255, 289, 290

L

Langer, S., 23, 24, 27
Large language models (llms), 257
Lattices, 17, 57, 60, 61, 86, 87, 132–134, 136, 139, 141, 143, 144, 175, 197, 199–201, 210, 211, 216, 217, 225–227, 231, 232, 242, 255, 355, 575, 579, 581, 596
Law, 17, 373, 585
Layout generation, 276–279, 347
Le Corbusier, 39, 422, 429, 594
Lecun Y., 257, 258
Leeuwenberg, E., 222, 223, 255
Lego, 17, 570
Leonardo da Vinci, 44
Lernaean hydra, 514, 515
Leyton, M., 222
LeWitt, S., 101
Li, A. I., 74, 172, 175, 177, 378, 434, 523–525, 529, 533, 534, 575
Liang, S., 523, 525–527, 530–534
Liew, H., 263–265, 274, 278, 282, 285
Lines + cells, 58
Logic of inquiry, 18
Lucretian swerve, ix

M

Machine learning, 4, 8, 23, 188, 292, 308, 322, 353, 485, 495
Madafa, 547, 553, 554, 557, 558, 560
Magic, 13, 70
Magical, 3, 15, 19, 27, 70, 163
Maher, M. L., 457, 459, 462, 467
Making grammars, 99, 100, 103, 106, 108, 121, 127, 246
Mappings, 85, 199, 207, 257, 329, 366, 370, 386–390, 500, 542, 593
March, L., 33–44, 46, 48, 55–57, 59, 61, 63–65, 70, 224, 256, 343, 355, 430, 463, 591–594, 596
Marquis, J. P, 292
Martini glass, 178, 179
Martin, L., 33, 34, 40, 42, 56, 59, 61, 63, 85, 88, 100, 131, 133, 144, 178, 179, 343, 592
Material
  shapes, 125, 126, 128–132, 134, 136, 139, 141, 143–146
  specification, vii
  system, 132
Maximal
  elements, 25, 79, 175, 196, 225, 256, 341, 485
  representation of shapes, 76, 81, 227, 256, 341
Maximal elements, 25, 79, 175, 196, 225, 256, 341, 485
McCarthy, J., 7, 257, 497
McCormac, J. P., 74, 165, 168, 176, 187, 377, 378, 400
McCulloch, W. S., 497
McKay, A., 74, 163, 172, 178–180, 356, 377, 378
McCullough, M., 157, 339, 347, 411, 412
Mechanical
  artifices, 88
  descriptions, 529
Memory, 33, 44, 61, 88, 186, 222, 245, 257, 290, 484
Mereology, 493, 494, 502, 503, 511, 512
Merge, 8, 26, 79, 109, 188, 189, 506
Metaphor, 29, 329, 457–459, 462, 465, 512, 514
Meta
  shape rules, 204, 205, 207, 218
  shape grammar, 195, 205, 207, 218
Metzger, W., 498, 509, 510
Michelangelo, 326
Midnight sun cocktail bar, 397

Mies van der Rohe, 350, 351
Minimalist design, 364, 368
Minimum space, 224, 256
Minsky, M., 7, 488, 497
Mitchell, W. J., 22, 57, 58
Mixed reality, 150, 152, 153, 459
Mondrian, P., 27, 40, 41, 104, 322, 575
Moore, G. E., 353
Morris, R., 101, 102
Muslimin, R., 127, 575, 577, 585
Mutt, R., 8

N
Names, 5, 13, 24, 33, 214, 266, 368–371, 569
Navigation rules, 463–465, 467, 473, 474
Negative capability, 27
Neural nets, 8, 189, 258, 328, 372, 484, 485, 486, 489, 497
Newman, M. H., 64, 102
Noise, 153, 503, 513
NP-hard, 483, 487
Number
    fields, 88
    systems, 88
NURBS curve, 174

O
Object-oriented, 74, 91, 345, 369, 370, 459, 461, 462, 571
Objets trouvés, 3
Oblique divisions, 104–107, 114, 117, 120
Ockham's razor, 4
Oliver, P., 585
Optimization
    approach, 318
    method, 188, 189
    problems, 318, 323, 487
Oracle, 372, 373
Organic
    and individual, 53
    development, 61
    growth, 61, 542, 563
Ornamental friezes, 240
Orsborn, S., 74, 188, 189
Orthogonal divisions, 104–106, 109, 110, 121
Overlap, 6, 37, 39, 61, 67, 81, 83, 134, 225, 245, 274, 340, 414
Ozkar, M., 81, 125–129, 145, 157, 174, 493

P
Pacioli, L., 44
Pakan, L., 576, 577
Palladian Canon, 47, 49
Palladian plans, 22, 57, 58, 330
Palladio, A., 34, 42, 44, 46, 57, 187, 267, 270, 330, 367, 409, 429, 489, 594
Paradigm shift, 348
Parallel
    coordinates, 363, 365
    grammars, 202, 434
    rules, 416, 418, 419, 421, 424, 425
Parametric
    model, 365
    representation, 357
    shape, 22, 57, 85, 86, 165, 175, 177, 343, 434, 445, 541, 549
    shape grammar, 86, 177, 343, 434, 445, 541, 549
    variation, 442
Parsing, 103, 583, 595
Part
    boundaries, 75
    equivalent, 297, 298
    relation, 75, 78, 290
Partial order, 196, 197, 201, 225, 290
Participation, xv
Partis, x
Passura, 127, 575–577, 579, 581–585, 587, 588
Patterson, S., 571
Perceptual change, 127, 341, 500
Preservation, 163, 164, 452
Price, C., 151
Principal Component Analysis, 188
Perception and cognition, 222, 510
Perception without prediction, 16
Perceive and implement, 221
Permutation, 13, 35, 39, 63
Phenomenology, 101, 488, 489, 576
Physical
    action, 127, 128, 130
    construction, 221–224, 243, 246, 254–257
    elements, 27
    evaluation, 189
    information, 127
    meaning, 303
    making, 99, 121, 130, 257
    object, 26, 126, 146, 223, 255, 257
    process, 100
    production, 231, 256, 257
    system, 125, 126, 146

Picasso, P., 4
Pictures and poems, 9, 15
Pinwheel, 62
Plagiarism, 8, 16
Platonic, 9, 44, 46
Play, 16, 17, 64, 110, 125, 139, 195, 222, 353, 422, 424, 547, 567–572, 594
Playground, 154–156
Point-line arrangement, 289, 291–309, 311–313
Polyominos, 86
Pope, A., 70
Portman, J., 349, 397, 410, 411, 422, 423, 429, 431
Portm-Ino, 422–430
Prairie house, 322
Problem solving, 186, 188, 258, 339, 348, 368, 510
Principles and proportions, 523, 525–527
Product-focused shape grammars, 187
Projection, 26, 201, 202, 440, 445, 506, 508
Proportional systems, 34
Proto-implementation, 409, 412–422, 424
Protracted refugee camps, 537–539, 541
Pulli kolam, 245, 246
Pythagorean, 88

Q
Q-phrase, 324, 325
Qualitative representation, 317, 324, 328
Queen Anne grammar, 90

R
Randomness, 263, 347, 478
Rational numbers, 88, 175, 380
Reading, 3, 81, 100, 339, 351, 355, 407, 409, 440, 591, 597
Readymade, 3, 4, 6, 8, 10, 11, 15
Reasoning about shapes, 320, 324
Receptors and effectors, 256, 257, 498
Recognition and manipulation, 273
Reconstruction and deconstruction, 152
Recursion and embedding, 258
Reduction, 340, 341, 510, 598
Refinement of perception, 255
Refugee camps, 537–543, 563
Registration points, 175–177, 379, 394–397, 399, 400, 401
Regular expressions, 261, 265–271, 285, 261, 265–271, 285
Rembrandt, 4
Renaissance arithmetic, 33, 34, 42, 43
Revision, 418, 419, 422, 424, 598
Rewriting shape rules, 410, 421, 428, 430
Rhythmic composition, 53
Rhythmic measures, 36, 38
Richards, I. A., 8, 17, 18
Rietveld, G.T., 27, 28
Rittel, H., 483, 489
Rorschach test, 6, 7, 9, 11, 14, 16, 18
Rowe, C., 429
Rubin vase, 11
Rules
    algebra, 75, 195, 201–203, 205, 209, 218
    application, 57, 58, 60, 65, 66, 73, 75, 81, 87, 89–91, 109, 110, 167, 176, 186, 190, 261–264, 269, 273, 274, 284, 285, 341, 345, 347, 351, 364, 366, 368, 369, 408, 412, 415, 421, 461, 464, 468, 478, 483, 487, 524, 558
    compositional, 27
    decomposition, 195, 210–212, 217, 219
    description, 273, 274, 496
    interaction, 463–465, 467, 474
    sequencing, 262, 263, 285, 345
Ruskin, J., 254, 255

S
Said, A. A., 576
Schemas
    addition, 54, 65
    grids as, 54, 63
    identity, 11
    inverse, 15, 67, 208, 209, 218
    part, 504, 508
    summation, 11, 13, 14, 15
Schindler House, 29
Schön, D., 39, 129, 157, 363, 373, 430
Schröder House, 27, 28
Schulze, F., 350
Scripting, 339, 460
Scruton, R., 29
Search space, 73, 89, 90
See-do cycle, 256
Seeing and doing, 7, 18, 19, 25, 223, 246, 257, 363–366, 368, 494, 498, 499, 502, 516, 595
Seeing and looking in art and design, 11
Seeing and making decisions, 106
Seeing in new ways, vii
Segmentation, 103, 504
Sequence of shape rules, 224, 478
Serbian medieval churches, 202

Index   609

Serialism, 39
Set
   intersection, 300, 302
   of shape rules, 90, 170, 188, 195, 351, 353, 357, 478
   union, 300, 301, 303
Shape
   algebra, 54, 55, 73–75, 84, 90, 202, 218, 224, 255, 256, 595
   annealing, 188–190
   boundary, 75, 174, 175
   comparison, 262, 289, 292, 338, 412
   derivation, 196, 198–200, 204, 206, 211, 212, 219
   equivalence, 299
   embedding, 289–291, 298, 339, 340, 343–345, 347–351, 356, 357, 368, 377–380, 394, 396, 397, 399, 401, 402
   extrapolation, 328, 329
   interpolation, 328
   interpretation, 128, 219, 222, 317, 329, 496
   interpreter, 90, 187, 356, 366, 495, 496, 501
   machine, 56, 60, 64, 70, 74, 89, 172, 337, 340–345, 347–358, 363–372, 374, 377–379, 381, 391, 394, 397, 399, 402, 409, 410, 412, 414, 421, 422, 424, 426, 427, 430, 433, 435, 438, 442, 452, 487, 488
   operations, 75, 76, 79, 81, 83, 84, 127, 240, 300, 464
   optimization, 188, 189
   recognition, 81, 85, 86, 175, 326, 483, 487, 488, 495
   representation, 73, 86, 87, 90, 321–324, 328, 330, 487, 489
   specification, vii
   transformation, 126, 130, 174, 176, 222, 578
Shape-based search, 337, 338, 340
Shape grammar implementation, 86, 90, 163, 172, 175, 177, 179, 180, 410, 435, 483–487
Shape grammar interpreter, 30, 73, 74, 76, 78, 81, 85–88, 91, 286, 339, 341, 343, 356, 377, 378, 397, 399, 400, 435, 483, 484, 487, 496, 524, 533
Shapely symbol, 54
Shape machine, 56, 60, 64, 70, 74, 89, 172, 337, 340–345, 347–358, 363–372, 374, 377–379, 381, 391, 394, 397, 399, 402, 409, 410, 412, 421, 422, 424, 426, 427, 430, 433, 435, 438, 442, 452, 487, 488
Shape machine interface, 363–372, 374
Shapes as unanalyzed entities, 222
Shea, K., 188
Shelley, P. B., 17
Shult, E. E., 311, 313
Sibat, 547, 553, 558, 559, 560
Simon, H. A., 7, 185, 189, 190
Situated learning, 317, 327
Sketching, 25, 102, 130, 218, 363
Smith, K., 29, 459
Social and cultural behaviors, 458
Sortal GI, 377, 487, 488
Spatial
   analysis, 542
   change, 76–78, 84, 86, 542
   composition, 53, 579
   relation, 6, 53, 54, 56–59, 60, 62–68, 70, 77, 133, 145, 232, 320, 325, 337, 352, 379, 380, 385, 408, 417, 421, 478, 509, 549, 551, 571, 594, 595, 597
Standard of taste, 11
Statistics, 8, 488
Steadman, P., 34, 62, 593, 594, 596
Steinhardt, N. S, 530
STEM and STEAM education, 567
Stewart, I., 343
Stillwell, J., 292
Stiny, G., 7, 22, 23, 25–27, 29, 30, 34, 44, 54, 55, 57, 58, 60, 65, 67, 70, 73–76, 78, 80, 84, 86, 87, 90, 91, 100, 106, 125, 126, 128, 134, 149, 152, 163, 164, 166, 172, 173–176, 178, 181, 185, 187, 190, 195, 197, 199, 208–210, 214, 217, 222–225, 227, 228, 255, 256, 258, 261–263, 267, 273, 289, 290, 317, 318, 321, 338, 339, 341, 343, 348, 349, 355–357, 363, 368, 378, 400, 407–409, 414, 430, 431, 435, 461, 463, 479, 484–489, 493–499, 501–505, 513–515, 524, 525, 527, 541, 572, 575, 585, 594–597
Stouffs, R., 74, 75, 78, 80–82, 84–86, 91, 172, 177, 273, 274, 276, 277, 283, 284, 339, 341, 343, 377, 378, 484, 487, 488, 496, 524, 533
Structural rhythm, 53, 232
Structure sharing, 87
Studio teaching, x
Sub-shape
   detection, 173–175, 496

recognition, 85, 86, 483, 487, 488
Subsymbolic AI, 485, 486, 488, 489
Sum and difference, 83
Sutherland, I., 7, 25, 257, 258
Swastika, 230, 231, 237, 238, 239
Sweep process, 84
Symbol, 4, 6, 54, 126, 208, 249, 261–263, 266, 270, 324, 337, 338, 345, 366, 402, 467, 468, 495, 512, 594, 595, 597, 598
Symbolic
  calculating, 483, 485
  logic, 320
  reasoning, 320, 486, 488
Symmetric difference, 84, 197, 201, 203, 206
Symmetry, 4, 12, 13, 38, 39, 42, 55, 56, 176, 199, 222, 223, 230, 240, 245, 246, 255, 275, 324, 326, 327, 355, 397, 440, 585, 593, 594
Syntax without semantics, 29
Synthetic, analytic, and descriptive, 524, 527

T
Tacit design/making, 577, 588
Takeshima, T., 525
Taleb, N. N., 501
Tapia, M., 86, 174, 175, 339, 343, 378, 394, 524
Taxonomy, 484, 592
Techne (τέχνη), 222
Technique, 36, 74, 100, 101, 126, 127, 131, 133, 145, 155, 222–224, 228, 229, 246, 254–256, 263, 347, 367, 451, 499, 508, 509, 514, 516, 539
Temporality of making, 99, 100, 103, 106, 121, 131
Thesis/antithesis and synthesis/ indifference, 12
3D virtual worlds, 457, 459–465, 476, 478–480
Topology, 17, 64, 86, 91, 176, 199, 211, 214, 216, 217, 221, 223, 226, 232, 255, 258, 289, 290, 328, 593
Total meaning, 8, 9, 18
Tractable shape rule application, 81
Training set, 4, 8, 23
Transparent, 506, 533, 534, 568
Trees (hierarchies), 11
Tree-trunks and clods of earth, 4, 15, 18
Triaxial weave, 136, 138–144

Truth in building construction, physical performance, and technology, xi
Truth in logic, xi
Turing, A. M., xiii
Turing machine, 4, 7, 15, 16, 290, 372, 595, 597
Turner, J. M. W., 3, 539
Two-dimensional patterns, 136, 221, 223, 254

U
Ullman, S., 509, 510
Unconscious, 255
Units, 17, 18, 39, 46, 111, 112, 199, 277, 320, 324, 391, 392, 394, 411, 412, 496, 539, 543, 545, 547, 553–555, 557, 560, 562, 570, 598
Unity, 10, 11, 289
Unmaking, 149, 150–152, 154, 158, 159
Unreachable symbols, 5
Urpflanze, 17
Useless rules, 214, 216, 218

V
Van der Helm, P., 222, 223, 255
van Doesburg, T., 40
Vantongerloo, G., 41, 42
Vickers, S., 200
Virtual gallery, 457, 458, 460, 461, 466–469, 473–478
Visual
  analogy, 6, 7, 9–11, 15, 16, 18, 19, 329, 488
  calculating, 4, 7, 223, 289–291, 314, 483, 488
  complexity, 109–111, 113, 116, 117, 121, 514
  recognition, 223
  representation, 407, 408, 414, 501, 502
  style, 328, 595
  thinking, 24, 25
  making, 128, 499, 501, 502, 506
Visual features
  abstract, 223, 224, 230, 240, 245, 255
  figural, 223, 226, 230, 255
Vitruvian
  canon, 18
  pairs, 55, 56
  principles, 55, 61
Vocabularies of counters, 17
Vocabulary and visual analogies, 18
Vocabulary of shapes, 21, 30, 246

Index

von Neumann, J., 6, 7, 18, 19

W

Waring, T. M., 246
Weaving grammar, 127, 131, 146
Wedge border divisions, 106, 119
Weighted shape, 74, 78–81
Weights, 12, 53, 63, 78–81, 127, 131, 177, 189, 435, 446, 448, 485, 486, 500
Weiser, M., 459
Wellek, R., 8
Wertheim, M., 458
Wescoat, B., 434, 436–438, 442, 449
Whitehead, A. N., 101, 102, 318
Wicked problem, 502, 503, 505, 509, 511
Wiener, N., 8

Wilde, O., 7, 13, 15–19
Wilde's critical formula, 13, 15, 16, 17
Winged eye, 498
Winston, P. H., 497
Wittgenstein, L., 222
Wittkower, R., 42
Wonder (θαύμα), 222
Wondering (θαυμάζειν), 222
Woodbury, R., 89, 357, 363, 412, 593
Woolley, B., 458
Wortmann, T., 81, 82, 85, 484, 486–488, 496
Woven textiles, 180

Y

Yingzao fashi, 434, 523, 525–534, 575

The manufacturer's authorised representative in the EU is Springer Nature Customer Service Centre GmbH, Europaplatz 3, 69115 Heidelberg, Germany. If you have any concerns regarding our products, please contact ProductSafety@springernature.com

Printed and bound by CPI Group (UK) Ltd, Croydon, CR0 4YY
23/03/2026
02076723-0001